P9-CDU-874

Probability Models and Applications

SECOND EDITION

Probability Models and Applications

Ingram Olkin Stanford University

Leon J. Gleser University of Pittsburgh

Cyrus Derman Columbia University

Macmillan College Publishing Company New York

Maxwell Macmillan Canada Toronto

Maxwell Macmillan International
New York Oxford Singapore Sydney

Editor: Robert W. Pirtle
Production Supervisor: Elisabeth Belfer
Production Manager: Paul Smolenski
Text and Cover Designer: Eileen Burke

This book was set in Times Roman by Bi-Comp, Incorporated, and was printed and bound by Book Press, Inc.
The cover was printed by Phoenix Color Corp.

Printed in the United States of America

Macmillan College Publishing Company
866 Third Avenue, New York, New York 10022

Macmillan College Publishing Company is part
of the Maxwell Communication Group of Companies.

Maxwell Macmillan Canada, Inc.
1200 Eglinton Avenue East
Suite 200
Don Mills, Ontario M3C 3N1

Library of Congress Cataloging in Publication Data

Olkin, Ingram.
Probability models and applications / Ingram Olkin, Leon J.
Gleser, Cyrus Derman. -- 2nd ed.
p. cm.
Includes bibliographical references and indexes.
ISBN 0-02-389220-X
1. Probabilities. I. Gleser, Leon Jay. II. Derman, Cyrus.
III. Title.
QA273.04 1994
519.2--dc20 93-22412
CIP

Printing: 1 2 3 4 5 6 7 8 Year: 4 5 6 7 8 9 0 1 2 3

To Leah and Jared;
Noah and Sophia;
Rachel and Jeremy

Ingram Olkin

To Kimberly

Leon J. Gleser

In memory of Adam

Cyrus Derman

Preface

The primary goal of the first edition of this textbook was to bridge the gap between the mathematical theory of probability and the actual construction, use, and interpretation of probability models in a variety of disciplines. Rather than a treatise that presented the axioms, calculations, and main concepts of probability abstractly, divorced from applications, our desire was to promote cross-disciplinary research into the modeling of the increasingly complex data arising in scientific and technological research by describing the role probability concepts and models play in scientific and practical applications. Numerous examples, many with actual data, were provided of situations where probability models have been used in practice. Detailed descriptions were given of the properties and uses of probability models, other than the normal distribution, that have successfully modeled real phenomena. Because judgments whether a probability model successfully "fits" data involve the use of statistical inference, a firm foundation, particularly in techniques for deriving distributions, was built to prepare for subsequent study of statistics.

Although the first edition successfully met its goals, the need for substantial changes in a new edition has become evident. In particular, treatment of fitting probability models was incomplete without a description of methods for testing goodness of fit, a topic usually left to books on statistics. We therefore describe these methods but without a major digression to statistical theory and practice.

Changes in the Second Edition

The most apparent change is the addition of a new chapter (Chapter 14) on testing goodness of fit. Markov chains, which is normally treated in more advanced probability modeling courses, has been deleted. Material on the distributions, moments, and moment generating functions of sums of independent random variables has now been gathered into a separate chapter (Chapter 7) and is presented before individual families of distributions (e.g., binomial distributions) are discussed. This makes unnecessary complicated summation and integral derivations of means and variances for such distributional families. Other pedagogical changes include (a) placing conditional probability and independence of events in a separate chapter, (b) concentrating almost exclusively on bivariate distributions as examples of multivariate models, and (c) isolating the more theoretical mathematical derivations and proofs into subsections that can be treated separately by those who wish to emphasize formal mathematical arguments.

For the convenience of students, glossaries of new terminology and summaries of important new results have been included at the end of each chapter. Also, important formulas and results are highlighted by boxes in the text. Finally, numerous new exercises and examples have been added.

Prerequisites

Because of the changes made in the presentation, the material in the unstarred[1] sections of the second edition should be easily accessible to readers who have had a one-semester undergraduate (service) course in calculus. Exposure to partial derivatives and integrals in two variables will be helpful for understanding Chapter 11, "Bivariate Distributions." The proofs in starred subsections and in Chapter 13 (transformations of more than one variable) require somewhat greater sophistication; exposure to a one-semester course in advanced calculus (limits, sequences, and series) is recommended. Such a mathematical background, however, is not essential to the comprehension and use of the results given in Chapter 13.

How to Use This Book

This book is intended for either a one-semester or a two-quarter course on probability models and concepts. Two main classes of user are envisioned: (1) those taking a probability course for its own sake, particularly those who want to learn about available models and techniques that can be used in scientific or technological areas of interest to them, and (2) those who intend to use this material as background for more advanced probability courses or courses in statistical theory and methods.

[1] Sections, chapters, and exercises that involve more technical mathematical material are marked with a star in the text. If desired by the instructor, such material can be made optional for students.

Chapters 1–7 should be covered by all users. It is also recommended that Chapter 9 and at least Sections 1–4 of Chapter 8 and Sections 1, 2, and 4 of Chapter 10 be discussed, with students being invited to skim omitted sections in Chapters 8 and 10 on their own. (The material in Chapters 8–10, and also 12, can serve as a useful handbook of models for later use in applications, as well as providing examples of distributions for use in later courses.) Students in group 1 can complete their introduction to probability models by exposure to Sections 1–3 and 5 of Chapter 11, Sections 1 and 2 of Chapter 12, and Sections 1 and 2 of Chapter 14. Students in group 2 can most benefit from Chapters 11 and 13; for those interested in statistics, Chapter 14 can serve as a useful introduction.

The exercises were chosen to illustrate ideas and to provide additional learning material. More technical or theoretical exercises have been separated from those exercises that only require applications of material presented in the text.

Necessary tables appear in Appendix B. Blank entries in tables are either equal to 0 or correspond to undefined quantities. Because of roundoff errors, sums of probabilities over all possible outcomes need not exactly equal the theoretical value of 1. Where these discrepancies are minor and recognizable, we have made no attempt to correct them. For readers unfamiliar with the technique of linear interpolation in tables, a brief introduction is given at the beginning of Appendix B.

Acknowledgments

Both editions of this book have gone through numerous drafts and thus required the dedication and patience of many typists, too many to list individually. We are indebted to all of these individuals, without whom we would have been unable to fulfill our goals for this book. We are also grateful to the many colleagues who took the time to make helpful suggestions and correct errors in the first edition and in drafts of the current edition—and to the Macmillan reviewers Lynne Billard, University of Georgia; Roger Carlson, University of Missouri; Arthur Cohen, Rutgers University; Jay Devore, California Polytechnic State University; and Prem Goel, The Ohio State University. We, of course, take sole responsibility for any errors or flaws that may remain and would appreciate having these brought to our attention.

Finally, we are grateful to our families who put up with absences, short tempers, and a tendency to talk business at all times.

I. O.
L. J. G.
C. D.

Contents

3 *Finite Probability Models and Random Sampling* *54*

4 *Conditional Probability and Probabilistic Independence* *85*

5 *Random Variables* *144*

6 *Descriptive Properties of Distributions* *215*

Probability Models and Applications

1 Introduction

Everyone encounters uncertainty. Unpredictable results are said to be "random" or to "happen by chance." The news media often discuss the "odds" or probability of a certain occurrence: for example, a favorite team winning a crucial game, or mortgage rates declining. Such forecasts may help us in choosing actions. Thus a weather prediction of a "50% chance of rain" may suggest carrying an umbrella. Businesses and governments also take risks in the face of uncertainty. Scientists and engineers encounter phenomena whose results cannot be predicted. Out of the need to find an order or pattern in unpredictable phenomena and to measure risks precisely, mathematical models of uncertainty, called **probability models,** have been developed. The present book is intended to introduce modern mathematical probability theory and the various methods used to calculate odds or probabilities. Some widely used probability models are described, along with examples of how such models are developed and used in science, technology, business, and everyday life.

1 History

Archeological evidence shows that games of chance were played in the earliest human civilizations. Chance was seen as being controlled by the whims of the gods; consequently, disputes were often settled by drawing lots, and future events were forecast by casting bones (precursors of dice). One of the synonyms of probability, the word **stochastic,** is derived from the ancient Greek word for a person who forecast the future.

Gambling was discouraged by the medieval church but revived in popularity in the late Middle Ages and early Renaissance. In the search for winning strategies, crude calculations were made of odds for games of chance. However, little progress was made until the middle of the seventeenth century. In 1654, a French nobleman, Antoine Gombaud, Chevalier de Meré (1607–1684), consulted his countryman Blaise Pascal (1623–1662) about two problems arising in games of chance. This started a celebrated correspondence between Pascal and Pierre de Fermat (1601–1665). The interest shown by these two great mathematicians stimulated other European mathematicians to work on similar problems. The honor of publishing the first treatise on probability fell to the Dutch physicist Christian Huygens (1629–1695), whose *De Ratiociinis in Alea Ludo* appeared in 1657. By the middle of the eighteenth century, the efforts of Jakob (James) Bernoulli (1654–1705), Abraham de Moivre (1667–1754), and others had led to the development of methods of considerable generality for calculating odds in games of chance.

The seventeenth century also saw the rise of probability's sister discipline, **statistics.** States began collecting demographic facts (births, deaths, marriages) about their citizens. In 1662, the Englishman John Graunt (1620–1674) published the first "statistical" treatment of birth and death records, illustrating thereby the value of gathering such data. His countryman Sir William Petty (1623–1687) advocated "the art of reasoning by figures upon things related to government" (political economy). The interest of merchants in protecting themselves against such risks as shipwrecks led to the growth of insurance companies and the actuarial profession. Actuaries compiled life and accident tables, calculated risks, and set premiums for insurance policies.

As statistical data were gathered, demographers and actuaries soon noted parallels between patterns of variability in their records and the results of games of chance. For example, the records of successive male or female births closely resembled the succession of heads and tails in repeated tosses of a coin. This suggested modeling variability in natural phenomena (such as the sex of a newborn child) as if it resulted from a game of chance played by nature. Knowing the game that nature played allowed one to predict the results of such phenomena (e.g., the percentage of male births). Conversely, since certain results would appear more often in some games than others, observations would permit the investigator to infer what "game" nature was playing (statistical inference). Consequently, the developing theory of probability became a basis for statistical reasoning—a role that probability continues to play today.

The eighteenth and nineteenth centuries saw further application of probabilistic calculations and models in statistical reasoning. The great German scientist and mathematician Carl Friedrich Gauss (1777–1855) developed a "Law of Errors" to deal with variability in astronomical and physical measurements. This same "law" was later championed by the Belgian scientist Adolphe Quetelet (1796–1874) as a way of describing human variability, both social and biological.

In 1773, the French mathematician Pierre Simon Laplace (1749–1827) introduced a form of probabilistic reasoning called "inverse probability" for using data to decide the truth of scientific theories. Apparently unknown to Laplace, this form of reasoning had earlier been proposed by the Englishman Thomas Bayes (1702–1761), whose ideas had largely gone unnoticed. Two kinds of probabilities were used by this method of reasoning: those governed by nature and appearing in data, and the "odds" that individuals might assign to various competing theories.

Despite probability's contributions to statistical reasoning, most physical scientists of the eighteenth and early nineteenth centuries did not regard probability or chance as fundamental to the structure of the physical universe. However, the English physicist James Clerk Maxwell (1831–1879) derived laws for the behavior of gases on the basis of the probabilities of molecular velocities, and the German physicist Max Planck (1858–1947) described radiation in terms of probabilities for energy states of atoms and molecules (quantum mechanics). Such advances led many scientists and mathematicians to search for a theory that would unify all the ways in which probability was applied to model and reason about variation in nature.

The German mathematician Richard von Mises (1883–1953) gave an empirical definition of probability in terms of the "stability properties of relative frequencies" (Section 3). This definition clarified the relationship between patterns of variability in repeated observations of natural phenomena and probabilities assigned to various possible happenings in such observations. However, it did not apply to the "odds" used in inverse probability reasoning (or other odds that individuals apply to unique one-time future occurrences). This definition also was criticized as being unverifiable.

In 1933, the Russian mathematician Andrei Nikolaevich Kolmogorov (1903–1987) published a formal axiomatic definition of probability [see Kolmogorov (1956)]. This definition gave the basic rules (axioms) that all types of probabilities or odds must follow, and provided justification for the stability of relative frequencies. Equally important, it showed the basic relationship of probability calculations to general techniques of mathematical analysis that had been developed in the nineteenth and twentieth centuries. Probability theory could now serve as a bridge between abstract mathematical theory and scientific and technological applications, to the enrichment of all of these fields of inquiry. New applications of probability models would suggest new questions for mathematical analysis. Advances in mathematical analysis could both answer these questions and suggest new models and insights to contribute to greater understanding of the variable and changing world around us.

2 Probability Models

A **scientific model** is a simplified description of a given phenomenon stated in terms of an analogy with some other phenomenon or structure for which more knowledge is available. We have already noted that games of chance served as models for the variability of such natural phenomena as the determination of sex at conception. Similarly, social systems have been modeled as biological organisms [e.g., see Spencer (1877)], economic systems have been analyzed in terms of thermodynamic concepts, and psychological behavior has been described in terms of games [e.g., see Tversky and Kahneman (1982)]. If the analogy on which a model is based is valid, known facts about the phenomenon or structure used as the model can suggest new ideas, interpretations, or explanations concerning the phenomenon being modeled (see Exhibit 2.1).

Most modern scientific models are mathematical. That is, important elements of a natural phenomenon (e.g., the position, mass, and charge of a particle) are identified with basic entities of a mathematical structure. Certain fundamental facts concerning these elements are expressed as rules (axioms) relating the analogous mathematical entities. The logical consequences (theorems) of these axioms in the mathematical system then correspond to assertions (laws) concerning more complex relationships among the elements of the natural phenomenon. These relationships can then be experimentally verified. A bonus gained by using mathematical models is that such models help us focus and simplify our knowledge. Rather than having to recall many facts concerning a phenomenon, we can derive these findings as logical consequences of a few basic rules.

Probabilistic (stochastic) models are mathematical models for variability or uncertainty. During the eighteenth and nineteenth centuries, it was believed that variability or uncertainty in an observed phenomenon could be attributed to a failure to identify and control its causes. Given sufficient information, any phenomenon could be predicted exactly, even the face on a tossed coin. In short, nature was seen as being a fully predictable machine. Such **mechanistic models** proved very fruitful in physics, chemistry, and engineering, and their success led scientists to question whether models involving chance or uncertainty were needed, or realistic.

However, as more complex phenomena were studied, it became impossible to measure or control the numerous factors affecting such phenomena.

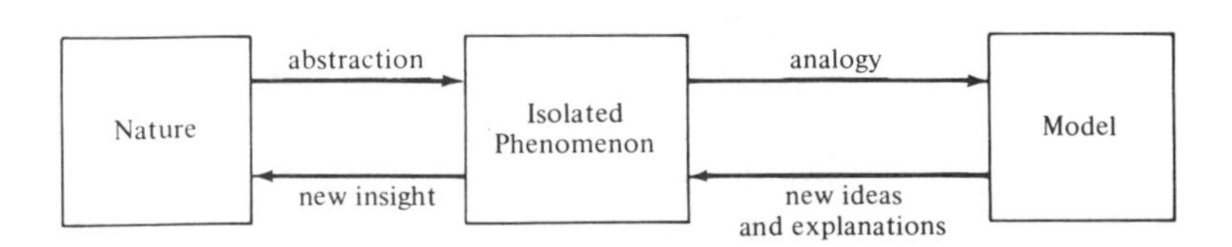

EXHIBIT 2.1 *Use and development of a scientific model.*

Further, as argued by the German physicist Werner Heisenberg (1901–1976), in some contexts the act of measurement itself had the potential to produce unpredictable variations in observations. Finally, statisticians demonstrated that a deliberate introduction of uncertainty in some experiments helped to eliminate subconscious biases on the part of both experimenters and subjects, and often substantially reduced the cost of obtaining needed information.

Thus it became useful to describe the natural phenomena by probabilistic models. In science, the decisive criteria for choosing a model are practical, not metaphysical. Because probabilistic models often provide a simple, yet comprehensive explanation of known phenomena, randomness has become an intrinsic component of many scientific models.

3 The Empirical Basis of Probability Theory

Probability theory is useful because many chance phenomena exhibit statistical regularities. A clear example of this fact is the ancient game of dice. Although we cannot say what the result of any given throw of a die will be, in a large number of throws of a fair die the proportion of throws that result in (say) a six is predictable. We use this information to place our bets for each throw. A more practical application of the statistical regularity of chance phenomena occurs in the field of insurance. Although it is nearly impossible to predict the life span of a given individual, precise statements concerning longevity can be made about large populations. Such a statement may be, for example, that 80% of the present population will live past the age of 65. These statements are adequate to enable the insurance company to "place its bet"—that is, to decide what premium to charge for a given amount of insurance. In offering a policy to a large group of people, it is not important to the company to know what happens to any single person but only what happens to the group as a whole.

Let us discuss this subject in a slightly more formal way. Basic to all discussions of chance phenomena is the notion of a random experiment. An experiment is a clearly specified procedure for obtaining an observation of some phenomenon. It should be possible to repeat, or replicate, an experiment, so that observations and conclusions obtained by one investigator can be checked by others. In many of the classical experiments of physics and chemistry, one knows in advance what will be observed. For example, in Galileo's (1564–1642) famous gravitational experiment, a rock and a feather dropped simultaneously from a tall tower will hit the ground at the same time. In contrast, a random experiment can yield any one of many distinct possible observations, called **outcomes.** One, and only one, of these outcomes can occur when the experiment is performed, but that outcome cannot be predicted with certainty in advance of running the experiment.

EXAMPLE 3.1

Sample Survey

A telephone respondent chosen by lottery is asked whether he or she is listening to a certain radio station. The possible outcomes (responses) are "yes," "no," and perhaps "refuses to respond." Before the phone call, we cannot say for sure what the person's response will be. ♦

EXAMPLE 3.2

Physical Measurement

A physicist attempts to find the weight of a given object. Here the outcomes are all possible weights. The experiment is random because minute changes in moisture content, air pressure, and temperature in the room affect both the scale and the actual weight of the object in unknown and unpredictable ways, and because there may be slight visual or adjustment errors on the part of the physicist. ♦

Each repetition of a (random) experiment is called a **trial.** Before making a trial of a random experiment, it is a good idea to list all of the outcomes that might possibly occur. Such a list of all possible outcomes is called a **sample space** for the random experiment. Thus in Example 3.1 a sample space is the list {yes, no, refuses to respond}. Because weights are nonnegative numbers, a sample space for the physicist's measurement in Example 3.2 is all nonnegative numbers—that is, the interval $[0, \infty)$. If we knew more about the object weighed, or the scale being used to weigh the object, a smaller sample space could be given.

EXAMPLE 3.3

Three skiers A, B, C choose straws to decide the order in which they will jump. An outcome of this random experiment is a description of the order in which the skiers jump. One possible outcome is that C jumps first, A jumps second, and B jumps third, which we could represent notationally as CAB. The sample space of this experiment is the list $\{ABC, ACB, BAC, BCA, CAB, CBA\}$. ♦

EXAMPLE 3.4

At a cafeteria, customers can choose to purchase one of four brands A, B, C, D of soft drink, and select either a small or large cup of that beverage. Since customers' preferences vary unpredictably, the choice made by a customer can be regarded as a trial of a random experiment. The outcomes of this experiment can be described by indicating the brand chosen and the size of cup selected. One possible outcome, Q, is that the customer does not purchase a beverage. The outcome sA denotes choice of a small cup of brand A, and lB denotes choice of a large cup of brand B. A sample space for this random experiment is thus the following list of 9 possible outcomes: $\{Q, sA, lA, sB, lB, sC, lC, sD, lD\}$. ◆

Statements that relate to, or describe, the outcome of a trial of a random experiment can now be made. In Example 3.3, a statement might be "A jumps first." In Example 3.4, a statement might be "the customer chooses brand A." Each trial of the experiment yields a single outcome that may or may not satisfy the given statement.

To decide whether an outcome satisfies the given statement, we could list in advance all outcomes in the sample space that satisfy the statement. Denote this list or collection of outcomes by E. In order to have a name for E, let us call E the **event** defined by the given statement. If a trial yields an outcome belonging to the collection E, we say that the **event E has occurred;** otherwise, we say that E has not occurred.

EXAMPLE 3.3 *(continued)*

In Example 3.3, the event E defined by the statement "A jumps first" contains the outcomes ABC and ACB; that is, $E = \{ABC, ACB\}$. If the outcome ACB is observed, E has occurred (skier A jumps first). If the outcome BAC is observed, E has not occurred. ◆

EXAMPLE 3.4 *(continued)*

In Example 3.4, the event E defined by "the customer chooses brand A" contains the outcomes sA and lA. Thus $E = \{sA, lA\}$. ◆

We are now ready to discuss statistical regularity. Suppose that N repetitions **(repeated trials)** of a random experiment are made, with the outcomes of the trials having no influence on one another. We record the outcome observed at each trial. Because the experiment is random, the outcome of each trial is unpredictable, and in a new series of N repeated trials a different list of outcomes would be obtained.

Among the N trials that we observe, we can count the number, $\#E$, of times that a particular event E occurs. This number is the **frequency** of E in the N trials. The **relative frequency** of E,

$$\text{r.f.}(E) = \frac{\#E}{N},$$

is found by dividing the frequency, $\#E$, of E by the number N of trials.

EXAMPLE 3.4 *(continued)*

Provided that customers are not influenced by choices made by other customers, Example 3.4 provides a good example of repeated trials. Each customer's choice is a trial. The choices made by N customers are then N repeated trials of this experiment. Observation of the choices made by $N = 10$ customers might produce the following table of outcomes:

Trial (*customer*)	1	2	3	4	5	6	7	8	9	10
Result	*lA*	*sB*	*Q*	*sA*	*sD*	*lC*	*lB*	*lC*	*sA*	*Q*

The event E defined by "the customer chooses brand A" occurs on a given trial if either the outcome sA or the outcome lA is observed. Thus the frequency of E in these 10 trials is $\#E = 3$ and the relative frequency of E is r.f.$(E) = \frac{3}{10} = 0.3$.

As we continue to observe trials, we might construct a table showing how the relative frequency of r.f.(E) of the event E changes as a function of the number N of trials, and see what happens to this relative frequency as the number of trials becomes large. Thus the 10 trials described in Example 3.4 would produce the following table:

Trial, N	1	2	3	4	5	6	7	8	9	10
r.f.(E)	1	$\frac{1}{2}$	$\frac{1}{3}$	$\frac{2}{4}$	$\frac{2}{5}$	$\frac{2}{6}$	$\frac{2}{7}$	$\frac{2}{8}$	$\frac{3}{9}$	$\frac{3}{10}$

Continuing this table as we obtain more and more trials would be difficult, so instead let us graph the relative frequency r.f.(E) against the number N of trials, as in Exhibit 3.1. Note that in Exhibit 3.1 a logarithmic scale for N is used to exhibit a large number of trials more easily. As the number N of trials grows larger and larger, the number r.f.(E) should move closer to a certain value, which we shall call p. In Exhibit 3.1, $p = \frac{2}{5} = 0.4$. ◆

For small N, the relative frequencies of the event E may vary widely over the range of numbers from 0 to 1; but as N gets large, r.f.(E) approaches p. This phenomenon of convergence is what we call the **statistical regularity of chance phenomena;** it is also known as the **stability of relative frequencies.** Such stability may be all that we need to describe and utilize chance phenomena (a point noted earlier in connection with life insurance policies). When the property of stability of relative frequencies is assumed to hold for all events (statements) of interest in a given experiment, we can use probability theory as a mathematical model for this chance phenomenon.

We do not always test the stability of relative frequencies to justify the use of probability theory. In many random experiments it is impractical to

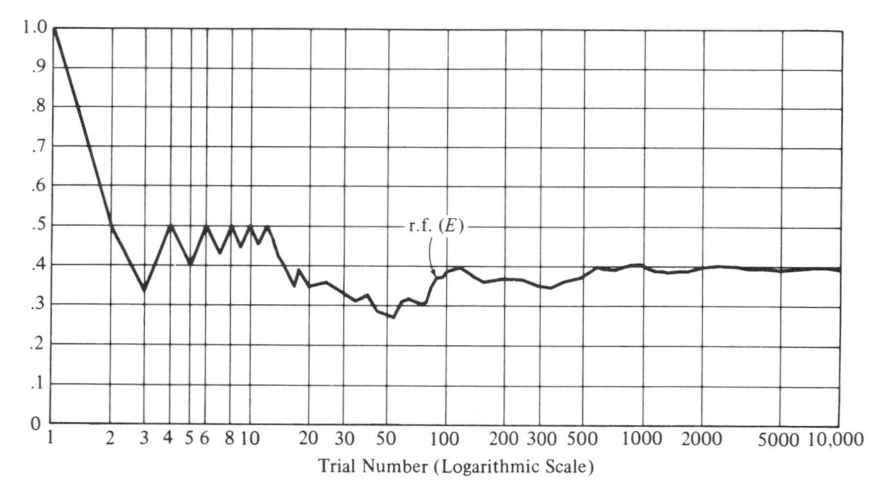

EXHIBIT 3.1 *Relative frequency of the result E.*

perform many repeated trials (although we can perhaps imagine such trials). However, even when repeated trials are not performed, we may have theoretical reasons for assuming the relative frequencies stabilize. The only test of this assumption then becomes the success of the model in describing and explaining the phenomenon.

Notes and References

The reader interested in the history of probability and in the use of scientific models may find the books by Todhunter (1865), Ore (1953), David (1962), Kline (1962), Pearson and Kendall (1970), Kendall and Plackett (1977), Hacking (1975), and Stigler (1986) and the articles by Ore (1960) and Kendall (1963), to be of interest. A series of original essays on probability by Pierre Laplace, Charles S. Peirce, John Maynard Keynes, Henri Poincaré, and Ernest Nagel is reproduced in Newman (1956). Another interesting original source is an English translation of Laplace's popular exposition of probability theory [Laplace (1951)].

Glossary and Summary

Experiment: A clearly specified procedure (intended to be replicable) for obtaining an observation of some phenomenon.

Trial: A single replication of an experiment.

Repeated trials: Repetitions of an experiment under the specified conditions that define the experiment.

Random experiment: An experiment having many possible results, with the result that occurs on any trial being unpredictable with certainty.

Outcome: Any single possible result of a random experiment, defined so that one and only one outcome can be observed at any trial of the experiment.

Event: A collection of outcomes usually described by a statement or property.

Sample space: The collection of all possible outcomes.

Occurrence of an event E: If an outcome is observed that belongs to the event E.

Frequency of an event E: The number of times that the event E occurs in repeated trials of a random experiment (denoted $\#E$).

Relative frequency of the event E: The ratio, $\text{r.f.}(E) = \#E/N$, of the frequency of the event to the total number N of trials.

Stability of relative frequencies of an event E: When the relative frequencies of E vary less and less about a limiting value p as the number of repeated trials of a random experiment increases.

Probability: A number chosen to correspond to the limit or stable value of the relative frequency of an event in a large number of trials; also, the odds assigned subjectively to the truth of an assertion or to the occurrence of an event by an individual.

Scientific model: A simplified description of a phenomenon, or class of phenomena, based on analogy with a known phenomenon or structure.

Mathematical model: A scientific model based on a mathematical structure (axiomatic system).

Probability model: A mathematical model of uncertainty or variation.

Mechanistic model: Natural phenomena viewed as predictable machines once all factors affecting those phenomena are known.

Statistical regularity: When relative frequencies of all possible events stabilize.

Scientists use models as a way of summarizing available information and to obtain new insights and verifiable hypotheses. Models are used more for these ends than because they are "true." Probabilistic models have proved useful, and thus are used, even by scientists who view nature as mechanistic. One important use of probability models is as a basis for statistical inference, where one uses partial information (data) about a phenomenon to infer what probability model describes the variability of the outcomes.

Exercises

1. In each of the following experiments, list all of the outcomes in an appropriate sample space.

(a) An investor follows the Dow Jones, Standard and Poor's, and over-the-counter stock averages for one year. At the end of the year, the investigator determines for each stock average whether its price has gone up, gone down, or remained the same.

(b) A market researcher asks a consumer the following questions:

(i) What is your sex: male or female?

(ii) Which soft drink do you prefer: Pepsi, New Coke, Classic Coke?

(iii) How many bottles of soft drink do you drink a day: 0, 1, more than 1?

2. The plans of a college student for spring break can be regarded as a random experiment. Such a student might choose to go to a Florida resort, skiing in Colorado, on a boat trip to save the whales, or home to visit parents. If the student goes to Florida, he or she could choose to camp out or to stay with friends. Similar choices of lodging are available in Colorado. However, the student only sleeps on the boat for the boat trip, and can only stay with family at home. Finally, the student may or may not decide to bring along textbooks to study. The outcomes of this experiment must indicate the location and type of lodging chosen by the student and whether or not the student brings along textbooks. List all such outcomes. (It may help you to draw a tree in which the branches are the choice of location, limbs off each branch designate the choice of lodging, and twigs off each limb indicate the student's decision about textbooks. Each path along a branch, limb, and twig is an outcome. Such *tree diagrams* are often useful in visualizing the outcomes of an experiment.)

3. In some random experiments, there may be too many outcomes in the sample space to list. In this case we can describe the form of a typical outcome and give some indication of what the sample space would look like. For example, if the random experiment is to count the number of defects in a 100-yard bolt of cloth, the possible outcomes are nonnegative integers x, and the sample space is $\{0, 1, 2, 3, \ldots\}$. In each of the following experiments, describe both a typical outcome and the sample space of the experiment.

(a) The net income (losses are negative incomes) to the nearest dollar is observed for a business at the end of the fiscal year 1993.

(b) The weight in kilograms of iron extracted from a shipment of ore is measured very precisely (to an arbitrary number of decimal places) on a scale whose largest possible reading is 1000 kg.

(c) Both the length of time (in minutes) that a customer waits in line at the deposit window at a bank and the size of the customer's deposit (in dollars) are observed.

4. Exhibit E.1 gives the frequency with which the letters A through Z of the English alphabet occur as the first letter of the nouns used in a sample

EXHIBIT E.1 *Frequency of Occurrence of the Initial Letters in a Sample of 2276 Nouns*

Initial Letter	*Frequency*	*Initial Letter*	*Frequency*
A	111	N	40
B	147	O	41
C	210	P	188
D	153	Q	7
E	69	R	133
F	112	S	256
G	72	T	112
H	110	U	16
I	72	V	43
J	22	W	100
K	18	X	0
L	84	Y	5
M	124	Z	1

Source: G. Herdan, *Quantitative Linguistics* (London: Butterworth & Company Ltd., 1964).

from the works of the English author John Bunyan (1628–1688).

(a) What is the relative frequency of the event E defined by "the letter chosen is a vowel"? (The vowels are A, E, I, O, and U.)

(b) What letter has the largest relative frequency?

5. Consider the following experiment. With a tape measure, a person measures the length of the longest wall in a given room to the nearest $\frac{1}{8}$ inch.

(a) What are the possible outcomes of this experiment?

(b) Twenty-five people were asked to perform this experiment. The outcomes obtained are shown in Exhibit E.2. Regard each person's measurement as a repeated trial of the measuring experiment described above. Find the negative frequency of each outcome actually observed in these repeated trials.

6. A standard deck of cards is well shuffled, and the top card is turned up. The suit (clubs, spades, hearts, or diamonds) and the value (ace, deuce, . . . , jack, queen, king) of the card are noted. The result of this process is one trial of a certain random experiment.

(a) List the outcomes of this experiment.

(b) List the outcomes belonging to the event E_1 defined by "the card drawn is a heart."

(c) If this experiment were repeated many times, at what number p would the relative frequency r.f.(E_1) stabilize?

(d) List the outcomes belonging to the event E_2 defined by "the card drawn is a 'face card' (king, queen, jack)."

(e) If you conducted many trials of this experiment, to what number p would r.f.(E_2) tend?

EXHIBIT E.2 *Measured Lengths of a Wall Obtained by 25 People*

Person No.	*Length Obtained*	*Person No.*	*Length Obtained*
1	20 ft $8\frac{1}{8}$ in.	14	20 ft $7\frac{3}{8}$ in.
2	20 ft $8\frac{1}{2}$ in.	15	20 ft $8\frac{1}{8}$ in.
3	20 ft $8\frac{3}{8}$ in.	16	20 ft 8 in.
4	20 ft $7\frac{7}{8}$ in.	17	20 ft $8\frac{1}{4}$ in.
5	20 ft 8 in.	18	20 ft $7\frac{7}{8}$ in.
6	20 ft $8\frac{1}{8}$ in.	19	20 ft $7\frac{3}{4}$ in.
7	20 ft $7\frac{3}{4}$ in.	20	20 ft 8 in.
8	20 ft $7\frac{7}{8}$ in.	21	20 ft $8\frac{1}{8}$ in.
9	20 ft $8\frac{1}{2}$ in.	22	20 ft $7\frac{1}{2}$ in.
10	20 ft $8\frac{1}{4}$ in.	23	20 ft $8\frac{1}{4}$ in.
11	20 ft $7\frac{5}{8}$ in.	24	20 ft 8 in.
12	20 ft 8 in.	25	20 ft $8\frac{3}{8}$ in.
13	20 ft $7\frac{3}{4}$ in.		

7. In a cloud-seeding experiment, clouds are randomly chosen and categorized as follows:

Category A. The cloud dissipates before producing rain.
Category B. The cloud produces rain on the area over which it was first observed.
Category C. The cloud produces rain somewhere else but not on the area over which it was first observed.

Exhibit E.3 gives the outcomes of observations on 100 clouds, listed in the order in which these clouds were observed.

(a) Consider the event E_1 defined by "the observed cloud produces rain somewhere." List the outcomes in E_1. Construct a table showing how the relative frequency of E_1 varies with the number of trials.

EXHIBIT E.3 *Outcomes of Observation of 100 Clouds*[a]

A	A	A	A	B	B	C	B	A	B	C	A	C	C	B	B	A	A	B	C
B	A	A	A	A	B	A	A	A	C	A	A	B	C	A	B	C	A	A	B
A	B	B	A	B	A	A	C	C	A	C	A	A	A	B	B	B	A	A	A
C	C	A	B	A	A	A	B	C	C	C	A	A	A	A	B	A	A	A	B
C	B	A	A	B	A	B	A	B	B	C	A	B	B	A	B	A	C	A	B

[a] Read across.

Graph the relative frequencies so obtained against the number of trials (see Exhibit 3.1). Around what number p do the relative frequencies stabilize?

(b) Consider the event E_2 defined by "the cloud does not produce rain on the area over which it was first observed." List the outcomes in E_2. Construct a table and a graph for r.f.(E_2) similar to the table and graph constructed in part (a). Around what number p do the relative frequencies stabilize?

8. Use of the "random number generator" of a personal computer to produce a random digit can be thought of as a "trial" of an experiment in which an integer is randomly chosen from among the integers $0, 1, 2, \ldots, 9$.

(a) Consider the event E defined by "the integer generated is less than or equal to 2." What outcomes are contained in E? At about what number p would the relative frequencies of E stabilize in a large number of trials?

(b) A personal computer was made to generate 100 random digits. The outcomes are listed in Exhibit E.4. Construct a table showing how the relative frequency of the event E defined in part (a) varies with the number N of trials. [It is enough to find r.f.(E) for $N = 1, 2, 5, 10, 20, 40, 60, 80, 90, 100$.] Graph the relative frequencies of E against the number of trials N. Around what number p do the relative frequencies appear to stabilize? Compare to your answer in part (a).

EXHIBIT E.4 *Generation of Random Digits*

Trials	*Results*
1–20	0, 4, 5, 5, 0, 3, 7, 6, 7, 6, 0, 5, 2, 3, 6, 3, 5, 6, 6, 1
21–40	9, 7, 8, 5, 0, 5, 5, 0, 7, 9, 3, 4, 5, 7, 7, 5, 2, 7, 1, 4
41–60	9, 5, 8, 4, 9, 1, 0, 9, 9, 5, 9, 6, 3, 7, 5, 9, 1, 9, 7, 8
61–80	9, 8, 1, 3, 6, 3, 1, 8, 4, 7, 3, 0, 4, 6, 3, 0, 0, 4, 2, 4
81–100	8, 0, 4, 8, 5, 9, 1, 4, 0, 4, 9, 0, 6, 6, 1, 4, 2, 7, 2, 9

9. A diagnostic computer receives a medical history and a list of symptoms from a patient, from which it makes a diagnosis of the patient's problem. For each patient, a team of physicians also makes a diagnosis. If the diagnoses agree, an A is recorded; otherwise, a D is recorded. The outcomes for 80 patients are shown in Exhibit E.5.

(a) The table can be regarded as giving the outcomes of 80 trials of a certain random experiment. What is the sample space of that experiment?

EXHIBIT E.5 *Results of the Diagnosis of 80 Patients*[a]

A	A	A	D	A	A	A	A	A	D	A	A	A	A	A	A	A	D	D	A
A	A	A	A	A	A	A	A	A	D	A	A	A	A	A	A	A	A	A	D
A	A	A	A	D	A	A	D	A	A	A	A	A	A	D	A	A	A	A	A
A	A	A	A	A	A	D	A	A	A	A	A	D	A	A	A	D	A	A	A

[a] Read across.

(b) For each of the events $E_1 = \{A\}$, $E_2 = \{D\}$, $E_3 = \{A, D\}$, determine whether or not the relative frequencies stabilize. These three events, plus the impossible event E_4 that contains no outcomes, comprise all the events that can be defined for this random experiment. Based on your results for these data, can the random experiment described in part (a) be modeled using probability theory?

2 The Elements of Probability Theory

1 Introduction

Our discussion in Chapter 1 identified three basic elements in all random experiments: **outcomes, events,** and **relative frequencies.** We now seek a mathematical structure with entities that correspond to, and have the properties of, those basic elements. Corresponding to outcomes of a random experiment, we define basic mathematical entities in our model which, for convenience, we also call outcomes. We denote a typical outcome by the Greek lowercase letter omega (ω). When more than one outcome is discussed, we may use numerical subscripts to identify the various outcomes: that is, ω_1, ω_2, ω_3, and so on. Whenever possible, however, we use a notation for outcomes that helps us remember the context of whatever random experiment is being modeled. For example, if the experiment is to toss a coin and record the upturned face, we may denote the outcomes by H (for heads) and T (for tails) rather than using the notation ω_1, ω_2.

Events of a random experiment are collections of outcomes; similarly, events in our model are defined to be collections of the basic mathematical entities corresponding to outcomes. Because of this correspondence, we do not distinguish between the outcomes (or events) of a random experiment and the outcomes (or events) of our model.

EXAMPLE 1.1 In Exercise 4 of Chapter 1, we discussed a random experiment whose outcomes are the letters of the English alphabet. If we wished to maintain the distinction between experiment and model, we would need to define outcomes $\omega_1, \omega_2, \ldots, \omega_{26}$ for our model to correspond to the letters A, B, . . . , Z in our experiment. The event in the experiment described by the statement "the letter observed is a vowel, that is, A, E, I, O, U," would then correspond to the event in our model that is the collection of the outcomes ω_1, ω_5, ω_9, ω_{15}, and ω_{21}. Because this is a cumbersome and unnecessary distinction to maintain, we instead say that the outcomes of our model are A, B, . . . , Z, and that the event "the letter observed is a vowel" in our model is the collection of the outcomes A, E, I, O, U. ♦

A **set** is a well-defined collection of mathematical entities. Thus an event is a set of outcomes. In the following section, we provide a brief outline of some of the more useful concepts, rules, and methods of analysis of the mathematical theory of sets, described in terms of events and outcomes.

2 A Brief Outline of Set Theory

We have seen that events E are collections of outcomes ω. We can define an event E by giving a proposition or property $\mathcal{P}$ that is to be satisfied by every outcome in the event. In such a case, we write[1]

$$E = \{\omega : \omega \text{ satisfies property } \mathcal{P}\},$$

or, more succinctly,

$$E = \{\omega \text{ satisfies property } \mathcal{P}\}.$$

Another way to describe an event E is by enumeration of all the outcomes that belong to E. Thus if E consists of the outcomes ω_1, ω_2, and ω_6, we write

$$E = \{\omega_1, \omega_2, \omega_6\}. \tag{2.1}$$

The event consisting of all the outcomes defined for a given model is called the **sample space** for that model and is denoted by Ω. The event Ω is often called the **sure event** (or certain event) because Ω contains all the outcomes of the random experiment and thus is sure to occur when the experiment is performed. The event $\varnothing$ that has no outcomes is called the **impossible** or the **null event,** because $\varnothing$ cannot occur on any trial of the experiment.

[1] The braces in $\{\omega : \omega$ satisfies property $\mathcal{P}\}$ mean "collection of" and the colon means "such that." Thus $\{\omega : \omega$ satisfies property $\mathcal{P}\}$ is the collection of all ω such that ω satisfies property $\mathcal{P}$.

EXAMPLE 2.1

In the social sciences, a subject is often asked to assign one of the ranks 1, 2, 3, ... , 10 to an object, where higher ranks correspond to greater desirability. One event E_1 of potential interest is "the subject gives a rank greater than 5 to the object." We can describe E_1 in any of the following equivalent ways:

$$\begin{aligned} E_1 &= \{\omega : \omega \text{ is a rank greater than } 5\} \\ &= \{\omega : \omega > 5\}. \\ &= \{6, 7, 8, 9, 10\}. \end{aligned}$$

Another event E_2 consists of ranks (outcomes) which are even integers:

$$\begin{aligned} E_2 &= \{\omega : \omega \text{ is an even integer}\} \\ &= \{2, 4, 6, 8, 10\}. \end{aligned}$$

The sure event Ω is $\{1, 2, 3, 4, \ldots, 10\}$. Alterntively, Ω can be described as $\{\omega : \omega$ is either an even integer or an odd integer$\}$. The null event $\varnothing$ can be described as $\{\omega : \omega$ is both an even and an odd integer$\}$ or as $\{\omega : \omega$ is less than 5 and is greater than 5$\}$. ◆

EXAMPLE 2.2

Three toys A, B, C are placed in the playpen of an infant. The interest of the experimenter is in the *order* with which the infant picks up and plays with the toys. The possible outcomes of this random experiment are

$$\begin{matrix} (A, B, C), & (B, A, C), & (C, A, B), \\ (A, C, B), & (B, C, A), & (C, B, A). \end{matrix}$$

Note the convenient shorthand notation that we have used to give the order—the first toy listed is the first chosen by the infant, the second listed is the second toy chosen, and so on.

Whenever a list is given in parentheses, it is always understood that the entries are being given in a definite order. In contrast, a list given in braces [such as the list of outcomes contained in an event; see equation (2.1)] is simply a collection, with no order specified. For example, when we list the outcomes in the event E_1 defined by the statement "toy B was chosen first by the infant," we write

$$E_1 = \{(B, A, C), (B, C, A)\}.$$

The event E_1 is a collection of two outcomes (B, A, C) and (B, C, A) and the braces tell us that no ordering of the outcomes is intended. Hence it is also the case that

$$E_1 = \{(B, C, A), (B, A, C)\}.$$

On the other hand, (B, A, C) is not the same outcome as (B, C, A) because the parentheses indicate that the order in which the toys are listed is important. A second event of possible interest is the event E_2: "the infant plays with toy A before toy C." Here

$$E_2 = \{(A, B, C), (A, C, B), (B, A, C)\}.$$ ◆

Inclusion of Events

An **event** E_1 *is said to* **include** *another event* E_2 *if every outcome belonging to* E_2 *also belongs to* E_1. This relation is illustrated in Exhibit 2.1 [called a **Venn diagram** in honor of the English logician John Venn (1834–1923)]. In the figure, the dots represent outcomes ω, the event E_1 is represented by the larger oval, and the event E_2 is represented by the smaller oval (contained in E_1). Notice that every dot (outcome) contained in E_2 is also contained in E_1.

If the event E_1 includes the event E_2, we write $E_2 \subset E_1$ or $E_1 \supset E_2$. If two events E_1 and E_2 each include the other, these events must have the same outcomes as members, and so are *equal* to one another ($E_1 = E_2$). Thus

(2.2) $$E_1 = E_2 \quad \text{if and only if} \quad E_1 \supset E_2 \text{ and } E_2 \supset E_1.$$

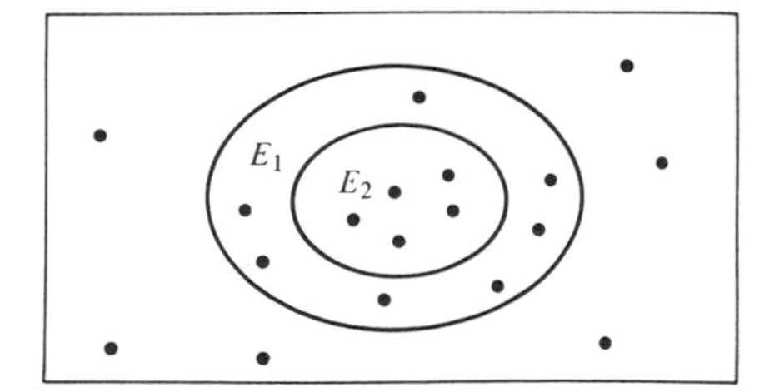

EXHIBIT 2.1 *Venn diagram illustrating inclusion.*

The Complement of an Event

The **complement** E^c of an event E is the event consisting of all outcomes in the sample Ω that are not members of E (see Exhibit 2.2). That is,

$$E^c = \{\omega : \omega \text{ is not a member of } E\}.$$

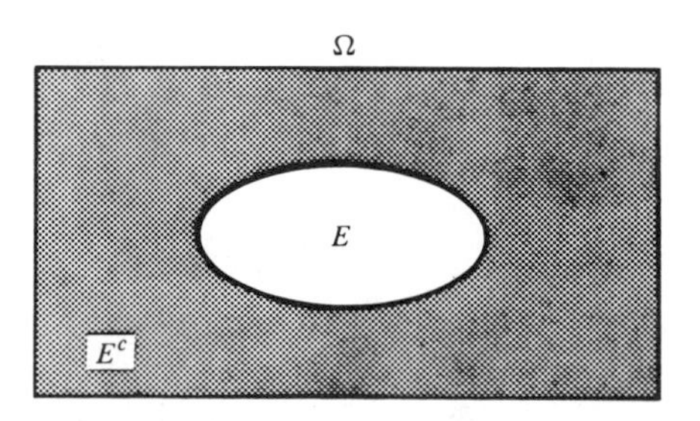

EXHIBIT 2.2 *Venn diagram illustrating the complement of an event.*

The Union of Two Events

The **union** of two events E_1 and E_2, denoted $E_1 \cup E_2$, consists of all outcomes that are members of E_1, of E_2, or of both E_1 and E_2 (represented by all shaded regions in Exhibit 2.3); that is,

$$E_1 \cup E_2 = \{(\omega : \omega \text{ is a member of } E_1, \text{ of } E_2, \text{ or of both } E_1 \text{ and } E_2\}.$$

Another way to describe the union $E_1 \cup E_2$ of two events is to say that $E_1 \cup E_2$ is the collection of all outcomes that are members of *at least one* of the two events E_1, E_2.

It follows directly from the definition of $E_1 \cup E_2$ that

$$E_1 \subset (E_1 \cup E_2), E_2 \subset (E_1 \cup E_2).$$

Note also that

$$E_1 \cup E_2 = E_2 \cup E_1,$$

as is clear from either the definition or Exhibit 2.3.

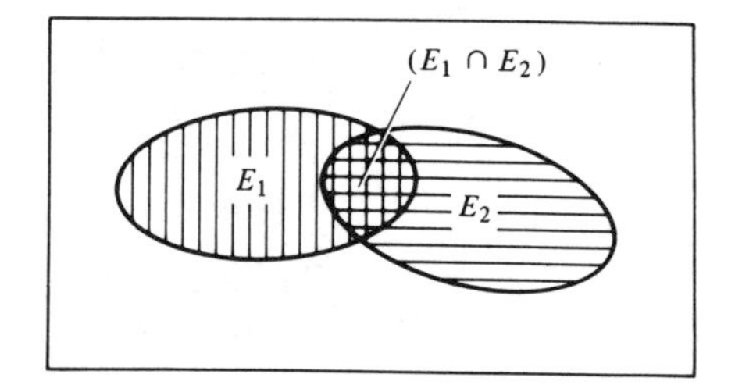

EXHIBIT 2.3 *Venn diagram illustrating the union and intersection of two events.*

The Intersection of Two Events

The **intersection** of two events E_1 and E_2, denoted $E_1 \cap E_2$, consists of outcomes that are members of *both* E_1 *and* E_2 (see Exhibit 2.3); that is,

$$E_1 \cap E_2 = \{\omega : \omega \text{ is a member of } E_1 \textit{ and } \text{of } E_2\}.$$

It follows from the definition of $E_1 \cap E_2$ that

$$(E_1 \cap E_2) \subset E_1, \qquad (E_1 \cap E_2) \subset E_2,$$

and also

$$E_1 \cap E_2 = E_2 \cap E_1.$$

We can describe the intersection $E_1 \cap E_2$ as the collection of all outcomes that are in *exactly two* of the events E_1, E_2.

EXAMPLE 2.1 *(continued)*

In this example, a subject is asked to assign one of the ranks 1 to 10 to an object. The operations of union and intersection on the events $E_1 = \{\omega : \omega \text{ exceeds } 5\}$ and $E_2 = \{\omega : \omega \text{ is even}\}$ yield

$$\begin{aligned} E_1 \cup E_2 &= \{\omega : \omega \text{ exceeds 5 } \textit{or } \omega \text{ is even}\} \\ &= \{2, 4, 6, 7, 8, 9, 10\}, \\ E_1 \cap E_2 &= \{\omega : \omega \text{ exceeds 5 } \textit{and } \omega \text{ is even}\} \\ &= \{6, 8, 10\}. \end{aligned}$$

The complement of E_1 is

$$\begin{aligned} E_1^c &= \{\omega : \omega \text{ does not exceed } 5\} = \{\omega : \omega \leq 5\} \\ &= \{1, 2, 3, 4, 5\}, \end{aligned}$$

whereas the complement of E_2 is

$$E_2^c = \{\omega : \omega \text{ is not even}\} = \{\omega : \omega \text{ is odd}\} = \{1, 3, 5, 7, 9\}. \quad ◆$$

Translating from Words to Set Notation and Back

Events of interest in random experiments are usually described in words or phrases. The translation of these descriptions into the notation used for events in our mathematical model sometimes requires considerable care. There is a direct relationship between the logical connectives "not," "or," "and" and the operations of complementation, union, and intersection on events. An example of such a translation may be helpful.

EXAMPLE 2.3

In a marketing survey, a subject is asked "which of the magazines A, B do you read?" Two events of interest are "the subject reads magazine A" (denoted A) and "the subject reads magazine B" (denoted B).[2]

How do we represent the event "the subject does *not* read magazine A"? The word "not" in the description of this event tells us that the complement A^c of the event A is being described. Similarly, the word "or" in the description of the event "the subject reads either magazine A *or* magazine B" tells us that this event is the union $A \cup B$ of the events A, B. Another way that the event $A \cup B$ might be described is "the subject reads *at least one* of the magazines A, B." Three ways in which the intersection $A \cap B$ of the events A, B can be described are (a)

[2] In examples, it is often helpful, for mnemonic purposes, to use other capital letters besides E (or subscripted E's) to represent events.

"the subject reads magazine A *and* magazine B," (b) "the subject reads *both* of the magazines A, B," and (3) "the subject reads *exactly two* of the magazines A, B." In statement (a), the word "and" clearly indicates that the intersection of events A, B is desired. In statements (b) and (c), we must recognize that the given statements are logically equivalent to statement (a).

A slightly more complicated statement is "the subject reads magazine A but not magazine B." The phrase "but not" is a stylistically preferred way of saying "and not" in English. Thus the event in question can be restated as "the subject reads magazine A *and* does *not* read magazine B," which is the event $A \cap B^c$.

Consider the event E: "the subject reads *neither* magazine A *nor* B." Although the word "nor" is related to the word "or," the event E is not a union. In fact, the connectives "neither-nor" convey the same meaning as "not-and-not." Thus an alternative description of E is "the subject does *not* read magazine A *and* does *not* read magazine B," from which it is apparent that E is actually the intersection $A^c \cap B^c$ of the complements A^c, B^c of the events A, B. The event $A^c \cap B^c$ can also be described by "the subject reads *none* of the magazines A, B."

Finally, the event F: "the subject reads *exactly one* of the magazines A, B" can be written in many equivalent ways in terms of the events A, B. For example, an alternative description of the event F is "the subject reads *at least one* of the magazines A, B *but not both*." This statement is translated into set notation by writing

$$F = (A \cup B) \cap (A \cap B)^c.$$

Alternatively, we can think of two events whose union is the event F:

1. Where the subject reads magazine A but not magazine B.
2. Where the subject reads magazine B but not magazine A.

Event 1 is $A \cap B^c$; similarly, event 2 is $B \cap A^c$. Thus we conclude that

$$F = (A \cap B^c) \cup (B \cap A^c) = (A \cap B^c) \cup (A^c \cap B). \tag{2.3}$$

Mutually Exclusive Events

If two events have no outcomes in common, we say that they are **mutually exclusive** or **disjoint.** The events E_1 and E_2 are mutually exclusive events if, and only if,

$$E_1 \cap E_2 = \varnothing,$$

because asserting that E_1 and E_2 have no outcomes in common is equivalent to asserting that their intersection $E_1 \cap E_2$ contains no outcomes as members, and thus is the null event.

EXAMPLE 2.4

A space satellite is equipped with two radio transmitters but only one antenna. The National Space Agency wishes to construct a probability model whose outcomes list the operating status, at a given time, of these three components. For example, one such outcome is

$$\omega = \text{(antenna working, first radio is not working, second radio working).}$$

Only when at least one radio and the antenna are working can one communicate with the satellite. Let E_1 be the event that "the antenna is not working," E_2 be the event "first radio not working" and E_3 be the event that "one can communicate with the satellite." Then E_1 and E_3 are mutually exclusive because if the antenna is not working (E_1), communication with the satellite (E_3) is impossible; thus E_1 and E_3 have no outcomes in common. On the other hand, the events E_2 and E_3 are not mutually exclusive, because they have the outcome ω given above in common. ◆

Union and Intersection of Three or More Events

We can define the union of three or more events. For example, the union $E_1 \cup E_2 \cup E_3$ of the events E_1, E_2, E_3 is the event whose outcomes each belong to at least one of the events E_1, E_2, E_3 (see Exhibit 2.4); that is,

$$E_1 \cup E_2 \cup E_3 = \{\omega : \omega \text{ is a member of } E_1, \text{ of } E_2, \textit{ or } \text{of } E_3\}.$$

In general, the union $E_1 \cup E_2 \cup \cdots \cup E_k$ of k events is the event whose outcomes belong to *at least one* of the events $E_1, E_2, \ldots, E_k$.

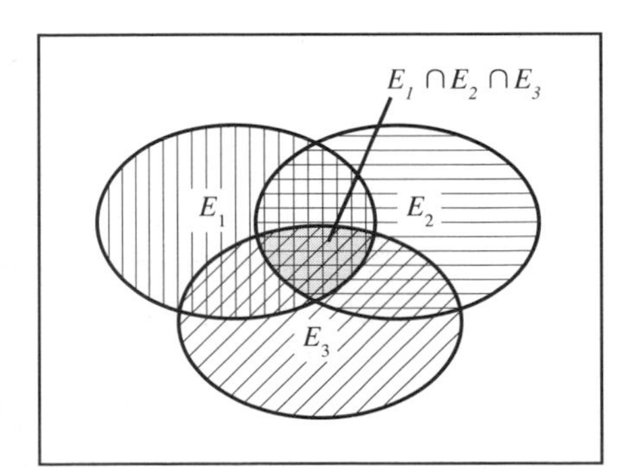

EXHIBIT 2.4 *Venn diagram illustrating the union and intersection of three events.*

Analogously, we can define the intersection of three or more events. Given E_1, E_2, E_3, their intersection $E_1 \cap E_2 \cap E_3$ is the event whose outcomes are members of all the events E_1, E_2, E_3 (see Exhibit 2.4); that is

$$E_1 \cap E_2 \cap E_3 = \{\omega : \omega \text{ is a member of } E_1, \text{ of } E_2, \textit{ and } \text{of } E_3\}.$$

In general, the intersection $E_1 \cap E_2 \cap \ldots \cap E_k$ of k events is the event whose outcomes belong to all of the events $E_1, E_2, \ldots, E_k$.

Some Useful Rules

The following useful identities are direct consequences of the definitions of union, intersection, and complementation:

Rule 2.1. $\Omega^c = \varnothing, \quad \varnothing^c = \Omega.$

Rule 2.2. For any event E:

(a) $E \cap E^c = \varnothing$. (b) $E \cup E^c = \Omega$. (c) $E \cap \varnothing = \varnothing$.
(d) $E \cup \varnothing = E$. (e) $E \cap \Omega = E$. (f) $E \cup \Omega = \Omega$.
(g) $(E^c)^c = E$.

Note that Rule 2.2(a) states that any event E and its complement E^c are mutually exclusive. This result, and the remainder of the assertions in Rules 2.1 and 2.2, are readily apparent from a Venn diagram. These results can also be established by formal logical argument (see Exercises T.1 and T.2).

It follows directly from the definitions of the union and intersection of three events that

$$\begin{aligned} E_1 \cup (E_2 \cup E_3) &= (E_1 \cup E_2) \cup E_3 = E_1 \cup E_2 \cup E_3, \\ E_1 \cap (E_2 \cap E_3) &= (E_1 \cap E_2) \cap E_3 = E_1 \cap E_2 \cap E_3. \end{aligned} \tag{2.4}$$

That is, we can form unions and intersections stepwise in any order. On the other hand, when unions and intersections appear together in the representation of an event, the order in which the operations are performed is important. For example, $E_1 \cup (E_2 \cap E_3)$ is not the same event as $(E_1 \cup E_2) \cap E_3$ (see Exhibit 2.5).

The following rule relates the operations of union and intersection to one another.

Rule 2.3. For any events E_1, E_2, E_3:

(a) $E_1 \cup (E_2 \cap E_3) = (E_1 \cup E_2) \cap (E_1 \cup E_3)$.
(b) $E_1 \cap (E_2 \cup E_3) = (E_1 \cap E_2) \cup (E_1 \cap E_3)$.

Rule 2.3 is one of the basic facts about the algebra of sets (Boolean algebra).[3] Using such rules, one can often rewrite very complex representations of events into simpler and more usable form. Computer programs that do logical analyses of very complex statements (stated in terms of complements, unions, and intersections of large numbers of events) make heavy use of the rules of Boolean algebra. However, when complements, unions, and intersections of only two or three events are under consideration, it is often simpler to refer to a Venn diagram. (Venn diagrams become unwieldy when more than three events are being combined.) Since most of the examples in this book are of sufficient simplicity that a Venn diagram can be of help to us, we will not make extensive use of Rule 2.3.

One rule of Boolean algebra that will be of frequent use to us is named for the English mathematician Augustus de Morgan (1806–1871), and relates the operation of complementation to the operations of union and intersection.

Rule 2.4 (de Morgan's Laws). For any two events E_1, E_2:

(a) $(E_1 \cup E_2)^c = E_1^c \cap E_2^c$.
(b) $(E_1 \cap E_2)^c = E_1^c \cup E_2^c$.

[3] The name Boolean algebra is in honor of the English mathematician George Boole (1815–1864), who pioneered in the area of logic.

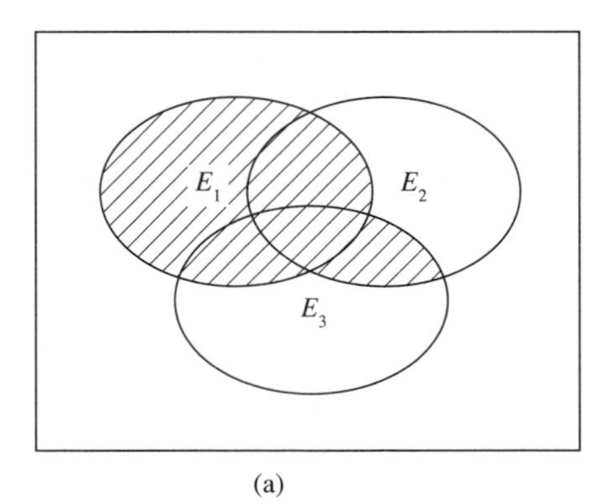

(a)

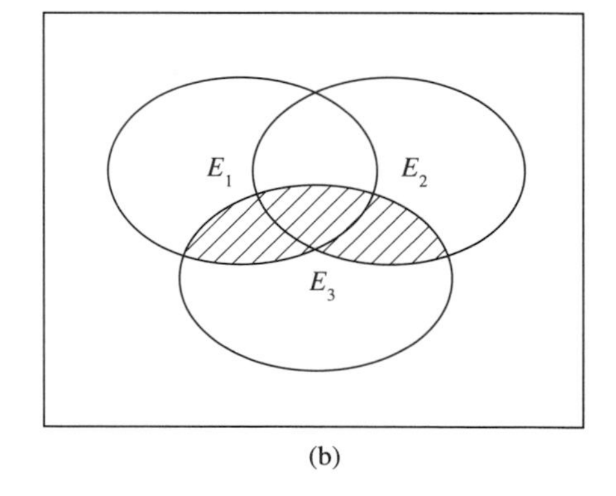

(b)

EXHIBIT 2.5 *Venn diagrams illustrating (a)* $E_1 \cup (E_2 \cap E_3)$; *(b)* $(E_1 \cup E_2) \cap E_3$.

Remember that the union $E_1 \cup E_2$ of E_1, E_2 can be described as the event "at least one of the events E_1, E_2 occur." Since not "at least one" is the same as "none," the complement $(E_1 \cup E_2)^c$ of this event is the event "none of the events E_1, E_2 occur," which we have seen previously (Example 2.4) is $E_1^c \cap E_2^c$. This argument verifies Rule 2.4(a). Similarly, $E_1^c \cup E_2^c$ is the event that "at least one of the events E_1, E_2, does not occur." This is equivalent to saying that the events E_1, E_2 do *not both* occur. Thus $E_1^c \cup E_2^c = (E_1 \cap E_2)^c$, and Rule 2.4(b) is established. [It is also possible to obtain Rule 2.4(b) as a formal consequence of Rules 2.4(a) and 2.2(g).]

More Tricks of Translation

We have noted that Venn diagrams provide a helpful way to simplify complicated expressions involving three or fewer events. Such diagrams are also useful in translating from verbal description to set notation. If we draw a Venn diagram of three events E_1, E_2, E_3, such as Exhibit 2.6, then for any verbal description of a new event E defined in terms of the original events E_1, E_2, E_3, we can mark on the diagram the pieces that satisfy the verbal description. The union of these pieces (events) is then a way of representing the new event E in set notation.

EXAMPLE 2.3 *(continued)*

Suppose that we are interested in whether or not subjects read a certain consumer magazine C (event C), as well as the two weekly news magazines A, B considered previously. Denote the event "the subject reads *exactly one* of the magazines A, B, C" by the letter G. Note that each of the events $A \cap B^c \cap C^c$, $A^c \cap B \cap C^c$, $A^c \cap B^c \cap C$ contains outcomes in which exactly one magazine is read. Thus

$$G = (A \cap B^c \cap C^c) \cup (A^c \cap B \cap C^c) \cup (A^c \cap B^c \cap C).$$

The event H: "the subject reads magazine C and exactly one of the magazines A, B" can be translated as follows. We saw earlier that

$$\begin{aligned} F &= \{\omega : \text{subject reads exactly one of magazines } A, B\} \\ &= (A \cap B^c) \cup (A^c \cap B). \end{aligned}$$

Consequently,

$$H = C \cap F = C \cap [(A \cap B^c) \cup (A^c \cap B)],$$

which from Rule 2.3b is

$$H = (A \cap B^c \cap C) \cup (A^c \cap B \cap C).$$

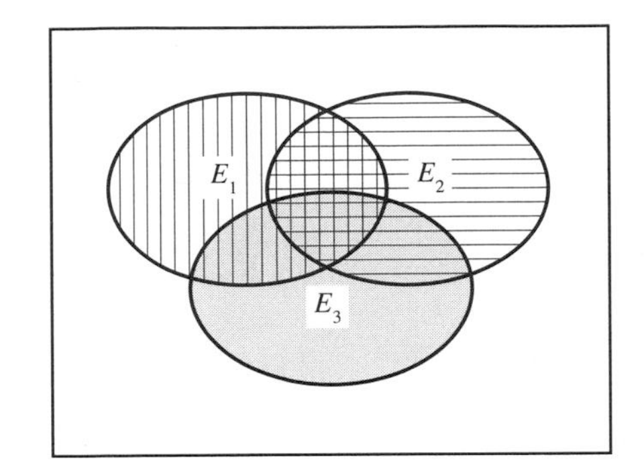

EXHIBIT 2.6 *Venn diagram of three events.*

Note that the two representations for H just obtained could also be obtained as a consequence of Rule 2.3(b) and equation (2.4); that is,

$$\begin{aligned} C \cap [(A \cap B^c) \cup (A^c \cap B)] &= [C \cap (A \cap B^c)] \cup [C \cap (A^c \cap B)] \\ &= (A \cap B^c \cap C) \cup (A^c \cap B \cap C). \end{aligned} \quad \blacklozenge$$

Inductive Derivations★

Rules 2.3 and 2.4 and equation (2.4) can be generalized to cover any number of events $E_1, E_2, \ldots, E_k$. For example, it follows from the definitions of union and intersection for k events that

$$E_1 \cup E_2 \cup \cdots \cup E_k = E_1 \cup (E_2 \cup \cdots \cup E_k) \tag{2.5}$$

and

$$E_1 \cap E_2 \cap \cdots \cap E_k = E_1 \cap (E_2 \cap \cdots \cap E_k). \tag{2.6}$$

Rule 2.3(a) has the following generalization: For any events $E, E_1, E_2, \ldots, E_k$,

$$E \cup (E_1 \cap E_2 \cap \cdots \cap E_k) = (E \cup E_1) \cap (E \cup E_2) \cap \cdots \cap (E \cup E_k). \tag{2.7}$$

We can verify equation (2.7) by making repeated use of Rule 2.3(a) and equation (2.5):

$$\begin{aligned} &E \cup (E_1 \cap E_2 \cap \cdots \cap E_k) \\ &\quad = E \cup [E_1 \cap (E_2 \cap \cdots \cap E_k)] \\ &\quad = (E \cup E_1) \cap [E \cup (E_2 \cap \cdots \cap E_k)] \\ &\quad = (E \cup E_1) \cap (E \cup E_2) \cap [E \cup (E_3 \cap \cdots \cap E_k)], \end{aligned}$$

and so on, proceeding as shown above until (2.7) is obtained. This same kind of **recursive argument** can be used to obtain the generalization

$$E \cap (E_1 \cup \cdots \cup E_k) = (E \cap E_1) \cup (E \cap E_2) \cup \cdots \cup (E \cap E_k) \tag{2.8}$$

of Rule 2.3(b) and also the generalizations

$$\begin{aligned} (E_1 \cup E_2 \cup \cdots \cup E_k)^c &= E_1^c \cap E_2^c \cap \cdots \cap E_k^c, \\ (E_1 \cap E_2 \cap \cdots \cap E_k)^c &= E_1^c \cup E_2^c \cup \cdots \cup E_k^c \end{aligned} \tag{2.9}$$

of de Morgan's laws.

3 Formal Structure: Probabilities

We have assumed that in a large number of repeated trials of our random experiment the relative frequency of every event stabilizes to a certain number, the long-run relative frequency of that event. Correspondingly, we assume that for every event E in our model, there exists a number $P(E)$, called the **probability** of the event E, that is the idealization of the long-run relative frequency of E. In order to make the analogy between the concepts of probability and relative frequency useful, we would like the probability of an event to have all the important properties of relative frequencies.

Recall that the relative frequency r.f.(E) of an event E in N trials is $\#E/N$, where $\#E$ is the number of times that the event E occurs in the N trials. Because $\#E$ and N are nonnegative numbers, and $\#E$ cannot exceed N, we must have $0 \leq \text{r.f.}(E) \leq 1$. We have already remarked that the sure event Ω is certain to occur (has relative frequency 1), and that the impossible event $\varnothing$ has relative frequency 0. Finally, if two events E_1 and E_2 cannot happen together on the same trial (they have no outcomes in common, they are mutually exclusive), then

$$\#(E_1 \text{ or } E_2) = \#E_1 + \#E_2$$

and it follows that

$$\text{r.f.}(E_1 \cup E_2) = \text{r.f.}(E_1) + \text{r.f.}(E_2).$$

The three properties just described are basic properties obeyed by relative frequencies. Corresponding to these basic properties, we require probabilities of events to obey the following axioms.

> ***AXIOM 1.*** For every event E, $0 \le P(E) \le 1$.
>
> ***AXIOM 2.*** $P(\Omega) = 1$, $P(\varnothing) = 0$.
>
> ***AXIOM 3.*** If E_1 and E_2 are mutually exclusive events, then
>
> $$P(E_1 \cup E_2) = P(E_1) + P(E_2).$$

The null event $\varnothing$ need not be the only event that has probability 0. Certain events may be so rare that they "happen only once in a lifetime." The relative frequencies of such events in a large number of trials will be almost indistinguishable from 0. The conceptual importance of this fact will become apparent when we discuss random variables in Chapter 5.

4 Probability Calculations Using Venn Diagrams

A Venn diagram represents events as regions. The area of any such region cannot be greater than the area of the entire Venn diagram nor less than zero. Further, when two regions A, B are nonoverlapping (Exhibit 4.1), the union $A \cup B$ of these regions has area equal to the sum of the areas of regions A and B. Hence, if we assign an area of 1 to the entire Venn diagram (the box marked Ω in Exhibit 4.1), the areas of regions in a Venn diagram satisfy Axioms 1, 2, 3 for probabilities.

Because both probabilities of events and areas of regions in a Venn diagram obey the same axioms, consequences of those axioms can also be represented by Venn diagrams. In particular, we regard each region in a Venn diagram as having area equal to the probability of the event which that

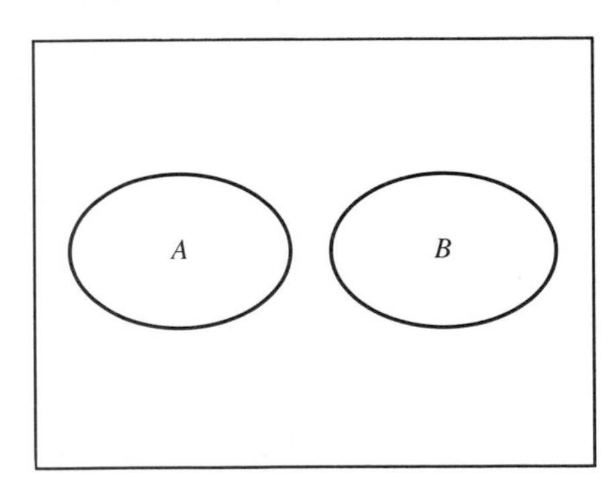

EXHIBIT 4.1 *Venn diagram with nonoverlapping regions.*

region represents. (The Venn diagram need not be drawn so that regions have the correct areas—the concept is enough.) We proceed as follows:

1. Identify regions on the Venn diagram that correspond to events whose probabilities are known and assign to these regions areas equal to the given probabilities.
2. Find the region that corresponds to the event E whose probability is desired. Use the Venn diagram to see how to obtain the area of this region from the given areas. The area of the region corresponding to event E is then the probability $P(E)$ of E.

EXAMPLE 4.1

Two events E_1, E_2 are known to have the following probabilities: $P(E_1) = 0.6$, $P(E_2) = 0.4$, $P(E_1 \cap E_2) = 0.1$. What is the probability of the event $E_1 \cap E_2^c$?

Because the intersection $E_1 \cap E_2$ of the events E_1, E_2 has a probability greater than 0, the Venn diagram used for this problem must represent the events E_1, E_2 as overlapping regions, as in Exhibit 4.2. Observe from Exhibit 4.2 that the area of the region marked $E_1 \cap E_2^c$ equals the difference between the area of the region E_1 and the area of the region $E_1 \cap E_2$. Thus

$$\begin{aligned} P(E_1 \cap E_2^c) &= \text{area of } E_1 \cap E_2^c \\ &= (\text{area of } E_1) - (\text{area of } E_1 \cap E_2) \\ &= 0.6 - 0.1 = 0.5. \end{aligned}$$ ◆

The argument used in Example 4.1 yields the formula

$$(4.1) \qquad P(E_1 \cap E_2^c) = P(E_1) - P(E_1 \cap E_2),$$

which is true for any two events E_1, E_2. If $E_1 = \Omega$, $E_2 = E$, it follows from Rule 2.2(e) that $E_1 \cap E_2^c = \Omega \cap E^c = E^c$ and $E_1 \cap E_2 = \Omega \cap E = E$.

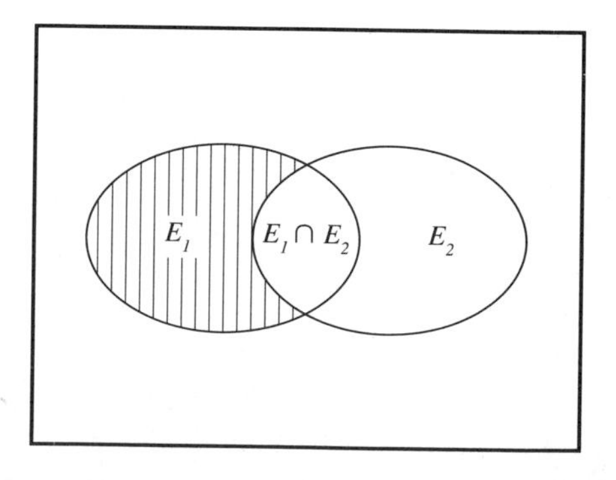

EXHIBIT 4.2 *Venn diagram for Example 4.1.*

Substitution of these facts into equation (4.1) yields the following well-known result:

> **Law of Complementation.**[4] For any event E, $P(E^c) = 1 - P(E)$.

EXAMPLE 4.1 *(continued)*

To find the probability of the union $E_1 \cup E_2$ of the events E_1, E_2, we see from Exhibit 4.2 that the area of the region $E_1 \cup E_2$ is not the sum of the areas of the regions E_1, E_2. Such a sum adds the area of the overlapped region $E_1 \cap E_2$ twice, whereas it should only be added once. However, if we subtract the area of the overlap $E_1 \cap E_2$ from the sum of the areas of E_1, E_2 we correct for this error. Consequently, we have shown that

$$P(E_1 \cup E_2) = P(E_1) + P(E_2) - P(E_1 \cap E_2) = 0.6 + 0.4 - 0.1 = 0.9.$$

Alternatively, Exhibit 4.2 reveals that the area of the region $E_1 \cup E_2$ is the sum of the areas of the three nonoverlapping regions $E_1 \cap E_2^c$, $E_1 \cap E_2$, $E_1^c \cap E_2$. Although we do not know the area of $E_1^c \cap E_2$, we can find this area by subtracting the area of $E_1 \cap E_2$ from the area of E_2. Thus the area of $E_1^c \cap E_2$ is $0.4 - 0.1 = 0.3$, and because we earlier found that $P(E_1 \cap E_2^c) = 0.5$,

$$\begin{aligned} P(E_1 \cup E_2) &= P(E_1 \cap E_2^c) + P(E_1 \cap E_2) + P(E_1^c \cap E_2) \\ &= 0.5 + 0.1 + 0.3 = 0.9, \end{aligned}$$

which agrees with the answer for $P(E_1 \cup E_2)$ already obtained.

To find $P(E_1^c \cap E_2^c)$ note from Exhibit 4.2 that the area of $E_1^c \cap E_2^c$ is the difference between the area of the whole Venn diagram Ω and the area of the union $E_1 \cup E_2$ of the regions E_1, E_2. Because the Venn diagram has area 1, and the union $E_1 \cup E_2$ has area 0.9, the region $E_1^c \cap E_2^c$ has area $1 - 0.9 = 0.1$. That is,

$$P(E_1^c \cap E_2^c) = 1 - P(E_1 \cup E_2) = 1 - 0.9 = 0.1.$$

Actually, this result is just an application of the law of complementation, because by de Morgan's law, $E_1^c \cap E_2^c = (E_1 \cup E_2)^c$ and thus

$$P(E_1^c \cap E_2^c) = P((E_1 \cup E_2)^c) = 1 - P(E_1 \cup E_2) = 1 - 0.9 = 0.1. \quad \blacklozenge$$

[4] A formal proof of this result and equation (4.1) is given in Section 5.

Once again, the arguments in this particular example, when applied generally, yield useful formulas.[5] For example, we have demonstrated the following useful result:

> **General Law for Addition of Two Events.** For any two events E_1, E_2,
>
> $$P(E_1 \cup E_2) = P(E_1) + P(E_2) - P(E_1 \cap E_2).$$

Notice that this general law of addition includes Axiom 3 for probabilities as a special case. That is, when the events E_1, E_2 are mutually exclusive, their intersection $E_1 \cap E_2$ is the impossible event $\varnothing$ (which has probability zero) and

$$\begin{aligned} P(E_1 \cup E_2) &= P(E_1) + P(E_2) - P(E_1 \cap E_2) = P(E_1) + P(E_2) - P(\phi) \\ &= P(E_1) + P(E_2) - 0 = P(E_1) + P(E_2), \end{aligned}$$

as is asserted by Axiom 3.

Because the area of the union of nonoverlapping regions is the sum of the areas of these regions, we obtain the following (corresponding) fact about probabilities:

> **Special Law of Addition.** If $E_1, E_2, \ldots, E_k$ are k events, no two of which have outcomes in common (i.e., all pairs of events E_i, E_j are mutually exclusive), then
>
> $$\begin{aligned} P(E_1 \cup E_2 \cup \cdots \cup E_k) &= P(E_1) + P(E_2) + \cdots P(E_k) \\ &= \sum_{i=1}^{k} P(E_i). \end{aligned}$$

The Special Law of Addition extends Axiom 3 (which is the case $k = 2$ of this law) to cover the union of more than two mutually exclusive events. This law has numerous applications. [We made use of this law in finding $P(E_1 \cup E_2)$ in Example 4.1.] When we apply this law, we must always check that the events whose union we want are indeed mutually exclusive.

EXAMPLE 4.2

Police statistics in a certain city indicate that of all cars stolen, 25% are less than 1 year old, 20% are at least 1 year old but less than 2 years old,

[5] Formal proofs of the following general and special laws are given in Section 5.

20% are between 2 and 3 years old, 15% are between 3 and 5 years old, 11% are between 5 and 10 years old, and 9% are 10 years old or older. What percentage of stolen cars are 2 years old or older?

The percentages given us can be converted to proportions by dividing by 100. Thus the proportion of stolen cars less than 1 year old is 0.25. These proportions in turn can be regarded as being relative frequencies obtained in a random experiment in which the age of a stolen car is observed. Because relative frequencies and probabilities obey the same axioms, the question asked can be treated as if it were a probability problem in which we are given the probabilities of six events, as listed in the following table:

Age of car (years)	Less than 1	1–2	2–3	3–5	5–10	10 or more
Event	E_1	E_2	E_3	E_4	E_5	E_6
Probability	0.25	0.20	0.20	0.15	0.11	0.09

The event of interest is the union

$$E = E_3 \cup E_4 \cup E_5 \cup E_6$$

of the four events E_3, E_4, E_5, E_6. From their definitions, we see that no two of these events have outcomes in common. Hence, by the Special Law of Addition,

$$\begin{aligned} P\{\text{stolen car is 2 years old or older}\} &= P(E_3 \cup E_4 \cup E_5 \cup E_6) \\ &= P(E_3) + P(E_4) + P(E_5) + P(E_6) \\ &= 0.20 + 0.15 + 0.11 + 0.09 \\ &= 0.55. \end{aligned}$$

Converting back to percentages, we conclude that 100(0.55) = 55% of all stolen cars are 2 years old or older. ◆

It would be helpful to have a formula to calculate the probability of the union of k events that have outcomes in common (are not mutually exclusive). We can find such a law for $k = 3$ events with the aid of a Venn diagram. We illustrate this with an example.

EXAMPLE 4.3 Suppose that three events A, B, C have probabilities:

$$\begin{array}{lll} P(A) = 0.5, & P(A \cap B) = 0.2, & P(A \cap B \cap C) = 0.01. \\ P(B) = 0.3, & P(A \cap C) = 0.1, & \\ P(C) = 0.4, & P(B \cap C) = 0.1, & \end{array}$$

We wish to find the probability of the union $A \cup B \cup C$ of these events.

From Exhibit 4.3 we see that the sum $P(A) + P(B) + P(C)$ does not give the correct answer, because the areas of the regions $A \cap B \cap C^c$, $A \cap B^c \cap C$, $A^c \cap B \cap C$ are counted twice and the area of the region $A \cap B \cap C$ is counted three times. If we subtract $P(A \cap B)$, $P(A \cap C)$, and $P(B \cap C)$ from this sum, we correct for the overcount of the regions $A \cap B \cap C^c$, $A \cap B^c \cap C$, $A^c \cap B \cap C$, but overcorrect for the three-way region $A \cap B \cap C$. Adding back $P(A \cap B \cap A)$ corrects for this. Consequently, we arrive at the result

$$\begin{aligned} P(A \cup B \cup C) &= P(A) + P(B) + P(C) - P(A \cap B) - P(A \cap C) \\ &\quad - P(B \cap C) + P(A \cap B \cap C) \\ &= 0.5 + 0.3 + 0.4 - 0.2 - 0.1 - 0.1 + 0.01 = 0.81. \end{aligned}$$

Alternatively, we can use the Special Law of Addition to obtain

$$\begin{aligned} P(A \cup B \cup C) &= P(A \cap B^c \cap C^c) + P(A^c \cap B \cap C^c) \\ &\quad + P(A^c \cap B^c \cap C) + P(A \cap B \cap C^c) \\ &\quad + P(A \cap B^c \cap C) + P(A^c \cap B \cap C) \\ &\quad + P(A \cap B \cap C), \end{aligned} \tag{4.2}$$

because $A \cup B \cup C$ is the union of the seven events whose probabilities are added in equation (4.2), and no two of these seven events have outcomes in common. We now need to find the probabilities of these seven events. From Exhibit 4.3,

$$\begin{aligned} P(A \cap B) &= P((A \cap B \cap C) \cup (A \cap B \cap C^c)) \\ &= P(A \cap B \cap C) + P(A \cap B \cap C^c), \end{aligned}$$

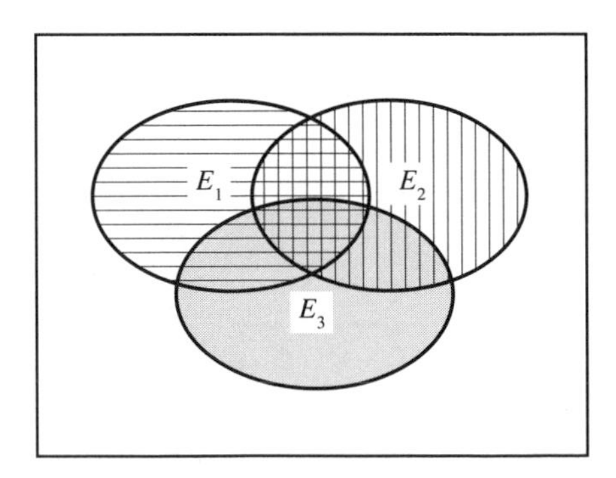

EXHIBIT 4.3 *Venn diagram relating probabilities for three events.*

because the regions $A \cap B \cap C$ and $A \cap B \cap C^c$ are nonoverlapping. Thus

$$P(A \cap B \cap C^c) = P(A \cap B) - P(A \cap B \cap C) = 0.2 - 0.01 = 0.19.$$

In similar fashion,

$$P(A \cap B^c \cap C) = P(A \cap C) - P(A \cap B \cap C) = 0.1 - 0.01 = 0.09,$$
$$P(A^c \cap B \cap C) = P(B \cap C) - P(A \cap B \cap C) = 0.1 - 0.01 = 0.09.$$

Next, note from Exhibit 4.3 that

$$P(A) = P(A \cap B^c \cap C^c) + P(A \cap B \cap C^c) + P(A \cap B^c \cap C) + P(A \cap B \cap C),$$

so that

$$\begin{aligned} P(A \cap B^c \cap C^c) &= P(A) - P(A \cap B \cap C^c) - P(A \cap B^c \cap C) - P(A \cap B \cap C) \\ &= 0.5 - 0.19 - 0.09 - 0.01 = 0.21. \end{aligned}$$

Similarly,

$$P(A^c \cap B \cap C^c) = 0.3 - 0.19 - 0.09 - 0.01 = 0.01,$$
$$P(A^c \cap B^c \cap C) = 0.4 - 0.09 - 0.09 - 0.01 = 0.21.$$

Substituting these results into equation (4.2) yields

$$\begin{aligned} P(A \cup B \cup C) &= 0.21 + 0.01 + 0.21 + 0.19 + 0.09 + 0.09 + 0.01 \\ &= 0.81, \end{aligned}$$

which is the same result obtained previously. Although these calculations were more cumbersome, we have learned methods to obtain probabilities of events such as $A \cap B \cap C^c$ and $A \cap B^c \cap C^c$.

Indeed, consider the event E defined by "exactly one of the events A, B, C occurs." Arguing as in Example 2.3, E is the union of cases where one of the events A, B, C occurs and the other two do not; that is,

$$E = (A \cap B^c \cap C^c) \cup (A^c \cap B \cap C^c) \cup (A^c \cap B^c \cap C).$$

From the Special Law of Addition

$$\begin{aligned} P(E) &= P(A \cap B^c \cap C^c) + P(A^c \cap B \cap C^c) + P(A^c \cap B^c \cap C) \\ &= 0.21 + 0.01 + 0.21 = 0.43. \end{aligned}$$

Similarly, the probability of the event F that "exactly two of the events A, B, C occur" is

$$\begin{aligned} P(F) &= P(A \cap B \cap C^c) + P(A \cap B^c \cap C) + P(A^c \cap B \cap C) \\ &= 0.19 + 0.09 + 0.09 = 0.37. \end{aligned}$$

For completeness, note that the event "exactly three of the events A, B, C occur" is $A \cap B \cap C$, which has probability 0.01. Also, the event "none of the events A, B, C occur" is $A^c \cap B^c \cap C^c$, which by the generalization of de Morgan's law is the complement of the event $A \cup B \cup C$. Thus

$$\begin{aligned} P(A^c \cap B^c \cap C^c) &= P((A \cup B \cup C)^c) = 1 - P(A \cup B \cup C) \\ &= 1 - 0.81 = 0.19. \end{aligned}$$ ♦

The arguments used to find the probability of the union of three events in Example 4.3 apply in general. In particular, we have the following useful result.

General Law of Addition for Three Events.[6] For any three events E_1, E_2, E_3,

$$\begin{aligned} P(E_1 \cup E_2 \cup E_3) = P(E_1) + P(E_2) + P(E_3) - P(E_1 \cap E_2) \\ - P(E_1 \cap E_3) - P(E_2 \cap E_3) + P(E_1 \cap E_2 \cap E_3). \end{aligned}$$

As noted in Section 2, Venn diagrams are unwieldy for displaying more than three overlapping (not mutually exclusive) events. This limits the use of Venn diagrams in probability calculations, and has forced the development of a proability calculus to handle problems too complicated for the use of Venn diagrams. This probability calculus includes some of the general formulas obtained in this section, plus various generalizations. Many of these generalizations are obtained recursively from simple cases. For example, from equation (4.1),

$$P(E_1 \cap E_2^c) = P(E_1) - P(E_1 \cap E_2).$$

If we let $E_1 = A \cap B$, $E_2 = C$, then

$$P(A \cap B \cap C^c) = P(A \cap B) - P(A \cap B \cap C),$$

[6] A formal proof of this result is given in Section 5.

a result that could have been used in Example 4.3. We can also let $E_1 = A \cap B \cap C$, $E_2 = D$, and obtain

$$P(A \cap B \cap C \cap D^c) = P(A \cap B \cap C) - P(A \cap B \cap C \cap D),$$

and so on.

Also, in laws that hold for two or three events, we sometimes can find a pattern that allows us to extend these laws to more than three events. For example, the idea of correcting for overlap used to obtain the general laws of addition for two and three events suggests the following generalization to four events:

$$\begin{aligned} P(E_1 \cup E_2 \cup E_3 \cup E_4) = &\sum_{i=1}^{4} P(E_i) - \sum_{i<j} P(E_i \cap E_j) \\ &+ \sum_{i<j<k} P(E_i \cap E_j \cap E_k) \\ &- P(E_1 \cap E_2 \cap E_3 \cap E_4). \end{aligned} \tag{4.3}$$

Experience shows that arguments from diagrams can be erroneous. For example, one may draw a diagram that is not sufficiently general and thus obtain a result that is not always true. Consequently, it may be preferable to derive results of the probability calculus as rigorous logical consequences of the axioms of probability. In Section 5 we give logical derivations for some key formulas obtained in this section in order to illustrate the kinds of arguments commonly used.

In deference to mathematical rigor, we note that Axioms 1, 2, 3 are not strong enough to support many of the calculations and models used later in this book. For example, suppose that we want to consider a probability model in which the outcomes ω are positive integers 1, 2, ... If we want to determine the probability of the event $\{\omega$ is an even integer$\}$ and we know the probabilities of the mutually exclusive events $E_i = \{\omega = i\}$, $i = 1, 2, 3, \ldots$, it seems reasonable that

$$\begin{aligned} P\{\omega \text{ is an even integer}\} &= P(\omega = 2 \text{ or } 4 \text{ or } 6 \text{ or } 8 \text{ or } \cdots) \\ &= P(E_2 \cup E_4 \cup E_6 \cup E_8 \cup \cdots) \end{aligned}$$

should be equal to the sum $\sum_{i=1}^{\infty} P(E_{2i})$ of the probabilities of the events E_2, E_4, E_6, E_8, ... Unfortunately, this result cannot be proved using Axioms 1, 2, and 3. In particular, the Special Law of Addition applies only to a union of a *finite* number k of mutually exclusive events, whereas the event $\{\omega$ is an even integer$\}$ is a union of an infinite number of events E_i.

A mathematically correct and completely general theory of probability requires that we replace Axiom 3 by the following:

AXIOM 3′. For any *countable* collection $E_1, E_2, E_3, \ldots$ of events, no two of which have outcomes in common,

$$P(E_1 \cup E_2 \cup E_3 \cup \cdots) = \sum_{i=1}^{\infty} P(E_i).$$

This axiom corresponds to dropping the requirement in the Special Law of Addition that k be finite.

5 Formal Proofs

In this section we derive some of the basic results of Section 4 as logical consequences of Axioms 1, 2 and 3.

Derivation of Equation (4.1) and the Law of Complementation

We start by noting that *for any two events* E_1, E_2,

(5.1) $$E_1 \cap E_2 \text{ and } E_1 \cap E_2^c \text{ are mutually exclusive.}$$

To see this, observe that by properties of the operation of intersection, and Rules 2.2(a), 2.2(c),

$$\begin{aligned}(E_1 \cap E_2) \cap (E_1 \cap E_2^c) &= E_1 \cap E_2 \cap E_1 \cap E_2^c = (E_1 \cap E_1) \cap (E_2 \cap E_2^c),\\ &= E_1 \cap \varnothing = \varnothing,\end{aligned}$$

so that no outcomes are common to $E_1 \cap E_2$ and $E_1 \cap E_2^c$.

Further, by Rules 2.2(b), 2.2(e), and 2.3(a),

(5.2) $$E_1 = E_1 \cap \Omega = E_1 \cap (E_2 \cup E_2^c) = (E_1 \cap E_2) \cup (E_1 \cap E_2^c).$$

Thus, by Axiom 3,

(5.3) $$P(E_1) = P((E_1 \cap E_2) \cup (E_1 \cap E_2^c)) = P(E_1 \cap E_2) + P(E_1 \cap E_2^c).$$

The formula (4.1) now follows by subtracting $P(E_1 \cap E_2)$ from both sides of equation (5.3). As already noted in Section 4, the Law of Complementation is the special case of equation (4.1) when $E_1 = \Omega$, $E_2 = E$.

Derivation of the Special Law of Addition

The Special Law of Addition can be derived recursively or by mathematical induction (see "Inductive Derivations" at the end of Section 2). The first step in either derivation is to note that the case $k = 2$ of this law is Axiom 3. A proof by recursion would then derive the law for $k = 3$ events from the law for $k = 2$ events, the law for $k = 4$ events from the law for $k = 3$ events, and so on. Mathematical induction requires a demonstration that the truth of the law for k events implies the truth of the law for $k + 1$ events, $k = 2, 3, 4, \ldots$. Using the latter approach, suppose that the Special Law of Addition holds for any k mutually exclusive events $E_1, E_2, \ldots, E_k$, and that E_{k+1} is still another event having no outcomes in common with $E_1, E_2, \ldots, E_k$. Consequently, by (2.8),

$$\begin{aligned} &E_{k+1} \cap (E_1 \cup E_2 \cup \cdots \cup E_k) \\ &\quad = (E_{k+1} \cap E_1) \cup (E_{k+1} \cap E_2) \cup \cdots \cup (E_{k+1} \cap E_k) \\ &\quad = \varnothing \cup \varnothing \cup \cdots \cup \varnothing = \varnothing, \end{aligned}$$

so that E_{k+1} and $(E_1 \cup E_2 \cup \cdots \cup E_k)$ have no outcomes in common. By (2.5) and Axiom 3,

$$\begin{aligned} P(E_1 \cup E_2 \cup \cdots \cup E_k \cup E_{k+1}) &= P((E_1 \cup E_2 \cup \cdots \cup E_k) \cup E_{k+1}) \\ &= P(E_1 \cup E_2 \cup \cdots \cup E_k) + P(E_{k+1}) \\ &= \sum_{i=1}^{k} P(E_i) + P(E_{k+1}), \end{aligned}$$

where the third equality follows from the induction hypothesis. Consequently, the truth of the Special Law of Addition has been verified for $k + 1$ events, and the induction proof is completed.

Derivation of the General Law of Addition for Two Events

From equation (5.3),

$$P(E_1) = P(E_1 \cap E_2) + P(E_1 \cap E_2^c). \tag{5.4}$$

Reversing the roles of E_1 and E_2 in (5.2) yields

$$P(E_2) = P(E_2 \cap E_1) + P(E_2 \cap E_1^c) = P(E_1 \cap E_2) + P(E_1^c \cap E_2). \tag{5.5}$$

Thus from (5.4) and (5.5),

(5.6)

$$\begin{aligned} &P(E_1) + P(E_2) - P(E_1 \cap E_2) \\ &\quad = [P(E_1 \cap E_2) + P(E_1 \cap E_2^c)] + [P(E_1 \cap E_2) + P(E_1^c \cap E_2)] \\ &\qquad - P(E_1 \cap E_2) \\ &\quad = P(E_1 \cap E_2) + P(E_1 \cap E_2^c) + P(E_1^c \cap E_2). \end{aligned}$$

From Exercise T.1(e),

$$E_1 \cup E_2 = E_1 \cup (E_1^c \cap E_2).$$

Further, E_1 and $(E_1^c \cap E_2)$ are mutually exclusive events. Thus by Axiom 3, (5.4), and (5.6),

$$\begin{aligned} P(E_1 \cup E_2) &= P(E_1) + P(E_1^c \cap E_2) \\ &= P(E_1 \cap E_2) + P(E_1 \cap E_2^c) + P(E_1^c \cap E_2) \\ &= P(E_1) + P(E_2) - P(E_1 \cap E_2), \end{aligned}$$

which completes the argument.

Proof of the General Law of Addition for Three Events

By (2.4) and the General Law of Addition for Two Events (with $E_2 \cap E_3$ substituted for E_2),

(5.7)
$$\begin{aligned} P(E_1 \cup E_2 \cup E_3) &= P(E_1 \cup (E_2 \cup E_3)) \\ &= P(E_1) + P(E_2 \cup E_3) - P(E_1 \cap (E_2 \cup E_3)). \end{aligned}$$

However,

(5.8)
$$P(E_2 \cup E_3) = P(E_2) + P(E_3) - P(E_2 \cap E_3).$$

Also by Rule 2.3(b),

$$E_1 \cap (E_2 \cup E_3) = (E_1 \cap E_2) \cup (E_1 \cap E_3).$$

Applying the General Law of Addition one final time gives

(5.9)
$$\begin{aligned} &P(E_1 \cap (E_2 \cup E_3)) \\ &\quad = P((E_1 \cap E_2) \cup (E_1 \cap E_3)) \\ &\quad = P(E_1 \cap E_2) + P(E_1 \cap E_3) - P((E_1 \cap E_2) \cap (E_1 \cap E_3)) \\ &\quad = P(E_1 \cap E_2) + P(E_1 \cap E_2) - P(E_1 \cap E_2 \cap E_3), \end{aligned}$$

because

$$(E_1 \cap E_2) \cap (E_1 \cap E_3) = E_1 \cap E_2 \cap E_1 \cap E_3 = (E_1 \cap E_1) \cap E_2 \cap E_3 = E_1 \cap E_2 \cap E_3.$$

Substituting (5.8) and (5.9) into (5.7) yields

$$\begin{aligned} P(E_1 \cup E_2 \cup E_3) &= P(E_1) + P(E_2) + P(E_3) - P(E_2 \cap E_3) \\ &\quad - [P(E_1 \cap E_2) + P(E_1 \cap E_3) - P(E_1 \cap E_2 \cap E_3)] \\ &= P(E_1) + P(E_2) + P(E_3) - P(E_1 \cap E_2) - P(E_1 \cap E_3) \\ &\quad - P(E_2 \cap E_3) + P(E_1 \cap E_2 \cap E_3), \end{aligned}$$

which is the desired result.

6 Fitting a Probability Model

To describe a random experiment in terms of a probability model, we need to determine a rule or function that assigns a probability to every event. This function is called the **probability measure** for the model.

If the sample space Ω consists of 15 outcomes, it can be shown that there are $2^{15} = 32{,}768$ distinct events in the probability model. Thus, even in a random experiment with relatively few outcomes, the task of constructing a probability measure appears quite formidable. Fortunately, because of the formulas of the probability calculus (such as those discussed in Section 4), only probabilities for certain basic events need to be determined. In practice, such probabilities are determined in one of two different ways. We may actually take a large number of trials of the random experiment, calculate the resulting relative frequencies, and set the probabilities equal to these relative frequencies. Alternatively, we may employ theoretical arguments to specify that certain mathematical relationships are to hold between the probabilities of the basic events. We give below an example of each of these two methods for determining the probability measure of a random experiment.

EXAMPLE 6.1

Experimental Methods

Insurance companies have traditionally been concerned with records of human life spans. These are usually summarized in **life tables,** the earliest example of which appears to have been given by John Graunt (1620–1674) in 1662. The foundation of a theory of life annuities is due to the English astronomer Edmund Halley (1656–1742) in 1693. Halley's famous life table for the city of Breslau, Poland, is given in Exhibit 6.1. The outcomes are the ages at death (last birthday); the probabilities are actually relative frequencies calculated over thousands of observed life spans. ♦

EXHIBIT 6.1 *Life Table for People Living in the City of Breslau (Wroclaw), Poland, 1693*

Age	Probability	Age	Probability	Age	Probability	Age	Probability
0	0.145	21	0.007	42	0.010	63	0.010
1	0.057	22	0.006	43	0.010	64	0.010
2	0.038	23	0.006	44	0.010	65	0.010
3	0.028	24	0.007	45	0.010	66	0.010
4	0.022	25	0.007	46	0.010	67	0.010
5	0.018	26	0.007	47	0.010	68	0.010
6	0.012	27	0.007	48	0.011	69	0.011
7	0.010	28	0.008	49	0.011	70	0.011
8	0.009	29	0.008	50	0.011	71	0.011
9	0.008	30	0.008	51	0.011	72	0.011
10	0.007	31	0.008	52	0.011	73	0.010
11	0.006	32	0.008	53	0.010	74	0.010
12	0.006	33	0.009	54	0.010	75	0.010
13	0.006	34	0.009	55	0.010	76	0.010
14	0.006	35	0.009	56	0.010	77	0.009
15	0.006	36	0.009	57	0.010	78	0.008
16	0.006	37	0.009	58	0.010	79	0.007
17	0.006	38	0.009	59	0.010	80	0.006
18	0.006	39	0.009	60	0.010	81	0.005
19	0.006	40	0.009	61	0.010	≥82	0.023
20	0.006	41	0.010	62	0.010		

EXAMPLE 6.2

Theoretical Approach

Suppose that we are concerned with the number of fatal automobile accidents that occur over a typical weekend. This is a random experiment whose outcomes are nonnegative integers 0, 1, 2, ... Using certain theoretical assumptions about the nature of automobile accidents, the probability of d deaths on the weekend can be shown (Chapter 8, Section 3) to have the form

$$P\{d \text{ deaths}\} = \frac{e^{-\lambda}\lambda^d}{d!},$$

where λ is a positive number whose value depends on the rate at which accidents occur, and $d! = (d)(d-1)(d-2)\cdots(2)(1)$.

To determine whether this probability measure provides a good model for the number of fatal automobile accidents, we would take experimental data and see whether the probabilities computed from our model agree to a close enough approximation with the observed relative frequencies. ♦

Notes and References

One of the advantages of the mathematical conceptualization of a real phenomenon is that the same conceptualization may serve, with suitable translations in language, as a model for more than one reality. Probabilities were conceptualized to correspond to relative frequencies in repeated trials of a random experiment. Such probabilities are sometimes called **objective probabilities.** However, in many cases people are willing to express judgments about the truth of a statement or the outcome of some random process in terms of personal probabilities. Thus a person might say that "there is a 50% probability that it will rain today," or "I'll give you 2 : 1 odds (that is, probability of $\frac{2}{3}$) that the Sox will win the pennant," or "my probability is 0.25 that human beings will someday travel faster than the speed of light." In each of these instances, the stated probability or odds are based on that person's beliefs and information about the subject under discussion. Another person might state different probabilities based on different information. Hence such probabilities have been called **subjective.**

Although the probabilities that people assign to events or statements may fail to satisfy the requirements imposed by the axioms of probability, there are certain criteria of consistency which, if satisfied, uniquely determine a probability measure for the events or statements under consideration. Such **subjective probability** measures have been shown to be useful in managerial decision making. Many scientists and philosophers also advocate the use of such subjective probabilities in assessing scientific evidence. They point out that the act of assessing and announcing personal probabilities brings out into the open the subjective assessments that have been made. Whether one uses objective or subjective probabilities is tangential to the topics treated in this book because the probability calculus developed in this chapter, and in succeeding chapters, is valid for *either type of probability used.*

An example may help to clarify concepts. Suppose that you are shown the white pages of a city phone directory of 300 pages. What odds would you give to the assertion E_1 that "the first person listed on page 130 of this directory has last name beginning with the letter H"? Note that by opening the phone book to page 130, you could determine with certainty the truth of this assertion. Such an "experiment" can only be run once; the relative frequency of assertion E_1 being true in this experiment is either 1 or 0. However, you are not given the phone book, so you do not know for certain which is the case. You do have some information that may be helpful. For example, the letter H is the seventh letter of the 26 letters of the alphabet, while page 130 is close to the middle of the phone book. You also know that some letters appear more frequently at the beginnings of last names than do others. You may even know something about the names that are likely to appear in the phone book of this city. On this basis, you might arrive at a numerical measure (odds) of belief in assertion E_1—say, for example, odds of 1 out of 100. In similar fashion you could arrive at odds for the assertion E_2 that "the last person on page 230 has last name beginning with the letter S," and for the assertion "E_1 and E_2" that both of these statements are true. Consistency of these odds would require, for example, that the odds of "E_1 and E_2" be less than the odds of E_1 and also of the odds of E_2 (because "E_1 and E_2" being true implies that both E_1 and E_2 are true). Each of the assertions E_1, E_2, "E_1 and E_2" is either true or false (relative frequency 1 or 0). What you are modeling is not relative frequency but your personal knowledge of (belief in) the truth of these assertions. Other people, with different information and willingness to make guesses, may arrive at different odds (subjective probabilities) than you do.

The reader interested in further discussion of the meaning, construction, and uses of probability models is encouraged to read the *Scientific American* survey articles by Ayer (1965), Kac (1964), and Weaver (1950, 1952). The remaining chapters in this book emphasize the use of probability in modeling relative frequencies for observations (data) encountered in practical and scientific work, in the spirit of the classic but more advanced textbook of Feller (1968). Among textbooks which treat the modeling of subjective probabilities (odds), and the combination of these with probabilities for data, are those of Jeffreys (1961), DeGroot (1970), Winkler (1972), de Finetti (1975), Good (1983), and Jaynes (1983).

Glossary and Summary

We summarize below some of the key ingredients that arise repeatedly in the development of a probability model:

Outcomes: The basic elements of the model corresponding to the outcomes of a random experiment.
Sample space: The collection of all possible outcomes; the sure event.
Event: A collection of outcomes.
Null event: The event that has no outcomes; the impossible event.
Probability of an event E: Idealization of the long-run relative frequency of E; also, idealization of the subjective odds assigned to a possibility or assertion.

The following are operations and relationships defined for events and some basic rules satisfied by these operations:

Inclusion ($E_2 \subset E_1$): E_2 is included in E_1 if every outcome in E_2 is also included in E_1.

Equality ($E_1 = E_2$): Two events are equal if they contain the same outcomes ($E_1 \subset E_2$ and also $E_2 \subset E_1$).

Complement (E^c): The collection of outcomes not included in E.

Intersection ($E_1 \cap E_2 \cap \cdots \cap E_k$): The collection of outcomes common to all k events $E_1, E_2, \ldots, E_k$ (for $k = 2$, all outcomes in both E_1 and E_2).

Union ($E_1 \cup E_2 \cup \cdots \cup E_k$): The collection of all outcomes belonging to at least one of $E_1, E_2, \ldots, E_k$ (for $k = 2$, all outcomes in E_1, in E_2, or in both E_1 and E_2).

Mutually exclusive: Two events E_1 E_2 are mutually exclusive if $E_1 \cap E_2 = \varnothing$; that is, if they have no outcomes in common.

Rule 2.1. $\Omega^c = \varnothing$, $\varnothing^c = \Omega$.

Rule 2.2. For any event E: (a) $E \cap E^c = \varnothing$; (b) $E \cup E^c = \Omega$; (c) $E \cap \varnothing = \varnothing$; (d) $E \cup \varnothing = E$; (e) $E \cap \Omega = E$; (f) $E \cup \Omega = \Omega$; (g) $(E^c)^c = E$.

Rule 2.3. For any events E_1, E_2, E_3:

$$\text{(a) } E_1 \cup (E_2 \cap E_3) = (E_1 \cup E_2) \cap (E_1 \cup E_3).$$
$$\text{(b) } E_1 \cap (E_2 \cup E_3) = (E_1 \cap E_2) \cup (E_1 \cap E_3).$$

For events $E, E_1, \ldots, E_k$:

$$E \cup (E_1 \cap E_2 \cap \cdots \cap E_k) = (E \cup E_1) \cap (E \cup E_2) \cap \cdots \cap (E \cup E_k),$$
$$E \cap (E_1 \cup E_2 \cup \cdots \cup E_k) = (E \cap E_1) \cup (E \cap E_2) \cup \cdots \cup (E \cap E_k).$$

Rule 2.4 (de Morgan's laws). For any events $E_1, \ldots, E_k$:

$$(E_1 \cup E_2 \cup \cdots \cup E_k)^c = E_1^c \cap E_2^c \cap \cdots \cap E_k^c$$
$$(E_1 \cap E_2 \cap \cdots \cap E_k)^c = E_1^c \cup E_2^c \cup \cdots \cup E_k^c$$

Thus $(E_1 \cup E_2)^c = E_1^c \cap E_2^c$, $(E_1 \cap E_2)^c = E_1^c \cup E_2^c$.

The following are the axioms (basic requirements) satisfied by probabilities:

Axiom 1. For every event E, $0 \le P(E) \le 1$.

Axiom 2. $P(\Omega) = 1$, $P(\varnothing) = 0$.

Axiom 3. If E_1 and E_2 are mutually exclusive events, then

$$P(E_1 \cup E_2) = P(E_1) + P(E_2).$$

Axiom 3′. For any countable collection $E_1, E_2, E_3, \ldots$ of events, no two of which have outcomes in common, $P(E_1 \cup E_2 \cup E_3 \cup \cdots) = \sum_{i=1}^{\infty} P(E_i)$.

The following are important rules of the probability calculus:

1. $P(E_1 \cap E_2^c) = P(E_1) - P(E_1 \cap E_2)$.
2. **Law of Complementation.** $P(E^c) = 1 - P(E)$.
3. **General Laws of Addition.**

$$P(E_1 \cup E_2) = P(E_1) + P(E_2) - P(E_1 \cap E_2),$$

$$\begin{aligned} P(E_1 \cup E_2 \cup E_3) = {} & P(E_1) + P(E_2) + P(E_3) - P(E_1 \cap E_2) \\ & - P(E_1 \cap E_3) - P(E_2 \cap E_3) \\ & + P(E_1 \cap E_2 \cap E_3), \end{aligned}$$

$$\begin{aligned} P(E_1 \cup E_2 \cup E_3 \cup E_4) = {} & \sum_{i=1}^{4} P(E_i) - \sum_{1 \le i < j \le 4} P(E_i \cap E_j) \\ & + \sum_{1 \le i < j < k \le 4} P(E_i \cap E_j \cap E_k) \\ & - P(E_1 \cap E_2 \cap E_3 \cap E_4). \end{aligned}$$

4. **Special Law of Addition.** If $E_1, E_2, \ldots, E_k$ are k events, no two of which have outcomes in common, then

$$P(E_1 \cup E_2 \cup \cdots \cup E_k) = \sum_{i=1}^{k} P(E_i).$$

5. $P(E_1^c \cap E_2^c) = 1 - P(E_1 \cup E_2)$.

Venn diagrams exhibit events as ovals. Intersections of the ovals correspond to intersections of events. The sample space Ω is the entire diagram (the box). Venn diagrams can be used to solve probability problems by visualizing probabilities of events as areas of the ovals representing the events.

Finally, a **probability measure** is the rule or function that assigns a probability value $P(E)$ to each event E in the probability model. When probability measures involve subjective probabilities, care needs to be taken that these probabilities are assigned in such a way as not to violate the axioms of probability. Objective probability measures are established either from observed relative frequencies (empirically) or from theoretical arguments.

Exercises

1. In Example 2.1, list the outcomes that belong to each of the following events.
(a) $E_1^c \cap E_2$. (b) $E_1 \cup E_2^c$. (c) $E_1^c \cap E_2^c$.

(d) $E_1^c \cup E_2^c$. (e) $(E_1^c \cap E_2) \cup E_2^c$. (f) $(E_1 \cup E_2^c)^c$.
(g) The event "exactly one of the events E_1, E_2 occurs,"
(h) The event "at most one of the events E_1, E_2 occurs."

2. In Example 2.2, list the outcomes that belong to each of the following events.
(a) $E_1 \cap E_2$. (b) $E_1 \cup E_2$. (c) E_2^c.
(d) $E_1 \cap E_2^c$. (e) $(E_1^c \cup E_2) \cup E_2^c$. (f) $E_2 \cap (E_1 \cap E_2)^c$.
(g) The event "exactly one of the events E_1, E_2 occurs."
(h) The event "the infant chooses toy A before toy B, or toy B before toy C."

3. A family with exactly three children is selected by lottery, and the sexes (M = male, F = female) are recorded in sequence of birth. One possible outcome of this random experiment is MFM, in which the first-born child is male, the second-born is female, and the third-born is male.
(a) List all outcomes of this random experiment.
(b) List the outcomes in the event E_1 defined by "exactly two of the three children are male."
(c) List the outcomes in the event E_2 defined by "the first-born child is female."
(d) List the outcomes in the following events: (i) $E_1 \cap E_2$; (ii) $E_1 \cup E_2$; (iii) $E_1^c \cap E_2$; (iv) $E_1 \cup E_2^c$.
(e) Suppose that in this random experiment one only counts the *number* of male children. List the outcomes of this experiment.

4. Customers in a donut shop can purchase up to 3 donuts at one time, and also have the chance to purchase a single beverage (coffee, orange juice, tea, or milk). Customers can purchase donuts without purchasing a beverage, or a beverage without purchasing donuts, but must purchase one or the other. As a random experiment, we observe the purchases of a customer at a given time. The number of donuts and the beverage (if any) that are purchased are recorded.
(a) List the outcomes of this experiment.
(b) List the outcomes contained in the event E_1 defined by "the customer does not purchase a donut."
(c) List the outcomes contained in the event E_2 defined by "the customer does not purchase a beverage."
(d) Are E_1 and E_2 mutually exclusive events?
(e) In terms of the events E_1, E_2 defined in parts (b) and (c), describe the following events in set notation:
 (i) "The customer purchases at least one donut."
 (ii) "The customer purchases a beverage and at least one donut."

5. Let E_1, E_2, E_3 be three events associated with a random experiment. Express the following verbal statements in set notation (see Example 2.3).

(a) At most one of the events E_1, E_2, E_3 occurs.
(b) At least two of the events E_1, E_2, E_3 occur.
(c) E_1 occurs, and either E_2 occurs or E_3 occurs.
(d) At least one of the events E_1, E_2, E_3 does not occur.
(e) E_1 occurs, but neither E_2 nor E_3 occurs.
(f) Exactly two of the events occur, but E_1 does not occur.

6. Which of the following pairs of events A and B are mutually exclusive?
(a) A: being the daughter of a lawyer, B: being born in Chicago.
(b) A: being under 18 years of age, B: voting in a presidential election.
(c) A: owning a Chevrolet, B: owning a Ford.

7. An economic forecaster is trying to assign subjective probabilities to the events E_1 that "the national economy will experience a recession" and E_2 that "the rate of inflation will increase." The forecaster assigns odds of 5 to 3 to event E_1 and 2 to 1 to event E_2; these correspond to subjective probabilities of $P(E_1) = 5/(5 + 3) = 0.625$, $P(E_2) = 2/(2 + 1) = 0.667$.
(a) If the forecaster also assigns subjective probability $P(E_1 \cap E_2) = 0.500$ to the event that "a recession will occur and the rate of inflation will rise," what probabilities should be assigned to the following events?
 (i) $E_1^c \cap E_2$ that "no recession will occur but the rate of inflation will rise."
 (ii) $E_1^c \cap E_2^c$ that "neither recession nor a rise in inflation will occur."
 (iii) "Exactly one of the possibilities E_1, E_2 will occur."
(b) Suppose, instead, that the forecaster decides to assign probability 0.250 to the event $E_1 \cap E_2$. Is this choice consistent with the probabilities 0.625, 0.667 assigned to the events E_1, E_2? Explain your answer.

8. As reported in the 1989 *Statistical Abstracts* (p. 550), 47.7% of all adults (18 years or older) are males (event M), 21.9% of all adults are single (event S) and 42.5% of all adults are neither male nor single (event $M^c \cap S^c$). What percentage of all adults are (a) females; (b) single males; (c) single females?

9. In a survey about income and investments, two events of interest are "the investor owns common stock" (event A) and "the investor owns municipal bonds" (event B). The following probabilities are found: $P(A) = 0.68$, $P(B) = 0.23$, $P(A \cap B) = 0.15$. Find the probability that:
(a) An investor owns common stock but does not own municipal bonds.
(b) An investor owns neither common stock nor municipal bonds.
(c) An investor owns common stock or municipal bonds, but not both.

10. Suppose that we are given a probability model in which two mutually

exclusive events A and B have the respective probabilities $P(A) = 0.30$ and $P(B) = 0.40$. Find the probability that:
(a) Either A or B occurs.
(b) Both A and B occur.
(c) Exactly one of the events A and B occurs.
(d) Neither A nor B occurs.
(e) B occurs but A does not occur.
(f) Either B does not occur or A does not occur.

11. The Venn diagram in Exhibit E.1 provides probabilities for different events in a random experiment. For example, the diagram states that $P(A \cap B \cap C) = 0.10$. Using this information, determine the following probabilities:
(a) $P(A \cap B)$; (b) $P(B \cap C)$; (c) $P\{\text{only } C \text{ happens}\}$;
(d) $P\{\text{exactly one of the events } A, B, C \text{ happens}\}$;
(e) $P\{\text{none of the events } A, B, C \text{ happens}\}$.

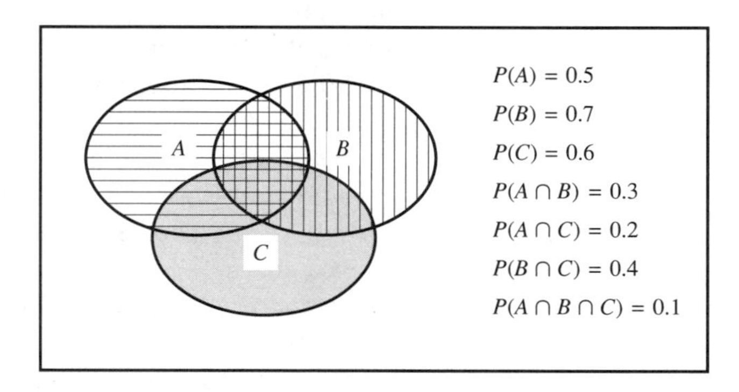

EXHIBIT E.1 *Venn diagram providing probabilities for different events.*

12. A survey of the sources of caffeine intake among college students reported that 65% of the students regularly used coffee, 75% of the students regularly used cola beverages, and 20% of the students regularly used tea. Also, 45% regularly used both coffee and cola beverages, 12% regularly used both coffee and tea, 8% regularly used both cola beverages and tea, and 5% regularly used coffee, cola beverages, and tea. What percentage of the college students:
(a) Regularly use at least one of these beverages?
(b) Regularly use exactly one of these beverages?
(c) Regularly use none of these beverages?
(d) Regularly use cola beverages but neither coffee or tea?
(e) Regularly use only tea?

13. In a certain city of population 100,000 the proportions of people who read the most popular magazines are:

A: 0.15	B: 0.10	C: 0.08
A and B: 0.08	A and C: 0.02	B and C: 0.04
A and B and C: 0.01		

(a) Find the proportion and the number of people who read exactly one magazine.
(b) How many people read at least two magazines?
(c) If A and C arrive on Fridays and B arrives on Monday, how many people read at least one Friday magazine and the Monday magazine?
(d) How many people read only *one* Friday magazine and the Monday magazine?

14. Refer to Example 4.2.
(a) Find the probability that a stolen car is 3 years old or older.
(b) Find the probability that a stolen car is 5 years old or younger.
(c) Find the probability that a stolen car is at least 1 year old but not older than 3 years of age.

15. Suppose that in 1963 a person was randomly chosen from the city of Breslau, Poland. Use the life table (Exhibit 6.1) to answer the following questions.
(a) What is the probability that this person lived past a 60th birthday?
(b) What is the probability that this person did not live to reach an 18th birthday?
(c) What is the probability that this person lived past a 15th birthday, but died before a 70th birthday?

16. A study of the choice of elective courses by seventh-grade students showed that 60% took home economics, 30% took band, 50% took music appreciation, 20% took both home economics and band, and 30% took both home economics and music appreciation. It was not possible for a student to take both band and music appreciation because these courses were scheduled at the same hour. What percentage of students:
(a) Took all three of these courses:
(b) Took exactly two of these courses?
(c) Took at least one of these courses?
(d) Took none of these courses?
(e) Took home economics but neither band nor music appreciation?
(Draw your Venn diagram carefully.)

17. Let the sample space of a probability model have as outcomes all schools in the United States. According to the 1987 *Statistical Abstract of the United States* (p. 124), there were 57,471 public elementary schools, 15,631 private elementary schools, 22,336 public secondary schools, 2621 private secondary schools, 935 public junior colleges, and 371 private junior colleges.
(a) How many schools were public schools?
(b) How many schools were elementary schools?
(c) How many schools were either private schools or junior colleges?

(d) How many schools were either private schools or elementary schools, but not both?
(e) How many schools were *not* secondary schools?
(f) How many schools had no more than one of the following characteristics: private, junior college, elementary school?
(g) How many schools had *none* of the following characteristics: private, elementary, secondary? (*Hint:* Use Venn diagrams.)

18. A distress signal is sent by a ship at sea. There are three coast guard stations A, B, C within receiving distance of the signal. Each station has probability 0.3 of receiving the signal. The probability that station A is the *only* one to receive the signal is 0.20, while the probabilities of being the only one to receive the signal for stations B, C are 0.15 and 0.18, respectively. Finally, the probability that station C is the only one *not* to receive the signal is 0.05. Find the probability that at least one station receives the distress signal, so that the ship is rescued. (Use a Venn diagram.)

19. State the General Law of Addition for five events; that is, give a formula for $P(E_1 \cup E_2 \cup E_3 \cup E_4 \cup E_5)$ (see Section 4).

Theoretical Exercises

Starting in this chapter, exercises are given which ask readers to provide logical justifications of assertions made in the text. In addition, results are presented that supplement and extend theory in the text; readers are also asked to verify these new results. Most of the verifications are straightforward.

T.1. One way to prove that two events E_1, E_2 are equal ($E_1 = E_2$) is to show that $E_1 \subset E_2$ and that $E_2 \subset E_1$. That is, we show that any outcome ω belonging to E_1 also belongs to E_2, and that any outcome ω belonging to E_2 also belongs to E_1. Use this method to verify the following assertions.
(a) Rule 2.2(b). (b) Rule 2.2(g). (c) Rule 2.3(a).
(d) Rule 2.3(b). (e) $E_1 \cup E_2 = E_1 \cup (E_1^c \cap E_2)$.

T.2. (a) Show by formal proof the following facts about inclusion of events.
(i) $\varnothing \subset E \subset \Omega$.
(ii) If $E_1 \subset E_2$, then $E_1 \cap E_2 = E_1$ and $E_1 \cup E_2 = E_2$.
(iii) $E_1 \subset E_2$ implies that $E_2^c \subset E_1^c$.
(iv) $E_1 \subset E_2$ and $E_2 \subset E_3$ implies that $E_1 \subset E_3$.
(v) $(E_1 \cap E_2) \subset E_i \subset (E_1 \cup E_2)$, $i = 1, 2$.
(b) Use (i) and (ii) of part (a) to verify Rules 2.2(c), 2.2(d), 2.2(e), and 2.2(f).

(c) Use (iii) of part (a) to verify the assertion that $E_1 = E_2$ if and only if $E_1^c = E_2^c$.

T.3. (a) Use Rules 2.4(a) and 2.2(g) to verify Rule 2.4(b). [*Hint:* Apply Rule 2.4(a) to E_1^c, E_2^c. Take complements of both sides of the resulting equality.]

(b) Use Rules 2.3(a), 2.2(g), and 2.4 to verify Rule 2.3(b). [*Hint:* Apply Rule 2.3(a) to E_1^c, E_2^c, E_3^c, take complements of both sides of the resulting equality and use Rules 2.4(a) and 2.4(b).]

T.4. Use recursive arguments to verify equations (2.8) and (2.9).

T.5. Show that if two events E_1, E_2 satisfy $E_1 \subset E_2$, then $P(E_1) \le P(E_2)$. Thus show that for any two events E_1, E_2,

$$P(E_1 \cap E_2) \le P(E_i) \le P(E_1 \cup E_2), \qquad i = 1, 2.$$

T.6. (a) Prove the inequality

$$P(E_1 \cup E_2) \le P(E_1) + P(E_2).$$

(b) Now use recursive arguments to show that

$$P(E_1 \cup E_2 \cup \cdots \cup E_k) \le P(E_1) + P(E_2) + \cdots + P(E_k)$$

for $k = 2, 3, 4, \ldots$. This inequality is known as **Boole's Inequality,** after George Boole (see Section 2). It is one of a series of inequalities for the probability $P(E_1 \cup E_2 \cup \cdots \cup E_k)$ obtained by the Italian mathematician Carlo Emilio Bonferroni (1892–1960).

(c) Another of Bonferroni's inequalities is

$$P(E_1 \cup E_2 \cup \cdots \cup E_k) \ge S_k(1) - S_k(2), \qquad k \ge 2,$$

where

$$\begin{aligned} S_k(1) &= P(E_1) + P(E_2) + \cdots + P(E_k), \\ S_k(2) &= \text{sum of all probabilities of intersections } E_i \cap E_j, \\ & \ i < j. \end{aligned}$$

For $k = 2$ this inequality is actually an equality $[P(E_1 \cup E_2) = P(E_1) + P(E_2) - P(E_1 \cap E_2)]$. For $k = 3$, the inequality asserts that

$$\begin{aligned} P(E_1 \cup E_2 \cup E_3) \ge\ & P(E_1) + P(E_2) + P(E_3) \\ & - P(E_1 \cap E_2) - P(E_1 \cap E_3) - P(E_2 \cap E_3). \end{aligned}$$

Prove this inequality ($k = 3$).

(d) Prove the Bonferroni inequality of part (c) for $k = 4$. (This proof is more difficult.)

T.7. Note that

$$P(E_1 \cup E_2) = P(E_1) + P(E_2) \quad \text{if and only if} \quad P(E_1 \cap E_2) = 0.$$

(a) Use this fact to prove that $P(E_1 \cup E_2 \cup E_3) = P(E_1) + P(E_2) + P(E_3)$ if and only if $P(E_1 \cap E_2) = P(E_1 \cap E_3) = P(E_2 \cap E_3) = 0$.

(b) Use recursive arguments (mathematical induction) to show that

$$P(E_1 \cup E_2 \cup \cdots \cup E_k) = P(E_1) + P(E_2) + \cdots + P(E_k)$$

if and only if $P(E_i \cap E_j) = 0$ for all $i \neq j$.

T.8. (a) For any two events E_1, E_2, show that

$$P\{\text{exactly one of } E_1, E_2 \text{ occur}\} = P(E_1) + P(E_2) - 2P(E_1 \cap E_2).$$

(b) For any three events E_1, E_2, E_3 show that

$$P\{\text{exactly two of } E_1, E_2, E_3 \text{ occur}\}$$
$$= P(E_1 \cap E_2) + P(E_1 \cap E_3) + P(E_2 \cap E_3) - 3P(E_1 \cap E_2 \cap E_3).$$

(c) Verify the General Law of Addition for four events given in equation (4.3).

3

Finite Probability Models and Random Sampling

In this chapter we discuss probability models whose sample spaces contain only a finite number of outcomes. When every outcome is equally probable (the **uniform probability model**), the probabilities of events can be determined by counting outcomes. In Section 2 we show how uniform probability models arise in statistical sampling problems. Rules of combinational analysis useful for counting outcomes for probability calculations are developed in Section 3.

1 Finite Probability Models

There are a variety of random experiments whose sample spaces are finite. Examples include population sampling, experiments in the social and biological sciences, experiments in genetics, and physical experiments dealing with microscopic structures. Most popular games of chance also have sample spaces that are finite.

If the sample space Ω of a probability model contains a finite number M of outcomes $\omega_1, \omega_2, \ldots, \omega_M$ the model is said to be a **finite probability model.** We define the **simple events** $S_i = \{\omega_i\}$, $i = 1,2, \ldots, M$, of such a probability model to be the events consisting of exactly one outcome. Exhibit 1.1 illustrates the sample space Ω of a finite probability model in which there are $M = 7$ outcomes $\omega_1, \omega_2, \ldots, \omega_7$ (represented by points in Exhibit 1.1) and seven simple events $S_1, S_2, \ldots, S_7$, represented by the small circles in Exhibit 1.1. The entire box in Exhibit 1.1 represents Ω, whereas one particular event E, represented by the large circle, contains the outcomes ω_1, ω_2, and ω_4. Note that

$$E = \{\omega_1, \omega_2, \omega_4\} = S_1 \cup S_2 \cup S_4.$$

Exhibit 1.1 illustrates the following facts about finite probability models.

1. Any two simple events S_i and S_j, $i \neq j$, are mutually exclusive.
2. Every event E is the union of simple events corresponding to the outcomes contained in E.

These two facts, together with the Special Law of Addition (Chapter 2), lead to a fundamental rule for computing probabilities in finite probability models.

> **Rule 1.1.** The probability $P(E)$ of any event E of a finite probability model is equal to the sum of the probabilities of the simple events whose union is E.

From Rule 1.1 it follows that any finite probability model an be completely described by a table such as

Simple event	S_1	S_2	$\cdots$	S_{M-1}	S_M
Probability	p_1	p_2	$\cdots$	p_{M-1}	p_M

where $p_j = P(S_j), j = 1, 2, \ldots, M$. Note that because $\Omega = S_1 \cup S_2 \cup \cdots \cup S_M$ and $P(\Omega) = 1$, Rule 1.1 implies that

$$\sum_{j=1}^{M} p_j = P(\Omega) = 1. \tag{1.1}$$

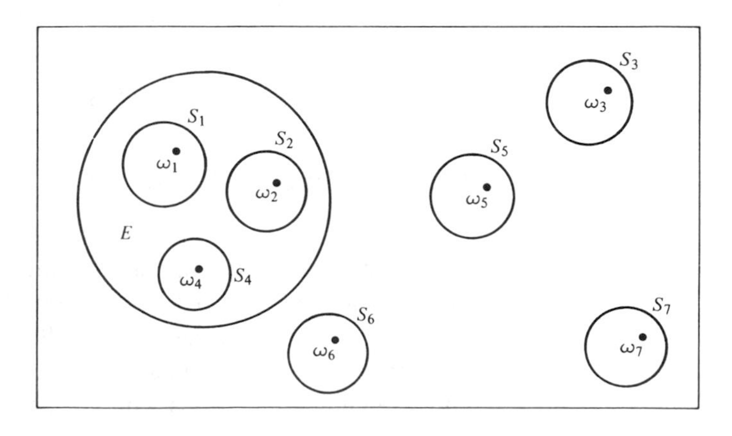

EXHIBIT 1.1 *Sample space of a finite probability model.*

Rule 1.2. The sum of the probabilities of all simple events is equal to 1.

EXAMPLE 1.1

For the sample space illustrated by Figure 1.1, suppose that a probability model is defined by the following table of probabilities.

Simple event	S_1	S_2	S_3	S_4	S_5	S_6	S_7
Probability	0.08	0.25	0.33	0.04	0.16	0.05	0.09

Because the event $E = S_1 \cup S_2 \cup S_4$, Rule 1.1 implies that

$$P(E) = P(S_1) + P(S_2) + P(S_4) = 0.08 + 0.25 + 0.04 = 0.37. \quad \blacklozenge$$

EXAMPLE 1.2

A random experiment consists of spinning a roulette wheel. The outcomes of this experiment are the numbers 1,2, . . . , 35, 36, 0, 00 printed on the wheel; there are $M = 38$ outcomes in all. The simple events are $S_j = \{j \text{ is observed}\}$, $j = 1, \ldots, 36$, and $S_{37} = \{0 \text{ is observed}\}$, $S_{38} = \{00 \text{ is observed}\}$. Assuming that every number on the wheel has an equal probability of being observed when the wheel is spun, we have the following table of probabilities:

Simple event	S_1	S_2	$\cdots$	S_{36}	S_{37}	S_{38}
Probability	$\frac{1}{38}$	$\frac{1}{38}$	$\cdots$	$\frac{1}{38}$	$\frac{1}{38}$	$\frac{1}{38}$

If E is the event consisting of outcomes that are even numbers (excluding the zeros), then $E = S_2 \cup S_4 \cup \cdots \cup S_{36}$, and

$$P(E) = P(S_2) + P(S_4) + \cdots + P(S_{36}) = \frac{18}{38} = 0.474. \quad \blacklozenge$$

Uniform Probability Models

In many random experiments, such as Example 1.2, there is a symmetry (or interchangeability) among the outcomes. There is a symmetry between the two faces of a tossed coin, among the numbers on a balanced roulette wheel, and among the cards in a well-shuffled deck. Such symmetry among outcomes is often assumed for the choice of sex at birth, for guessing in multiple-choice situations, for the choice of magnetic polarity in a given electron, and in many other experiments in science and technology. It is a *logical* consequence of this assumption of symmetry that the simple events of the probability model have *equal* probabilities. Any finite model for which the simple events S_i have equal probabilities is called a **uniform probability model.**

Because each of the M simple events S_i in a uniform probability model has the same probability p, it follows from Equation (1.1) that $p = 1/M$. If an event E consists of K outcomes, then E is the union of K simple events, each having probability $1/M$. From Rule 1.1, it follows that $P(E) = K/M$.

Rule 1.3. In a uniform probability model, the probability of any simple event S_i is $1/M$, and the probability of any event E is $P(E) = K/M$, where K is the number of outcomes in E and M is the number of outcomes in Ω.

EXAMPLE 1.3

A bottling company manager claims to be able to taste the difference between two flavors (an original and new flavor) of a cola beverage. She is given four unmarked cups of cola beverage to drink: two of the new flavor and two of the original flavor. The cups are presented in order, and in each case it is noted whether or not she answers correctly. A typical outcome of this experiment is (C, I, C, I), in which correct answers are given on the first and third cups and incorrect answers on the second and fourth cups. The 16 possible outcomes are

$$\begin{array}{llll}(C, C, C, C), & (I, C, C, C), & (C, I, I, C), & (I, I, C, I), \\ (C, C, C, I), & (C, C, I, I), & (I, C, C, I), & (I, C, I, I), \\ (C, C, I, C), & (C, I, C, I), & (I, I, C, C), & (C, I, I, I), \\ (C, I, C, C), & (I, C, I, C), & (I, I, I, C), & (I, I, I, I).\end{array}$$

Suppose that the manager really cannot tell the difference between the flavors, and just guesses. As long as her guesses do not depend on each other in any way, there is a symmetry among the outcomes, implying that a uniform probability model may be appropriate. If so, the probability of each simple event is $\frac{1}{16}$, because there are $M = 16$ outcomes in the sample space. By Rule 1.2 the probability that she makes exactly three correct answers is $\frac{4}{16}$, because there are $K = 4$ outcomes, (C, C, C, I), (C, C, I, C), (C, I, C, C), and (I, C, C, C), belonging to the event "the manager makes exactly three correct answers." ◆

It is clear from Rule 1.2 and Example 1.3 that probability calculations in uniform probability models reduce to problems of counting outcomes. Because it is not always possible to list all outcomes in the sample space, as we did in Example 1.3, counting rules can be helpful. A few of the most useful rules of **combinatorial analysis,** which is the mathematical theory of counting, are given in Section 3.

2 Simple Random Sampling

Perhaps the single most important use of uniform probability models is in survey sampling. Here the randomness in the experiment is deliberately introduced by the experimenter. This permits the experimenter to sample members of a population in a controlled random way, examine these chosen members, and then make inferences about the entire population through the use of probability theory.

The basic probability model for survey sampling is that of **random sampling.** Suppose that we have a population of N units $u_1, u_2, \ldots, u_N$. This population may consist of people, domiciles, radio tubes, chemical distillates, items on an examination, numbers, and so on. In choosing a random sample of n units from the totality of N units, we perform a random experiment in which the outcomes are **samples** of n units and in which each particular sample (outcome) is chosen with a specified probability.

Formally, a **sample** is an array of units listed in a definite order. For example, one possible sample of $n = 5$ units from the population consisting of the units $u_1, u_2, \ldots, u_N$ is the sample $(u_1, u_4, u_5, u_7, u_3)$. This sample would be distinguished from the sample $(u_5, u_1, u_4, u_7, u_3)$, even though the same units appear in both samples, because the *order* in which these units appear is different.[1] The definition of a sample should be interpreted to allow a given unit (say, u_2) to appear $0, 1, 2, \ldots, n$ times in an (ordered) sample of size n. For example, the ordered array of units $(u_3, u_2, u_2, u_7, u_1)$ and the ordered array of units $(u_1, u_1, u_1, u_2, u_2)$ are both permissible samples of 5 units.

EXAMPLE 2.1

In the population consisting of the 3 units u_1, u_2, u_3, the possible ordered samples of 2 units are

$$\begin{array}{lll}
\omega_1 = (u_1, u_1), & \omega_4 = (u_2, u_1), & \omega_7 = (u_3, u_1), \\
\omega_2 = (u_1, u_2), & \omega_5 = (u_2, u_2), & \omega_8 = (u_3, u_2), \\
\omega_3 = (u_1, u_3), & \omega_6 = (u_2, u_3), & \omega_9 = (u_3, u_3),
\end{array}$$ ◆

Simple random sampling with replacement occurs when the sample space Ω consists of all possible ordered samples and when each sample has

[1] The notational convention that a list given between parentheses is assumed to be ordered was discussed earlier in Chapter 2. Remember that lists given between braces are not ordered; thus $\{u_1, u_2\}$ and $\{u_2, u_1\}$ are the same, but (u_1, u_2) is not the same as (u_2, u_1). An ordered list of n items is called an *n-tuple*.

equal probability of being selected (a uniform probability model over ordered samples). **Simple random sampling without replacement** occurs when the random experiment is described by a probability model that gives zero probability to all ordered samples in which any unit appears more than once, and equal probability to all other ordered samples. This latter sampling yields a uniform model over those ordered samples in which no unit appears more than once.

EXAMPLE 2.1
(continued)

For the population of 3 units discussed in Example 2.1, the probability model for simple random sampling with replacement is

Simple event	$\{\omega_1\}$	$\{\omega_2\}$	$\{\omega_3\}$	$\{\omega_4\}$	$\{\omega_5\}$	$\{\omega_6\}$	$\{\omega_7\}$	$\{\omega_8\}$	$\{\omega_9\}$
Probability	$\frac{1}{9}$	$\frac{1}{9}$	$\frac{1}{9}$	$\frac{1}{9}$	$\frac{1}{9}$	$\frac{1}{9}$	$\frac{1}{9}$	$\frac{1}{9}$	$\frac{1}{9}$

and the probability model for simple random sampling without replacement is

Simple event	$\{\omega_1\}$	$\{\omega_2\}$	$\{\omega_3\}$	$\{\omega_4\}$	$\{\omega_5\}$	$\{\omega_6\}$	$\{\omega_7\}$	$\{\omega_8\}$	$\{\omega_9\}$
Probability	0	$\frac{1}{6}$	$\frac{1}{6}$	$\frac{1}{6}$	0	$\frac{1}{6}$	$\frac{1}{6}$	$\frac{1}{6}$	0

◆

Simple random sampling is commonly performed with the aid of a **table of random numbers,** an example of which appears as Exhibit B.1 in Appendix B. A random number table is an array of digits with the property that each digit in the array has equal probability $\frac{1}{10}$ of being any of the numbers 0, 1, 2, . . . , 9; that each pair of adjacent digits has equal probability $\frac{1}{100}$ of being 00, 01, 02, . . . , 98, 99; that each triple of adjacent digits has equal probability $\frac{1}{1000}$ of being 000, 001, 002, . . . , 998, 999; and so on. Such random number tables are typically created by computer programs called **random number generators.** If available, these computer programs can be used directly as a source of random digits.

To carry out simple random sampling, every unit in the population is first assigned a distinct number. Digits in the random number table are grouped so that they correspond to the numbers assigned to the population units. Thus a population of 75 units assigned index numbers 01, . . . , 74, 75 requires digits in the random number table to be grouped in pairs. On the other hand, triples of random digits are needed to draw samples from a population of 563 units indexed by the numbers 001, 002, . . . , 562, 563.

Once units in the population have been indexed, numbers are drawn from the random number table one at a time to select units for the sample.

(Numbers not corresponding to units in the population are discarded.) For **sampling with replacement,** numbers are drawn and units with the corresponding index are selected one at a time until n units have been selected. The number n of units selected for the sample is called the **sample size.** For **sampling without replacement,** the process of drawing numbers and selecting units with corresponding index is similar to that of sampling with replacement, except that when a random number is drawn corresponding to a unit that has already been selected for the sample, this number is not used (it is discarded). Samples (outcomes) are represented as ordered lists of units to reflect this sampling process. That is, the first unit listed is the first unit chosen for the sample, the second unit listed is the second chosen, and so on.

EXAMPLE 2.2

Suppose that a population consists of $N = 75$ units, $u_1, u_2, \ldots, u_{75}$. We wish to select a simple random sample of $n = 3$ units. From a random number table or a random number generator, pairs of random digits are obtained as follows:

$$23 \mid 07 \mid 00 \mid 23 \mid 34 \mid 15 \mid \cdots \mid .$$

If we sample *with replacement,* the first unit taken for our sample is unit u_{23}, and the second unit taken is u_7. After discarding the number 00, because it does not index any unit in our population, we take unit u_{23} again as the third unit for our sample. Thus our sample is (u_{23}, u_7, u_{23}).

If we sample *without replacement*, we take unit u_{23} first and unit u_7 second. We then discard 00 because it does not index any unit in our population, and we also discard 23 because it was previously drawn. Because the next pair of digits (34) in the string has not previously been drawn, we take unit u_{34} as the third unit for our sample. Thus our sample is (u_{23}, u_7, u_{34}). ◆

Other methods for obtaining a simple random sample can be used. In raffles or lotteries, for example, balls or tickets with numbers on them are often placed in an urn and one ball (or ticket) is selected at a time, with the urn being vigorously shaken before each draw. Although simple random sampling *with replacement* can only be performed in the one-at-a-time fashion we have described, a simple random sample *without replacement* can also be taken by selecting all n units simultaneously. For example, a simple random sample of 5 pieces of candy could be drawn without replacement from a bag of candies either 1 piece at a time, or by taking a handful of 5 pieces all at once. Consequently, the outcomes of simple random sampling without replacement could be represented as unordered lists of the units drawn for the sample. We have chosen not to do this in order to have a common framework for discussing the two kinds (with replacement, without replacement) of simple random sampling.

Probability Calculations

Because both simple random sampling with replacement and simple random sampling without replacement define uniform probability models over their respective sample spaces, in order to find the probability of any event E defined by some property $\mathcal{P}$ we need only compute

M = the total number of possible samples (outcomes) in the sample space Ω for the given type of sampling,

and

K = the number of samples (outcomes) in Ω which have the property $\mathcal{P}$.

Rule 1.2 then tells us that $P(E) = K/M$.

The following facts are established in Section 3.

Rule 2.1. If there are T_1 ways to do task 1, and for each such way T_2 ways to do task 2, the total number of ways in which the two tasks can be done together is $T = (T_1)(T_2)$.

Rule 2.2. The total number of ordered samples of size n drawn with replacement from a population of N units is N^n.

Rule 2.3. The total number of ordered samples of size n from a population of N units in which no unit appears more than once in the sample (i.e., drawn without replacement) is

$$N(N - 1) \cdot \cdot \cdot (N - n + 1).$$

EXAMPLE 2.3

A quality control engineer selects a simple random sample of $n = 2$ parts *without replacement* from a lot of $N = 100$ parts produced by a factory, and tests each of the parts selected to see whether or not it is defective. Suppose that $Q = 15$ of the $N = 100$ parts in the lot are, in fact, defective (and the remaining $N - Q = 85$ parts are good). What is the probability that exactly q of the parts selected are defective (where $q = 0, 1, 2$)?

By Rule 2.3, the total number of outcomes (samples) in the sample space Ω for this problem is $M = (100)(99) = 9900$. Thinking of the sample of $n = 2$ units as having been drawn one unit at a time, let E_1 be the event "the first unit drawn for the sample is defective" and let E_2 be the event "the second unit drawn for the sample is defective." Note that the event "$q = 2$ (both) of the parts selected are defective" is the same as the event $E_1 \cap E_2$. To obtain two defective parts for the sample, we would have to select them from the $Q = 15$ defective parts in the population (lot). By Rule 2.3, this can be done in $K = (15)(14) = 210$ ways. Thus

$$P\{q = 2 \text{ of the parts selected are defective}\}$$
$$= P(E_1 \cap E_2) = \frac{(15)(14)}{(100)(99)} = \frac{210}{9900} = 0.021.$$

Similarly, the event "$q = 0$ (none) of the parts selected are defective" is the same as the event $E_1^c \cap E_2^c$. If there are no defective parts in the sample, both parts in the sample must be good. By Rule 2.3 these two good parts can be selected from the $N - Q = 85$ good parts in the population in $K = (85)(84) = 7140$ ways. Thus

$$P\{q = 0 \text{ of the parts selected are defective}\}$$
$$= P(E_1^c \cap E_2^c) = \frac{(85)(84)}{(100)(99)} = \frac{7140}{9900} = 0.721.$$

Finally, note that the event $\{q = 1$ of the parts selected are defective$\}$ is the complement of the union $(E_1 \cap E_2) \cup (E_1^c \cap E_2^c)$ of the two mutually exclusive events whose probabilities we have just calculated. Thus by the Law of Complementation and Axiom 3,

$$P\,\{q = 1 \text{ of the parts elected are defective}\}$$
$$= 1 - P((E_1 \cap E_2) \cup (E_1^c \cap E_2^c)) = 1 - P(E_1 \cap E_2) - P(E_1^c \cap E_2^c)$$
$$= 1 - 0.021 - 0.721 = 0.258.$$

However, it is instructive to obtain this probability by direct means. If exactly one defective part has been obtained in our sample of $n = 2$ parts, the other part in the sample must be good. The defective part could be selected first, followed by the good part, namely $(E_1 \cap E_2^c)$, *or* the good part could be selected and then the defective part, namely $(E_1^c \cap E_2)$. Thus

(2.1)
$$\{q = 1 \text{ of the parts elected are defective}\} = (E_1 \cap E_2^c) \cup (E_1^c \cap E_2).$$

This result is not surprising because another way of defining the event "$q = 1$ of the parts selected are defective" is to say "*exactly one* of the events E_1, E_2 occur" (see Chapter 2). Recall from Chapter 2 that the events $(E_1 \cap E_2^c)$, $(E_1^c \cap E_2)$ are mutually exclusive. Therefore, by equation (2.1) and Axiom 3,

(2.2)
$$P\{q = 1 \text{ of the parts elected are defective}\} = P(E_1 \cap E_2^c) + P(E_1^c \cap E_2).$$

To obtain an outcome (sample) belonging to event $E_1 \cap E_2^c$, we need to select a defective part from among the $Q = 15$ defective parts in the lot (population) and we need to select a good part from among the $N - Q = 85$ good parts in the lot. The first task can be done in $T_1 = 15$ ways, and the second task in $T_2 = 85$ ways. Thus, by Rule 2.1, there are

$$K = (T_1)(T_2) = (15)(85)$$

outcomes in the event $E_1 \cap E_2^c$, and

$$P(E_1 \cap E_2^c) = \frac{(15)(85)}{(100)(99)} = \frac{1275}{9900} = 0.129.$$

Similarly, there are $K = (85)(15)$ outcomes in event $E_1^c \cap E_2$, so that

$$P(E_1^c \cap E_2) = \frac{(85)(15)}{(100)(99)} = 0.129.$$

Thus from equation (2.2),

$$\begin{aligned} P\{q = 1 \text{ of the parts elected are defective}\} &= P(E_1 \cap E_2^c) + P(E_1^c \cap E_2). \\ &= 0.129 + 0.129 = 0.258, \end{aligned}$$

which is the result already obtained by the Law of Complementation.

◆

Notice that both $E_1 \cap E_2^c$ and $E_1^c \cap E_2$ in Example 2.3 have the same probability. This suggests the possibility of a pattern that we may be able to exploit in more complicated problems of this sort.

Although the event notation used in Example 2.3 is helpful in pointing out the relationship of the probability calculations in sampling problems to the concepts discussed in Chapter 2, this notation can become cumbersome. The next example illustrates a convenient shorthand for intersections of events.

EXAMPLE 2.4

A committee consists of $n = 3$ people randomly chosen from a population of $N = 60$, in which 35 persons favor proposal A and the remaining 25 favor a competing proposal B. What is the probability that exactly 2 of the 3 persons chosen for the committee favor proposal A?

Because the committee must be chosen without replacement, we use Rule 2.3 to find the number M of outcomes (ordered samples) in the sample space Ω; thus

$$M = (60)(59)(58) = 205,320.$$

The 2 supporters of proposal A (and 1 supporter of proposal B) can be chosen for the committee in three distinct orders: AAB, ABA, BAA. In each such order, the two A's are drawn without replacement from among the 35 supporters of proposal A. Rule 2.3 shows that this task can be done in $T_1 = (35)(34) = 1190$ ways. For every such choice, there are $T_2 = 25$ ways of choosing the single supporter of proposal B. Thus there are, by Rule 2.1, a total of $T_1T_2 = [(35)(34)](25)$ samples exhibiting each kind of order. Hence the total number of committees in which 2 supporters of proposal A appear is

$$\begin{aligned} K &= \text{number of samples of type AAB} \\ &\quad + \text{number of type ABA} + \text{number of type BAA} \\ &= (35)(34)(25) + (35)(34)(25) + (35)(34)(25) = 89,250. \end{aligned}$$

By Rule 1.2,

$$P\{\text{exactly 2 members of the committee favor proposal A}\} = \frac{K}{M} = \frac{89,250}{205,320} = 0.435.$$

By use of Rule 2.3 in both numerator and denominator, we can also determine that

$$P\{\text{all 3 committee members favor proposal A}\} = \frac{\text{number of samples of type AAA}}{M} = \frac{(35)(34)(33)}{(60)(59)(58)} = 0.191.$$

and

$$P\{\text{no member favors proposal A}\} = \frac{\text{number of samples of type BBB}}{M} = \frac{(25)(24)(23)}{(60)(59)(58)} = 0.067.$$

Finally, using Rules 2.1 and 2.3, we have

$$\begin{aligned} P\{\text{exactly 1 member favors proposal A}\} &= \frac{(35)(25)(24) + (35)(25)(24) + (35)(25)(24)}{(60)(59)(58)} \\ &= \frac{(3)(35)(25)(24)}{(60)(59)(58)} = 0.307. \end{aligned}$$

◆

EXAMPLE 2.5

An ecologist studying grazing patterns of deer in a particular area notes that there are 50 deer, 10 of which are male (and 40 of which are female). At three separate times the identities of the deer grazing closest to a predetermined point are recorded. If deer graze at random, the sample so obtained can be regarded as a simple random sample of size $n = 3$ *with replacement* from the population of $N = 50$ deer. (Note that the same deer can be observed more than once.) Following the arguments used in Example 2.4 (with A representing male deer, B representing female deer), but with Rule 2.3 replaced by Rule 2.2 (because sampling is with replacement), we can show that

$$P\{\text{3 male deer are observed}\} = \frac{(10)^3}{(50)^3} = 0.008,$$

$$P\{\text{exactly 2 male deer are observed}\} = \frac{(3)(10)^2(40)}{(50)^3} = 0.096,$$

$$P\{\text{exactly 1 male deer is observed}\} = \frac{(3)(10)(40)^2}{(50)^3} = 0.384,$$

$$P\{\text{no male deer are observed}\} = \frac{(40)^3}{(50)^3} = 0.512.$$

◆

Although Examples 2.3 to 2.5 are typical of probability calculations that arise in random sampling contexts, other kinds of problems can be encountered. The following example makes use of the distinction between with replacement and without replacement to obtain an intriguing result.

EXAMPLE 2.6

Suppose that the dates of birth of n persons assembled in a room are a simple random sample *with replacement* of size n from among the $N = 365$ days of the year (excluding February 29). What is the probability of the event E that no two people in the room have the same birthday? To obtain the probability of this event, we count the number K of ordered samples of size n from among the $N = 365$ days of the year in which no day (birthdate) appears more than once, and divide this number K by the total number M of ordered samples of size n with replacement from among the $N = 365$ days of the year. From Rule 2.2, $M = 365^n$ and from Rule 2.3, $K = (365)(364) \cdots (365 - n + 1)$. For example, if $n = 2$,

$$P(E) = \frac{K}{M} = \frac{(365)(364)}{(365)^2} = \frac{364}{365} = 0.997,$$

whereas if $n = 5$,

$$P(E) = \frac{K}{M} = \frac{(365)(364)(363)(362)(361)}{(365)^5} = 0.973,$$

and when $n = 22$,

$$P(E) = \frac{K}{M} = \frac{(365)(364) \cdots (344)}{(365)^{22}} = 0.525.$$

Interestingly enough, when $n = 23$,

$$P(E) = \frac{K}{M} = \frac{(365)(364) \cdots (344)(343)}{(365)^{23}} = 0.493.$$

Thus if the assumptions of this example hold, the probability that at least two people have the same birthdate at a party of $n = 23$ persons is

$$P(E^c) = 1 - P(E) = 1 - 0.493 = 0.507.$$

Considering that there are 365 days to choose from, the result that there is better than a 50–50 chance of a match in birthdays among 23 people appears counterintuitive. ◆

The next example extends the calculations in Examples 2.3 to 2.5 to samples from populations that have units belonging to more than two categories.

EXAMPLE 2.7

Suppose that there are 4 black socks, 8 green socks, and 6 yellow socks all mixed up in a dresser drawer. If one selects 2 socks without looking, what is the probability that the socks will have the same color?

Note that there are a total of $N = 18$ socks in the drawer. Because 2 socks are drawn simultaneously from the drawer, the socks are being selected by simple random sampling without replacement. Thus the total number of possible outcomes (samples) is, by Rule 2.3,

$$M = (18)(17) = 306.$$

The 2 socks selected will have the same color if both are black (BB), both are green (GG), or both are yellow(YY). For both socks to be black, they must be taken from among the 4 black socks in the drawer; by Rule 2.3 there are $(4)(3) = 12$ ways to do this. Similarly, there are $(8)(7) = 56$ ways to obtain samples of the type GG, and $(6)(5) = 30$ ways to obtain samples of the type YY. Hence because these three types of samples are mutually exclusive, there are

$$K = 12 + 56 + 30 = 98$$

outcomes (samples) in the event "both socks have the same color." By Rule 1.2,

$$P\{\text{both socks have the same color}\} = \frac{K}{M} = \frac{98}{306} = 0.320. \quad \blacklozenge$$

Lotteries

Lotteries furnish an application of simple random sampling. The use of a lottery as a fair method of choosing among rival candidates is quite old. The Old Testament contains a number of examples in which lotteries were used to determine the division of land, the choice of a scapegoat as a sacrifice, the allocation of duties in the temple, as well as in the division of an estate between brothers. [For details concerning these examples, see Hasofer (1970) and Rabinovitch (1973).]

More recently, the lottery has been used in a variety of circumstances. Its use in gambling is tradiitonal. Various states now use lotteries as a source of income. Lotteries are also used to select random samples of units, individuals, and so on, for various purposes in scientific research. Jury lists are often selected by a lottery from among all eligible voters in a given community. In the state of Arizona there is an annual hunt of buffalo. The U.S. Department of the Interior determines how thin the herd should be and designates certain animals for the hunt. Hunters register for the hunt and are then chosen by lottery (*New York Times,* October 12, 1971, p. 45).

Perhaps the most controversial lotteries in the United States in this century have been the draft lotteries of 1917, 1940, and 1970. The big issue in the controversy was not whether a lottery is a fair procedure, but rather whether equal probabilities were actually guaranteed to all the possible outcomes by the method used to run the lottery. [For details, see Fienberg (1971).]

3 Combinatorial Analysis

Combinatorial analysis is the calculus of counting. Counting rules are useful when the number of items or possibilities is very large, so that a simple listing is impractical. As seen in Section 2, counting rules can be useful in dealing with uniform probability models. However, such rules are merely shortcuts to computation and are not intrinsic to the understanding of the theory of probability.

Combining Choices

Many counting problems involve counting the number of distinct ways in which one can combine the results of a series of choices. For example, a restaurant may provide 4 possible choices of an appetizer, 3 choices of an entrée, 4 choices of a dessert, and 5 choices of beverage. We may be interested in the number of meals we can select by choosing one item from each of the foregoing categories.

Let us start by asking how many combinations of appetizers and entrées we can choose. We can represent the possibilities in a **counting tree,** as shown in Exhibit 3.1. Here the 4 limbs of the tree represent our choices of appetizer, and the 3 branches on each limb represent our choices of entrée. Each limb–branch combination represents a different choice available to us. We see that there are $3 + 3 + 3 + 3 = (4)(3) = 12$ possible choices. The argument we have used here is easily generalized to verify Rule 2.1 in Section 2.

Returning to our problem of choosing a meal, we see that for every one of the $T_1 = (4)(3) = 12$ choices of appetizer and entrée, there are $T_2 = 4$ choices of a dessert. Thus by Rule 2.1 there are

$$T_1T_2 = [(4)(3)](4) = (4)(3)(4) = 48$$

available combinations of appetizer, entrée, and dessert. Again, for every one of these $T_1 = (4)(3)(4) = 48$ choices, there are $T_2 = 5$ choices of beverage. Consequently, another application of Rule 2.1 shows that there are a total of

$$[(4)(3)(4)](5) = (4)(3)(4)(5) = 240$$

possible meals that we can choose.

The recursive type of argument just used enables us to apply Rule 2.1 whenever we have a series of choices to make, and we wish to count the number of distinct combinations of choices available to us.

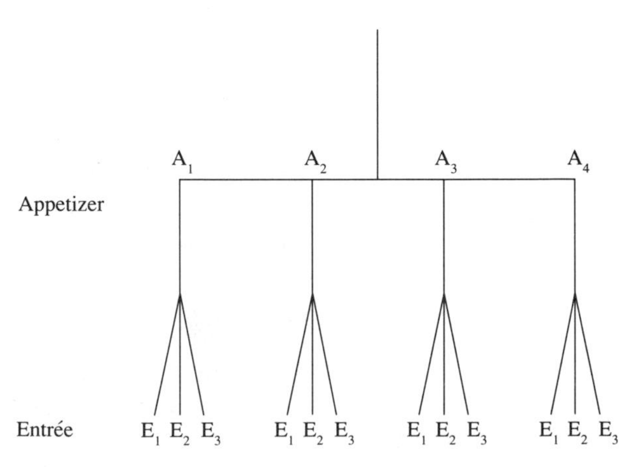

EXHIBIT 3.1 *A counting tree.*

EXAMPLE 3.1

Suppose that 50 individually tagged fish are in a pond. If we catch 3 fish, one at a time, and record the identity of each fish as it is caught, how many possible catches are there?

The answer depends on whether or not fish are returned to the pond after they are caught. In either case, there are 50 ways to catch the first fish. If the first fish is returned to the pond after being caught, 50 fish are still available in the pond when we make our second catch. Thus, by Rule 2.1, there are $(50)(50) = (50)^2$ ways to catch the first two fish. Again, if we also return the second fish caught to the pond, there are 50 fish available for our third catch. Thus there are a total of

$$(50)^2(50) = (50)^3 = 125{,}000$$

distinct ways to catch the three fish.

On the other hand, if fish are not returned to the pond after they are caught, only 49 fish remain in the pond after the first fish is caught. Thus there are (50)(49) ways to catch the first two fish. Since both of these fish are not returned to the pond, there are only 48 fish remaining in the pond for the third catch. Hence when caught fish are not returned to the pond, there are a total of

$$[(50)(49)](48) = (50)(49)(48) = 117{,}600$$

distinct (ordered) ways to catch 3 fish. ♦

Example 3.1 is an example of (ordered) sampling. By the reasoning used in this example, we see that there are

$$(N)(N) \cdots (N) = N^n$$

possible ordered samples of size n that can result from sampling with replacement from a population of N units. This fact is stated as Rule 2.2 in Section 2. Similarly, there are

$$(N)(N-1) \cdots (N-n+1)$$

possible ordered samples of size n that can result from sampling without replacement from the given population. This fact is stated as Rule 2.3 in Section 2.

Because the product $(N)(N-1) \cdots (N-n+1)$ is frequently used in sampling contexts, it is helpful to have a notation for this product. Thus we write

$$N_{(n)} = (N)(N-1) \cdots (N-n+1). \tag{3.1}$$

For example,

$$5_{(2)} = (5)(4) = 20, \qquad 6_{(3)} = (6)(5)(4) = 120,$$
$$10_{(4)} = (10)(9)(8)(7) = 5040,$$

and so on.

Permutations

A special case of sampling without replacement is when the number n of units sampled equals the size of the population. Because units sampled from the population are not returned and are listed in the order in which they are selected, the samples obtained in this way represent an ordering of the entire population. From equation (3.1), the number of ways in which we can order a population of n distinct items (units, objects) is

$$(3.2) \quad n_{(n)} = (n)(n-1)\cdots(n-n+1) = (n)(n-1)(n-2)\cdots(2)(1),$$

the product of all of the numbers from 1 to n. Instead of using the symbol $n_{(n)}$, it is more common to use the factorial symbol $n!$ (read "n factorial"). An arrangement of n distinct objects in a given order is called a **permutation.**

> **Rule 3.1.** The number of distinguishable permutations (arrangements, orderings) of n distinct items is
>
> $$n! = (n)(n-1)\cdots(1).$$

Thus

$$1! = 1, \quad 2! = 2, \quad 3! = 6, \quad 4! = 24, \quad 5! = 120, \quad 6! = 720, \quad 7! = 5040,$$

and so on. For mathematical convenience in writing formulas, we define $0!$ to be 1.

EXAMPLE 3.2

Six different wines are to be used in a taste-testing experiment. The number of distinct ways in which the wines can be ordered is

$$6! = 720.$$

◆

A useful formula is obtained by noting that if $1 \le n \le N$,

$$
\begin{aligned}
N_{(n)} &= (N)(N-1)\cdots(N-n+1) \\
&= \frac{[(N)(N-1)\cdots(N-n+1)][N-n)(N-n-1)\cdots(2)(1)]}{(N-n)(N-n-1)\cdots(2)(1)} \qquad (3.3)\\
&= \frac{N!}{(N-n)!}.
\end{aligned}
$$

Unordered Samples

A sample of n items taken without replacement does not have to be taken one at a time. One can select all n items at once. For example, in Example 3.1 we could have caught the fish we wanted in a net. Consequently, the order in which the items were selected for our sample may not be relevant. We may only be interested in *which* items were chosen.

In Chapter 2 we noted the distinction between the notation in which a list of items is given in parentheses [e.g., (u_3, u_1, u_4)], which indicates that the items on the list appear in order, and the set notation where the list appears in braces (e.g., $\{u_3, u_1, u_4\}$), where no order is intended and only membership in the collection is shown. The quantity $N_{(n)}$ tells us how many distinguishable *ordered* lists of n items can be constructed as samples from a population of N distinct items. Now, however, we are interested in counting how many distinguishable *unordered* lists (collections) of n items can be constructed from a population of N distinct items. Clearly, there are more possible ordered samples than unordered samples, because every distinct unordered sample of n items can be ordered in $n!$ ways.

Indeed, we can think of obtaining an ordered sample of n items in two steps: (1) choose an unordered sample of n items, and (2) order this sample. Denote by X the number of distinguishable ways in which we can choose an unordered sample of n items. There are then X ways to complete step 1. For every choice of an unordered sample of size n in step 1, there are $n!$ ways of ordering this sample. Thus by Rule 3.1,

$$N_{(n)} = \text{number of ordered samples} = (X)(n!),$$

so that

$$X = \frac{N_{(n)}}{n!} = \frac{N!/(N-n)!}{n!} = \frac{N!}{n!\,(N-n)!}.$$

Rule 3.2. The number of distinguishable *unordered* samples of size n that can be drawn without replacement from a population of N distinct items is given by the expression

$$\binom{N}{n} \equiv \frac{N!}{n!\,(N-n)!} = \frac{N_{(n)}}{n!} = \frac{(N)(N-1)\cdots(N-n+1)}{(n)(n-1)\cdots(2)(1)}.$$

EXAMPLE 3.1 (continued)

In this example, there were 50 tagged fish in a pond. If instead of catching 3 fish one at a time (without replacement), we netted the 3 fish all at once, the number of different collections (unordered samples) of fish we could obtain is

$$\binom{50}{3} = \frac{50!}{3!\,47!} = \frac{(50(49)(48)}{(3)(2)(1)} = \frac{117{,}600}{6} = 19{,}600$$

Notice how small (19,600 versus 117,600) the number of unordered samples of 3 fish is when compared to the number of ordered samples of 3 fish. ♦

The following example may help further clarify the distinction between ordered and unordered samples.

EXAMPLE 3.3

An ice cream store sells 33 flavors of ice cream. Suppose that a two-scoop cone contains different flavored scoops. Because the top scoop is eaten before the bottom scoop, we can distinguish a cone with chocolate on top of vanilla from a cone with vanilla on top of chocolate. Thus every two-scoop cone is an ordered sample without replacement of $n = 2$ flavors from among $N = 33$ possibilities. There are $33_{(2)} = (33)(32) = 1056$ distinguishable samples (two-scoop cones). If, instead, both scoops were combined into a cup, you could only distinguish which flavors were present. Since order would be irrelevant, there would be only

$$\binom{33}{2} = \frac{33_{(2)}}{2!} = \frac{1056}{2} = 528$$

distinguishable two-scoop cups. ♦

One of the most important uses of Rule 3.2 is in arranging n items belonging to two distinct types, where there are k items of type A and $n - k$ items of type B. For example, we may have k black tiles and $n - k$ white tiles to arrange in a row. Items of the same type cannot be told apart. Thus we can only distinguish arrangements by the patterns that they display. Exhibit 3.2 shows the 10 distinguishable arrangements (designs) formed by 3 black tiles and 2 white tiles.

To determine any such arrangement, we need only select k locations in which to place the items of type A. The $n - k$ items of type B will then be placed into the remaining locations. Because we only want to know *which* locations are to be chosen for the type A items (the order in which the choice is made is irrelevant to us), Rule 3.2 applies. Consequently, there are

$$(3.4) \qquad \binom{n}{k} = \frac{n!}{k!\,(n-k)!} = \frac{(n)(n-1)\cdots(n-k+1)}{(k)(k-1)\cdots(2)(1)}$$

possible arrangements.

Rule 3.3. The number of distinguishable arrangements (orderings) of n items, of which k are of type A and $n - k$ are of type B (type A and type B being distinct) is $\binom{n}{k}$.

EXAMPLE 3.4

The number of distinguishable designs that can be made with 3 black tiles and 2 white tiles is

$$\binom{3+2}{3} = \binom{5}{3} = \frac{(5)(4)(3)}{(3)(2)(1)} = 10,$$

as we can also see from Exhibit 3.2. ◆

Notice that we could just as easily have chosen $n - k$ locations in which to place the items of type B, instead of choosing locations for the type A items. It follows that

$$(3.5) \qquad \binom{n}{k} = \binom{n}{n-k}.$$

Equation (3.5) is easily verified algebraically by noting that from equation (3.4),

$$\binom{n}{k} = \frac{n!}{k!\,(n-k)!} = \frac{n!}{(n-k)!\,k!} = \binom{n}{n-k}.$$

BBBWW	BWBWB
BBWBW	WBBWB
BWBBW	BWWBB
WBBBW	WBWBB
BBWWB	WWBBB

EXHIBIT 3.2 *All arrangements of 3 black and 2 white tiles.*

Binomial Coefficients

The expression

$$\binom{n}{k} = \frac{n!}{k!\,(n-k)!} = \binom{n}{n-k}$$

is called the kth **binomial coefficient.** The terminology "binomial coefficient" arose historically because $\binom{n}{k}$ is the coefficient of $a^k b^{n-k}$ in the binomial expansion

$$(a + b)^n = \sum_{j=0}^{n} \binom{n}{j} a^j b^{n-j}.$$

Values of $\binom{n}{k}$ for $n = 1, 2, \ldots, 10$ and various values of k are presented in Exhibit 3.3. The binomial coefficients can be arranged to form a triangle-like figure called **Pascal's triangle,** where each coefficient is obtained as the sum of the two coefficients directly above it (see Exhibit 3.4).

Multinomial Coefficients*

Rule 3.3 can be extended to situations where the n items to be arranged belong to r distinct types and consist of n_j items of type j, $j = 1, 2, \cdots, r$.

EXHIBIT 3.3 *Values of the kth Binomial Coefficient* $\binom{n}{k}$ *for* $k \leq n$, *1, 2, · · ·, 10*

	k										
n	0	1	2	3	4	5	6	7	8	9	10
1	1	1									
2	1	2	1								
3	1	3	3	1							
4	1	4	6	4	1						
5	1	5	10	10	5	1					
6	1	6	15	20	15	6	1				
7	1	7	21	35	35	21	7	1			
8	1	8	28	56	70	56	28	8	1		
9	1	9	36	84	126	126	84	36	9	1	
10	1	10	45	120	210	252	210	120	45	10	1

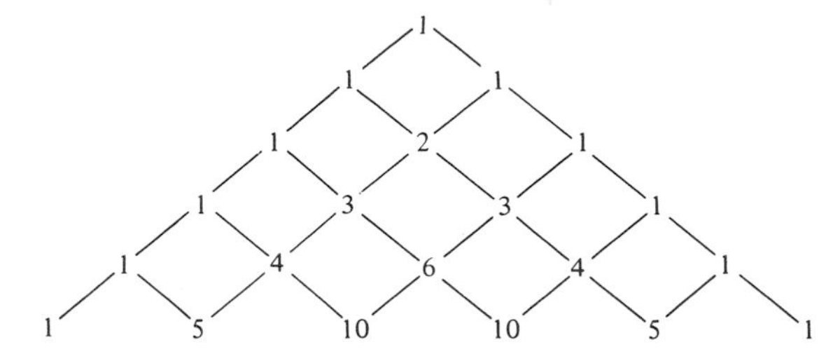

EXHIBIT 3.4 *Pascal's triangle.*

> **Rule 3.4.** The number of distinguishable arrangement of n items, n_1 of which are of type 1, n_2 of type 2, . . . , and n_r of type r, is
>
> $$\binom{n}{n_1, n_2, \ldots, n_r} = \frac{n!}{n_1!\, n_2! \cdots n_r!}.$$
>
> Here $n_1 + n_2 + \cdots + n_r = n$ and the types $1, 2, \ldots, r$ are distinct.

The quantity $\binom{n}{n_1, n_2, \ldots, n_r}$ is known as a **multinomial coefficient** because it is the coefficient of $(a_1^{n_1} a_2^{n_2} \cdots a_r^{n_r})$ in the multinomial expansion

$$(a_1 + a_2 + \cdots + a_r)^n = \sum_{\substack{0 \leq n_i \leq n \\ n_1 + \cdots + n_r = n}} \binom{n}{n_1, n_2, \ldots, n_r} a_1^{n_1} a_2^{n_2} \cdots a_r^{n_r}.$$

EXAMPLE 3.5

We wish to find a pleasing abstract design by placing patterned tiles in 20 positions in a row. Seven of the tiles are of pattern A, 5 are of pattern B, 4 are of pattern C, and 4 are of pattern D. From Rule 3.4, we thus have a total of 20!/(7! 5! 4! 4!) = 6,983,776,800 designs from which to choose. ◆

EXAMPLE 3.6

Card games are particularly well suited as a basis for examples of use of the multinomial coefficient because a deck of 52 cards contains 4 suits, each of 13 cards: hearts, diamonds, spades, clubs.

In dealing 52 cards to 4 players the number of ways that the first receives 7 hearts, the second receives 3 hearts, the third receives 2 hearts, and the fourth receives 1 heart is obtained directly from the multinomial coefficient:

$$\frac{13!}{7!\, 3!\, 2!\, 1!} = 102{,}960.$$

The number of ways that one particular player receives more than three times as many hearts as any other player is

$$\frac{13!}{13!\,0!\,0!\,0!} + \frac{13!}{12!\,1!\,0!\,0!} + \frac{13!}{11!\,2!\,0!\,0!} + \frac{13!}{11!\,1!\,1!\,0!} + \frac{13!}{10!\,3!\,0!\,0!} + \frac{13!}{10!\,2!\,1!\,0!} + \frac{13!}{10!\,1!\,1!\,1!} + \frac{13!}{9!\,2!\,2!\,0!} + \frac{13!}{9!\,2!\,1!\,1!} + \frac{13!}{8!\,2!\,2!\,1!} + \frac{13!}{7!\,2!\,2!\,2!} = 209{,}028. \quad \blacklozenge$$

Notes and References

A number of books contain extensive tables of binomial coefficients. For example, in the *Handbook of Mathematical Functions,* edited by Abramowitz and Stegun (1965), binomial coefficients are provided for $n = 1(1)50$. The notation $n = 1(1)50$ means that the first value of n is 1, the last value of n is 50, and the steps are in units of 1. A table of the factorials is given in *Tables of $n!$ and $\Gamma(n + \frac{1}{2})$ for the First Thousand Values of n* [National Bureau of Standards (1951)].

Glossary and Summary

Finite probability model: A probability model whose sample space contains a finite number of outcomes.

Simple events: Events containing only one outcome.

Uniform probability model: A finite probability model in which the simple events all have the same probability.

Unit: A member of a population.

Population size N: The number of units in a population.

Sample size n: The number of units to be sampled; the number of observations.

Random sampling: A method of choosing samples in which each sample has a specified probability of being drawn.

Ordered sample: A collection of units distinguished not only by which units appear in the collection, but also by the order in which these units are listed.

Unordered sample: A collection of units distinguished only by which units appear in the collection (order not important).

Simple random sampling with replacement: Random sampling in which every ordered sample has the same probability of being selected.

Simple random sampling without replacement: Random sampling where zero probability is given to samples in which the same unit appears more than once, and equal probability is given to all other ordered samples.
Table of random numbers: An array of digits generated by a process (a random number generator) such that every digit has probability $\frac{1}{10}$ of being 0, 1, 2, . . . , 8 or 9; every pair of digits has probability $\frac{1}{100}$ of being 00, 01, . . . , 98, 99; and so on.
Lottery: An application of simple random sampling to make choices or for gambling.
Combinatorial analysis: The arithmetic of counting; counting rules.
Permutation: An ordering or arrangement of distinct objects.

Rules and Formulas

Rule 1.1 The probability of any event E is the sum of the probabilities of the simple events whose union is E (assuming a finite probability model).

Rule 1.2 In a finite probability model, the sum of the probabilities of all simple events is equal to 1.

Rule 1.3 In a uniform probability model, the probability of any simple event is $1/M$ and the probability of any event E is K/M, where M is the number of outcomes in the sample space and K is the number of outcomes in the event E.

Rule 2.1 If there are T_1 ways to do task 1 and T_2 ways to do task 2 (once task 1 has been done), the total number of ways in which the two tasks can be done together is $R = T_1T_2$. The total number of ways to do k tasks together, if there are T_i ways to do task i, $1 \leq i \leq k$, is $T = T_1T_2 \cdots T_{k-1}T_k$.

Rule 2.2 The total number of ordered samples of size n drawn with replacement from a population of N units is N^n.

Rule 2.3 The total number of ordered samples of size n from a population of N units in which no unit appears more than once in the sample (without replacement) is $N_{(n)} = N(N-1) \cdots (N-n+1)$.

Rule 3.1 The number of distinguishable permutations (orderings, arrangements) of n distinct items is $n! = n(n-1) \cdots (2)(1)$. (*Note:* $0! = 1$.)

Rule 3.2 The number of distinguishable unordered samples of size n that can be drawn without replacement from a population of N distinct units is

$$\binom{N}{n} = \frac{N!}{n!\,(N-n)!} = \frac{N_{(n)}}{n!} = \frac{N(N-1)\cdots(N-n+1)}{n(n-1)\cdots(2)(1)}.$$

Rule 3.3 The number of distinguishable arrangements of n items of which k are of type A and $n - k$ of type B (type A and type B being distinct) is given by the **binomial coefficient:**

$$\binom{n}{k} = \frac{n!}{k!\,(n-k)!} = \binom{n}{n-k}.$$

Rule 3.4. The number of distinguishable arrangements of n items, n_1 of which are of type 1, n_1 of type 2, . . . , and n_r of type r (the types being distinct and $n_1 + n_2 + \cdots + n_r = n$) is given by the **multinomial coefficient:**

$$\binom{n}{n_1, n_2, \ldots, n_r} = \frac{n!}{n_1!\, n_2! \cdots n_r!}.$$

Exercises

1. Each of two players separately chooses to hold up one, two, or three fingers. Player I wins if the sum of the number of fingers held up by the two players is an even integer. Player II wins if the sum is an odd integer.
   (a) List the possible outcomes for one play of this game.
   (b) If the players make their choices at random, each of the possible outcomes of the game has the same probability. What is the probability that player I will win this game?
   (c) Under the assumptions of part (b), what is the probability that the sum of the number of fingers held up by the two players is equal to 4?
2. In the game of roulette described in Example 1.2, find probabilities of the following events.
   (a) E_1: the wheel stops at a number that strictly exceeds 18.
   (b) E_2: the wheel stops at a number (other than 0 and 00) that is divisible by 3.
   (c) $E_1 \cap E_2$.
   (d) $E_1 \cup E_2$.
3. Families with three sons are studied with respect to color blindness.
   (a) List the possible outcomes of the experiment in which a family is randomly drawn, and for each son (in order of birth) it is recorded whether or not he is color blind.
   (b) Suppose that we have a uniform probability model for this experiment.
      (i) Find the probability that the oldest son is color blind.

(ii) Find the probability that exactly one son is color blind.
(iii) Find the probability that at least one son is color blind.

4. An urn contains 5 red balls and 5 black balls. Person A takes a simple random sample of $n = 2$ balls *with* replacement from the urn, and then replaces the balls in the urn. Then Person B takes a simple random sample *without* replacement of $n = 2$ balls. Let E be the event "the sample contains 1 red and 1 black ball." Find $P(E)$ for: (a) person A's sample; (b) person B's sample.

5. A botanist studying the process of fertilization in a certain species of plant chooses a random sample of $n = 2$ plants without replacement from a greenhouse containing 30 plants of the desired species. Unknown to the botanist, 3 of the 30 plants have damaged buds and cannot flower. Find the probability that:
(a) None of the chosen plants flower.
(b) All of the chosen plants flower.
(c) Exactly 1 of the chosen plants flowers.

6. Repeat Exercise 5 assuming that $n = 3$ plants are sampled.

7. Before accepting a shipment of 20 new dishwashers, an appliance dealer randomly samples $n = 3$ dishwashers *without* replacement from among the 20 and has a service person inspect these dishwashers thoroughly for defects. Suppose that 2 of the 20 new dishwashers in the shipment are defective. Find the probability that *at least one* defective dishwasher is obtained in the sample.

8. When a personal computer generates successive random digits, the probability model is that of a simple random sample *with replacement* from the population $\{0, 1, 2, 3, 4, 5, 6, 7, 8, 9\}$.
(a) How many outcomes are there in the sample space for this experiment?
(b) What is the probability that the computer produces the outcome $(1, 2, 3, 4)$?
(c) What is the probability that the 4 integers generated are all different?
(d) What is the probability that the 4 integers are all the same?
(e) What is the probability that 3 of the integers are the same and the fourth integer is different? (Be careful!)

9. At a charity carnival, 60 people each buy one ticket for a raffle. Three prizes are to be awarded (there will be 3 draws), and a person is allowed to win more than one prize (the winning ticket is replaced after each draw).
(a) How many possible outcomes are there?
(b) What is the probability that three different persons win prizes?
(c) What is the probability that one person wins at least two prizes?
(d) Ten of the 60 ticket holders belong to the same club. What is the probability that at least one member of the club wins a prize?

10. A dentist has a unique way of deciding how much to charge patients. When it comes time to present the bill, the dentist asks the patient to sample $n = 3$ balls randomly without replacement from an urn in which there are 20 balls. Of the balls in the urn, 9 are marked \$5, 7 are marked \$10, and 4 are marked \$15. The patient pays the *sum* of the amounts shown on the balls, unless all 3 balls chosen have the *same* amount marked on them—in which case, the patient pays nothing.
(a) What is the probability that the patient pays nothing?
(b) What is the probability that the patient pays \$40?
(c) What is the probability that the patient pays \$45?

11. A mechanic has a jar containing 6 screws of the straight-slot type and 3 screws of the Phillips type. She needs 2 screws of each type for a job she is doing, but because of her position she cannot see into the jar. Thus she blindly reaches into the jar and randomly draws out 4 screws. What is the probability that she gets the combination of screws (two of each type) that she needs?

12. A child takes a handful of 3 cookies at random from a jar of cookies containing 10 chocolate chip cookies, 8 vanilla wafers, and 5 sugar cookies. (Note that this sample must be a simple random *without* replacement. Very few children will replace cookies without at least a bite!) Find the probability that:
(a) All 3 of the cookies picked by the child are of the same type.
(b) The 3 cookies are each of different type.

13. You are going on a trip and want to take 3 books with you to read on the plane. In your library, there are 20 novels, 15 nonfiction books, and 5 poetry books. You cannot decide what books to take, so you put the titles of the books on slips of paper, shake the slips of paper up in a box, and draw 3 slips at random without replacement from the box. The books whose titles are on the slips of paper you draw are the ones you take on your trip.
(a) If you pay attention to the order in which the books are drawn, how many possible outcomes are there?
(b) What is the probability that the three books you take on your trip are all of the same type?
(c) What is the probability that the three books you take are of different types (i.e., one novel, one nonfiction, one poetry)?

14. A small college has 25 full professors, 20 associate professors, and 55 assistant professors. The college also has an unusual method for choosing commencement speakers—it is done by lottery. The names of all the college faculty are put in an urn and a random sample of $n = 3$ names is drawn *without replacement* by the president of the board of trustees. The first name drawn gives the greeting, the second name gives the main speech, and the last name drawn gives the closing.
(a) What is the probability that an associate professor gives the greet-

ing, a full professor gives the main speech, and an associate professor gives the closing?

(b) What is the probability that the three speeches are given by two associate professors and one full professor?

(c) What is the probability that only the main speech and closing speeches are given by associate professors? (*Hint:* Who gives the greeting? List cases.)

15. At an office Christmas party attended by 30 people, 3 door prizes are to be awarded. Two schemes are being considered for awarding prizes.

Scheme 1. All names of people attending the party are placed in a hat, and 3 names are selected in one draw.

Scheme 2. All names are placed in a hat, and the 3 names are selected one at a time with winning names replaced in the hat after they are drawn. (The hat is shaken vigorously before each draw.)

(a) How many possible outcomes are there for Scheme 1? Scheme 2?

(b) If 5 of the 30 people at the party are managers, what is the probability that at least one manager wins a prize:

(i) If scheme 1 is used?

(ii) If scheme 2 is used?

In each of Exercises 16 to 29, state explicitly the counting rules that you use to solve the problem.

16. The emblems on six flags are respectively a stripe, a dot, a triangle, a rectangle, a bar, and a circle. A signal is obtained by showing 2 different flags, one after the other. How many different signals are possible?

17. Bicycle locks frequently are arranged with 4 adjacent disks, each disk having 5 numbers. The correct combination is an ordered list of 4 numbers.

(a) If we wish to open the lock, how many combinations might need to be tried?

(b) Suppose that the digit on the third disk represents a geographical area where the lock is made. Thus 1 may denote the northeastern states, 2 the southeastern states, 3 the midwest, 4 the northwest, and 5 the southwest. If we know the bicycle comes from a western state (northwest or southwest), how many combinations do we have to try?

18. A dating service chooses one of their 12 male clients and one of their 8 female clients for a prize to attend one of 9 Broadway shows and one of 7 restaurants. How many possible combinations of male, female, show, and restaurant are there?

19. A telephone dial has a finger hole for each of the 10 digits.

(a) How many telephone numbers, each with 7 digits but with no digit repeated, are possible?

(b) How many telephone numbers, each with 7 digits ending in one or more zeros, are possible?

20. (a) How many five-place numbers can be made using each of the following groups of digits?

(i) 3, 4, 5, 8, 9.

(ii) 0, 4, 5, 8, 9.

(iii) 3, 4, 5, 5, 9.

(b) How many *even* five-place numbers can be made using each of the digits 2, 3, 5, 7, and 9?

21. If 13 international diplomats are asked to line up for a group picture with the host diplomat always in the center, in how many distinguishable ways can they be arranged?

22. If 7 types of meat are available for dinners, a different meat for every day of the week, can we vary (permute) our choices of meats so that the meats appear in a different order each week of the year?

23. A wine taster claims to be able to discriminate among 5 different varieties of wine by taste. The taster is blindfolded and given the wine varieties one at a time.

(a) In how many different possible orders could the 5 wine varieties be presented to the wine taster?

(b) The taster was really boasting. When the actual experiment is run, the order in which the 5 varieties are named is chosen at random from among all possible orders. What is the probability that the wine taster guesses correctly?

(c) What is the probability that the wine taster guesses exactly 3 right out of 5?

24. Of 7 chest x-rays, 2 show a disease and the others are normal. The radiologist only pays attention to the distinction between diseased and normal x-rays. In how many possible distinguishable orders could the x-rays be presented to the radiologist?

25. A child has 3 black blocks and 3 white blocks to arrange in one line.

(a) How many possible distinguishable patterns can be made?

(b) If the child arranges the blocks in 2 rows of 3 blocks each, how many distinguishable patterns can be made?

26. A family has a choice of 5 vacation spots. They decide to visit 2 of these spots on their vacation, spending part of their time at each. How many different choices of vacations do they have under each of the following conditions?

(a) We distinguish the order in which they visit the vacation spots that they choose.

(b) We do not distinguish the order in which they visit the chosen vacation spots but name only which vacation spots were chosen.

27. Nine skaters—3 from the United States, 3 from Russia, and 3 from China—compete. At the end of the contest, the skaters will be ranked from best to worst (no ties are permitted), but the scoring will only take account of the countries the skaters represent, not their individual identities.

(a) For the purpose of scoring, how many possible distinguishable outcomes are there to this contest?

(b) How many outcomes correspond to results in which the U.S. skaters are ranked 1, 2, 3?

28. A stained-glass window is made of 25 panes (5 rows of 5 panes each). The artist who constructs the window has 7 blue panes, 5 red panes, 6 green panes, and 7 yellow panes. In how many different distinguishable ways can the panes be arranged to make an abstract design for the window?

29. What is the probability that no 2 people in a room of 4 have the same birth month? (Assume birth months are randomly selected with replacement from the 12 months of the year.)

Theoretical Exercises

T.1. Show that:

(a) $n! = n[(n - 1)!]$.

(b) $N_{(n)} = N[(N - 1)_{(n)}] = (N - n + 1)[N_{(n-1)}]$.

(c) $\binom{n}{k} = \left[\frac{n}{n-k}\right]\binom{n-1}{k} = \left[\frac{n-k+1}{n}\right]\binom{n}{k-1}$.

T.2. (a) Prove that $\binom{n-1}{k-1} + \binom{n-1}{k} = \binom{n}{k}$. This is the basis for Pascal's triangle.

(b) Use mathematical induction on n and part (a) to verify the binomial expansion

$$(a + b)^n = \sum_{i=0}^{n} \binom{n}{i} a^i b^{n-i}.$$

(c) Show that $\sum_{i=0}^{n} \binom{n}{i} = 2^n$. [Use part (b).]

(d) Show that $\sum_{i=0}^{n} \binom{n}{i} a^i = (1 + a)^n$.

T.3. (a) Show that

$$\binom{n-1}{i-1, j, n-i-j+1} + \binom{n-1}{i, j-1, n-i-j+1} + \binom{n-1}{i, j, n-i-j-1} = \binom{n}{i, j, n-i-j}.$$

(b) Use mathematical induction on n and part (a) to verify the **trinomial expansion:**

$$(a + b + c)^n = \sum_{\substack{i,j=0,1,2,\ldots,n \\ i+j\leq n}} \binom{n}{i,\ j,\ n-i-j} a^i b^j c^{n-i-j}.$$

T.4 If a sample space Ω contains T outcomes, show that the number of possible events E (including the null event $\varnothing$ and the sure event Ω) is 2^T. [*Hint:* Any event E can be described by deciding for each outcome ω_i whether or not ω_i belongs to E, $i = 1, \ldots, T$. Alternatively, Exercise T.2(c) may be helpful.]

T.5 In statistical mechanics in physics, there are a variety of probability models used to describe the location of particles. What is called the "phase space" is divided into a large number N of small regions or cells. The models describe how n particles are assigned randomly to these cells.

(a) Suppose that every assignment of particles to cells, where a cell can contain $0, 1, 2, \ldots, n$ particles, is equally likely (Maxwell–Boltzmann "statistics"). If particles can be distinguished from one another, what is the probability that all n particles fall into one cell? (To what kind of simple random sampling does this model correspond?)

(b) Instead, suppose that no cell can contain more than one particle, but otherwise all assignments of particles to cells are equally likely (this requires n to be less than N). If particles can be distinguished from one another, what is the probability of any single outcome? (To what kind of simple random sampling does this correspond?)

(c) Fermi–Dirac "statistics" correspond to the model of part (b), except that particles are indistinguishable. Thus we do not know what particles fall into what cells but only which cells are occupied. All such identifiable possibilities of cell choices for the n particles are equally likely. What is the common probability of any such choice (outcome)?

(d) Bose–Einstein "statistics" correspond to the model of part (a) in that more than one particle can be assigned to a cell. However, particles are indistinguishable. Thus we know which cells are occupied and by how many particles, but not what particles are in what cells. All such possibilities have the same probability. Show that this common probability is

$$1\Big/\binom{N+n-1}{n}.$$

[This calculation is more difficult than parts (a) to (c). See pp. 38–39 of Feller (1968).]

4

Conditional Probability and Probabilistic Independence

In this chapter we introduce the important concept of **conditional probability** and illustrate how knowledge of conditional probabilities aids both in computing probabilities of complex results and in prediction. The closely related and very useful concept of **probabilistic independence** is then discussed. Finally, experiments whose outcomes are n-tuples (pairs, for example) resulting from recording the outcomes of n component experiments are introduced, and it is shown how these **composite experiments** can be modeled using the concept of probabilistic independence.

1 Conditional Probability

In some situations partial knowledge concerning the outcome of a random experiment may become available before the complete result is known. By taking account of such partial information, we may wish to reevaluate the probabilities previously assigned to the outcomes of the experiment.

EXAMPLE 1.1

A woman has breast cancer, for which surgery has been recommended. A magazine article reports that the probability of recurrence of cancer after surgery is 0.30. Her doctor points out that this probability is based on tumors of all sizes. However, the woman's tumor was discovered while it was still small, and the chance of recurrence of cancer after surgery for such tumors is only 0.09. Because of the additional information on the size of the tumor, the woman's chances of complete recovery are much higher than the magazine article had led her to believe. ♦

EXAMPLE 1.2

A high school counselor is approached by a student who plans to drop out of school. The student points to friends who have dropped out of school and gotten jobs, and questions the need to obtain a high school degree. The counselor shows the student the following table (based on 1988 employment data reported on p. 150 of the 1989 *Statistical Abstract of the United States*):

Educational Attainment	*Proportion Employed*
High school graduate	0.770
High school dropout	0.528
All persons (ages 16–24)	0.693

The proportions provide probabilities for a random experiment in which the possible outcomes specify both "educational attainment" and "employment status." The student's decision about continuing in school will determine "educational attainment" and thus will provide partial information about the outcome. When this partial information is available, the proportion, 0.693, of all individuals employed is not particularly relevant to the student's chances of employment. Instead, the appropriate probability of employment is 0.528 if the student drops out of high school and is 0.770 if the student completes a high school degree. ◆

The process of reevaluating probabilities in the light of partial information is made precise by the concept of **conditional probability**. Formally, let A be an event such that $P(A) > 0$, and let B be another event. The conditional probability, $P(B \mid A)$, of B occurring *given* that A has occurred is defined by the equation

$$P(B \mid A) = \frac{P(A \cap B)}{P(A)}. \tag{1.1}$$

If $P(A) = 0$, the conditional probability of B given A is undefined.

To distinguish the conditional probability $P(B \mid A)$ of the event B *given* the event A from the probability $P(B)$ of B evaluated before the experiment begins, we sometimes speak of $P(B)$ as being the **unconditional, marginal,** or **absolute** probability of B. Of course, all probabilities are conditional in the sense that we assign (or calculate) these probabilities in the light of our present knowledge of the properties of whatever random experiment we are observing. Thus even the unconditional probability $P(B)$ could be written $P(B \mid \Omega)$.

EXAMPLE 1.3

A study performed for a local television station asks people (1) whether they regularly watch the station's evening news show, and (2) whether they are familiar with a certain product advertised on that news show. Let B be the event that a person is familiar with the product, and let A be the event that the person regularly watches the news show. The study finds that the following are approximate probabilities for the events A, B and their intersection $A \cap B$:

$$P(A) = 0.40, \qquad P(B) = 0.50, \qquad P(A \cap B) = 0.24.$$

Given that a person regularly watches the news show, the conditional probability that this person is familiar with the product is

$$P(B \mid A) = \frac{P(A \cap B)}{P(A)} = \frac{0.24}{0.40} = 0.60.$$

Note that $P(A^c) = 1 - P(A) = 1 - 0.40 = 0.60$ and that

$$P(B \cap A^c) = P(B) - P(A \cap B) = 0.50 - 0.24 = 0.26.$$

Thus the probability that a person who does not regularly watch the news show is familiar with the product is

$$P(B \mid A^c) = \frac{P(B \cap A^c)}{P(A^c)} = \frac{0.26}{0.60} = 0.43.$$

Because nonviewers of the news program are less likely to be familiar with the product than viewers, the television station argues that advertisement of the product on their news show is an effective way to bring the product to the attention of potential buyers. ◆

Frequency Interpretation of Conditional Probability

The frequency interpretation for the probability $P(E)$ of an event E states that the proportion r.f.(E) of trials in which E occurs in a large number of repetitions of a random experiment tends to be close to $P(E)$. In a similar manner, a frequency interpretation can be given for $P(B \mid A)$. Let $\#A$ and $\#(A \cap B)$ be the number of times that the events A and $A \cap B$, respectively, occur in N repetitions of a random experiment. Then, when N is large,

$$\frac{\#(A \cap B)}{\#A} = \frac{\#(A \cap B)/N}{\#A/N} = \frac{\text{r.f.}(A \cap B)}{\text{r.f.}(A)},$$

which by the frequency interpretation of the measure $P(E)$ tends to be approximately $P(A \cap B)/P(A)$. *Thus the conditional probability of B given A is* *the abstraction of the long-run proportion of times that B occurs among those trials for which A occurs.*

EXAMPLE 1.2 *(continued)*

The table that the high school counselor showed to the student who was thinking of dropping out of high school was derived from Exhibit 1.1. Let B represent "employed" and A represent "high school graduate." The counselor calculated the proportion of all persons who are employed by dividing the total number (9,167,000) of employed people by the population size $N = 13{,}230{,}000$:

$$\text{r.f.}(B) = \frac{\#B}{N} = \frac{9{,}167{,}000}{13{,}230{,}000} = 0.693.$$

The proportion *of high school graduates* who are employed is

$$\frac{\#(A \cap B)}{\#A} = \frac{6{,}932{,}000}{8{,}999{,}000} = 0.770.$$

The italicized phrase "of high school graduates" restricts the population under consideration to that represented by the first row of Exhibit 1.1, for which the population size is $\#A = 8{,}999{,}000$ and the number who are employed is $\#A \cap B = 6{,}932{,}000$. Similarly, the counselor calculated the proportion of high school dropouts who are employed by

$$\frac{\#(A^c \cap B)}{\#A^c} = \frac{2{,}235{,}000}{4{,}231{,}000} = 0.528,$$

where the population of interest is the second row of Exhibit 1.1.

The calculation of conditional probabilities mimics the steps above, but with Exhibit 1.1 replaced by a corresponding table of relative frequencies (which are treated as probabilities). To obtain such a table,

EXHIBIT 1.1 *Employment Status of Americans Aged 16–24*[a]

High School Educational Attainment	Work Status		Total
	Employed	*Not Employed*	
Graduate	6,932,000	2,067,000	8,999,000
Dropout	2,235,000	1,996,000	4,231,000
Total	9,167,000	4,063,000	13,230,000

[a] Excludes high school graduates who continued to college.

every entry in Exhibit 1.1 is divided by $N = 13{,}230{,}000$. The result is shown in Exhibit 1.2. Note that the unconditional probability $P(B)$ of being employed replaces $\#B$, the unconditional probability $P(A)$ of being a high school graduate replaces $\#A$, and the unconditional probability $P(A \cap B)$ of being a high school graduate and employed replaces $\#A \cap B$. The conditional probability, $P(B \mid A)$, that a person is employed *given* that the person is a high school graduate is given by a formula similar to that used to obtain the relative frequency, $\#(A \cap B)/\#A$, from Exhibit 1.1—that is,

$$P(B \mid A) = \frac{P(A \cap B)}{P(A)} = \frac{0.524}{0.680} = 0.770.$$

Similarly,

$$P(B \mid A^c) = \frac{P(A^c \cap B)}{P(A^c)} = \frac{0.169}{0.320} = 0.528$$

corresponds to the earlier calculation of $\#(A^c \cap B)/\#A^c$. Of course, this is to be expected because of the way we defined the probabilities:

$$\frac{P(A \cap B)}{P(A)} = \frac{(\#A \cap B)/N}{(\#A)/N} = \frac{\#(A \cap B)}{\#A}.$$

The crucial points are that the conditional probability calculations are similar to the way the proportions were earlier obtained from Exhibit 1.1, and that the result does not depend on the population size N. Any population yielding the same table of relative frequencies (probabilities) will result in the same proportions (conditional probabilities).

The counselor might also want to tell the student the proportion of employed persons who are high school graduates. Rather than use Exhibit 1.1, this answer can be obtained from conditional probability calculations:

EXHIBIT 1.2 *Probabilities of Employment*

High School Educational Attainment	Work Status		
	Employed, B	*Not Employed, B^c*	*Total*
Graduate, A	$P(A \cap B) = 0.524$	$P(A \cap B^c) = 0.156$	$P(A) = 0.680$
Dropout, A^c	$P(A^c \cap B) = 0.169$	$P(A^c \cap B^c) = 0.151$	$P(A^c) = 0.320$
Total	$P(B) = 0.693$	$P(B^c) = 0.307$	$P(\Omega) = 1.000$

$$P(A \mid B) = \frac{P(A \cap B)}{P(B)} = \frac{0.524}{0.693} = 0.756.$$

Thus approximately $\frac{3}{4}$ of those people who are employed are high school graduates. ◆

The Probability Calculus for Conditional Probabilities

There is another useful way of thinking of conditional probabilities. As already noted, knowing that event A has occurred changes the sample space of the experiment. Rather than expecting the outcome ω of the random experiment to belong to the sample space Ω, it now is known that ω must belong to the event A, which thus becomes the sample space of a new probability model.

In Exhibit 1.3, the old sample space Ω is replaced by a new sample space A, and each event B is replaced by the event $B \cap A$. Because the relative frequency of the event B given that event A has occurred is proportional to r.f.$(B \cap A)$ = r.f.$(A \cap B)$, the new model must have a probability measure P^* that assigns a probability to the event B proportional to the probability of the event $A \cap B$ under the old (unconditional) model; that is,

$$P^*(B) = cP(A \cap B),$$

where c is a constant. Because A is the new sample space, $P^*(A)$ must equal 1, so that

$$1 = P^*(A) = cP(A \cap A) = cP(A).$$

Solving for the constant c, yields $c = 1/P(A)$. Hence for any event B, $P^*(B) = P(A \cap B)/P(A) = P(B \mid A)$.

Before $P(B \mid A)$ can be regarded as a probability measure, it must be shown that $P(B \mid A)$, as defined by (1.1), satisfies Axioms 1, 2, 3 for probabilities. This is shown is Exercise T.1 at the end of the chapter.

Because $P(B \mid A)$ is a legitimate probability measure, it must obey the following rules:

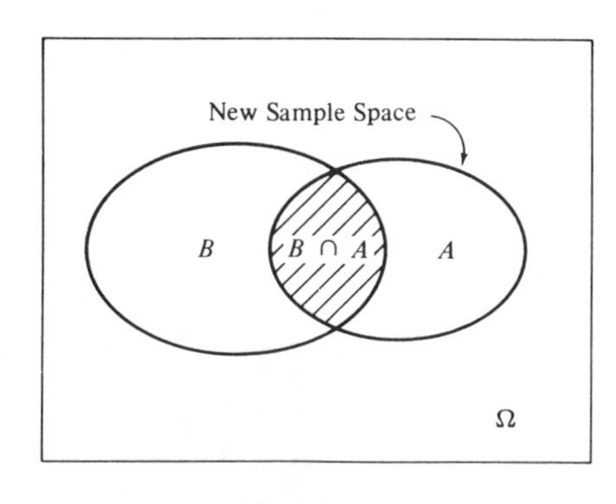

FIGURE 1.3 *Given that A occurs, the event A replaces the sample space Ω, and B ∩ A replaces B.*

Law of Complementation for Conditional Probabilities. For any event E,

$$P(E^c \mid A) = 1 - P(E \mid A).$$

General Law of Addition (Two Events) for Conditional Probabilities. For any two events E_1, E_2,

$$P(E_1 \cup E_2 \mid A) = P(E_1 \mid A) + P(E_2 \mid A) - P(E_1 \cap E_2 \mid A).$$

Other results of the probability calculus for unconditional probabilities (Section 4 of Chapter 2) can be converted to corresponding results for conditional probabilities by simply replacing unconditional probabilities $P(E)$ by corresponding conditional probabilities $P(E \mid A)$ in the relevant formulas. For example, the general law of addition for three events becomes

$$\begin{aligned} P(E_1 \cup E_2 \cup E_3 \mid A) &= P(E_1 \mid A) + P(E_2 \mid A) + P(E_3 \mid A) \\ &\quad - P(E_1 \cap E_2 \mid A) - P(E_1 \cap E_3 \mid A) \\ &\quad - P(E_2 \cap E_3 \mid A) + P(E_1 \cap E_2 \cap E_3 \mid A). \end{aligned} \tag{1.2}$$

EXAMPLE 1.4

In a probability model concerning investments, A is the event that common stock prices fall, E_1 is the event that bond prices rise, and E_2 is the event that commodity prices rise. The model states that

$$\begin{aligned} P(E_1 \mid A) &= P\{\text{bond prices rise} \mid \text{stock prices fall}\} = 0.63, \\ P(E_2 \mid A) &= P\{\text{commodity prices rise} \mid \text{stock prices fall}\} = 0.57, \\ P(E_1 \cap E_2 \mid A) &= P\{\text{bond } \textit{and} \text{ commodity prices rise} \mid \text{stock prices fall}\} \\ &= 0.50. \end{aligned}$$

From this information, the conditional probability that either bond prices or commodity prices rise *given* a drop in stock prices is

$$\begin{aligned} P(E_1 \cup E_2 \mid A) &= P(E_1 \mid A) + P(E_2 \mid A) - P(E_1 \cap E_2 \mid A) \\ &= 0.63 + 0.57 - 0.50 = 0.70, \end{aligned}$$

whereas the conditional probability of a drop in both bond prices and commodity prices *given* a drop in stock prices is

$$\begin{aligned} P(E_1^c \cap E_2^c \mid A) &= P((E_1 \cup E_2)^c \mid A) = 1 - P(E_1 \cup E_2 \mid A) \\ &= 1 - 0.70 = 0.30. \end{aligned}$$

◆

Notation and Translation

It is not always easy to distinguish between conditional and unconditional probabilities in the statement of a problem. It helps to think of a probability as a fraction (a relative frequency) whose denominator depends on the sample space given. For unconditional probabilities $P(E)$, $P(A)$, $P(E \cap A)$, the sample space is understood to be that of all outcomes (Ω). In the notation for conditional probability, the event listed to the right of the vertical slash is the given sample space. Thus for $P(E \mid A)$ the sample space is A, whereas for $P(A \mid E)$ the sample space is E. Consequently, $P(E \mid A)$ and $P(A \mid E)$ have different interpretations. To distinguish among the various kinds of probabilities, it is important to determine what sample space is given.

For example, two events of interest might be ''is married'' (event A) and ''is employed'' (event E). The sample space Ω might be adult males (men). The statement ''50% of *all men* are married'' is translated as the unconditional probability $P(A) = 0.50$. The statement ''40% *of all males* are employed *and married*'' is translated as the unconditional probability: $P(A \cap E) = 0.40$. On the other hand, the statement ''80% *of all married men* are employed'' is translated as the *conditional* probability $P(E \mid A) = 0.80$. Finally, the question: ''What is the probability that an *employed* man is married?'' can be restated as: ''*Among* all employed men, what fraction are married? or ''*Given that* a man is employed, what is the probability that he is married?'' This question thus asks for the value of the conditional probability $P(A \mid E)$. Because the wording of these statements appear very similar, it is important to be clear about what meaning is intended.

Rules for combining conditional probabilities are often misused. For example, the statement

$$P(E \mid A) + P(E^c \mid A) = 1$$

is correct, but the statement

$$P(E \mid A) + P(E \mid A^c) = 1$$

is incorrect. The key point to note is that the two fractions $P(E \mid A)$ and $P(E^c \mid A)$ have the same denominator, whereas the fractions $P(E \mid A)$ and $P(E \mid A^c)$ do not. Adding two conditional probabilities with different given sample spaces is counterintuitive and may not yield meaningful results.

EXAMPLE 1.5

Among registered voters in a certain town, 60% vote in municipal elections, 80% vote in national elections, and 50% of adults vote in both kinds of election. What proportion of those registered voters who vote in national elections do not vote in municipal elections?

The information given can be reexpressed as unconditional probabilities for the events "vote in municipal elections" (event M), "vote in national elections" (event N), and of the intersection $M \cap N$ of these two events. That is,

$$P(M) = 0.6, \qquad P(N) = 0.8, \qquad P(M \cap N) = 0.5.$$

The phrase "of those registered who vote in national elections" restricts the outcomes to the event N. Thus the question asks for the conditional probability $P(M^c \mid N)$. Using equation (4.1) of Chapter 2 yields

$$P(M^c \cap N) = P(N) - P(M \cap N) = 0.8 - 0.5 = 0.3.$$

Consequently, applying the definition (1.1) gives

$$P(M^c \mid N) = \frac{P(M^c \cap N)}{P(N)} = \frac{0.3}{0.8} = 0.375,$$

which is the desired answer.

Note that $P(M \mid N) = P(M \cap N)/P(N) = 0.5/0.8 = 0.625$, so that

$$P(M \mid N) + P(M^c \mid N) = 1,$$

in agreement with the Law of Complements for Conditional Probabilities. On the other hand,

$$P(N^c) = 1 - P(N) = 1 - 0.8 = 0.2,$$
$$P(M \cap N^c) = P(M) - P(M \cap N) = 0.6 - 0.5 = 0.1,$$

so that

$$P(M \mid N^c) = \frac{P(M \cap N^c)}{P(N^c)} = \frac{0.1}{0.2} = 0.5.$$

The sum $P(M \mid N) + P(M \mid N^c) = 0.625 + 0.50 = 1.125$ is not equal to 1 or is even a possible probability. This demonstrates that we cannot add conditional probabilities with different given sample spaces and expect to obtain a meaningful result. ◆

Law of Multiplication

If we know $P(A)$ and $P(B \mid A)$, or instead know $P(B)$ and $P(A \mid B)$, the definition (1.1) of conditional probability tells us how to obtain $P(A \cap B)$:

Law of Multiplication

$$P(A \cap B) = P(A)P(B \mid A) = P(B)P(A \mid B).$$

EXAMPLE 1.6

Long-term records of the performance of a particular type of electric generator show that the probability, $P(A)$, of failure is 0.25. Given that a generator has failed, the conditional probability $P(B \mid A)$ that the failure cannot be repaired is 0.40. Thus the probability that a generator will experience an irreparable failure is

$$P(A \cap B) = P(A)P(B \mid A) = (0.25)(0.40) = 0.10. \quad \blacklozenge$$

The Law of Multiplication can be generalized, using recursive arguments, to any number of events. For example, for three events E_1, E_2, E_3, use of the Law of Multiplication first with $A = E_1 \cap E_2$, $B = E_3$, and again with $A = E_1$, $B = E_2$, yields

$$\begin{aligned} P(E_1 \cap E_2 \cap E_3) = P((E_1 \cap E_2) \cap E_3) &= P(E_1 \cap E_2)P(E_3 \mid E_1 \cap E_2) \\ &= P(E_1)P(E_2 \mid E_1)P(E_3 \mid E_1 \cap E_2). \end{aligned}$$

Because $E_1 \cap E_2 \cap E_3 = E_1 \cap E_1 \cap E_3 = E_3 \cap E_1 \cap E_2$, and so on, we obtain alternative representations

$$\begin{aligned} P(E_1 \cap E_2 \cap E_3) &= P(E_2)P(E_1 \mid E_2)P(E_3 \mid E_1 \cap E_2) \\ &= P(E_3)P(E_3 \mid E_1)P(E_2 \mid E_1 \cap E_3), \end{aligned}$$

say. What is important is to label events in such a way that the required probabilities are available.

A similar result for k events $E_1, E_2, \ldots, E_k$ is

$$\begin{aligned} P(E_1 \cap E_2 \cap \cdots \cap E_k) = P(E_1)&P(E_2 \mid E_1)P(E_3 \mid E_1 \cap E_2) \cdots \\ &P(E_k \mid E_1 \cap E_2 \cap \cdots \cap E_{k-1}). \end{aligned} \tag{1.3}$$

EXAMPLE 1.7

A high school's records reveal that 65% of their graduates enter college. Of those who enter college, 50% receive a college degree. Of those who enter college and receive a college degree, 10% go on to receive a professional or graduate degree. What percentage of all graduates of the high school eventually complete both undergraduate and advanced degrees?

Let E_1 be the event "enters college," E_2 be the event "receives college degree," and E_3 be the event "receives professional or graduate degree." The facts presented can be restated in terms of probabilities: $P(E_1) = 0.65$, $P(E_2 \mid E_1) = 0.50$, and $P(E_3 \mid E_1 \cap E_2) = 0.10$. We are asked for $P(E_2 \cap E_3)$. However, because one cannot receive a college degree without entering a college,

$$E_2 \cap E_3 = E_1 \cap E_2 \cap E_3.$$

Thus, by (1.3),

$$\begin{aligned} P(E_2 \cap E_3) &= P(E_1 \cap E_2 \cap E_3) \\ &= P(E_1)P(E_2 \mid E_1)P(E_3 \mid E_1 \cap E_2) \\ &= (0.65)(0.50)(0.10) = 0.0325. \end{aligned}$$

We conclude that 3.25% of all graduates of the high school eventually complete both undergraduate and advanced degrees. ◆

In Chapter 3, counting rules were used to obtain probabilities connected with a simple random sampling situation. However, any random sample of size n can be conceived of as arising from n consecutive draws from the population. Before each draw is observed, we have the opportunity to reassess our probabilities for that draw and all future draws in the light of the results of the draws already observed.

EXAMPLE 1.8

Example 2.4 of Chapter 3 deals with a population of 60 people, 35 of whom favor proposal A, and 25 who favor proposal B. A committee (sample) of $n = 3$ members is randomly drawn *without replacement* from the 60 people. Let E_j be the event that the jth person drawn for the committee favors proposal A, $j = 1, 2, 3$. Because each person has an equal chance to be drawn on the first draw, and because 35 of these 60 people favor proposal A, $P(E_1) = 35/60$. Given that a person favoring proposal A has been drawn on the first draw (E_1 occurs), each of the $60 - 1 = 59$ remaining people has an equal *conditional probability* to be drawn on the second draw, and $35 - 1 = 34$ of these people favor proposal A. Thus $P(E_2 \mid E_1 = 34/59$. Finally, *given* that the first two people drawn favor proposal A ($E_1 \cap E_2$ occurs), the *conditional* probability $P(E_3 \mid E_1 \cap E_2)$ that the third person drawn favors proposal A is

$$P(E_3 \mid E_1 \cap E_2) = \frac{35 - 2}{60 - 2} = \frac{33}{58}.$$

We conclude from (1.3) that

$$\begin{aligned}&P\{\text{all 3 members favor proposal A}\}\\&\quad = P(E_1 \cap E_2 \cap E_3) = P(E_1)P(E_2 \mid E_1)P(E_3 \mid E_1 \cap E_2)\\&\quad = \left(\frac{35}{60}\right)\left(\frac{34}{59}\right)\left(\frac{33}{58}\right) = 0.191,\end{aligned}$$

which is the result we obtained in Chapter 3 by counting. Similarly,

$$\begin{aligned}P(E_1^c) &= \frac{25}{60},\\ P(E_2^c \mid E_1^c) &= \frac{25-1}{60-1} = \frac{24}{59},\\ P(E_3^c \mid E_1^c \cap E_2^c) &= \frac{25-2}{60-2} = \frac{23}{58},\end{aligned}$$

so that

$$\begin{aligned}&P\{\text{no members favor proposal A}\}\\&\quad = P(E_1^c \cap E_2^c \cap E_3^c) = P(E_1^c)P(E_2^c \mid E_1^c)P(E_3^c \mid E_1^c \cap E_2^c)\\&\quad = \left(\frac{25}{50}\right)\left(\frac{24}{59}\right)\left(\frac{23}{58}\right) = 0.067.\end{aligned}$$

Finally,

$$\begin{aligned}&P\{\text{2 members favor proposal } A \text{ and 1 member favors } \textit{proposal } B\}\\&\quad = P((E_1 \cap E_2 \cap E_3^c) \cup (E_1 \cap E_2^c \cap E_3) \cup (E_1^c \cap E_2 \cap E_3))\\&\quad = P(E_1 \cap E_1 \cap E_3^c) + P(E_1 \cap E_2^c \cap E_3) + P(E_1^c \cap E_2 \cap E_3)\\&\quad = P(E_1)P(E_2 \mid E_1)P(E_3^c \mid E_1 \cap E_2)\\&\qquad + P(E_1)P(E_2^c \mid E_1)P(E_3 \mid E_1 \cap E_2^c)\\&\qquad + P(E_1^c)P(E_2 \mid E_1^c)P(E_3 \mid E_1^c \cap E_2)\\&\quad = \left(\frac{35}{60}\right)\left(\frac{34}{59}\right)\left(\frac{25}{58}\right) + \left(\frac{35}{60}\right)\left(\frac{25}{59}\right)\left(\frac{34}{58}\right) + \left(\frac{25}{60}\right)\left(\frac{35}{59}\right)\left(\frac{34}{58}\right)\\&\quad = \frac{(3)(35)(34)(25)}{(60)(59)(58)} = 0.435.\end{aligned}$$

◆

EXAMPLE 1.9

An auditor has 500 receipts, of which, unknown to the auditor, 10 have errors. The auditor randomly draws one receipt at a time, putting each receipt aside after it is inspected. What is the probability that the auditor draws 3 receipts that have no errors before drawing one that does have an error on it?

Here, sampling is clearly without replacement. Let E_i be the event that the ith receipt drawn contains an error, $i = 1, 2, 3$, and so on. The question asks for the probability of the event $E_1^c \cap E_2^c \cap E_3^c \cap E_4$. Because 490 of the 500 receipts have no error, $P(E_1^c) = 490/500$. Given that the first receipt drawn has no error, there are 499 receipts left, of which 489 are without errors. Thus $P(E_2^c \mid E_1^c) = 489/499$. Similarly, $P(E_3^c \mid E_1^c \cap E_2^c) = 488/498$. Given that 3 receipts without error have been drawn, there are 497 receipts left and 10 of these have errors. Consequently, $P(E_4 \mid E_1^c \cap E_2^c \cap E_3^c) = 10/497$. Applying (1.3) yields

$$\begin{aligned} &P(E_1^c \cap E_2^c \cap E_3^c \cap E_4) \\ &\quad = P(E_1^c)P(E_2^c \mid E_1^c)P(E_3^c \mid E_1^c \cap E_2^c)P(E_4 \mid E_1^c \cap E_2^c \cap E_3^c) \\ &\quad = \left(\frac{490}{500}\right)\left(\frac{489}{499}\right)\left(\frac{488}{498}\right)\left(\frac{10}{497}\right) \\ &\quad = 0.019. \end{aligned}$$

◆

We will have more to say about the use of conditional probabilities for calculating probabilities in simple random sampling in Section 2.

Partitions and Bayes' Rule

Events $E_1, E_2, \ldots, E_k$ are said to **partition** the sample space Ω if no two of these events have outcomes in common (they are mutually exclusive), and their union is Ω. Exhibit 1.4 illustrates a partition of the sample space by four events E_1, E_2, E_3, E_4. The shaded oval represents an event B split up by this partition into nonoverlapping pieces $E_1 \cap B$, $E_2 \cap B$, $E_3 \cap B$, $E_4 \cap B$. That is,

$$B = (E_1 \cap B) \cup (E_2 \cap B) \cup (E_3 \cap B) \cup (E_4 \cap B),$$

where $E_1 \cap B$, $E_2 \cap B$, $E_3 \cap B$, $E_4 \cap B$ are (pairwise) mutually exclusive.

It follows from the Special Law of Addition that

$$P(B) = P(E_1 \cap B) + P(E_2 \cap B) + P(E_3 \cap B) + P(E_4 \cap B) = \sum_{i=1}^{4} P(E_i \cap B).$$

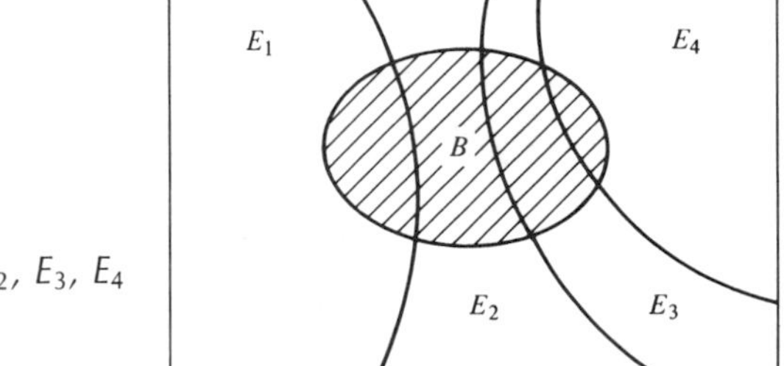

EXHIBIT 1.4 *The events E_1, E_2, E_3, E_4 partition the sample space.*

However, the Law of Multiplication yields

$$P(E_i \cap B) = P(E_i)P(B \mid E_i), \qquad i = 1, 2, 3, 4.$$

Consequently,

$$P(B) = \sum_{i=1}^{4} P(E_i \cap B) = \sum_{i=1}^{4} P(E_i)P(B \mid E_i).$$

The same reasoning can be used to verify the following result.

Law of Partitions. If the events $E_1, E_2, \ldots, E_k$ partition the sample space Ω, the probability of any event B can be calculated as follows:

$$P(B) = \sum_{i=1}^{k} P(E_i \cap B) = \sum_{i=1}^{k} P(E_i)P(B \mid E_i).$$

This law is useful in determining the overall rate of occurrence of a trait in a population that is divided into distinct subpopulations, each of which is a known proportion of the total population and has a known rate of occurrence for the trait in question.

EXAMPLE 1.10

A manufacturer produces motors in three factories. Factory 1 produces 30% of all motors produced, factory 2 produces 50%, and factory 3 produces the remaining 20%. An inspection has shown that 5% of the motors produced by factory 1 are unsatisfactory, 6% of the motors produced by factory 2 are unsatisfactory, and 2% of the motors produced by factory 3 are unsatisfactory. What percentage of all motors produced by the manufacturer are unsatisfactory?

Let E_i be the event "the motor is produced by factory i," $i = 1, 2, 3$, and let B be the event "the motor is unsatisfactory." Note that the sample space Ω consists of all motors (outcomes) produced by the manufacturer, and that events E_1, E_2, E_3 partition Ω. Converting the given information into probabilities, we have

$$\begin{aligned} P(E_1) &= 0.30, & P(E_2) &= 0.50, & P(E_3) &= 0.20, \\ P(B \mid E_1) &= 0.05, & P(B \mid E_2) &= 0.06, & P(B \mid E_3) &= 0.02. \end{aligned}$$

Consequently, by the Law of Partitions,

$$\begin{aligned}P(B) &= P(E_1)P(B \mid E_1) + P(E_2)P(B \mid E_2) + P(E_3)P(B \mid E_3)\\ &= (0.30)(0.05) + (0.50)(0.06) + (0.20)(0.02)\\ &= 0.049.\end{aligned}$$

Thus, converting back to percentages, we see that 4.9% of all motors produced are unsatisfactory. ◆

It is worth noting that an event D and its complement D^c always partition the sample space Ω. The usefulness of this partition is shown in the next example.

EXAMPLE 1.11

A medical test is designed to give a response when a patient has a certain disease. Previous experience shows that the conditional probability of a response (event B) *given* that the patient has the disease (event D) is $P(B \mid D) = 0.80$, whereas the conditional probability of a response *given* that the patient does not have the disease is $P(B \mid D^c) = 0.15$. The probability that a randomly chosen patient has the disease is $P(D) = 0.25$. What is the probability $P(B)$ that the medical test, when applied to a randomly chosen patient, yields a response? Because D and D^c partition the sample space and $P(D^c) = 1 - P(D) = 0.75$, we can apply the Law of Partitions to obtain

$$\begin{aligned}P(B) &= P(D)P(B \mid D) + P(D^c)P(B \mid D^c)\\ &= (0.25)(0.80) + (0.75)(0.15) = 0.3125.\end{aligned}$$ ◆

In Example 1.11, a physician may ask: ''Given that there is a response on the medical test, what is the (conditional) probability $P(D \mid B)$ that the patient has the disease?'' This is a reassessment of the probability that the patient has the disease in the light of the fact that a response has been obtained on the medical test.

By the definition (1.1) of conditional probability,

$$P(D \mid B) = \frac{P(D \cap B)}{P(B)} = \frac{P(D)P(B \mid D)}{P(B)} = \frac{(0.25)(0.80)}{0.3125} = 0.64.$$

Thus if a response has been obtained on the medical test, the probability that the patient has the disease has increased from 0.25 to 0.64.

The method for reassessing probabilities in the light of new information that has just been illustrated has been used since the seventeenth century, when the Reverend Thomas Bayes (1702–1761) first employed the rule that now bears his name.

Bayes' Rule. If k events $E_1, E_2, \ldots, E_k$ partition the sample space Ω, and if B is any other event for which $P(B) > 0$, the conditional probability of any particular partitioning event E_i given that event B has occurred is

$$P(E_i \mid B) = \frac{P(E_i \cap B)}{P(B)} = \frac{P(E_i)P(B \mid E_i)}{\sum_{i=1}^{k} P(E_i)P(B \mid E_i)}, \qquad i = 1, 2, \ldots, k.$$

Observe that Bayes' rule is derived by (1) applying the definition (1.1) of the conditional probability $P(E_i \mid B)$ in terms of $P(E_i \cap B)$ and $P(B)$, (2) using the Law of Partitions to calculate $P(B)$, and (3) using the Law of Multiplication to calculate $P(E_i \cap B)$.

EXAMPLE 1.12

A randomly chosen student takes an examination that is passed by 80% of all "good" students, 60% of all "average" students, and 30% of all "poor" students. In the population, 25% of all students are good students (event E_1), 50% are average (event E_2), and 25% are poor (event E_3). Given that the chosen student passes the exam (event B), what is the conditional probability that the student is a good student?

The information given is that $P(B \mid E_1) = 0.8$, $P(B \mid E_2) = 0.6$, $P(B \mid E_3) = 0.3$, and that $P(E_1) = P(E_3) = 0.25$, $P(E_2) = 0.50$. Thus we can use Bayes' rule to find

$$\begin{aligned}
&P\{\text{student is good} \mid \text{passes exam}\} \\
&\quad = P(E_1 \mid B) \\
&\quad = \frac{P(B \mid E_1)P(E_1)}{\sum_{i=1}^{3} P(B \mid E_i)P(E_i)} \\
&\quad = \frac{(0.8)(0.25)}{(0.8)(0.25) + (0.6)(0.5) + (0.3)(0.25)} = 0.35.
\end{aligned}$$

Because the student passed the exam, the odds have increased that the student is "good." ◆

EXAMPLE 1.13

A manufacturer produces items of which 6% have defect A, another 4% have defect B, and the remainder are nondefective. (No item has both types of defect.) A quality detector correctly identifies as defective 95% of those items with defect A, 75% of those items with defect B, and incorrectly identifies 10% of the nondefective items as defective. All items identified as defective by the detector are rejected; the remainder are packaged for sale. What proportion of the items packaged for sale are defective?

This example illustrates a standard quality control assessment problem. Let A, B, C denote the events "has defect A," "has defect B," "is nondefective," respectively, and let E be the event that an item is packaged. We know that

$$P(A) = 0.06, \qquad P(B) = 0.04, \qquad P(C) = 1 - 0.06 - 0.04 = 0.90.$$

Also, because an item is packaged if the detector does not discard it as defective,

$$P(E \mid A) = 1 - 0.95 = 0.05, \qquad P(E \mid B) = 1 - 0.75 = 0.25,$$
$$P(E \mid C) = 1 - 0.10 = 0.90.$$

We want to know

$$\begin{aligned} &P\{\text{item is defective} \mid \text{item is packaged for sale}\} \\ &\quad = P\{\text{item has defect } A \text{ } or \text{ defect } B \mid \text{item is packaged for sale}\} \\ &\quad = P(A \cup B \mid E) = P(A \mid E) + P(B \mid E), \end{aligned}$$

where the last equality follows from Axiom 3 applied to conditional probabilities (the events A, B are mutually exclusive). Thus, by two applications of Bayes' rule, we have

$$\begin{aligned} P(A \mid E) &= \frac{P(A)P(E \mid A)}{P(A)P(E \mid A) + P(B)P(E \mid B) + P(C)P(E \mid C)} \\ &= \frac{(0.06)(0.05)}{(0.06)(0.05) + (0.04)(0.05) + (0.90)(0.90)} \\ &= 0.0036, \\ P(B \mid E) &= \frac{P(B)P(E \mid B)}{P(A)P(E \mid A) + P(B)P(E \mid B) + P(C)P(E \mid C)} \\ &= \frac{(0.04)(0.25)}{(0.06)(0.05) + (0.04)(0.25) + (0.90)(0.90)} \\ &= 0.0122, \end{aligned}$$

and the desired proportion is

$$P(A \mid E) + P(B \mid E) = 0.0036 + 0.0122 = 0.0158.$$

The manufacturer notes that if the detector had not been used, the proportion of items packaged for sale that are defective would have been equal to

$$P(A) + P(B) = 0.06 + 0.04 = 0.10.$$

Thus the use of a detector has materially improved the quality of the items packaged, even though the detector is not infallible. The proportion of defective items among those packaged for sale has decreased from 0.10 to 0.0158. ◆

It is worth pointing out that an alternative solution can be given to the problem in Example 1.13. Because C is the event that an item is nondefective, it is easier to find $P(C \mid E)$ by Bayes' rule and then subtract this value from 1. That is, by the Law of Complementation for Conditional Probability,

$$\begin{aligned}&P\{\text{the item is defective} \mid \text{the item is packaged}\}\\&\quad = 1 - P\{\text{the item is nondefective} \mid \text{the item is packaged}\}\\&\quad = 1 - P(C \mid E).\end{aligned}$$

By Bayes' rule,

$$\begin{aligned}P(C \mid E) &= \frac{P(C)P(E \mid C)}{P(A)P(E \mid A) + P(B)P(E \mid B) + P(C)P(E \mid C)}\\&= \frac{(0.90)(0.90)}{(0.06)(0.05) + (0.04)(0.25) + (0.90)(0.90)}\\&= 0.9842,\end{aligned}$$

and the desired answer is

$$1 - P(C \mid E) = 1 - 0.9842 = 0.0158,$$

as obtained previously.

Conditional Probability Trees

The use of conditional probability trees can be helpful in organizing information in conditional probability problems and applying the Law of Multiplication, the Law of Partitions, and Bayes' rule. These trees have some resemblance to the counting trees used in Chapter 3 but are used for a different purpose. We illustrate the use of such trees using Example 1.10. In that example, a motor was produced by factory 1 with probability $P(E_1) = 0.30$, by factory 2 with probability $P(E_2) = 0.50$, or by factory 3 with probability $P(E_3) = 0.20$. Thus we start by drawing three branches labeled E_1, E_2, and E_3 (see Exhibit 1.5), and writing the respective probabilities of the events corresponding to these branches on the branches.

Now, from each branch we draw two limbs labeled B and B^c, corresponding to the events "motor is unsatisfactory" and "motor is satisfactory," respectively. On these limbs, we write the conditional probability of that limb given the branch from which it is drawn. Thus on the limb labeled B

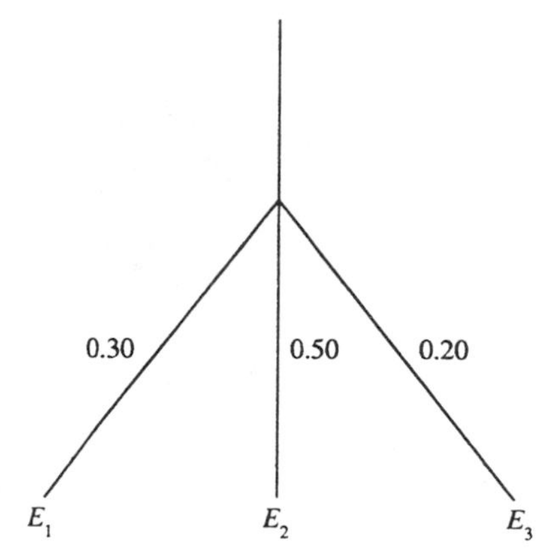

EXHIBIT 1.5 *Branches of a conditional probability tree.*

drawn from the branch labeled E_1, we write the conditional probability $P(B \mid E_1) = 0.05$, whereas on the limb marked B^c coming from the branch E_1 we write the conditional probability $P(B^c \mid E_1) = 1 - P(B \mid E_1) = 0.95$ (see Exhibit 1.6).

The path along the branch labeled E_1 and the limb labeled B corresponds to the intersection $E_1 \cap B$ of these two events. By the Law of Multiplication, the unconditional probability of this path equals

$$P(E_1 \cap B) = P(E_1)P(B \mid E_1) = (0.30)(0.05) = 0.015.$$

Thus we find the unconditional probability of any path (or of the corresponding intersection of events) by multiplying the numbers written on the paths. As another example, the probability of the event $E_3 \cap B^c$ is found by multiplying the probabilities written on the branch E_2 and on the limb B^c extending from the branch:

$$P(E_2 \cap B^c) = (0.20)(0.98) = 0.196.$$

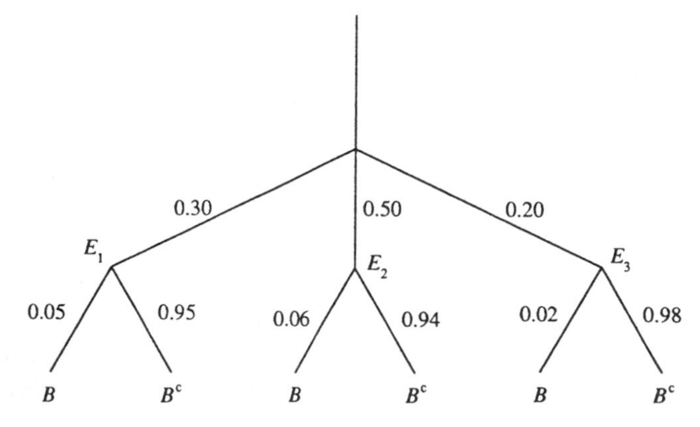

EXHIBIT 1.6 *A conditional probability tree for Example 1.10.*

If we want to find the unconditional probability of the event B, we simply add the probabilities of all paths ending in a limb labeled B. Because each path has a probability given by a product of numbers, we thus add products:

$$\begin{aligned} P(B) &= \{\text{probability of path } E_1 \to B\} + \{\text{probability of path } E_2 \to B\} \\ &\quad + \{\text{probability of path } E_3 \to B\} \\ &= (0.30)(0.05) + (0.50)(0.06) + (0.20)(0.02) \\ &= 0.049. \end{aligned}$$

This is the same result we obtained in Example 1.10 using the Law of Partitions.

Suppose that we want to find the conditional probability that a motor is made by factory 1 given that it is unsatisfactory. That is, we want to find $P(E_1 \mid B)$. Because

$$P(E_1 \mid B) = \frac{P(E_1 \cap B)}{P(B)}$$

is the ratio of the probability of path $E_1 \to B$ to the probability of all paths ending in a limb marked B, we find that

$$\begin{aligned} P(E_1 \mid B) &= \frac{\text{probability of path } E_1 \to B}{\text{sum of all probabilities of paths ending in } B} \\ &= \frac{(0.30)(0.05)}{(0.30)(0.05) + (0.50)(0.06) + (0.20)(0.02)} \\ &= \frac{0.015}{0.049} = 0.306, \end{aligned}$$

which is the result that would be given by Bayes' rule.

Conditional probability trees can also be applied to more complicated problems, as the following example illustrates.

EXAMPLE 1.14

In a crucial ball game, the home team has the bases loaded with one out in the last of the ninth inning with the score tied. The team at bat needs one run to win the game. The player at bat has a batting average of 0.250, so he has probability 0.250 of getting a hit (which will win the game). If, on the other hand, he makes an out, he can do this by striking out, grounding out, or flying out. These events, given that the player makes an out, have respective conditional probabilities 0.10, 0.50, 0.40. If the player strikes out, the conditional probability that the runner on third will score (because the catcher mishandles the ball) is 0.01. If the player grounds out, the conditional probability that a run will score is

0.30. Finally, if the player flies out, the conditional probability that a run will score is 0.60. The manager wants to calculate the unconditional probability that a run will score.

To solve this problem, we draw the following conditional probability tree (Exhibit 1.7). Note that the branch labeled "hit" has one limb marked "run" at its end. This limb has 1.00 marked on it because a run scores with conditional probability 1.00 if there is a hit. The other branch (labeled "out") has three limbs labeled "strike out," "ground out," and "fly out," and each limb is marked with the conditional probability of its corresponding event. From each such limb are two "twigs," labeled "run" and "no run." On these twigs are marked the conditional probabilities given the events corresponding to the branch and limb leading to that twig. For example,

$$P\{\text{run} \mid \text{out} \cap \text{ground out}\} = 0.30.$$

The general law of multiplication (Equation 1.3) justifies the assertion that the probability of any branch–limb–twig path is the product of the numbers written on the branch, the limb and the twig. Thus

$$P\{\text{out} \cap \text{ground out} \cap \text{run}\} = (0.700)(0.50)(0.30) = 0.105,$$

and

$$P\{\text{out} \cap \text{ground out} \cap \text{no run}\} = (0.700)(0.50)(0.70) = 0.245.$$

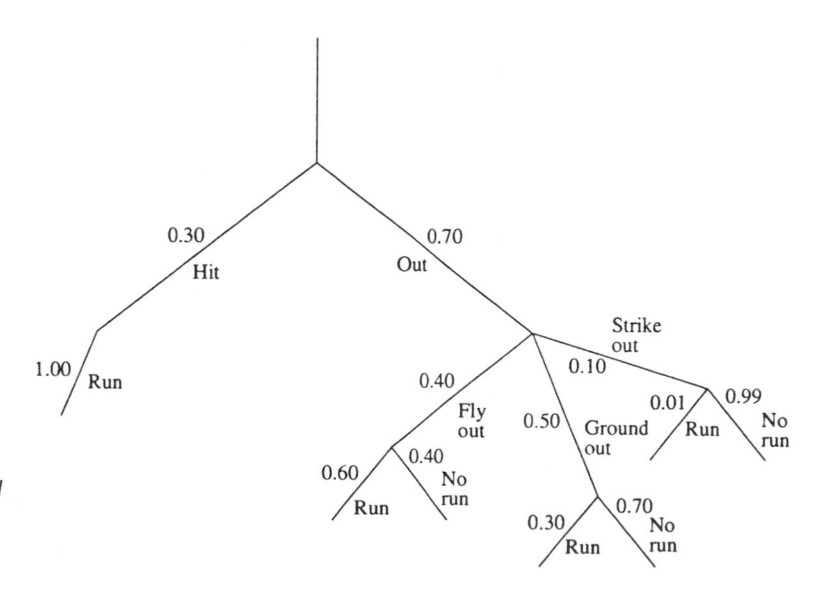

EXHIBIT 1.7 *A conditional probability tree for Example 1.14.*

The unconditional probability that a run will score is the sum of the probabilities of all paths ending in a twig (or limb) labeled "run." Thus

$$\begin{aligned} P(\text{run}) &= (0.300)(1.00) + (0.700)(0.10)(0.01) + (0.700)(0.50)(0.30) \\ &\quad + (0.700)(0.40)(0.60) \\ &= 0.5737. \end{aligned}$$

◆

2 Probabilistic Independence

Partial information about the outcome of a random experiment permits us to reassess the probabilities of events of interest to us. Frequently, the conditional probability $P(B \mid A)$ assigned to an event B after an event A has occurred is different from the (unconditional) probability $P(B)$ assigned to that event prior to the experiment. Thus, in this sense, the probability of the event B depends on our knowledge concerning the event A. If, for example, B denotes the event that "a person is wealthy" and A denotes "a person has completed 4 years of college," information that $P(B) \neq P(B \mid A)$ suggests that there is a statistical relationship between financial success and educational preparation.

Suppose, however, that

$$P(B \mid A) = P(B). \tag{2.1}$$

Here the information that event A has occurred does not change the probability assigned to event B. In this case we would say that event B is **probabilistically** or **statistically independent** of the event A.

Another intuitive way to describe probabilistic independence is to say that the (conditional) probability of B is the same whether A does or does not occur. That is,

$$P(B \mid A) = P(B \mid A^c). \tag{2.2}$$

That (2.1) and (2.2) are equivalent can be shown as follows. Note that by the Law of Partitions,

$$P(B) = P(A)P(B \mid A) + P(A^c)P(B \mid A^c). \tag{2.3}$$

If (2.2) holds, then from (2.3), using $P(A) + P(A^c) = 1$, we obtain

$$\begin{aligned} P(B) &= P(A)P(B \mid A) + P(A^c)P(B \mid A) \\ &= [P(A) + P(A^c)]\, P(B \mid A) = P(B \mid A), \end{aligned}$$

so that (2.1) holds. On the other hand, if (2.1) holds, then from (2.3),

$$P(B \mid A) = P(B) = P(A)P(B \mid A) + P(A^c)P(B \mid A^c).$$

Subtracting $P(A)P(B \mid A)$ from both sides of this equality, and noting that $1 - P(A) = P(A^c)$, yields

$$P(A^c)P(B \mid A) = P(A^c)P(B \mid A^c).$$

Now, $P(B \mid A^c)$ is not defined unless $P(A^c) \neq 0$. Thus, dividing both sides of the foregoing equality by $P(A^c)$, we obtain (2.2).

We have thus shown that (2.1) and (2.2) both define the *same* concept of probabilistic independence. We have also shown that *event B is probabilistically independent of event A if and only if event B is probabilistically independent of the complement A^c of event A*. It is also the case that *the event B is probabilistically independent of the event A if and only if the event A is probabilistically independent of the event B*. That is, equality (2.1) implies that

$$P(A \mid B) = P(A), \tag{2.4}$$

and vice versa.

To demonstrate this assertion, note that by the Law of Multiplication,

$$P(A \cap B) = P(A)P(B \mid A) = P(B)P(A \mid B).$$

If (2.1) holds, then $P(B \mid A) = P(B)$ and

$$P(A \mid B) = \frac{P(A \cap B)}{P(B)} = \frac{P(A)P(B \mid A)}{P(B)} = \frac{P(A)P(B)}{P(B)} = P(A).$$

Thus (2.4) is true. On the other hand, if (2.4) holds, then $P(A \mid B) = P(A)$ and

$$P(B \mid A) = \frac{P(A \cap B)}{P(A)} = \frac{P(B)P(A \mid B)}{P(A)} = \frac{P(B)P(A)}{P(A)} = P(B),$$

so that (2.1) is true.

We now have given three equivalent definitions of probabilistic independence, each of which is based on conditional probabilities. These definitions state in different ways the basic concept underlying the notion of probabilistic independence: namely, that knowledge of whether or not one of the events (say, A) has occurred does not affect the probability assigned to the other event (B).

EXAMPLE 2.1

If a dry summer is anticipated, wheat farmers need to switch to strains of wheat that require less water. However, using drought-resistant strains of wheat in a summer that is not dry is inefficient (less crop for the same effort). Thus wheat farmers would like to be able to predict dry summers. About 10% of all summers in a certain wheat-growing region are dry.

The farmers in the region once used an almanac to predict dry summers. If the almanac predicted a dry summer, the farmers would switch crops to drought-resistant wheat. This practice stopped after one farmer checked past records and found that dry summers had occurred in 10% of those years for which the almanac had predicted a dry summer. Hence, the event "a dry summer occurs" (event B) is probabilistically independent of the event "the almanac predicts a dry summer" (event A) because $P(B) = 0.10 = P(B \mid A)$. In consequence, the almanac's prediction of a dry summer does not provide useful information; the chance of a dry summer is the same whether we know the almanac's prediction or not. [It also follows from the independence of events A and B that if the almanac predicts a normal, non-dry, summer, the probability that a dry summer will actually occur is still 0.10; that is, $P(B \mid A^c) = 0.10 = P(B \mid A)$.]

It should be noted that although we can conclude from a probabilistic (or statistical) perspective that no relationship exists between the almanac's prediction and the event that a dry summer occurs, this probabilistic assertion does not rule out the possibility that the almanac can make a correct prediction in a given year. The lack of relationship asserted by "probabilistic independence" is between relative frequencies (probabilities) calculated over many years (trials) and does not rule out a successful prediction in any one particular year.

The farmers also considered making use of a saying of the early settlers ("a dry summer is preceded by a snowy winter") to predict dry summers. A check of past dry summers found that 70% were preceded by snowy winters (event C). Thus, $P(C \mid B) = 0.70$. This excited the farmers until they discovered that 70% of all normal summers were also preceded by snowy winters; thus, $P(C \mid B^c) = 0.70$. Because $P(C \mid B) = P(C \mid B^c)$, the event C is probabilistically independent of the event B, which implies that event B is probabilistically independent of the event C. Consequently, the odds of a dry summer are the same (0.10) whether the preceding winter is snowy or not. (It also follows that $P(C) = P(C \mid B) = 0.70$, so that 70% of all winters in that region are snowy.) Knowledge of whether or not the previous winter was snowy does not improve the farmer's ability to predict a dry summer. ♦

In verifying that (2.1) and (2.4) each implies the other, we also established one last characterization of probabilistic independence: *The event B is probabilistically independent of the event A if and only if*

$$P(A \cap B) = P(A)P(B). \tag{2.5}$$

If event B is probabilistically independent of event A, then by definition (2.1), $P(B \mid A) = P(B)$. Hence by the Law of Multiplication,

$$P(A \cap B) = P(A)P(B \mid A) = P(A)P(B),$$

and (2.5) is true. On the other hand, if (2.5) is true, then

$$P(B \mid A) = \frac{P(A \cap B)}{P(A)} = \frac{P(A)P(B)}{P(A)} = P(B),$$

so that event B is probabilistically independent of event A.

Although (2.5) is a less conceptually meaningful definition of probabilistic independence than either (2.1), (2.2), or (2.4), it is often computationally more useful. It also has the advantage that it clearly reveals the symmetry of the events A and B in the definition of probabilistic independence. Because of this symmetry, it makes sense to talk of "probabilistically independent events."

Definition of the Probabilistic Independence of Two Events. Two events A, B are probabilistically independent if any *one* of the following equivalent assertions is true:

1. B is probabilistically independent of A [i.e., $P(B) = P(B \mid A)$].
2. A is probabilistically independent of B [i.e., $P(A) = P(A \mid B)$].
3. $P(B \mid A) = P(B \mid A^c)$.
4. $P(A \mid B) = P(A \mid B^c)$.
5. $P(A \cap B) = P(A)P(B)$.

To repeat, if any *one* of the foregoing characterizations of the probabilistical independence of the events A, B is true, then *all* of these characterizations hold. To determine whether or not the events A, B are probabilistically independent, we pick that characterization—(1), (2), (3), (4), or (5)—that is easiest to check.

If we are told that $P(A \mid B) = 0.6$ and $P(A \mid B^c) = 0.5$, it is easiest to check characterization (4). Because

$$P(A \mid B) = 0.6 \neq 0.5 = P(A \mid B^c),$$

we see that events A, B are not probabilistically independent (they are *probabilistically dependent*). Consequently, we know (without the need to check further) that $P(A \cap B) \neq P(A)P(B)$, $P(A \mid B) \neq P(A)$, and so on.

On the other hand, we might be given

$$P(A) = 0.6, \qquad P(B) = 0.4, \qquad P(A \cap B) = 0.24.$$

Here characterization (5) is easiest to use. Because

$$P(A \cap B) = 0.24 = (0.6)(0.4) = P(A)P(B),$$

we conclude that the events A, B are probabilistically independent. Consequently, we can infer from characterizations (1), (2) that

$$P(A \mid B) = P(A) = 0.6, \qquad P(B \mid A) = P(B) = 0.4,$$

and also from characterizations (3), (4) that

$$P(A \mid B^c) = P(A \mid B) = 0.6, \qquad P(B \mid A^c) = P(B \mid A) = 0.4.$$

Further, we know from (1) and (3) that A and B^c are independent, so that

$$P(A \cap B^c) = P(A)P(B^c).$$

Similarly, A^c and B are independent, so that

$$P(A^c \cap B) = P(A^c)P(B).$$

These last results are worth summarizing as follows:

Law of Multiplication for Two Probabilistically Independent Events. If the events A, B are probabilistically independent, then

$$P(A \cap B) = P(A)P(B), \qquad P(A \cap B^c) = P(A)P(B^c),$$
$$P(A^c \cap B) = P(A^c)P(B), \qquad P(A^c \cap B^c) = P(A^c)P(B^c).$$

EXAMPLE 2.2

A charity sells 500 raffle tickets. There are two prizes to be awarded; the same ticket is allowed to win both prizes. Consequently, the raffle is a simple random sample of size $n = 2$ *with replacement*. What is the probability that a customer who has purchased 50 tickets wins at least one prize?

Let the events A, B be "the customer wins the first prize awarded" and "the customer wins the second prize awarded," respectively. We need to find the probability $P(A \cup B)$ that "at least one of the events A, B occurs." Because the winning ticket on the first draw is replaced before the second winning ticket is drawn, the probability that the customer wins the second prize awarded is the same whether the customer wins the first prize awarded (event A) or does not win that prize (event A^c). Hence

$$P(B \mid A) = P(B \mid A^c),$$

and we see by characterization 3 of the probabilistic independence of two events that the events A, B are probabilistically independent. Note that by the General Law of Addition for two events,

$$P(A \cup B) = P(A) + P(B) - P(A \cap B).$$

Further, because on each draw the customer has 50 chances to win out of 500, we know that

$$P(A) = P(B) = \tfrac{50}{500} = 0.10.$$

Finally, because the events A, B are probabilistically independent,

$$P(A \cap B) = P(A)P(B) = (0.10)(0.10) = 0.01.$$

Consequently,

$$\begin{aligned} &P\{\text{customer wins at least one prize}\} \\ &\quad = P(A \cup B) \\ &\quad = P(A) + P(B) - P(A \cap B) = 0.10 + 0.10 - 0.01 = 0.19. \end{aligned}$$

Alternatively, it may be easier to note that

$$P((A \cup B)^c) = P(A^c \cap B^c) = P(A^c)P(B^c) = (1 - 0.1)(1 - 0.1) = 0.81,$$

because A, B are probabilistically independent events. Consequently,

$$P(A \cup B) = 1 - P(A^c \cap B^c) = 1 - 0.81 = 0.19,$$

which is the result already obtained. ◆

In Example 2.2, a heuristic argument was given to explain why events A and B are independent. Note that event A refers to the first draw and event B

refers to the second draw. Because the sampling is with replacement, knowledge of the result of either draw yields no information as to the probabilities governing the other draw. This lack of information is what is reflected in the mathematical notion of probabilistic independence.

In contrast, if the winning ticket on the first draw in the raffle was not permitted to win the prize on the second draw, the raffle would be a simple random sample without replacement. In this case, if the customer wins the prize on the first draw (event A occurs), the conditional probability that the customer wins on the second draw is $P(B \mid A) = 49/499$, because $50 - 1 = 49$ of the customer's tickets would remain among the $500 - 1 = 499$ tickets still eligible for the prize. If the customer does not win the prize on the first draw (A^c), the customer still has 50 eligible tickets, and consequently has conditional probability $P(B \mid A^c) = 50/499$ of winning a prize on the second draw. Because $P(B \mid A \neq P(B \mid A^c)$, the events A, B are *not* probabilistically independent. (They are probabilistically dependent.)

Mutually Exclusive Events and Probabilistically Independent Events—A Fallacy

There is a tendency to equate the concepts "mutually exclusive" and "probabilistically independent." This is a fallacy. Indeed, unless one of the events has probability zero, mutually exclusive events can never be probabilistically independent, and vice versa. To illustrate, suppose that A, B are events for which $P(A) = 0.4$, $P(B) = 0.3$. If A, B are mutually exclusive, then $A \cap B = \varnothing$, and

$$P(A \cap B) = P(\varnothing) = 0.$$

On the other hand, if A, B are probabilistically independent, then

$$P(A \cap B) = P(A)P(B) = (0.4)(0.3) = 0.12 \neq 0.$$

Clearly, both of these equations cannot be true simultaneously. If A, B are mutually exclusive, they are not probabilistically independent. If A, B are probabilistically independent, they are not mutually exclusive. [Note that it could actually be the case that $P(A \cap B) = 0.10$. In this case, the events A, B are neither mutually exclusive nor independent!]

A related fallacy is that one can see probabilistic independence in a Venn diagram. Again, this confuses mutually exclusive events with probabilistically independent events. Mutually exclusive events are represented on a Venn diagram as nonoverlapping regions; thus one can easily visualize the property of being mutually exclusive. To see probabilistic independence of two events A, B, one has to be able to see either that the ratio of $P(A \cap B)$ to

$P(A)$ is equal to the ratio of $P(B)$ to $P(\Omega) = 1$, or that $P(A \cap B)$ is the product of $P(A)$ and $P(B)$. Even if Venn diagrams are drawn so that areas of regions equal the probabilities of events represented by these regions, such relationships will not be easy to ascertain.

Terminology

The adverbs "statistically" and "stochastically" are often used in place of "probabilistically" when discussing probabilistic independence or dependence of events. On the other hand, it is customary in the literature to omit these adverbs entirely whenever it is clear what kind of "independence" or "dependence" is being discussed. We adopt this last custom in the remainder of this book. Thus if we say that two events A and B "are independent," we mean that these events are probabilistically independent as defined in Section 2.

In many practical problems, independence (probabilistic independence) of two events A, B is made part of the statement of the model, and interest is directed toward the consequences of such an assumption.

EXAMPLE 2.3

For the same cost, a communications engineer can either (a) send two identical bits (units of information) *independently* through a noisy system in which the probability that any particular bit is successfully received (identified) is 0.7, or (b) send one bit through a clean system in which the probability that the bit is successfully received is 0.9. If the goal is to transmit a bit with highest probability of successful receipt, which method should be used?

For method (b), the probability of success is 0.9. Thus we concentrate on method (a). Let A be the event that one bit (which we arbitrarily call "bit 1") is successfully received, and let B be the event that the other bit (bit 2) is successfully received. If either bit 1 or bit 2 is successfully received, the desired information has been transmitted. Thus the probability of success for method (a) is $P(A \cup B)$. The probability model states that events A and B are probabilistically independent. Consequently, the probability of success for method (a) is

$$\begin{aligned} P(A \cup B) &= 1 - P(A^c \cap B^c) = 1 - P(A^c)P(B^c) \\ &= 1 - (1 - 0.7)(1 - 0.7) = 1 - 0.09 = 0.91. \end{aligned}$$

Because this probability of success exceeds that of method (b), method (a) should be used to transmit information.

The probability that *exactly one* of the two bits sent by method (a) is successfully received is

$$
\begin{aligned}
&P\{\text{exactly one bit is successfully received}\}\\
&\quad = P((A \cap B^c) \cup (A^c \cap B)) = P(A \cap B^c) + P(A^c \cap B)\\
&\quad = P(A)P(B^c) + P(A^c)P(B) = (0.7)(1 - 0.7) + (1 - 0.7)(0.7)\\
&\quad = 0.42,
\end{aligned}
$$

because $A \cap B^c$ and $A^c \cap B$ are mutually exclusive events, and the events A and B are assumed to be independent. This computation illustrates the respective proper uses of the often confused concepts "mutually exclusive" and "probabilistic independence." ◆

Mutually Independent Events

We have given two equivalent types of definition of the independence of two events E_1, E_2: (1) an intuitive definition based on conditional probabilities, and (2) the computational definition based on equalities such as $P(E_1 \cap E_2) = P(E_1)P(E_2)$, $P(E_1 \cap E_2^c) = P(E_1)P(E_2^c)$, and so on. Both kinds of definition can be generalized to provide definitions of the **mutual (probabilistic) independence** of n events.

Intuitively, n events $E_1, E_2, \ldots, E_n$ *are mutually independent if information about any* $n - 1$ *of these events is probabilistically independent of the remaining event*. That is, for any $k = 1, 2, \ldots, n$ consider E_k and the collection of events $E_1, \ldots, E_{k-1}, E_{k+1}, \ldots, E_n$. Let A_k be any event constructed from $E_1, \ldots, E_{k-1}, E_{k+1}, \ldots, E_n$ by the operations of complementation, union, and intersection; such an event A_k expresses information about the events $E_1, \ldots, E_{k-1}, E_{k+1}, \ldots, E_n$. We require that E_k be probabilistically independent of A_k for every choice of E_k and A_k.

Although the definition above certainly expresses the intuitive notion of probabilistic independence, it is difficult to verify this definition in practice because of the large numbers of event pairs, E_k, A_k, that must be considered.

Alternatively, we can define $E_1, E_2, \ldots, E_n$ to be mutually independent by generalizing the computational definition of independence for two events.

The events $E_1, E_2, \ldots, E_n$ *are mutually independent if*

$$P(E_{i_1} \cap E_{i_2} \cap \cdots \cap E_{i_k}) = P(E_{i_1})P(E_{i_2}) \cdots P(E_{i_k}) \tag{2.6}$$

for all subcollections $E_{i_1}, E_{i_2}, \ldots, E_{i_k}$ *of size* k for $k = 2, 3, \ldots, n$, *of the collection* $E_1, E_2, \ldots, E_n$.

For example, the events E_1, E_2, and E_3 are mutually independent if *all* the following equalities hold:

$$\begin{aligned} P(E_1 \cap E_2) &= P(E_1)P(E_2), \\ P(E_1 \cap E_3) &= P(E_1)P(E_3), \\ P(E_2 \cap E_3) &= P(E_2)P(E_3), \\ P(E_1 \cap E_2 \cap E_3) &= P(E_1)P(E_2)P(E_3). \end{aligned}$$

It can be shown (see Exercise 26 for the case of $n = 3$ events) that the intuitive conditional definition and the computational definition (2.6) are equivalent.

Pairwise and Mutual Independence

It is clear from either of the definitions of mutual independence above that *every subcollection* $E_{i_1}, E_{i_2}, \ldots, E_{i_k}$, $2 \leq k \leq n$, *of a collection* $E_1, E_2, \ldots, E_n$ *of mutually independent events is itself mutually independent*. In particular, *every pair of events* E_i, E_j, $i \neq j$ *in a collection* $E_1, E_2, \ldots, E_n$ *of mutually independent events is independent*. For example, if the events E_1, E_2, E_3 are mutually independent, so are the pairs E_1 and E_2, E_1 and E_3, and E_2 and E_3.

Although mutual independence of the events $E_1, E_2, \ldots, E_n$ implies the independence of every pair E_i, E_j, $i \neq j$, the converse of this assertion is not true in general; that is, *pairwise independence of all pairs* E_i, E_j, $i \neq j$, $i, j = 1, 2, \ldots, n$ *does not imply mutual independence* of the events $E_1, E_2, \ldots, E_n$, except trivially when $n = 2$. For example, suppose that the sample space ω of a random experiment has four outcomes $\omega_1, \omega_2, \omega_3, \omega_4$, and that $P(\{\omega_i\}) = \frac{1}{4}$, $i = 1, 2, 3, 4$. Suppose that

$$E_i = \{\omega_i, \omega_4\}, \qquad i = 1, 2, 3.$$

Then

$$P(E_i \cap E_j) = P(\{\omega_4\}) = \tfrac{1}{4} = (\tfrac{1}{2})(\tfrac{1}{2}) = P(E_i)P(E_j)$$

for all $i \neq j$, $i, j = 1, 2, 3$. Thus E_i and E_j are pairwise independent events for all $i \neq j$. However,

$$P(E_1 \cap E_2 \cap E_3) = P(\{\omega_4\}) = \tfrac{1}{4} \neq \tfrac{1}{8} = P(E_1)P(E_2)P(E_3),$$

so that (2.6) fails to hold, and E_1, E_2, E_3 are not mutually independent events.

Because of the fact that every pair E_i, E_j, $i \neq j$, of the events $E_1, E_2, \ldots, E_n$ can be independent, even when $E_1, E_2, \ldots, E_n$ are not mutually independent, we must be careful in establishing terminology for the case when n events are not mutually independent. Thus if $E_1, E_2, \ldots, E_n$ are not mutually independent, we say that they are **mutually dependent.**

Mutual Independence as a Modeling Assumption

Even the computational definition (2.6) of the mutual independence of n events is difficult to verify in practice, particularly if the number n of events is large. Consequently, such verification is rarely attempted in practice (except, of course, for the special case $n = 2$ of pairwise independence).

The concept of mutual independence is most often used in constructive probability models for random experiments. If we can *model* events $E_1, E_2, \ldots, E_n$ as being mutually independent, the calculation of probabilities of complex events based on $E_1, E_2, \ldots, E_n$ is greatly simplified. For example, the following generalization of the Law of Multiplication for two independent events is frequently useful.

General Law of Multiplication for Mutually Independent Events. If $E_1, E_2, \ldots, E_n$ are mutually independent events, then

$$P(F_1 \cap F_2 \cap \cdots \cap F_n) = P(F_1)P(F_2) \cdots P(F_n),$$

where each F_i can be either E_i or E_i^c or Ω, $i = 1, 2, \ldots, n$.

For example, the General Law of Multiplication for $n = 3$ mutually independent events states that

$$\begin{aligned}
P(E_1 \cap E_2 \cap E_3) &= P(E_1)P(E_2)P(E_3),\\
P(E_1 \cap E_2 \cap E_3^c) &= P(E_1)P(E_2)P(E_3^c),\\
P(E_1 \cap E_2^c \cap E_3) &= P(E_1)P(E_2^c)P(E_3),\\
P(E_1^c \cap E_2 \cap E_3) &= P(E_1^c)P(E_2)P(E_3),\\
P(E_1 \cap E_2^c \cap E_3^c) &= P(E_1)P(E_2^c)P(E_3^c),\\
P(E_1^c \cap E_2 \cap E_3^c) &= P(E_1^c)P(E_2)P(E_3^c),\\
P(E_1^c \cap E_2^c \cap E_3) &= P(E_1^c)P(E_2^c)P(E_3),\\
P(E_1^c \cap E_2^c \cap E_3^c) &= P(E_1^c)P(E_2^c)P(E_3^c),
\end{aligned}$$

and also that

$$\begin{aligned}
P(E_1 \cap E_2) &= P(E_1 \cap E_2 \cap \Omega) = P(E_1)P(E_2)P(\Omega) = P(E_1)P(E_2),\\
P(E_1 \cap E_3^c) &= P(E_1 \cap \Omega \cap E_3^c) = P(E_1)P(\Omega)P(E_3^c) = P(E_1)P(E_3^c),
\end{aligned}$$

and so on.

EXAMPLE 2.4

A radio with 3 batteries will get good reception if at least 2 of the batteries function. The probability that any particular battery will continue to function is 0.80. If the batteries operate independently, what is the probability that the radio continues to get good reception?

Let E_i = {battery i functions}, $i = 1, 2, 3$. The probability model states that $P(E_i) = 0.80$, $i = 1, 2, 3$, and that the events E_1, E_2, E_3 are mutually independent. Because 2 or more batteries must function for the radio to work, we can write the event A "the radio continues to get good reception" as

$$A = (E_1 \cap E_2 \cap E_3^c) \cup (E_1 \cap E_2^c \cap E_3) \cup (E_1^c \cap E_2 \cap E_3) \\ \cup (E_1 \cap E_2 \cap E_3).$$

Because A is a union of mutually exclusive events (draw a Venn diagram of E_1, E_2, E_3), and the events E_1, E_2, E_3 are mutually independent,

$$\begin{aligned} P(A) &= P(E_1 \cap E_2 \cap E_3^c) + P(E_1 \cap E_2^c \cap E_3) + P(E_1^c \cap E_2 \cap E_3) \\ &\quad + P(E_1 \cap E_2 \cap E_3) \\ &= P(E_1)P(E_2)P(E_3^c) + P(E_1)P(E_2^c)P(E_3) + P(E_1^c)P(E_2)P(E_3) \\ &\quad + P(E_1)P(E_2)P(E_3) \\ &= (0.8)(0.8)(1 - 0.8) + (0.8)(1 - 0.8)(0.8) + (1 - 0.8)(0.8)(0.8) \\ &\quad + (0.8)(0.8)(0.8) \\ &= 0.896. \end{aligned}$$

◆

EXAMPLE 2.5

An investor following the Dow Jones index records daily whether this index increased or did not increase. One proposed model for common stocks states that stocks have a probability of 0.50 of increasing from day to day, and that daily stock fluctuations are mutually independent. Over a 4-day period, what is the probability that the Dow Jones index goes up at least once?

Let E_i = {Dow Jones index goes up on day i}, $i = 1, 2, 3, 4$. The probability model given to us states that $P(E_i) = 0.50$, $i = 1, 2, 3, 4$, and that E_1, E_2, E_3, E_4 are mutually independent events. Further, the event A that "the Dow Jones index goes up at least once" can be written as

$$A = E_1 \cup E_2 \cup E_3 \cup E_4.$$

Note that

$$A^c = (E_1 \cup E_2 \cup E_3 \cup E_4)^c = E_1^c \cap E_2^c \cap E_3^c \cap E_4^c.$$

That is, the complement of "at least once" is "none." Consequently, by the Law of Complementation and the General Law of Multiplication for mutually independent events,

$$\begin{aligned} P(A) = 1 - P(A^c) &= 1 - P(E_1^c \cap E_2^c \cap E_3^c \cap E_4^c) \\ &= 1 - P(E_1^c)P(E_2^c)P(E_3^c)P(E_4^c) \\ &= 1 - (1 - 0.50)(1 - 0.50)(1 - 0.50)(1 - 0.50) \\ &= 0.938. \end{aligned}$$

Suppose that the investor has watched the Dow Jones index increase for 99 consecutive days. What is the (conditional) probability that the index will increase on the 100th day? Many people would say that "the law of averages" makes it nearly certain that the Dow Jones index will fall. [That is, $P(E_{100}^c \mid E_1 \cap E_2 \cap \cdots \cap E_{99})$ is close to 1.] However, if the assumed probability model holds, and the events $E_1, E_2, \ldots, E_{100}$ are mutually independent, the intuitive definition of mutual independence shows that

$$P(E_{100}^c \mid E_1 \cap E_2 \cap \cdots \cap E_{99}) = P(E_{100}^c) = 1 - 0.50 = 0.50. \quad \blacklozenge$$

3 Composite Random Experiments

In many cases, the outcomes of a random experiment are the consequence of combining the outcomes of two or more **component random experiments.** For example, if we measure the number of years of schooling and the index of social status of a randomly selected person, our experiment $\mathscr{E}$ combines two **component random experiments** $\mathscr{E}^{(1)}$ and $\mathscr{E}^{(2)}$. In experiment $\mathscr{E}^{(1)}$, the person's years of schooling are measured, while in experiment $\mathscr{E}^{(2)}$ we measure the index of social status of the person. As another example, each unit in a random sample of n units can be thought of as a sample of size 1 from a population of units. The n draws that create the sample are then component random experiments $\mathscr{E}^{(1)}, \mathscr{E}^{(2)}, \ldots, \mathscr{E}^{(n)}$ whose outcomes are combined to form the outcome of the composite random experiment, $\mathscr{E}$, in which the entire sample of n units is drawn. This last example is a special case of a composite experiment consisting of n repeated trials of the same random experiment (see Chapter 1).

An event E defined for a composite random experiment $\mathscr{E}$ with component random experiments (*components*) $\mathscr{E}^{(1)}, \mathscr{E}^{(2)}, \ldots, \mathscr{E}^{(n)}$ is said to be determined by the component $\mathscr{E}^{(j)}$ if the occurrence or nonoccurrence of the event E can be demonstrated by observing only the component $\mathscr{E}^{(j)}$. For example, if we observe the years of education and index of social status of a person chosen at random, the event E = {the person had 16 years of schooling} is determined by the component $\mathscr{E}^{(1)}$, in which the person's years of schooling is observed. Similarly, if we observe n trials of a certain random experiment and E is the event that a certain happening occurs on the fifth trial, the event E is determined by the fifth trial (component) $\mathscr{E}^{(5)}$ of the composite random experiment.

The assumption that the components of a composite random experiment are mutually independent can greatly simplify computations of probabilities of events. Two components $\mathscr{E}^{(1)}$ and $\mathscr{E}^{(2)}$ of a composite random experiment $\mathscr{E}$ are said to be **independent** if every event E_1 determined by component $\mathscr{E}^{(1)}$ is independent of every event E_2 determined by component $\mathscr{E}^{(2)}$. The k components $\mathscr{E}^{(1)}, \mathscr{E}^{(2)}, \ldots, \mathscr{E}^{(k)}$ of a composite random experiment $\mathscr{E}$ are said to be **mutually independent** if for every collection $E_1, E_2, \ldots, E_k$ of events, where E_i is determined by $\mathscr{E}^{(i)}$, $i = 1, 2, \ldots, k$, the events $E_1, E_2, \ldots, E_k$ are mutually independent. A composite random experiment consisting of k mutually independent repetitions of the same experiment is said to consist of **independent and identically distributed** trials.

If we are given a probability model for a composite random experiment $\mathscr{E}$, we can attempt to verify that the components $\mathscr{E}^{(1)}, \mathscr{E}^{(2)}, \ldots, \mathscr{E}^{(n)}$ are mutually independent. However, unless there is some simple way of describing probabilities for the random experiment $\mathscr{E}$, the task of establishing mutual independence from a probability model for $\mathscr{E}$ can be difficult. Instead, we frequently construct the probability model for $\mathscr{E}$ on the *assumption* that its components are mutually independent, basing that assumption on the nature of the process that gives rise to the joint results of these component experiments.

For example, every draw in a random sample *with replacement* of n items from a population of N items is unaffected by the results of other (previous) draws, because all previous draws are replaced in the population before each new draw. Thus the draws in a random sample of size n with replacement from a given population are independent and identically distributed trials of the random experiment of drawing a sample of size 1 from that population.

EXAMPLE 3.1

Suppose that there are $N = 100$ fish in a pond, of which 30 fish are of legal size. Think of the fish caught as a random sample *with replacement* from the fish in the pond. If we catch 3 fish, what is the probability that all 3 fish caught are of legal size?

Let E_i equal the event that the ith fish caught (ith draw) is of legal size, $i = 1, 2, 3$. Then $P(E_i) = \frac{30}{100} = 0.3$. Because we sample with replacement, our draws (catches) are mutually independent, and thus

$$\begin{aligned} P\{\text{all 3 fish are of legal size}\} &= P(E_1 \cap E_2 \cap E_3) \\ &= P(E_1)P(E_2)P(E_3) \\ &= (0.3)(0.3)(0.3) = 0.027. \end{aligned}$$

Similarly,

$$\begin{aligned} P\{&\text{at least 1 fish is of legal size}\} \\ &= P(E_1 \cup E_2 \cup E_3) \\ &= P((E_1^c \cap E_2^c \cap E_3^c)^c) = 1 - P(E_1^c \cap E_2^c \cap E_3^c) \\ &= 1 - P(E_1^c)P(E_2^c)P(E_3^c) = 1 - (0.7)^3 = 0.657. \end{aligned}$$ ♦

Other examples of mutually independent components of a composite random experiment occur when the component experiments are physically separated (isolated). For example, if n people are asked to solve a puzzle, and if they work separately, the success (or lack of success) of some of them should not statistically affect the chance of success of the remaining people in solving the puzzle. Thus each person would be a component of the overall random experiment "n people are asked to solve a puzzle," and the random experiment could be modeled as having mutually independent components (persons). Similarly, if the same person is asked to solve n puzzles, and no learning effect is assumed (doing one puzzle does not improve skill in later puzzles), the person's attempts at the n puzzles could be regarded as n independent trials of a component experiment "solve a puzzle."

EXAMPLE 3.2

A building has 3 elevators run by separate machinery. Each elevator is a component of the random experiment in which we check the functioning of all three elevators. Assume that these components are mutually probabilistically independent. Suppose that E_i is the event "elevator i is functioning," $i = 1, 2, 3$, and that $P(E_1) = 0.7$, $P(E_2) = 0.8$, $P(E_3) = 0.9$. What is the probability that on a given day, *exactly one* elevator is not functioning?

The event A "exactly one elevator is not functioning" can be written

$$A = (E_1^c \cap E_2 \cap E_3) \cup (E_1 \cap E_2^c \cap E_3) \cup (E_1 \cap E_2 \cap E_3^c).$$

Because this is a union of mutually exclusive events,

$$P(A) = P(E_1^c \cap E_2 \cap E_3) + P(E_1 \cap E_2^c \cap E_3) + P(E_1 \cap E_2 \cap E_3^c).$$

Because the events E_1, E_2, E_3 are determined, respectively, by the mutually independent components $\mathscr{E}_1$ = elevator 1, $\mathscr{E}_2$ = elevator 2, $\mathscr{E}_3$ = elevator 3, these events are mutually independent. Thus by the General Law of Multiplication for mutually independent events,

$$\begin{aligned} P(E_1^c \cap E_2 \cap E_3) &= P(E_1^c)P(E_2)P(E_3) = (1 - 0.7)(0.8)(0.9) = 0.216, \\ P(E_1 \cap E_2^c \cap E_3) &= P(E_1)P(E_2^c)P(E_3) = (0.7)(1 - 0.8)(0.9) = 0.126, \\ P(E_1 \cap E_2 \cap E_3^c) &= P(E_1)P(E_2)P(E_3^c) = (0.7)(0.8)(1 - 0.9) = 0.056, \end{aligned}$$

and

$$P\{\text{exactly one elevator is not functioning}\} = P(A) = 0.216 + 0.126 + 0.056 = 0.398. \quad \blacklozenge$$

In working with mutually independent trials or components, one must be careful to remember that not all collections of events defined for such experiments are mutually independent. Only those collections of events for which each event is determined by a different component are known to be mutually independent.

EXAMPLE 3.2 *(continued)*

We are told that exactly one elevator is not functioning, but are not informed which. What is the probability that the malfunctioning elevator is elevator 1? The naive answer would be $P(E_1^c) = 0.3$, but this ignores the information that exactly one elevator is malfunctioning.

Instead, we need to find $P(E_1^c \mid A)$, where A is the event "exactly one elevator is not functioning." It might be thought that events E_1^c and A are probabilistically independent. If this were the case, $P(E_1^c \mid A) = P(E_1^c) = 0.3$. However, event A depends on the results of all three components (elevators). Thus one cannot assume that E_1^c and A are independent events. Indeed,

$$P(E_1^c \mid A) = \frac{P(E_1^c \cap A)}{P(A)} = \frac{P(E_1^c \cap E_2 \cap E_3)}{P(A)} = \frac{0.216}{0.398} = 0.543,$$

because the only way for an outcome to be in both event A ("exactly one elevator is not functioning") and event E_1^c ("elevator 1 is not functioning") is for elevator 1 not to function and the other two elevators to be functioning. Note that

$$P(E_1^c \mid A) = 0.543 > 0.3 = P(E_1^c),$$

so that the events E_1^c, A are *not* probabilistically independent. ◆

Checking Probabilistic Independence of Two Component Experiments: Contingency Tables

There is one situation, frequently arising in scientific practice, where from the probabilities assigned to events of a composite random experiment, we can easily verify whether the components of this experiment are probabilistically independent. This situation is one in which there are two component experiments, $\mathscr{E}^{(1)}$, $\mathscr{E}^{(2)}$, each of which has a finite number of possible outcomes. For example, in biology we might wish to determine whether the color of an insect's eyes is independent (statistically) of the shape of its wings. The demonstration of such independence may indicate that these characteristics are determined by different chromosomes. Here "color of eyes" is one component $\mathscr{E}^{(1)}$ of the experiment in which we observe eye color and wing shape of an insect, and "wing shape" is the other component $\mathscr{E}^{(2)}$.

Suppose that there are 4 possible eye colors, C_1, C_2, C_3, C_4, and 3 possible wing shapes S_1, S_2, S_3. We observe numerous insects of the given type and construct a table of probabilities (Exhibit 3.1). Such a table is called a **contingency table.** It can be shown (see Exercise T.4) that the components $\mathscr{E}^{(1)}$ = "eye color," $\mathscr{E}^{(2)}$ = "wing shape" are probabilistically independent if and only if

$$P(S_i \cap C_j) = P(S_i)P(C_j), \qquad \text{all } i, \text{ all } j, \tag{3.1}$$

where

$$P(S_i) = P(S_i \cap C_1) + P(S_i \cap C_2) + P(S_i \cap C_3) + P(S_i \cap C_4) \tag{3.2}$$

are given by the row sums of the contingency table ($i = 1, 2, 3$) and

$$P(C_j) = P(S_1 \cap C_j) + P(S_2 \cap C_j) + P(S_3 \cap C_j) \tag{3.3}$$

are given by the column sums ($j = 1, 2, 3, 4$). That is, *the components are probabilistically independent if and only if every entry in the table is the product of the corresponding row and column sums.* Note that to show probabilistic independence, we must verify that $P(S_i \cap C_j) = P(S_i)P(C_j)$ for all i, j; whereas if we can show for *any* pair (i, j) that

$$P(S_i \cap C_j) \neq P(S_i)P(C_j),$$

we can conclude that the components are probabilistically *dependent* (there is a statistical relationship connecting some eye color with some wing shape).

Perhaps the easiest way to check (3.1) is to construct a new contingency table whose entries are the products $P(S_i)P(C_j)$ of the row and column sums

EXHIBIT 3.1 *A Two-Component Contingency Table*

	Eye Color			
Wing Shape	C_1	C_2	C_3	C_4
S_1	$P(S_1 \cap C_1)$	$P(S_1 \cap C_2)$	$P(S_1 \cap C_3)$	$P(S_1 \cap C_4)$
S_2	$P(S_2 \cap C_1)$	$P(S_2 \cap C_2)$	$P(S_2 \cap C_3)$	$P(S_2 \cap C_4)$
S_3	$P(S_3 \cap C_1)$	$P(S_3 \cap C_2)$	$P(S_3 \cap C_3)$	$P(S_3 \cap C_4)$

of the original contingency table, and compare the entries of the new table to those of the original table. If all entries match, the components are independent; if there is at least one nonmatch, the components are probabilistically dependent.

EXAMPLE 3.3

Suppose that the contingency table shown in Exhibit 3.2 is observed for a given type of insect. Because

$$P(S_1 \cap C_1) = 0.05 \neq (0.30)(0.28) = P(C_1)P(S_1),$$

we conclude that eye color and wing shape are probabilistically dependent for this type of insect. If eye color and wing length had been probabilistically independent but the row and column sums in the table remained unchanged, we would have obtained the table shown in Exhibit 3.3. It is interesting to note that although the components "eye color" and "wing shape" are dependent, the eye color C_1 is independent (probabilistically) of the wing shape S_2, because the entries agree for $C_1 \cap S_2$ in Exhibits 3.2 and 3.3.

EXHIBIT 3.2 *Probabilities for a Two-Component Contingency Table*

	Eye Color				
Wing Shape	C_1	C_2	C_3	C_4	*Row Sum*
S_1	0.05	0.12	0.07	0.04	0.28
S_2	0.15	0.14	0.11	0.10	0.50
S_3	0.10	0.09	0.02	0.01	0.22
Column sum	0.30	0.35	0.20	0.15	

EXHIBIT 3.3 *Probabilities for the Independence of Two Components*

	Eye Color				
Wing Shape	C_1	C_2	C_3	C_4	*Row Sum*
S_1	0.084	0.098	0.056	0.042	0.28
S_2	0.150	0.175	0.100	0.075	0.50
S_3	0.066	0.077	0.044	0.033	0.22
Column sum	0.30	0.35	0.20	0.15	

◆

Notes and References

Intuition can be a poor guide for determining conditional probabilities. Even quantitatively sophisticated people are sometimes led astray in apparently straightforward situations, as illustrated by the following example.

A Question of Information

It is rumored that a conglomerate plans a takeover of one of three comparable computer software companies: A, B, or C. Each company is viewed as having the same chance, $\frac{1}{3}$, of being chosen, with the consequence that the stock price of each company has risen. An investor who purchased stock in company A before the rumors began is now considering selling her stock for a profit. Doing so, of course, would lose her the chance to make a much larger profit if company A is chosen for takeover. Unless the odds that company A will be chosen are somehow increased, however, she feels that the profit made by selling now is worth taking.

Subsequent information becomes available that negotiations with company C have failed. The investor argues that this new information reduces the number of possibilities to two (companies A and B) and thus increases the odds in favor of company A from $\frac{1}{3}$ to $\frac{1}{2}$. Consequently, she decides not to sell company A's stock.

Was the subsequent announcement about company C really informative? A counterargument that the announcement is not informative is as follows: Because it was known in advance that at least one company other than company A would not be chosen, naming this particular company should not change the odds in favor of company A.

Clearly, one of these two arguments is faulty. The investor is incorrect. Her mistake comes in her conceptualization of an appropriate sample space. She modeled this sample space as having three equally likely outcomes A, B, C. When the new information eliminated outcome C, she believed that the conditional odds in favor of outcome A became

$$\begin{aligned} P\{A \text{ chosen} \mid A \text{ or } B \text{ chosen}\} &= \frac{P\{\text{``}A \text{ chosen''} \cap \text{``}A \text{ or } B \text{ chosen''}\}}{P(A \text{ or } B \text{ chosen})} \\ &= \frac{P(A \text{ chosen})}{P(A \text{ or } B \text{ chosen})} = \frac{\frac{1}{3}}{\frac{1}{3}+\frac{1}{3}} = \frac{1}{2}. \end{aligned}$$

What she forgot was that the subsequent announcement about failed negotiations is part of the outcome. This is not important when company B or company C has been chosen for takeover, because then the announcement is determined. Note that if a company has been chosen for takeover, the subsequent announcement cannot say that negotiations have failed with that company. Further, an announcement that company A has not been chosen for takeover would end any question as to whether the investor should sell her holdings in that company. It follows that if company B has been chosen, the announcement of a failure of negotiations must necessarily refer to company C. Similarly, if company C has been chosen, the announcement could only refer to company B. However, if the investor's company (company A) is the one chosen, the announcement could have been either that company B was not chosen or that company C was not chosen. Assuming that either of these two companies was equally likely to have been mentioned in the announcement, the possible outcomes and their probabilities are the following:

Outcome	*Company Chosen for Takeover*	*Company Announced as Not Chosen*	*Probability*
1	A	B	$(\frac{1}{3})(\frac{1}{2}) = \frac{1}{6}$
2	A	C	$(\frac{1}{3})(\frac{1}{2}) = \frac{1}{6}$
3	B	C	$(\frac{1}{3})(1) = \frac{1}{3}$
4	C	B	$(\frac{1}{3})(1) = \frac{1}{3}$

Thus

$$P\{\text{company } A \text{ chosen} \mid \text{announcement that company } C \text{ not chosen}\}$$
$$= \frac{P(\text{outcome } 2)}{P(\text{outcome 2 or outcome 3})} = \frac{\frac{1}{6}}{\frac{1}{6} + \frac{1}{3}} = \frac{1}{3},$$

and not $\frac{1}{2}$ as the investor thought.

Although the announcement concerning company C did not alter the odds in favor of company A, it is incorrect that no important information had been given to the investor. By similar reasoning.

$$P\{\text{company } B \text{ chosen} \mid \text{announcement that company } C \text{ not chosen}\}$$
$$= 1 - P\{\text{company } A \text{ chosen} \mid \text{company } C \text{ not chosen}\}$$
$$= 1 - \tfrac{1}{3} = \tfrac{2}{3}.$$

Thus, if the investor had not been misled by her intuition, on the basis of the announcement she could sell her stock in company A and buy stock in company B, with an excellent chance of making profits in both investments!

The example above illustrates the rule that information used in determining conditional probabilities must be modeled in the outcomes. It is partial information from the *outcomes* of a random experiment that allows one to compute conditional probabilities of such outcomes. The example "A Question of Information" is illustrative of a class of problems in which ignoring this rule can produce apparently paradoxical results. Other similar examples include "The Prisoner's Dilemma" [Mosteller (1965)] and "The Monty Hall Problem" [Selvin (1975); see also Morgan, Chanty, Dahiya, and Doviak (1991)]. The latter of these two examples was published in a *Parade Magazine* column in 1990 by the columnist Marilyn vos Savant. Her solution to the problem, which was correct, was disputed by many mathematicians and scientists. This controversy became so heated that it was discussed on the front page of the July 21, 1991 *New York Times*. (See also letters, August 11, 1991.)

The Monty Hall Problem

In the popular television game show "Let's Make a Deal," Monty Hall is the master of ceremonies. At certain times during the show, a contestant is allowed to choose one of three identical doors A, B, C, behind only one of which is a valuable prize (a new car). After the contestant picks a door (say, door A), Monty Hall opens another door and shows the contestant that there is no prize behind that door. (Monty Hall knows where the prize is and always chooses a door where there is no prize.) He then asks the contestant whether he or she wants to stick with their choice of door or switch to the remaining unopened door. Should the contestant switch doors? Does it matter?

Glossary and Summary

Conditional probability of B given A: The proportion of trials in which A occurs that B also occurs (frequency definition); a probability for event B taking into account the fact that only outcomes in A are possible (so that A is the new sample space); $P(B \mid A) = P(A \cap B)/P(A)$ for $P(A) > 0$ (formal definition).

Partition of a sample space: Mutually exclusive events $E_1, E_2, \ldots, E_k$ whose union $E_1 \cup E_2 \cup \cdots \cup E_k$ is the sample space Ω.

Independent events: A and B are independent events if the probability that B occurs knowing that A has occurred is the same as the probability that B occurs without knowing whether or not A has occurred. That is, the odds of B do not depend upon the occurrence or nonoccurrence of A (intuitive definition); A and B are independent events if any one of the following is true:

(i) $P(A \mid B) = P(A)$.
(ii) $P(A \mid B) = P(A \mid B^c)$.
(iii) $P(B \mid A) = P(B)$.
(iv) $P(B \mid A) = P(B \mid A^c)$.
(v) $P(A \cap B) = P(A)P(B)$.

Dependent events: A and B are dependent events if A and B are *not* independent events.

Mutually independent events: $A_1, \ldots, A_k$ are mutually independent events if A_j is independent of any event formed by union, intersection, and complementation from the remaining events $A_1, \ldots, A_{j-1}, A_{j+1}, \ldots, A_k$, for every choice of $j = 1, 2, \ldots, k$. That is, the occurrence or nonoccurrence of any event based on $A_1, \ldots, A_{j-1}, A_{j+1}, \ldots, A_k$ does not change the odds for A_j, $j = 1, 2, \ldots, k$.

Pairwise independent events: A collection $A_1, \ldots, A_k$ of events such that every pair of events A_i, A_j, $1 \leq i \neq j \leq k$, taken from this collection are independent.

Mutually dependent events: Events that are not mutually independent.

Composite random experiment: An experiment whose outcomes are k-tuples formed from the outcomes of k component experiments; for example, one can observe the height and weight of a person—in this case, the outcomes of the composite experiment formed from the component experiments "observe height," "observe weight" are the pairs (height, weight).

Component experiments: The separate observations that are combined to make a composite experiment.

Independent component experiments: Two experiments $\mathscr{E}_1$ and $\mathscr{E}_2$ are independent if every event E_1 whose occurrence or nonoccurrence can be observed entirely from experiment $\mathscr{E}_1$ is independent of every event E_2 depending in a similar way on experiment $\mathscr{E}_2$.

Contingency table: A table of joint probabilities for a composite experiment in which the rows of the table are defined by outcomes of one component experiment and the columns are defined by the outcomes of the other component experiment.

Mutually independent experiments: Experiments (viewed as components of a larger experiment) having the property that any choice of events $E_1, E_2, \ldots, E_k$, one from each experiment, are mutually independent.

Independent identically distributed trials: Repetitions (trials) of a common experiment that are mutually independent of one another.

Rules and Results

1. Conditional probabilities (*given* an event A) obey Axioms 1 to 3 and all other rules for probability measures:
   (a) **Axiom 1.** $0 \leq P(E \mid A) \leq 1$ for all events E.
   (b) **Axiom 2.** $P(\varnothing \mid A) = 1$, $P(\Omega \mid A) = P(A \mid A) = 1$.
   (c) **Axiom 3.** For any two mutually exclusive events E_1, E_2,

$$P(E_1 \cup E_2 \mid A) = P(E_1 \mid A) + P(E_2 \mid A).$$

(d) **Law of Complements.** $P(E^c \mid A) = 1 - P(E \mid A)$.
(e) **General Law of Addition.**

$$P(E_1 \cup E_2 \mid A) = P(E_1 \mid A) + P(E_2 \mid A) - P(E_1 \cap E_2 \mid A).$$

2. In general, combining conditional probabilities with different "given" events makes no sense. For example, $P(E \mid A^c) \neq 1 - P(E \mid A)$, in general, and $P(E \mid A) + P(E \mid B)$ has no meaning (and can exceed 1).
3. **Law of Multiplication**

$$\begin{aligned} P(A \cap B) &= P(A)P(B \mid A), \\ P(A \cap B \cap C) &= P(A)P(B \mid A)P(C \mid A \cap B), \\ P(E_1 \cap E_2 \cap \cdots \cap E_k) &= P(E_1)P(E_2 \mid E_1)P(E_3 \mid E_1 \cap E_2) \\ &\quad \cdots P(E_k \mid E_1 \cap E_2 \cap \cdots \cap E_{k-1}). \end{aligned}$$

4. **Law of Multiplication for Independent Events**
(a) If A and B are independent events,

$$\begin{aligned} P(A \cap B) &= P(A)P(B), & P(A \cap B^c) &= P(A)P(B^c), \\ P(A^c \cap B) &= P(A^c)P(B), & P(A^c \cap B^c) &= P(A^c)P(B^c). \end{aligned}$$

(b) If $E_1, \ldots, E_k$ are mutually independent events, then

$$P(F_2 \cap F_2 \cap \cdots \cap F_k) = P(F_1)P(F_2) \cdots P(F_k),$$

where each F_i is either E_i or E_i^c or Ω, $i = 1, \ldots, k$. For example, if E_1, E_2, E_3, E_4 are mutually independent events,

$$\begin{aligned} P(E_1 \cap E_2^c \cap E_3) &= P(E_1)P(E_2^c)P(E_3)P(\Omega) \\ &= P(E_1)P(E_2^c)P(E_3). \end{aligned}$$

5. **Law of Partitions.** For any partition $E_1, E_2, \ldots, E_k$ of the sample space and any event A,

$$P(A) = \sum_{i=1}^{k} P(E_i)P(A \mid E_i).$$

6. **Bayes' rule.** For any partition $E_1, E_2, \ldots, E_k$ of the sample space and any event A,

$$P(E_j \mid A) = \frac{P(E_j)P(A \mid E_j)}{\sum_{i=1}^{k} P(E_i)P(A \mid E_i)}, \qquad j = 1, 2, \ldots, k.$$

7. Unless $P(A) = 0$ or $P(B) = 0$, mutually exclusive events A, B cannot be independent, and independent events A, B cannot be mutually exclusive.

Exercises

1. A census of small businesses in a midwestern state showed that:
   45% had raised capital through state development bonds.
   70% had obtained reduced tax rates from their local communities.
   26% had taken advantage of *both* of the incentives listed above.
   (a) What percentage of all small businesses in the state took advantage of only (exactly) one of the incentives above? (Draw a Venn diagram.)
   (b) What percentage of those small businesses that obtained reduced tax rates raised capital through state development bonds?
   (c) What percentage of those small business that did not raise capital through state development bonds were able to obtain reduced tax rates?
2. A woman has two children. Assume that all possible outcomes for the sexes of these children (e.g., MM, MF, FM, FF) are equally likely. Given that at least one of the woman's children is a male, what is the conditional probability that both of her children are males? (The answer is not 0.5; can you explain why?)
3. A pebble is selected at random from the shore of Lake Michigan, and the color (brown, black, green) of this pebble is observed. Based on a group of 1000 pebbles studied by Miller and Kahn (1962), an appropriate probability model for this experiment is the following:

Simple Event	{brown}	{black}	{green}
Probability	0.852	0.093	0.055

   (a) Find the probability that a randomly selected pebble is either black or brown.
   (b) Find the conditional probability that the pebble is brown *given* that the pebble is not green. Compare the conditional probability to the unconditional probability that the pebble is brown.
4. Based on studies described by Gregory (1963, p. 78), the following probability model describes the number of floods in a given wet season.
   (i) The outcomes are the numbers 0, 1, 2, 3, . . . of floods. The simple events are {0}, {1}, {2}, {3}, and so on.
   (ii) The probability model is determined by

Simple Event	{0}	{1}	{2}	{3}	{4}	{5}	{6}
Probability	0.2466	0.3452	0.2417	0.1127	0.0395	0.0110	0.0033

Simple events corresponding to more than 6 floods have probability 0.

(a) Find the probability of 2 or more floods in a wet season.

(b) Find the conditional probability of 4 or more floods in a wet season given that 2 floods have already been observed.

(c) Find the conditional probability of at least 2 *additional* floods in a wet season given that a flood has already been observed.

(*Hint:* If k floods have already been observed, you know that k or more floods will occur.)

5. A study of magazine readership habits show that for the three major sports magazines A, B, C: 8.0% of the population read A, 7.5% of the population read B, 7.0% of the population read C, 3.5% of the population read A and B, 3.0% of the population read A and C, 3.0% of the population read B and C, and 1.0% of the population read all three magazines.
   (a) What percent of the population reads *exactly one* of the three sports magazines?
   (b) What percent of all people who read magazine A also read magazine C? (This is a conditional probability question.)
   (c) What percent of all people who read exactly one of the three sports magazines read magazine C?
6. Exhibit E.1 shows data obtained by Lazarsfeld and Thielens (1958) on the relationships among age, productivity, and party vote in 1952 for a sample of 2117 social scientists studied in 1954–1955. Calculate the following percentages.
   (a) Scientists who voted Democratic in 1952.
   (b) Scientists in the age group 41–50 who are classified in the middle of the productivity score.
   (c) Scientists 51 years or older and with a low productivity score among all who voted Democratic in 1952.

EXHIBIT E.1: *Social Scientists, Classified by Age, Productivity Score, and Party Vote in 1952*

	Productivity Score					
	Low		Middle		High	
Age	*Democrats*	*Others*	*Democrats*	*Others*	*Democrats*	*Others*
40 or younger	260	118	226	60	224	60
41–50	60	60	78	46	231	91
51 or older	43	60	59	60	206	175

(d) Scientists 40 years or younger among all who did not vote Democratic in 1952.
(e) Scientists 51 years or older and with a low productivity score among all who did not vote Democratic in 1952.
(f) Scientists low on the productivity scale and voting Democratic in 1952 among all who are in the age group 41–50 years.

7. In 1982, 43% of the U.S. Labor force consisted of women. Of the women who worked, 7% held managerial positions, whereas 14% of men who worked held such positions.
(a) What percentage of all workers were women managers?
(b) What percentage of all workers were managers?

8. Glass and Hall (1954) obtained occupational status data on male residents of England and Wales and their fathers. Three occupation states were distinguished: upper level (professional or executive), middle level (supervisory, nonmanual, and skilled manual) and lower level (semiskilled and unskilled manual). Let S_1, F_1 be the events that sons, fathers, respectively, are upper level; S_2, F_2 be the events that sons, fathers are middle level; S_3, F_3 be the events that sons, fathers are lower level. The following conditional probabilities were found:

$$P(S_1 \mid F_1) = 0.45, \quad P(S_1 \mid F_2) = 0.05, \quad P(S_1 \mid F_3) = 0.01,$$
$$P(S_2 \mid F_1) = 0.48, \quad P(S_2 \mid F_2) = 0.70, \quad P(S_2 \mid F_3) = 0.50,$$
$$P(S_3 \mid F_1) = 0.07, \quad P(S_3 \mid F_2) = 0.25, \quad P(S_3 \mid F_3) = 0.49.$$

Suppose that $P(F_1) = 0.50$, $P(F_2) = 0.40$, $P(F_3) = 0.10$.
(a) Find the probability that both father and son have upper-level occupations.
(b) Find the probability that father and son have the *same* occupational level.
(c) Given that a father has a middle-level occupation, what is the conditional probability that the son's occupational level is at least as high as his father [i.e., what is $P(S_1 \text{ or } S_2 \mid F_2)$]?
(d) Find the unconditional probability that a son's occupational state is at least as high as his father's.

9. A study has been made of salary disputes in a certain professional sport. Of all such disputes, 60% were settled without arbitration and 40% required arbitration. Of those disputes settled without arbitration, 35% resulted in final salaries closer to the player's original demand, 20% resulted in a final salary midway between the player's demand and the owners' offer, and 45% resulted in a final salary closest to the owners' offer. Of those disputes requiring arbitration, 50% resulted in final salaries closer to the player's demand, 30% resulted in a final salary midway between the player's demand and the owners' offer, and 20% resulted in a final salary closest to the owners' offer.

(a) What percentage of all salary disputes were settled with a final salary midway between the player's demand and the owners' offer?
(b) What percentage of all cases in which the final salary was closer to the player's original demand were settled by arbitration?

10. A marketing study of TV viewing habits concentrated on regular watchers of two types of program: news programs and sports programs. Suppose that it was found that 80% of college-educated adults regularly watched the news, 50% watched sports, and 40% watched both news and sports. On the other hand, 65% of adults who had not completed college watched news, 85% watched sports, and 55% watched both. Finally, assume that 35% of all adults are college educated.
(a) What percent of all college-educated adults watch neither news nor sports on TV?
(b) What percent of all adults watch neither news nor sports on TV?
(c) What percent of all non-college-educated adults watch sports but not news on TV?
(d) What percent of all adults watch sports but not news on TV?

11. Three boys A, B, C are throwing a ball to one another. If A has the ball, he throws it to B with probability 0.7 and to C with probability 0.3. If B has the ball, he throws it to A with probability 0.6 and to C with probability 0.4. If C has the ball, he throws it to A with probability 0.5 and to B with probability 0.5. Note that the probabilities for who gets the ball at the kth throw are conditional, depending on which boy got the ball on the $(k - 1)$st throw, but not on who had the ball previous to that. Such a probability model for a series of trials over time (in which conditional probabilities for the future depend only on the present and not on the past) is called a *Markov Chain* model. To start the game, suppose that the boys draw straws for who goes first; thus A, B, C all have probability $\frac{1}{3}$ for being the first to have the ball.
(a) What is the (unconditional) probability that the ball starts with A and then is thrown to B?
(b) What is the (unconditional) probability that the ball goes from A to B to C?
(c) What is the (unconditional) probability that the ball goes from A to B and back to A?
(d) What is the probability that B gets the ball on the second toss?

12. A company classifies potential employees as superior, acceptable, or unsatisfactory. They estimate that 10% of all persons who apply for a job are superior, 40% are acceptable, and 50% are unsatisfactory. Every applicant takes a screening test that yields three possible summary scores: E (excellent), G (good), P (poor). It is known that 80% of superior applicants score E, and 10% of such applicants score G, 60% of acceptable applicants score G, and 20% score E; 50% of unsatisfactory applicants score P and 30% score G.

(a) What percentage of superior applicants score P?

(b) What percentage of all applicants score E? What percentage of all applicants score P?

(c) Given that an applicant scores E on the screening test, what is the conditional probability that the applicant is superior ?

(d) Given that an applicant scores G on the screening test, what is the conditional probability that the applicant is either superior or acceptable?

13. A study of middle management in a large corporation classifies the educational background of managers as follows.

Class A: technical degree plus MBA
Class B: nontechnical degree plus MBA
Class C: technical degree, no MBA
Class D: nontechnical degree, no MBA

Of all middle managers, 30% are in class A, 25% in class B, 20% in class C, and 25% in class D. Promotion records show that 5% of class A managers are eventually promoted to top management, along with 4% of class B, 4% of class C, and 2% of class D.

(a) What percent of all middle managers are eventually promoted to top management?

(b) Of all those promoted, what percent are in class A?

(c) Of all those promoted, what percent have an MBA?

14. A diagnostic test is used by a clinic to determine whether or not a person has a certain disease. If the test is positive (event T), the person is assumed to have the disease, while if the test is negative (event T^c), the person is assumed not to have the disease. Let D be the event that a person has the disease. Suppose that $P(D) = 0.10$, $P(T \mid D) = 0.90$, and $P(T^c \mid D^c) = 0.95$.

(a) Find $P(T)$.

(b) A mistaken diagnosis occurs when either the test is negative and the person has the disease, or the test is positive and the person does not have the disease. What is the probability that the clinic will make a mistaken diagnosis?

(c) Find $P(D \mid T)$ and $P(D^c \mid T^c)$.

15. A company has an option to buy thorium rights at a mining site. The company knows that 10% of all such sites are rich in thorium, 30% of all such sites are marginally profitable (but not rich), 60% of all such sites have no commercially useful deposits. The company can gather geological data with the following properties:

(i) If the site is really rich in thorium, the data will indicate rich deposits in 80% of cases, marginally profitable deposits in 15% of cases, and no useful deposits in 5% of cases.

(ii) If the site really has marginally profitable deposits, the data will indicate rich deposits in 20% of cases, marginally profitable deposits in 50% of cases, and no useful deposits in 30% of cases.

(iii) If the site really has no useful deposits, the data will indicate rich deposits in 5% of cases, marginally profitable deposits in 25% of cases, and no useful deposits in 70% of cases.

Answer the following questions.

(a) If the geological data indicate that a site has rich deposits, what is the probability that the site really has no useful deposits?

(b) If the geological data indicate that a site has no useful deposits, what is the probability that the site really does have rich deposits?

(c) In what proportion of all sites do the geological data *correctly* indicate the true nature of the thorium deposits?

16. Police records in a certain city show that 70% of all homicides are committed by acquaintances of the victim (and the remaining 30% by strangers). Murders by acquaintances are done with guns in 40% of cases, with knives or fists in 50% of cases, and by chemicals (poisons) in 10% of cases. Murders by strangers are done with guns in 60% of cases, with knives or fists in 35% of cases, and by chemicals in 5% of cases.

(a) A murder is being investigated and the coroner states that the victim was killed by poison. What is the probability that the crime was committed by an acquaintance of the victim?

(b) If the victim was knifed to death, what is the probability that the murderer was an acquaintance of the victim?

17. A box contains three balls numbered 1, 2, 3. In addition, there are two balls outside the box numbered 4, 5. Balls 1 and 5 are black, whereas balls 2, 3, 4 are white. A random experiment consists of the following two steps:

Step 1. One ball is drawn at random from the box and replaced by the ball outside the box that has the opposite color. (Thus if ball 2 is chosen, it is replaced by ball 5.)

Step 2. A second ball is then drawn at random from the box. The numbers of the two balls drawn from the box are recorded.

(a) List all outcomes in the sample space Ω of this experiment.

(b) Find the probabilities of the outcomes in the sample space Ω.

(c) Find the probability that the two balls drawn have the same color.

18. A job placement center at Midwestern University is studying the results of job interviews by their students. An interview (together with the actions and decisions that follow) can be regarded as one trial of a random experiment. The center records the following facts about each interview: (1) whether or not the student is invited for a plant trip, (2) whether or not the student is eventually offered a job, and (3) whether or not a student who is offered the job accepts the job. Note that a student not invited for a plant trip can still be offered a job, but a student not offered a job cannot make a decision about whether or not to take a job.

(a) List all possible outcomes of the experiment.

(b) The center's records show that 30% of all students interviewed are invited for plant trips. Of all students invited for plant trips, 60% are offered jobs; whereas of all students not invited for plant trips, only 20% are offered jobs. Further, 50% of students offered jobs after taking a plant trip accept the job offered, while 90% of those offered jobs but not invited for plant trip accept the job offered.

 (i) What percentage of all students interviewed are offered jobs?

 (ii) What percentage of all students who are offered jobs accept the job offered?

19. Robin Hood had finally been caught by the sheriff of Nottingham and was scheduled for execution. Because Robin was a popular hero and the sheriff wanted to seem generous, he offered Robin a chance to go free: "Here is a box with 8 black balls and 2 white balls. You will pick one ball from the box while blindfolded, and if you pick a white ball you go free. If you pick a black ball, you die." Robin, who had taken a probability course, proposed instead that he (Robin) be allowed to sort the balls into two boxes. The sheriff would then choose one of the two boxes at random and Robin would select from that box while blindfolded. The sheriff thought that Robin's suggestion would make no difference (the odds that Robin would go free would still be only $\frac{2}{10} = 0.20$). Hence he agreed to Robin's proposal. The sheriff should have taken a probability course. Below are three ways in which Robin could arrange the 8 black balls and 2 white balls in the two boxes. In each case, find the probability that Robin goes free (picks a white ball).

(a) | ○ ○ | | ● ● ● ● ● ● ● ● |.

(b) | ○ ● | | ○ ● ● ● ● ● ● ● |.

(c) | ○ | | ○ ● ● ● ● ● ● ● ● |.

Which of these ways is best for Robin? (That is, which way gives Robin the highest chance to go free?)

20. The following is a joint probability model for the population of Stateville, U.S.A.:

Gender	*Juvenile* (*age* < 21)	*Adult* ($21 \leq$ *age* < 65)	*Senior Citizen* (*age* ≥ 65)
Male	0.13	0.27	0.06
Female	?	0.30	0.09

(a) What is the probability that a person randomly chosen from the population of Stateville is both a female *and* a juvenile?

(b) What is the probability that a person randomly chosen from the population of Stateville is at least 21 years old?

(c) Given that a person randomly chosen from the population of Stateville is at least 21 years old, what is the conditional probability that this person is a male?

(d) For the town of Stateville, U.S.A., are the events "the individual is a male" and "the individual is at least 21 years old" independent events? Briefly justify your answer.

(e) Are the two events described in part (d) mutually exclusive? Briefly justify your answer.

21. In the land of Phaze, 40% of the population are human, 40% of the population are mythical beasts, and 20% of the population are spirits. The most prized ability in Phaze is to be able to do magic. Only 12% of the humans can do magic, whereas 8% of the beasts and 10% of the spirits can do magic.

(a) What is the proportion of individuals in Phaze who can do magic?

(b) Are the events "the individual is a spirit" and "the individual can do magic" statistically independent in the land of Phaze? Support your assertion by factual evidence.

22. The proportions of people falling into blood groups O, A, B, and AB are as follows:

Blood Group	O	A	B	AB
Proportion	0.44	0.42	0.10	0.04

Assume that choice of a marriage partner is probabilistically independent of blood group. A married couple is chosen by a simple random sample and the blood groups of the spouses are observed.

(a) What is the probability that the wife has blood group A and her husband has blood group B?

(b) What is the probability that one spouse has blood group A and the other spouse has blood group B?

(c) What is the probability that both spouses have the same blood group?

23. The probability that a package will be lost by the U.S. Postal Service is 0.05, and by an express company is 0.08. To make sure that a package reaches its destination, you mail 2 packages, one by each of the delivery services. Assume that the packages you mail are handled *independently*.

(a) What is the probability that at least one package reaches its destination?

(b) What is the probability that exactly one package reaches its destination?

(c) Given that exactly one package reaches its destination, what is the conditional probability that it was the one mailed through the U.S. Postal Service?

24. In electrical systems, components may be inserted in "parallel" or in "series." If 2 or more components appear in parallel, electric current will pass through if *at least one* component is capable of conducting it. If 2 or more components appear in series, *all* components must conduct current if electricity is to pass through. In the following, all components are assumed to be mutually independent of one another and the following are the probabilities that components of various types will conduct electricity at any given time:

Type	*Probability*
A	0.7
B	0.9
C	0.6
D	0.9
E	0.9

Find the probabilities that the following units conduct electricity.

(a) → [A] / [A] → in parallel.

(b) →[D]→[E]→ in series.

(c) 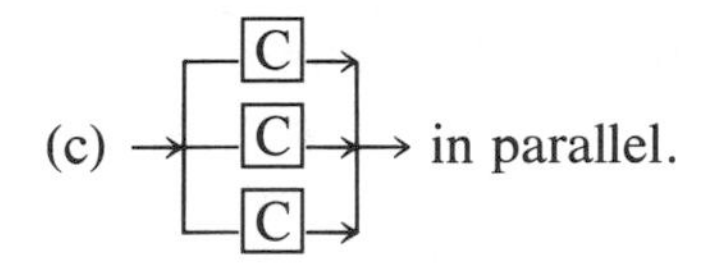

in parallel.

(d) 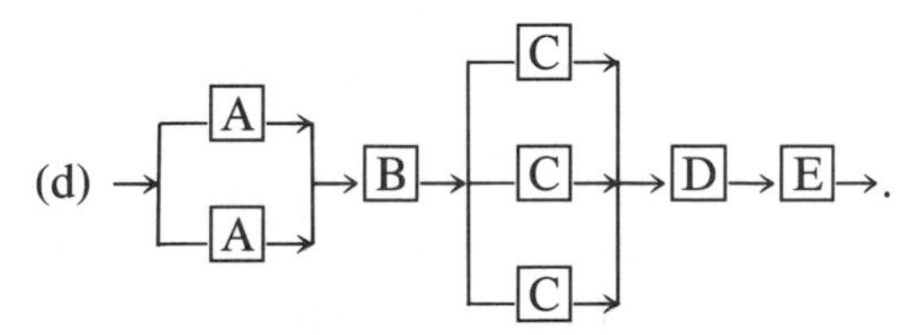

Note that the units in parts (a), (b), (c) appear in series (along with component B) in part (d). Consequently, you can use your answers in parts (a), (b), (c) to help you answer part (d). Now, answer the following.

(e) Given that the system in part (d) fails to conduct electricity, what is the *conditional* probability that component B is not conducting electricity?

25. You redial a long-distance call every 5 minutes *until you obtain an open line*. Your chance of getting an open line on any call is 0.40, and attempts are probabilistically independent.

(a) What is the probability that you first succeed in obtaining an open line on your fifth attempt?

(b) What is the probability that you require more than 4 tries before you place your call?

(c) Given that you have tried 99 times to obtain an open line without success, what is the (conditional) probability that you succeed on the 100th try?

26. In a study of marriages, both husbands and wives were classified as being either good tempered (G) or bad tempered (B). The following probability model was found to hold.

Temper of Husband	Temper of Wife	
	G	B
G	0.22	0.24
B	0.31	0.23

(a) What is the conditional probability that a husband is good tempered given that his wife is good tempered?

(b) What is the conditional probability that a husband is good tempered given that his wife is bad tempered?

(c) Are the tempers of husband and wife probabilistically independent of one another? Based on the foregoing table of probabilities, if you want to find a good-tempered wife, should you look for a good-tempered husband?

27. A particle moves on the vertices of a square. At any point in time it has probability 0.40 of moving to the vertex on its left, 0.40 of moving to the vertex on its right, and 0.20 of moving to the vertex diagonally across from it. What the particle does at any one time is statistically independent of what it did previously. What is the probability that the particle returns to the vertex at which it started (at time 0) at (a) time 2, (b) time 3, (c) time 4? Note that because a particle always moves away from the vertex where it presently is, it cannot return at time 1 to where it started. (*Hint:* Draw a square and label the vertices.)

28. Two baseball teams, A and B, meet in a short playoff. The winner is the first team to win 2 games.

(a) List all possible outcomes of this playoff series. Define your notation clearly.

(b) Suppose that the result of any game is statistically independent of the results of all previous games, and that team A has probability 0.6 of winning any particular game. What is the probability that the playoff lasts 3 games?

(c) If the playoff lasts only 2 games, what is the conditional probability that team A wins the playoff?

29. An investor will try to seize control of three companies: A, B, C. The probability that she succeeds in controlling company A is 0.4, that she succeeds in controlling company B is 0.3, and the probability that she succeeds in controlling company C is 0.5. She believes that the events "controls A," "controls B," "controls C" are mutually statistically independent.

(a) List the possible outcomes of the investor's attempt. How many outcomes are there?

(b) What is the probability that the investor succeeds in controlling exactly one company?

(c) *Given* that the investor succeeds in controlling exactly one company, what is the *conditional* probability that the company controlled is company A?

30. In a given probability model, three events A, B, C are of special interest. We are told that

$$P(A) = 0.6,$$
$$P(B) = 0.5,$$
$$P(C) = 0.1,$$

that A, B are independent events, and that C has no outcomes in common with either A or B. Using a Venn diagram, find

(a) $P(A \cap B \cap C)$.

(b) $P(A \cup B)$.

(c) $P(A \cup C)$.

(d) The probability that *exactly one* of the events A, B, C occurs.

31. In a certain random experiment three events A, B, C are of interest. It is known that A and B are statistically independent events, that B and C are mutually exclusive events, that $P(A) = 0.7$, $P(B) = 0.2$, $P(C) = 0.5$, and that $P(A \cap C) = 0.30$.

(a) What is the probability that *none* of the events A, B, C occurs?

(b) Find the probability that *at most one* of the events A, B, C occurs.

(c) Given that event C occurs, what is the conditional probability that event A also occurs?

32. Miller and Kahn (1962, p. 6) report an experiment in which samples of rocks were collected and classified both according to type and according to modal size. Based on their data, the probability model shown in Exhibit E.2 has been constructed.

(a) Note that this experiment is a composite random experiment in which there are two component random experiments: the observation of the rock type of the chosen rock sample and the observation of the modal size of the rock sample. Find the probability models for each of the component experiments (considered separately) of this composite random experiment.

(b) Are the two component experiments probabilistically independent? That is, is rock type independent of rock size?

33. Francis Galton in his book *Finger Prints* (1892) discusses the prints of the right forefingers of 101 pairs of schoolchildren chosen at random from a large collection. In addition, 105 pairs of children having a fraternal relation are also chosen. The frequency tables for the two sets of data are given as Exhibit E.3. Convert these frequencies to relative frequencies, and in each table determine whether the right forefinger prints of A children and B children are probabilistically independent. If the measurements on the A and B children were actually independent components, what relative frequencies would we expect to see?

34. What is your answer to the "Monty Hall Problem" cited in "Notes and References"? Support your answer by exhibiting a sample space and calculating appropriate conditional probabilities.

35. Two hospitals, A and B, are comparing their recovery rates. Both hospitals treat three types of patients: minor illnesses, severe illnesses, and life-threatening illnesses. For each type of illness, hospital A has a higher recovery rate than hospital B; nevertheless, a higher proportion of all patients treated by hospital B recover than do patients treated by hospital A. Is this possible? The answer is "yes," as you are asked to show below. This seemingly paradoxical fact is known in the statistical

EXHIBIT E.2 *Rock Samples Classified by Type and Modal Size*

	Modal Size			
Rock Type	*Coarse*	*Medium*	*Fine*	*Very Fine*
Arkose	0.160	0.075	0.065	0.025
L.R. Gray Woeke	0.025	0.250	0.225	0.050
Quartzite	0.040	0.050	0.010	0.025

EXHIBIT E.3 *Characteristics of Fingerprints of Fraternal and Nonfraternal Schoolchildren*

(a) Nonfraternal children

B Children	A Children			
	Arches	*Loops*	*Whorls*	*Total*
Arches	5	12	8	25
Loops	8	18	8	34
Whorls	9	20	13	42
Total	22	50	29	101

(b) Fraternal children

B Children	A Children			
	Arches	*Loops*	*Whorls*	*Total*
Arches	5	12	2	19
Loops	4	42	15	61
Whorls	1	14	10	25
Total	10	68	27	105

literature as **Simpson's paradox**. The following are the recovery rates for the two hospitals:

Type of Patient	*Probability of Recovery*	
	Hospital A	*Hospital B*
Mild illness	0.60	0.50
Severe illness	0.30	0.25
Life-threatening illness	0.08	0.05

Note that these are conditional probabilities given the type of illness. Now, of all patients treated by hospital A, 30% had a mild illness, 40% had a severe illness, and the remainder had a life-threatening illness, whereas 50% of all patients treated by hospital B had a mild illness, 30% had a severe illness, and the remainder had a life-threatening illness. Note that these are unconditional probabilities.

(a) Find the unconditional probabilities of recovery for the two hospitals, and note that hospital B's overall rate of recovery is higher than hospital A's rate. Can you explain intuitively why this is the case?

(b) What proportion of all patients who recovered from illness at hospital A had a life-threatening disease? Answer the same question for hospital B.

36. Two baseball players A and B battled for the batting crown in their league. Player A had a higher batting average than player B before the All-Star break, and also had a higher batting average after the All-Star break. Yet player B won the batting crown. Is this possible? If so, construct an example to show that it is. (Is this an example of Simpson's paradox? See Exercise 35.)

Theoretical Exercises

T.1. Let A be an event with positive probability. In Section 1 it is asserted that the conditional probability measure $P(E \mid A)$ for events E satisfies Axioms 1, 2, and 3 for probability measures. Verify this assertion. [*Hints:* For Axiom 1, use Exercise T.5 of Chapter 2 to show that $0 \leq P(E \cap A) \leq P(A)$. For Axiom 2, use Rule 2.2(c) of Chapter 2 to show that $A \cap \varnothing = \varnothing$. For Axiom 3, show that if E_1, E_2 are mutually exclusive events, then $E_1 \cap A$, $E_2 \cap A$ are also mutually exclusive. Also note that by Rule 2.3(b) of Section 2, $(E_1 \cup E_2) \cap A$ equals $(E_1 \cap A) \cup (E_2 \cap A)$.]

T.2. In a probability model, events $E_1, E_2, \ldots, E_k$ are said to be decreasing or *nested* if $E_k \subset E_{k-1} \subset \cdots \subset E_2 \subset E_1$. Show that if $E_1, E_2, \ldots, E_k$ are nested events, then

$$P(E_k) = P(E_1)P(E_2 \mid E_1)P(E_3 \mid E_2) \cdots P(E_{k-1} \mid E_{k-2})P(E_k \mid E_{k-1}).$$

T.3. In the Pólya–Eggenberger urn model for a contagious disease, we begin with an urn consisting of r red balls and w white balls. A ball is drawn at random from the urn and replaced together with an additional s balls of the same color. This process is repeated twice. Let the event R_1 be that a red ball is drawn on draw 1, and let R_2, R_3 be defined similarly for draws 2, 3. Find $P(R_1)$, $P(R_2)$, $P(R_2^c \mid R_1)$, $P(R_2^c \mid R_1^c)$, $P(R_1 \cap R_2)$.

T.4. Suppose that a random experiment $\mathscr{E}$ has two component random experiments $\mathscr{E}^{(1)}$ and $\mathscr{E}^{(2)}$ and that the sample spaces $\Omega^{(1)}$ and $\Omega^{(2)}$ of $\mathscr{E}^{(1)}$ and $\mathscr{E}^{(2)}$ each have a finite number of outcomes. Show that $\mathscr{E}^{(1)}$ and $\mathscr{E}^{(2)}$

are independent if and only if

$$P(S_1 \cap S_2) = P(S_1)P(S_2)$$

for every simple event S_1 of $\mathscr{E}^{(1)}$ and every simple event S_2 of $\mathscr{E}^{(2)}$. (*Hint:* Every event in an experiment is the union of simple events.)

T.5. Show that events E_1, E_2, E_3 are mutually probabilistically independent if and only if

$$P(E_1 \cap E_2) = P(E_1)P(E_2), \qquad P(E_1 \cap E_3) = P(E_1)P(E_3),$$
$$P(E_2 \cap E_3) = P(E_2)P(E_3), \qquad P(E_1 \cap E_2 \cap E_3) = P(E_1)P(E_2)P(E_3).$$

Recall that E_1, E_2, E_3 are mutually probabilistically independent if for any one of these events E_i, E_i is probabilistically independent of any event A composed from the remaining two events E_j, E_k by the operations of complement, union, and intersection. (*Hint:* Suppose that $i = 1$. Use Exercise T.4 to show that it is only necessary to show that E_1 is independent of $A_1 = E_2 \cap E_3$, $A_2 = E_2 \cap E_3^c$, $A_3 = E_2^c \cap E_3$, $A_4 = E_2^c \cap E_3^c$.)

5

Random Variables

1 Introduction

Most random phenomena are complex. Detailed descriptions of the outcomes of such phenomena can sometimes be more confusing than informative. Consequently, investigators find it useful to focus on one or more numerical aspects of the outcomes. For example, for purposes of evaluating radiation therapy, investigators may decide to study only the change, X, in a tumor's diameter. Similarly, although answers to individual questions on a multiple-choice exam may have diagnostic importance, instructors typically record only the number, X, of correct answers.

If a random experiment has outcomes ω, but only a single numerical aspect X of that experiment is reported, then a rule is implicitly being used that assigns a number $x = X(\omega)$ to every outcome ω. Because there is uncertainty, the outcome ω is not predictable, and neither is the value $X(\omega)$. That is, X varies randomly as the random outcome ω of the experiment varies; consequently, X is a **random variable.** In mathematical language, a random variable X is a **function** whose domain is the sample space Ω of outcomes ω, and whose range is (some subset of) the real numbers.

EXAMPLE 1.1

In a study of child development, a child is given three puzzles A, B, D to solve. Originally, the possible outcomes of this random experiment may have been listed as follows:

$$\begin{aligned} \omega_1 &= (A, B, D), & \omega_5 &= (A^c, B^c, D), \\ \omega_2 &= (A, B, D^c), & \omega_6 &= (A^c, B, D^c), \\ \omega_3 &= (A, B^c, D), & \omega_7 &= (A, B^c, D^c), \\ \omega_4 &= (A^c, B, D), & \omega_8 &= (A^c, B^c, D^c), \end{aligned}$$

where, for example, (A, B^c, D) indicates that the child solved puzzles A and D, but did not solve puzzle B. The investigator, however, is interested primarily in the number, X, of puzzles solved. Note that

$$X(\omega_1) = 3, \quad X(\omega_2) = X(\omega_3) = X(\omega_4) = 2, \quad X(\omega_5) = X(\omega_6) = X(\omega_7) = 1, \quad X(\omega_8) = 0,$$

where $X(\omega)$ is the value of the random variable X for outcome ω. Thus the random variable X is a function with domain $\Omega = \{\omega_1, \omega_2, \ldots, \omega_8\}$ and range $\{0, 1, 2, 3\}$. The numbers 0, 1, 2, 3 are the **possible values** of X. ◆

If only the quantity X is of interest in a random experiment, the extra detail provided by the outcomes ω is superfluous. This suggests reformulating the model for the experiment to make the possible values x become the outcomes. Events in this new model are then collections of values (numbers). To complete the model, we need to specify how probability is assigned to (or distributed over) the possible values of X. Such an assignment of probabilities is called the **probability distribution** of X.

The determination of the distribution of X from the probabilities for events in the original model for the experiment is conceptually straightforward. The procedure is most easily illustrated by an example.

EXAMPLE 1.1 (continued)

Assume that the child makes independent attempts to solve puzzles A, B, D. Further, suppose that the probabilities of solving puzzles A, B, and D are 0.7, 0.6, and 0.2, respectively. Under these assumptions, the probability of outcome $\omega_1 = (A, B, D)$ is

$$P(\{\omega_1\}) = P(A)P(B)P(D) = (0.7)(0.6)(0.2) = 0.084.$$

Proceeding similarly, the following table of outcomes, corresponding values of X, and probabilities is constructed:

Outcome	ω_1	ω_2	ω_3	ω_4	ω_5	ω_6	ω_7	ω_8
Probability	0.084	0.336	0.056	0.036	0.024	0.144	0.224	0.096
Value x of X	3	2	2	2	1	1	1	0

The probability of the event $\{X = 1\}$ is now obtained by adding the probabilities of all outcomes ω that yield the value $x = 1$ for X. Thus

$$\begin{aligned} P(\{X = 1\}) &= P(\{\omega : X(\omega) = 1\}) = P(\{\omega_5, \omega_6, \omega_7\}) \\ &= 0.024 + 0.144 + 0.224 = 0.392. \end{aligned}$$

Similarly, $P(\{X = 3\}) = P(\{\omega_1\}) = 0.084$, $P(\{X = 0\}) = P(\{\omega_8\}) = 0.096$, and

$$P(\{X = 2\}) = P(\{\omega_2, \omega_3, \omega_4\}) = 0.336 + 0.056 + 0.036 = 0.428.$$

These results can be summarized in the following table of probabilities for the values of X:

Value x	0	1	2	3
Probability	0.096	0.392	0.428	0.084

Probabilities of more complicated events involving X can also be determined. For example, the probability of the event $\{X \le 1\}$ that the child solves no more than one puzzle is

$$\begin{aligned} P(\{X \le 1\}) &= P(\{\omega : X(\omega) \le 1\}) = P(\{\omega_5, \omega_6, \omega_7, \omega_8\}) \\ &= 0.024 + 0.144 + 0.224 + 0.096 = 0.488. \end{aligned}$$

However, it is easier to use the table of probabilities for the values of X:

$$P(\{X \le 1\}) = P(\{X = 0, 1\}) = 0.096 + 0.392 = 0.488.$$ ◆

In similar fashion, this table of probabilities can be used to find the probability of any event concerning (the values of) X, without the need to refer to the probabilities assigned to the original outcomes $\omega_1, \ldots, \omega_8$. Hence this table of probabilities defines the probability distribution of X. If only the number X of puzzles solved by the child is of interest, this probability distribution provides an adequate probability model for the random experiment.

Of course, this new probability model loses the more detailed information contained in the probability model for the original outcomes ω. For example, it is impossible to find the probability of the event "the child solves puzzle A, but fails to solve B" from the probability distribution of X. In agreeing to confine attention to the number X of puzzles solved, we indicated our lack of interest in determining such a probability.

Example 1.1 illustrates a situation where probabilities for more detailed outcomes ω, plus knowledge of the rule assigning values $X(\omega)$ to outcomes ω, determines the probability distribution for X. In other cases, it may be necessary to construct a probability distribution directly on the values x of X—usually by combining understanding of the processes affecting X with empirical evidence obtained from observed relative frequencies. Approaches to such modeling, and ways by which the probability distribution of

X can be described, depend on the type of random variable X under consideration.

2 Discrete and Continuous Random Variables

For the most part, random variables fall into two categories: discrete and continuous. **Discrete random variables** are random variables that take on only "isolated" values. That is, if the possible values are marked on the real line, gaps exist between the marks (see Exhibit 2.1). Examples of discrete random variables include (1) the number of puzzles solved by the child in Example 1.1, (2) the score of a student on a 50-question true–false test, (3) the proportion of employees in a company employing 100 employees that are late to work on a given day, and (4) the income to the nearest dollar of a randomly chosen family.

A **continuous random variable** is a random variable that (at least conceptually) can be measured to any desired degree of accuracy. The collection of all possible values of a continuous random variable consists of one or more intervals of real numbers. Note that no matter how close two numbers within an interval of numbers may be, there is always another number between these numbers which is also in the interval. Thus between the numbers 0.68 and 0.69 in the interval [0, 1], there is another number 0.685 in that interval; between 0.68 and 0.685 there is 0.6825; between 0.68 and 0.6825 there is 0.68125; and so on. Thus values of continuous variables are not isolated, but instead, have neighbors arbitrarily close to them.

The concept of a continuous random variable is a mathematical abstraction. Imagine a person being weighed on a scale that can give the weight to an arbitrary number of decimal places. Obviously, such accuracy can never be achieved, but its conception can provide mathematical simplification in many real problems. Examples of such problems are situations where weight, size, capacity, or time is measured—the amount of oxygen released in a chemical reaction, the tonnage of ore obtained from a mine, the distance a car can travel on 1 gallon of gas, the blood pressure of a patient. The intelligence (IQ) of a person and the monetary value that a person has for a given object or action are other examples of variables that, for mathematical convenience, are often regarded as being continuous.

One important distinction between discrete and continuous variables is worth noting. For discrete random variables, $P(\{X = x\})$ is not zero for some values x. However, for continuous random variables, it is necessary to assume that for all possible values x, $P(\{X = x\}) = 0$. This assertion seems

EXHIBIT 2.1 Values of a discrete random variable.

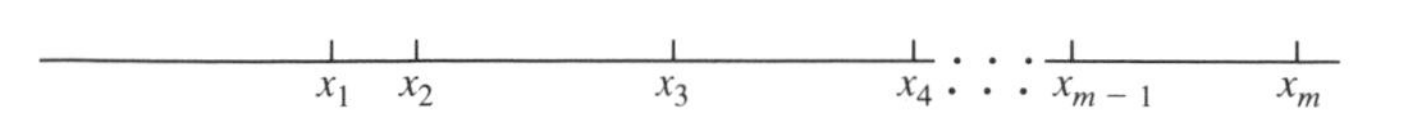

counterintuitive. How can a value of X be "possible" and yet be assigned probability zero?

To clarify this, consider repeated weighings of a person for which the weight X on a single such weighing is regarded as a continuous random variable. The repeated weighings constitute repeated trials of the random experiment in which X is observed. On the first weighing one may obtain a value x for X, but on the next weighing the measurement might differ by a very small amount from x. Indeed, it is unlikely that with a perfectly accurate scale one would ever obtain the same weight x more than once. Therefore, the relative frequency of any one value will approach zero as a limit as more and more weighings are made. Consequently, even though one may observe the value x once (x is possible), the limiting relative frequency interpretation of probability leads one to assign probability 0 to x.

On the other hand, if a person's weight is only measured to one decimal place of accuracy, the information that the person's weight is 160.1 pounds defines an **interval** [160.05, 160.15] of possible values. Certain such intervals (events) will be obtained fairly frequently in repeated weighings. Thus although every value x in such an interval has probability zero, the interval itself can have positive probability. A similar seemingly paradoxical fact is true in geometry, where points have zero length, but lines, which are made up of points, have nonzero length.

Because discrete and continuous random variables are so fundamentally different in nature, different methods are required to describe or model their probability distributions. In Section 3 it is shown how **probability mass functions** are used to describe the probability distributions of discrete random variables. For continuous random variables, **probability density functions** play a corresponding role (Section 4). It would be mathematically and conceptually convenient not to have to treat discrete and continuous random variables as separate special cases. This is accomplished by the **cumulative distribution function,** discussed in Section 5.

3 Probability Mass Functions of Discrete Random Variables

A useful way to describe the probability model of a discrete random variable X is through its probability mass function

$$p(x) = P(\{X = x\}) = P\{X = x\}, \qquad (3.1)$$

which assigns to each number x the probability of the event $\{X = x\}$ that x is observed. Consequently, this function conveys the same information as a table of probabilities for the possible values of X.

Note. The notation $P(\{X \text{ in } A\})$ for the probability of the event $\{X \text{ in } A\}$ that the observed value of X is contained in a collection of numbers, A, is correct, but cumbersome. As illustrated by Equation (3.1), henceforth the parentheses will be dropped, and $P\{X \text{ in } A\}$ will be understood to represent $P(\{X \text{ in } A\})$. Thus we will write $P\{X = 5\}$ instead of $P(\{X = 5\})$, and $P\{1 \le X \le 5\}$ instead of $P(\{1 \le X \le 5\})$.

EXAMPLE 1.1 *(continued)*

In Example 1.1 we obtained the following table of probabilities for the values of the random number, X, of puzzles solved by a child.

Value x	0	1	2	3
Probability	0.096	0.392	0.428	0.084

The corresponding probability mass function for X is

$$p(x) = \begin{cases} 0.096, & \text{if } x = 0, \\ 0.392, & \text{if } x = 1, \\ 0.428, & \text{if } x = 2, \\ 0.084, & \text{if } x = 3, \\ 0, & \text{other values of } x. \end{cases}$$ ♦

In Example 1.1 there appears to be little advantage to using a probability mass function, $p(x)$, in place of a table of probabilities. However, in other cases $p(x)$ can be expressed by a mathematical formula, which provides a compact summarization of the probabilities assigned to the values x of X.

EXAMPLE 3.1

The local baseball team has lost frequently on a road trip. Their owner is desperate for a win in order to bring fans back to the ballpark. Suppose that games are played independently, and that the local team has probability 0.3 of winning in any given game. Let X be the number of games that the team must play until they win a game (including the game that they win). Note that X is a discrete random variable, because the possible values of X are positive integers 1, 2, 3, ... Because it is conceivable that the local team could keep losing games for the indefinite future, no upper limit can be given to the possible values of X. That is, at least conceptually, X has an infinite number of possible values.

If the local team wins their first game, then $X = 1$. Thus $P\{X = 1\} = 0.3$. For X to equal 2, the team must lose the first game and win the second. The probability of this event is $(0.7)(0.3) = 0.21$. The probability that $X = 3$ is the probability of the event that the team loses the first two games and wins the third; that is, $P\{X = 3\} = (0.7)(0.7)(0.3) =$

0.147. In general, for any positive integer x, the event $\{X = x\}$ occurs if the local team loses the first $(x - 1)$ games and wins the xth game. Thus

$$p(x) = P\{X = x\} = (0.7)^{x-1}(0.3), \qquad x = 1, 2, 3, \ldots,$$

and $p(x) = 0$ for all other values of x. The probability mass function $p(x)$ summarizes compactly the probabilities for all possible values of X. ◆

The word *mass* in *probability mass function* derives historically from an analogy to classical mechanics, where the mass of a physical object is treated as if it is concentrated at a point. For discrete random variables, probabilities $p(x)$ of values x can be thought of as weights of probability masses located at points x on the real line. In a **bar graph** of the probability mass function $p(x)$, the magnitude of the weight $p(x)$ placed at x is represented by the height of the vertical line at x (see Exhibit 3.1). The analogy between probabilities $p(x)$ of discrete random variables and weights of masses at points x has provided useful intuition for analyzing and understanding probability distributions for discrete random variables.

Estimating a Probability Mass Function

In Example 3.1 the probability mass function $p(x)$ of a discrete random variable X was derived by a mathematical argument. In some applications, however, there may not be a mechanism that leads to the values of x. In such situations, empirical methods can be used to estimate $p(x)$.

Given N independent trials of the random experiment on X, the **sample probability mass function**

$$\hat{p}(x) = \text{r.f.}\{X = x\} \tag{3.2}$$

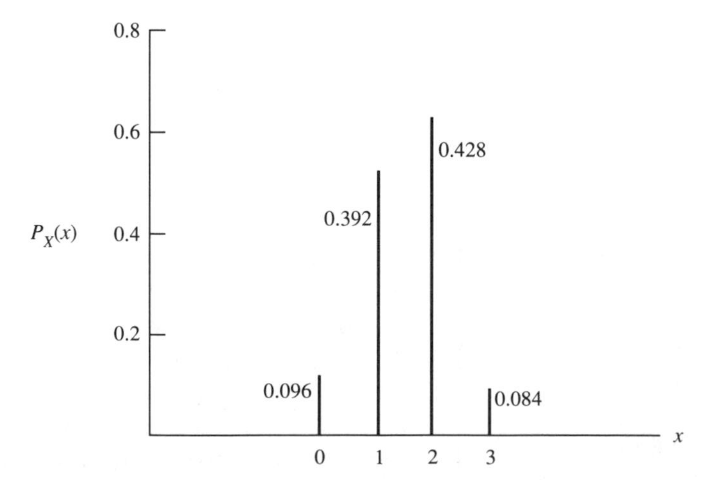

EXHIBIT 3.1 *A bar graph of the probability mass function of Example 1.1.*

provides an estimate of $p(x)$. If the number of trials, N, is sufficiently large, the sample probability mass function $\hat{p}(x)$ will be close to $p(x)$.

EXAMPLE 3.2

There have been numerous attempts to identify literary style statistically by means of sentence lengths [Yule (1939); Williams (1940)], number of unusual words used [Efron and Thisted (1976, 1987)], and number of syllables per word [Yule, unpublished; quoted in Williams (1956)]. These and more sophisticated techniques have been used to attempt to settle questions of disputed authorship [Mosteller and Wallace (1984); Efron and Thisted (1987)]. One of the earliest such studies was made by an American physicist, T. C. Mendenhall (1841–1922), who studied the probability distributions of the number of letters, X, in words used by such noted authors as Charles Dickens and William M. Thackeray [see Williams (1956)]. The intent was to use such probability distributions to compare and identify authors in much the same way that chemical compounds are identified by spectroscopic analysis. In his words: "It is proposed to analyse a composition by forming what may be called a 'word spectrum' . . . in which shall be a graphic representation of the arrangement of words according to their length and the relative frequency of their occurrence." Exhibit 3.2 gives the frequencies and sample probability mass function of the lengths (number of letters) X of words in Thackeray's *Vanity Fair*, based on a (nonrandom) sample of $N = 2000$ words. A bar graph of the sample probability mass function $\hat{p}(x) = \text{r.f.}\{X = x\}$ is given in Exhibit 3.3.

EXHIBIT 3.2 *Frequencies and Sample Probability Mass Function of Word Length X in 2000 Words from Vanity Fair*

Number X of Letters in the Word	*Frequency of x*	$\hat{p}(x)$	*Number X of Letters in the Word*	*Frequency of x*	$\hat{p}(x)$
1	58	0.0290	8	100	0.0500
2	315	0.1575	9	63	0.0315
3	480	0.2400	10	43	0.0215
4	351	0.1755	11	16	0.0080
5	244	0.1220	12	15	0.0075
6	154	0.0770	13	4	0.0020
7	152	0.0760	14	5	0.0025

Source: C. B. Williams, Studies in the history of probability and statistics. IV. A note on an early statistical study of literary style. *Biometrika*, vol. 43, pp. 248–56 (1956). ◆

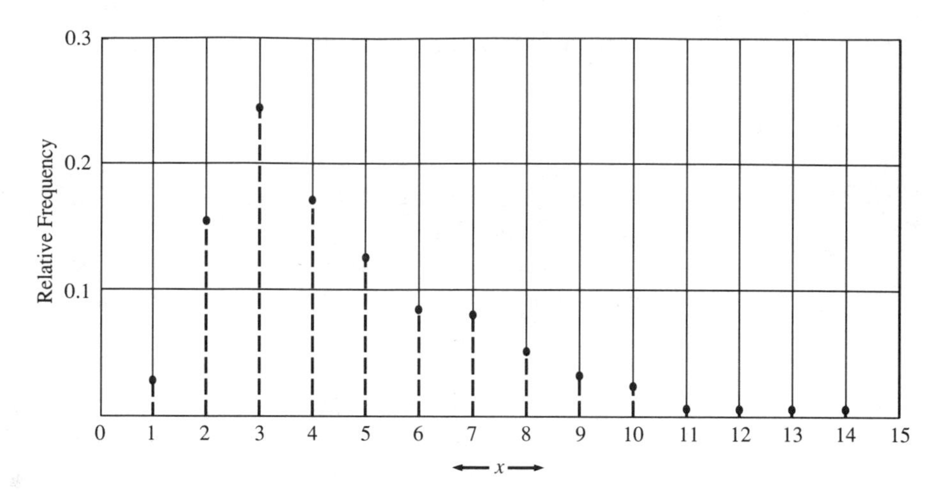

EXHIBIT 3.3 *Graph of the probability mass function of the number of letters in words from Thackey's Vanity Fair.*

Although the number of trials in Example 3.2 is fairly large ($N = 2000$), the sample probability mass function $\hat{p}(x)$ will not exactly equal the true probability mass function $p(x)$ at all values x. Keeping this in mind, one can compare $\hat{p}(x)$ with various probability mass functions suggested by assumptions concerning the mechanism (literary style) giving rise to X. Alternatively, these functions might be chosen solely for their convenient mathematical form. Probability modelers have at their disposal a large number of possible probability mass functions; some of the most important of these are discussed in Chapter 8. If any such function $p_*(x)$ has values in close agreement with the values of the sample probability mass function $\hat{p}(x)$, we say that $p_*(x)$ "fits the data." There is then a good possibility that $p_*(x)$ is either the true probability mass function $p(x)$ of X, or a very close approximation. [That is, the sample probability mass function $\hat{p}(x)$ should be "close" to the true probability mass function $p(x)$. Thus if $p_*(x)$ is "close" to $\hat{p}(x)$, it is also "close" to $p(x)$.] Statistical methods exist (see Chapter 14) to assess the "goodness of fit" of any hypothesized probability mass function $p_*(x)$.

EXAMPLE 3.2 *(continued)*

Williams (1956) suggests the function

$$p_*(x) = \begin{cases} 0.17307 \exp\left[-\frac{1}{2}\left(\frac{\log x - 1.224}{0.558}\right)^2\right], & x = 1, 2, \ldots, \\ 0, & \text{other values of } x, \end{cases}$$

as a possible probability mass function for the number X of letters in a

EXHIBIT 3.4 *Comparison of Hypothesized and Sample Probability Mass Functions for the Number X of Letters in a Random Word from Vanity Fair*

x	$\hat{p}(x)$	$p_*(x)$	x	$\hat{p}(x)$	$p_*(x)$
1	0.0290	0.0198	8	0.0500	0.0601
2	0.1575	0.1151	9	0.0315	0.0440
3	0.2400	0.1692	10	0.0215	0.0322
4	0.1755	0.1666	11	0.0080	0.0236
5	0.1220	0.1396	12	0.0075	0.0174
6	0.0770	0.1086	13	0.0020	0.0128
7	0.0760	0.0815	14	0.0025	0.0096

random word from *Vanity Fair*. Exhibit 3.4 compares $p_*(x)$ to the sample probability mass function $\hat{p}(x)$.

The "fit" of $p_*(x)$ to $\hat{p}(x)$ is not as close as might be desired. The maximum unsigned difference between $\hat{p}(x)$ and $p_*(x)$ is 0.0708 at $x = 3$, which is fairly substantial. Further, $p_*(x)$ appears to consistently overstate the probabilities of values of $x \geq 5$, and understate the probabilities of values $x \leq 4$. One is therefore led to seek alternative choices for $p_*(x)$ (see Exercise 11). ◆

Properties of Probability Mass Functions

Every probability mass function $p(x)$ must satisfy the following properties:

(a) $p(x) \geq 0$ for all numbers x.
(b) The sum $\sum_x p(x)$ of the values of $p(x)$ must equal 1.

The justification for these requirements is easily given. Property (a) holds because, by definition, $p(x)$ gives the probability of the event $\{X = x\}$, and probabilities are nonnegative numbers (Axiom 1, Chapter 2). For any collection of numbers A, the probability of the event that the value of X is a member of the collection A is the sum of the probabilities of all values x belonging to A. The notation $\{x \in A\}$ means $\{x \text{ is in } A\}$. Thus,

$$P\{X \text{ is in } A\} = \sum_{x \in A} P\{X = x\} = \sum_{x \in A} p(x). \quad (3.3)$$

Because all possible values of X are real numbers, X is certain to be included in the interval $(-\infty, \infty)$. Thus

$$1 = P\{X \text{ is in } (-\infty, \infty)\} = \sum_{x \in (-\infty,\infty)} p(x) = \sum_x p(x).$$

This verifies property (b).

It can also be shown that any function satisfying properties (a) and (b) is the probability mass function for some discrete random variable. Consequently, properties (a) and (b) *characterize* probability mass functions.

EXAMPLE 3.3

The function

$$q(x) = \begin{cases} \frac{1}{9}(x-2)^3, & x = 0, 1, 2, 3, 4, 5, \\ 0, & \text{other values of } x, \end{cases}$$

might be proposed as a possible probability mass function. Note that $q(x)$ satisfies property (b) of probability mass functions because

$$\begin{aligned}\sum_x q(x) &= \tfrac{1}{9}(0-2)^3 + \tfrac{1}{9}(1-2)^3 + \tfrac{1}{9}(2-2)^3 + \tfrac{1}{9}(3-2)^3 \\ &\quad + \tfrac{1}{9}(4-2)^3 + \tfrac{1}{9}(5-2)^3 \\ &= (-\tfrac{4}{9}) + (-\tfrac{1}{9}) + 0 + \tfrac{1}{9} + \tfrac{4}{9} + \tfrac{9}{9} = 1.\end{aligned}$$

However, the calculations above also show that $q(0)$ and $q(1)$ are negative numbers, violating property (a). Thus $q(x)$ cannot be a probability mass function for X.

On the other hand, the function

$$r(x) = \begin{cases} \frac{1}{9}(x-2)^2, & x = 0, 1, 2, 3, 4, 5, \\ 0, & \text{all other } x \text{ values}, \end{cases}$$

satisfies property (a) but is not a probability mass function because $\sum_x r(x) = 19/9 \neq 1$, in violation of property (b). Finally, the function

$$p(x) = \begin{cases} \frac{1}{19}(x-2)^2, & x = 0, 1, 2, 3, 4, 5, \\ 0, & \text{all other } x \text{ values}, \end{cases}$$

does satisfy both property (a) and property (b), and hence is a probability mass function. ◆

4 Probability Density Functions

For a continuous random variable, the function that plays a role analagous to that of the probability mass function is called the **probability density function,** or simply the **density function.** However, the probability density function $f(x)$ of a continuous random variable X does *not* give the probability that X equals the number x. (Recall that for a continuous random variable, $P\{X = x\} = 0$ for all numbers x.) Rather, $f(x)$ provides a means to calculate the probabilities of *intervals* of numbers. If $f(x)$ is graphed as in Exhibit 4.1, the area over the interval $[a, b]$ under the graph of $f(x)$ represents the probability, $P\{a \leq X \leq b\}$, that the observed value of X will belong to $[a, b]$. It thus follows that if $f(x)$ is smooth enough to permit integration,

$$P\{a \leq X \leq b\} = \int_a^b f(x)\, dx. \tag{4.1}$$

Because the probability is 1 that X is equal to some real number, it follows that *the area under the graph of $f(x)$ over the entire horizontal axis is always equal to 1*. That is,

$$\int_{-\infty}^{\infty} f(x)\, dx = 1. \tag{4.2}$$

As in the case of the probability mass function, there is a physical analogy. Think of the region bounded by the graph of $f(x)$ and the horizontal axis in Exhibit 4.1 as being a uniform sheet of metal of total weight 1. The value of $f(x)$ at x is then the density of the metal sheet at x. Observe that the weight of the sheet of metal over the interval $[a, b]$ is equal to the area of the sheet over that interval. Thus $P\{a \leq X \leq b\}$ can be regarded as a weight.

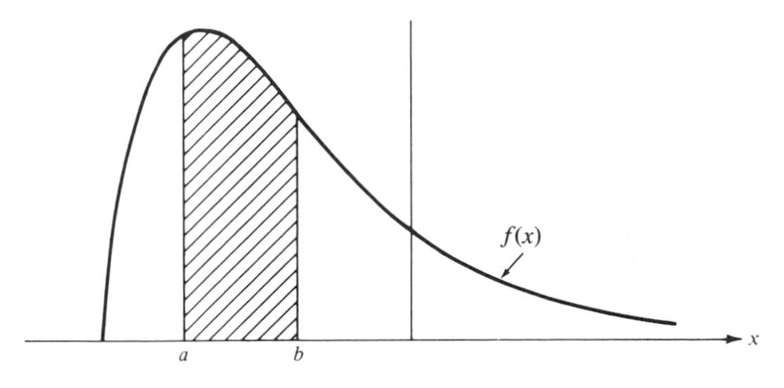

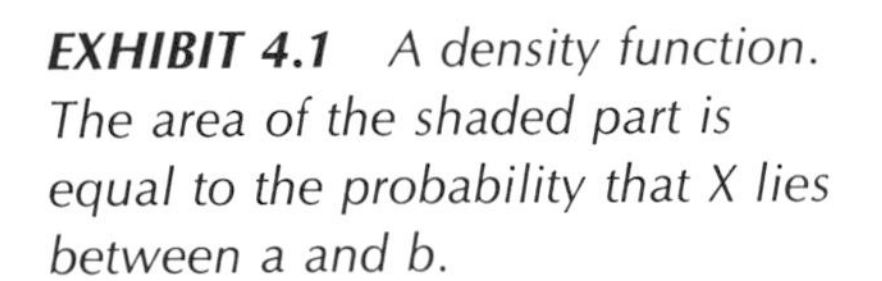
EXHIBIT 4.1 *A density function. The area of the shaded part is equal to the probability that X lies between a and b.*

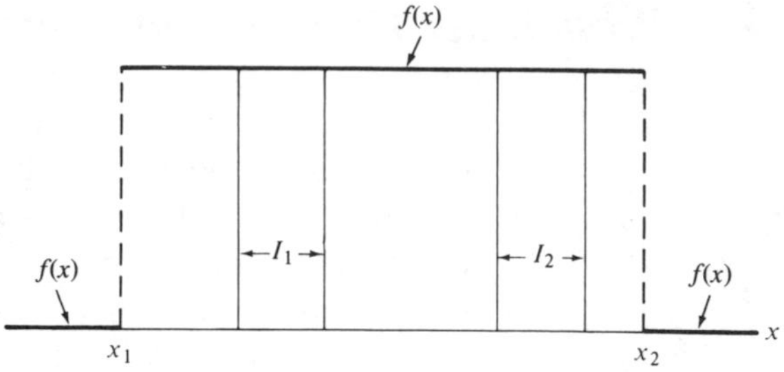

EXHIBIT 4.2 *Constant density between the values x_1 and x_2 implies equal probability for the intervals I_1 and I_2 of equal length.*

Note also that the weight of the sheet over any point x equals 0. This corresponds to the fact that $P\{X = x\} = 0$. Observe that the greater the height of the graph $f(x)$, the greater is the weight of the sheet in a small interval $[x - \Delta, x + \Delta]$ about x. (Here, Δ is a small positive number.) That is, although $f(x)$ is not itself a probability, it does provide information about the concentration of probability weight near the point x.

From the probability density function $f(x)$, certain qualitative conclusions concerning the variation of the random variable X over repeated independent and identical trials are evident. If the probability density function is constant over the line segment from x_1 to x_2, all subintervals of $[x_1, x_2]$ of equal length are equally probable (see Exhibit 4.2). On the other hand, if the probability density function is such that most of the area beneath it is concentrated within a very narrow range (see Exhibit 4.3), repeated observations on X tend to yield values mostly close to where the area is concentrated [where $f(x)$ is "spiked"].

Because $P\{X = a\} = 0$ and $P\{X = b\} = 0$ for any continuous random variable X, it follows that for continuous random variables,

(4.3)

$$P\{a \le X \le b\} = P\{a < X \le b\} = P\{a \le X < b\} = P\{a < X < b\}.$$

This result is, in general, not true for discrete random variables.

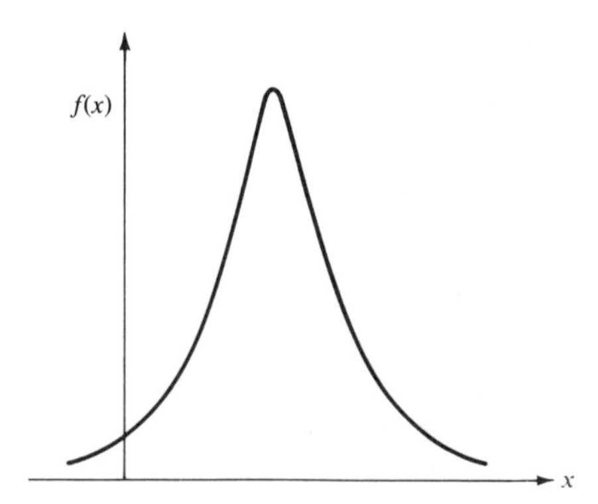

EXHIBIT 4.3 *A highly centralized density function.*

EXAMPLE 4.1

A delicatessen has 400 pounds of meat delivered every day. A study has shown that if X is the weight of the meat sold in a day (assumed measured exactly), the probability distribution (model) for the random variable X can be described by the probability density function

$$f(x) = \begin{cases} (1.25)10^{-5}x, & \text{if } 0 \le x \le 400, \\ 0, & \text{if } x < 0 \text{ or } x > 400. \end{cases}$$

This function is graphed in Exhibit 4.4.

The probability that between 200 and 300 pounds of meat are sold on a given day is

$$\begin{aligned} P\{200 \le X \le 300\} &= \int_{200}^{300} (1.25)10^{-5}x \, dx \\ &= (1.25)10^{-5} \frac{x^2}{2} \Bigg|_{200}^{300} = 0.3125. \end{aligned}$$

Similarly, the probability that at least 150 pounds of meat are sold in a day is

$$\begin{aligned} P\{150 \le X < \infty\} &= \int_{150}^{400} (1.25)10^{-5}x \, dx + \int_{400}^{\infty} 0 \, dx \\ &= (1.25)10^{-5} \frac{x^2}{2} \Bigg|_{150}^{400} + 0 \\ &= 0.8594. \end{aligned}$$

Observe that because the density $f(x)$ is defined differently over the intervals [150, 400] and [400, ∞), the probability was calculated as the sum of two integrals: one over the range $150 \le x \le 400$, and one over the range $400 \le x \le \infty$. ◆

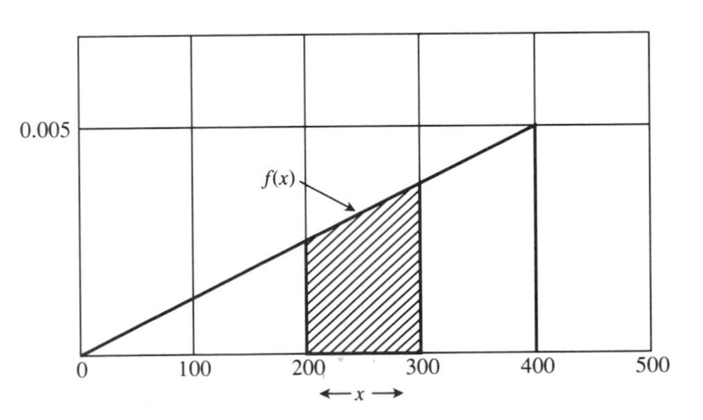

EXHIBIT 4.4 *Graph of the probability density function f(x) of the weight X of luncheon meat sold in a given morning. The shaded area gives the probability that between 200 and 300 pounds of luncheon meat is sold.*

EXAMPLE 4.2

A dispenser drops an amount, X, of coffee into an 8-ounce paper cup. Studies of the dispenser lead to modeling X as a continuous random variable with probability density function

$$f(x) = \begin{cases} 0, & \text{if } x < 7.7, \\ 10(x - 7.7), & \text{if } 7.7 \leq x < 8.1, \\ 40(8.2 - x), & \text{if } 8.1 \leq x < 8.2, \\ 0, & \text{if } 8.2 \leq x. \end{cases}$$

This density function is graphed in Exhibit 4.5. [Note that $f(8.1) = 4$. This again illustrates that $f(x)$ is not a probability, because a probability is never greater than 1.]

It is of obvious interest to find the probability, $P\{X > 8.0\}$, that more coffee is dispensed than an 8-ounce cup can hold. Because the density $f(x)$ is defined differently over different regions, this probability is calculated as a sum of integrals:

$$\begin{aligned} P\{8.0 < X < \infty\} &= \int_{8.0}^{8.1} 10(x - 7.7)\,dx + \int_{8.1}^{8.2} 40(8.2 - x)\,dx + \int_{8.2}^{\infty} 0\,dx \\ &= 10\left(\frac{x^2}{2} - 7.7x\right)\Bigg|_{8.0}^{8.2} + 40\left(8.2x - \frac{x^2}{2}\right)\Bigg|_{8.1}^{8.2} + 0 \\ &= 0.35 + 0.20 + 0 = 0.55. \end{aligned}$$

This calculation shows that more than 8 ounces of coffee will be dispensed in over 50% of all uses of the dispenser. ◆

Estimating a Probability Density Function: Histograms

Just as empirical methods were used to estimate probability mass functions for discrete random variables, data can be used to estimate probability

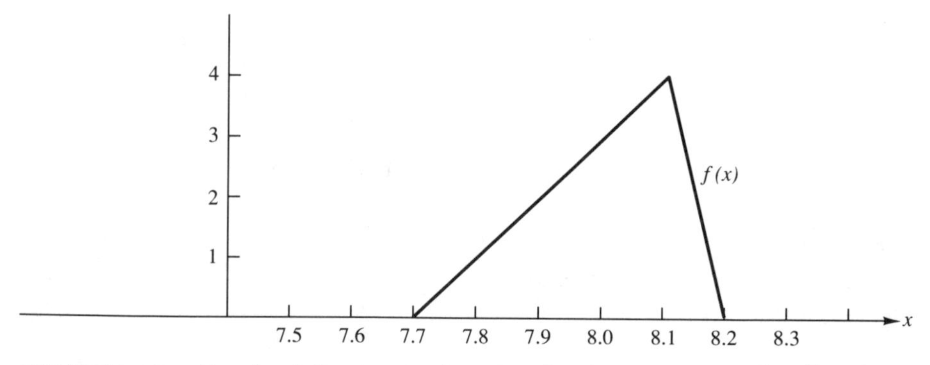

EXHIBIT 4.5 *Graph of the density function for the amount X of coffee dispensed into an 8 ounce cup.*

density functions for continuous random variables. (Due to the fact that the distribution of probability is described over an infinite number of possible values using only a finite amount of data, a certain amount of arbitrariness in the method of estimation is inevitable.) Because probability density functions are used to find probabilities of intervals, it seems natural to start by choosing a collection of intervals, and to use the relative frequencies with which these intervals occur in the data to estimate their probabilities. The choice of which intervals to use is one of the arbitrary aspects of the estimation procedure. It is generally agreed that these intervals should be chosen independently of the data, and that they should be (1) nonoverlapping, and (2) exhaustive of the data. For computational convenience, it is often desirable for all intervals, except perhaps the leftmost and rightmost intervals, to have the same length, and for there to be neither too few nor too many intervals. A rule of thumb frequently advocated is that between 5 and 15 intervals be used.

Suppose that 417 forty-watt, 110-volt internally frosted incandescent lamps are kept lit continuously until they burn out. The observed lifetimes, X, in hours, of the lamps are measured, and vary between a minimum of 225 hours and a maximum of 1690 hours [Davis (1952)]. Because of the large number of observations, we use 15 intervals. (With fewer observations, fewer and broader intervals might be more desirable.) If the left endpoint of the first interval is 200 hours and the right endpoint of the last interval is 1700 hours, all of the data are accounted for, and the common length of the intervals can be taken to be $l = (1700 - 200)/15 = 100$. The data of the lifetimes of the bulbs are now summarized in a **grouped relative frequency distribution**, as shown in Exhibit 4.6.

EXHIBIT 4.6 *Grouped Relative Frequency Distribution of the Lifetimes (in Hours) of 417 Forty-Watt Incandescent Lamps*[a]

Interval	*Frequency*	*Relative Frequency*	*Interval*	*Frequency*	*Relative Frequency*
200–300	1	0.0024	1000–1100	85	0.2038
300–400	0	0.0000	1100–1200	80	0.1918
400–500	0	0.0000	1200–1300	44	0.1055
500–600	3	0.0072	1300–1400	23	0.0552
600–700	10	0.0240	1400–1500	9	0.0216
700–800	21	0.0504	1500–1600	3	0.0072
800–900	45	0.1079	1600–1700	2	0.0048
900–1000	91	0.2182		417	

[a] Observations such as $X = 600$ are put into the 500–600 interval rather than the 600–700 interval.

From the relative frequency distribution in Exhibit 4.6, it is a direct process to form a **modified relative frequency histogram.** At the midpoint of each interval, a bar is constructed with height equal to the ratio of the relative frequency of the interval to the length of the interval. (Thus the height of the bar for the interval [200, 300] is r.f.$\{200 \leq X \leq 300\}/l = (1/417)/100 = 0.00024$.) The relative frequency of the interval is now "spread" over the interval by constructing a rectangle with base equal to the length of the interval and height equal to the height of the bar (see Exhibit 4.7). Notice that the area of such a rectangle gives the relative frequency of the interval. Thus the curve (graph) formed by the tops of the constructed rectangles has the property that the area under the curve over each interval equals the estimated probability of that interval. This curve or graph defines a probability density function that can serve as a crude estimate $\hat{f}(x)$ of the actual probability density function $f(x)$. For example, $\hat{f}(x) = 0.00192$ for $1100 \leq x < 1200$.

Alternative possibilities for the probability density function of X may be more accurate than $\hat{f}(x)$. For example, the shape of Exhibit 4.7 suggests that the interval (950, 1000] may have greater probability of containing X than does the interval (900, 950]. A more intuitively appealing estimate of the true probability density function $f(x)$ can be obtained by drawing a smooth curve through the top of each rectangle of the histogram at the midpoint of the interval (see Exhibit 4.8). If this curve is chosen carefully, the area under this curve can be made to equal 1, and the resulting graph defines a new estimate $\tilde{f}(x)$ of $f(x)$. However, it should be noted that many such curves [and corresponding estimators of $f(x)$] are possible, and no single choice is clearly more correct than another.

Frequently, the true probability density function is assumed to have a particular functional form. The most commonly used such functional forms are discussed in Chapters 9 and 10. For any particular probability density function $f(x)$, a table such as Exhibit 4.6 can be used to see how well the proposed probability density function "fits" the data. To do so, the theoreti-

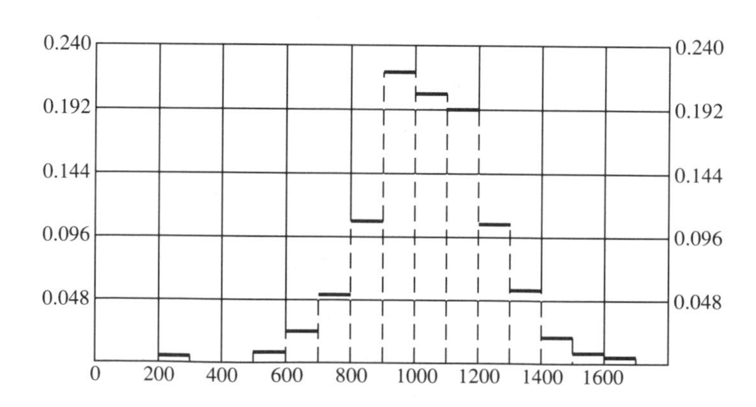

EXHIBIT 4.7 *Modified relative frequency histogram of data on life testing.*

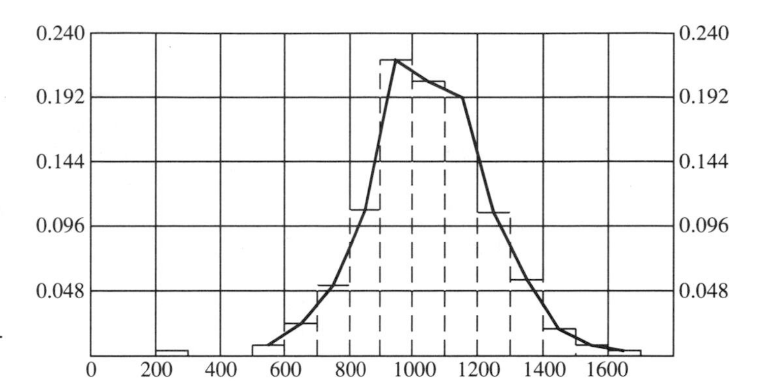

EXHIBIT 4.8 *Representation of life-testing data as a modified relative frequency curve.*

cal probability of each interval in the grouped relative frequency distribution is computed and is compared to the actual relative frequency of the interval.

The probability density functions that most life-testing specialists would use to fit the data in Exhibit 4.6 require special tables for calculation of probabilities. Instead, we use a less accurate, but simpler, choice of density function:

$$f(x) = \begin{cases} 0, & \text{if } x \leq 200, \\ (2.40)10^{-9}(x - 200)^2, & \text{if } 200 < x \leq 1000, \\ (5.16)10^{-9}(1700 - x)^2, & \text{if } 1000 < x \leq 1700, \\ 0, & \text{if } x > 1700. \end{cases} \tag{4.4}$$

Note that

$$\begin{aligned} \int_a^{a+100} (2.40)(10^{-9})(x - 200)^2\, dx &= (2.40)(10^{-9}) \left. \frac{(x - 200)^3}{3} \right|_a^{a+100} \\ &= (0.8)(10^{-9}) \left[(a - 100)^3 - (a - 200)^3\right] \end{aligned}$$

for $a = 200, 300, \ldots, 900$, and that

$$\begin{aligned} \int_a^{a+100} (5.16)(10^{-9})(1700 - x)^2\, dx &= (5.16)(10^{-9}) \left[-\frac{(1700 - x)^3}{3} \right] \Bigg|_a^{a+100} \\ &= (1.72)(10^{-9}) \left[(1700 - a)^3 - (1600 - a)^3\right] \end{aligned}$$

for $a = 1000, 1100, \ldots, 1600$. Using these results, theoretical probabilities based on the density (4.4) can be computed for the intervals in Exhibit 4.8. These, plus the corresponding relative frequencies, are given in Exhibit 4.9.

As expected, the fit of the probability density function $f(x)$ defined by equation (4.4) to the data is not particularly good. The calculations used to construct Exhibit 4.9 illustrate, however, how theoretical probabilities can be computed for more appropriate choices of $f(x)$.

EXHIBIT 4.9 *Comparison of Observed Relative Frequencies and Theoretical Probabilities for the Lifetimes of Forty-Watt Incandescent Lamps*

Interval	*Relative Frequency*	*Theoretical Probability*	*Interval*	*Relative Frequency*	*Theoretical Probability*
200–300	0.0024	0.0008	1000–1100	0.2038	0.2184
300–400	0.0000	0.0056	1100–1200	0.1918	0.1565
400–500	0.0000	0.0152	1200–1300	0.1055	0.1049
500–600	0.0072	0.0296	1300–1400	0.0552	0.0636
600–700	0.0240	0.0488	1400–1500	0.0216	0.0327
700–800	0.0504	0.0728	1500–1600	0.0072	0.0120
800–900	0.1079	0.1016	1600–1700	0.0048	0.0017
900–1000	0.2182	0.1352			

Properties of Probability Density Functions

A probability density function must satisfy the following properties:

$$(a)\ f(x) \geq 0 \text{ for all numbers } x.$$

$$(b)\ \int_{-\infty}^{\infty} f(x)\, dx = 1.$$

In addition, $f(x)$ must be smooth enough to permit integration over any interval $[a, b]$. We recognize property (b) as asserting that the probability of the sure event $\{-\infty < X < \infty\}$ must be 1. If $f(x)$ is negative over any interval of numbers $[a, b]$, no matter how small, the integral of $f(x)$ over $[a, b]$ will be negative. This contradicts the fact that for a probability density function,

$$\int_a^b f(x)\, dx = P\{a \leq X \leq b\}$$

must be a probability, and thus must be nonnegative. Taking the limit as b approaches a, and assuming that $f(x)$ is "smooth" shows that $f(a)$ must be nonnegative for any number a. [This argument is not entirely rigorous mathematically, because the value of $f(x)$ can be changed at any fixed value x_0 without affecting the area under $f(x)$ over any interval containing x_0. However, this argument does give the intuition behind a more rigorous proof.]

Any integrable function which satisfies properties (a) and (b) is the probability density function of some continuous random variable X. Thus properties (a) and (b) *characterize* probability density functions.

EXAMPLE 4.3 Consider the functions

$$g_1(x) = \begin{cases} 0, & \text{if } x < 0, \\ \frac{1}{2}(2x - 1), & \text{if } 0 \le x < 2, \\ 0, & \text{if } x \ge 2, \end{cases} \qquad g_2(x) = \begin{cases} 0, & \text{if } x < 0 \\ x, & \text{if } 0 \le x < 2, \\ 0, & \text{if } x \ge 2, \end{cases}$$

$$g_3(x) = \begin{cases} 0, & \text{if } x < 0, \\ \frac{1}{2}x, & \text{if } 0 \le x < 2, \\ 0, & \text{if } x \ge 2. \end{cases}$$

The function $g_1(x)$ satisfies property (b) because

$$\int_{-\infty}^{\infty} g_1(x)\,dx = \int_{-\infty}^{0} 0\,dx + \int_0^2 \tfrac{1}{2}(2x - 1)\,dx + \int_2^{\infty} 0\,dx$$
$$= 0 + \tfrac{1}{2}(x^2 - x)\Big|_0^2 + 0 = 1,$$

but fails to satisfy property (a) because $g_1(x) < 0$ for $0 \le x < \frac{1}{2}$. Thus $g_1(x)$ is not a probability density function. The function $g_2(x)$ satisfies property (a), but

$$\int_{-\infty}^{\infty} g_2(x)\,dx = \int_{-\infty}^{0} 0\,dx + \int_0^2 x\,dx + \int_2^{\infty} 0\,dx$$
$$= 0 + \frac{x^2}{2}\Big|_0^2 + 0 = 2,$$

so that property (b) is not satisfied. Because $g_3(x) = \frac{1}{2}g_2(x)$ for all x, $g_3(x)$ satisfies properties (a) and (b) and thus is a probability density function for some continuous random variable X. Note that

$$g_3(x) = \frac{g_2(x)}{\int_{-\infty}^{\infty} g_2(x)\,dx}.$$

As long as $\int_{-\infty}^{\infty} g_2(x)\,dx$ is finite, the equation above provides a way to convert a function $g_2(x)$ satisfying property (a) into a density function. ◆

5 Cumulative Distribution Functions

In this section we discuss an alternative way to define the distribution of a random variable which has the considerable advantage that it is applicable to both discrete and continuous variables.

Consider the function $F(x)$, which gives the probability of the event $\{X \leq x\}$ for every number x. That is, for each x,

$$F(x) = P\{X \leq x\} = P\{-\infty < X \leq x\}. \tag{5.1}$$

This function is called the **cumulative distribution function** of the random variable X.

Suppose that X is a discrete random variable with probability mass function $p(x)$. For each number x, the cumulative distribution function $F(x)$ of X is the sum of the probabilities of all numbers t which are less than or equal to x. That is,

$$F(x) = \sum_{t \leq x} p(t). \tag{5.2}$$

Equation (5.2) illustrates why $F(x)$ is called a cumulative distribution function; this function shows how probability accumulates as we move from smaller to larger values of x.

EXAMPLE 5.1

Let X be a discrete random variable with probability mass function

$$p(x) = \begin{cases} \dfrac{8}{15}\left(\dfrac{1}{2}\right)^x, & \text{if } x = 0, 1, 2, 3, \\ 0, & \text{other values of } x. \end{cases}$$

From the mass function $p(x)$, construct the following table of probabilities:

x	0	1	2	3
$p(x)$	$\frac{8}{15}$	$\frac{4}{15}$	$\frac{2}{15}$	$\frac{1}{15}$

To illustrate the calculation of the cumulative distribution function $F(x)$,

$$F(2.5) = \sum_{t \leq 2.5} p(t) = p(0) + p(1) + p(2) = \frac{8}{15} + \frac{4}{15} + \frac{2}{15} = \frac{14}{15}.$$

Specification of $F(x)$ for all numbers x proceeds by cases. First note that for any $x < 0$, there are no possible values of X less than or equal to

x. Thus

$$F(x) = 0, \qquad \text{for } x < 0.$$

For $x = 0$, and for any x between 0 and 1, the only possible value of X that is less than or equal to x is $t = 0$. Thus

$$F(x) = \sum_{t \le x} p(t) = p(0) = \frac{8}{15}, \qquad \text{if } 0 \le x < 1.$$

For $x = 1$, and for any x between 1 and 2, there are two possible values, 0 and 1, of X which are less than or equal to x. Therefore,

$$F(x) = \sum_{t \le x} p(t) = p(0) + p(1) = \frac{8}{15} + \frac{4}{15} = \frac{12}{15}, \qquad \text{if } 1 \le x < 2.$$

Similarly,

$$F(x) = \sum_{t \le x} p(t) = p(0) + p(1) + p(2) = \frac{14}{15}, \qquad \text{if } 2 \le x < 3,$$

$$F(x) = p(0) + p(1) + p(2) + p(3) = 1, \qquad \text{if } x \ge 3.$$

Hence the cumulative distribution function of X is

$$F(x) = \begin{cases} 0, & \text{if } x < 0, \\ \frac{8}{15}, & \text{if } 0 \le x < 1, \\ \frac{12}{15}, & \text{if } 1 \le x < 2, \\ \frac{14}{15}, & \text{if } 2 \le x < 3, \\ 1, & \text{if } 3 \le x. \end{cases}$$

This function is graphed in Exhibit 5.1. ♦

Note that the graph of the cumulative distribution function $F(x)$ in Exhibit 5.1 appears as a series of steps, and for this reason $F(x)$ is said to be a **step function.** In general, *the cumulative distribution function $F(x)$ for any discrete random variable X is a step function.*

Because $F(x) = P\{-\infty < X \le x\}$, the cumulative distribution function of a continuous random variable X with probability density function $f(x)$ is

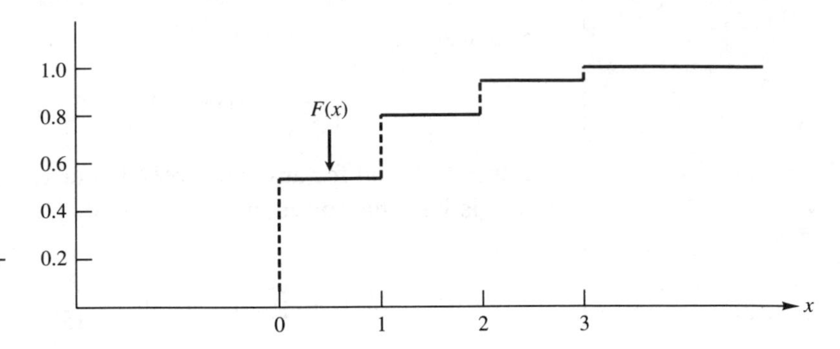

EXHIBIT 5.1 The cumulative distribution function for Example 5.1.

given by the definite integral

$$(5.3) \qquad F(x) = \int_{-\infty}^{x} f(t)\, dt.$$

EXAMPLE 5.2

Let X be a continuous random variable with probability density function

$$f(x) = \begin{cases} 0, & \text{if } x < 0, \\ e^{-x}, & \text{if } x \geq 0. \end{cases}$$

To find $F(2.5)$, calculate

$$\begin{aligned} F(2.5) &= \int_{-\infty}^{2.5} f(t)\, dt = \int_{-\infty}^{0} 0\, dt + \int_{0}^{2.5} e^{-t}\, dt = 0 + (-e^{-t})\Big|_{0}^{2.5} \\ &= -e^{-2.5} - (-e^{-0}) = -e^{-2.5} + 1 = 0.917915. \end{aligned}$$

To find $F(x)$ for general values of x, one again proceeds by cases:

$$F(x) = \int_{-\infty}^{x} f(t)\, dt = \int_{-\infty}^{x} 0\, dt = 0, \qquad \text{if } x < 0.$$

$$F(x) = \int_{-\infty}^{x} f(t)\, dt = \int_{-\infty}^{0} 0\, dt + \int_{0}^{x} e^{-t}\, dt = 1 - e^{-x}, \qquad \text{if } x \geq 0.$$

Consequently, the cumulative distribution function of X is

$$F(x) = \begin{cases} 0, & \text{if } x < 0, \\ 1 - e^{-x}, & \text{if } x \geq 0. \end{cases}$$

This function is graphed in Exhibit 5.2. ◆

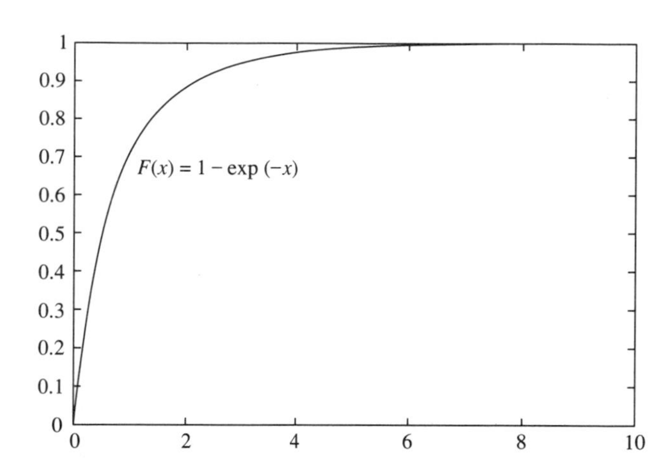

EXHIBIT 5.2 *The cumulative distribution function for Example 5.2.*

The graph of the cumulative distribution function $F(x)$ in Exhibit 5.2 is a smooth curve. In general, *the cumulative distribution function $F(x)$ of a continuous random variable X is always a continuous function of its argument x.*

It is important to remember that each value of the cumulative distribution function $F(x)$ of a continuous random variable is an *area*. Visualizing $F(x)$ as an area, perhaps by drawing a picture of the area desired, may help in obtaining formulas, and in avoiding common errors.

EXAMPLE 5.3

Let X be a continuous random variable with probability density function

$$f(x) = \begin{cases} 0, & \text{if } x < 0, \\ x, & \text{if } 0 \le x < 1, \\ \frac{1}{2}, & \text{if } 1 \le x < 2, \\ 0, & \text{if } 2 \le x. \end{cases}$$

The value of the cumulative distribution function $F(x)$ at any number x is the area over the interval $(-\infty, x]$ between the graph of the probability density function $f(x)$ and the horizontal axis. For example, the area required to obtain $F(0.8)$ is illustrated in Exhibit 5.3(a).

It is apparent from this figure that $F(0.8)$ is the area of a right triangle with base $[0, 0.8]$ of length 0.8 and with height $f(0.8) = 0.8$. Thus, by the formula for the area of a right triangle, $F(0.8) = \frac{1}{2}(0.8)(0.8) = 0.32$. Equation (5.3) confirms this assertion:

$$\begin{aligned} F(0.8) &= \int_{-\infty}^{0.8} f(t)\,dt = \int_{-\infty}^{0} 0\,dt + \int_{0}^{0.8} t\,dt \\ &= 0 + \tfrac{1}{2}t^2\Big|_0^{0.8} = \tfrac{1}{2}(0.8)^2 - \tfrac{1}{2}(0)^2 \\ &= 0.32. \end{aligned}$$

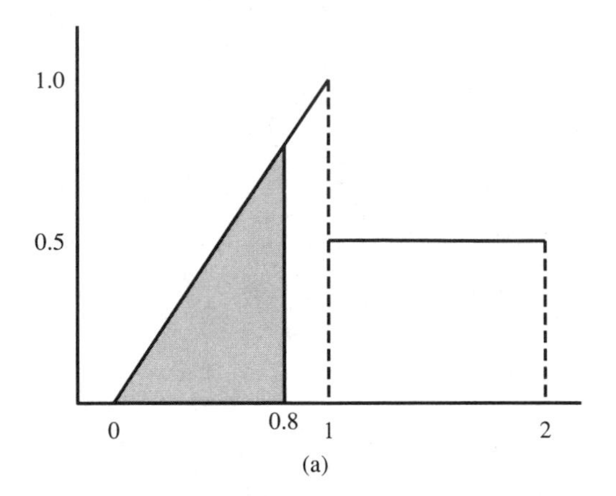

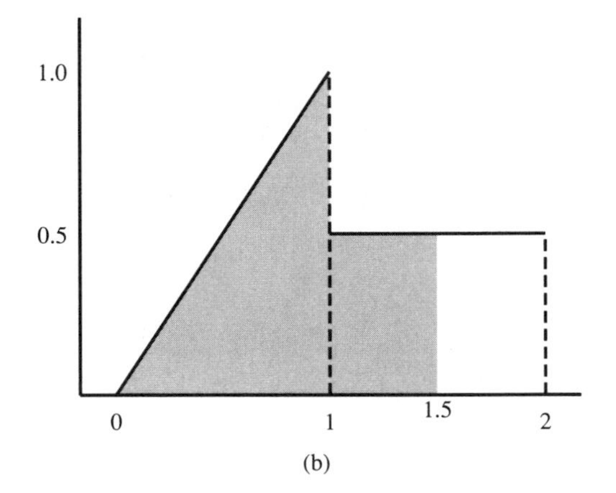

EXHIBIT 5.3 *Computation of probabilities for Example 5.3.*

In general, for $0 \le x < 1$, the area needed to calculate $F(x)$ is that of a right triangle with base $[0, x]$ of length x and height $f(x) = x$. Thus $F(x) = \frac{1}{2}x^2$ for $0 \le x < 1$. Alternatively,

$$F(x) = \int_{-\infty}^{x} f(t)\, dt = \int_{-\infty}^{0} 0\, dt + \int_{0}^{x} t\, dt = 0 + \tfrac{1}{2}t^2\Big|_0^x = \tfrac{1}{2}x^2,$$

if $0 \le x < 1$.

Now consider $F(1.5)$, which is the sum of the areas of the triangle and rectangle shown in Exhibit 5.3(b). Because the triangle has base $[0, 1]$ of length 1 and has height 1, its area is $\frac{1}{2}(1)^2 = 0.5$. The rectangle has base $[1, 1.5]$ of length $1.5 - 1 = 0.5$ and has height $\frac{1}{2}$; thus the area of this rectangle is $(0.5)(\frac{1}{2}) = 0.25$. Adding these areas together yields

$$F(1.5) = 0.5 + 0.25 = 0.75.$$

This result can be confirmed by using equation (5.3):

$$F(1.5) = \int_{-\infty}^{1.5} f(t)\,dt = \int_{-\infty}^{0} 0\,dt + \int_{0}^{1} t\,dt + \int_{1}^{1.5} \tfrac{1}{2}\,dt$$
$$= 0 + \tfrac{1}{2}t^2\Big|_0^1 + \tfrac{1}{2}t\Big|_1^{1.5}$$
$$= 0 + (\tfrac{1}{2} - 0) + [\tfrac{1}{2}(1.5) - \tfrac{1}{2}(1)] = 0.75.$$

In general, for $1 \le x < 2$,

$$F(x) = \int_{-\infty}^{x} f(t)\,dt = \int_{-\infty}^{0} 0\,dt + \int_{0}^{1} t\,dt + \int_{1}^{x} \tfrac{1}{2}\,dt$$
$$= 0 + \tfrac{1}{2} + \tfrac{1}{2}t\Big|_1^x$$
$$= \tfrac{1}{2} + \tfrac{1}{2}x - \tfrac{1}{2} = \tfrac{1}{2}x.$$

To complete the derivation of the cumulative distribution function $F(x)$, note that for $x < 0$,

$$F(x) = \int_{-\infty}^{x} f(t)\,dt = \int_{-\infty}^{x} 0\,dt = 0,$$

because $f(t) = 0$ for $t < 0$. Also when $x \ge 2$, all of the area under the graph of $f(t)$ is to the left of x. Because the total area under the graph of a probability density function equals 1, $F(x) = 1$ for $x \ge 2$. In summary,

$$F(x) = \begin{cases} 0, & \text{if } x < 0, \\ \tfrac{1}{2}x^2, & \text{if } 0 \le x < 1, \\ \tfrac{1}{2}x, & \text{if } 1 \le x < 2, \\ 1, & \text{if } 2 \le x. \end{cases}$$

The graph of $F(x)$, which is a smooth curve, is given in Exhibit 5.4. ♦

Random Variables with Mixed Distributions

One of the advantages of using a cumulative distribution function is that such a function can be defined for random variables that are discrete or continuous, or are neither wholly discrete nor wholly continuous.

EXAMPLE 5.4

The probability that it snows during a randomly chosen winter day at a certain ski resort is 0.6. If it snows, the amount X (in feet) of snow that falls during the day is a continuous random variable with probability density function

$$f(x) = \begin{cases} 0, & \text{if } x < 0 \text{ or } x > 10, \\ \tfrac{1}{10}, & \text{if } 0 \le x \le 10. \end{cases}$$

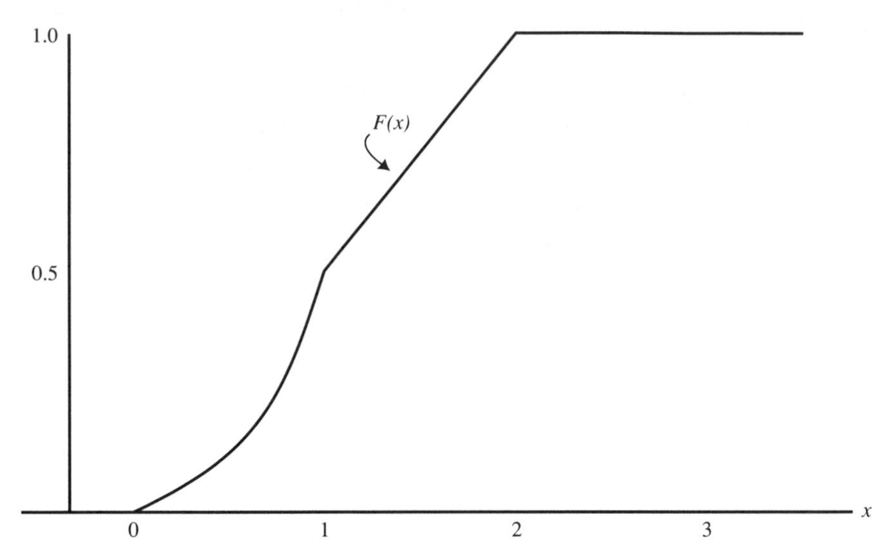

EXHIBIT 5.4 *The cumulative distribution function for Example 5.4.*

Furthermore,

$$P\{X = 0\} = P\{\text{no snow}\} = 1 - 0.6 = 0.4.$$

Thus there is probability mass at $x = 0$, and consequently, X cannot be a continuous random variable. On the other hand, positive values of X have probability density, rather than probability mass, so that X is not a discrete random variable. Because X is a mixture of discrete and continuous parts (values), we say that X has a *mixed* discrete–continuous distribution.

Because X is the amount of snow that falls, negative values for X are impossible. Thus for every $x < 0$,

$$F(x) = P\{-\infty < X \le x\} = 0.$$

On the other hand,

$$\begin{aligned} F(0) = P\{-\infty < X \le 0\} &= P\{-\infty < X < 0\} + P\{X = 0\} \\ &= 0 + 0.4 = 0.4. \end{aligned}$$

Given that it snows, the conditional probability of the event $\{-\infty < X \le x\}$ is

$$
\begin{aligned}
P(-\infty < X \le x \mid \text{snow}) &= \int_{-\infty}^{x} f(t)\,dt \\
&= \int_{-\infty}^{0} 0\,dt + \int_{0}^{x} \tfrac{1}{10}\,dt \\
&= 0 + \tfrac{1}{10}t\Big|_{0}^{x} = \tfrac{1}{10}x, \qquad 0 < x < 10,
\end{aligned}
$$

$$
P(-\infty < X \le x \mid \text{snow}) = \int_{-\infty}^{0} 0\,dt + \int_{0}^{10} \tfrac{1}{10}\,dt + \int_{10}^{x} 0\,dt = 1, \quad 10 \le x.
$$

On the other hand, for all $x > 0$,

$$
P(-\infty < X \le x \mid \text{no snow}) = 1,
$$

because when it doesn't snow, $X = 0$. Thus by the law of partitions,

$$
\begin{aligned}
F(x) &= P(\text{no snow})\, P(-\infty < X \le x \mid \text{no snow}) \\
&\quad + P(\text{snow})\, P(-\infty < X \le x \mid \text{snow}) \\
&= (0.4)(1) + (0.6)(\tfrac{1}{10}x)
\end{aligned}
$$

for $0 < x < 10$, and similarly,

$$
F(x) = (0.4)(1) + (0.6)(1) = 1
$$

for $x \ge 10$. Thus the cumulative distribution function of X is

$$
F(x) = \begin{cases} 0, & \text{if } x < 0, \\ 0.4, & \text{if } x = 0, \\ 0.4 + 0.06x, & \text{if } 0 < x < 10, \\ 1, & \text{if } x \ge 10. \end{cases}
$$

The graph of $F(x)$, given in Exhibit 5.5, has a jump at $x = 0$ but is otherwise smooth. This mixture of smooth parts with jumps at isolated values is what one would expect from the graph of the cumulative distribution function of a mixed discrete–continuous random variable. The jumps occur at values x where there is probability mass, whereas $F(x)$ is smooth over values x where there is probability density. ◆

Uses of the Cumulative Distribution Function

The cumulative distribution function provides a direct method for finding probabilities of intervals of numbers. Indeed, $F(x)$ itself is the probability of an interval: namely, the interval $(-\infty, x]$. For example, $F(5)$ is the probability of the interval $(-\infty, 5]$, or equivalently of the event $\{-\infty < X \le 5\}$.

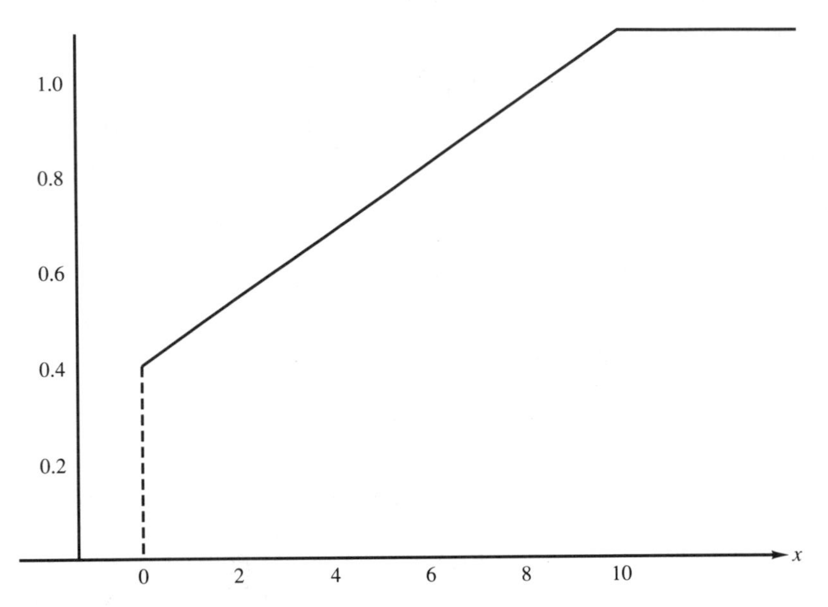

EXHIBIT 5.5 *A cumulative distribution function with a jump at 0 (Example 5.4).*

To find the probability $P\{5 < X < \infty\}$ of the interval $(5, \infty)$, note that the event $\{5 < X < \infty\}$ is the complement of the event $\{-\infty < X \leq 5\}$. It follows from the law of complementation and the definition (5.1) of $F(5)$ that

$$P\{5 < X\} = 1 - P\{-\infty < X \leq 5\} = 1 - F(5).$$

In general, for any number a,

$$(5.4) \qquad P\{a < X\} = 1 - F(a).$$

The calculation of $P\{3 < X \leq 5\}$ is obtained from

$$P\{3 < X \leq 5\} = P\{X \leq 5\} - P\{X \leq 3\} = F(5) - F(3),$$

as can be seen from Exhibit 5.6. More precisely,

$$\begin{aligned} P\{3 < X \leq 5\} &= P(\{X \leq 5\} \cap \{X \leq 3\}^c) \\ &= P\{X \leq 5\} - P(\{X \leq 3\} \cap \{X \leq 5\}) \\ &= P\{X \leq 5\} - P\{X \leq 3\} \\ &= F(5) - F(3). \end{aligned}$$

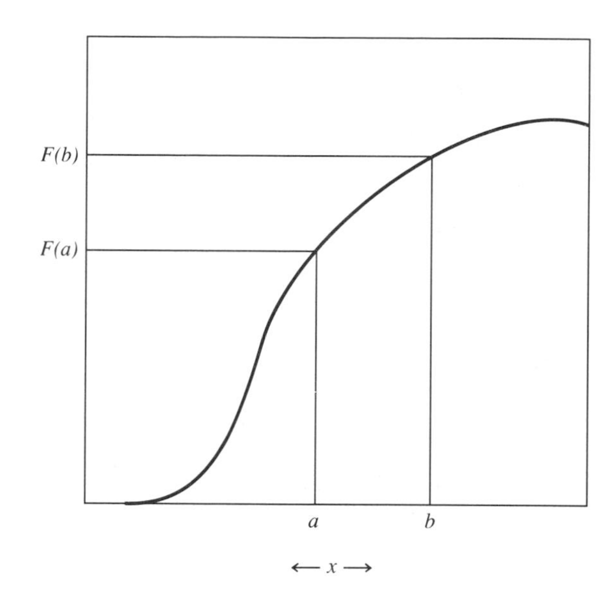

EXHIBIT 5.6 *Using the cumulative distribution function to determine the probability of falling in an interval.*

In general,

(5.5)	$P\{a < X \leq b\} = F(b) - F(a).$

Exhibit 5.7 gives formulas for calculating the probabilities of other events of interest for a random variable X using the cumulative distribution function $F(x)$.

EXHIBIT 5.7 *Formulas for Calculating Probabilities of Certain Events Using the Cumulative Distribution Function F(x)*

Event	*Formula for Probability of the Event*
$\{X = a\}$	Height of jump of graph of $F(x)$ at $x = a$
$\{a < X\}$	$1 - F(a)$
$\{a \leq X\}$	$1 - F(a) + P\{X = a\}$
$\{X \leq b\}$	$F(b)$
$\{X < b\}$	$F(b) - P\{X = b\}$
$\{a < X \leq b\}$	$F(b) - F(a)$
$\{a < X < b\}$	$F(b) - F(a) - P\{X = b\}$
$\{a \leq X \leq b\}$	$F(b) - F(a) + P\{X = a\}$
$\{a \leq X < b\}$	$F(b) - F(a) + P\{X = a\} - P\{X = b\}$

The reason that $P\{X = a\}$ equals the jump in the cumulative distribution function at $x = a$ has the following intuitive justification. Recall that the cumulative distribution function $F(x)$ gives the accumulation of probabilities for all values of the random variable X that are no greater than x. When x is less than a, the probability of the event $\{X = a\}$ is not a part of this accumulation of probabilities. This probability is added in for the first time (as x is increased) when $F(x)$ is computed for $x = a$; the height of the jump (if any) in $F(x)$ at $x = a$ reflects the addition of this new probability, and thus must equal $P\{X = a\}$. We can now see why the cumulative distribution functions of continuous random variables are continuous. Because $P\{X = a\} = 0$ for all numbers a, the cumulative distribution function $F(x)$ of a continuous random variable has no jumps, and hence must be smooth.

The argument above also suggests how to obtain the probability mass function $p(x)$ of a discrete random variable X from its cumulative distribution function $F(x)$:

(5.6) $\quad p(a) = P\{X = a\} =$ height of jump of $F(x)$ at $x = a$.

EXAMPLE 5.1 *(continued)*

In this example the cumulative distribution function of the discrete random variable X was

$$F(x) = \begin{cases} 0, & \text{if } x < 0 \\ \frac{8}{15}, & \text{if } 0 \le x < 1, \\ \frac{12}{15}, & \text{if } 1 \le x < 2, \\ \frac{14}{15}, & \text{if } 2 \le x < 3, \\ 1, & \text{if } 3 \le x. \end{cases}$$

Exhibit 5.1 reveals that $F(x)$ jumps at $x = 0, 1, 2,$ and 3. The jump at $x = 0$ has height equal to the difference between $F(0)$ and $F(x)$ for x "just less than 0." That is,

$$p(0) = \text{height of jump in } F(x) \text{ at } x = 0 = \frac{8}{15} - 0 = \frac{8}{15}.$$

Similarly, the jump at $x = 1$ has height equal to the difference between $F(1)$ and $F(x)$ for x "just less than 1," so that

$$p(1) = \frac{12}{15} - \frac{8}{15} = \frac{4}{15}.$$

In the same manner,

$$p(2) = \frac{14}{15} - \frac{12}{15} = \frac{2}{15}, \qquad p(3) = 1 - \frac{14}{15} = \frac{1}{15}.$$

Because $F(x)$ has no other jumps,

$$p(x) = 0 \text{ for all other values of } x.$$

Thus

$$p(x) = \begin{cases} \frac{8}{15}, & \text{if } x = 0, \\ \frac{4}{15}, & \text{if } x = 1, \\ \frac{2}{15}, & \text{if } x = 2, \\ \frac{1}{15}, & \text{if } x = 3, \\ 0, & \text{all other values of } x, \end{cases}$$

is the probability mass function. ◆

When $F(x)$ is the cumulative distribution function of a continuous random variable X, we can find the probability density function $f(x)$ of X by taking derivatives instead of differences. That is, using (5.3) and the Fundamental Theorem of Calculus,

$$\frac{d}{dx} F(x) = \frac{d}{dx} \int_{-\infty}^{x} f(t)\, dt = f(x), \tag{5.7}$$

for all values x.

EXAMPLE 5.2 *(continued)*

In Example 5.2 the continuous random variable X was shown to have the cumulative distribution function

$$F(x) = \begin{cases} 0, & \text{if } x < 0, \\ 1 - e^{-x}, & \text{if } x \geq 0. \end{cases}$$

Consequently, the probability density function of X is

$$f(x) = \frac{d}{dx} F(x) = \begin{cases} \frac{d}{dx}(0) = 0, & \text{if } x < 0, \\ \frac{d}{dx}(1 - e^{-x}) = e^{-x}, & \text{if } x \geq 0. \end{cases}$$ ◆

Equations (5.2), (5.3), (5.6), and (5.7) show that the cumulative distribution function provides an alternative, and equivalent, way of defining the distribution of probabilities for both discrete and continuous variables. From a probability mass function $p(x)$ or a probability density function $f(x)$ one can obtain the cumulative distribution function $F(x)$, and vice versa. However, Example 5.4 illustrates that the cumulative distribution function $F(x)$ can also be used to define the distribution of probabilities for random variables of mixed discrete–continuous type. It is for these reasons, and also because $F(x)$ can be used to find probabilities of events concerning X (see Exhibit 5.7), that the study of cumulative distribution functions is important.

It may be helpful at this point to illustrate some of the ways in which the cumulative distribution function $F(x)$ can be used to describe properties of a random variable X.

EXAMPLE 5.5

Suppose that a random variable X has cumulative distribution function

$$F(x) = \begin{cases} 0, & \text{if } x < 0, \\ \frac{1}{4}x^2, & \text{if } 0 \le x < 1, \\ \frac{1}{3}x, & \text{if } 1 \le x < 2, \\ 1, & \text{if } 2 \le x. \end{cases}$$

Note that 0, $\frac{1}{4}x^2$, $\frac{1}{3}x$, and 1 are all continuous functions of x. However, $F(x)$ may still have jumps at the points $x = 0$, $x = 1$, $x = 2$ where the formula for computing $F(x)$ changes. For x "just before" 0, $F(x) = 0$. Also $F(0) = \frac{1}{4}(0)^2 = 0$. Thus there is no jump in $F(x)$ at $x = 0$. For x just before 1, $F(x)$ has the value $\frac{1}{4}(1)^2 = \frac{1}{4}$. [Strictly speaking, $\lim_{x\uparrow 1} F(x) = \lim_{x\uparrow 1} \frac{1}{4}(x)^2 = \frac{1}{4}$.] On the other hand, $F(1) = \frac{1}{3}(1) = \frac{1}{3} > \frac{1}{4}$, so that there is a jump of

$$F(1) - \lim_{x\uparrow 1} F(x) = \tfrac{1}{3} - \tfrac{1}{4} = \tfrac{1}{12}$$

at $x = 1$. Similarly, there is a jump of

$$F(2) - \lim_{x\uparrow 2} F(x) = 1 - \lim_{x\uparrow 2} \tfrac{1}{3}x = 1 - \tfrac{1}{3}(2) = \tfrac{1}{3}$$

in $F(x)$ at $x = 2$. Consequently, X is a mixed random variable, with discrete part having probability masses $p(1) = \frac{1}{12}$, $p(2) = \frac{1}{3}$, and continuous part having probability density function

$$f(x) = \frac{d}{dx} F(x) = \begin{cases} \frac{d}{dx}\,0 \\ \frac{d}{dx}\,\frac{1}{4}x^2 \\ \frac{d}{dx}\,\frac{1}{3}x \\ \frac{d}{dx}\,1 \end{cases} = \begin{cases} 0, & \text{if } x < 0, \\ \frac{1}{2}x, & \text{if } 0 \le x < 1, \\ \frac{1}{3}, & \text{if } 1 \le x < 2, \\ 0, & \text{if } 2 \le x. \end{cases}$$

The possible values of X are where $F(x)$ is increasing: $0 \le x < 1$, at $x = 1$, $1 < x < 2$, and at $x = 2$. Thus the possible values of X are the numbers between, and including, 0 and 2. That is, $0 \le x \le 2$.

To find the probability of the event $\{1 \le X \le 1.5\}$, note from Exhibit 5.7 that

$$\begin{aligned} P\{1 \le X \le 1.5\} &= F(1.5) - F(1) + [\text{jump in } F(x) \text{ at } x = 1] \\ &= \tfrac{1}{3}(1.5) - \tfrac{1}{3}(1) + [F(1) - \lim_{x \uparrow 1} F(x)] \\ &= 0.5 - \tfrac{1}{3} + \tfrac{1}{12} \\ &= 0.25. \end{aligned}$$

On the other hand, from Exhibit 5.7,

$$\begin{aligned} P\{0.5 < X < 2\} &= F(2) - F(0.5) - [\text{jump in } F(x) \text{ at } x = 0.5] \\ &= 1 - \tfrac{1}{4}(0.5)^2 - 0 \\ &= 0.9375, \end{aligned}$$

because there is no jump in $F(x)$ at $x = 0.5$. ◆

The Sample Cumulative Distribution Function

For N independent trials (observations) of the random experiment in which the random variable X is observed, and for every number x, the probability $F(x) = P\{X \le x\}$ of the event $\{X = x\}$ can be estimated by the relative frequency of this event. Consequently, the function

$$\hat{F}(x) = \text{r.f.}\{X \le x\} = \frac{\text{number of observations} \le x}{N}$$

serves as an estimate of $F(x)$ for all values x. The function $\hat{F}(x)$ is called the **sample** (or **empirical**) **cumulative distribution function** of X.

For example, if

$$-3.1,\ -1.9,\ -0.8,\ 0.1,\ 0.3,\ 0.3,\ 0.5,\ 1.3,\ 1.3,\ 1.5$$

are the results of $N = 10$ trials, then

$$\hat{F}(x) = \begin{cases} 0, & \text{if } x < -3.1, \\ \frac{1}{10}, & \text{if } -3.1 \le x < -1.9, \\ \frac{2}{10}, & \text{if } -1.9 \le x < -0.8, \\ \frac{3}{10}, & \text{if } -0.8 \le x < 0.1, \\ \frac{4}{10}, & \text{if } 0.1 \le x < 0.3, \\ \frac{6}{10}, & \text{if } 0.3 \le x < 0.5, \\ \frac{7}{10}, & \text{if } 0.5 \le x < 1.3, \\ \frac{9}{10}, & \text{if } 1.3 \le x < 1.5, \\ 1, & \text{if } 1.5 \le x. \end{cases}$$

The graph of $\hat{F}(x)$ is shown in Exhibit 5.8. As more and more trials are taken $(N \to \infty)$, the frequency interpretation of probability implies that the graph of $\hat{F}(x)$ will more closely resemble the graph of the true cumulative distribution function $F(x)$—see Exhibit 5.9.

If a theoretical cumulative distribution function $F(x)$ of X has a certain functional form, a superimposed graph of $\hat{F}(x)$ and $F(x)$ can be used to check whether the theory "fits" the data. For some widely used cumulative distribution functions $F(x)$, special graph paper (**probability paper**) is used to make this comparison. On such graph paper, the horizontal and/or vertical axes are specially scaled so that if the function $F(x)$ "fits" the data, the graph of $\hat{F}(x)$ will be (nearly) a straight line. Use of such paper spares us the need to plot $F(x)$ along with $\hat{F}(x)$.

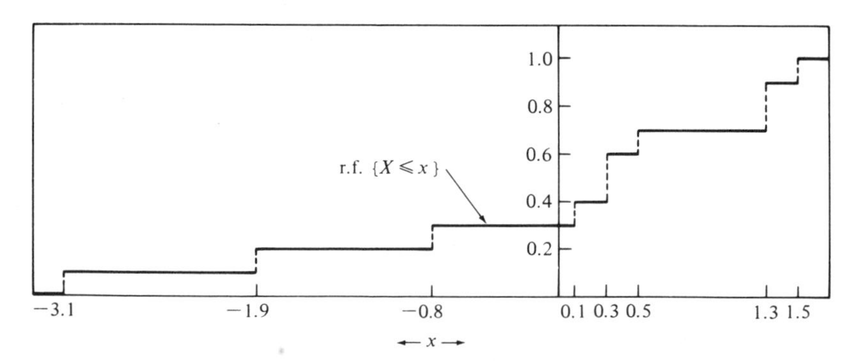

EXHIBIT 5.8 *Sample or empirical cumulative distribution function obtained from N = 10 trials.*

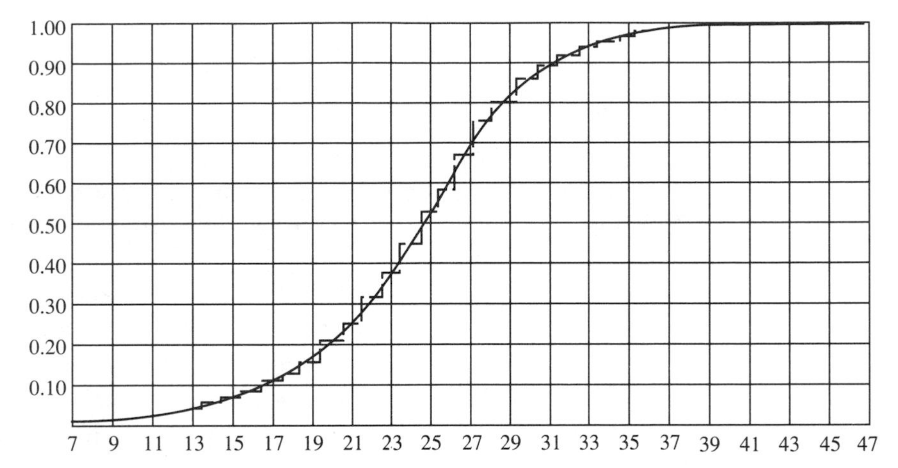

EXHIBIT 5.9 *The smooth curve is the theoretical cumulative distribution function. The step curve is the sample or empirical cumulative distribution function for large N.*

Properties of Cumulative Distribution Functions

Cumulative distribution functions $F(x)$ have the following properties:

Property 1. A cumulative distribution function $F(x)$ is always nondecreasing in x. That is,

$$F(a) \leq F(b)$$

whenever $a \leq b$.

Property 2. The values of $F(x)$ always lie between 0 and 1; that is, $0 \leq F(x) \leq 1$ for all x.

Property 3. $F(x)$ approaches 0 as x becomes arbitrarily small, and $F(x)$ approaches 1 when x becomes arbitrarily large. More formally,

$$\lim_{x \to -\infty} F(x) = 0, \qquad \lim_{x \to \infty} F(x) = 1.$$

Property 4. For any number a, as x takes values *decreasing* to a, the value of $F(x)$ will approach $F(a)$. That is, $\lim_{x \downarrow a} F(x) = F(a)$, and $F(x)$ is *continuous from the right* at every value $x = a$.

Property 1 is easily verified by noting that when $a \leq b$ all numbers (outcomes) in the event $\{X \leq a\}$ are included in the event $\{X \leq b\}$. Thus

$$F(a) = P\{X \leq a\} \leq P\{X \leq b\} = F(b).$$

Alternatively, note that if $a < b$ but $F(b) < F(a)$, then

$$P\{a < X \leq b\} = F(b) - F(a) < 0,$$

which is impossible.

Property 2 is a direct consequence of the facts that $F(x)$ is the probability of the event $\{X \leq x\}$ and that probabilities are always numbers between 0 and 1.

As $x \to \infty$, the event $\{X \leq x\}$ becomes larger, and eventually must contain all possible values of X. On the other hand, as $x \to -\infty$, the event $\{X \leq x\}$ becomes smaller, and eventually contains no outcomes. Property 3 follows immediately from these assertions.

Property 4 reflects the fact that $F(x)$ is an accumulation of probabilities. For any $x > a$,

$$F(x) = P\{X \leq x\} = P\{X \leq a\} + P\{a < X \leq x\}.$$

As x approaches a "from above" denoted $x \downarrow a$, the interval of values $(a, x]$, or equivalently, the event $\{a < X \leq x\}$, becomes empty. Thus

$$\begin{aligned}\lim_{x \downarrow a} F(x) &= \lim_{x \downarrow a} P\{X \leq a\} + \lim_{x \downarrow a} P\{a < X \leq x\} \\ &= P\{X \leq a\} + 0 = F(a).\end{aligned}$$

On the other hand, we have seen that when $F(x)$ has a jump at $x = a$, then the limit of $F(x)$ as x increases to a is not equal to $F(a)$. Indeed, the height of the jump in $F(x)$ at $x = a$ equals $F(a) - \lim_{x \uparrow a} F(x)$.

Every cumulative distribution function $F(x)$ of a random variable X must satisfy Properties 1 through 4. On the other hand, *any function $F(x)$ satisfying Properties 1 to 4 is the cumulative distribution function of some random variable X*. Consequently, Properties 1 to 4 characterize cumulative distribution functions.

EXAMPLE 5.6

Although $-e^{-x}$ is an antiderivative of e^{-x}, the function

$$G(x) = \begin{cases} 0, & \text{if } x < 0, \\ -e^{-x}, & \text{if } x \geq 0, \end{cases}$$

is not the cumulative distribution function of the random variable X in Example 5.2. Indeed, $G(x)$ is not a cumulative distribution function for *any* random variable X. To see this, note that $G(x) = -e^{-x}$ is strictly less than 0 when $x \geq 0$, violating Property 2. Further, because

$\lim_{x\to\infty} e^{-x} = 0$, it follows that

$$\lim_{x\to\infty} G(x) = 0,$$

violating Property 3. This example illustrates how Properties 1 to 4 can be used to check whether one has correctly specified or derived a cumulative distribution function. ◆

6 Transformations of a Random Variable

In many experiments, the variable of chief interest is not always the variable measured but rather, is a function of that variable. For example, an engineer might measure the time X that it takes a rocket to achieve a height of 5 miles above the ground but may be more interested in the average velocity $Y = 5/X$ achieved by the rocket. Similarly, the circumference X of a tree at a particular height may be measured, but it is the cross-sectional area $Y = (16\pi)^{-1}X^2$ that is usually of interest. In some experiments, measurements in the units of one scale can be converted into measurements in the units of another scale. Thus a measurement X of distance in feet can be converted to a corresponding measurement Y in meters, a temperature measurement X in degrees Fahrenheit can be converted to the corresponding temperature Y in degrees Celsius, and so on.

In each such case, the variable Y of interest is a function $Y = g(X)$, or **transformation,** of the variable X actually measured. In many cases, a probability model for X may be known. Because the values of Y are determined by the values of X, we should be able to determine probabilities for Y from the given probabilities for X. When X is a discrete random variable, this goal can be accomplished by the same steps used in section 1 to obtain probabilities of the values of a discrete variable from the probabilities of nonquantitative outcomes.

EXAMPLE 6.1

Suppose that X is a discrete random variable with probability mass function

$$p(x) = \begin{cases} 0.15, & \text{if } x = 0, 3, \\ 0.20, & \text{if } x = 1, 2, \\ 0.30, & \text{if } x = 4, \\ 0, & \text{all other values of } x, \end{cases}$$

and $Y = (X - 2)^2$. To find the probability mass function of Y, first list outcomes x, their associated probabilities $p(x)$, and the value y corresponding to each outcome x, as in the following table:

Outcome x	*Probability*	$y = (x - 2)^2$
0	0.15	4
1	0.20	1
2	0.20	0
3	0.15	1
4	0.30	4

For each possible value y of Y, the probability $P\{Y = y\}$ is obtained by adding the probabilities of all outcomes x for which $(x - 2)^2 = y$. For example,

$$P\{Y = 4\} = P\{X = 0 \text{ or } 4\} = 0.15 + 0.30 = 0.45.$$

Consequently,

$$p_Y(y) = P\{Y = y\} = \begin{cases} 0.20, & \text{if } y = 0, \\ 0.35, & \text{if } y = 1, \\ 0.45, & \text{if } y = 4, \\ 0, & \text{all other values of } y. \end{cases}$$ ◆

Note. Example 6.1 initiates a notational convention that is used henceforth in this book. When more than one random variable is being discussed, we distinguish the probability mass functions (or probability density functions, or cumulative distribution functions) of these variables by a subscript. Thus $p_X(\cdot)$ denotes the probability mass function of X, whereas $p_Y(\cdot)$ denotes the probability mass function of Y. For the sake of simplicity, when the context is clear, this subscripting convention is omitted.

In Example 6.1, more than one x-value could yield the same value y. When there is only one x-value for every value of y, the situation is simpler. In this case, a relabeling of outcomes yields the probability mass function of Y.

EXAMPLE 6.2

The number X of sales of farm equipment on a randomly selected business day has probability mass function $p(x)$ defined by the following table of probabilities:

Number of Sales x	0	1	2	3
Probability p(x)	0.512	0.384	0.096	0.008

A salesman who receives a commission of \$200 for each sale and has daily travel expenses of \$75 receives net earnings of $Y = 200X - 75$ dollars. What is the probability distribution of Y?

For every possible value y of Y there is only one value x for which $y = 200x - 75$, as can be seen from the following table:

x	0	1	2	3
y	-75	125	375	525

Thus y is just a relabeling of the outcomes x. In consequence,

$$p_Y(y) = \begin{cases} 0.512, & \text{if } y = -75, \\ 0.384, & \text{if } y = 125, \\ 0.096, & \text{if } y = 325, \\ 0.008, & \text{if } y = 525, \\ 0, & \text{all other values of } y. \end{cases}$$

A direct mathematical derivation of the probability mass function $p_Y(y)$ of Y proceeds in the following steps:

$$\begin{aligned} p_Y(y) = P\{Y = y\} &= P\{200X - 75 = y\} \\ &= P\{200X = y + 75\} \\ &= P\left\{X = \frac{y + 75}{200}\right\} = p_X\left(\frac{y + 75}{200}\right). \end{aligned}$$

The key step in this derivation is that, because $Y = 200X - 75$, the events $\{Y = y\}$ and $\{200X - 75 = y\}$ are identical and thus have the same probability. Applying the formula so obtained gives us

$$p_Y(125) = P\{Y = 125\} = p_X\left(\frac{125 + 75}{200}\right) = p_X(1) = 0.384,$$

in agreement with our earlier calculation. ◆

A function $g(x)$ is said to be **one-to-one** (or invertible) if for every value y in the range of values of $g(x)$, there is only one value x for which $y = g(x)$. If $g(x)$ is a one-to-one function, then $Y = g(X)$ is a one-to-one transformation. Linear functions $g(x) = ax + b$ ($a \neq 0$) are special cases of one-to-one functions; for example, the function $200x - 75$ used in Example 6.2 is linear and thus one-to-one. Other examples of one-to-one functions are $g(x) = e^x$, $g(x) = 1/x$, $g(x) = \sqrt{x}$, and $g(x) = \log x$; in the latter two examples, x can only be a nonnegative number. One-to-one transformations from a random variable X to a random variable Y simply relabel outcomes, as was the case in Example 6.2. Consequently, it is easy to give a general formula for finding the probability mass function of Y from the probability mass function of X.

For a one-to-one function $g(x)$, every distinct value y of $g(x)$ is obtained from a unique x, which is obtained by solving

$$y = g(x)$$

for x. Thus if $g(x) = e^x$, the solution of $y = e^x$ for x is

$$\log y = \log e^x = x.$$

The function $h(y)$ that assigns to every y the x-value satisfying $y = g(x)$ is called the **inverse function** for $g(x)$. Thus $h(y) = \log y$ is the inverse function for $g(x) = e^x$. Similarly, $h(y) = (y + 75)/200$ is the inverse function for $g(x) = 200x - 75$.

Suppose that $Y = g(X)$ is a function of a discrete random variable X, and that $g(x)$ is a one-to-one function of x with inverse function $h(y)$. Because $y = g(x)$ if and only if $x = h(y)$,

$$\{Y = y\} = \{g(X) = y\} = \{X = h(y)\}.$$

Consequently,

$$(6.1) \qquad p_Y(y) = P\{Y = y\} = P\{X = h(y)\} = p_X(h(y)).$$

EXAMPLE 6.3

Given a discrete random variable X with probability mass function defined by the following table of probabilities:

x	0	1	2	3	4
$p_X(x)$	$\frac{16}{31}$	$\frac{8}{31}$	$\frac{4}{31}$	$\frac{2}{31}$	$\frac{1}{31}$

what is the probability mass function of $Y = X^2$? Although the function $g(x) = x^2$ is not one-to-one over all numbers x, it is one-to-one over nonnegative numbers x. Because the possible values 0, 1, 2, 3, 4 of X are nonnegative, $g(x) = x^2$ can be treated as if it were a one-to-one function. Note that $y = x^2$ implies (because x must be nonnegative) that $x = \sqrt{y}$. Thus $h(y) = \sqrt{y}$ is the inverse function of $g(x) = x^2$.

From Equation (6.1),

$$p_Y(y) = p_X(h(y)) = p_X(\sqrt{y}).$$

Thus

$$p_Y(4) = p_X(\sqrt{4}) = p_X(2) = \tfrac{4}{31},$$
$$p_Y(16) = p_X(\sqrt{16}) = p_X(4) = \tfrac{1}{31},$$

and so on.

The probability mass function $p_X(x)$ of X can be written in the compact mathematical form

$$p_X(x) = \begin{cases} 2^{4-x}/31, & \text{if } x = 0, 1, 2, 3, 4, \\ 0, & \text{other values of } x. \end{cases}$$

Thus

$$p_Y(y) = p_X(\sqrt{y}) = \begin{cases} 2^{4-\sqrt{y}}/31, & \text{if } \sqrt{y} = 0, 1, 2, 3, 4, \\ 0, & \text{other values of } y. \end{cases}$$

$$= \begin{cases} 2^{4-\sqrt{y}}/31, & \text{if } y = 0, 1, 4, 9, 16, \\ 0, & \text{other values of } y. \end{cases}$$ ◆

Using Cumulative Distribution Functions

Individual values x of continuous random variables X have probability 0. Thus the methods used in Examples 6.1 to 6.3 are not applicable to continuous random variables. However, the basic concept that probabilities for values of $Y = g(X)$ come from probabilities for the values of X still applies. In particular, the cumulative distribution function $F_Y(y)$ of Y, which gives probabilities of the intervals $\{-\infty < Y \leq y\}$, can be obtained from knowledge of the cumulative distribution function $F_X(x)$ of X. To find the distribution of a transformed random variable $Y = g(X)$, it is sufficient to determine the cumulative distribution function $F_Y(y)$ of Y.

EXAMPLE 6.4

A rural electric company interested in whether it should supply electricity to an outlying district models the monthly demand X in thousands of kilowatt-hours for electricity. If X has cumulative distribution function

$$F_X(x) = \begin{cases} 0, & \text{if } x < 0, \\ 1 - e^{-1.1x}, & \text{if } x \geq 0, \end{cases}$$

and the company charges \$300 per thousand kilowatt-hours, their monthly income will be $Y = 300X$. To obtain the probability distribution

$F_Y(y)$ of Y, note that the possible values of Y are nonnegative numbers. Thus

$$F_Y(y) = P\{-\infty < Y \le y\} = 0, \qquad \text{if } y < 0.$$

Consider $y = 5$. Because $Y = 300\,X$,

$$\begin{aligned} F_Y(5) &= P\{-\infty < Y \le 5\} = P\{-\infty < 300\text{X} \le 5\} \\ &= P\{-\infty < X \le \tfrac{5}{300}\} = F_X(\tfrac{5}{300}) \\ &= 1 - e^{-1.1(5/300)} = 1 - 0.98183 \\ &= 0.01817. \end{aligned}$$

This same calculation can be carried out for any number $y \ge 0$; that is,

$$\begin{aligned} F_Y(y) &= P\{-\infty < Y \le y\} = P\{-\infty < 300X \le y\} \\ &= P\left\{-\infty < X \le \frac{y}{300}\right\} = F_X\left(\frac{y}{300}\right) = 1 - e^{-1.1(y/300)}. \end{aligned}$$

It follows that the cumulative distribution function $F_Y(y)$ of the monthly income Y is

$$F_Y(y) = \begin{cases} 0, & \text{if } y < 0, \\ 1 - e^{-1.1(y/300)}, & \text{if } y \ge 0. \end{cases}$$

Note that Y is a continuous random variable because $F_Y(y)$ is a continuous function of y.

To compute the probability that the company's income is between 400 and 600 dollars, we use $F_Y(y)$ to make this calculation:

$$\begin{aligned} P\{400 \le Y \le 600\} &= F_Y(600) - F_Y(400) + P\{Y = 400\} \\ &= (1 - e^{-1.1(600/300)}) - (1 - e^{-1.1(400/300)}) + 0 \\ &= 0.88920 - 0.76922 = 0.11998. \end{aligned}$$

However, this probability can be obtained directly from the distribution of X [without having to find $F_Y(y)$]:

$$\begin{aligned} P\{400 \le Y \le 600\} &= P\{400 \le 300X \le 600\} = P\{1.333 \le X \le 2.000\} \\ &= F_X(2.000) - F_X(1.333) \\ &= (1 - e^{-1.1(2.000)}) - (1 - e^{-1.1(1.333)}) \\ &= 0.88920 - 0.76922 = 0.11998. \end{aligned}$$

◆

An important class of functions $g(x)$ for which the cumulative distribution function of $Y = g(X)$ is straightforwardly obtained is the class of *strictly monotone functions*. A function $g(x)$ is strictly monotone if it is always strictly increasing as x increases, or if it is always strictly decreasing as x increases. For example, $g(x) = 300x$, $g(x) = e^x$, and $g(x) = \sqrt{x}$ (for $x \geq 0$) are examples of strictly increasing functions, whereas $g(x) = 1 - x$, $g(x) = 1/x$ (for $x > 0$), and $g(x) = e^{-x}$ are examples of strictly decreasing functions. All of these functions are strictly monotone. The following two examples illustrate how we can find the cumulative distribution function of Y.

EXAMPLE 6.5

Crystals of a certain mineral are cubes. The length X in millimeters of a side of a random crystal is a continuous random variable with cumulative distribution function

$$F_X(x) = \begin{cases} 0, & \text{if } x < 0, \\ \frac{1}{4}x^2, & \text{if } 0 \leq x < 2, \\ 1, & \text{if } x \geq 2. \end{cases}$$

A crystallographer might be interested in the volume $Y = X^3$ of the crystal. Note that $g(x) = x^3$ is a strictly increasing function of x. To find the probability that the volume Y of the crystal is no greater than 3.375, calculate

$$\begin{aligned} \mathrm{P}\{Y \leq 3.375\} &= P\{X^3 \leq 3.375\} = P\{X \leq (3.375)^{1/3} = 1.5\} \\ &= F_X(1.5) = \tfrac{1}{4}(1.5)^2 = 0.5625. \end{aligned}$$

The cumulative distribution function of Y is found by generalizing this calculation:

$$\begin{aligned} F_Y(y) = P\{Y \leq y\} &= P\{X^3 \leq y\} = P\{X \leq y^{1/3}\} \\ &= F_X(y^{1/3}), \end{aligned}$$

where

$$F_X(y^{1/3}) = \begin{cases} 0, & \text{if } y^{1/3} < 0, \\ \frac{1}{4}(y^{1/3})^2, & \text{if } 0 \leq y^{1/3} \leq 2, \\ 1, & \text{if } 1 \leq y^{1/3}. \end{cases}$$

Thus

$$F_Y(y) = \begin{cases} 0, & \text{if } y < 0, \\ \frac{1}{4}y^{2/3}, & \text{if } 0 \leq y < 8, \\ 1, & \text{if } 8 \leq y. \end{cases}$$

The probability density function of Y is

$$f_Y(y) = \frac{d}{dy} F_Y(y) = \begin{cases} 0, & \text{if } y < 0, \\ \frac{1}{6} y^{-1/3}, & \text{if } 0 \le y < 8, \\ 0, & \text{if } 8 \le y. \end{cases}$$

◆

EXAMPLE 6.6

The length of time X (in hours) that it takes an intercity bus to travel the 60 miles from city A to city B is a continuous random variable with cumulative distribution function

$$F_X(x) = \begin{cases} 0, & \text{if } x < 0.8, \\ 2.5(x - 0.8), & \text{if } 0.8 \le x < 1.2, \\ 1, & \text{if } 1.2 \le x. \end{cases}$$

The average velocity Y of the bus for this trip in miles per hour is $Y = 60/X$. Note that $g(x) = 60/x$ is a decreasing function of x for $x \ge 0$, and that all possible values of X are nonnegative. (The possible values of X are $0.8 \le x \le 1.2$.) Also note that the largest possible value of $Y = 60/X$ is $60/0.8 = 75$, and that the smallest possible value of Y is $60/1.2 = 50$. Thus $F_Y(y) = 0$ for $y < 50$ and $F_Y(y) = 1$ for $y \ge 75$.

The probability that the average velocity of the bus stays below the speed limit of 55 is

$$\begin{aligned} P\{Y \le 55\} &= P\left\{\frac{60}{X} \le 55\right\} = P\left\{\frac{60}{55} \le X\right\} \\ &= 1 - F_X\left(\frac{60}{55}\right) \\ &= 1 - (2.5)\left(\frac{60}{55} - 0.8\right) = 0.273. \end{aligned}$$

Similar calculations yield the cumulative distribution function $F_Y(y)$ of Y for $50 \le y < 75$:

$$\begin{aligned} F_Y(y) &= P\{Y \le y\} = P\left\{\frac{60}{X} \le y\right\} \\ &= P\left\{\frac{60}{y} \le X\right\} = 1 - F_X\left(\frac{60}{y}\right) \\ &= 1 - (2.5)\left(\frac{60}{y} - 0.8\right) = 3 - \frac{150}{y}. \end{aligned}$$

Thus

$$F_Y(y) = \begin{cases} 0, & \text{if } y < 50, \\ 3 - \dfrac{150}{y}, & \text{if } 50 \le y < 75, \\ 1, & \text{if } 75 \le y. \end{cases}$$

The probability density function of Y is

$$f_Y(y) = \frac{d}{dy} F_Y(y) = \begin{cases} 0, & \text{if } y < 50, \\ \dfrac{150}{y^2}, & \text{if } 50 \le y < 75, \\ 0, & \text{if } 75 \le y. \end{cases}$$ ◆

We now present general formulas for calculating the cumulative distribution function of Y when Y is a strictly monotone function $Y = g(X)$ of a random variable X. To begin with, note that every strictly monotone function $g(x)$ is one-to-one and hence has an inverse function $h(y)$, which is found by solving $y = h(x)$ for x in terms of y. If $g(x)$ is strictly increasing in x, then for every number y,

$$\{Y \le y\} = \{X \le h(y)\}.$$

This assertion is illustrated graphically in Exhibit 6.1. Consequently,

$$\text{(6.2)} \qquad F_Y(y) = P\{Y \le y\} = P\{X \le h(y)\} = F_X(h(y)).$$

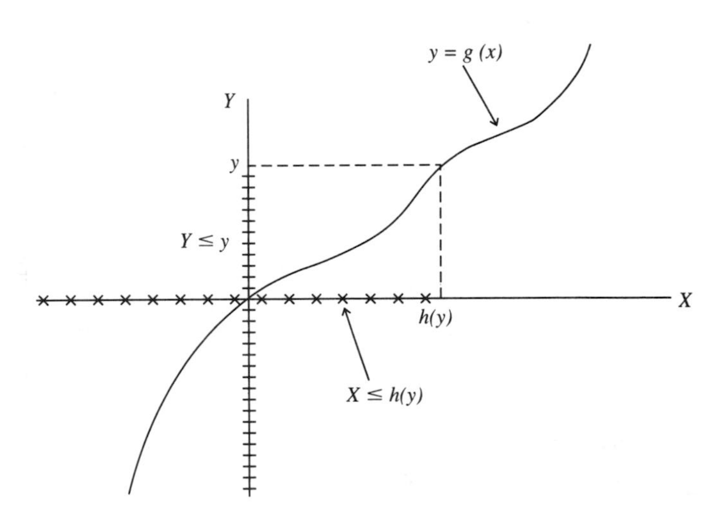

EXHIBIT 6.1 *For a strictly increasing function g, if Y = g(X) ≤ y, then X ≤ h(y).*

On the other hand, if $g(x)$ is strictly decreasing in x, then

$$\{Y \leq y\} = \{X \geq h(y)\},$$

as is illustrated in Exhibit 6.2. Consequently,

$$(6.3) \qquad \begin{aligned} F_Y(y) = P\{Y \leq y\} &= P\{X \geq h(y)\} \\ &= 1 - F_X(h(y)) + P\{X = h(y)\}, \end{aligned}$$

where we recall that $P\{X = h(y)\}$ is given by the jump in $F_X(x)$ at $x = h(y)$.

Suppose that the random variable X has cumulative distribution function

$$F_X(x) = \begin{cases} 0, & \text{if } x \leq 0, \\ 1 - e^{-x}, & \text{if } x \geq 0, \end{cases}$$

and that $Y = \sqrt{X}$. Note that $\sqrt{x}$ is a strictly increasing function of x for $x \geq 0$, with inverse function $h(y) = y^2$. From equation (6.2),

$$F_Y(2) = F_X(h(2)) = F_X((2)^2) = F_X(4) = 1 - e^{-4},$$

and in general

$$F_Y(y) = F_X(h(y)) = F_X(y^2) = 1 - e^{-y^2},$$

for all numbers $y \geq 0$. [Because $Y = \sqrt{X}$ can never be negative, $F_Y(y) = 0$ for all $y < 0$.]

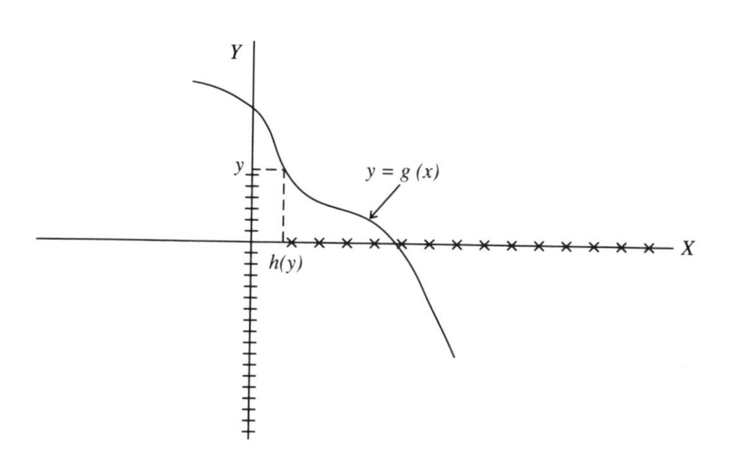

EXHIBIT 6.2 *For a strictly decreasing function g, if* $Y = g(X) \leq y$, *then* $X \geq h(y)$.

On the other hand, define $W = e^{-X}$. The function $g(x) = e^{-x}$ is strictly decreasing in x, with inverse function $h(w) = -\log w$. From equation (6.3),

$$\begin{aligned} F_W(0.5) &= 1 - F_X(h(0.5)) + P\{X = h(0.5)\} \\ &= 1 - F_X(-\log(0.5)) + P\{X = -\log(0.5)\} \\ &= 1 - (1 - e^{-(-\log(0.5))}) + 0 \\ &= e^{\log(0.5)} = 0.5, \end{aligned}$$

because X is a continuous random variable. In general,

$$\begin{aligned} F_W(w) &= 1 - F_X(-\log w) + P(\{X = -\log w\}) \\ &= 1 - (1 - e^{-(-\log w)}) + 0 \\ &= e^{\log w} = w \end{aligned}$$

for $0 \le w < 1$. Because $0 \le e^{-x} \le 1$ for all $x \ge 0$, the possible values of W lie in the interval $[0, 1]$. Thus $F_W(w) = 0$ for all $w < 0$, and $F_W(w) = 1$ for all $w \ge 1$. We conclude that

$$F_W(w) = \begin{cases} 0, & \text{if } w < 0, \\ w, & \text{if } 0 \le w < 1, \\ 1, & \text{if } 1 \le w. \end{cases}$$

Equations (6.2) and (6.3) are very useful, but the basic concept that probabilities for $Y = g(X)$ come from probabilities for X can be applied even when $g(x)$ is not a strictly monotone function. For example, suppose that the random variable X is continuous with cumulative distribution function

$$F_X(x) = \begin{cases} 0, & \text{if } x < 0, \\ \frac{1}{2}x, & \text{if } 0 \le x < 2, \\ 1, & \text{if } 2 \le x. \end{cases}$$

Suppose that $Y = (X - 1)^2$. Over the interval $0 \le x \le 2$ of possible values for X, the function $g(x) = (x - 1)^2$ is not strictly monotone (see Exhibit 6.3). In fact, for each possible value of Y other than $y = 0$, there are two x-values yielding $(x - 1)^2 = y$, namely $x = 1 + \sqrt{y}$ and $x = 1 - \sqrt{y}$. Nevertheless, Exhibit 6.3 shows that

$$\{Y \le y\} = \{(X - 1)^2 \le y\} = \{1 - \sqrt{y} \le X \le 1 + \sqrt{y}\}$$

for $y \ge 0$. Thus

$$\begin{aligned} F_Y(y) = P\{Y \le y\} &= P\{1 - \sqrt{y} \le X \le 1 + \sqrt{y}\} \\ &= F_X(1 + \sqrt{y}) - F_X(1 - \sqrt{y}) \end{aligned}$$

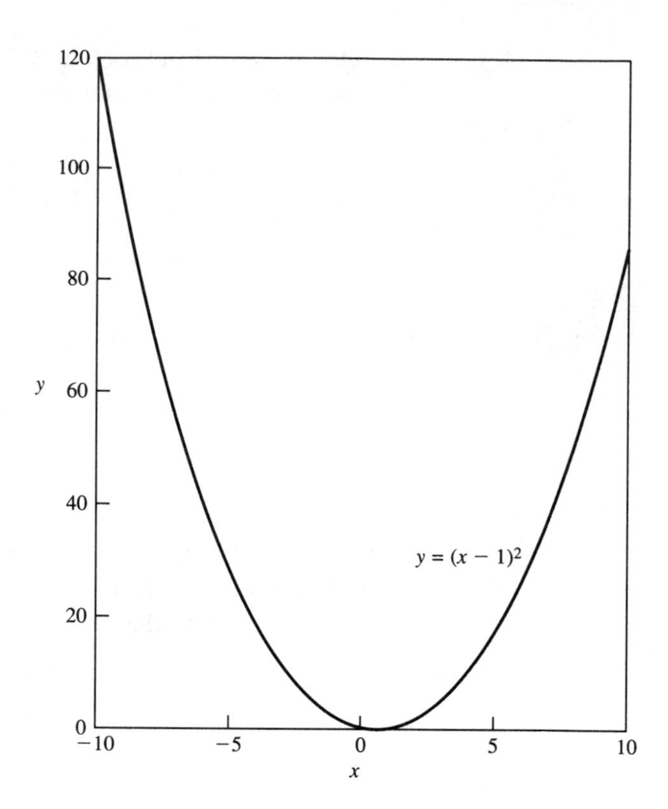

EXHIBIT 6.3 *Graph of the function $y = (x - 1)^2$ showing that each y value yields two x values.*

because X is a continuous random variable. Note that the smallest possible value of $(x - 1)^2$ over the interval $0 \le x \le 2$ of possible values of X is 0, and that the largest possible value is 1. Consequently, $F_Y(y) = 0$ for $y < 0$ and $F_Y(y) = 1$ for $y \ge 1$. For $y = \frac{1}{4}$, say,

$$\begin{aligned} F_Y(\tfrac{1}{4}) &= F_X(1 + \sqrt{1/4}) - F_X(1 - \sqrt{1/4}) = F_X(1 + \tfrac{1}{2}) - F_X(1 - \tfrac{1}{2}) \\ &= \tfrac{1}{2}(1 + \tfrac{1}{2}) - \tfrac{1}{2}(1 - \tfrac{1}{2}) = \tfrac{1}{2}. \end{aligned}$$

In general,

$$\begin{aligned} F_Y(y) &= F_X(1 + \sqrt{y}) - F_X(1 - \sqrt{y}) \\ &= \tfrac{1}{2}(1 + \sqrt{y}) - \tfrac{1}{2}(1 - \sqrt{y}) \\ &= \sqrt{y} \end{aligned}$$

for $0 \le y < 1$. Thus the cumulative distribution function of $Y = (X - 1)^2$ is

$$F_Y(y) = \begin{cases} 0, & \text{if } y < 0, \\ \sqrt{y}, & \text{if } 0 \le y < 1, \\ 1, & \text{if } 1 \le y. \end{cases}$$

The following example is another illustration of the basic principle used in this section.

EXAMPLE 6.7

In Example 4.2, a dispenser drops X ounces of coffee into an 8-ounce cup. The random variable X is continuous with probability density function

$$f(x) = \begin{cases} 0, & \text{if } x < 7.7, \\ 10(x - 7.7), & \text{if } 7.7 \le x < 8.1, \\ 40(8.2 - x), & \text{if } 8.1 \le x < 8.2, \\ 0, & \text{if } 8.2 \le x, \end{cases}$$

and cumulative distribution function

$$F_X(x) = \begin{cases} 0, & \text{if } x < 7.7, \\ 5(x - 7.7)^2, & \text{if } 7.7 \le x < 8.1, \\ 1 - 20(8.2 - x)^2, & \text{if } 8.1 \le x < 8.2, \\ 1, & \text{if } 8.2 \le x. \end{cases}$$

Because the cup can only hold 8 ounces of coffee, when $X \ge 8.0$ the cup overflows. Assume in this case that 8.0 ounces remains in the cup. Let Y be the amount of coffee in the cup after the dispenser has finished. Note that

$$Y = \begin{cases} X, & \text{if } 7.7 \le X \le 8.0, \\ 8.0, & \text{if } 8.0 < X \le 8.1, \end{cases}$$

and that the possible values of Y are $7.7 \le y \le 8.0$. Thus $F_Y(y) = 0$ if $y < 7.7$ and $F_Y(y) = 1$ if $y \ge 8.0$. For $7.7 \le y < 8.0$, the event $\{Y \le y\}$ is the same as the event $\{X \le y\}$, because no coffee spills out of the cup. Consequently,

$$F_Y(y) = P\{Y \le y\} = P\{X \le y\} = F_X(y) = 5(y - 7.7)^2$$

for $7.7 \le y < 8.0$. We conclude that

$$F_Y(y) = \begin{cases} 0, & \text{if } y < 7.7, \\ 5(y - 7.7)^2, & \text{if } 7.7 \le y < 8.0, \\ 1, & \text{if } 8.0 \le y. \end{cases}$$

Note that $F_Y(y)$ is continuous, except at $y = 8.0$, where there is a jump of size

$$F_Y(8.0) - \lim_{y \uparrow 8.0} F_Y(y) = 1 - 5(8.0 - 7.7)^2 = 0.55.$$

This jump is precisely the probability that the cup overflows, which was calculated in Example 4.2. Even though X is a continuous random variable, the random variable Y is a mixture of discrete and continuous parts. ◆

In general, random variables that are transformations of discrete random variables are themselves discrete. On the other hand, random variables Y obtained as transformations $Y = g(X)$ of continuous random variables X can be discrete [e.g., when $g(x) = 1$ for some values of x and 0 for the remaining values of x], mixed (see Example 6.7), or continuous (Examples 6.4 to 6.6).

Obtaining Probabilities

So far, we have concentrated on obtaining the cumulative distribution function $F_Y(y)$ of $Y = g(X)$ from knowledge of the cumulative distribution function $F_X(x)$ of X. We have done so because the cumulative distribution function of a random variable enables us to find probabilities of events concerning that random variable. Also, from $F_Y(y)$ we can find the probability mass function or probability density function of Y, depending on what type of random variable Y is.

In Example 6.4 we also noted that we do not have to obtain the cumulative distribution function $F_Y(y)$ of Y in order to find the probability of an event such as $P\{400 \le Y \le 600\}$, but instead can calculate such a probability directly from knowledge of the probability distribution of X. Earlier, we used the cumulative distribution function $F_X(x)$ of X to specify that distribution, but this is not really necessary. For example, we can make such a calculation directly from the probability mass function or probability density function of X, whichever is appropriate.

EXAMPLE 6.2 *(continued)*

In this example the variable originally measured was the number X of sales of farm equipment in a day. This is a discrete random variable with probability mass function $p(x)$ defined by

Number of Sales x	0	1	2	3
$p(x)$	0.512	0.384	0.096	0.008

The net earnings of a salesman is given by $Y = 200X - 75$. The probability that the salesman makes more than \$300 in a day is

$$
\begin{aligned}
P\{Y > 300\} &= P\{200X - 75 > 300\} \\
&= P\left\{X > \frac{300 + 75}{200} = 1.875\right\} \\
&= P\{X = 2 \text{ or } 3\} \\
&= p(2) + p(3) = 0.096 + 0.008 = 0.104.
\end{aligned}
$$
◆

EXAMPLE 6.8

The velocity X of a randomly chosen particle of unit mass is a continuous random variable with density function

$$
f(x) = \begin{cases} \frac{4}{9}x\left(1 - \frac{x^2}{16}\right), & \text{if } 2 \le x \le 4, \\ 0, & \text{all other values of } x. \end{cases}
$$

The kinetic energy Y of such a particle is related to its velocity X by $Y = \frac{1}{2}X^2$. Note that $g(x) = \frac{1}{2}x^2$ is a strictly increasing function over the possible values $2 \le x \le 4$ of X. To find $P\{2.5 \le Y \le 5\}$, note that

$$
\begin{aligned}
\{2.5 \le Y \le 5\} = \{2.5 \le \tfrac{1}{2}X^2 \le 5\} &= \{5 \le X^2 \le 10\} \\
&= \{\sqrt{5} \le X \le \sqrt{10}\},
\end{aligned}
$$

so that

$$
\begin{aligned}
P\{2.5 \le Y \le 5\} &= P\{\sqrt{5} \le X \le \sqrt{10}\} \\
&= \int_{\sqrt{5}}^{\sqrt{10}} f(x)\,dx = \int_{\sqrt{5}}^{\sqrt{10}} \frac{4}{9}x\left(1 - \frac{x^2}{16}\right)dx \\
&= \left(\frac{2}{9}x^2 - \frac{1}{144}x^4\right)\Bigg|_{\sqrt{5}}^{\sqrt{10}} \\
&= \left(\frac{20}{9} - \frac{100}{144}\right) - \left(\frac{10}{9} - \frac{25}{144}\right) = \frac{95}{144} = 0.65972.
\end{aligned}
$$

The cumulative distribution function $F_Y(y)$ of Y can be obtained without first finding the cumulative distribution function $F_X(x)$ of X. To do so, first note that because the possible values of X are $2 \le x \le 4$, the possible values of $Y = \frac{1}{2}X^2$ are $2 \le y \le 8$. Thus $F_Y(y) = 0$ for $y < 2$ and $F_Y(y) = 1$ for $y \ge 8$. For $2 \le y < 8$,

$$
\begin{aligned}
F_Y(y) &= P\{Y \le y\} = P\{\tfrac{1}{2}X^2 \le y\} = P\{X \le \sqrt{2y}\} \\
&= \int_{-\infty}^{\sqrt{2y}} f(x)\,dx = \int_{2}^{\sqrt{2y}} \frac{4}{9}x\left(1 - \frac{x^2}{16}\right)dx = \left(\frac{2}{9}x^2 - \frac{1}{144}x^4\right)\Bigg|_{2}^{\sqrt{2y}} \\
&= \frac{4}{9}y - \frac{1}{36}y^2 - \frac{7}{9} = 1 - \frac{1}{36}(y - 8)^2.
\end{aligned}
$$

Thus

$$F_Y(y) = \begin{cases} 0, & \text{if } y < 2, \\ 1 - \frac{1}{36}(8 - y)^2, & \text{if } 2 \le y < 8, \\ 1, & \text{if } 8 \le y, \end{cases}$$

from which the probability density function $f_Y(y)$ of Y is

$$f_Y(y) = \frac{d}{dy} F_Y(y) = \begin{cases} 0, & \text{if } y < 2, \\ \frac{8 - y}{18}, & \text{if } 2 \le y < 8, \\ 0, & \text{if } 8 \le y. \end{cases}$$

◆

Density Function of a Transformed Variable

Suppose that we know the probability density function $f_X(x)$ of a continuous random variable X and want to determine the probability density function $f_Y(y)$ of $Y = g(X)$, where $g(x)$ is a differentiable function of x. One method for obtaining $f_Y(y)$ has already been illustrated in Example 6.8—namely, use $f_X(x)$ to find $F_Y(y)$, and then take the derivative of $F_Y(y)$ with respect to y to obtain $f_Y(y) = \frac{d}{dy} F_Y(y)$.

If $g(x)$ is a strictly monotone function, a shortcut is available. If $g(x)$ is a strictly increasing function, we know from equation (6.2) that

$$F_Y(y) = F_X(h(y)),$$

where $h(y)$ is the inverse function of $g(x)$. Using the chain rule for differentiation gives

$$\begin{aligned} f_Y(y) = \frac{d}{dy} F_Y(y) &= \left(\frac{d}{dy} h(y)\right) \left(\frac{d}{dx} F_X(x)\Big|_{x=h(y)}\right) \\ &= \left(\frac{d}{dy} h(y)\right) f_X(h(y)) \end{aligned}$$

because $f_X(x) = \frac{d}{dx} F_X(x)$. Similarly, if $g(x)$ is a strictly decreasing function,

$$F_Y(y) = 1 - F_X(h(y)) + P\{X = h(y)\} = 1 - F_X(h(y)) + 0,$$

because X is a continuous random variable. Thus

$$f_Y(y) = \frac{d}{dy} F_Y(y) = \frac{d}{dy}(1 - F_X(h(y)))$$
$$= \left(-\frac{dh(y)}{dy}\right) f_X(h(y)).$$

However, the inverse function $h(y)$ of a strictly increasing (decreasing) function $g(x)$ is itself strictly increasing (decreasing). Consequently,

$$\frac{dh(y)}{dy} \text{ is } \begin{cases} >0 & \text{when } g(x) \text{ is strictly increasing,} \\ <0 & \text{when } g(x) \text{ is strictly decreasing.} \end{cases}$$

It then follows that

$$(6.4) \qquad f_Y(y) = \left|\frac{dh(y)}{dy}\right| f_X(h(y)),$$

whenever $Y = g(X)$ and $g(x)$ is a strictly monotone differentiable function of x over the range of possible values of X. (The notation $|x|$ means absolute value.)

In Example 6.8, $Y = \frac{1}{2}X^2$ and $g(x) = \frac{1}{2}x^2$ is strictly monotone (strictly increasing) over the range $2 \leq x \leq 4$ of possible values of X. The inverse function $h(y)$ of $g(x) = \frac{1}{2}x^2$ is $h(y) = \sqrt{2y}$. Applying equation (6.4), we obtain

$$f_Y(y) = \left|\frac{d\sqrt{2y}}{dy}\right| f_X(\sqrt{2y})$$
$$= \left|\frac{1}{\sqrt{2y}}\right| \begin{cases} \frac{4}{9}\sqrt{2y}\left(1 - \frac{(\sqrt{2y})^2}{16}\right), & \text{if } 2 \leq \sqrt{2y} \leq 4, \\ 0, & \text{otherwise} \end{cases}$$
$$= \begin{cases} \frac{8-y}{18}, & \text{if } 2 \leq y \leq 8, \\ 0, & \text{otherwise.} \end{cases}$$

EXAMPLE 6.9

The amount of time X (in hours) needed to repair a small electrical appliance is a continuous random variable whose probability density function is assumed to be

$$f(x) = \begin{cases} 20x^3(1 - x), & \text{if } 0 \leq x \leq 1, \\ 0, & \text{otherwise.} \end{cases}$$

Startup costs (tools and supplies) for the repair are \$10 and the person making the repair is paid \$15 an hour. Thus the cost of repair for the appliance is $Y = 15X + 10$. What is the probability density function of Y?

Here, $Y = g(X)$, where $g(x) = 15x + 10$ is a strictly increasing function of x. The inverse function $h(y)$ of $g(x)$ is $h(y) = (y - 10)/15$. Note that

$$\frac{d}{dy} h(y) = \frac{d}{dy}\left(\frac{y-10}{15}\right) = \frac{1}{15}.$$

Thus the probability density function of Y is

$$\begin{aligned}
f_Y(y) &= \left|\frac{1}{15}\right| f\left(\frac{y-10}{15}\right) \\
&= \begin{cases} \dfrac{20}{15}\left(\dfrac{y-10}{15}\right)^3\left(1 - \dfrac{y-10}{15}\right), & \text{if } 0 \le \dfrac{y-10}{15} \le 1, \\ 0, & \text{otherwise,} \end{cases} \\
&= \begin{cases} \dfrac{20}{(15)^4}(y-10)^3(25-y), & \text{if } 10 \le y \le 25, \\ 0, & \text{otherwise.} \end{cases}
\end{aligned}$$

◆

Notes and References

Survival and Hazard Functions

Suppose that a random variable X gives the length of life of a certain machine (or organism). The cumulative distribution function $F(x) = P\{X \le x\}$ of X gives the probability that the machine or organism will "die" (cease to function) before x units of time have passed. Thus the quantity

$$\bar{F}(x) = 1 - F(x) = P\{X > x\},$$

called the **survival function,** gives the probability that the machine survives more than x units of time. In studies of the reliabilities of machines (or electronic systems), in medical studies of lifetimes of diseased organisms under various treatments, and in actuarial studies, much of the investigators' efforts are devoted to modeling the survival function $\bar{F}(x)$. Frequently, this function (when identified) will be tabled in preference to tabling the cumulative distribution function $F(x)$. Of course, either function $F(x)$ or $\bar{F}(x)$ can be used to determine probabilities. For example,

$$\begin{aligned}
P\{a \le X \le b\} &= F(b) - F(a) \\
&= (1 - F(a)) - (1 - F(b)) \\
&= \bar{F}(a) - \bar{F}(b),
\end{aligned}$$

and

$$P\{X \le a\} = F(a) = 1 - \overline{F}(a).$$

The survival function is often called the **tail probability function.** Tables of the tail probability function, $\overline{F}(x)$, are used in evaluating the properties of statistical methods for testing hypotheses about probability models (Chapter 14).

When X is a continuous random variable giving the lifetime of a machine or organism, or more generally the waiting time until the occurrence of an event, it is often useful to determine the **hazard function** $h(x)$ of X. To motivate the definition of the hazard function, suppose that a machine has already survived x units of time; thus it is known that $X > x$. Given that $X > x$, the conditional probability that the machine "dies" before an additional δ units of time, $\delta > 0$, have passed is

$$\begin{aligned} P\{X \le x + \delta \mid X > x\} &= \frac{P\{X \le x + \delta \text{ and } X > x\}}{P\{X > x\}} \\ &= \frac{P\{x < X \le x + \delta\}}{P\{X > x\}} \\ &= \frac{F(x + \delta) - F(x)}{1 - F(x)}. \end{aligned}$$

The risk or rate of "death" per unit time, conditional upon the event that the machine or organism has survived x units of time, is

$$\frac{1}{\delta} P\{X \le x + \delta \mid X > x\} = \frac{[F(x + \delta) - F(x)]/\delta}{1 - F(x)}.$$

As δ becomes arbitrarily small, the *instantaneous risk of "death"* at time x given survival to time x is

$$\begin{aligned} h(x) = \lim_{\delta \to 0} \frac{1}{\delta} P\{X \le x + \delta \mid X > x\} &= \frac{\lim_{\delta \to 0} \dfrac{F(x + \delta) - F(x)}{\delta}}{1 - F(x)} \\ &= \frac{\dfrac{d}{dx} F(x)}{1 - F(x)} = \frac{f(x)}{1 - F(x)} = \frac{f(x)}{\overline{F}(x)}. \end{aligned}$$

The function $h(x)$ is called the hazard function. Note that

$$\frac{d}{dx}[-\log \overline{F}(x)] = \frac{\dfrac{d}{dx}F(x)}{\overline{F}(x)} = \frac{f(x)}{\overline{F}(x)}.$$

Thus an alternative formula for the hazard function $h(x)$ is

$$h(x) = \frac{d}{dx}[-\log \bar{F}(x)] = -\frac{d}{dx}\log \bar{F}(x).$$

EXAMPLE

Suppose that X has a density function $f(x)$ which has constant value over the interval $[0, a]$. That is,

$$f(x) = \begin{cases} \frac{1}{a}, & \text{if } 0 \le x \le a, \\ 0, & \text{otherwise.} \end{cases}$$

It follows that the tail probability function $\bar{F}(x)$ of X is

$$\bar{F}(x) = P\{X > x\} = \begin{cases} 1, & \text{if } x < 0, \\ \frac{a - x}{a}, & \text{if } 0 \le x < a, \\ 0, & \text{if } x \ge a, \end{cases}$$

and

$$h(x) = -\frac{d}{dx}\log \bar{F}(x) = \begin{cases} 0, & \text{if } x < 0, \\ \frac{1}{a - x}, & \text{if } 0 \le x < a, \\ \text{undefined}, & \text{if } x > a. \end{cases}$$

The fact that the hazard function is undefined for $x > a$ is not a problem, because the event $\{X > a\}$ has zero probability. ◆

The hazard function can be used in place of the probability density function to characterize the distribution of a continuous nonnegative random variable X. Define

$$H(x) = \int_0^x h(t)\, dt.$$

Then

$$\log \bar{F}(x) = -H(x)$$

and

$$\bar{F}(x) = e^{-H(x)}.$$

Consequently,

$$F(x) = 1 - \bar{F}(x) = 1 - e^{-H(x)}.$$

Because the cumulative distribution function $F(x)$ determines the probability distribution of X, it follows that knowledge of the hazard function $h(x)$ determines the distribution of X. Indeed, by the chain rule,

$$\begin{aligned} f(x) = \frac{d}{dx} F(x) &= \frac{d}{dx}\left[1 - e^{-H(x)}\right] \\ &= \left[\frac{d}{dx} H(x)\right] e^{-H(x)} \\ &= h(x) \exp\left[-\int_0^x h(t)\, dt\right], \end{aligned}$$

so that the probability density function $f(x)$ of X can be obtained from the hazard function $h(x)$.

EXAMPLE

What continuous nonnegative random lifetimes X have a constant hazard of death? That is, if

$$h(x) = c$$

for all positive values of x, what is the probability density function $f(x)$? For $x > 0$,

$$f(x) = h(x) \exp\left[-\int_0^x h(t)\, dt\right] = c \exp\left(-\int_0^x c\, dt\right) = ce^{-cx},$$

and otherwise (because X is nonnegative with probability 1), $f(x) = 0$.

Note that for $f(x)$ to be a density function, it is required that $f(x) \geq 0$ and $\int_0^\infty f(x)\, dx = 1$. Consequently, c must be a positive number. Random variables with probability density functions of the form $f(x) = c \exp^{-cx}$, $x > 0$, are said to have an **exponential distribution** (Chapter 10, Section 1). ◆

Further information on hazard functions and their use in modeling lifetimes of machines or organisms can be found in Barlow and Proschan (1981).

Glossary and Summary

Random variable: A function assigning real numbers to the outcomes of an experiment; a numerical aspect of a random experiment.

Probability distribution: The rule that assigns probabilities to values, or intervals of values, of a random variable X.

Discrete random variable: A random variable whose possible values are "isolated"; the number of possible values is either finite or countably infinite (can be indexed by integers).

Continuous random variable: A random variable whose possible values comprise one or more intervals of real numbers (and hence are uncountably infinite in number); an approximation that assumes that measurements could be made to any arbitrary accuracy.

Probability mass functions: A rule $p(x)$ that assigns probabilities to the possible values x of a discrete random variable; the probabilities play the role of masses placed on the real line.

Probability density function: A function $f(x)$ on the real line that defines probabilities for a continuous random variables by means of areas under the graph of the function $f(x)$.

Cumulative distribution function: A function $F(x)$ that gives for every number x the probability $F(x) = P\{X \leq x\}$ of all values equal to or smaller than x.

Sample probability mass function: A rule that gives the relative frequency $\hat{p}(x) = \text{r.f.}\{X = x\}$ in repeated trials for every value x of a discrete random variable X; the graph of such a function is called a bar gram.

Sample probability density function (or modified relative frequency histogram): A curve formed by the tops of rectangles constructed over intervals which partition the real line; the area of each such rectangle equals the relative frequency in repeated trials of the interval over which the rectangle is constructed.

Sample cumulative distribution function: A function $\hat{F}(x)$ that assigns to every value x the relative frequency $\hat{F}(x) = \text{r.f.}\{X \leq x\}$ in repeated trials of all values equal to or less than x.

"Fits the data": When a theoretical probability mass function, probability density function or cumulative distribution function is reasonably close (over all values x) to the corresponding sample probability mass function, probability density function, or cumulative distribution function.

Mixed random variables: Random variables whose distribution has both discrete and continuous parts; the graph of the cumulative distribution function $F(x)$ of such a random variable shows both smoothly rising parts and jumps.

Transformation of a random variable: A new random variable Y defined as a function $Y = g(X)$ of an original random variable X.

Monotone function: A function $g(x)$ that is either always rising (increasing) or always falling (decreasing).

One-to-one onto function: A function $g(x)$ having the property that for every number x there is one and only one number y for which $g(x) = y$.

Linear function: A function $g(x) = ax + b$ whose graph is a straight line.

Inverse function: If $g(x)$ is a one-to-one onto function of x, the function $h(y)$ that assigns to every value of y in the range of possible values of $g(x)$ the unique x for which $g(x) = y$.

Survival function: A function that assigns to every value x the probability $\bar{F}(x) = P\{X > x\}$ of all values larger than x; the tail probability function.

Hazard function: The instantaneous risk of "death"; assigns to each value of x the limit $\lim_{\delta \to 0} \delta^{-1} P\{X \le x + \delta \mid X > x\}$.

Useful Facts

1. Let $p(x)$ be the probability mass function of a discrete random variable X. Then for any collection A of values X, $P(X \in A) = \sum_{X \in A} p(x)$. For example, $P\{X = 1, 2, \text{ or } 3\} = p(1) + p(2) + p(3)$.
2. For $p(x)$ to be the probability mass function of a discrete random variable, it is necessary and sufficient that $p(x) \ge 0$ for every x and $\sum_{-\infty < x < \infty} p(x) = 1$.
3. Let $f(x)$ be the probability density function of a continuous random variable X. Then for any interval A of values of X, $P\{X \in A\} = \int_A f(x)\,dx$. For example, $P\{1 \le X \le 3\} = \int_1^3 f(x)\,dx$.
4. For $f(x)$ to be the density function of a continuous random variable it is necessary and sufficient that $f(x) \ge 0$ for every x and $\int_{-\infty}^{\infty} f(x)\,dx = 1$.
5. The cumulative distribution function $F(x)$ of X is given by

$$F(x) = \sum_{-\infty \le t \le x} p(t), \qquad \text{if } X \text{ is discrete.}$$

$$= \int_{-\infty}^{x} f(t)\,dt, \qquad \text{if } X \text{ is continuous.}$$

   If X is discrete, the probability mass function $p(x)$ of X is obtained from $F(x)$ by $p(x) =$ size of jump of $F(x)$ at x. If X is continuous, the probability density function $f(x)$ of X is obtained from $F(x)$ by $f(x) = (d/dx)F(x)$. Although $F(x)$ is an antiderivative of $f(x)$, it is not always the antiderivative one learns in a calculus course.
6. Probabilities of intervals of values can be obtained from the cumulative distribution function $F(x)$ by means of the formulas given in Exhibit 5.7. For example, $P\{X \le 1\} = F(1)$, $P\{1 < X \le 2\} = F(2) - F(1)$, $P\{1 \le X \le 2\} = F(2) - F(1) +$ jump in $F(x)$ at $x = 1$, and $P\{X > 2\} = 1 - F(2)$.
7. For $F(x)$ to be the cumulative distribution function of a random variable, it is necessary and sufficient that $0 \le F(x) \le 1$ for all values x, $F(x)$ is

nondecreasing in x, $\lim_{x\to-\infty} F(x) = 0$, $\lim_{x\to\infty} F(x) = 1$, and $F(x)$ is continuous from the right at all values x.

8. Let X be a random variable with cumulative distribution function $F(x)$. Let $g(x)$ be a monotone increasing function of x. Then the cumulative distribution function $F_Y(y)$ of $Y = g(X)$ is given by

$$F_Y(y) = F(h(y)),$$

where $h(y)$ is the inverse function of $g(x)$. If, instead, $g(x)$ is a monotone decreasing function of x, then

$$F_Y(y) = 1 - F(h(y)) + [\text{jump in } F(x) \text{ at } x = h(y)].$$

Also,

$$P\{a \le Y \le b\} = P\{h(a) \le X \le h(b)\}.$$

9. If X is a continuous random variable with probability density function $f(x)$ and $g(x)$ is a monotone differentiable function of x, the probability density function of $Y = g(X)$ is

$$f_Y(y) = \left|\frac{dh(y)}{dy}\right| f(h(y)),$$

where $h(y)$ is the inverse function of $g(x)$.

10. If X is a nonnegative random variable with cumulative distribution function $F(x)$ and probability density function $f(x)$, the hazard function $h(x)$ of X is given by

$$h(x) = \frac{d}{dx}[-\log(1 - F(x))] = \frac{f(x)}{1 - F(x)}.$$

Further,

$$F(x) = 1 - \exp\left[-\int_0^x h(t)\,dt\right]$$

and

$$f(x) = h(x)\exp\left[-\int_0^x h(t)\,dt\right].$$

Exercises

1. At a hospital, the progress of 4 patients suffering from a certain disease is observed. Let X be the number of patients who recover from the disease during the period of observation. Assume that the probability of recovery from the disease is 0.75, and that patients recover or fail to recover independently (in the statistical sense). Note that X is a discrete random variable.
   (a) Find the probability density function $p(x)$ of X.
   (b) Find the probability $P\{X \le 2\}$ that no more than 2 of the 4 patients recover.
   (c) Find the cumulative distribution function $F(x)$ of X.
2. On Halloween night trick-or-treaters are offered the choice of 3 hollow plastic balls from an urn containing 100 balls, of which 80 contain 10 cents and 20 contain 25 cents. (Balls picked from the urn by one person are replaced by similar balls before the next person makes a choice.) Let X be the amount of money that a random trick-or-treater gets. By listing outcomes and their probabilities, and the values of X corresponding to each outcome, find the probability mass function $p(x)$ of X.
3. A gardening company has 4 gardeners: Al, Ben, Carl, and his apprentice Dave. Because the pay is minimum wage, the gardeners often find a higher-paid job. On any day, Al will miss work with probability 0.2, Ben will miss work with probability 0.1, and Carl will miss work with probability 0.4. Al, Ben, and Carl act independently (statistically) of each other, but David comes to work on exactly the same days as Carl. Let X be the number of workers who skip work on a given day. Find the probability mass function $p(x)$ of X. (*Hint:* List possible outcomes, their probabilities, and corresponding values of X.)
4. In a large urban shopping mall, the stores are arranged as shown in the accompanying figure. Each store contains an underground garage. If a shopper wants to remain indoors, he or she can pass from store to store (and garage to garage) by the routes shown in the figure. The distances between adjacent stores are the same—say, 1 unit apart. A shopper parks at a garage at one store and finishes shopping at another store. Let X be the number of units of distance that the shopper has to walk to reach his or her car. Find the probability mass function $p(x)$ of X in each of the following cases.
   (a) When both the store under which the shopper parks and the store where the shopper stops shopping are randomly chosen (with replacement) from among the stores $A, B, \ldots, F$.
   (b) When the two stores are randomly chosen *without* replacement from among the stores $A, B, \ldots, F$.

(c) When the garages under stores C and D are closed for repair, so that the shopper can only park under stores A, B, E, F, but otherwise the store under which the shopper parks and the store where the shopper stops shopping are selected randomly and independently.

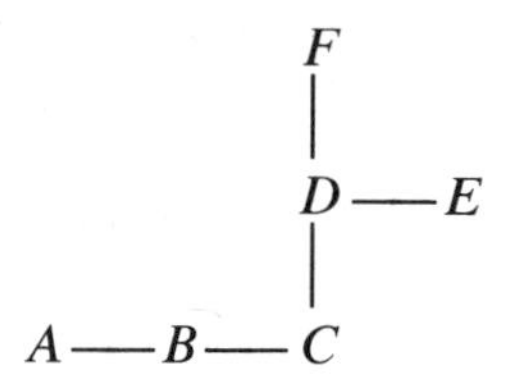

5. Consider the discrete random variable X that has possible values 0, 1, 2, . . . , 9 and for which $P\{X = j\} = \frac{1}{10}$, $j = 0, 1, 2, \ldots, 9$.
   (a) Let Y be the remainder obtained after dividing X^2 by 10 (e.g., 9^2 divided by 10 has remainder 1). Find the probability mass function of Y.
   (b) Let Z be the remainder obtained after dividing Y^2 by 10. Find the probability mass function of Z.
   (c) Let W be the remainder obtained after dividing Z^2 by 10. Find the probability mass function of W. Compare the probability mass functions of Z and W.
   (d) If, starting with the random variable W, we continue to define new random variables in terms of old random variables by the process described in parts (a), (b), and (c), do we ever obtain a probability mass function different from the probability mass function of W?
6. The probability mass function $p(x)$ of a discrete random variable X has the form

$$p(x) = \begin{cases} \left(\frac{1}{2}\right)^x\left(\frac{32}{31}\right), & \text{for } x = 1, 2, 3, 4, 5, \\ 0, & \text{otherwise.} \end{cases}$$

   (a) Show that $\sum_{x=1}^{5} p(x) = 1$.
   (b) Determine the cumulative distribution function $F(x)$ of X.
   (c) Find (i) $P\{X \geq 3\}$; (ii) $P\{2 \leq X \leq 4\}$; (iii) $P\{X < 4\}$.
7. A discrete random variable X has the following probability mass function:

$$p(x) = \begin{cases} \left(\frac{1}{4}\right)\left(\frac{3}{4}\right)^x, & \text{for } x = 0, 1, 2, 3, \\ \frac{81}{256}, & \text{for } x = 4, \\ 0, & \text{otherwise.} \end{cases}$$

   (a) Show that $\sum_{x=0}^{4} p(x) = 1$.
   (b) Find the cumulative distribution function $F(x)$ of X.
   (c) Find (i) $P\{X > 3\}$; (ii) $P\{1 \leq X \leq 5\}$; (iii) $P\{X < 3\}$.

8. A discrete random variable X has probability mass function of the following form:

$$p(x) = \begin{cases} cx^2, & \text{for } x = 1, 2, 3, 4, \\ 0, & \text{otherwise.} \end{cases}$$

(a) What is the value of the constant c?
(b) Find the cumulative distribution function $F(x)$ of X.
(c) Find $P\{1 < X \le 3\}$ and $P\{X \text{ is an even integer}\}$.

9. In a study of the incidence of dental caries in adults, a random sample of 100 adults was selected. These adults were offered free dental examinations at the beginning and end of a given year. A total of 80 adults accepted the offer. For each adult studied, the number X of new caries that developed over the year were counted. The data obtained from this experiment are given below.

2	1	3	0	1	2	2	1	1	1	2	2	3	1	3	0	1	0	2	1
0	0	1	2	2	1	0	2	1	5	0	0	1	2	1	3	1	4	1	1
0	1	0	1	0	0	0	0	1	1	4	0	0	1	1	0	1	3	2	0
0	0	1	0	1	1	2	0	0	1	0	1	2	0	0	1	0	1	0	1

(a) Find the sample probability mass function for X.
(b) Construct a bar graph for these data.
(c) Determine the sample cumulative distribution function for X.

10. In Exercise 9, determine how well the data (sample probability mass function) are fit by the probability mass function

$$p(x) = \begin{cases} \dfrac{e^{-1}}{x!}, & x = 0, 1, 2, 3, \ldots, \\ 0, & \text{other values of } x. \end{cases}$$

11. Exhibit 3.4 in Example 3.2 uses one possible probability mass function $p_*(x)$ to fit the sample probability mass function $\hat{p}(x)$ of the number X of letters in a random work from *Vanity Fair*. Instead, see how well the probability mass function

$$p_{**}(x) = \begin{cases} \dfrac{x^2 e^{-x}}{6}, & x = 0, 1, 2, 3, \ldots, \\ 0, & \text{other values of } x. \end{cases}$$

fits $\hat{p}(x)$. Compare the two fits. Which choice of theoretical probability mass function do you prefer?

12. The probability density function $f(x)$ of a continuous random variable X has the form

$$f(x) = \begin{cases} 0.4, & \text{if } 0 \leq x < 1, \\ 0.2, & \text{if } 1 \leq x < 2, \\ 0.3, & \text{if } 2 \leq x < 3, \\ 0.1, & \text{if } 3 \leq x < 4, \\ 0.0, & \text{if } x < 0 \text{ or } x \geq 4. \end{cases}$$

(a) Find the following probabilities.

(i) $P\{0 < X \leq 2\}$. (ii) $P\{0 \leq X \leq 2\}$. (iii) $P\{2 \leq X \leq 3\}$.
(iv) $P\{\frac{1}{2} \leq X < \frac{3}{2}\}$. (v) $P\{X \geq 1\}$. (vi) $P\{X < 2\}$.

[*Hint:* It will be helpful to graph $f(x)$ and compute probabilities by means of areas.]

(b) Find the cumulative distribution function $F(x)$ of X.

13. Suppose that the continuous random variable Y has a probability density function $f(y)$ of the form

$$f(y) = \begin{cases} 0, & \text{if } y < 0, \\ 3y^2, & \text{if } 0 \leq y \leq 1, \\ 0, & \text{if } y > 1. \end{cases}$$

(a) Find $P\{0 \leq Y \leq \frac{1}{2}\}$ and $P\{\frac{1}{4} \leq Y \leq \frac{3}{4}\}$.

(b) Find the cumulative distribution function $F(y)$ of Y.

14. The probability density function $f(z)$ of a continuous random variable Z has the form

$$f(z) = \begin{cases} 0, & \text{if } z < 1, \\ z - 1, & \text{if } 1 \leq z < 2, \\ 3 - z, & \text{if } 2 \leq z \leq 3, \\ 0, & \text{if } z > 3. \end{cases}$$

(a) Find $P\{Z \leq \frac{3}{2}\}$, $P\{\frac{3}{2} \leq Z \leq \frac{5}{2}\}$, and $P\{Z > \frac{5}{2}\}$.

(b) Find the cumulative distribution function $F(z)$ of Z.

15. Suppose that the lifetime X in years of a certain brand of light bulb has density

$$f_X(t) = \begin{cases} 0, & \text{if } t \leq 0, \\ (1 + t)^{-2}, & \text{if } t > 0. \end{cases}$$

(a) What is the probability that such a light bulb lasts *at least* 2 years?

(b) Find the conditional probability that the bulb lasts at least 7 years, given that it lasts at least 5 years. In other words, if you have a bulb that is exactly 5 years old and still working, what is the (conditional) probability that it will still be working after another 2 years?

16. A continuous random variable W has a probability density function of the form

$$f(w) = \begin{cases} ce^{-2w}, & \text{if } w \geq 0, \\ 0, & \text{otherwise.} \end{cases}$$

(a) What is the value of the constant c?
(b) Find the cumulative distribution function $F(w)$ of W.
(c) Find $P\{W \geq 1\}$ and $P\{2 \leq W < 5\}$.
(d) Find the number d for which $P\{W \leq d\} = 0.5$.

17. To show that a function $f(t)$ is a density function, you must show that $f(t) \geq 0$ for all t and that

$$\int_{-\infty}^{\infty} f(t)\, dt = 1.$$

Which of the following functions are density functions? Explain your answers.

(a) $f(t) = \begin{cases} 5t^4, & \text{if } 0 \leq t \leq 1, \\ 0, & \text{otherwise.} \end{cases}$

(b) $f(t) = \begin{cases} 2t, & \text{if } -1 \leq t \leq 2, \\ 0, & \text{otherwise.} \end{cases}$

(c) $f(t) = \begin{cases} 1/2, & \text{if } -1 \leq t \leq 1, \\ 0, & \text{otherwise.} \end{cases}$

(d) $f(t) = \begin{cases} t^{-1/2} & \text{if } \frac{1}{4} \leq t \leq 1, \\ 0, & \text{otherwise.} \end{cases}$

(e) $f(t) = e^{-|t|}$ if $-\infty < t < \infty$.

18. One measure of air pollution is the amount Y of beta radioactivity concentration in the air (measured in microcuries per cubic meter). Data for those states that have air sampling stations are collected by the federal government in the *Statistical Abstracts of the United States*. The results for 1965, listed alphabetically by states, are: 2.2, 6.5, 5.8, 3.9, 5.9, 4.0, 4.6, 4.9, 4.5, 2.6, 5.7, 4.7, 4.8, 3.5, 4.2, 5.0, 4.0, 4.1, 4.5, 5.0, 5.2, 3.9, 4.6, 4.1, 4.8, 3.9, 8.8, 5.9, 4.2, 4.8, 4.6, 5.5, 4.3, 5.0, 3.3, 5.2, 4.0, 4.6, 3.7, 5.7, 4.6, 5.6, 5.2, 4.4, 3.6, 4.9, 5.4, 6.2, 2.2. Prepare a modified relative frequency histogram from these data. Use 8 intervals of length 1.0 starting at $y = 1.45$.

19. Construct the sample cumulative distribution function of the data in Exercise 18.

20. A sixth-grade science class has been blowing soap bubbles as a class project. According to their textbook, the radius (in inches) of a soap bubble (which is approximately a sphere) is a random variable X having *cumulative* distribution function

$$F(x) = \begin{cases} 0, & \text{if } x < 0, \\ \dfrac{(1 + x)^2}{100}, & \text{if } 0 \le x < 9, \\ 1, & \text{if } 9 \le x. \end{cases}$$

(a) Is the random variable X discrete, continuous, or mixed?
(b) What are the possible values of X?
(c) The previous year, the sixth-grade class had blown a bubble 7.95 inches in radius. This year's class is going to try to break that record. Assuming that each try is statistically independent of all other tries, what is the probability that in 3 tries they break the record at least once?

21. A random variable X has the cumulative distribution function

$$F_X(x) = \begin{cases} 0, & x < 1, \\ \frac{1}{2}, & 1 \le x < 2, \\ \frac{7}{8}, & 2 \le x < 4, \\ 1, & x < 4. \end{cases}$$

A random variable Y has the cumulative distribution function

$$F_Y(y) = \begin{cases} 0, & y < 1, \\ \frac{1}{2}(y - 1), & 1 \le y < 2, \\ \frac{1}{4}y, & 2 \le y < 4, \\ 1, & y \ge 4. \end{cases}$$

(a) Which one of the following assertions is correct?
  (i) The random variables X and Y are both discrete.
  (ii) X is discrete and Y is continuous.
  (iii) Y is discrete and X is continuous.
  (iv) X and Y are both continuous.
  (v) None of the statements above are correct.
(b) What are the possible values of X? What are the possible values of Y?
(c) Consider the events $A = \{X \ge 2\}$, $B = \{Y \ge 2\}$. Find $P(A)$ and $P(B)$.
(d) If A and B in part (c) are independent events, find $P\{A \cup B\}$.

22. Let X have cumulative distribution function

$$F(x) = \begin{cases} 0, & x < -1, \\ 0.3(x + 1), & -1 \le x < 0, \\ 0.6x + 0.7, & 0 \le x < \frac{1}{2}, \\ 1, & x \ge \frac{1}{2}. \end{cases}$$

(a) Find $P\{-0.5 \le X < 0\}$.
(b) What are the possible values of X?
(c) Note that the distribution of X is *mixed*. Thus there are some values x of X where there is nonzero probability mass and some values x where there is density. Find those values x of X where there is nonzero probability mass, and for each such value of x find

$$p(x) = P\{X = x\}.$$

For all other values x of X, there is a density function $f(x)$ that helps determine probabilities for X. Find $f(x)$.
(d) Show that $\sum_{-\infty<x<\infty} p(x) = 0.4$ and $\int_{-\infty}^{\infty} f(x)\,dx = 0.6$. However, the total probability of all values of X is 1.

23. Gemstones (particularly diamonds) are weighed in terms of units called *carats* [1 carat = 200 mg (roughly)]. The weight X of a gem-quality diamond is a continuous random variable having probability density function

$$f(x) = \begin{cases} 0, & \text{if } x < 0 \text{ or } x \ge 2, \\ 2x^2, & \text{if } 0 \le x < 1, \\ \frac{2}{3}(2 - x), & \text{if } 1 \le x < 2. \end{cases}$$

The value V of a gem-quality diamond in dollars is related to the weight X of the diamond (in carats) by the formula

$$V = (3000)X^2.$$

(a) What is the probability that a gem-quality diamond is worth more than \$750?
(b) Find the cumulative distribution function, $F_V(v)$, of V.

24. The kinetic energy X of a randomly chosen atomic particle of unit mass is a continuous random variable with density function:

$$f(x) = \begin{cases} 2(1 + x)^{-3}, & \text{if } x > 0, \\ 0, & \text{if } x \le 0. \end{cases}$$

(a) Find the probability, $P\{1 \le X \le 5\}$, that a randomly chosen particle of unit mass has kinetic energy between 1 and 5.

(b) Find the cumulative distribution function $F(x)$ of X.

(c) The velocity Y of a particle of unit mass is related to the kinetic energy by the formula

$$Y = \sqrt{2X}.$$

Find the probability density function $f_Y(y)$ of Y.

25. A certain rocket fuel will deliver a constant vertical acceleration of 60 miles per second per second to a rocket for a random period X of time (and then will be exhausted). The time X, measured in seconds, until the fuel is exhausted is a continuous random variable with density function

$$f(x) = \begin{cases} 24x^{-4}, & \text{if } 2 \le x < \infty, \\ 0, & \text{if } x < 2. \end{cases}$$

The distance in miles that the rocket travels until the fuel is exhausted is $Y = \frac{1}{2}(60)X^2 = 30X^2$.

(a) Find the probability $P\{Y \ge 200\}$ that the rocket travels at least 200 miles.

(b) Find the cumulative distribution function $F_Y(y)$ of Y.

(c) Find the probability density function $f_Y(y)$ of Y.

26. The duration, T, of a business executive's long-distance phone call in minutes has density function

$$f_T(t) = \begin{cases} t, & \text{if } 0 < t \le 1, \\ \frac{3}{10}, & \text{if } 1 < t \le 2, \\ \frac{2}{10}, & \text{if } 2 < t \le 3, \\ 0, & \text{otherwise.} \end{cases}$$

When the phone company charges for a call, they first compute

$$X = \text{"}T \text{ rounded to nearest whole number."}$$

(a) Note that X is a discrete random variable. Find the probability mass function of X.

(b) The telephone company charges $Y = 0.25X$ dollars for the call. Thus the call is free if $T < \frac{1}{2}$ (because $X = 0$) and costs a quarter if $\frac{1}{2} \le T \le \frac{3}{2}$ (because $X = 1$). What is the probability that a random call will cost 50 cents or more?

(c) Suppose that the telephone company bases its charges directly on the measured time T, so that the cost Y of a call equals $0.25T$. What now is the probability that a random call will cost 50 cents or more?

27. A tavern has a dartboard that is a circle of radius 6 inches. When a player hits the board with a dart, the probability that the dart lands in a region A on the face of the dartboard is proportional to the area of A. (Because it is assumed that the dart hits the board, the probability of hitting the entire board is 1.) Let X be the distance in inches from the center of the board (the bull's-eye) to where the dart hits. Assuming that the dart hits the board, find the cumulative distribution function of X. Is X a discrete or a continuous variable, or is it mixed discrete–continuous?

Theoretical Exercises

T.1. The following problem is a prototype of a set of problems known as "matches," "coincidences," or "recontre." It has many variants, some of which date from the early eighteenth century. Suppose that we have N pairs of distinguishable tickets, divided into 2 decks each of which contains 1 ticket from each pair. Both decks of N tickets are shuffled, and 1 ticket at a time is taken from the top of each deck and matched against a corresponding ticket from the top of the other deck. If the tickets agree, we say that a *match* has occurred. Among the N tickets compared, we count the number X_N of matches. Assuming that both decks of tickets are well shuffled, find the probability mass function of X_N when (a) $N = 2$, (b) $N = 3$, (c) $N = 4$, and (d) $N = 5$. Can you give a general formula for this mass function (i.e., one that holds for arbitrary $N \geq 1$)?

T.2. (a) Write down the Taylor's expansion of e^{λ} about $\lambda = 0$.
(b) Use your result in part (a) to show that the following function is a probability mass function

$$p(t) = \begin{cases} \dfrac{\lambda^t}{t!} e^{-\lambda}, & \text{if } t = 0, 1, 2, \ldots, \\ 0, & \text{otherwise,} \end{cases}$$

where λ is a given positive constant. [You must show that $p(t) \geq 0$, all t, and that $\sum_{t=0}^{\infty} p(t) = 1$.]
(c) If $\lambda < 0$, explain why $p(t)$ is not a probability mass function.

T.3. Suppose that a continuous random variable X has a probability density function $f(x)$. *If* $Y = aX + b$, $a \neq 0$, is a linear transformation of X, show that the probability density function of Y is

$$f_Y(y) = \frac{1}{|a|} f\left(\frac{y - b}{a}\right).$$

T.4. Suppose that a discrete random variable X has a probability mass function $p(x)$ and that $Y = aX + b$, $a \neq 0$. Show that the probability mass function of Y is $p_Y(y) = p((y - b)/a)$.

T.5. Suppose that a continuous random variable X has cumulative distribution function $F(x)$. Show that $Y = F(X)$, the **probability integral transform** of X, has:

(a) Cumulative distribution function

$$F_Y(y) = \begin{cases} 0, & \text{if } y < 0, \\ y, & \text{if } 0 \leq y < 1, \\ 1, & \text{if } 1 \leq y. \end{cases}$$

(b) Probability density function

$$f_Y(y) = \begin{cases} 1, & \text{if } 0 \leq y < 1, \\ 0, & \text{otherwise.} \end{cases}$$

6

Descriptive Properties of Distributions

1 Introduction

Probability mass functions and probability density functions specify how probabilities are distributed over the values x of a random variable X. For this reason, such functions are said to give the **distribution** of X. The graphs of these functions provide profiles of the population of x-values.

For many practical and scientific purposes, it is unnecessary to have complete probabilistic information about X. Instead, a few descriptive indices of the distribution of X may suffice. Two of the most frequently useful types of descriptor of a distribution are its **location** and its **dispersion.** A measure of location seeks to identify where the graph (profile) of a distribution is located on the horizontal axis, whereas a measure of dispersion indicates how widely spread the graph is about its point of location. Such indices are particularly useful for comparing distributions (populations). For example, we may be interested in comparing the distribution of incomes X and Y of coal miners in 1980 and 1990. Comparison of measures of location for the distributions of X and Y gives us a crude idea of how incomes changed over that time, whereas comparison of measures of dispersion for these distributions would indicate whether incomes among workers were more or less variable in 1990 than in 1980.

Exhibit 1.1 illustrates what we mean by the location of a distribution. Here, the two densities shown have identical graphical shape, but differ in where they are placed on the horizontal axis. To identify the locations of the graphs of the densities, we have arbitrarily used the values x_0, x_1 where the densities $f(x)$, $f^*(x)$ have maximum height (the modes). Any other value of x that can be identified from the **shape** of the graph of a density would do

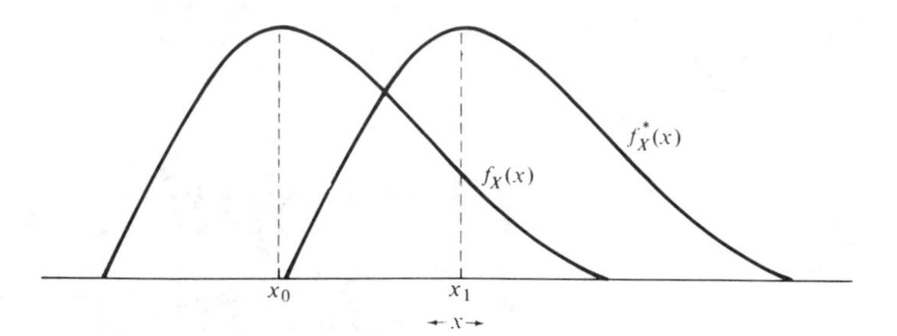

EXHIBIT 1.1 *Two densities differing in location.*

equally well to locate that density—for example, the point x that acts as the center of gravity or balance point for the graph (the mean), or the point that divides the area under the graph in half (the median). When the graph moves d units along the horizontal line, any such measure will change its value by d units. This property will be used to define a measure of location: *A descriptive measure that changes by an amount d whenever a distribution moves d units along the horizontal axis is called a measure of location.*

Suppose that we have a random variable X which has a distribution defined by a probability mass function $p_X(x)$ or a probability density function $f_X(x)$. For example, X might be the income of a randomly chosen coal miner in 1980. If Y is the income of that coal miner in 1990, we might believe that Y represents an increase (or decrease) of d dollars over X; that is, $Y = X + d$. In this ideal case, the graph of the distribution [$p_Y(y)$ or $f_Y(y)$] of Y has the same shape as that of X, but is displaced by d units along the horizontal axis relative to the graph of the distribution of X. (This assertion may be verified using Exercise T.3 of Chapter 5.) Consequently, comparison of corresponding measures of location for X and Y will reflect this change. If $M(X)$, $M(Y)$ are comparable measures of location for X, Y, respectively, then

$$M(Y) = M(X + d) = M(X) + d. \tag{1.1}$$

Thus a change in income from 1980 to 1990 can be identified from the measure of location $M(\cdot)$ without the need to determine the complete distributions of X and Y.

Measures of dispersion of a distribution reflect how strongly the distribution concentrates about a particular central value c on the horizontal axis. A measure of dispersion is small or large, depending on whether the spread of the distribution about c is small or large, and equals zero when all the probability is concentrated at the single point c. For example, the distribution graphed in Exhibit 1.2 has a smaller spread about c than the distribution graphed in Exhibit 1.3, and thus its measure of dispersion should be smaller. Similarly, the distributions graphed in Exhibit 1.4 have different dispersions. Note that the dispersion or spread of the graph of the distribution of X reflects the **variability** of X. Random variables whose distributions have large

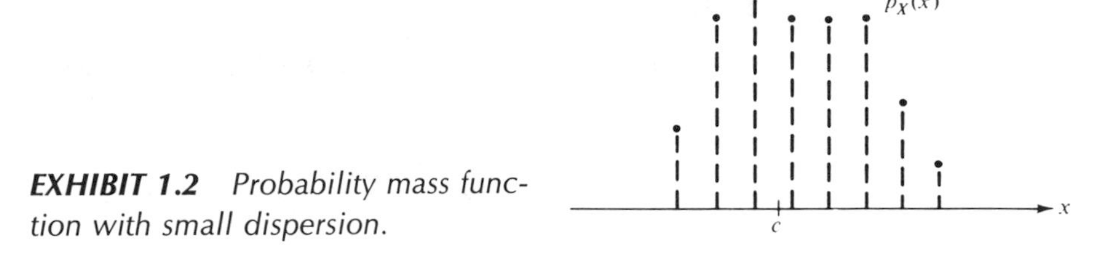

EXHIBIT 1.2 Probability mass function with small dispersion.

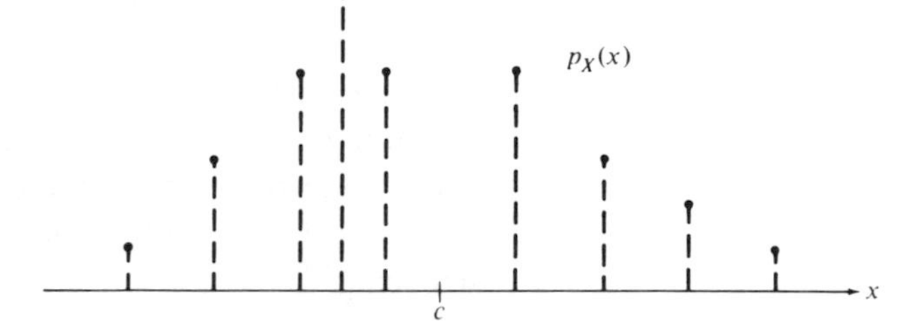

EXHIBIT 1.3 Probability mass function with a larger dispersion than that of the mass function in Exhibit 1.2.

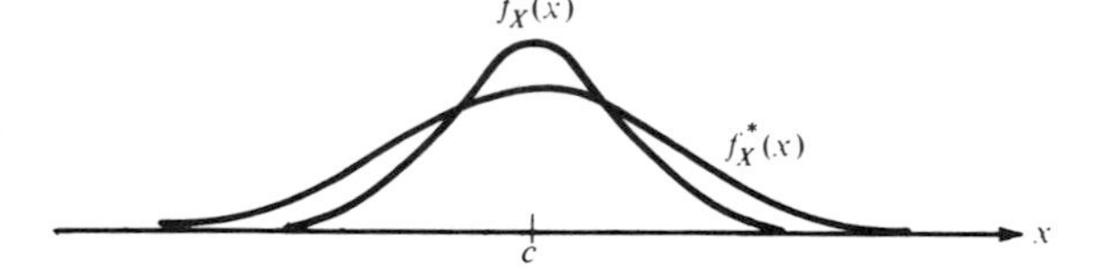

EXHIBIT 1.4 Two density functions with different dispersions.

spread will vary (from trial to trial) more than random variables whose distributions are less widely dispersed.

In this chapter we introduce and compare several useful measures of location for a distribution (mean, median, quantile, mode). We also discuss measures of dispersion, one of which (the variance) plays a prominent role in statistical analysis. Other descriptive measures and descriptive terminology (skewness, peakedness or kurtosis, and so on) are briefly mentioned. Finally, we show how a knowledge of a measure of location and a measure of dispersion of a random variable X allows us to approximate probabilities of events, even when little else is known about the distribution of X.

2 Measures of Location for a Distribution: The Mean

One of the most useful measures of location of a distribution is its **mean** (also called the **expected value** or the **expectation**). The mean can be viewed as being the fulcrum, the center of gravity, or the balance point of the distribution on the real line.

Definition of the Mean

For a discrete random variable X with possible values $x_1, \ldots, x_k$ and corresponding probabilities

$$p_i = p(x_i) = P\{X = x_i\}, \qquad i = 1, \ldots, k,$$

the graph (see Exhibit 2.1) of the probability mass function $p(x)$ of X can be viewed as representing masses p_i placed at the points x_i on the horizontal axis. The center of gravity of these masses is defined in physics as the point μ_X on the horizontal axis (thought of as a seesaw) at which a fulcrum could be placed to balance the seesaw. The forces acting on the seesaw are given by the products, $(x_i - \mu_X)p_i$, of the masses p_i and their signed distances $x_i - \mu_X$ from the fulcrum; positive products $(x_i - \mu_X)p_i$ push the seesaw clockwise, whereas negative values push it counterclockwise. Thus the seesaw balances if

$$\sum_{i=1}^{k} (x_i - \mu_X)p_i = 0.$$

However,

$$0 = \sum_{i=1}^{k} (x_i - \mu_X)p_i = \sum_{i=1}^{k} x_i p_i - \mu_X \sum_{i=1}^{k} p_i = \sum_{i=1}^{k} x_i p_i - \mu_X,$$

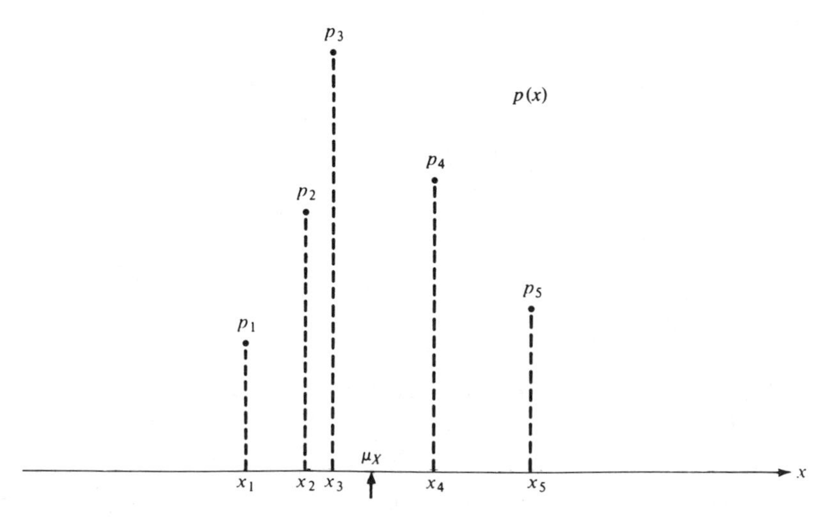

EXHIBIT 2.1 *The probability mass function $p(x)$ of a discrete random variable X having possible values x_1, x_2, x_3, x_4, x_5 with respective probabilities p_1, p_2, p_3, p_4, p_5, and with mean μ_X.*

because the sum of the probabilities $\sum_{i=1}^{k} p_i$ equals 1. Thus the center of gravity is given by $\mu_X = \sum_{i=1}^{k} x_i p_i$, leading to the following definition for the mean.

If X has a probability mass function $p(x)$ that places probabilities $p_1, \ldots, p_k$ at values $x_1, \ldots, x_k$, then the mean of X is defined by

$$\mu_X = \sum_{i=1}^{k} x_i p_i = \sum x_i p(x_i) \equiv \sum_{x} x p(x). \tag{2.1}$$

EXAMPLE 2.1

A sociologist is interested in comparing the sizes of households in the United States and in a less industrially developed country. Let X and Y be the numbers of people in a randomly chosen household in the United States and the less developed country, respectively. (*Note*: A household consists of all persons who live together in a dwelling unit.) From recent census figures for the two countries, the sociologist obtains the following probability models for X and Y:

Number of Individuals	1	2	3	4	5	6	7	8	9	10
X Probability	0.237	0.317	0.178	0.157	0.070	0.026	0.015	0.000	0.000	0.000
Y Probability	0.015	0.146	0.242	0.191	0.156	0.164	0.050	0.021	0.010	0.005

The means of X and Y are

$$\begin{aligned}\mu_X &= (1)(0.237) + (2)(0.178) + \cdots + (7)(0.015) + (8)(0.000) \\ &\quad + (9)(0.000) + (10)(0.000) \\ &= 2.644,\end{aligned}$$

$$\begin{aligned}\mu_Y &= (1)(0.015) + (2)(0.146) + \cdots + (9)(0.010) + (10)(0.005) \\ &= 4.219.\end{aligned}$$

These means provide a crude comparison of how many people live in households in the respective countries. Comparing $\mu_X = 2.644$ and $\mu_Y = 4.132$ suggests that U.S. households tend to have approximately 1.5 fewer persons than do households in the less industrially developed country. The sociologist then may seek to explain this difference. Of course, comparing only the means does not yield a complete picture of the differences between the distributions of X and Y. (For example, the distribution of X has a single peak, whereas the distribution of Y has two peaks—at $y = 3$ and $y = 6$.) A similar criticism can be directed at the use of any other single index in place of the entire distribution of a random variable. ◆

Note. In some problems, a discrete random variable X may have an infinite number of possible values $x_1, x_2, \ldots$. In this case, the definition (2.1) of the mean of X can be extended as follows:

$$\mu_X = \sum_{i=1}^{\infty} x_i p(x_i), \tag{2.2}$$

provided that the infinite sum on the right side of (2.2) is well defined. A necessary and sufficient condition for the sum (2.2) to be well defined is that $\sum_{i=1}^{\infty} |x_i| p(x_i)$ is finite. This need not always be the case (see Exercise T.8).

The center of gravity for a smooth distribution of physical mass on a line is defined as the integral of the product of the density of mass at each point with the distance of that point from a given origin. Correspondingly, the mean μ_X of a continuous random variable X with probability density function $f(x)$ is defined to be

$$\mu_X = \int_{-\infty}^{\infty} x f(x)\, dx. \tag{2.3}$$

EXAMPLE 2.2

Suppose that the continuous random variable X has density function

$$f(x) = \begin{cases} \frac{1}{2}, & \text{for } -1 \le x \le 1, \\ 0, & \text{otherwise.} \end{cases}$$

This density is graphed in Exhibit 2.2. From equation (2.3),

$$\begin{aligned} \mu_X &= \int_{-\infty}^{-1} (x)(0)\, dx + \int_{-1}^{1} (x)\left(\frac{1}{2}\right) dx + \int_{1}^{\infty} (x)(0)\, dx \\ &= 0 + \frac{x^2}{4}\bigg|_{-1}^{1} + 0 = 0. \end{aligned}$$

The result, $\mu_X = 0$, is intuitively reasonable because of the concept of the mean as the center of gravity of the distribution of probability, and the symmetry of the graph of $f(x)$ about the point $x = 0$. ◆

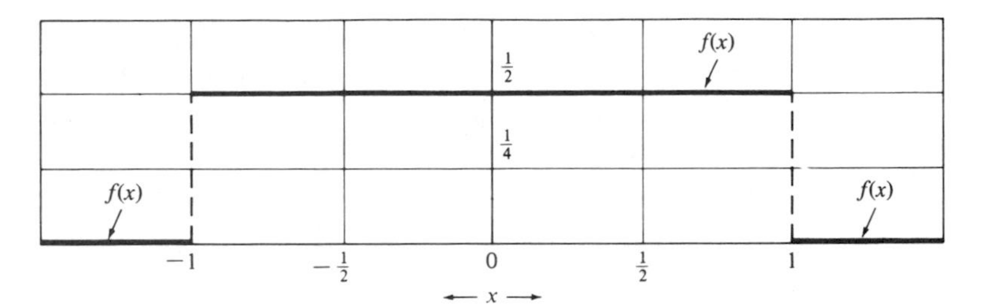

EXHIBIT 2.2 *Graph of the density function f(x), which is flat from x = −1 to x = 1 and is zero elsewhere.*

The concept that the mean μ_X of a random variable X is the center of gravity of the distribution of probability mass or density can be helpful in checking calculations. In Example 2.1, for example, a misplaced decimal point might have yielded the result 26.44 for μ_X. This result is clearly impossible because 26.44 is greater than the largest possible value (10) for X. On the other hand, if we had transposed the first two digits, obtaining a mean of 6.244, a rough graph of the probability mass function $p(x)$ of X would reveal that 6.244 is too far to the right of the main concentration of probability mass to be the center of gravity for the distribution of X.

EXAMPLE 2.3

A delicatessen receives an order of 400 pounds of meat every day. The total weight X of meat sold in a day is a continuous random variable with probability density function

$$f(x) = \begin{cases} (1.25)(10^{-5})x, & \text{for } 0 \le x \le 400, \\ 0, & \text{otherwise.} \end{cases}$$

A graph of this density would reveal that it forms a triangle with the x-axis. The "heavy" end of the triangle is to the right, so that we would expect the mean or balance point μ_X to be bigger than the midpoint $x = 200$ of the interval [0, 400] of possible values of x. Thus we expect that $\mu_X \ge 200$. In fact, the mean amount of meat that the delicatessen sells per day is

$$\begin{aligned} \mu_X &= \int_{-\infty}^{0} (x)(0)\,dx + \int_{0}^{400} (x)(1.25)(10^{-5})x\,dx + \int_{400}^{\infty} (x)(0)\,dx \\ &= (1.25)10^{-5} \int_{0}^{400} x^2\,dx \\ &= \frac{(1.25)10^{-5}}{3} x^3 \Big|_{0}^{400} = 266.67 \text{ pounds.} \end{aligned}$$

◆

The mean μ_X is often called the **expected value** of X. However, this terminology does not imply that μ_X is the value that we expect to obtain when we observe X. Indeed, the mean of X need not even be a possible value of X. For example, the means $\mu_X = 2.644$ and $\mu_Y = 4.219$ of the random variables X, Y in Example 2.1 are not possible values.

EXAMPLE 2.4

In an investigation of the theory of Brownian motion in physics, the Swedish physicist T. Svedberg (1912) counted the number of particles, X, in an optically isolated volume of a colloidal solution of gold at a certain instance of time and obtained the following table of probabilities:

Number of Particles, x	0	1	2	3	4	5	6	7
$P\{X = x\}$	0.182	0.310	0.264	0.150	0.064	0.022	0.006	0.002

for which the mean of the number of particles is $\mu_X = 1.704$. Clearly, one can never observe $X = 1.704$ particles. ◆

Instead, we expect the mean μ_X of X to be very close to the **arithmetic average** of the values of X observed in a large number, N, of trials. Suppose that we obtain N observations $t_1, t_2, \ldots, t_N$ of a discrete random variable X. The arithmetic average of the observed values is

$$\bar{t} = \frac{t_1 + t_2 + \cdots + t_N}{N}.$$

However, an alternative way to calculate the average $\bar{t}$ is by the formula

$$\bar{t} = \sum_t [t \times \text{r.f.}\{X = t\}],$$

where the sum is over all distinct values t of X that are observed. To see this, suppose that $N = 7$, and that we observe the values 1, 1, 2, 2, 2, 5, 6 for X. Then

$$\bar{t} = \frac{1 + 1 + 2 + 2 + 2 + 5 + 6}{7} = (1)\left(\frac{2}{7}\right) + (2)\left(\frac{3}{7}\right) + (5)\left(\frac{1}{7}\right) + (6)\left(\frac{1}{7}\right)$$

$$= \sum_t t \times \text{r.f.}\{X = t\}.$$

As the number N of trials becomes very large, the relative frequency interpretation of probability (Chapter 1) implies that r.f.$\{X = t\}$ should be close to $P\{X = t\}$. Thus when the number of trials is large,

$$(2.4) \qquad \bar{t} = \sum_t [t \times \text{r.f.}\{X = t\}] \simeq \sum_t [t \times P\{X = t\}] = \mu_X,$$

where the symbol "$\simeq$" means "approximately equal to." Consequently, in a large number of trials the arithmetic average (also called **sample average** or **sample mean**) of the observed values of X is approximately equal to μ_X. This discussion also indicates why μ_X is often called the population average of X.

EXAMPLE 2.5

The approximation given by equation (2.4) is frequently used in practice to predict the magnitude of the sum $\Sigma_{i=1}^N t_i = N\bar{t}$ of the observed values of X before these values are actually observed. For example, suppose that a city council learns that a developer plans to construct housing for $N = 1000$ new households in their city. The council is concerned about the increase in population this will create. Assuming that the numbers $t_1, t_2, \ldots, t_{1000}$ of people in these households are a random sample of U.S. households, and using the probability model for X given in Example 2.1, they can expect the average household size,

$$\bar{t} = \frac{t_1 + \cdots + t_{1000}}{1000},$$

to approximately equal the mean $\mu_X = 2.644$ of X. Consequently, the population of their city will increase by approximately

$$1000\bar{t} \simeq 1000\mu_X = 2644 \text{ persons.}$$

◆

EXAMPLE 2.6

A 44-year-old man buys a 1-year term life insurance policy of \$25,000 for a premium of \$100. The insurance company will pay this man's heirs \$25,000 if he dies during the year. From a standard mortality table (1982 Life Insurance Fact Book, American Council of Life Insurance, Washington, D.C.), the probability that a man of age 44 will live to be age 45 is $p = 0.99787$. If X denotes the amount of profit (or loss) in dollars that the insurance company makes from the transaction, the company earns $X = \$100$ if the man lives, whereas if he dies $X = \$100 - \$25{,}000$. Consequently, the mean μ_X of the random variable X is

$$\begin{aligned}\mu_X &= (\$100)p + (\$100 - \$25{,}000)(1 - p)\\ &= (\$100)(0.99787) + (-\$24{,}900)(0.00213)\\ &= \$46.75.\end{aligned}$$

Notice that $\mu_X = \$46.75$ does not refer to the profit that the insurance company makes on the given transaction (this profit is either \$100 if the man lives, or $-\$24{,}900$ if he dies). Instead, the value of μ_X refers to the *average* profit that the company can expect over a great many transactions of this type. The larger the number N of such transactions, the more accurately μ_X indicates what the average profit per transaction will be. ◆

The Mean as a Measure of Location

The mean is a measure of location for a probability distribution. The interpretation of the mean as the center of gravity of a probability distribution provides intuitive support to this assertion. To verify that μ_X satisfies the formal definition of a measure of location, we find what happens to the mean when we add a constant d to X.

If X is a discrete random variable with possible values $x_1, x_2, \ldots$ and corresponding probabilities $p_1, p_2, \ldots$, and $Z = X + d$, then

$$p_i = P\{X = x_i\} = P\{Z = x_i + d\},$$

because $X = x_i$ if and only if $Z = x_i + d$. Thus, noting that the possible values of Z are $x_1 + d, x_2 + d, \ldots,$

$$\begin{aligned}\mu_Z &= \sum_i z_i P\{Z = z_i\} = \sum_i (x_i + d)P\{Z = x_i + d\} \\ &= \sum_i (x_i + d)p_i = \sum_i x_i p_i + d \sum_i p_i \\ &= \mu_X + d,\end{aligned}$$

because $\sum_i p_i = 1$. A similar result holds when X is a continuous random variable (see Exercise T.9).

Indeed, for $a \neq 0$, and any number c, a similar argument can be used to show that:

> The mean of $aX + c$ is
>
> (2.5) $$\mu_{aX+c} = a\mu_X + c.$$

Our previous result, $\mu_{X+d} = \mu_X + d$, is the special case of (2.5) with $a = 1$, $c = d$.

EXAMPLE 2.7

A towing firm is paid \$50 by the state for every abandoned vehicle towed off state highways. The mean number of vehicles towed per week is $\mu_X = 8.3$. To determine the mean amount of money received per week, let X be the number of abandoned vehicles towed in a week and Y be the amount of money (in dollars) received for such towing. Then $Y = 50X$, and from equation (2.5),

$$\mu_Y = \mu_{50X} = 50\mu_X = 50(8.3) = \$415. \quad \blacklozenge$$

3 Other Measures of Location for a Distribution: Median, Quantile, Mode

The mean μ_X is not the only measure of location for a distribution. Other measures of location that are often useful are medians, quantiles, and modes.

Quantiles and the Median

Let p be a fixed proportion, $0 < p < 1$. If X is a continuous random variable, we can find a value $x = x^*$ such that the area under the probability density function, $f(x)$, of X to the left of x^* equals p (see Exhibit 3.1). This point x^* is called the pth **quantile** (or 100pth **percentile**) of X, and is denoted $Q_X(p)$. Formally, $Q_X(p)$ satisfies

$$P\{X \leq Q_X(p)\} = p,$$

or

$$F(Q_X(p)) = p, \tag{3.1}$$

where $F(x) = P\{X \leq x\}$ is the cumulative distribution function of X.

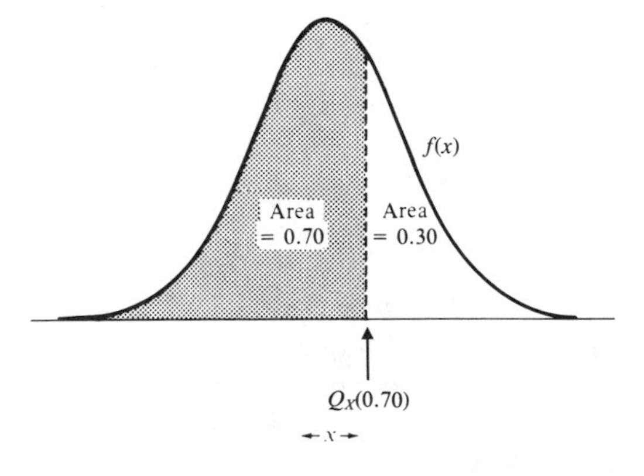

EXHIBIT 3.1 *Graph of the probability density function f(x) of a continuous random variable X showing how the 0.70th quantile $Q_X(0.70)$ divides the total area into two parts.*

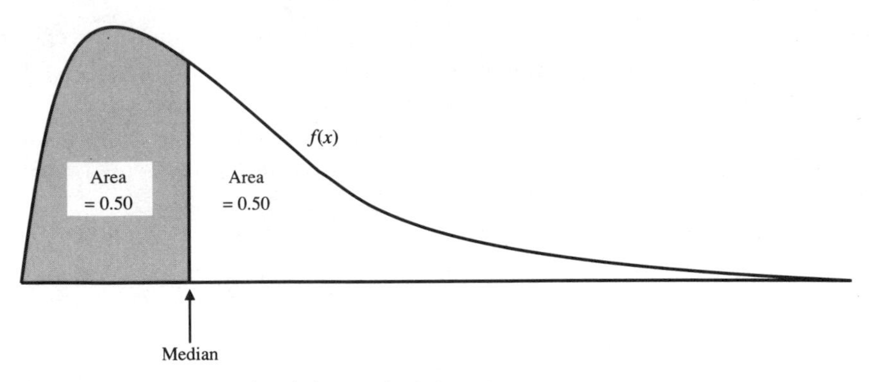

EXHIBIT 3.2 *Graph of the probability density function f(x) of a continuous random variable X showing how the median Med(x) divides the total area into two equal parts.*

As its name suggests, the **median** is a value x^* so that the area under $f(x)$ to the left of x^* is $\frac{1}{2}$ (Exhibit 3.2). That is, the median is the middle value of X in the sense that 50% of the probability is to the left of the median and 50% to the right. It immediately follows that the median is the 0.50th quantile $Q_X(0.50)$ of X. We denote the median by either $Q_X(0.50)$ or Med(X).

The pth quantile, $Q_X(p)$, of a continuous random variable X is most easily obtained analytically by solving the equation $F(x) = p$ for x. For example, if the cumulative distribution function of X is

$$F(x) = \begin{cases} 0, & \text{if } x < 0, \\ x^2, & \text{if } 0 \leq x < 1, \\ 1, & \text{if } x \geq 1, \end{cases}$$

then we solve $x^2 = p$ for x, obtaining $x = Q_X(p) = p^{1/2}$. Thus $Q_X(0.50) = \text{Med}(X) = (0.50)^{1/2} = 0.707$.

Exhibit 3.3 shows how to find the pth quantile graphically (for $p = 0.50$, 0.75). Such a graphical solution is usually less convenient than the analytical

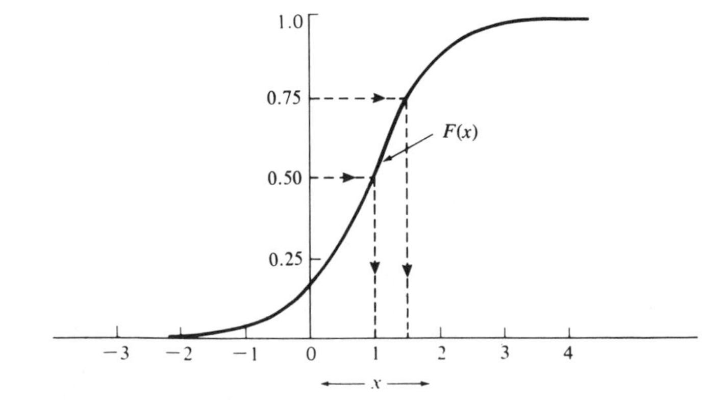

EXHIBIT 3.3 *Finding a median, Med(x), and 0.75th quantile, $Q_X(0.75)$, of a continuous random variable. Here Med(x) = 1 and $Q_X(0.75) = 1.5$.*

solution obtained by solving $F(x) = p$ for x but does help to conceptualize what the analytic solution means.

Because the cumulative distribution function $F(x)$ of a continuous random variable X is a continuous function of x, and $\lim_{x\to-\infty} F(x) = 0$, $\lim_{x\to\infty} F(x) = 1$, a solution x^* of the equation $F(x) = p$ always exists $(0 < p < 1)$.

EXAMPLE 3.1

Suppose that X has cumulative distribution

$$F(x) = \begin{cases} 0, & \text{if } x < 0, \\ \frac{1}{4}x, & \text{if } 0 \le x < 1, \\ \frac{1}{4}x^2, & \text{if } 1 \le x < 2, \\ 1, & \text{if } x \ge 2. \end{cases}$$

To find the median, $Q_X(0.50)$, of X, we need to find a value x^* satisfying $F(x^*) = 0.50$. Do we solve $\frac{1}{4}x = 0.50$ or $\frac{1}{4}x^2 = 0.50$ for x? As x varies within the limits $0 \le x \le 1$, $F(x)$ varies between 0 and $\frac{1}{4}$. [Here, graphing $F(x)$ might help.] Thus the solution of $F(x) = 0.50$ cannot be in this interval of x-values. Consequently, to find the median we solve $\frac{1}{4}x^2 = 0.50$ for x, obtaining $x = Q_X(0.50) = \sqrt{2} = 1.414$. On the other hand, to find the 0.25th quantile $Q_{0.25}(X)$, we would solve $\frac{1}{4}x = 0.25$ for x, finding that $x = Q_X(0.25) = 1.00$. ◆

If X is a discrete random variable, its cumulative distribution function moves only in jumps. For some values of p there may be no value of x for which $F(x) = p$. Instead, $F(x)$ may "jump through" p at some value x^*. (This is shown in Exhibit 3.4 in the case $p = 0.50$.) If $F(x)$ jumps through p

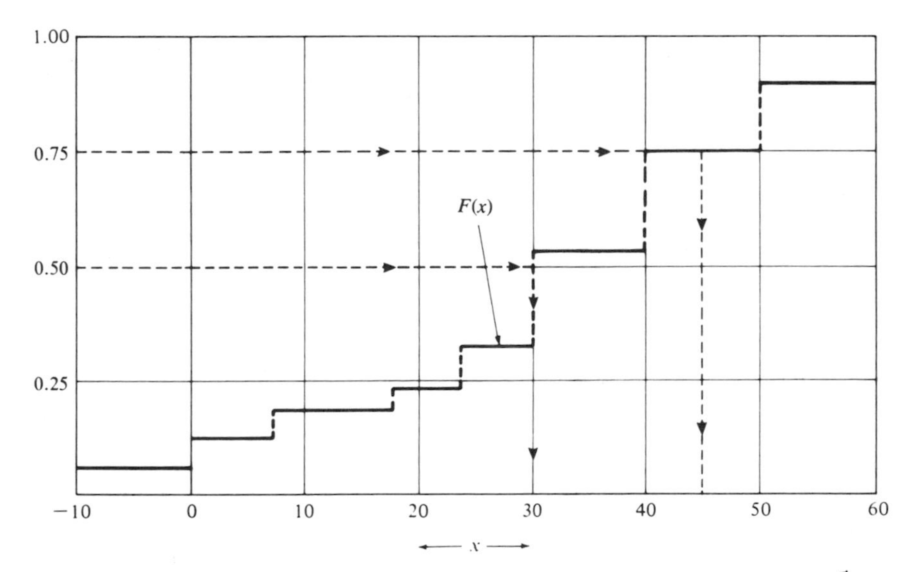

EXHIBIT 3.4 *Finding a median, Med(x), and 0.75th quantile, $Q_X(0.75)$, of a discrete random variable, X. Here Med(x) = 30, $Q_X(0.75) = 0.45$.*

when $x = x^*$, then x^* is defined to be the value of the pth quantile $Q_X(p)$. The only other possibility is that there is an interval of x-values for which $F(x) = p$. (This is shown in Exhibit 3.4 in the case $p = 0.75$.) The midpoint of this interval of x-values is then taken to be the value of $Q_X(p)$.

When X is a discrete random variable, obtaining $Q_X(p)$ through a graph of the cumulative distribution function $F(x)$ can be awkward. Instead, it is often easier to work with a **table of cumulated probabilities** for X. Use of this method is illustrated by the following example.

EXAMPLE 3.2

Suppose that the distribution of the discrete random variable X is given by the following table of probabilities:

x	-2	-1	0	1	2
$P\{X = x\}$	0.12	0.13	0.21	0.30	0.24

To find $Q_X(0.90)$, the 0.90th quantile of X, convert this table to a table of cumulative probabilities:

x	-2	-1	0	1	2
$F(x) = P\{X \leq x\}$	0.12	0.25	0.46	0.76	1.00

where, for example, $F(0) = P\{X \leq 0\}$ is computed by summing 0.12, 0.13, and 0.21. Because the jumps of $F(x)$ occur only at $x = -2, -1, 0, 1, 2$, and the graph of $F(x)$ is flat between jumps, this table could be used to graph $F(x)$. However, this is unnecessary; rather, the entries in the table are used to find the value of x at which $F(x)$ either equals 0.90 or jumps from under 0.90 to over 0.90. Here this value is $x = 2$, because $F(1) = 0.76 < 0.90$ whereas $F(2) = 1.00$. Hence the 0.90th quantile of X is $Q_X(0.90) = 2$. Similarly, because $F(-1) = 0.25$, all numbers $-1 \leq x < 0$ are 0.25th quantiles of X. In this case, the midpoint $Q_X(0.25) = \frac{1}{2}(-1 + 0) = -0.5$ of this interval is reported as the value of $Q_X(0.25)$. ♦

EXAMPLE 2.1 (continued)

In Example 2.1 hypothetical probability models were presented for the number X of persons in a randomly chosen U.S. household and the number Y of people in a random household from a less industrially developed country. From the table of probabilities given in Example 2.1, we calculate the following table of cumulated probabilities $F_X(a) = P\{X \leq a\}$, $F_Y(a) = P\{Y \leq a\}$ for X, Y, respectively:

Number of People, a	1	2	3	4	5	6	7	8	9	10
$F_X(a)$	0.237	0.554	0.732	0.889	0.959	0.985	1.000	1.000	1.000	1.000
$F_Y(a)$	0.015	0.161	0.403	0.594	0.750	0.914	0.964	0.985	0.995	1.000

Using this table and the method illustrated in Example 3.1, a few quantiles of the distributions of X and Y have been obtained:

Quantile	$Q(0.10)$	$Q(0.25)$	$Q(0.50) = \text{Med}$	$Q(0.75)$	$Q(0.90)$
X	1.0	2.0	2.0	4.0	5.0
Y	2.0	3.0	4.0	5.5	6.0

Notice that all of these quantiles satisfy $Q_Y(p) \geq Q_X(p)$, consistent with the ordering $\mu_Y \geq \mu_X$ of the means observed in Example 2.1. However, the differences between corresponding measures of location for X and Y are not all the same. Thus $Q_Y(0.10) - Q_X(0.10) = 1.0$, whereas $Q_Y(0.75) - Q_X(0.75) = 1.5$. ◆

Quantiles and Medians as Measures of Location

A pth quantile is a measure of location for a probability distribution in the sense that $Q_X(p)$ divides the graph of the distribution of probability into two well-defined parts. For continuous random variables, the division is into exact proportions p to the left, $1 - p$ to the right (see Exhibit 3.1). For discrete variables, some probability mass may be located at $x = Q_X(p)$. Consequently, the mass is divided into a proportion less than or equal to p to the left, less than or equal to $1 - p$ to the right, and perhaps some mass at $Q_X(p)$ (see Exhibit 3.5).

If $Y = aX + c$, $a > 0$, then

$$F_Y(y) = F_X\left(\frac{y - c}{a}\right),$$

and a solution $x = Q_X(p)$ of $F_X(x) = p$ yields a solution $y = Q_Y(p)$ of $F_Y(y) = p$ for which

$$\frac{Q_Y(p) - c}{a} = Q_X(p),$$

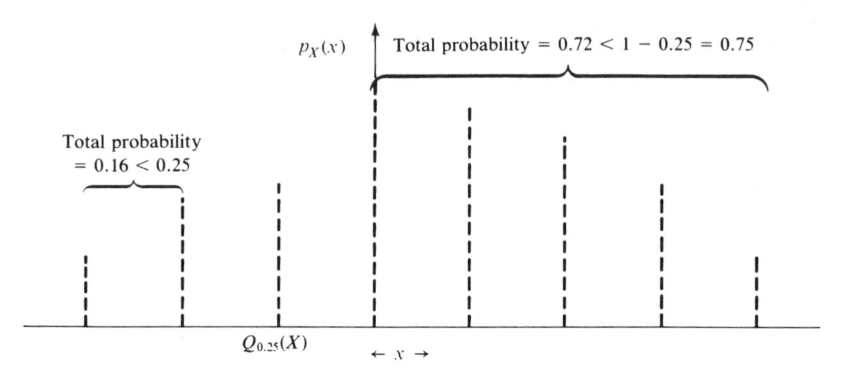

EXHIBIT 3.5 *Graph of the probability mass function $p_X(x)$ of a discrete random variable, X, showing how the 0.25th quantile $Q_X(0.25)$ divides the probability mass into two parts.*

or

$$Q_Y(p) = aQ_X(p) + c.$$

This result is also true when the cumulative distributions $F_X(x)$, $F_Y(y)$ jump through p at $x = Q_X(p)$, $y = Q_Y(p)$, respectively.

Consequently, it has been shown that for all $0 < p < 1$, all $a > 0$, and all c,

$$Q_{aX+c}(p) = aQ_X(p) + c. \tag{3.2}$$

The special case $a = 1$ of equation (3.2) verifies that $Q_X(p)$ and $\text{Med}(X) = Q_X(0.50)$ are measures of location:

$$Q_{X+c}(p) = Q_X(p) + c, \qquad \text{Med}(X + c) = \text{Med}(X) + c.$$

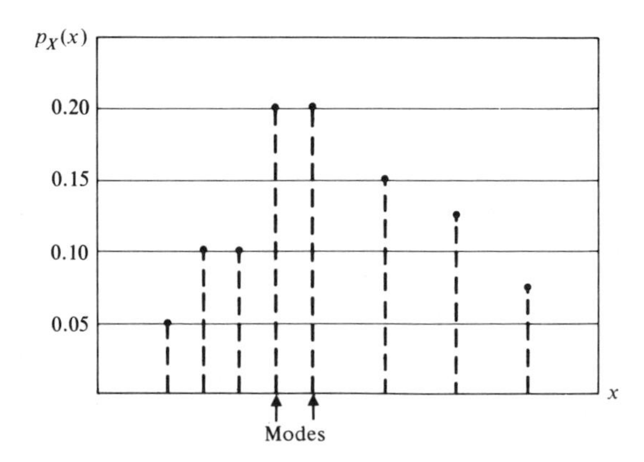

EXHIBIT 3.6 *A discrete unimodal distribution.*

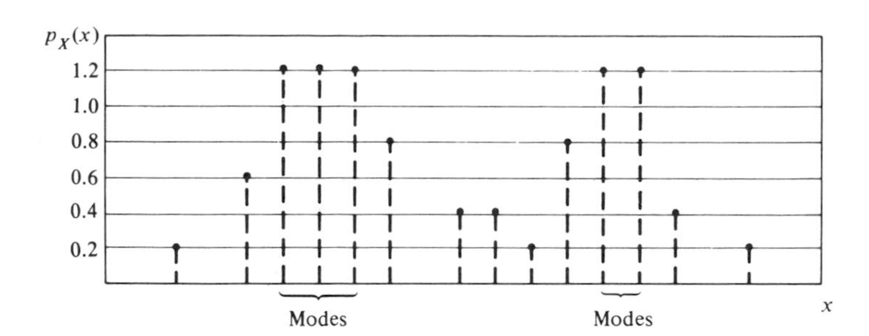

EXHIBIT 3.7 *A discrete bimodal distribution.*

Modes

A **mode** of a discrete random variable X is the "most probable" value of X. The mode, Mode (X), of X is the value x where the probability mass function $p(x)$ of X has its largest value. There can be one, or more than one, mode of the distribution of a discrete random variable (see Exhibits 3.6 and 3.7).

Because every possible value x of a continuous random variable X has zero probability of occurrence, the mode of a continuous random variable X requires a different definition. Instead of maximizing the probability of observing a value x, one maximizes the probability density $f(x)$ of x. That is, the mode of a continuous random variable X with probability density function $f(x)$ is defined as that value of x at which $f(x)$ is a maximum. As in the discrete case, the distribution of a continuous random variable can have one, or more than one, mode (see Exhibits 3.8 and 3.9).

A **relative mode** of a discrete (continuous) random variable X is any possible value of X for which the probability mass function (probability density function) is *locally* the highest. Examples of this concept are shown in Figures 3.6 to 3.9. A mode, of course, is always a relative mode.

The terms *unimodal* and *bimodal* are often used to describe distributions. A **unimodal distribution** has a single "hump" (see Exhibits 3.6 and 3.8). In contrast, a **bimodal distribution** has two "humps" (see Exhibits 3.7 and 3.9). A bimodal distribution has two different separated relative modes but may have a single unique mode.

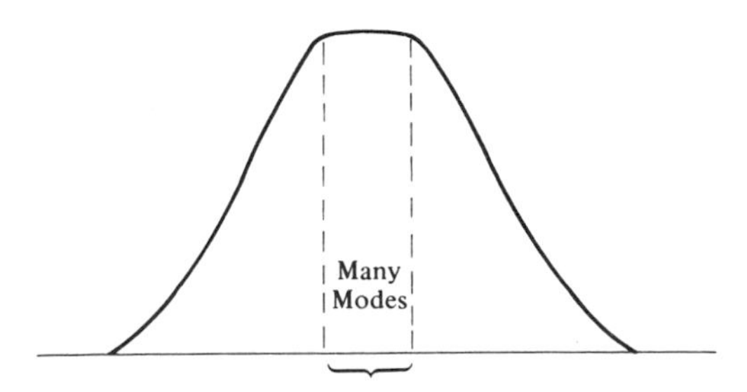

EXHIBIT 3.8 *A continuous unimodal distribution.*

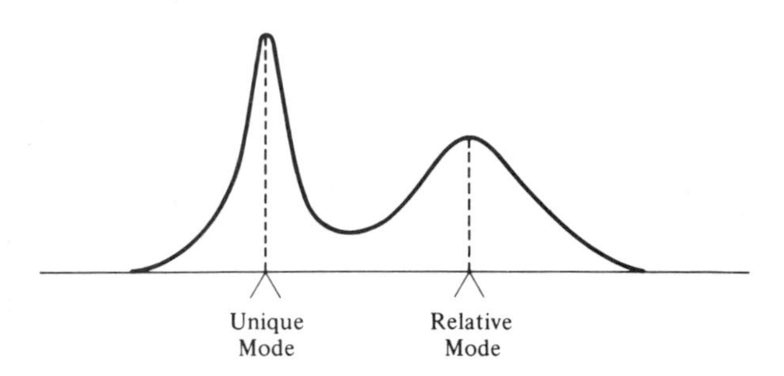

EXHIBIT 3.9 *A continuous bimodal distribution.*

Choice of a Measure of Location

There are a number of potential measures of location for a distribution: the mean μ_X, the median Med(X), the quantiles $Q_X(p)$, and the mode. Each measure of location has both advantages and disadvantages, and there is no single best measure of location. The following example illustrates how the information that a measure of location provides about a distribution is influenced by the shape of the distribution, and why the choice of a measure of location is largely determined by the goals of the investigation.

EXAMPLE 3.3

Economists A and B studying the distribution of income are concerned with poverty-oriented legislation and with market research, respectively. The data available yield an estimate of the probability distribution of total pretax family income X (in dollars). Although pretax income is discrete, the number of possible values of this variable is sufficiently large that it can be approximated by a continuous random variable with a probability density function $f(x)$. The shape of the density function $f(x)$ of family income X in the United States resembles that of the function graphed in Exhibit 3.10. Important features of this graph

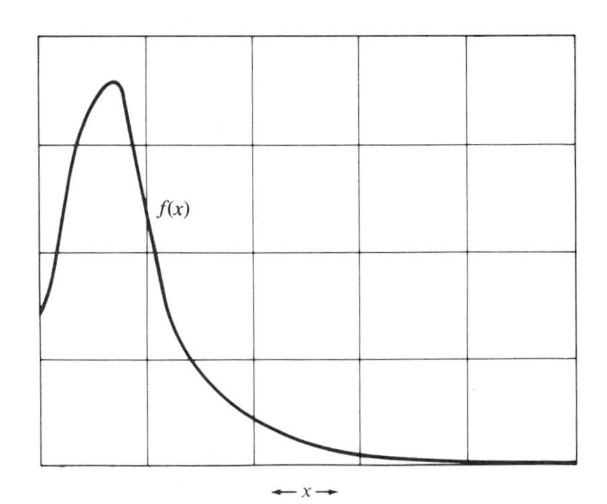

EXHIBIT 3.10 *Probability density function of the income of a randomly chosen family.*

include a long "tail" to the right, indicating incomes of the relatively few very wealthy families, and a single "hump" to the left, representing the low-to-moderate incomes of most families.

Economist A would not use the mean μ_X to describe the location of the distribution of family incomes X, because the large incomes of the few wealthy families can inflate the value of μ_X. As an extreme illustration, if 99% of all families earned \$10,000 and the other 1% earned \$1,000,000, then $\mu_X = (10{,}000)(0.99) + (1{,}000{,}000)(0.01) = 19{,}900$ dollars. Thus the mean μ_X would overstate (by nearly twice) the income of 99% of the population. On the other hand, the value of Med(X) is not influenced by large incomes occurring with small probability. Knowing that Med(X) = \$10,000 tells us that 50% of all families have incomes less than or equal to \$10,000 and provides an indication of the financial well-being of the majority of the population. Similarly, as long as p is not too close to 1, $Q_X(p)$ is not affected by extremely large incomes that have small probability. The values of $Q_X(0.10)$ and $Q_X(0.25)$, for example, might be used to indicate the financial status of the poorest 10% and 25%, respectively, among families. Consequently, economist A would report one or more quantiles of the distribution of X.

Economist B is concerned with predicting family consumer purchases for any given year based on knowledge of family income in the immediately preceding year. If economist B reports Med(X) instead of μ_X as the "average" or "typical" family incomes, this will ignore the incomes of the very rich families, leading to an underprediction of consumer spending for the coming year. Indeed, it was shown in Section 2 that the mean μ_X allows us to estimate the total income for *all* families, and thus permits prediction of total consumer purchases for the following year.

Neither economist A nor B would give much consideration to the mode of X as a measure of location. The mode tells little about how low the incomes of the lower-income families actually are, nor does it help in calculating, even approximately, the total income of all families.[1] ◆

4 The Expected Value of a Function of a Random Variable

Given a probability model for a random variable X, one might want to predict (approximately) the average value of some transformation $g(X)$ of X. For example, if X represents measures of velocities of randomly selected

[1]In fact, the mode is rarely the preferred measure of location in practical problems. The mode is useful to predict a possible value (or interval of possible values) for a future observation of a random variable X with the highest probability of making a correct prediction. In that case the mode of X (or an interval containing the mode) may be the proper predictor to use.

atomic particles of known mass m, one might want to predict the average kinetic energy $Y = \frac{1}{2}mX^2$ of these particles. As remarked in Chapter 5, Y is then a random variable with a probability distribution of its own, which can be found from the probability distribution of X.

The average of the values of $Y = g(X)$ in a large sample of values $t_1, t_2, \ldots, t_N$ observed on X can be approximated by the mean μ_Y of the distribution of Y. Because the formulas of Section 2 for calculating μ_Y were expressed in terms of the probability distribution of Y, it appears that one must determine the probability distribution of Y. Fortunately, as suggested by the fact that

$$\text{(4.1)} \qquad \text{average of } g(t_i) \text{ values} = \frac{1}{N}\sum_{i=1}^{N} g(t_i) = \sum_{t} g(t) \text{ r.f.}\{X = t\},$$

μ_Y can be found directly from the distribution of X. To do so, it is convenient to define $E[g(X)]$, called the **expected value** of $g(X)$.

Expected value of $g(X)$. As suggested by (4.1),

$$\text{(4.2)} \qquad E[g(X)] = \sum_{x} g(x)p(x)$$

if X is a discrete random variable with probability mass function $p(x)$. Analogously, when X is a continuous random variable with probability density function $f(x)$,

$$\text{(4.3)} \qquad E[g(X)] = \int_{-\infty}^{\infty} g(x)f(x)\,dx.$$

If the sum in (4.2) or the integral in (4.3) equals $-\infty$ or ∞, or is not well defined, the expected value $E[g(X)]$ does not exist.

Note from the definition of $E[g(X)]$ in the case $g(x) = x$ that

$$\text{(4.4)} \qquad \mu_X = E[X].$$

The notations μ_X and $E[X]$, however, are intended to play different roles in our vocabulary. The mean μ_X of X is an **index** of the distribution of X that helps locate this distribution on the real line. On the other hand, $E[X]$ and, more generally, $E[g(X)]$ denote certain *computations* based on the probability distribution of X. This distinction is analogous to the distinction commonly made between the derivative as being a measure of rate of change, and differentiation as the operation by which the derivative is obtained.

We have asserted that if $Y = g(X)$, the mean μ_Y of Y can be computed directly from the probability distribution of X, without the need to obtain the probability distribution of Y. Indeed,

(4.5) $$\mu_Y = \mu_{g(X)} = E[g(X)].$$

A general proof of this assertion would require us to discuss mathematical technicalities not needed elsewhere in the book. To illustrate the basic idea of the general proof, we use the example $Y = (X - 3)^2$, where X is a discrete random variable with probability distribution given by

x	0	1	2	3	4	5	6
$p(x)$	$\frac{1}{16}$	$\frac{2}{16}$	$\frac{3}{16}$	$\frac{4}{16}$	$\frac{4}{16}$	$\frac{1}{16}$	$\frac{1}{16}$

Note that the possible values of Y are 0, 1, 4, and 9, and that $Y = 0$ when $X = 3$, $Y = 1$ when $X = 2$ or 4, $Y = 4$ when $X = 1$ or 5, $Y = 9$ when $X = 0$ or 6. Consequently, the probability mass function of Y is given by

y	0	1	4	9
$p_Y(y)$	$p(3) = \frac{4}{16}$	$p(2) + p(4) = \frac{7}{16}$	$p(1) + p(5) = \frac{3}{16}$	$p(0) + p(6) = \frac{2}{16}$

It follows from equation (2.1) that

$$\mu_Y = \sum y p_Y(y) = (0)\left(\tfrac{4}{16}\right) + (1)\left(\tfrac{7}{16}\right) + (4)\left(\tfrac{3}{16}\right) + (9)\left(\tfrac{2}{16}\right) = \tfrac{37}{16}.$$

On the other hand, using equation (4.1) and grouping terms with similar values of $(x - 3)^2$,

$$\begin{aligned} E[(X - 3)^2] &= \sum_{x=0}^{3} (x - 3)^2 p(x) \\ &= (0)p(3) + (1)[p(2) + p(4)] + (4)[p(1) + p(5)] \\ &\quad + (9)[p(0) + p(6)] \\ &= (0)\left(\tfrac{4}{16}\right) + (1)\left(\tfrac{7}{16}\right) + (4)\left(\tfrac{3}{16}\right) + (9)\left(\tfrac{2}{16}\right) \\ &= \tfrac{37}{16} = \mu_Y. \end{aligned}$$

Thus we have verified equation (4.5) in this special case.

EXAMPLE 4.1

Suppose that the probability density for the velocity X of a randomly chosen atomic particle of known mass m is

$$f(x) = \begin{cases} \frac{1}{V}, & \text{if } 0 \le x \le V, \\ 0, & \text{otherwise.} \end{cases}$$

The kinetic energy of the particle is

$$Y = \tfrac{1}{2}mX^2.$$

Thus the mean kinetic energy μ_Y of particles of mass m can be computed from equation (4.5) as

$$\mu_Y = E\left[\frac{m}{2}X^2\right] = \int_0^V \left(\frac{m}{2}x^2\right)\frac{1}{V}\,dx = m\frac{V^2}{6}.$$ ♦

One advantage of the expected value notation is that we do not have to distinguish between discrete and continuous random variables. This advantage results from two important properties of the expected value.

Linearity Property of the Expected Value. If X is any random variable, $g(\cdot)$ and $h(\cdot)$ are any two functions, and a and b are any two constants, then

$$E[ag(X) + bh(X)] = aE[g(X)] + bE[h(X)]$$

provided that the expected values exist.

Order Preserving Property of the Expected Value. If X is any random variable, and $g(X)$ is nonnegative with probability equal to 1, then $E[g(X)] \ge 0$.

The proof of linearity of the expected value in the case when X is a discrete random variable follows from noting that

$$\sum_x [ag(x) + bh(x)]p(x) = a\sum_x g(x)p(x) + b\sum_x h(x)p(x).$$

Similarly, if X is a continuous random variable,

$$\int_{-\infty}^{\infty} [ag(x) + bh(x)]f(x)\,dx = a\int_{-\infty}^{\infty} g(x)f(x)\,dx + b\int_{-\infty}^{\infty} h(x)f(x)\,dx.$$

The order-preserving property of expected values results from the fact that $\sum_x g(x)p(x)$ and $\int_{-\infty}^{\infty} g(x)f(x)\,dx$ are a sum and an integral of nonnegative quantities, respectively.

Remark. It is worth noting that the probability $P(A)$ of any event A concerning X can be expressed as is the expected value of the **indicator function**

$$I_A(X) = \begin{cases} 1, & \text{if } X \text{ satisfies the property defining event } A, \\ 0, & \text{otherwise.} \end{cases}$$

That is,

$$P(A) = E[I_A(X)]. \tag{4.6}$$

In deriving probabilities of events, use of equation (4.6) permits us to take advantage of the properties of the expected value. Equation (4.6) is discussed in greater detail in Exercise T.11.

5 Measures of Dispersion for a Distribution: The Variance and the Standard Deviation

The **variance** σ_X^2 of a random variable X is a measure of the dispersion of the distribution of X around the mean μ_X. We use the function $(X - \mu_X)^2$ to measure how close X is to its mean μ_X, and define

$$\sigma_X^2 = E[(X - \mu_X)^2]. \tag{5.1}$$

In a large number of observations $t_1, t_2, \ldots, t_N$ upon X, σ_X^2 would be approximately equal to the *average squared distance* of the observations $t_1, t_2, \ldots, t_N$ from the mean μ_X. Thus the variance σ_X^2 quantifies how extensively X varies "on the average" about its mean μ_X. Large values of the variance are associated with large dispersions of X about μ_X, and small values of the variance are associated with small dispersions. If the variance of X is equal to zero, X does not vary at all about μ_X, and $P\{X = \mu_X\} = 1$.

The term **standard deviation** of X refers to the square root, σ_X, of the variance of X. The standard deviation is expressed in terms of the same units of measurement as X, whereas the variance is not. For this reason, the standard deviation is frequently used as a measure of dispersion in preference to the variance.

Using equations (5.1) (4.2), and (4.3), σ_X^2 is calculated by

(5.2)

$$\sigma_X^2 = \begin{cases} \sum_x (x - \mu_X)^2 p(x), & \text{if } X \text{ is a discrete random variable,} \\ \int_{-\infty}^{\infty} (x - \mu_X)^2 f(x)\, dx, & \text{if } X \text{ is a continuous random variable.} \end{cases}$$

EXAMPLE 5.1

An investor has a choice of two stock investments. Stock A is thought to have probability $\frac{1}{2}$ of doubling in value within one year, but has a probability of $\frac{1}{4}$ of losing half of its value in that time, and otherwise will end the year at the same price as it was purchased. Stock B is supposed to have probability $\frac{3}{4}$ of increasing in value by 50% in the year, and otherwise will end the year at the price of purchase. Thus if the investor invests \$1000 in stock A, the investor's profit X is a discrete random variable with probability mass function

$$p_X(x) = \begin{cases} \frac{1}{2}, & \text{if } x = 1000, \\ \frac{1}{4}, & \text{if } x = -500, \\ \frac{1}{4}, & \text{if } x = 0, \\ 0, & \text{otherwise.} \end{cases}$$

If the investor instead invests \$1000 in stock B, the investor's profit Y has probability mass function

$$p_Y(y) = \begin{cases} \frac{3}{4}, & \text{if } y = 500, \\ \frac{1}{4}, & \text{if } y = 0, \\ 0, & \text{otherwise.} \end{cases}$$

The mean profits for the two investments are the same:

$$\begin{aligned} \mu_X &= \sum_x x p_X(x) = (1000)\left(\tfrac{1}{2}\right) + (-500)\left(\tfrac{1}{4}\right) + (0)\left(\tfrac{1}{4}\right) \\ &= \frac{1500}{4} = \uparrow 375.00 \end{aligned}$$

$$\mu_Y = \sum_y y p_Y(y) = (500)\left(\tfrac{3}{4}\right) + (0)\left(\tfrac{1}{4}\right) = \$375.00.$$

If the mass functions of X and Y are graphed as in Exhibit 5.1, it is apparent that the distribution of X is more spread out (dispersed) than the distribution of Y. This assertion is supported by a comparison of the variances

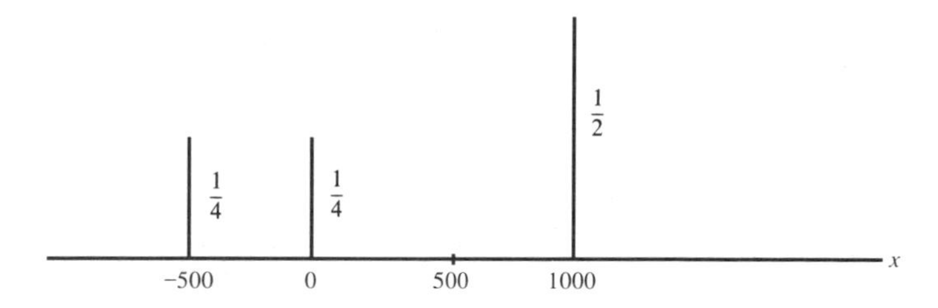

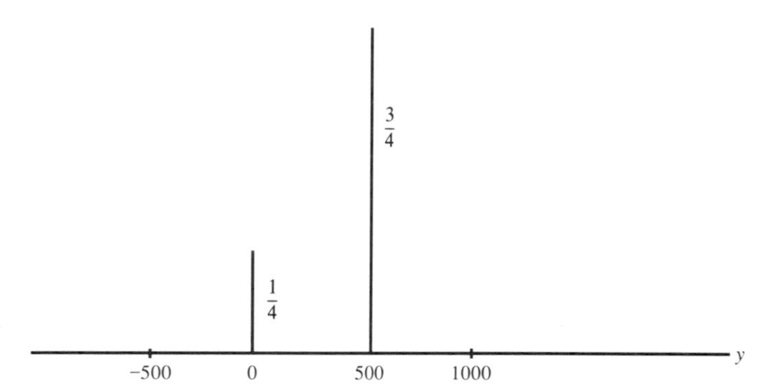

EXHIBIT 5.1 *Comparison of the mass functions of the profits from two stock investments.*

$$\begin{aligned}\sigma_X^2 &= \sum_x (x - 375)^2 p_X(x)\\ &= (1000 - 375)^2\left(\tfrac{1}{2}\right) + (-500 - 375)^2\left(\tfrac{1}{4}\right) + (0 - 375)^2\left(\tfrac{1}{4}\right)\\ &= 421{,}875,\end{aligned}$$

$$\begin{aligned}\sigma_Y^2 &= \sum_y (y - 375)^2 p_Y(y) = (500 - 375)^2\left(\tfrac{3}{4}\right) + (0 - 375)^2\left(\tfrac{1}{4}\right)\\ &= 46{,}875.\end{aligned}$$

Because $\sigma_X^2 > \sigma_Y^2$, the profit X from investing in stock A is more variable than the profit Y from investing in stock B, and thus is more uncertain (risky). Because the two stock investments have the same mean profit, a conservative investor will favor the less risky investment (stock B). ◆

EXAMPLE 5.2

The law requires boxes of cereal to contain at least the amount of cereal (16 ounces) stated on the carton. If the company underfills too many boxes, they may be in legal trouble, but if they overfill their boxes, they lose money. A study of the machines used to fill the boxes indicates that the probability density function of the amount X of cereal placed in a randomly chosen box is given by

$$f(x) = \begin{cases} 2.5, & 15.8 \le x \le 16.2,\\ 0, & \text{otherwise.}\end{cases}$$

The mean of X is

$$\mu_X = \int_{-\infty}^{\infty} xf(x)\,dx = \int_{15.8}^{16.2} x(2.5)\,dx = 1.25x^2\Big|_{15.8}^{16.2} = 16.0,$$

so that the machines fill the correct amount "on the average." However, there is a substantial probability of under- and overfilling the boxes. The variance of X is

$$\sigma_X^2 = \int_{-\infty}^{\infty} (x-16)^2 f(x)dx = \int_{15.8}^{16.2} (x-16)^2(2.5)\,dx$$
$$= \frac{(x-16)^3}{3}(2.5)\Big|_{15.8}^{16.2} = 0.0133.$$

To decrease the dispersion, the machines are overhauled, yielding a new probability density for X:

$$f(x) = \begin{cases} 25(x-16)+5, & 15.8 \le x < 16, \\ 25(16-x)+5, & 16 \le x < 16.2, \\ 0, & \text{otherwise.} \end{cases}$$

Now it is still the case that $\mu_X = 16.0$, so that the machines fill the correct amount "on the average," but the new variance is only half of the former variance:

$$\sigma_X^2 = \int_{-\infty}^{\infty} (x-16)^2 f(x)\,dx = \int_{15.8}^{16.0} (x-16)^2[25(x-16)+5]\,dx$$
$$+ \int_{16.0}^{16.2} (x-16)^2[25(16-x)+5]\,dx$$
$$= 5\int_{15.8}^{16.2} (x-16)^2\,dx + 25\left[\int_{15.8}^{16.0} (x-16)^3\,dx - \int_{16.0}^{16.2} (x-16)^3\,dx\right]$$
$$= \tfrac{5}{3}(x-16)^3\Big|_{15.8}^{16.2} + \tfrac{25}{4}(x-16)^4\Big|_{15.8}^{16.0} - \tfrac{25}{4}(x-16)^4\Big|_{16.0}^{16.2}$$
$$= 0.026\overline{6} - 0.01 - 0.01 = 0.006\overline{6}.$$

This lower variance indicates less variation about the mean, which is due to the reduction in the probabilities of under- and overfilling under the new distribution of X. [Compare Exhibit 5.2(a) to Exhibit 5.2(b).]

◆

Some caution should be noted with respect to the use of σ_X^2 as a measure of variation. The relationship of the degree of clustering of the distribu-

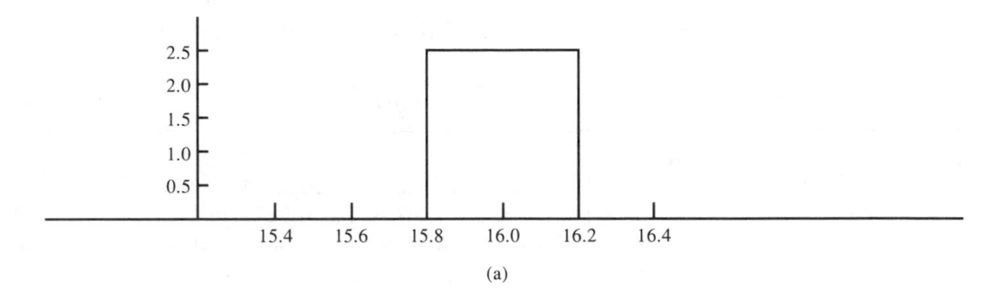

(a)

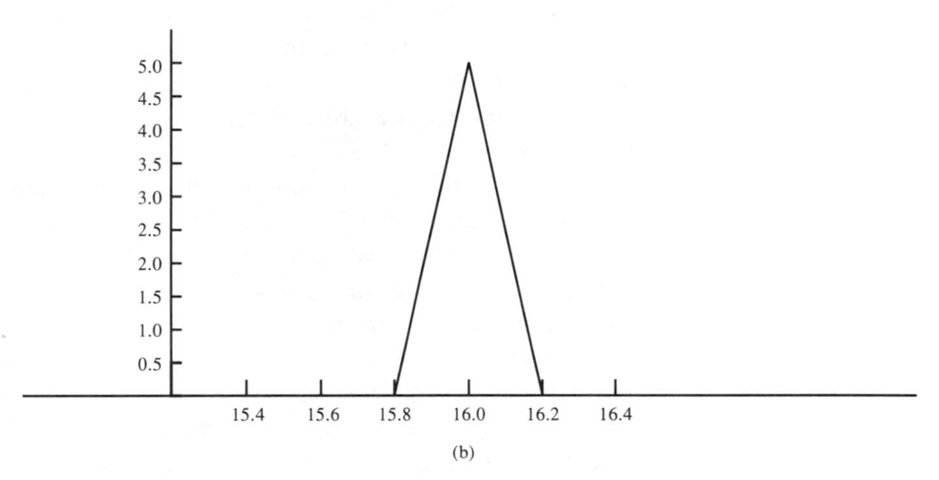

(b)

EXHIBIT 5.2 *Probability density function of X (a) before and (b) after overhaul of machines used to fill cereal boxes.*

tion of X to the size of σ_X^2 holds in most practical cases; however, examples can be given where the variance is not a particularly good indicator of the spread (variability) of the distribution. For example, if the random variable X has the probability model

x	0	d
$P\{X = x\}$	$1 - \dfrac{1}{d}$	$\dfrac{1}{d}$

where $d > 1$ is a constant, then the mean and variance of X are

$$\mu_X = (0)\left(1 - \frac{1}{d}\right) + (d)\left(\frac{1}{d}\right) = 1,$$

$$\sigma_X^2 = (0 - 1)^2\left(1 - \frac{1}{d}\right) + (d - 1)^2\left(\frac{1}{d}\right) = d - 1.$$

When $d = 10$, 100, and 1000, the probability models for X are:

$d = 10$

x	0	10
$P\{X = x\}$	0.9	0.1

$d = 100$

x	0	100
$P\{X = x\}$	0.99	0.01

$d = 1000$

x	0	1000
$P\{X = x\}$	0.999	0.001

The mean, in each case, is 1, but $\sigma_X^2 = 9$, 99, and 999 for $d = 10$, 100, and 1000, respectively. Thus, as d becomes larger, almost all of the probability mass of X becomes located at zero, with less and less probability being placed at d, and thus X is less and less variable. On the other hand, the variance σ_X^2 increases as d grows larger, falsely suggesting that X is more and more dispersed.

The reason for this apparent contradiction is that the variance is unduly sensitive to a small amount of probability placed away from most of the probability mass of the distribution. In Section 3 we noted that the mean μ_X has a similar sensitivity to extreme values assumed with small probability, and that because of this sensitivity it is sometimes preferable to use one of the quantiles $Q_X(p)$ in place of μ_X as a measure of location. Similarly, one might use a measure of dispersion based on the quantiles in place of the variance when there are extreme values with small probabilities.

One measure of dispersion based on the quantiles is the **interquartile range** IQR $= Q_X(0.75) - Q_X(0.25)$. Its justification is based on the fact that the interval of numbers between and including $Q_X(0.25)$ and $Q_X(0.75)$ includes at least 50% of the total probability mass of the distribution of X.

An important property of the variance σ_X^2 and standard deviation σ_X of a random variable X is that for any two numbers a and c,

$$(5.3) \qquad \sigma_{aX+c}^2 = a^2\sigma_X^2, \qquad \sigma_{aX+c} = |a|\sigma_X.$$

Verification of this property will be given later in this section.

Property (5.3) is most useful when we wish to change units of measurement. For example, to convert a measurement, X, of temperature in degrees Fahrenheit to a measurement $Z = (5/9)X - 160/9$ in degrees Celsius, $a = 5/9$ and $c = -160/9$. Similarly, to convert a measurement X of height in feet to the corresponding measurement $Z = 12X$ in inches, $a = 12$, $c = 0$. We could, of course, obtain the distribution of Z in each case from knowledge of the distribution of X, and then compute the variance σ_Z^2 of Z from the distribution of Z. However, (5.3) provides an easier way, provided that the variance σ_X^2 of X is known.

EXAMPLE 5.3

In Florida, temperatures X during the winter season have a mean of $\mu_X = 77$ degrees Fahrenheit (°F) and a standard deviation of $\sigma_X = 5$ °F. To convert these indices to the Celsius scale, let $Z = \frac{5}{9}X - \frac{160}{9}$. Using equation (2.5), Florida's mean winter season temperature in degrees Celsius is

$$\mu_Z = \frac{5}{9}(\mu_X) - \frac{160}{9} = \frac{5}{9}(77) - \frac{160}{9} = 25 \text{ °C}.$$

Using equation (5.3), the standard deviation in degrees Celsius of Florida's winter season temperatures is

$$\sigma_Z = \left(\frac{5}{9}\right)\sigma_X = \left(\frac{5}{9}\right)5 = 2.78 \text{ °C}.$$

◆

Notice that the constant c appears in the formula (2.5) for μ_{aX+c}, but does not appear in the formulas (5.3) for σ^2_{aX+c} and σ_{aX+c}. This is intuitively reasonable, because adding a constant to a random variable changes the location of the distribution of that variable (and thus the mean) but does not affect the spread or dispersion of the distribution.

The variance σ^2_X has been defined as the expected (or mean) squared distance from X to its mean μ_X. Instead of using the mean to compute this distance, any other constant c could have been chosen. That is, $E(X - c)^2$ could also be used as a measure of dispersion for X. Justification for choosing $c = \mu_X$ comes partly from the physical interpretation of σ^2_X as the second moment of inertia of the probability mass about its center of gravity; and partly as a consequence the usefulness of μ_X as a measure of location for the distribution of X. However, there is also an important mathematical justification for choosing $c = \mu_X$. Among all quantities of the form $E(X - c)^2$, choosing $c = \mu_X$ results in the smallest value. That is,

$$E(X - c)^2 \geq E(X - \mu_X)^2 = \sigma^2_X \tag{5.4}$$

for all constants c. Inequality (5.4) follows as a consequence of the useful formula

$$E(X - c)^2 = \sigma^2_X + (\mu_X - c)^2. \tag{5.5}$$

Applying equation (5.5) with $c = 0$ yields

$$\sigma^2_X = E(X^2) - (\mu_X)^2, \tag{5.6}$$

which often provides a convenient shortcut for computing the variance.

EXAMPLE 5.2 *(continued)*

In Example 5.2 we considered the continuous random variable X, which was the amount of cereal placed in a randomly chosen box of cereal. Before the machines were overhauled, X had the probability density function

$$f(x) = \begin{cases} 2.5, & \text{if } 15.8 \le x \le 16.2, \\ 0, & \text{otherwise.} \end{cases}$$

For this distribution $\mu_X = 16.0$, and

$$E(X^2) = \int_{-\infty}^{\infty} x^2 f(x)\, dx = \int_{15.8}^{16.2} x^2(2.5)\, dx = \frac{2.5}{3} x^3 \Big|_{15.8}^{16.2} = 256.013\bar{3},$$

so that

$$\sigma_X^2 = E(X^2) - \mu_X^2 = 256.013\bar{3} - (16.0)^2 = 0.013\bar{3}.$$

which agrees with our previous computation. ♦

EXAMPLE 5.4

Suppose that the continuous random variable X has density function

$$f(x) = \begin{cases} \dfrac{1.4427}{x}, & \text{if } 1 \le x \le 2, \\ 0, & \text{if } x < 1 \text{ or } x > 2. \end{cases}$$

A graph of $f(x)$ appears in Exhibit 5.3. We can compute

$$\mu_X = \int_{-\infty}^{\infty} x f(x)\, dx = 0 + \int_1^2 x\left(\frac{1.4427}{x}\right) dx + 0 = 1.4427,$$

$$E(X^2) = \int_{-\infty}^{\infty} x^2 f(x)\, dx = 0 + \int_1^2 x^2\left(\frac{1.4427}{x}\right) dx + 0 = \int_1^2 1.4427x\, dx$$

$$= \frac{1.4427}{2} x^2 \Big|_1^2 = 2.16405,$$

so that equation (5.7) yields

$$\sigma_X^2 = E(X^2) - (\mu_X)^2 = 2.16405 - (1.4427)^2 = 0.08267.$$ ♦

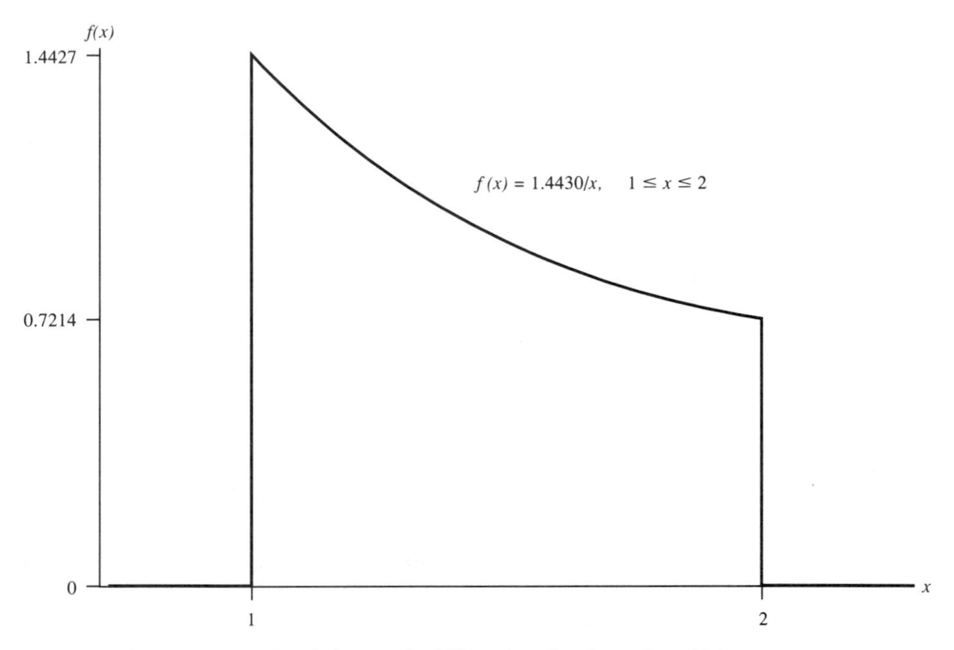

EXHIBIT 5.3 *Graph of the probability density function f(x).*

Verification of Equations (5.3) and (5.5)

To verify the assertion in equation (5.3) that $\sigma^2_{aX+c} = a^2\sigma^2_X$, let $Z = aX + c$. From equation (2.5),

$$\mu_Z = \mu_{aX+c} = a\mu_X + c,$$

and thus

$$\begin{aligned}\sigma^2_Z &= E(Z - \mu_Z)^2 = E[(aX + c) - (a\mu_X + c)]^2 \\ &= E[a(X - \mu_X)]^2 = a^2E[(X - \mu_X)^2] = a^2\sigma^2_X.\end{aligned}$$

The result stated in (5.3) concerning the standard deviation then follows because $\sigma_{aX+c} = (\sigma^2_{aX+c})^{1/2} = (a^2\sigma^2_X)^{1/2} = |a|\sigma_X$.

To verify the assertion of equation (5.5) that $E(X - c)^2 = \sigma^2_X + (\mu_X - c)^2$, use the linearity property of expected values:

$$\begin{aligned}E(X - c)^2 &= E(X - \mu_X + \mu_X - c)^2 \\ &= E[(X - \mu_X)^2 + 2(\mu_X - c)(X - \mu_X) + (\mu_X - c)^2] \\ &= E(X - \mu_X)^2 + 2E[(\mu_X - c)(X - \mu_X)] + E(\mu_X - c)^2 \\ &= \sigma^2_X + 2(\mu_X - c)E(X - \mu_X) + (\mu_X - c)^2.\end{aligned}$$

Because $E(X - \mu_X) = 0$, the asserted result (5.5) follows.

6 Higher Moments

The mean μ_X of the random variable X is sometimes referred to as the **first moment** of the distribution of X. This terminology comes from physics, where the center of gravity (mean) is the first moment of inertia of a collection of masses. A generalization of the moment concept is $E(X - c)^r$, $r = 1, 2, \ldots,$ which is referred to as the rth **moment about the point** c of the distribution of X. The rth moment about the mean μ_X of X,

$$\mu_X^{(r)} = E(X - \mu_X)^r, \qquad r = 1, 2, \ldots, \tag{6.1}$$

is called the rth **central moment** of X. Notice that the variance σ_X^2 of X is the second central moment; that is,

$$\sigma_X^2 = \mu_X^{(2)} = E(X - \mu_X)^2.$$

The third and fourth central moments, $\mu_X^{(3)}$ and $\mu_X^{(4)}$, are also often used as indices of the distribution of X.

The third central moment $\mu_X^{(3)}$ measures the symmetry of the distribution of X about its mean. If X is measured in feet, then σ_X will also be measured in feet, whereas $\mu_X^{(3)}$ and σ_X^3 are expressible in units of $(\text{feet})^3$. A measure of symmetry of a distribution whose magnitudes do not depend on the units of measurement of X is the **measure of skewness,**

$$\nu_1 = \frac{\mu_X^{(3)}}{\sigma_X^3}. \tag{6.2}$$

The index ν_1 is zero when the distribution of X is symmetric about the mean μ_X (for an example, see Exhibit 6.1). Negative values of ν_1 are usually found when the distribution of X is **skewed to the left** (or negatively skewed; see Exhibit 6.2), whereas ν_1 tends to be positive when the distribution of X is **skewed to the right** (positively skewed; see Exhibit 6.3). Skewness to the left or right implies a long tail on the left or right, respectively.

The fourth central moment, $\mu_X^{(4)} = E(X - \mu_X)^4$, gives an indication of the "peakedness" or "kurtosis" of a distribution. As in the case of the measure ν_1 of symmetry, a dimension-free **measure of kurtosis** is

$$\nu_2 = \frac{\mu_X^{(4)}}{\sigma_X^4}. \tag{6.3}$$

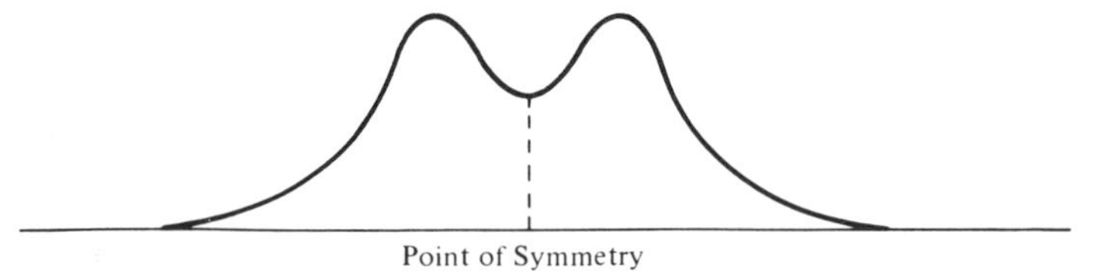

EXHIBIT 6.1 *Graph of a symmetric probability density function.*

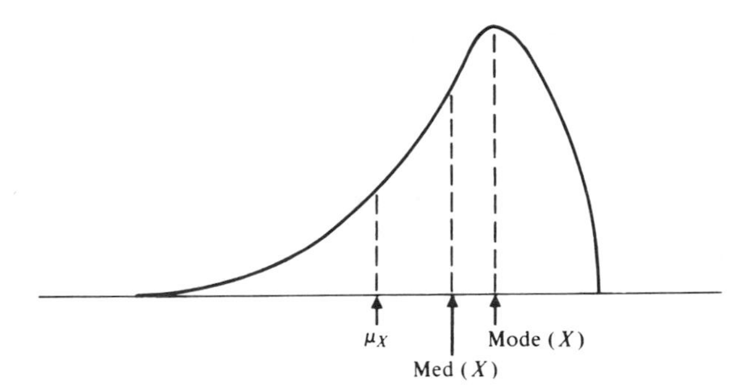

EXHIBIT 6.2 *Graph of a probability density function skewed to the left.*

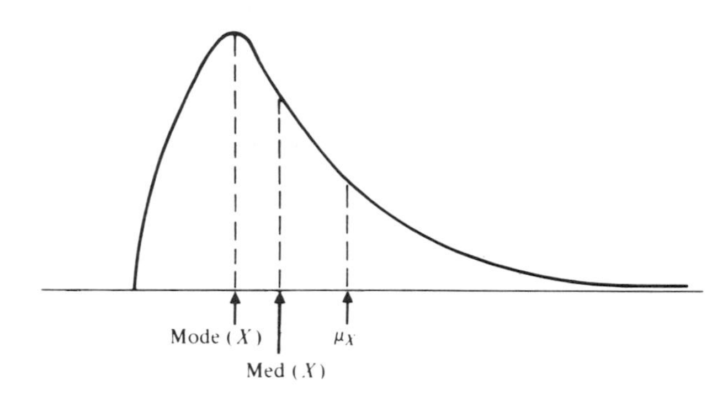

EXHIBIT 6.3 *Graph of a probability density function skewed to the right.*

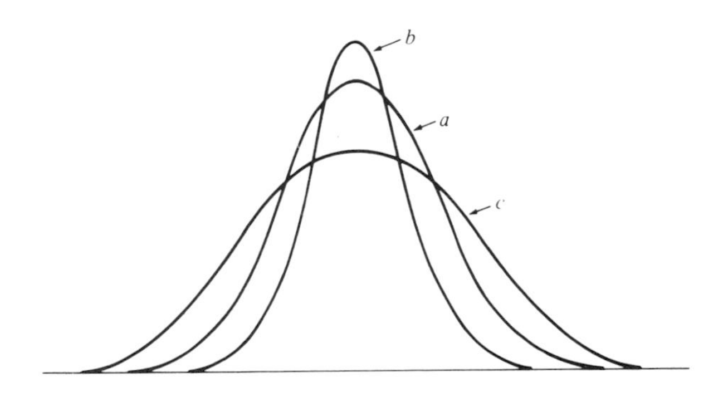

EXHIBIT 6.4 *Graph of the probability density functions of the normal distribution (a), a more peaked distribution (b), and a less peaked distribution (c).*

To determine the "peakedness" of a distribution using ν_2, the famous **bell-shaped normal distribution** (Chapter 9) is used as a standard. For a normal distribution, $\nu_2 = 3$. A distribution is "less peaked" (platykurtic) than the normal distribution if $\nu_2 < 3$, and is "more peaked" (leptokurtic) than the normal distribution if $\nu_2 > 3$ (see Exhibit 6.4).[2] An alternative interpretation of the index ν_2 is given in Exercise T.10. Knowledge of the location, dispersion, skewness, and peakedness of a distribution is usually enough to give a coarse, but useful picture of the distribution.

[2] The prefix *platy* means broad; the prefix *lepto* means slender or narrow.

7 Some Probability Inequalities

Various descriptive indices of distributions were discussed in previous sections. We now focus on determining how much information about a probability distribution can be obtained from knowledge of only a few such indices.

If only the mean $\mu_X = \mu$ of the distribution is given, we cannot in general make any nontrivial statements about the probability distribution of X. However, if it is known that X is nonnegative, then we can obtain information from the mean μ_X about the probabilities of certain events. The source of this information is the *Markov inequality*, named after the Russian mathematician A. A. Markov (1856–1922), who first obtained the result.

Markov Inequality. If X is a nonnegative random variable, and if $\mu_X = \mu$ is positive and finite, then for any positive constant c,

$$P\{X \geq c\mu\} \leq \frac{1}{c}.$$

Proof of the Markov Inequality*

An intuitive justification of the Markov inequality comes from considering the random variable Y formed from X by replacing all values of X greater than or equal to $c\mu$ by $c\mu$, and all values of X less than $c\mu$ by 0. That is,

$$Y = g(X) = \begin{cases} c\mu, & \text{if } X \geq c\mu, \\ 0, & \text{if } 0 \leq X < c\mu. \end{cases}$$

Because Y equals $c\mu$ with probability $P\{X \geq c\mu\}$ and equals 0 with probability $P\{X < c\mu\}$,

$$\mu_Y = (c\mu)P\{X \geq c\mu\} + (0)P\{X < c\mu\} = c\mu P\{X \geq c\mu\}.$$

By construction, Y is never greater than X; hence the same should be true of the means of Y and X:

$$\mu_Y = c\mu P\{X \geq c\mu\} \leq \mu_X = \mu.$$

It follows, dividing all sides of this last inequality by the positive number $c\mu$, that $P\{X \geq c\mu\} \leq 1/c$, as asserted by the Markov inequality.

EXHIBIT 7.1 *Largest Possible Values 1/c of the Probability $P\{X \geq c\mu\}$ in the Tail of the Distribution of a Nonnegative Random Variable X*

c	Largest Possible Value of $P\{X \geq c\mu\}$	c	Largest Possible Value of $P\{X \geq c\mu\}$
1.0	1.000	4.5	0.222
1.5	0.667	5.0	0.200
2.0	0.500	10.0	0.100
2.5	0.400	30.0	0.033
3.0	0.333	50.0	0.020
3.5	0.286	100.0	0.010
4.0	0.250		

Use of the Markov Inequality

The Markov inequality is uninformative if $1/c$ exceeds 1, because it is already known that $P\{X \geq c\mu\} \leq 1$. However, for $1/c$ less than 1, the information supplied by this inequality is useful. Exhibit 7.1 shows the largest possible probabilities of the event $\{X \geq c\mu\}$ for various values of c (see also Exhibit 7.2).

EXAMPLE 7.1

A car rental company advertises that the mean waiting time to pick up a car is $\mu = 5$ minutes. What is the probability that a customer will be delayed more than 60 minutes?

Noting that $60 = 12(5) = 12\mu$, the value of c is 12. Regardless of the probability distribution of the waiting time, the Markov inequality shows that the probability that the customer waits longer than 60 minutes for a car cannot be larger than $1/c = \frac{1}{12} = 0.083$. That is, on the average at least 92% of the customers will wait less than 60 minutes, and at most 8% will wait longer. ◆

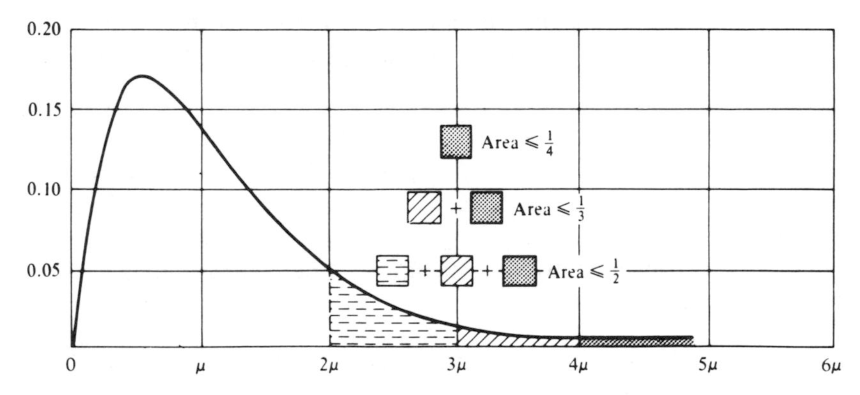

EXHIBIT 7.2 *Inequalities for areas based on the value of μ.*

The Bienaymé–Chebychev Inequality

The Markov inequality requires that X be nonnegative. However, not all random variables are nonnegative. If both the mean μ and variance σ^2 of a random variable X are known, the following inequality holds whether or not X is nonnegative.

> **Bienaymé–Chebychev Inequality.** For any number $d > 0$, and for any random variable X with mean μ and variance σ^2,
>
> $$P\{X \leq \mu - d\sigma \text{ or } X \geq \mu + d\sigma\} \leq \frac{1}{d^2}.$$

The Bienaymé–Chebyshev inequality is named after the French mathematician I. J. Bienaymé (1796–1878) and the Russian mathematician P. L. Chebychev (Čebyšev) (1821–1894), who discovered the result independently. Note that $Y = (X - \mu)^2$ is a nonnegative random variable with mean

$$\mu_Y = E(X - \mu)^2 = \sigma^2,$$

and that $Y \geq d^2\sigma^2$ if and only if $(X - \mu)^2 \geq d^2\sigma^2$, which in turn is true if and only if $X \leq \mu - d\sigma$ or $X \geq \mu + d\sigma$. Thus, applying the Markov inequality to Y, with $c = d^2$,

$$P\{X \leq \mu - d\sigma \text{ or } X \geq \mu + d\sigma\} = P\{Y \geq d^2\sigma^2\} \leq \frac{1}{d^2},$$

which establishes the desired inequality.

The event $\{\mu - d\sigma < X < \mu d\sigma\}$ is the complement of the event $E = \{X \leq \mu - d\sigma \text{ or } X \geq \mu + d\sigma\}$. Because we know that $P(E) \leq 1/d^2$, it follows that

$$P(E^c) = 1 - P(E) \geq 1 - \frac{1}{d^2}.$$

Consequently, the Bienaymé–Chebychev inequality implies that

$$P\{\mu - d\sigma < X < \mu + d\sigma\} \geq 1 - \frac{1}{d^2}. \quad (7.1)$$

Note that it is also true that

$$P\{\mu - d\sigma \le X \le \mu + d\sigma\} \ge 1 - \frac{1}{d^2}$$

because adding the values $x = \mu - d\sigma$, $x = \mu + d\sigma$ to the event $\{\mu - d\sigma < X < \mu + d\sigma\}$ can only increase the probability.

Use of the Bienaymé–Chebychev Inequality

Only when $d > 1$ do we gain useful information about probabilities of events of the form $\{X \le \mu - d\sigma \text{ or } X \ge \mu + d\sigma\}$ or their complements from the Bienaymé–Chebychev inequality. Exhibit 7.3 gives values for the largest possible probability of the event $\{X \le \mu - d\sigma \textit{ or } X \ge \mu + d\sigma\}$ and the smallest possible probability of the event $\{\mu - d\sigma < X < \mu + d\sigma\}$ for various values of $d \ge 1$. Note that the probability is nearly 1 (bounded below by 0.9996) that X will be within $d = 50$ standard deviations σ of its mean μ. Indeed, regardless of the distribution of X, the event $\{\mu - 4.5\sigma < X < \mu + 4.5\sigma\}$ always has probability at least $1 - (1/4.5)^2 = 0.951$. Exhibit 7.4 illustrates the inequality (7.1) for $d = 1, 2, 3$ and X a continuous random variable.

EXHIBIT 7.3 *Largest Possible Value (Upper Bound) U for* $P\{X \le \mu - d\sigma \text{ or } X \ge \mu + d\sigma\}$ *and Smallest Possible Value (Lower Bound) L for* $P\{\mu - d\sigma < X < \mu + d\sigma\}$

d	1.0	1.5	2.0	3.0	3.5	4.0	4.5	5.0	10.0	30	50	100
U	1.000	0.444	0.250	0.111	0.082	0.063	0.049	0.040	0.010	0.0011	0.0004	0.0001
L	0.000	0.556	0.750	0.889	0.918	0.937	0.951	0.960	0.990	0.9989	0.9996	0.9999

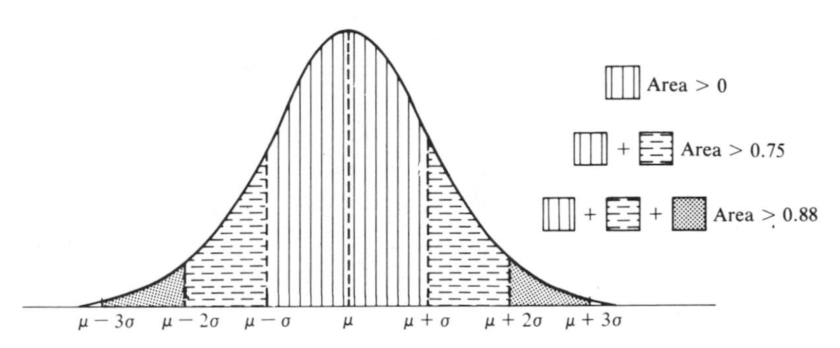

EXHIBIT 7.4 *Inequalities for areas based on the values of* μ *and* σ.

EXAMPLE 7.2

The score X on a multiple-choice reading comprehension test (in which incorrect answers receive negative scores) has mean $\mu_X = 0$ and variance $\sigma^2 = 100$. What can we say about a student whose score is 30? Using the Bienaymé–Chebychev inequality,

$$\begin{aligned} P\{X \geq 30\} &\leq P\{X \leq -30 \text{ or } X \geq 30\} \\ &= P\{X \leq \mu - 3\sigma \text{ or } X \geq \mu + 3\sigma\} \\ &\leq \frac{1}{(3)^2} = 0.111. \end{aligned}$$

Thus the student has a reading ability that is exceeded by at most 11.1% of those taking the test. ◆

Other Probability Inequalities

There are other inequalities that make use of the values of the mean and variance of a random variable X. As an example, when X is a nonnegative random variable, with mean μ and variance σ^2, then for any $b > 0$,

$$P\{X \geq \mu + b\sigma\} \leq \frac{1}{1 + b^2}. \tag{7.2}$$

This inequality generally gives smaller bounds (more precise bounds) for probabilities of events of the form $\{X \geq a\}$ than does the Markov inequality, but requires more information (namely, the value of the variance σ^2 of X).

Notes: Estimation of the Mean and Variance

The mean μ_X and variance σ_X^2 are indices of the probability distribution of X that are frequently useful for describing this distribution. In situations where the mean and variance of X are unknown, but observations on X are available, we can estimate these indices.

Suppose that N probabilistically independent observations $t_1, t_2, \ldots, t_N$ have been obtained on X. As remarked in Section 2, μ_X should be approximately equal to the average

$$\bar{t} = \frac{\sum_{i=1}^{N} t_i}{N} = \sum_t [t \times \text{r.f.}\{X = t\}]$$

of the observations $t_1, t_2, \ldots, t_N$ when N is large. This suggests estimating μ_X by $\hat{\mu}_X = \bar{t}$.

Remark. A notational convention in the statistical literature, which we have adopted here, is to denote an estimate of an unknown quantity, say θ, by the symbol $\hat{\theta}$. Thus $\hat{\mu}_X$ is an estimate of μ_X.

The variance $\sigma_X^2 = E(X - \mu_X)^2$ of X will be approximately equal to the average $(1/N) \sum_{i=1}^N (t_i - \mu_X)^2$ of the squared distances of the observations from the mean μ_X when N is large. Because $\hat{\mu}_X$ should be close to μ_X, this suggests estimating σ_X^2 by

$$\hat{\sigma}_X^2 = \frac{1}{N} \sum_{i=1}^{N} (t_i - \hat{\mu}_X)^2 = \sum_t (t - \hat{\mu}_X)^2 \text{ r.f.}\{X = t\}.$$

A convenient shortcut formula for calculating $\hat{\sigma}_X^2$, analogous to Equation (5.6), is

$$\hat{\sigma}_X^2 = \frac{1}{N} \sum_{i=1}^{N} (t_i^2 - (\hat{\mu}_X)^2 = \sum_t t^2 \text{ r.f.}\{X = t\} - (\hat{\mu}_X)^2.$$

EXAMPLE 1

Suppose that a sample of $N = 10$ people are asked to give a rank X of desirability for a new brand of deodorant on a scale of 1 to 10. If the observations on X are: 2, 1, 1, 2, 3, 4, 5, 9, 1, 1, we obtain

$$\hat{\mu}_X = \frac{2 + 1 + 1 + 2 + 3 + 4 + 5 + 9 + 1 + 1}{10} = 2.9,$$

$$\hat{\sigma}_X^2 = \frac{(2)^2 + (1)^2 + (1)^2 + (2)^2 + (3)^2 + (4)^2 + (5)^2 + (9)^2 + (1)^2 + (1)^2}{10} - (2.9)^2 = 5.89.$$

Alternatively, we might have constructed an empirical probability mass function from these observations (Exhibit 1). In this case,

$$\begin{aligned}\hat{\mu}_X = \sum_t t \text{ r.f.}\{X = t\} &= (1)(\tfrac{4}{10}) + (2)(\tfrac{1}{10}) + (3)(\tfrac{1}{10}) + (4)(\tfrac{1}{10}) \\ &\quad + (5)(\tfrac{1}{10}) + (9)(\tfrac{1}{10}) \\ &= 2.9,\end{aligned}$$

$$\begin{aligned}\hat{\sigma}_X^2 &= \sum_t t^2 \text{ r.f.}\{X = t\} - \hat{\mu}_X^2 \\ &= [(1)^2(\tfrac{4}{10}) + (2)^2(\tfrac{2}{10}) + (3)^2(\tfrac{1}{10}) + (4)^2(\tfrac{1}{10}) + (5)^2(\tfrac{1}{10}) + (9)^2(\tfrac{1}{10})] \\ &\quad - (2.9)^2 \\ &= 5.89.\end{aligned}$$

◆

EXHIBIT 1 *Empirical Probability Mass Function Based on the Observations $t_1 = 2$, $t_2 = 4$, etc.*

t	1	2	3	4	5	9
r.f.$\{X = t\}$	$\frac{4}{10}$	$\frac{2}{10}$	$\frac{1}{10}$	$\frac{1}{10}$	$\frac{1}{10}$	$\frac{1}{10}$

The individual observations of X are called the **raw data**. In the preceding example, X was a discrete random variable, and Exhibit 1 provided the same information as the original raw data for purposes of estimating the mean and variance of X. The result of taking N observations upon a continuous random variable X may, however, be reported to us in the form of Exhibit 2, where the data have been grouped in k intervals of equal length. (For example, such grouping is done when constructing modified relative frequency histograms from observations.) In this case the calculations for $\hat{\mu}_X$ and $\hat{\sigma}_X^2$ are carried out as if all observations in an interval were placed at the midpoint of that interval. That is,

$$\hat{\mu}_X = \sum_{i=1}^{k} (\text{midpoint of interval } i)(\text{relative frequency of interval } i),$$

$$\hat{\sigma}_X^2 = \sum_{i=1}^{k} (\text{midpoint})^2(\text{relative frequency of interval } i) - (\hat{\mu}_X)^2.$$

For example, the computations for $\hat{\mu}_X$ and $\hat{\sigma}_X^2$ based on the observations in Exhibit 2 are

$$\begin{aligned}\hat{\mu}_X &= (1.0)(\tfrac{1}{40}) + (2.0)(\tfrac{2}{40}) + (3.0)(\tfrac{4}{40}) + (4.0)(\tfrac{7}{40}) + (5.0)(\tfrac{12}{40}) \\ &\quad + (6.0)(\tfrac{8}{40}) + (7.0)(\tfrac{5}{40}) + (8.0)(\tfrac{1}{40}) \\ &= \tfrac{196}{40} = 4.90\end{aligned}$$

and

$$\hat{\sigma}_X^2 = (1.0)^2(\tfrac{1}{40}) + (2.0)^2(\tfrac{2}{40}) + \cdots + (8.0)^2(\tfrac{1}{40}) - (4.90)^2 = 2.34.$$

Note that the rule for calculating $\hat{\mu}_X$ from data grouped in intervals is almost certain to produce a different value for $\hat{\mu}_X$ than would have been obtained if $\hat{\mu}_X$ had been calculated from the raw data. (It is unlikely that every observation in an interval will equal the midpoint of that interval.) However, the error made by assuming that each observation lies at the midpoint of its interval is never greater than half the common length L of the intervals. It follows that the discrepancy between $\hat{\mu}_X$, as calculated from the raw data, and $\hat{\mu}_X$ as calculated from the grouped data *can never be greater than L/2*. Hence the smaller is the common length L of each interval used to construct the relative frequency histogram, the smaller is the discrepancy between the two estimates of μ_X.

EXHIBIT 2 *Relative Frequency Distribution Based on $N = 40$ Observations of the Random Variable X*

Midpoint, x, of Interval of Values	*Observed Relative Frequency of the Event* $\{x - 0.5 \leq X < x + 0.5\}$
1.0	$\frac{1}{40}$
2.0	$\frac{2}{40}$
3.0	$\frac{4}{40}$
4.0	$\frac{7}{40}$
5.0	$\frac{12}{40}$
6.0	$\frac{8}{40}$
7.0	$\frac{5}{40}$
8.0	$\frac{1}{40}$

Glossary and Summary

Distribution of a random variable: A rule (function) specifying how probabilities are distributed over the values of the random variable. Examples are probability mass functions, probability density functions, and cumulative distribution functions.

Measure of location: An index of a distribution that specifies where on the horizontal axis the graph or profile of the distribution is located (*synonym:* measure of central tendency). This measure changes by an amount d whenever the graph of the distribution moves d units along the horizontal axis.

Measure of dispersion: An index of a distribution that measures how widely spread the graph of the distribution is about a specified point c on the horizontal axis (*synonyms:* measure of variation, measure of spread). The larger the value of the index, the larger the spread.

Mean: The center of gravity, balance point, or first moment of inertia of a distribution (viewing probability as weight) (*synonyms:* expected value, expectation, population average). For a discrete random variable X with probability mass function $p(x)$, the mean μ_X of X is defined by

$$\mu_X = \sum_x x\, p(x),$$

whereas if X is a continuous random variable with probability density function $f(x)$, then

$$\mu_X = \int_{-\infty}^{\infty} xf(x)\,dx.$$

pth Quantile: For a continuous random variable X, a number $Q_X(p)$ on the horizontal axis that divides the total area (probability) under the graph of the probability density function $f(x)$ of X into a proportion p to the left and $1 - p$ to the right. That is,

$$\int_{-\infty}^{Q_X(p)} f(x)dx = p, \qquad \int_{Q_X(p)}^{\infty} f(x)\,dx = 1 - p.$$

This number can most easily be found by solving the equality

$$F(x) = p$$

for x, where $F(x)$ is the cumulative distribution function of x. (If more than one solution exists, take the midpoint of the interval of solutions.) For a discrete random variable X with probability mass function $p(x)$, the definition of the pth quantile has to take account of the clumping of probability at values of X, so that $Q_X(p)$ satisfies

$$\sum_{x<Q_X(p)} p(x) < p \leq \sum_{x\leq Q_X(p)} p(x).$$

That is, $Q_X(p)$ is the x-value at which the cumulative probability "jumps through" p. Consequently, $Q_X(p)$ does not necessarily divide the probability distribution into a proportion p to the left and $1 - p$ to the right [because there may be some probability mass at $x = Q_X(p)$].

100pth Percentile: A synonym for the pth quantile, $Q_X(p)$.

Median: The 0.50th quantile or 50th percentile, $\text{Med}(X) = Q_X(0.50)$. The median, Med($X$) of a continuous random variable X, divides the total area under the probability density function of X into two equal parts. In this sense, Med (X) is the middle or median value of X. When X is discrete, some probability mass may fall on $x = \text{Med}(X)$, so that the probabilities are not divided into two equal parts. Instead,

$$\sum_{x<\text{Med}(X)} p(x) < \tfrac{1}{2} \leq \sum_{x\leq\text{Med}(X)} p(x).$$

Quartiles: The first quartile of X is $Q_X(0.25)$, the second quartile of X is Med(X), and the third quartile of X is $Q_X(0.75)$. When X is a continuous random variable, the quartiles break up the distribution of probability into four equal parts: $\frac{1}{4}$ to the left of the first quartile, $\frac{1}{4}$ between the first quartile and the median, $\frac{1}{4}$ between the median and the third quartile, and $\frac{1}{4}$ to the right of the third quartile.

Mode: A mode of a discrete random variable X is the most probable value of X; that is, the value x where the probability mass function $p(x)$ of X has its largest value. For a continuous random variable X, the mode is the value x where the probability density function $f(x)$ of x has its largest value. There can be one, or more than one, mode of a distribution.

Relative mode: A value of x at which the probability mass function $p(x)$ or probability density function $f(x)$ is locally the highest (has a ''peak''). A distribution function (mass function, density function) with one relative mode (which is then the mode of the distribution) is called **unimodal**. If a distribution has two separated relative modes, it is called **bimodal,** and the mode of the distribution is the relative mode at which the mass function or density function has the largest value.

Expected value of a function: If $g(x)$ is a function defined over the values x of a random variable X, the expected value of $g(X)$ is defined to be

$$\begin{aligned} E[g(x)] &= \sum_x g(x)p(x), && \text{if } X \text{ is discrete,} \\ &= \int_{-\infty}^{\infty} g(x)f(x)\,dx, && \text{if } X \text{ is continuous.} \end{aligned}$$

The expected value of $g(X)$ provides an easy way to calculate the mean, μ_Y, of the transformed random variable $Y = g(X)$. In particular, the mean of X is given by the computation

$$\mu_X = E[X].$$

Intuitively, $E[g(X)]$ is the value that we expect the average $\sum_{i=1}^{n} g(x_i)/n$ of a large number of observations of $g(X_i)$ to be.

Variance: The variance σ_X^2 is the expected value of the squared deviation $(X - \mu_X)^2$ of a random variable X from its mean μ_X; that is,

$$\sigma_X^2 = E[(X - \mu_X)^2] \begin{cases} \sum_x (x - \mu_X)^2 p(x), & \text{if } X \text{ is discrete,} \\ \int_{-\infty}^{\infty} (x - \mu_X)^2 f(x)\,dx, & \text{if } X \text{ is continuous,} \end{cases}$$

Large values of σ_X^2 imply that X is widely dispersed about its mean; small values imply small dispersion.

Standard deviation: The standard deviation σ_X is the square root of the variance σ_X^2 of X. That is,

$$\sigma_X = \sqrt{\sigma_X^2}.$$

An advantage of the standard deviation over the variance is that the standard deviation is expressed in the same units of measurement (e.g., feet) as is X, whereas the variance is expressed in squared units (e.g., squared feet).

Interquartile range: The interquartile range IQR of a random variable X is defined to be the distance between the first and third quartiles of X; that is,

$$\text{IQR} = Q_X(0.75) - Q_X(0.25).$$

The IQR serves as a measure of dispersion for X in that it gives the length of an interval that contains at least 50% of the probability of the distribution of X.

Moments about a point c: $E[(X - c)^r]$ is the rth moment of X about the point c, $r = 1, 2, \ldots$.

Central moments: $\mu_X^{(r)} = E[(X - \mu_X)^r]$ is the rth central moment of X, $r = 1, 2, \ldots$. The first central moment is 0; the second central moment, $\mu_X^{(2)}$, equals the variance σ_X^2; the third central moment, $\mu_X^{(3)} = E[(X - \mu_X)^3]$, measures the symmetry of X about its mean through the measure of skewness; and the fourth central moment of X, $\mu_X^{(4)} = E[(X - \mu_X)^4]$, measures "peakedness" through the measure of kurtosis.

Measure of skewness: $\nu_1 = \mu_X^{(3)}/\sigma_X^3$. Negative values of ν_1 are usually found when the distribution of X has a long "tail" to the left (**skewed to the left, negatively skewed;** see Exhibit 6.2), whereas ν_1 tends to be positive when the distribution of X has a long tail to the right (**skewed to the right, positively skewed;** see Exhibit 6.3). When the distribution of X is symmetric about its mean, $\nu_1 = 0$.

Measure of kurtosis: $\nu_2 = \mu_X^{(4)}/\sigma_X^4$. The value of this index is 3 for the famous **normal distribution** (Chapter 9). If $\nu_2 < 3$, the distribution of X is said to be **platykurtic** and is often "less peaked" than the normal distribution. If $\nu_2 > 3$, the distribution of X is said to be **leptokurtic** and is often "more peaked" than the normal distribution. However, alternative interpretations of this index are possible (Exercise T.10).

Raw data: The individual data values obtained from repeated observations on a random variable X.

Grouped data: Data grouped in intervals and summarized by giving the relative frequency of each such interval.

Sample mean and variance: The sample mean of the raw data $t_1, t_2, \ldots, t_n$ from observing a random variable X is

$$\hat{\mu}_X = \bar{t} = \frac{1}{n}\sum_{i=1}^{n} t_i$$

and serves as an estimate of the (population) mean μ_X of X. The sample mean is also called the (sample) **average**. If the data are grouped, the formula for the sample mean is

$$\hat{\mu}_X = \sum_{i=1}^{k} (\text{midpoint of interval } i)\ (\text{relative frequency of interval } i).$$

The sample variance $\hat{\sigma}_x^2$ is given by

$$\hat{\sigma}_x^2 = \frac{1}{n} \sum_{i=1}^{n} (t_i - \hat{\mu}_x)^2$$

or

$$\hat{\sigma}_X^2 = \sum_{i=1}^{k} (\text{midpoint of interval } i - \hat{\mu}_x)^2\ (\text{relative frequency of interval } i),$$

depending on whether the data appear in raw or grouped form. The sample variance can be used as an estimate of the population variance, σ_X^2, of X. (However, in statistical practice, use of the estimate $s^2 = \frac{n}{n-1} \hat{\sigma}_X^2$ is preferred, for reasons given in Chapter 14.)

Useful Facts

1. The following indices are measures of location for the distribution of a random variable X:
   (a) The mean μ_X.
   (b) The median Med(X).
   (c) Every pth quantile $Q_X(p)$, including the first and third quartiles.
   (d) The mode of X, provided that it is unique (the distribution is unimodal).

   Each of these indices satisfies the defining property of a measure of location $M(X)$, namely that $M(X + d) = M(X) + d$. Thus

$$\mu_{X+d} = \mu_X + d, \quad \text{Med}(X + d) = \text{Med}(X) + d, \quad Q_{X+d}(p) = Q_X(p) + d.$$

2. For any two numbers a and c,

$$\mu_{aX+c} = a\mu_X + c, \qquad \text{Med}(aX + c) = a\,\text{Med}(X) + c,$$
$$\sigma_{aX+c}^2 = a^2\sigma_X^2, \qquad \sigma_{ax+c} = |a|\sigma_X.$$

Also, provided that $a > 0$.

$$Q_{aX+c}(p) = aQ_X(p) + c.$$

3. It is often easier to compute the variance by the computation formula:

$$\sigma_X^2 = E[X^2] - (\mu_X)^2.$$

Similarly, computational formulas for the sample variance are

$$\hat{\sigma}_X^2 = \frac{1}{n}\sum t_i^2 - (\hat{\mu}_X)^2,$$

$$\hat{\sigma}_X^2 = \sum_{i=1}^{k} (\text{midpoint of interval } i)^2(\text{relative frequency of interval } i) - (\hat{\mu}_X)^2,$$

depending on whether the data are given in raw or grouped form, respectively.

4. The useful formula $E[X - c]^2 = \sigma_X^2 + (\mu_X - c)^2$ shows that choosing $c = \mu_X$ minimizes the second moment of X about c.
5. The expected value of a function $g(X)$ satisfies the following properties:
(a) **Linearity.** For any two functions $g(X)$, $h(X)$ and any numbers a, b,

$$E[ag(X) + bh(X)] = aE[g(x)] + bE[h(X)].$$

(b) **Order preserving.** If $P\{g(X) \geq 0\} = 1$, then $E[g(X)] \geq 0$.
In general, $E[g(X)] \neq g(E(X))$, although equality does hold if $g(x)$ is linear, that is, if $g(x) = ax + c$ (see Exercise T.3).
6. The mean μ_X and standard deviation σ_X of a random variable X yield probabilistic information about the probability distribution of X through the following inequalities:
(a) **Markov inequality.** If $P\{X \geq 0\} = 1$ and $\mu_X = \mu$, where $\mu > 0$, then for any positive number c,

$$P\{X \geq c\mu\} \leq \frac{1}{c}.$$

(b) **Bienaymé–Chebychev.** For any random variable X with mean $\mu_X = \mu$ and standard deviation $\sigma_X = \sigma$, and for any positive number d,

$$P\{X \leq \mu - d\sigma \text{ or } X \geq \mu + d\sigma\} \leq \frac{1}{d^2}$$

and

$$P\{\mu - d\sigma \le X \le \mu + d\sigma\} \ge 1 - \frac{1}{d^2}.$$

7. If grouping intervals all have length L, the absolute difference between the sample mean calculated from such grouped data and the sample mean calculated from the raw (ungrouped) data is never greater than $\frac{1}{2}L$.

Exercises

1. Let X have the following probability model:

x	0	1	2
Probability	$\frac{1}{3}$	$\frac{1}{2}$	$\frac{1}{6}$

(a) Find the mean, median, and mode of X. Which of these measures of location are possible values of X?
(b) Find the variance of X using equations (5.2) and (5.6); the results obtained by the two formulas should be the same.

2. A toothbrush manufacturer has introduced a new toothpaste. They publish the following probability model for the number X of new cavities found in the mouth of a child who uses their toothpaste for six months:

$$p(x) = \begin{cases} 0.07, & \text{if } x = 0, \\ 0.40, & \text{if } x = 1, \\ 0.36, & \text{if } x = 2, \\ 0.15, & \text{if } x = 3, \\ 0.01, & \text{if } x = 4, \text{ or } x = 5, \\ 0, & \text{otherwise.} \end{cases}$$

(a) What is the *mean* number of new cavities per child?
(b) What is the *median* number of new cavities per child?
(c) Suppose that you want to predict the number of cavities that will be found in the mouth of a child after six months of use of the new toothpaste, and want the highest probability possible of being exactly right. What measure of location for the distribution of X gives you that prediction? What number of cavities should you predict?

3. An integer is chosen at random from among the integers 1, 2, 3, 4, 5, 6, 7, 8, 9, and 10. Let X be the number of different positive integers that divide the chosen integer without remainder. (For example, if the integer 6 is chosen, then $X = 4$ since 6 can be divided without remainder by 1, 2, 3, and 6.)

(a) The possible values of X are 1, 2, 3, and 4. Find the probability mass function $p(x)$ of X.

(b) Find the mean, median, and mode of the distribution of X.

(c) If you were to play the "Guess the Number of Divisors Game," in which your opponent chooses an integer at random from among the integers 1, 2, . . . , 9, 10, and you have to guess the number X of divisors that this integer has, which value would you guess for X? What is the probability that your guess would be exactly correct?

(d) Find the variance of X.

4. An insurance company wishes to establish the premium C for selling a 1-year term insurance policy of \$1000 to a person of age 45. Suppose that the probability that a 45-year-old person will die during the next year is 0.00363. Let the random variable X denote the gain (or loss) to the insurance company as a result of selling a 1-year term insurance policy of \$1000 to a 45-year-old person. Thus the probability model for X is

x	C	$C - 1000$
Probability	0.99637	0.00363

(a) Determine C so that the company's expected gain is equal to 0.

(b) What premiums allow the company to make a profit "on the average"?

(c) The company sets a premium of \$5 for every \$1000 of term insurance and sells 100 \$1000 term insurance policies to 45-year-old people; this yields the company \$500 in premiums. Approximately what total net profit (or loss) will they have earned at the end of the year?

5. A special government loan program awards emergency loans of \$2000 to students. Past records show that 60% of all students repay the loan completely. Of those students who fail to repay all of their loans, 40% pay back $\frac{3}{4}$ of the loan, 30% pay back $\frac{1}{2}$ the loan, 20% pay back $\frac{1}{4}$ of the loan, and 10% fail to repay any of the loan.

(a) What is the mean loss (default) per loan for the loans under this special government program?

(b) A university has made loans to 1000 students under this program.

Approximately what *total* amount of money can they expect to have defaulted on these loans?

6. A gambler plays a sequence of games that he either wins (W) or loses (L). The outcomes of the games are mutually statistically independent, and the probability that the gambler wins any game is $\frac{2}{3}$. The gambler stops playing as soon as he either has won a total of 2 games or has lost a total of 3 games. Let T be the number of games played by the gambler.

(a) Find the probability mass function of T. (*Hint:* List all possible outcomes, the probabilities of these outcomes, and the value of T corresponding to each outcome. What are the possible values of T?)

(b) Find the mean, μ_T, and variance σ_T^2, of T.

(c) Find the median and mode of T.

7. Let X be the yearly income (in thousands of dollars) of a person who earns more than K thousand dollars and who pays some federal income tax. Although X is actually discrete, it is more convenient in economic applications to treat X as if it were a continuous random variable. The density function $f(x)$ of the random variable X is assumed to be of the form

$$f(x) = \begin{cases} \dfrac{aK^a}{x^{a+1}}, & \text{for } x \geq K, \\ 0, & \text{for } x < K. \end{cases}$$

Suppose that $K = 2$ and $a = 3$.

(a) Graph the probability density function $f(x)$. Find the mode of X.

(b) Find the mean of X.

(c) Find the cumulative distribution function $F(x)$ of X. Then find the median Med(X) of X, $Q_X(0.25)$, and $Q_X(0.75)$.

(d) What measure of location would you use to describe yearly income if you were interested in predicting tax revenues? What measure of location would you use if you wished to describe the economic well-being of the typical American who earns more than $2000 a year?

(e) Find the variance of X. [*Note:* The distribution of X is a special case of a *Pareto distribution*. This application of Pareto distributions is discussed by Hagstroem (1960) and Arnold (1983, Chapter 2).]

8. Let X be a continuous random variable with a probability density function given by

$$f(x) = \begin{cases} \frac{3}{8}(2 - x)^2, & \text{for } 0 \leq x \leq 2, \\ 0, & \text{for } x < 0 \text{ or } x > 2. \end{cases}$$

(a) Find the mean μ_X of X.
(b) Find the variance σ_X^2 of X.
(c) Show that the cumulative distribution function of X is given by

$$F(x) = \begin{cases} 0, & \text{if } x < 0, \\ 1 - \dfrac{(2-x)^3}{8}, & \text{if } 0 \le x \le 2, \\ 1, & \text{if } 2 < x. \end{cases}$$

Find the median of X and $Q_X(0.90)$.

9. Suppose that X is a *discrete* random variable with cumulative distribution function

$$F(x) = \begin{cases} 0, & x < 1, \\ 0.2, & 1 \le x < 2, \\ 0.6, & 2 \le x < 3, \\ 0.8, & 3 \le x < 4, \\ 0.9, & 4 < x < 5, \\ 1.0, & 4 < x. \end{cases}$$

(a) Find the median, Med(X), of X.
(b) Find $Q_X(0.90)$.
(c) Find the mean, μ_X, of X.
(*Hint*: What are the possible values of X and the corresponding probabilities?)
(d) Find the mode of X.
(e) Find the variance of X.

10. A continuous random variable X has the following cumulative distribution function:

$$F(x) = P\{X \le x\} = \begin{cases} 0, & \text{if } x < 1, \\ 1 - x^{-4}, & \text{if } x \ge 1. \end{cases}$$

(a) Find the median, Med(X), of X.
(b) Find $Q_X(0.10)$.
(c) Find the mean, μ_X, of X. [*Hint:* You first need to find the probability density function $f(x)$ of X.]
(d) Find the variance, σ_X^2, of X.

11. Let X have probability mass function

$$p(x) = \begin{cases} \frac{1}{8}, & \text{if } x = -2, -1, 1, 2, \\ \frac{1}{4}, & \text{if } x = 0, 3, \\ 0, & \text{otherwise.} \end{cases}$$

(a) Find the mean μ_X, of X.

(b) Find the probability mass function $p_Y(y)$ of $Y = (X - 1)^2$ and then find the mean, μ_Y, of Y.

(c) Find $E(X - 1)^2$ directly from the mass function $p(x)$ of X. Show that your answer is equal to the mean, μ_Y, of $Y = (X - 1)^2$ obtained in part (b).

12. The profit (in millions of dollars) from a capital investment is a continuous random variable X with probability density function

$$f(x) = \begin{cases} 2x^{-2}, & \text{if } 1 \le x < 2, \\ 0, & \text{otherwise.} \end{cases}$$

(a) Find the mean, μ_X, of X.

(b) Find the median of X.

(c) Find the variance of X.

(d) The firm owes their source of capital \$200,000 plus 10% of the profits X, so that after repaying their source of capital, the firm retains a profit

$$Y = (0.9)X - 0.2$$

from the investment. Find the mean and variance of Y directly from the mean and variance of X.

13. The probability of finding oil in any hole is 0.1, and the result at one hole is independent of what happens at any other hole. A drilling company drills one hole at a time. If they hit oil, they stop drilling. However, the company has only enough capital to drill 5 dry holes.

(a) Let X be the number of holes drilled. What is the mean of X? (What are the possible values of X and their associated probabilities?)

(b) Find the median of X. Find $Q_{0.25}(X)$.

(c) The drilling company has borrowed money (\$100,000) to buy equipment at a rate of 30% per drilling period, and cannot afford to pay any money back until they hit oil. Thus the amount owed (in thousands of dollars) at stopping is

$$Y = 100(1.30)^X.$$

Find the mean of Y. [Note that $\mu_Y = E(100(1.30)^X)$ and not $100(1.30)^{E(X)}$.]

14. An eccentric artist paints only square paintings. Each new painting has a random width (in feet), W, which is a continuous random variable with probability density function

$$f_W(w) = \begin{cases} \frac{1}{2}, & \text{if } 2 \le w \le 4, \\ 0, & \text{otherwise.} \end{cases}$$

It costs the artist \$5 per square foot in materials to create a new painting. He sells his paintings for \$50, regardless of their size. Let Y be the artist's profit on the sale of a painting. Find the mean, median, and variance of Y.

15. The velocity X of an atomic particle of unit mass in a certain physical system is a continuous random variable with density function

$$f(x) = \begin{cases} \left(\frac{6}{5}\right)\left(1 - \frac{x^2}{16}\right), & \text{if } 2 \le x \le 4, \\ 0, & \text{otherwise.} \end{cases}$$

(a) Find the mean velocity, μ_X, of a particle in this physical system.
(b) Find a velocity v such that 80% of the particles in this system have velocity greater than v [i.e., $P(X > v) = 0.80$]. (This velocity v is a pth quantile of the distribution of X. What is p?)
(c) The kinetic energy Y of a particle of unit mass is $Y = \frac{1}{2}X^2$, where X is the velocity. Find the mean kinetic energy, $\mu_Y = E(\frac{1}{2}X^2)$, of a particle of unit mass in this system.

16. A random variable X has the probability density function

$$f(x) = \begin{cases} \frac{3}{16}x^2, & \text{if } -2 < x < 2, \\ 0, & \text{otherwise.} \end{cases}$$

Find (a) μ_X; (b) $\text{Med}(X)$; (c) $Q_X(0.90)$; (d) σ_X^2; (e) $E[X^3]$.

17. If $Y = 3X - 5$ is a transformation from the random variable X to the new random variable Y, and if $\mu_Y = 0$, $\sigma_Y^2 = 9$, find μ_X and σ_X^2.

18. If X is a random variable with $\mu_X = -1$, $\sigma_X^2 = 4$, find $a > 0$ and b so that $Y = aX + b$ has mean 0 and variance 1.

19. For each of the following distributions, find $\mu_X^{(3)}$, $\mu_X^{(4)}$, ν_1, and ν_2. Interpret your findings in terms of the skewness and kurtosis (peakedness) of the distribution, and compare your conclusions to a rough graph of the probability density function.

(a) $f(x) = \begin{cases} 2x, & \text{if } 0 \le x \le 1, \\ 0, & \text{if } x < 0 \text{ or } x > 1. \end{cases}$

(b) $f(x) = \begin{cases} 2(1 - x), & \text{if } 0 \le x \le 1, \\ 0, & \text{if } x < 0 \text{ or } x > 1. \end{cases}$

(c) $f(x) = \begin{cases} 1, & \text{if } 0 \le x \le 1, \\ 0, & \text{if } x < 0 \text{ or } x > 1. \end{cases}$

(d) $f(x) = \begin{cases} 4x, & \text{if } 0 \le x \le \frac{1}{2}, \\ 4(1 - x), & \text{if } \frac{1}{2} < x \le 1, \\ 0, & \text{otherwise.} \end{cases}$

20. Scores X on the verbal scale of the Stochastic Aptitude Test (SAT Verbal) are nonnegative and have a mean $\mu_X = 650$.

(a) Without making any other assumptions about the distribution of SAT Verbal scores X, find the largest possible values of the following probabilities: (i) $P\{X \geq 715\}$; (ii) $P\{X \geq 780\}$.

(b) Suppose that we also know that the standard deviation σ_X of X is 64. Use inequality (7.2) to determine the largest possible value of $P\{X \geq 780\}$. Do you believe the claim that 30% of all SAT scores exceed 780? Do you believe the claim that 80% of all SAT scores exceed 715?

(c) Use the Bienaymé–Chebychev inequalities to find the smallest values for $P\{650 - 64d < X < 650 + 64d\}$ for $d = 1.1, 1.5, 1.8, 2.0$, and 2.2, and the largest values for $P\{X \geq 650 + 64b\}$ for $b = 0.5$, 0.7, 1.0, 1.2, and 1.5.

21. Suppose that X is a discrete random variable with the following probability mass function:

x	$\mu - k\sigma$	μ	$\mu + k\sigma$
Probability	$\frac{1}{2k^2}$	$1 - \frac{1}{k_2}$	$\frac{1}{2k^2}$

where μ is any number, σ is any positive number, and $k \geq 1$.

(a) Show that the mean of X is μ.

(b) Show that the variance of X is σ^2.

(c) Show that $P\{X \geq \mu + k\sigma \text{ or } X \leq \mu - k\sigma\} = 1/k^2$.

22. Linguistic studies have been made of the syllabication of words by various authors. For example, let X be the number of syllables in a word chosen at random from the works of the English author and lexicographer Samuel Johnson. In 2000 such words, the following sample probability mass function was obtained for X:

Number of syllables, x	1	2	3	4	5	6
Observed relative frequency of the event $\{X = x\}$	$\frac{1268}{2000}$	$\frac{423}{2000}$	$\frac{195}{2000}$	$\frac{77}{2000}$	$\frac{29}{2000}$	$\frac{8}{2000}$

(a) Find the sample mean $\hat{\mu}_X$ and sample variance $\hat{\sigma}_X^2$ of X.

(b) The number Y of syllables used in a random word taken from the works of an American author, A, had a population mean $\mu_Y = 2.1$ and a population variance $\sigma_Y^2 = 0.75$. Assume that the sample mean and variance of X found in part (a) equal the population mean

μ_X and variance σ_X^2 of X, respectively. Which author, Samuel Johnson or A, used more syllables on the average? Which author used a greater variety of word lengths (as measured in terms of the number of syllables)?

Theoretical Exercises

T.1. Suppose the continuous random variable X satisfies $P\{0 \le X \le a\} = 1$ for some number a, $0 < a < \infty$. (That is, X is a nonnegative and bounded random variable.) Let $F(x)$ be the cumulative distribution function of X. Use integration by parts to show that

$$\mu_X = \int_0^a [1 - F(x)]\, dx.$$

Apply this result to obtain μ_X in Exercise 8.

T.2. (a) Extend the result of T.1 to the case where $P\{0 \le X < \infty\} = 1$ and $E(X) < \infty$. That is, prove that

$$\mu_X = E(X) = \int_0^\infty [1 - F(x)]\, dx$$

by making use of integration by parts. What use do you make of the fact that $E(X) < \infty$?

(b) Apply the result of part (a) to obtain μ_X in Exercise 7.

(c) If X is any continuous random variable for which $E(|X|) < \infty$, show that

$$\mu_X = E(X) = -\int_{-\infty}^0 F(x)\, dx + \int_0^\infty [1 - F(x)]\, dx.$$

[*Hint*: Note that $E(X) = \int_{-\infty}^0 xf(x)\, dx + \int_0^\infty xf(x)\, dx$. Handle each integral separately using integration by parts.]

T.3. When $g(x) = ax + c$, $a > 0$, it has been shown that

$$\mu_{g(X)} = \mu_{aX+c} = a\mu_X + c = g(\mu_X).$$

However, the result $\mu_{g(X)} = g(\mu_X)$ is not generally true. To see this, let X be a continuous random variable with probability density function

$$f(x) = \begin{cases} 1, & \text{if } 0 \le x \le 1, \\ 0, & \text{if } x < 0 \text{ or } x > 1, \end{cases}$$

and let $g(x) = e^x$. Find $E[e^X]$ and show that it is not equal to $e^{\mu_X} = e^{1/2}$.

T.4. In contrast to what was shown in Exercise T.3, show that for any random variable X with a unique pth quantile $Q_X(p)$ and for any strictly increasing function $g(x)$, it is the case that

$$Q_{g(X)}(p) = g(Q_X(p)).$$

[*Hint:* Use the fact that $g(x)$ has an inverse function $h(y)$ that is nondecreasing, $h(g(x)) = x$, and apply the definition of $Q_{g(X)}(p)$.]

T.5. Let X be a random variable and let $g_1(x)$, and $g_2(x)$ be two functions of x for which $g_1(x) \geq g_2(x)$ for all x. Use the properties of the expected value to show that

$$E[g_1(X)] \geq E[g_2(X)].$$

T.6. Use the linearity property of expected values to show that:
(a) $E[X - \mu_X]^3 = E(X^3) - 3\sigma_X^2\mu_X - \mu_X^3$.
(b) $E[X - \mu_X]^4 = E(X^4) - 4\mu_X E(X - \mu_X)^3 - 6\mu_X^2 \sigma_X^2 - \mu_X^4$.

T.7. Let X be a nonnegative random variable (i.e., $P\{X \geq 0\} = 1$) with finite mean and variance $\mu_X = \mu$ and $\sigma_X^2 = \sigma^2$, respectively. Prove inequality (7.2), which states that $P\{X \geq \mu + b\sigma\} \leq (1 + b^2)^{-1}$ for any $b > 0$, using the following steps of proof.
(a) Let

$$I(x) = \begin{cases} 1, & \text{if } x \geq \mu + b\sigma, \\ 0, & \text{otherwise,} \end{cases}$$

and

$$g(x) = \frac{[(x - \mu)b + \sigma]^2}{\sigma^2(1 + b^2)^2}.$$

Show that $E[(X - \mu)b + \sigma]^2 = \sigma^2(b^2 + 1)$ and thus that

$$E[g(X)] = \frac{1}{1 + b^2}.$$

(b) Show that $g(X) - I(X) \geq 0$ if $X \geq 0$.
(c) Thus show that

$$0 \leq E[g(X) - I(X)] = \frac{1}{1 + b^2} - P\{X \geq \mu + b\sigma\},$$

which establishes inequality (7.2).

T.8. Suppose that X has probability mass function

$$p(x) = \begin{cases} \dfrac{6}{\pi i^2}, & \text{if } x = i^2, \quad i = 1, 2, \ldots, \\ 0, & \text{otherwise.} \end{cases}$$

Here the possible values of X are squares i^2 of positive integers. Show that

$$\mu_X = \sum_{i=1}^{\infty} x_i p(x_i) = \infty.$$

Hence μ_X is not well defined.

T.9. Suppose that X has density $f(x)$ and that $Z = aX + c$, $a > 0$.

(a) Show that $F_Z(z) = F_X\left(\dfrac{z - c}{a}\right)$.

(b) Use the chain rule to show that $f_Z(z) = \dfrac{1}{a} f_X\left(\dfrac{z - c}{a}\right)$.

(c) Thus

$$\mu_Z = \int_{-\infty}^{\infty} z f_Z(z)\, dz = \int_{-\infty}^{\infty} \frac{z}{a} f_X\left(\frac{z - c}{a}\right) dz.$$

Change variable of integration from Z to $X = (Z - c)/a$ and use the fact that $\int_{-\infty}^{\infty} f_X(x)\, dx = 1$ to show that $\mu_Z = a\mu_X + c$.

(d) Suppose that $a < 0$. Show that $F_Z(z) = 1 - F_X\left(\dfrac{z - c}{a}\right)$ and that $f_Z(z) = \dfrac{1}{|a|} f_X\left(\dfrac{z - c}{a}\right)$. Again show that $\mu_Z = a\mu_X + c$.

T.10. Let X be a random variable having mean $\mu_X = \mu$, variance $\sigma_X^2 = \sigma^2$, and measure of kurtosis ν_2. Let $Y = (X - \mu)^2/\sigma^2$.

(a) Show that the mean of Y is 1; that is, $\mu_Y = 1$.

(b) Show that the variance of Y is $\sigma_Y^2 = \nu_2 + 1$. Hence conclude that for any random variable X, the measure of kurtosis ν_2 for X always is -1 or greater.

(c) What insight does the result of part (b) give to the interpretation of the measure of kurtosis ν_2 as a descriptor of the distribution of X? [Recall that the variability of Y measures the dispersion of Y about its mean and that $Y = (X - \mu)^2/\sigma^2$.]

T.11. The **indicator function** $I_A(x)$ of a collection of values A is defined by

$$I_A(x) = \begin{cases} 1, & \text{if } x \text{ is in } A, \\ 0, & \text{if } x \text{ is not in } A. \end{cases}$$

For any random variable X, the following fact is true:

$$P\{X \in A\} = E[I_A(X)].$$

(a) Prove this result for the case where X is a discrete random variable with probability mass function $p(x)$.

(b) Prove this result when A is the interval $[a, b]$, $a \leq b$, and X is a continuous random variable with probability density function $f(x)$.

(c) Suppose that X is a nonnegative random variable. Use indicator functions and the properties of expected values to prove the Markov inequality. (The intuitive argument for the Markov inequality in Section 7 may suggest a proof.)

(d) Show that for any random variable X and any collection of values A, the variance of $Y = I_A(X)$ cannot exceed 0.25.

7

Sums and Averages of Independent Random Variables

1 Introduction

In many random experiments, attention may focus on more than one random variable. For example, both years of schooling and current income may be measured for people, both the temperature of the reaction and the amount of product yielded may be measured for chemical reactions, both the tensile strength and resistance to shear may be measured for bars of steel, and so on.

If the outcome of a random experiment is summarized by r random variables, $r \geq 2$, an outcome of such an experiment is given by a list of the observed values of these variables. For example, a typical outcome for two variables X and Y is the pair (x, y), where x is the value observed for X, and y is the value observed for Y. In general, if r random variables $X_1, X_2, \ldots, X_r$ are measured, a typical outcome is the r-tuple $(x_1, x_2, \ldots, x_r)$.

To construct a probability model for r random variables $X_1, X_2, \ldots,$ X_r, we need to describe how probability is distributed across the possible outcomes $(x_1, x_2, \ldots, x_r)$. Any such description defines a **joint** (or **multivariate**) **probability distribution** for $X_1, X_2, \ldots, X_r$. When $X_1, X_2, \ldots, X_r$ are all discrete variables, such a distribution can be defined by the function $p(x_1, \ldots, x_r)$ which gives the probability of the event that simultaneously $X_1 = x_1$, $X_2 = x_2$, and $X_r = x_r$; that is,

$$p(x_1, x_2, \ldots x_r) = P\{X_1 = x_1 \text{ and } X_2 = x_2 \text{ and } \ldots \text{ and } X_r = x_r\}.$$

Such a function, $p(x_1, x_2, \ldots, x_r)$, is called a **joint probability mass function.** When $X_1, X_2, \ldots, X_r$ are all continuous variables, and additional conditions hold, the corresponding function defining the joint probability distribution is the **joint probability density function** $f(x_1, x_2, \ldots, x_r)$. Alternatively, the **joint cumulative distribution function**

$$F(x_1, x_2, \ldots, x_r) = P\{X_1 \le x_1 \text{ and } X_2 \le x_2 \text{ and } \ldots \text{ and } X_r \le x_r\}$$

can always be used to define the joint distribution of $X_1, X_2, \ldots, X_r$.

Joint probability distributions allow us to model the probabilistical (statistical) dependence among measured variables, and thus are essential tools in studying the relationships among variables. In particular, such models help us to predict the values of one variable when we know the values of other variables (using the concepts of conditional probability—see Chapter 4). Note that a random experiment in which we observe random variables $X_1, X_2, \ldots, X_r$ is a composite random experiment, and that any particular variable X_i defines a component experiment of this composite random experiment.

Because more than one variable is considered at a time, the study of joint probability distributions is more complicated than the study of the distributions of individual random variables. New concepts and new descriptive indices are required, and more sophisticated tools of mathematical analysis are needed. In this chapter we concentrate on a very special type of joint probability distribution where the random variables $X_1, X_2, \ldots, X_r$ are mutually probabilistically independent. Discussion of more general joint probability models (multivariate distributions) is postponed to Chapter 11.

2 Mutually Independent Random Variables

In Chapter 4 we showed how the assumption of mutual probabilistic independence of the components of a composite random experiment can greatly simplify the modeling of probabilities. If the component experiments $\mathscr{E}_1, \mathscr{E}_2, \ldots, \mathscr{E}_r$ are mutually independent, then for any events $E_1, E_2, \ldots, E_r$, where each E_i is determined by the component experiment $\mathscr{E}_i$, $i = 1, 2, \ldots, r$,

$$P(E_1 \cap E_2 \cap \cdots \cap E_r) = P(E_1)\, P(E_2) \cdots P(E_r).$$

Consequently, probabilities of events for the entire composite experiment can be calculated using knowledge of the probabilities of the events determined by the component experiments.

Here, the individual variables $X_1, X_2, \ldots, X_r$ define component experiments of the composite experiment in which these variables are jointly measured. Events E_i determined by the random variable X_i are of the form

$$E_i = \{X_i \text{ is in } A_i\},$$

where A_i is any collection of numerical values, $i = 1, 2, \ldots, r$.

> The random variables $X_1, X_2, \ldots X_r$ are defined to be **mutually probabilistically independent** (mutually independent) if the events $\{X_i \text{ in } A_i\}$, $i = 1, 2, \ldots, r$, are mutually probabilistically independent for all choices of the collections $A_1, A_2, \ldots, A_r$.

That is, $X_1, X_2, \ldots, X_r$ are mutually probabilistically independent if (and only if)

$$\begin{aligned} P\{X_1 \text{ in } A_1, \text{ and } X_2 \text{ in } A_2, \text{ and } \ldots \text{ and } X_r \text{ in } A_r\} \\ = P\{X_1 \text{ in } A_1\}\, P\{X_2 \text{ in } A_2\} \cdots P\{X_r \text{ in } A_r\} \end{aligned} \tag{2.1}$$

for all collections $A_1, A_2, \ldots, A_r$. Note that by taking $A_{k+1}, A_{k+2}, \ldots, A_r$ to be the whole line, we obtain

$$P\{X_1 \text{ in } A_1, \ldots, X_k \text{ in } A_k\} = P\{X_1 \text{ in } A_1\} \cdots P\{X_k \text{ in } A_k\}$$

for all $k = 1, 2, \ldots, r$.

It follows from the definition (2.1) that if $X_1, X_2, \ldots, X_r$ are mutually independent discrete variables, their joint probability mass function

$$\begin{aligned} p(x_1, \ldots, x_r) &= P\{X_1 = x_1, \text{ and } X_2 = x_2, \text{ and } \ldots \text{ and } X_r = x_r\} \\ &= P\{X_1 = x_1\}\, P\{X_2 = x_2\} \cdots P\{X_r = x_r\} \\ &= p_{X_1}(x_1)\, p_{X_2}(x_2) \cdots p_{X_r}(x_r) \end{aligned} \tag{2.2}$$

is the product of the individual mass functions, $p_{X_i}(x_i)$, of these random variables. For any mutually independent random variables $X_1, X_2, \ldots, X_r$, their joint cumulative distribution function

$$\begin{aligned} F(x_1, \ldots, x_r) &= P\{X_1 \le x_1, \text{ and } X_2 \le x_2, \text{ and } \ldots \text{ and } X_r \le x_r\} \\ &= P\{X_1 \le x_1\}\, P\{X_2 \le x_2\} \cdots P\{X_r \le x_r\} \\ &= F_{X_1}(x_1)\, F_{X_2}(x_2) \cdots F_{X_r}(x_r) \end{aligned} \tag{2.3}$$

is the product of the individual cumulative distribution functions, $F_{X_i}(x_i)$, of these random variables. Finally, if $X_1, \ldots, X_r$ are mutually independent continuous variables with a joint probability density function $f(x_1, \ldots, x_r)$, it is shown in Chapter 11 that

$$f(x_1, \ldots, x_r) = f_{X_1}(x_1) f_{X_2}(x_2) \cdots f_{X_r}(x_r), \tag{2.4}$$

where $f_{X_i}(x_i)$ is the probability density function of X_i, $i = 1, 2, \ldots, r$.

Mutually independent random variables arise in many contexts. For example, every unit in a population of units can possess a certain quantitative characteristic (height, weight, number of defects, income, and so on). Suppose that we take a simple random sample of r units **with replacement** from such a population and record the value of the quantitative characteristic for each unit sampled. The result of this experiment is then an r-tuple $(x_1, x_2, \ldots, x_r)$ of observed measurements on the random variables $X_1, X_2, \ldots, X_r$, where X_i is the value of the quantitative characteristic measured on the ith unit drawn. Because the r draws are mutually independent components (Chapter 4) of this random experiment, $X_1, X_2, \ldots, X_r$ are mutually independent random variables.

EXAMPLE 2.1

Fish and Wildlife Service employees catch fish one at a time, measure the length of each fish, and return the fish to the pond. If fish are sampled with replacement, the lengths $X_1, X_2, \ldots, X_r$ of the r fish caught are mutually independent random variables. ◆

Example 2.1 is a special case of a more general situation in which r mutually independent observations are made of a particular quantitative phenomenon—for example, the length of the fish. If the phenomenon is unchanged while the observations are being made (the population of fish and their lengths remain constant), each such quantitative observation X_i has the *same probability distribution*. In this case we say that the r observations $X_1, X_2, \ldots, X_r$ are **independent, identically distributed trials** of the experiment of observing the quantitative characteristic X. If X is described by a probability mass function $p(x)$, then $X_1, X_2, \ldots, X_r$ all have this same mass function $p(x)$, and

$$p(x_1, \ldots, x_r) = p(x_1)\, p(x_2) \cdots p(x_r).$$

If X is described by a probability density function $f(x)$, then $f(x)$ is also the density function for each of the variables $X_1, X_2, \ldots, X_r$, and

$$f(x_1, x_2, \ldots, x_r) = f(x_1)\, f(x_2) \cdots f(x_r).$$

In the statistical literature, mutually independent, identically distributed trials (observations) $X_1, X_2, \ldots, X_r$ of a random variable X are said to be a **random sample of size r from the distribution of X**, even if no actual sampling from a population has been done.

EXAMPLE 2.2

The duration X of time in which a car battery maintains an electrical charge is a continuous random variable with probability density function $f(x)$. The durations X_1, X_2, X_3 of time that the batteries in 3 cars maintain electrical charge is a random sample of size $r = 3$ from the distribution of X. Consequently, the joint density $f(x_1, x_2, x_3)$ of X_1, X_2, X_3 is given by the product

$$f(x_1, x_2, x_3) = f(x_1)f(x_2)f(x_3).$$ ◆

EXAMPLE 2.3

The number, X, of dots on the upturned face of a tossed die is a discrete random variable with probability mass function $p(x) = \frac{1}{6}$ if $x = 1, 2, 3, 4, 5, 6$, and $p(x) = 0$ otherwise. If $r = 2$ dice are tossed, the numbers X_1, X_2 of dots on the upturned faces of these two dice are a random sample of size $r = 2$ from the distribution of X. Note that

$$\begin{aligned} p(x_1, x_2) &= P\{X_1 = x_1 \text{ and } X_2 = x_2\} = p(x_1)p(x_2) \\ &= \begin{cases} \dfrac{1}{36}, & \text{if } x_1, x_2 = 1, 2, 3, 4, 5, 6, \\ 0, & \text{otherwise.} \end{cases} \end{aligned}$$ ◆

Mutually independent random variables can arise in other contexts besides that of independent, identically distributed trials.

EXAMPLE 2.4

The numbers X, Y of units produced in a week by two factories of an industrial firm are discrete random variables that might have quite different probability mass functions, $p_X(x)$ and $p_Y(y)$. For example, one factory may have older equipment than the other, so that the amount X produced by the factory with older equipment may tend to be smaller than the amount Y produced by the more modern factory. Consequently, the probability mass function $p_X(x)$ of X will differ from the probability mass function $p_Y(y)$ of Y. If the two factories are physically separated and have distinctly different employees and suppliers of raw material, it can be argued that X and Y are probabilistically independent random variables. ◆

EXAMPLE 2.5

The car batteries for the three cars in Example 2.2 may be made by different manufacturers. In this case, the density functions $f_{X_1}(x_1)$, $f_{X_2}(x_2)$, $f_{X_3}(x_3)$, of the lifetimes X_1, X_2, X_3 of the batteries are likely to be different. Because the batteries are placed in separate cars, it is reasonable to assume that X_1, X_2, X_3 are mutually probabilistically independent. ◆

When mutually independent random variables $X_1, X_2, \ldots, X_r$ are observed in a random experiment, attention is often focused on some function

$$Y = g(X_1, X_2, \ldots, X_r)$$

of these variables. In this chapter we concentrate on the *sum* of the observations:

$$S = X_1 + X_2 + \cdots + X_r = \sum_{i=1}^{r} X_i, \tag{2.5}$$

because this function is so frequently of interest in random experiments.

3 Sums of Independent Random Variables

Because the sum (2.5) of $X_1, X_2, \ldots, X_r$ is a function of random variables, it is itself a random variable and hence has a probability distribution. This probability distribution can be determined from the joint probability distribution of the variables $X_1, X_2, \ldots, X_r$, as the following example illustrates.

EXAMPLE 3.1

Suppose that X and Y are two probabilistically independent discrete random variables, with respective probability mass functions:

x	0	1
$p_X(x)$	$\frac{1}{3}$	$\frac{2}{3}$

y	0	1	2
$p_Y(y)$	$\frac{1}{4}$	$\frac{1}{2}$	$\frac{1}{4}$

Then the joint probability mass function of X and Y is given by

$$p(x, y) = p_X(x)p_Y(y)$$

For example,

$$p(1, 0) = P\{X = 1 \text{ and } Y = 0\} = p_X(1)\, p_Y(0) = (\tfrac{2}{3})\,(\tfrac{1}{4}) = \tfrac{1}{6},$$
$$p(1, 1) = P\{X = 1 \text{ and } Y = 1\} = p_X(1)\, p_Y(1) = (\tfrac{2}{3})\,(\tfrac{1}{2}) = \tfrac{1}{3},$$

and so on. This mass function, $p(x, y)$, can be given in the following tabular display (see Chapter 4):

x	y			$p_X(x)$
	0	1	2	
0	$\frac{1}{12}$	$\frac{1}{6}$	$\frac{1}{12}$	$\frac{1}{3}$
1	$\frac{1}{6}$	$\frac{1}{3}$	$\frac{1}{5}$	$\frac{2}{3}$
$p_Y(y)$	$\frac{1}{4}$	$\frac{1}{2}$	$\frac{1}{4}$	

Note that $S = X + Y$ has possible values 0, 1, 2, 3. To find the probability that $S = s$, for any one of these values s, one simply adds the probabilities of all pairs (x, y) for which $x + y = s$. Thus

$$\begin{aligned} P\{S = 0\} &= p(0, 0) = \tfrac{1}{12}, \\ P\{S = 1\} &= p(0, 1) + p(1, 0) = \tfrac{1}{6} + \tfrac{1}{6} = \tfrac{1}{3}, \\ P\{S = 2\} &= p(0, 2) + p(1, 1) = \tfrac{1}{12} + \tfrac{1}{3} = \tfrac{5}{12}, \\ P\{S = 3\} &= p(1, 2) = \tfrac{1}{6}. \end{aligned}$$

Consequently, the probability mass function of S is

s	0	1	2	3
$p_S(s)$	$\frac{1}{12}$	$\frac{1}{3}$	$\frac{5}{12}$	$\frac{1}{6}$

◆

In general, if $X_1, X_2, \ldots, X_r$ are mutually independent discrete random variables, the probability mass function of the sum S is given by

$$\begin{aligned} p_S(s) &= \sum_{x_1+x_2+\cdots+x_r=s} p(x_1, x_2, \ldots, x_r) \\ &= \sum_{x_1+x_2+\cdots+x_r=s} p_{X_1}(x_1)\, p_{X_2}(x_2) \cdots p_{X_r}(x_r). \end{aligned} \tag{3.1}$$

There is a similar formula for the probability density function of the sum $S = X_1 + X_2 + \cdots + X_r$ of mutually independent continuous random variables $X_1, X_2, \ldots, X_r$ in terms of integrals over their joint density function (2.4); see Chapter 13. Despite the relative simplicity of Example 3.1, it is not easy in general to obtain the distribution of the sum S using (3.1) or its density function analog. One alternative method is the method of generating functions, which is discussed in Section 4.

Mean and Variance of Weighted Sums

As remarked in Chapter 6, for many purposes it is enough to provide certain descriptive indices (such as the mean and variance) of the distribution of the sum $S = \sum_{i=1}^{r} X_i$. The following very useful facts are verified in Chapter 11:

$$\mu_S = \sum_{i=1}^{r} \mu_{X_i}, \tag{3.2}$$

$$\sigma_S^2 = \sum_{i=1}^{r} \sigma_{X_i}^2. \tag{3.3}$$

It should be noted that equation (3.2) is true *whether or not the random variables* $X_1, X_2, \ldots, X_r$ *are mutually probabilistically independent*. On the other hand, equation (3.3) is not generally true unless the random variables $X_1, X_2, \ldots, X_r$ are mutually independent.

A simple example shows why this last assertion is true. Suppose that we have two random variables X, Y for which $X + Y = 0$ (i.e., $Y = -X$). Note that $S = X + Y$ has, in this case, only one possible value (namely, 0); consequently, $\sigma_S^2 = 0$, because S does not vary. On the other hand, if equation (3.3) were true,

$$\sigma_S^2 = \sigma_X^2 + \sigma_Y^2 = \sigma_X^2 + \sigma_{-X}^2 = \sigma_X^2 + (-1)^2\sigma_X^2 = 2\sigma_X^2 > 0,$$

if σ_X^2 is not zero. The explanation for this contradiction is that X and Y are **probabilistically dependent** (knowing the value of X, we know the value of $Y = -X$), and equation (3.3) does not hold in this case.

Consequently, *when applying equation* (3.3), *we need to make certain that the random variables* $X_1, X_2, \ldots, X_r$ *are mutually probabilistically independent.*

Equations (3.2) and (3.3) also apply to **weighted sums** $S = \sum_{i=1}^{r} a_i X_i$. Perhaps the most widely used weighted sum is the arithmetic average

$$\overline{X} = \frac{1}{r} \sum_{i=1}^{r} X_i \tag{3.4}$$

for which the weights a_i all equal $1/r$. As noted in Chapter 6, the arithmetic average $\overline{X}$ of a random sample $X_1, X_2, \ldots, X_r$ of r observations from the distribution of a random variable X is used to estimate the mean μ_X of the distribution of X. In considering weighted sums, we make the assumption that $a_1, a_2, \ldots, a_r$ are all not equal to 0, because if any $a_i = 0$, the corresponding variable X_i does not affect the value of the sum $\sum_{i=1}^{r} a_i X_i$.

To see how equations (3.2) and (3.3) apply to weighted averages, note first that any information that we have about a_iX_i is also information about X_i, and vice versa (assuming that $a_i \neq 0$). Computing a_iX_i simply relabels the outcomes (values) of X_i. Thus, knowing that $X_i = x$ tells us that $a_iX_i = a_ix$, whereas knowing that $a_iX_i = x^*$ tells us that $X_i = x^*/a_i$. Consequently, it follows from the definition of mutual probabilistic independent random variables that *$a_1X_1, a_2X_2, \ldots, a_rX_r$ are mutually probabilistically independent if and only if $X_1, X_2, \ldots, X_r$ are mutually probabilistically independent.* Because $Y_i = a_iX_i$ are themselves random variables, and are mutually probabilistically independent when $X_1, \ldots, X_r$ are mutually independent, equations (3.2) and (3.3) imply that

$$\mu_S = \mu_{\sum_{i=1}^r Y_i} = \sum_{i=1}^{r} \mu_{Y_i} = \sum_{i=1}^{r} \mu_{a_iX_i} = \sum_{i=1}^{r} a_i\mu_{X_i}, \tag{3.5}$$

$$\sigma_S^2 = \sigma^2_{\sum_{i=1}^r Y_i} = \sum_{i=1}^{r} \sigma^2_{Y_i} = \sum_{i=1}^{r} \sigma^2_{a_iX_i} = \sum_{i=1}^{r} a_i^2\sigma^2_{X_i}, \tag{3.6}$$

where $S = \sum_{i=1}^r a_iX_i$. Here we have made use of the fact that $\mu_{aX} = a\mu_X$ and $\sigma^2_{aX} = a^2\sigma^2_X$ (see Chapter 6).

Note that when $a_1 = a_2 = \cdots = a_r = 1$, (3.5) and (3.6) reduce to (3.2) and (3.3), respectively.

EXAMPLE 3.2

In Example 2.4 the total number of units produced in each week by the two factories is the sum $X + Y$ of their outputs, and the (random) outputs X, Y of the factories are mutually probabilistically independent. Suppose that the production X of the factory with older equipment has mean $\mu_X = 9600$ and variance $\sigma_X^2 = 8650$, and the production Y of the other factory has mean $\mu_Y = 12{,}500$ and variance $\sigma_Y^2 = 10{,}000$. Then the mean weekly total production of the firm is

$$\mu_{X+Y} = \mu_X + \mu_Y = 9600 + 12{,}500 = 22{,}100 \text{ units},$$

and the variance of the total weekly production is

$$\sigma^2_{X+Y} = \sigma^2_X + \sigma^2_Y = 8650 + 10{,}000 = 18{,}650.$$ ◆

EXAMPLE 3.3

A biologist is interested in whether a chemical retards the growth of mold. The biologist gives the chemical to one of two mold cultures chosen at random. The other culture is allowed to grow as usual. Let X and Y denote the growth (measured in terms of units of area of spread) of the chemically treated and untreated cultures, respectively. Let $D =$

$Y - X$ denote the difference in growth between the two cultures. Suppose that the chemical does "on the average" retard growth, and that

$$\begin{aligned} \mu_X &= 1.2, & \mu_Y &= 3.3, \\ \sigma_X^2 &= 0.044, & \sigma_Y^2 &= 0.042. \end{aligned}$$

Assuming that X and Y are probabilistically independent, the mean and variance of the growth difference $D = Y - X$ are

$$\mu_D = \mu_Y - \mu_X = 3.3 - 1.2 = 2.1,$$

$$\sigma_D^2 = (1)^2\sigma_Y^2 + (-1)^2\sigma_X^2 = \sigma_Y^2 + \sigma_X^2 = 0.042 + 0.044 = 0.086,$$

respectively. Notice that although we subtracted means, we *added* variances. ◆

EXAMPLE 3.4 Because his company requires him to obey all speed limits, a bus driver travels X_1 hours at 55 miles per hour (mph), X_2 hours at 35 mph, and X_3 hours at 25 mph. If X_1, X_2, X_3 are mutually independent random variables, and

$$\begin{aligned} \mu_{X_1} &= 4.1, & \mu_{X_2} &= 1.2, & \mu_{X_3} &= 1.7, \\ \sigma_{X_1}^2 &= 0.16, & \sigma_{X_2}^2 &= 0.25, & \sigma_{X_3}^2 &= 0.18, \end{aligned}$$

then the mean and variance of the total number $S = 55X_1 + 35X_2 + 25X_3$ of miles traveled by the bus driver are obtained by applying (3.5) and (3.6):

$$\begin{aligned} \mu_S &= 55\mu_{X_1} + 35\mu_{X_2} + 25\mu_{X_3} \\ &= (55)(4.1) + (35)(1.2) + (25)(1.7) \\ &= 310 \text{ miles}, \end{aligned}$$

$$\begin{aligned} \sigma_S^2 &= (55)^2\sigma_{X_1}^2 + (35)^2\sigma_{X_2}^2 + (25)^2\sigma_{X_3}^2 \\ &= (55)^2(0.16) + (35)^2(0.25) + (25)^2(0.18) \\ &= 902.75 \text{ (miles)}^2. \end{aligned}$$

◆

Mean and Variance of Totals and Averages of Mutually Independent, Identically Distributed Random Variables

If $X_1, X_2, \ldots, X_n$ are mutually independent, identically distributed trials of a random variable X (a random sample of size n from the distribution of X), then each of $X_1, X_2, \ldots, X_n$ has the same distribution as X, and hence each has the same mean $\mu_X = \mu$, and variance, $\sigma_X^2 = \sigma^2$, as X. In this case (3.5) and (3.6) yield the following results for the mean and variance of the sum $S_n = \Sigma_{i=1}^n X_i$ and average $\overline{X} = \frac{1}{n}\Sigma_{i=1}^n X_i$:

$$\mu_S = \sum_{i=1}^{n} \mu_{X_i} = \sum_{i=1}^{n} \mu = n\mu, \qquad \sigma_S^2 = \sum_{i=1}^{n} \sigma_{X_i}^2 = \sum_{i=1}^{n} \sigma^2 = n\sigma^2,$$

$$\mu_{\overline{X}} = \sum_{i=1}^{n} \frac{1}{n}\mu_{X_i} = \frac{1}{n}\sum_{i=1}^{n} \mu = \mu, \qquad \sigma_{\overline{X}}^2 = \sum_{i=1}^{n} \left(\frac{1}{n}\right)^2 \sigma_{X_i}^2 = \left(\frac{1}{n}\right)^2 \sum_{i=1}^{n} \sigma^2 = \frac{\sigma^2}{n}. \tag{3.7}$$

Note that the mean of the average $\overline{X}$ of $X_1, X_2, \ldots, X_n$ is equal to the common mean μ of these variables, but that the variance of $\overline{X}$ is smaller than the variance σ^2 of any single observation. Indeed, the variance of the average $\overline{X}$ decreases to zero as the sample size n increases.

The Law of Large Numbers

We can now partly confirm an assertion made in Chapter 6 concerning the relationship between the sample average $\overline{X}$ and the population mean μ of X in large samples. Namely, we show that as the sample size n increases, the probability that $\overline{X}$ lies in the interval $[\mu - c, \mu + c]$ approaches zero, no matter how small is the positive number c. To see this, note that by the Bienaymé–Chebychev inequality (Chapter 6):

$$\begin{aligned} P\{\mu - c \le \overline{X} \le \mu + c\} &= P\{-c \le \overline{X} - \mu \le c\} \\ &= P\{-d\sigma_{\overline{X}} \le \overline{X} - \mu \le d\sigma_{\overline{X}}\} \\ &\ge 1 - \frac{1}{d^2}, \end{aligned} \tag{3.8}$$

where $d\sigma_{\overline{X}} = c$. Consequently,

$$d = \frac{c}{\sigma_{\overline{X}}} = \frac{c}{\sqrt{\sigma^2/n}} = \frac{c\sqrt{n}}{\sigma}$$

and

$$P\{\mu - c \le \overline{X} \le \mu + c\} \ge 1 - \frac{1}{d^2} = 1 - \frac{\sigma^2}{nc^2}. \tag{3.9}$$

Because $P\{\mu - c \leq \overline{X} \leq \mu + c\}$ is always bounded above by 1 and is bounded below by $1 - (\sigma^2/nc^2)$, which tends to 1 as $n \to \infty$, we conclude that for any $c > 0$,

$$\lim_{n \to \infty} P\{\mu - c \leq \overline{X} \leq \mu + c\} = 1. \tag{3.10}$$

To summarize, **as the sample size n becomes large, the sample average $\overline{X}$ concentrates more and closely around its mean μ.** Using inequality (3.8), Exhibit 3.1 gives values of the sample size n required to guarantee with probability at least P (for $P = 0.95, 0.99$) that $\overline{X}$ is within various distances c from its mean μ when $\sigma^2 = 1$.

The conclusion of this analysis is that by making n large enough, we ensure a high probability of having $\overline{X}$ fall as close as we please to its mean μ, regardless of the common distribution of the observations $X_1, X_2, \ldots, X_n$. This assertion, which is a somewhat informal statement of the **Law of Large Numbers,** can be derived from the axioms of probability theory; it corresponds to the frequency interpretation of the population mean given in Chapter 6, Section 2.

A Popular Misconception

A popular misconception concerning the Law of Large Numbers is that it is a statement about the sample total, $X_1 + X_2 + \cdots + X_n$, rather than about the sample average $(1/n)(X_1 + X_2 + \cdots + X_n)$. For example, in tossing a fair coin, it is often stated that the number of heads observed should equal the number of tails. Thus if 50 heads in a row are observed,

EXHIBIT 3.1: *Values of the Sample Size n Required to Assure at Least a Probability of P that $\overline{X}$ Falls Within a Specified Number c of Units from Its Mean μ When $\sigma^2 = 1$*

Number of Units c	n	
	$P = 0.95$	$P = 0.99$
0.01	200,000	1,000,000
0.10	2,000	10,000
0.50	80	400
1.00	20	100
1.50	9	45
2.00	5	25
2.50	4	16
3.00	3	12

there is a belief that a tail will appear shortly thereafter in subsequent tosses, on the grounds that the "law of averages" says that the total number of heads must equal the total number of tails. But if the coin tosses are independent, as assumed, observing 50 heads on the first 50 tosses tells us nothing new about the probability that the 51st toss results in a tail: the probability of a tail is still $\frac{1}{2}$. The law of large numbers does not say that the *total number* of heads must eventually equal the *total number* of tails; rather, it says that the *proportion* of heads eventually equals the *proportion* of tails.

What does this have to do with sums? Consider a random variable X_i which equals 1 if the ith toss is heads, and 0 if the ith toss is tails. Note that in n tosses of the coin, the number of heads observed is $H = \sum_{i=1}^{n} X_i$ and the number of tails observed is $T = n - H$. To say that the number of heads equals the number of tails is the same as saying that

$$H = \sum_{i=1}^{n} X_i = T = n - \sum_{i=1}^{n} X_i$$

or, solving for $\sum_{i=1}^{n} X_i$, that $\sum_{i=1}^{n} X_i = n/2$. Thus, in $n = 500$ tosses, the misconception is that we should observe $\frac{1}{2}n = 250$ heads, whereas the correct assertion is that

$$\overline{X} = \frac{1}{n}\sum_{i=1}^{n} X_i = \frac{\text{number of heads}}{n} = \text{proportion of heads}$$

should approximately equal 1/2, if n is sufficiently large.

The difference between these assertions can be clarified by assuming that the first 10 tosses are heads ($X_1 = X_2 = \cdots = X_{10} = 1$, $\sum_{i=1}^{10} X_i = 10$), but that after the first 10 tosses, the number of heads does equal the number of tails. Thus in 1000 tosses we have 505 heads and 495 tails, in 10,000 tosses we have 5005 heads and 4995 tails, and so on. Always, the number of heads is 10 more than the number of tails, but the proportions of heads are 0.505 as against 0.495 (after 1000 tosses), 0.5005 as against 0.4995 (after 10,000 tosses), 0.50005 as against 0.49995 (after 100,000 tosses), and so on. That is when we compute proportions ($\overline{X}$) rather than totals ($\sum_{i=1}^{n} X_i$), the excess of heads is overwhelmed by the total number of tosses. The difference (10) between the total numbers of heads and tails observed becomes a smaller and smaller fraction of the total number, n, of tosses as n becomes large.

4 The Moment Generating Function

As noted in Section 3, it is not always easy to obtain the distributions of sums of random variables directly. Moment generating functions offer an indirect but often simpler method of obtaining such distributions. In addition, moment generating functions allow one to obtain important descriptive indices (the moments—such as the mean and variance) of a distribution. Indeed, when the moment generating function is well defined, this function uniquely characterizes or identifies its distribution.

The moment generating function of any random variable X is formally defined to be

$$M(t) = E(e^{tX}) \tag{4.1}$$

for all values t for which the expected value on the right side of (4.1) exists. Thus if X is a discrete random variable with probability mass function $p(x)$, then

$$M(t) = \sum_x e^{tx}\, p(x), \tag{4.2}$$

and if X is a continuous random variable with probability density function $f(x)$,

$$M(t) = \int_{-\infty}^{\infty} e^{tx} f(x)\, dx. \tag{4.3}$$

EXAMPLE 4.1

For the discrete random variable X with probability mass function

$$p(x) = \begin{cases} p, & \text{if } x = 1, \\ 1 - p, & \text{if } x = 0, \\ 0, & \text{otherwise,} \end{cases}$$

the moment generating function is

$$M(t) = \sum_{x=0}^{1} e^{tx} p(x) = e^{(t)(1)}p + e^{(t)(0)}(1 - p) = pe^t + (1 - p),$$

for all values of t. ◆

EXAMPLE 4.2

For the continuous random variable X with probability density function

$$f(x) = \begin{cases} 1, & \text{if } 0 \le x < 1, \\ 0, & \text{otherwise,} \end{cases}$$

the moment generating function is

$$M(t) = \int_0^1 e^{tx}(1)\,dx = \frac{1}{t}e^{tx}\Big|_0^1 = \frac{1}{t}(e^{(t)(1)} - e^{(t)(0)}) = \frac{e^t - 1}{t},$$

for all t. ◆

Because $e^0 = 1$, the moment generating function $M(t)$ is always defined at $t = 0$, and

$$M(0) = E(e^{0X}) = E(1) = 1.$$

When $M(t)$ is defined for values of t in an open interval (t_L, t_U) such that $t_L < 0 < t_U$, the moment generating function has the following important properties.

1. **Uniqueness.** *There is one and only one distribution that has the moment generating function $M(t)$.* Because the cumulative distribution function $F(x) = P\{X \le x\}$ determines the distribution of any random variable X, discrete or continuous, this property is equivalent to the assertion that *to each moment generating function $M(t)$ there corresponds a unique cumulative distribution function $F(x)$.*
2. **Determines moments.** Moments are obtained by taking derivatives evaluated at $t = 0$. That is,

$$E(X^k) = \left[\frac{d^k}{(dt)^k} M(t)\right]_{t=0}, \qquad k = 1, 2, \ldots,$$

3. **Moment generating function of $aX + b$.** If $M(t)$ is the moment generating function of X, the moment generating function of $Y = aX + b$ is

$$M_{aX+b}(t) = e^{bt}M(at). \tag{4.4}$$

4. **Determines distributions of sums of independent variables.** If X, Y are probabilistically independent and have moment generating functions $M_X(t)$, $M_Y(t)$, respectively, then

$$M_{X+Y}(t) = M_X(t)M_Y(t), \tag{4.5}$$

where $M_{X+Y}(t)$ is the moment generating function of $X + Y$.
5. **Preservation of limits.** For each $n = 1, 2, \ldots$, let $F^{(n)}(x)$ be a cumulative distribution function, with corresponding moment generating func-

tion $M^{(n)}(t)$ defined for t in an open interval (t_L, t_U), $t_L < 0 < t_U$. If there exists a moment generating function $M^*(t)$ defined for t in (t_L, t_U) for which

$$\lim_{n\to\infty} M^{(n)}(t) = M^*(t), \qquad \text{all } t \text{ in } (t_L, t_U), \tag{4.6}$$

then if $F^*(x)$ is the cumulative distribution function corresponding to $M^*(t)$,

$$\lim_{n\to\infty} F^{(n)}(x) \text{ exists}, \qquad \lim_{n\to\infty} F^{(n)}(x) = F^*(x), \tag{4.7}$$

for x such that $F^*(x)$ is continuous at x. [*Note*: It can also be shown that if (4.7) is true, then (4.6) is true.]

We now illustrate applications of these properties.

EXAMPLE 4.1 *(continued)*

In this example, X has mass function $p(x) = p$ if $x = 1$, $p(x) = 1 - p$ if $x = 0$, and has moment generating function $pe^t + (1 - p)$. Suppose that a random variable Y has moment generating function $M(t) = pe^t + (1 - p)$. In this case, the uniqueness property of moment generating functions implies that X and Y have the same cumulative distribution function and hence the same probability mass function. (The mass function can be obtained from the cumulative distribution function—see Chapter 5.) Thus we know that $P\{Y = 1\} = p$ and that $P\{Y = 0\} = 1 - p$.

It is easily seen that

$$\mu_X = \sum_{x=0}^{1} xp(x) = (0)(1 - p) + (1)(p) = p.$$

Alternatively, by Property 2,

$$\mu_X = \left[\frac{d}{dt} M(t)\right]_{t=0} = \frac{d}{dt}(pe^t + 1 - p)\bigg|_{t=0} = pe^t\bigg|_{t=0} = p.$$

Similarly,

$$E(X^2) = \left[\frac{d^2}{dt^2} M(t)\right]_{t=0} = \left[\frac{d}{dt}(pe^t)\right]_{t=0} = pe^t\bigg|_{t=0} = p,$$

so that

$$\sigma_X^2 = E(X)^2 - (\mu_X)^2 = p - p^2 = p(1 - p).$$

In the same way, $E(X^h) = p$ for $h = 3, 4, 5, 6, \ldots$. ◆

EXAMPLE 4.2 *(continued)*

Recall that in this example the random variable X has moment generating function $M(t) = (e^t - 1)/t$. Suppose that $Y = 5X + 7$. Then, by (4.4), the moment generating function $M_Y(t)$ of Y is

$$M_Y(t) = e^{7t}\, M(5t) = e^{7t}\left(\frac{e^{5t} - 1}{5t}\right)$$
$$= \frac{e^{12t} - e^{7t}}{5t}.$$

Consider a continuous random variable W with probability density function

$$f_W(w) = \begin{cases} \dfrac{1}{5}, & \text{if } 7 \le w \le 12, \\ 0, & \text{otherwise.} \end{cases}$$

Then W has moment generating function

$$M_W(t) = \int_7^{12} e^{tw}\left(\frac{1}{5}\right) dw = \frac{1}{5t}e^{tw}\Big|_7^{12} = \frac{e^{12t} - e^{7t}}{5t}.$$

Thus W and Y have the same moment generating function and hence (by the uniqueness property) have the same cumulative distribution function. Consequently, W and Y have the same density function. We conclude that the density function of Y is

$$f_Y(y) = \begin{cases} \dfrac{1}{5}, & \text{if } 7 \le y \le 12, \\ 0, & \text{otherwise.} \end{cases}$$ ◆

EXAMPLE 4.3

Suppose that each of the probabilistically independent continuous random variables X_1, X_2 has probability density function

$$f(x) = \begin{cases} e^{-x}, & \text{if } x \ge 0. \\ 0, & \text{is } x < 0. \end{cases}$$

Then the moment generating functions $M_{X_1}(t)$, $M_{X_2}(t)$ of X_1, X_2 equal

$$\begin{aligned} M(t) &= \int_0^\infty e^{tx}e^{-x}\,dx = \int_0^\infty e^{-(1-t)x}\,dx \\ &= -\frac{1}{1-t}e^{-(1-t)x}\bigg|_0^\infty = \frac{1}{1-t}(-e^{-(1-t)\infty} + e^{-(1-t)0}) \\ &= \frac{1}{1-t}(-0+1) = \frac{1}{1-t}, \end{aligned}$$

as long as $t < 1$. Because $M_{X_1}(t) = M_{X_2}(t) = M(t)$ exists for t in an interval $(-\infty, 1)$ surrounding $t = 0$, Property 4 allows us to obtain the moment generating function of $X_1 + X_2$:

$$\begin{aligned} M_{X_1+X_2}(t) &= M_{X_1}(t)M_{X_2}(t) = [M(t)]^2 \\ &= \left(\frac{1}{1-t}\right)^2 = \frac{1}{(1-t)^2}. \end{aligned}$$

Property 2 now yields the moments of $X_1 + X_2$. Thus

$$\begin{aligned} \mu_{X_1+X_2} &= \frac{d}{dt} M_{X_1+X_2}(t)\bigg|_{t=0} = \frac{d}{dt}\frac{1}{(1-t)^2}\bigg|_{t=0} \\ &= \frac{2}{(1-t)^3}\bigg|_{t=0} = 2. \end{aligned}$$

Also,

$$\begin{aligned} E(X_1 + X_2)^2 &= \frac{d^2}{dt^2} M_{X_1+X_2}(t)\bigg|_{t=0} \\ &= \frac{d}{dt}\frac{2}{(1-t)^3}\bigg|_{t=0} = \frac{6}{(1-t)^4}\bigg|_{t=0} = 6, \end{aligned}$$

and thus,

$$\begin{aligned} \sigma^2_{X_2+X_2} &= E(X_1 + X_2)^2 - (\mu_{X_1+X_2})^2 \\ &= 6 - (2)^2 = 2. \end{aligned}$$

Suppose that the continuous random variable V has probability density function

$$f_V(v) = \begin{cases} ve^{-v}, & \text{if } v \geq 0, \\ 0, & \text{if } v < 0. \end{cases}$$

Then V has moment generating function

$$\begin{aligned} M_V(t) &= \int_0^\infty e^{tv} v e^{-v}\, dv = \int_0^\infty v e^{-(1-t)v}\, dv \\ &= v\left(-\frac{1}{1-t} e^{-(1-t)v}\right)\Bigg|_0^\infty + \int_0^\infty \frac{1}{1-t} e^{-(1-t)v}\, dv \\ &= 0 + \frac{-1}{(1-t)^2} e^{-(1-t)v}\Bigg|_0^\infty \\ &= \frac{1}{(1-t)^2}, \qquad -\infty < t < 1, \end{aligned}$$

where we have integrated by parts and used the fact that $ve^{-(1-t)v}\big|_{v=\infty} = 0$ for $t < 1$. We now observe that $M_V(t) = M_{X_1+X_2}(t)$. Hence by the uniqueness property of moment generating functions, $X_1 + X_2$ has the probability density function $f_V(v)$. This demonstrates how moment generating functions can be used to obtain the distribution of sums of independent random variables. ◆

Property 5, preservation of limits, is used in the Appendix to Chapter 9 to prove the famous Central Limit Theorem.

Verification

Demonstrations of Properties 1, 2, and 5 of moment generating functions require introduction of considerable mathematical theory not needed elsewhere in this book. Consequently, we omit such demonstrations. We remark that verification of Property 1 is accomplished by exhibiting *inversion formulas* that show theoretically how to obtain the cumulative distribution function $F(x)$ corresponding to a moment generating function $M(t)$. (The methods for deriving such formulas are more advanced and are treated in books on Laplace transforms.) Some insight into the justification for Property 2 can be gained by noting that because

$$\frac{d^k}{dt^k} e^{tx} = x^k e^{tx},$$

we have

$$\frac{d^k}{dt^k}[M(t)] = \frac{d^k}{dt^k} E(e^{tX}) = E\left(\frac{d^k}{dt^k} e^{tX}\right) = E[X^k e^{tX}], \tag{4.8}$$

provided that the order in which the derivative d^k/dt^k and the expected value $E(e^{tX})$ are taken can be interchanged. The formula for $E[X^k]$ in Property 2 then follows by setting $t = 0$ in (4.8).

Verification of Property 3 is straightforward:

$$M_{aX+b}(t) = E(e^{t(aX+b)}) = E(e^{tb}e^{taX}) = e^{tb}E(e^{atX}) = e^{tb}M(at).$$

Property 4 can be derived in the case where X and Y are independent discrete random variables. Let $V = X + Y$. Then

$$\begin{aligned} M_V(t) = E(e^{tV}) &= \sum_v e^{tv}p_V(v) = \sum_v e^{tv} \sum_{x+y=v} p(x, y) \\ &= \sum_v \sum_{x+y=v} e^{t(x+y)}p_X(x)p_Y(y) \\ &= \left[\sum_x e^{tx}p_X(x)\right]\left[\sum_y e^{ty}p_Y(y)\right] \\ &= M_X(t)M_Y(t) \end{aligned}$$

Here use has been made of the fact that $e^{a+b} = e^a e^b$, and also the fact that

$$\sum_v \sum_{x+y=v} g(x, y) = \sum_x \sum_y g(x, y)$$

for any function $g(x, y)$. Verification of Property 4 in the case of continuous variables X, Y follows similarly but requires using line integrals of the form $\int_{x+y=v} f(x, y)\, dx\, dy$.

Property 4 can be generalized as follows. Let $X_1, X_2, \ldots, X_r$ be mutually independent random variables with moment generating functions $M_{X_1}(t), M_{X_2}(t), \ldots, M_{X_r}(t)$, respectively. Also let $a_1, a_2, \ldots, a_r$ be constants. The moment generating function $M_S(t)$ of the weighted sum $S = \sum_{i=1}^r a_iX_i$ is given by the formula

$$M_Y(t) = M_{X_1}(a_1t)M_{X_2}(a_2t) \cdots M_{X_r}(a_rt). \tag{4.9}$$

In particular, if $X_1, \ldots, X_n$ all have the same distribution, and thus the same moment generating function $M(t)$, the moment generating functions of their sum $S_n = \sum_{i=1}^n X_i$ and average $\overline{X} = (1/n)\sum_{i=1}^n X_i$ are

$$M_{S_n}(t) = [M(t)]^n, \qquad M_{\overline{X}}(t) = \left[M\left(\frac{t}{n}\right)\right]^n \tag{4.10}$$

respectively.

Verification of (4.9) proceeds recursively using Property 4 and the following fact: *If $X_1, X_2, \ldots, X_r$ are mutually independent, then a_kX_k is independent of $a_1x_1 + a_2X_2 + \cdots + a_{k-1}X_{k-1}$ for all k, $2 \le k \le r$.* This fact follows directly from the definition of mutual independence—namely, that any information about $X_1, X_2, \ldots, X_{k-1}, X_{k+1}, \ldots, X_r$ does not change probabilities for the remaining variable, X_k. Here the information about $X_1, X_2, \ldots, X_{k-1}, X_{k+1}, \ldots, X_r$ is a statement about the value of $a_1X_1 + a_2X_2 + \cdots + a_{k-1}X_{k-1}$. Because the conditional probabilities of events defined for a_kX_k given this statement are the same as the unconditional probabilities, it follows that every event defined for $a_1X_1 + a_2X_2 + \cdots + a_{k-1}X_{k-1}$ is independent of every event for a_kX_k, and hence $a_1X_1 + a_2X_2 + \cdots + a_{k-1}X_{k-1}$ and a_kX_k are probabilistically independent random variables. Consequently, repeatedly applying Property 4 yields

$$\begin{aligned} M_{\Sigma^r a_iX_i}(t) &= M_{\Sigma^r a_iX_i}(t)M_{a_rX_r}(t) \\ &= M_{\Sigma^r a_iX_i}(t)M_{a_{r-1}X_r}(t)M_{a_rX_r}(t) \\ &= \cdots \\ &= M_{a_1X_1}(t)M_{a_2X_2}(t) \cdots M_{a_rX_r}(t), \end{aligned}$$

and (4.9) now follows from Property 3.

The Characteristic Function

A major drawback to the use of the moment generating function is that such a function is not *well defined* [exists for all t in an interval (t_L, t_U) containing 0] for all random variables X. In contrast, the **characteristic function** (or Fourier transform) has properties parallel to Properties 1 to 5 of moment generating functions and is well defined for *all* random variables. With the use of the imaginary (complex) number $i = \sqrt{-1}$, the characteristic function $C(t)$ of a random variable X is defined to be

$$C(t) = E(e^{itX}) = \begin{cases} \sum_x e^{itx}p(x), & \text{if } X \text{ is discrete,} \\ \int_{-\infty}^{\infty} e^{itx}f(x)\,dx, & \text{if } X \text{ is continuous,} \end{cases}$$

for all t, $-\infty < t < \infty$. Understanding and use of this kind of generating function requires knowledge of the mathematical theory of complex variables. For this reason, the study of characteristic functions is left to more advanced books on probability theory.

Notes and References

Equations (3.5) and (3.6) give formulas for the mean and variance of a weighted sum $S = \sum_{i=1}^r a_iS_i$ of independent random variables $X_1, \ldots, X_r$ in terms of the means and variances of these variables. An obvious question to ask is whether similar results exist for the case when $X_1, \ldots, X_r$ are statistically dependent.

For the case of the mean of S, it has already been mentioned that equation (3.5) continues to be valid regardless of whether the X_i's are independent or dependent. The result $\mu_S = \sum_{i=1}^r a_i \mu_{X_i}$ is verified in Chapter 11. On the other hand, when the random variables are statistically dependent, the formula for the variance of S involves additional terms that reflect this statistical dependence among the variables.

As its name imples, the **covariance** σ_{XY} between two random variables X and Y measures how these variables vary together. Large positive values of σ_{XY} indicate that the two variables X, Y tend to increase together in a nearly linear fashion. Large negative values of σ_{XY} indicate an nearly inverse relationship—as X increases, Y tends to decrease. Values of σ_{XY} near 0 indicate that X is not linearly well predicted by Y, or vice versa. In particular, $\sigma_{XY} = 0$ when X and Y are probabilistically independent. However, it is possible for X and Y to be probabilistically dependent, yet $\sigma_{XY} = 0$. That is, independence of X, Y implies that $\sigma_{XY} = 0$ but *not* vice versa. An example of a case where $\sigma_{XY} = 0$ but X and Y are probabilistically dependent is given in Chapter 11.

A difficulty with the covariance is that it is expressed in terms of the product of the units of measurement of X and Y. That is, if X is measured in feet and Y in pounds, σ_{XY} is in foot $\times$ pounds units. Instead, for purposes of measuring the strength of (linear) dependence between X and Y, the **correlation**

$$\rho_{XY} = \frac{\sigma_{XY}}{\sqrt{\sigma_X^2 \sigma_Y^2}} = \frac{\sigma_{XY}}{\sigma_X \sigma_Y}$$

is often used. This index is dimension-free and takes values in the interval $[-1, 1]$. Either σ_{XY} or ρ_{XY} may be given in problems, so that one needs to have formulas in terms of either index.

The formula for the variance of $S = \sum_{i=1}^r a_i X_i$ can be given in terms of the variances of the individual variables X_i and either covariances $\sigma_{X_i X_j}$ or correlations $\rho_{X_i X_j}$ between pairs of these variables. Thus

$$\begin{aligned}\sigma_S^2 &= \sum_{i=1}^r a_i^2 \sigma_{X_i}^2 + 2 \sum_{i<j} a_i a_j \sigma_{X_i X_j} \\ &= \sum_{i=1}^r a_i^2 \sigma_{X_i}^2 + 2 \sum_{i<j} a_i a_j \sigma_{X_i} \sigma_{X_j} \rho_{X_i X_j}\end{aligned}$$

as verified in Chapter 11.

EXAMPLE

If we invest 5 thousand dollars in stocks and 3 thousand dollars in bonds, and the random rates of return for these investments are X, Y, respectively, the total return from these investments will be

$$S = 5X + 3Y$$

thousands of dollars.

Suppose that X has a mean $\mu_X = 0.20$ (20% return) and variance $\sigma_X^2 = 0.25$, whereas Y has mean $\mu_Y = 0.09$ and variance $\sigma_Y^2 = 0.16$. Because returns on both stocks and bonds depend on the state of the economy, X and Y are probabilistically dependent. Suppose that the correlation between X and Y is $\rho_{XY} = 0.5$. The mean and variance of the total return $S = 5X + 3Y$ are

$$\begin{aligned} \mu_S &= 5\mu_X + 3\mu_Y = 5(0.20) + 3(0.09) \\ &= 1.27 \text{ thousand dollars,} \end{aligned}$$

$$\begin{aligned} \sigma_S^2 &= (5)^2\sigma_X^2 + (3)^2\sigma_Y^2 + 2(5)(3)\sigma_X\sigma_Y\rho_{XY} \\ &= (25)(0.25) + (9)(0.16) + 30(\sqrt{0.25})(\sqrt{0.16})(0.5) \\ &= 6.25 + 1.44 + 3.00 = 10.69. \end{aligned}$$

Suppose instead that the bond market commonly goes up when the stock market goes down. A negative correlation of $\rho_{XY} = -0.5$ would then be more appropriate. Changing from a positive correlation to a negative correlation does not affect the value of the mean of S, which remains equal to 1.27 thousand dollars. However, the variance of S becomes

$$\begin{aligned} \sigma_S^2 &= (25)(0.25) + (9)(0.16) + 30(\sqrt{0.25})(\sqrt{0.16})(-0.50) \\ &= 6.25 + 1.44 - 3.00 = 4.69. \end{aligned}$$

Note that there is less variation in the total returns when the correlation is negative than when it is positive. Diversification into two investments that each act to counteract movements in the other (rates of return of one go up when rates of return of the other go down) results in less variation in total return than investing in similar investments whose movements reinforce each other. ◆

Glossary and Summary

Joint probability distribution: A rule assigning probabilities to the outcomes $(x_1, \ldots, x_r)$ obtained from jointly observing r random variables $X_1, \ldots, X_r$.

Joint probability mass function: A function that assigns to each r-tuple $(x_1, \ldots, x_r)$ the probability $p(x_1, \ldots, x_r)$ of the event $\{X_1 = x_1$ and $X_2 = x_2$ and $\cdots$ and $X_r = x_r\}$.

Joint probability density function: A function that plays the role for joint distributions of continuous random variables $X_1, \ldots, X_r$ that the probability density function plays for a single continuous variable X; thus probabilities are found by taking volumes under the graph of the density function.

Joint cumulative distribution function: A function that assigns to each r-tuple $(x_1, \ldots, x_r)$ the probability $F(x_1, \ldots, x_r)$ of the event $\{X_1 \le x_1$ and $X_2 \le x_2$ and $\cdots$ and $X_r \le x_r\}$.

Mutual probabilistic independence of random variables: The random variables $X_1, X_2, \ldots, X_r$ are mutually probabilistically independent if the events $\{X_i$ is in $A_i\}$, $i = 1, 2, \ldots, r$ are mutually probabilistically independent for all choices of the collections of values $A_1, A_2, \ldots, A_r$.

Independent identically distributed random variables: Mutually probabilistically independent random variables $X_1, X_2, \ldots, X_r$, each of which has the same probability distribution.

Random sample of size *r*: r independent identically distributed random variables. (A definition often used in the statistical literature.)

Moment generating function: A function $M(t)$ which for each real number t gives the expected value of e^{tX}. If this expected value exists (is well defined and finite), derivatives of $M(t)$ evaluated at $t = 0$ give the moments of the distribution of the random variable X.

Characteristic function: A function that is similar to the moment generating function except that it involves the expected value of e^{itX}, where i is the imaginary number $\sqrt{-1}$.

Useful Facts

If $X_1, \ldots, X_r$ are mutually probabilistically independent random variables, then:

1. When $X_1, \ldots, X_r$ are discrete, their joint probability mass function $p(x_1, \ldots, x_r)$ is the product of their individual (marginal) probability mass functions $p_{X_i}(x_i)$, $i = 1, 2, \ldots, r$; that is,

$$p(x_1, \ldots, x_r) = p_{X_1}(x_1)p_{X_2}(x_2) \cdots p_{X_r}(x_r).$$

2. Their joint cumulative distribution function $F(x_1, \ldots, x_r)$ is the product of their individual cumulative distribution functions $F_{X_i}(x_i)$, $i = 1, \ldots, r$; that is,

$$F(x_1, \ldots, x_r) = F_{X_1}(x_1)F_{X_2}(x_2) \cdots F_{X_r}(x_r).$$

3. When $X_1, \ldots, X_r$ are continuous and have a joint probability density function $f(x_1, \ldots, x_r)$, this density function is the product

$$f(x_1, \ldots, x_r) = f_{X_1}(x_1) f_{X_2}(x_2) \cdots f_{X_r}(x_r)$$

of the individual (marginal) probability density functions of $f_{X_i}(x_i)$ of X_i, $i = 1, \ldots, r$.

4. The mean and variance of the weighted sum $S = \sum_{i=1}^{r} a_i X_i$ are given in terms of the means μ_{X_i} and variances $\sigma_{X_i}^2$ of the individual variables X_i, $i = 1, \ldots, r$, by

$$\mu_S = \sum_{i=1}^{r} a_i \mu_{X_i}, \qquad \sigma_S^2 = \sum_{i=1}^{r} a_i^2 \sigma_{X_i}^2$$

respectively.

5. The moment generating function $M_S(t)$ of the weighted sum $S = \sum_{i=1}^{r} a_i X_i$ is given in terms of the individual moment generating functions $M_{X_i}(t)$ of the X_i, $i = 1, \ldots, r$, by

$$M_S(t) = M_{X_1}(a_1 t) M_{X_2}(a_2 t) \cdots M_{X_r}(a_r t).$$

6. If $X_1, X_2, \ldots, X_n$ are independent and identically distributed random variables whose common distribution has mean μ, variance σ^2, and moment generating function $M(t)$, then:
   (a) The mean, variance, and moment generating function of the sum $S = \sum_{i=1}^{n} X_i$ are given by

$$\mu_S = n\mu, \qquad \sigma_S^2 = n\sigma^2, \qquad M_S(t) = [M(t)]^n,$$

   respectively.
   (b) The mean, variance, and moment generating function of the average $\overline{X} = n^{-1}\sum_{i=1}^{n} X_i$ are given by

$$\mu_{\overline{X}} = \mu, \qquad \sigma_{\overline{X}}^2 = \frac{\sigma^2}{n}, \qquad M_{\overline{X}}(t) = \left[M\left(\frac{t}{n}\right)\right]^n,$$

   respectively.
   (c) **Law of Large Numbers.** As $n \to \infty$, the distribution of $\overline{X}$ concentrates more and more closely about its mean $\mu_{\overline{X}} = \mu$; that is, for any positive constant c (no matter how small),

$$\lim_{n\to\infty} P\{\mu - c \leq \overline{X} \leq \mu + c\} = 1.$$

7. The moment generating function $M(t)$ of a random variable X (assumed to exist for an internal of values of t containing 0) has the following properties:
   (a) It uniquely defines the distribution of X in the sense that there is one and only one distribution, or cumulative distribution function $F(x)$, corresponding to each moment generating function $M(t)$. Thus if X and Y both have the same moment generating function $M(t)$, they have the same distribution (probability mass function, probability density function, or cumulative distribution function).
   (b) Moments $E[X^k]$ of X are found by taking the kth derivative of $M(t)$ evaluated at $t = 0$; that is,

$$E[X^k] = \left[\frac{d^k}{(dt)^k} M(t)\right]_{t=0}.$$

   (c) The moment generating function $M_{aX+b}(t)$ of $Y = aX + b$ is given by

$$M_{aX+b}(t) = e^{bt}M(at).$$

In addition, the moment generating function has a certain preservation of limits property (see Section 4), and it is sometimes useful to remember that $M(0)$ is always equal to 1.

Exercises

1. In Example 2.3, find the probability mass function $p_S(s)$ of the sum $S = X_1 + X_2$ of the number of dots X_1, X_2 on the upturned faces of two independently tossed dice. Then find the mean μ_S and variance σ_S^2 of S from $p_S(s)$ and show that

$$\mu_S = \mu_{X_1} + \mu_{X_2}, \qquad \sigma_S^2 = \sigma_{X_1}^2 + \sigma_{X_2}^2,$$

by calculating μ_{X_1}, μ_{X_2}, $\sigma_{X_1}^2$, $\sigma_{X_2}^2$.
2. In Example 3.1, find the probability mass function $p_V(v)$ of $V = X - Y$. From $p_V(v)$ determine μ_V and σ_V^2. Show that

$$\mu_V = \mu_X - \mu_Y, \qquad \sigma_V^2 = \sigma_X^2 + \sigma_Y^2,$$

by calculating μ_X, μ_Y, σ_X^2, σ_Y^2.
3. Two prize pedigreed dogs are each about to produce their first litter of puppies. Puppies of the first dog's breed can be sold for $200 each.

From breeding records the distribution of the number X of puppies per litter produced by dogs of this breed is the following:

x	3	4	5	6	7
Probability	0.20	0.25	0.20	0.20	0.15

The second dog's breed is less popular, and puppies sell for $100 each. The distribution of the number Y of puppies produced per litter by this breed is the following:

y	3	4	5	6	7
Probability	0.10	0.25	0.35	0.20	0.10

(a) Find the means μ_X, μ_Y of X and Y. Which breed of dog tends to produce the largest litters?

(b) Find the variances σ_X^2, σ_Y^2 of X and Y. For which breed is the litter size most variable?

(c) The total proceeds, in hundreds of dollars, for selling the puppies will be $2X + Y$. Assuming that X and Y are independent, find the mean and variance of the total proceeds.

4. A contestant on a quiz show is asked 3 questions from each of 4 categories: Movies, History, Sports, and Politics. Assume that all 12 questions are answered independently by the contestant. Let $X_i = 1$ if the contestant answers the ith question correctly, and $X_i = 0$ otherwise, $i = 1, 2, \ldots, 12$. Then $T = \sum_{i=1}^{12} X_i$ is the total number of questions answered correctly. In each category, one question is easy, one question is of moderate difficulty, and one question is hard. Easy questions are answered correctly with probability 0.8, moderately difficult questions with probability 0.5, and hard questions with probability 0.1.

(a) Note that if $P\{X = 1\} = p$, $P\{X = 0\} = 1 - p$, then $\mu_X = p$, $\sigma_X^2 = p(1 - p)$. Using this fact, find the mean, μ_T, and variance, σ_T^2, of the total number T of questions answered correctly.

(b) Contestants are paid $10 for correct answers to easy questions (say, questions 1, 2, 3, 4), $25 for correct answers to moderately difficult questions (questions 5, 6, 7, 8), and $100 for correct answers to each hard question (questions 9, 10, 11, 12). Find the mean and variance of a contestant's winnings:

$$W = 10 \sum_{i=1}^{4} X_i + 25 \sum_{i=5}^{8} X_i + 100 \sum_{i=9}^{12} X_i.$$

5. The average price X (in thousands of dollars) at which a manufacturer will be able to sell a drill press is uncertain but has mean $\mu_X = 10$ and variance $\sigma_X^2 = 4$. The average cost Y in thousands of dollars to produce this product is also uncertain, and is assumed to be a random variable with mean $\mu_Y = 5$ and variance $\sigma_Y^2 = 36$. There is also a fixed cost of 10 thousand dollars for initiating production. Consequently, if the firm produces (and sells) 25 drill presses, the total profit is

$$T = 25X - 25Y - 10$$

thousand dollars.

(a) Find μ_T, the mean total profit.

(b) If X and Y are independent, find the variance, σ_T^2, of T.

(c) If X and Y have correlation $\rho_{X,Y} = 0.1$, find the variance of T (see "Notes and References").

6. The amount of a certain chemical that is precipitated in a chemical reaction is a continuous random variable X with probability density function

$$f(x) = \begin{cases} 0.003x^2, & \text{if } 0 \le x < 10, \\ 0, & \text{otherwise.} \end{cases}$$

(a) Find the mean, μ_X, and variance, σ_X^2, of X.

(b) The chemical reaction is independently repeated 10 times, yielding precipitation amounts $X_1, X_2, \ldots, X_{10}$ all having the same distribution as X. Find the values of the mean, $\mu_{\overline{X}}$, and variance, $\sigma_{\overline{X}}^2$, of the sample average

$$\overline{X} = \frac{1}{10} \sum_{i=1}^{10} x_i.$$

(c) Under the assumptions of part (b), find the values of the mean, μ_S, and variance, σ_S^2, of the total amount, $S = \sum_{i=1}^{10} X_i$, of chemical precipitated.

7. The diameter (in centimeters) of a ball bearing produced by a firm is a continuous random variable X with density function

$$f_X(x) = \begin{cases} 12(x - 1.5)^2, & \text{if } 1 \le x \le 2, \\ 0, & \text{otherwise.} \end{cases}$$

The diameter (in centimeters) of a circular hole drilled to receive the ball bearing is a continuous random variable Y with density function

$$f_Y(y) = \begin{cases} 8(2.5 - y), & \text{if } 2 \le y \le 2.5, \\ 0, & \text{otherwise.} \end{cases}$$

The ball bearing is manufactured independently (probabilistically) of the drilling process that produces the circular hole. The difference $L = Y - X$ describes the looseness of fit (in centimeters) of the ball bearing in the hole. (Notice that $Y - X \ge 0$ always, so the ball bearing will always fit the hole.)

(a) Find μ_X, μ_Y, σ_X^2, σ_Y^2.

(b) Then find the mean, μ_L, and variance, σ_L^2, of L.

8. Two randomly selected groups of people are independently given a reaction time test, the first group after taking a particular drug, and the second group after taking a placebo. The reaction times $X_1, X_2, \ldots, X_n$ of the people taking the drug are assumed to be mutually independent, and each X_i has mean μ and variance σ^2. The reaction times $Y_1, Y_2, \ldots, Y_m$ of the people taking the placebo are assumed to be mutually independent, and each Y_j has mean ν and variance σ^2.

(a) To evaluate the effect on reaction time of taking the drug, we estimate $\Delta = \mu - \nu$. Because $\overline{X} = (X_1 + X_2 + \cdots + X_n)/n$ is a reasonable approximation to the mean μ of any X_i when n is large, and $\overline{Y} = (Y_1 + Y_2 + \cdots + Y_m)/m$ is a reasonable approximation to the mean ν of any Y_j when m is large, we estimate Δ by $\overline{X} - \overline{Y}$. Find the mean and variance of $\overline{X} - \overline{Y}$.

(b) Use the Bienaymé–Chebychev inequality (see Chapter 6) and the answer to part (a) to show that the probability distribution of $\overline{X} - \overline{Y}$ concentrates more and more probability around the value Δ as the sample sizes n and m both become large.

(c) Assume that $n = m$ and that $\sigma^2 = 4$. How many people must take the drug (and how many people must take the placebo) in order that

$$P\{-1 \le \overline{X} - \overline{Y} - \Delta \le 1\} \ge 0.95?$$

Answer using the Bienaymé–Chebychev inequality.

9. (a) Find the moment generating functions corresponding to the following probability mass functions:

$$\text{(i)}\quad p(x) = \begin{cases} 1/2, & \text{if } x = -1, \\ 1/2, & \text{if } x = 1, \\ 0, & \text{otherwise.} \end{cases}$$

$$\text{(ii)}\quad p(x) = \begin{cases} \frac{1}{4}, & \text{if } y = -2, \\ \frac{1}{2}, & \text{if } y = 0, \\ \frac{1}{4}, & \text{if } y = 2, \\ 0, & \text{otherwise.} \end{cases}$$

(iii) The probability mass function of $T = X_1 + X_2$, where X_1, X_2 are probabilistically independent, each having the probability mass function $p(x)$ of (i). (Use Property 4 of moment generating functions. Do not obtain the probability mass function of T).

(b) Using the moment generating functions found in part (a), show that $T = X_1 + X_2$ in (iii) has the probability mass function $p(x)$ given in (ii).

10. (a) Find the moment generating functions corresponding to the following probability density functions:

$$\text{(i)}\quad f(x) = \begin{cases} \dfrac{e^x}{e-1}, & \text{if } 0 \le x \le 1, \\ 0, & \text{otherwise.} \end{cases}$$

$$\text{(ii)}\quad f(x) = \begin{cases} \dfrac{xe^x}{(e-1)^2}, & \text{if } 0 \le x < 1, \\ \dfrac{(2-x)e^x}{(e-1)^2}, & \text{if } 1 \le x \le 2, \\ 0, & \text{otherwise.} \end{cases}$$

(*Hint:* Using integration by parts, $\int_a^b xe^{cx}\,dx = c^{-1}xe^{cx}\big|_a^b - \int_a^b c^{-1}e^{cx}\,dx$.)

(iii) The probability density function of $T = X_1 + X_2$, where X_1, X_2 are probabilistically independent, each having the probabiliity density function $f(X)$ of (i). (Use Property 4 of moment generating functions. Don't try to obtain the probability density function of T.)

(b) Using the results in parts (ii) and (iii), show that $T = X_1 + X_2$ in (iii) has the probability density function given in (ii).

11. The probability density function

$$f(x) = \tfrac{1}{2}e^{-|x|}, \qquad -\infty < x < \infty,$$

defines a **double exponential distribution.**

(a) Find the moment generating function of the double exponential distribution. (*Hint:* Break up the integral from $-\infty$ to ∞ into the part from $-\infty$ to 0, and the part from 0 to ∞.)

(b) Find the moment generating function of $Y = X - 2$ where X has the double exponential distribution.

(c) Find the first four moments of the double exponential distribution.

(d) Let W_1, W_2 be independent continuous random variables, each with probability density function

$$f_W(w) = \begin{cases} e^{-w}, & \text{if } w \geq 0, \\ 0, & \text{if } w < 0. \end{cases}$$

Show that $W_1 - W_2$ has the double exponential distribution.

12. The moment generating function of a certain random variable X is

$$M(t) = \exp(\tfrac{1}{2}t^2).$$

(a) Find the first four moments, $E[X^k]$, $k = 1, 2, 3, 4$, of X. Find the variance of X.

(b) If X and Y are independent random variables each with the moment generating function $M(t)$, find the moment generating function $M_{X+Y}(t)$ of $X + Y$.

(c) Use the moment generating function $M_{X+Y}(t)$ of $X + Y$ to find the first four moments of $X + Y$ and the variance of $X + Y$. Check your results for the mean and variance by using the formulas

$$\mu_{X+Y} = \mu_X + \mu_Y, \qquad \sigma^2_{X+Y} = \sigma^2_X + \sigma^2_Y$$

to find the mean and variance of $X + Y$.

(d) Under the assumptions of part (b) find the moment generating function of $X - Y$ and show that $X + Y$ and $X - Y$ have the same distribution.

13. An automobile dealer can sell a car at a profit X in dollars that is a continuous random variable with probability density function

$$f(x) = \begin{cases} (0.01)e^{-(0.01)(x-100)}, & \text{if } x \geq 100, \\ 0, & \text{otherwise.} \end{cases}$$

(a) Find the moment generating function $M(t)$ of X and use $M(t)$ to find the mean and variance of X.

(b) Check your calculations by finding the mean of X directly from the density function $f(x)$.

(c) This type of car has probability (0.1) of having a certain defect, which must be corrected at a cost of \$50 to the dealer. The cost Y to the dealer is thus a discrete random variable with probability mass function

$$p(y) = \begin{cases} 0.1, & \text{if } y = 50, \\ 0.9, & \text{if } y = 0, \\ 0, & \text{otherwise.} \end{cases}$$

Find the mean, variance, and moment generating function of Y.

(d) The dealer believes that X and Y are independent random variables. (The dealer's profit on the original sale should have no relationship to the presence of the defect, of which both dealer and customer will be unaware.) The cost Y of correcting the defect will reduce the dealer's initial profit X on the car. Thus the dealer's final profit is $X - Y$. Find the mean, variance, and moment generating function of $X - Y$.

Theoretical Exercises

T.1. A random variable X is bounded if there exist numbers a, b with $-\infty < a \leq b < \infty$ for which $P\{a \leq X \leq b\} = 1$. Show that if X is a bounded random variable, the moment generating function $M(t)$ of X is defined for all t, $-\infty < t < \infty$.

T.2. Show that if X has density function,

$$f(x) = \frac{1}{\pi(1 + x^2)}, \qquad -\infty < x < \infty,$$

then the moment generating function $M(t)$ of X is not defined except at $t = 0$.

T.3. The **cumulant generating function** $K(t)$ of a random variable X is defined by

$$K(t) = \log M(t).$$

The quantities

$$\kappa_r = \left[\frac{d^r}{(dt)^r} K(t)\right]_{t=0}, \qquad r = 1, 2, \ldots$$

are known as the rth **cumulants** of the random variable X.

(a) Show that $\kappa_1 = \mu_X$, $\kappa_2 = \sigma_X^2$. Find κ_3 and κ_4 in terms of the moments of X.

(b) If X and Y are independent random variables, show that

$$K_{X+Y}(t) = K_X(t) + K_Y(t).$$

(c) If X and Y are independent random variables, show that the rth cumulant of $X + Y$ is the sum of the rth cumulants for X and Y.

8

Special Distributions: Discrete Case

Certain basic probability distributions are applicable in a wide variety of contexts, and thus arise repeatedly in practice. In this and in the next two chapters, we describe some frequently encountered discrete and continuous distributions.

1 Bernoulli Trials and the Binomial Distribution

A random experiment with two possible outcomes is called a **Bernoulli trial.** A Bernoulli trial occurs, for example, when we observe whether or not a person contracts a certain disease, whether the price of a stock does or does not rise, or whether a student does or does not answer a question correctly. In each such trial, we pick one outcome of the two possible outcomes and count how often this outcome is observed in repetitions of the experiment (trial). The outcome whose occurrences are counted is called a **success,** and the remaining outcome is called a **failure.** A ''success'' need not be a desirable outcome; for example, we may count illnesses or accidents. Rather, the terminology ''success'' is used simply to identify the outcome of a Bernoulli trial that is of chief interest to us.

More complex random experiments can also yield Bernoulli trials when interest focuses on the occurrence or nonoccurrence of a particular event. For example, we may observe the time T, in minutes, required to complete a certain task. Although the random variable T will generally have more than two possible values, we may be interested only in whether the task takes more than 5 minutes to complete. In this case, the event $E = \{T > 5\}$ can be called a success, the complementary event $E^c = \{T \leqslant 5\}$ is then called a failure, and we can regard the experiment as being a Bernoulli trial. Note that in repeated trials of the experiment, the number of successes observed is equal to the frequency, $\#E$, of the event E, so that in this manner Bernoulli trials can provide a framework for the process (Chapters 1 and 5) of fitting probability models to data.

The Bernoulli Distribution

Associated with a Bernoulli trial is the indicator X of success. That is, $X(\omega) = 1$ if the outcome ω is a success, and $X(\omega) = 0$ if the outcome ω is a failure. If p is the probability of the outcome called a success, then X is a discrete random variable having probability mass function

$$(1.1) \qquad p(x) = \begin{cases} 1 - p, & \text{if } x = 0 \\ p, & \text{if } x = 1 \\ 0, & \text{otherwise} \end{cases} = \begin{cases} p^x(1 - p)^{1-x}, & \text{if } x = 0, 1, \\ 0, & \text{otherwise.} \end{cases}$$

Any random variable having the probability mass function (1.1) for some p, $0 \le p \le 1$, is said to be a **Bernoulli random variable** and the distribution (1.1) is called a **Bernoulli distribution.** The form of the probability mass function (1.1) depends on the constant p, which is called the **parameter** of the Bernoulli distribution.

A straightforward computation shows the following:

The mean μ and variance σ^2 of the Bernoulli distribution (1.1) are

$$\mu = p, \qquad \sigma^2 = p(1 - p),$$

respectively.

Indeed,

$$\mu = \sum_{x=0}^{1} xp(x) = (0)(1 - p) + (1)p = p,$$

$$\sigma^2 = \sum_{x=0}^{1} (x - \mu)^2 p(x) = (0 - p)^2(1 - p) + (1 - p)^2 p = p(1 - p).$$

Binomial Distribution

As mentioned earlier, one often observes n independent replications of a Bernoulli trial and counts the number Z of successes that occur. The possible values of Z are 0, 1, 2, . . . , n. Note that Z is the frequency with which a success occurs in the n trials, and that

$$\hat{p} = \frac{Z}{n}$$

is the relative frequency of success. If n is large, the Law of Large Numbers states that $\hat{p}$ will be close to the true probability of success p. When n is of small or moderate size, it was seen in Chapter 1 that the values of $\hat{p}$ fluctuate. It is thus important to find the distribution of $\hat{p}$. Because the values of $\hat{p}$ are fractions, whereas the values of Z are integers, it is more convenient to find the distribution of Z and then use the fact that

$$P\left\{\hat{p} = \frac{k}{n}\right\} = P\{Z = k\}, \qquad k = 0, 1, \ldots, n,$$

to find the distribution of $\hat{p}$. Further, as will be seen, the distribution of Z has other important applications besides its use in determining the distribution of $\hat{p}$.

To illustrate how the distribution of Z is determined, consider the case where $n = 3$ independent replications of a Bernoulli trial are observed. Here, any of the following eight outcomes may be obtained: SSS, SSF, SFS, FSS, SFF, FSF, FFS, FFF, where for example the outcome SFS represents the situation where the first and third trials were successes and the second trial was a failure. To find probabilities for the values of Z, we first find probabilities for these outcomes, and then add the probabilities of all outcomes that correspond to a particular value z of Z to find the probability $P\{Z = z\}$, $z = 0$, 1, 2, 3.

Probabilities for the outcomes are calculated from the given facts that (1) trials are independent and (2) each trial has the same probability p of resulting in a success. For example, the probability of the outcome SFS is found as follows:

$$\begin{aligned} p(SFS) &= P(\text{first trial is } S \textit{ and } \text{second trial is } F \textit{ and } \text{third trial is } S) \\ &= P(\text{first trial is } S)\, P(\text{second trial is } F)\, P(\text{third trial is } S) \\ &= p(1 - p)p = p^2(1 - p) \end{aligned}$$

using first the fact that the trials are independent, and then the fact that the probability of a success S on any trial is equal to p. Exhibit 1.1 gives the outcomes, their probabilities, and the value of Z corresponding to each outcome.

EXHIBIT 1.1 *Probabilities of the Outcomes for an Experiment Consisting of $n = 3$ Independent Bernoulli Trials with Common Probability p of Success S*

Outcome	*Probability*	z
SSS	$(p)(p)(p) = p^3$	3
SSF	$(p)(p)(1-p) = p^2(1-p)$	2
SFS	$(p)(1-p)(p) = p^2(1-p)$	2
FSS	$(1-p)(p)(p) = p^2(1-p)$	2
SFF	$(p)(1-p)(1-p) = p(1-p)^2$	1
FSF	$(1-p)(p)(1-p) = p(1-p)^2$	1
FFS	$(1-p)(1-p)(p) = p(1-p)^2$	1
FFF	$(1-p)(1-p)(1-p) = (1-p)^3$	0

Notice from Exhibit 1.1 that all outcomes associated with the same value of z have the same probability. Thus the probability that $Z = z$ equals the product of the number of outcomes that correspond to the value z and the common probability of each such outcome. For example, there are 3 outcomes (SSF, SFS, FSS) that correspond to the value $z = 2$ and each such outcome has probability $p^2(1 - p)$. Therefore

$$P\{Z = z\} = (3)\,[p^2(1 - p)].$$

In this way, we obtain the following probability model for Z:

z	0	1	2	3
$p(z) = P\{Z = z\}$	$(1-p)^3$	$3p(1-p)^2$	$3p^2(1-p)$	p^3

In general, if Z is the number of successes in n independent replications of a Bernoulli trial having probability of success p, the probability distribution of Z is found by noting that for any possible value z of Z,

$$\begin{aligned} P\{Z = z\} &= \begin{bmatrix}\text{number of outcomes}\\ \text{with } z \text{ successes}\end{bmatrix}\begin{bmatrix}\text{probability of any outcome}\\ \text{with } z \text{ successes}\end{bmatrix} \\ &= \begin{bmatrix}\text{number of outcomes}\\ \text{with } z \text{ successes}\end{bmatrix}\left[p^z(1-p)^{n-z}\right]. \end{aligned}$$

The number of outcomes having z successes (and $n - z$ failures) is the same as the number of distinguishable arrangements of z S's and $n - z$ F's. By Rule 3.3 of Chapter 3, this number is given by the binomial coefficient

$$\binom{n}{z} = \frac{n!}{z!\,(n-z)!}.$$

Thus the probability mass function for Z is given by

$$(1.2)\quad p(z) = P\{Z = z\} = \begin{cases} \binom{n}{z} p^z(1-p)^{n-z}, & z = 0, 1, \ldots, n, \\ 0, & \text{otherwise.} \end{cases}$$

For example, if $n = 6$, we obtain the following probability mass function:

z	0	1	2	3	4	5	6
$p(z)$	$(1-p)^6$	$6(1-p)^5p$	$15(1-p)4p^4$	$20(1-p)^3p^3$	$15(1-p)^2p^4$	$6(1-p)p^5$	p^6

The distribution defined by the probability mass function (1.2) is called the **binomial distribution.** The functional form of $p(z)$ is determined by two **parameters** n and p, where n is the number of repeated Bernoulli trials and p is the probability of success on each one of these trials. When we wish to show the dependence of the probability mass function $p(z)$ on the parameters n and p, we write $p(z; n, p)$ instead of $p(z)$.

Exhibit 1.2 illustrates how values of the parameter p influence the probabilities for Z in the particular case $n = 10$. Observe that for any fixed value of p, the probabilities $p(z) = P\{Z = z\}$ increase, reach a maximum, and then decrease as z increases from $z = 0$ to $z = 10$. The value of z where the maximum of $p(z)$ occurs (the mode) depends on p; for $p = 0.1$, the mode is $z = 1$, whereas for $p = 0.9$, the mode is $z = 9$.

EXHIBIT 1.2. *Value of $p(z)$ for the Binomial Distribution When $n = 10$ and p Varies*

z	p								
	0.1	0.2	0.3	0.4	0.5	0.6	0.7	0.8	0.9
0	0.3487	0.1074	0.0282	0.0060	0.0010	0.0001	0.0000	0.0000	0.0000
1	0.3874	0.2684	0.1211	0.0403	0.0098	0.0016	0.0001	0.0000	0.0000
2	0.1937	0.3020	0.2335	0.1209	0.0439	0.0106	0.0014	0.0001	0.0000
3	0.0574	0.2013	0.2668	0.2150	0.1172	0.0425	0.0090	0.0008	0.0000
4	0.0112	0.0881	0.2001	0.2508	0.2051	0.1115	0.0368	0.0055	0.0001
5	0.0015	0.0264	0.1029	0.2007	0.2461	0.2007	0.1029	0.0264	0.0015
6	0.0001	0.0055	0.0368	0.1115	0.2051	0.2508	0.2001	0.0881	0.0112
7	0.0000	0.0008	0.0090	0.0425	0.1174	0.2150	0.2668	0.2013	0.0574
8	0.0000	0.0001	0.0014	0.0106	0.0439	0.1209	0.2335	0.3020	0.1937
9	0.0000	0.0000	0.0001	0.0016	0.0098	0.0403	0.1211	0.2684	0.3874
10	0.0000	0.0000	0.0000	0.0001	0.0010	0.0060	0.0282	0.1074	0.3487

Recognizing the Binomial Distribution

The main advantage of knowing that a random variable has a certain named distribution (e.g., the binomial distribution) is that it is not necessary to derive the probabilties for this random variable from first principles. Rather, one can go directly to formulas or tables to compute probabilities and descriptive indices (mean, variance, median) of the distribution. While the derivation of the binomial distribution is still fresh in our minds, let us list the facts that were used in deriving this distribution. This will permit one to recognize situations in which a discrete random variable Z has a binomial distribution.

1. First, Z was a count of the number of successes in a **fixed** number, n, of Bernoulli trials.
2. Second, the fact that the Bernoulli trials were **mutually independent** was used in deriving the probabilities of the outcomes.
3. Finally, use was made of the fact that each trial had the **same probability of success,** p.

Whenever these three requirements are met, the random variable Z has a binomial distribution with parameters n and p.

EXAMPLE 1.1

A professional baseball player has a batting average of 0.300. In a crucial series, this player bats 10 times. Let Z be the number of hits made by the player in the series. What is the probability mass function of Z? What is the probability that the player gets 5 or more hits?

To answer these questions, we first note that Z is a count of hits (success) in a fixed number, $n = 10$, of Bernoulli trials. (We assume that at every "at bat" the player either makes a hit or an out; walks and sacrifices are ignored because these are not counted as "at bats" in determining a player's batting average.) Can we assume that each time at bat is independent of all other times at bat? There is disagreement among baseball experts on this point (there may be "hitting streaks"), but statistical studies of baseball records seem to indicate that this assumption is a good approximation to reality. Of greater concern is the question of whether the probability of getting a hit is the same at each at bat. Most batters have different rates of success against different pitchers. Here, however, we are given no information about the pitchers faced by the player, so it is reasonable to assume that the player's probability of getting a hit, given by the player's batting average of $p = 0.300$, is the same for each time at bat. Thus, at least approximately, the key facts required to show that Z has a binomial distribution have been established. We conclude that Z has a binomial distribution with parameters $n = 10$ and $p = 0.300$. Using the column of Exhibit 1.2 headed by $p = 0.300$, we find that

$$\begin{aligned} P\{5 \text{ or more hits}\} \\ &= P\{Z \geq 5\} = \sum_{z=5}^{10} p(z) \\ &= 0.1029 + 0.0368 + 0.0090 + 0.0014 + 0.0001 + 0.0000 \\ &= 0.1502. \end{aligned}$$ ◆

EXAMPLE 1.2

A student takes an exam in which there are 5 hard questions and 5 easy questions. The student answers the questions independently. Let Z be the number of questions which the student answers correctly. Does Z have a binomial distribution?

Here, Z is a count of questions answered correctly (successes) in a fixed number, $n = 5 + 5 = 10$, of trials (questions). Further, we are told that the questions are answered independently. However, the wording of the example suggests that the probabilities of answering the questions correctly are not the same, presumably being smaller for hard questions than for easy questions. Thus Z does **not** have a binomial distribution. On the other hand, if every hard question has the same probability p of being answered correctly, the number of hard questions that the student answers correctly would have a binomial distribution with parameters $n = 5$ and p. ◆

EXAMPLE 1.3

Players A and B are evenly matched in ability for playing handball. They agree to continue to play games of handball until player A wins a game. Each game is played independently of all other games. Let Z be the number of games they play. Does Z have a binomial distribution?

Because A and B are said to be evenly matched, we can assume that the probability that A wins any given games is $p = 0.50$. Further, games are played independently. However, the number of trials is not fixed and we are not counting successes. (In fact, the number of successes must be 1 because a success is when A wins and A and B play until A wins exactly one game.) Instead, the variable Z counts trials, not successes. Thus Z does not have a binomial distribution. The distribution of Z for this example is discussed in Section 4. ◆

EXAMPLE 1.4

A pond is stocked with 100 fish of which 40 are of legal size. You catch 3 fish. Does the number Z of fish of legal size that you catch have a binomial distribution?

One can think of the process of catching fish as if it were a random sample of $n = 3$ fish taken from the pond. Each fish caught is a Bernoulli trial, with "success" being when the fish is of legal size. Note that a fish has probability $p = \frac{40}{100} = 0.4$ of being of legal size. Thus the only fact that remains to be verified is whether the trails (fish) are mutually independent. If each fish is returned to the pond after it is caught, and a caught fish fails to learn from its experience, the fish are being sampled with replacement. In this case the trials are independent, as demonstrated in Chapter 7, and Z has a binomial distribution with parameters $n = 3$ and $p = 0.4$.

On the other hand, most fishermen do not return caught fish to the pond; in this case, the fish are sampled without replacement. If the first 2 fish you catch are of legal size, the third fish is sampled from a population of 98 fish to whom 38 are of legal size. Hence

$$P(\text{third fish of legal size} \mid \text{first 2 fish are of legal size}) = \frac{38}{98} = 0.388 \neq 0.400 = P(\text{third fish of legal size}),$$

so that the trials (fish) are dependent. Thus Z does not have a binomial distribution. The correct distribution of Z for this case is discussed in Section 2. ♦

Calculating Binomial Probabilities

Exhibit B.2 of Appendix B gives the probability mass functions $p(z) = p(z; n, p)$, $z = 0, 1, \ldots, n$, of binomially distributed random variables for $p = 0.01(0.01)0.05(0.05)0.50$, $n = 1(1)10$, 12, 15, and 20. Such probabilities can be calculated by means of the recursion relationship:

$$\begin{aligned} p(0; n, p) &= (1 - p)^n \\ p(z; n, p) &= \frac{(n - z + 1)p}{z(1 - p)} p(z - 1; n, p), \qquad z = 1, 2, \ldots, n, \end{aligned} \tag{1.3}$$

which avoids the need repeatedly to take powers of p and $1 - p$. For example, if $n = 3$ and $p = 0.40$, we start by calculating $p(0) = (1 - 0.4)^3 = 0.216$. Then

$$\begin{aligned} p(1) &= \frac{(3 - 1 + 1)(0.4)}{(1)(1 - 0.4)} p(0) = 0.432, \\ p(2) &= \frac{(3 - 2 + 1)(0.4)}{(2)(1 - 0.4)} p(1) = 0.288, \\ p(3) &= \frac{(3 - 3 + 1)(0.4)}{(3)(1 - 0.4)} p(2) = 0.064. \end{aligned}$$

Exhibit B.2 does not give probability mass functions $p(z; n, p)$ for $p > 0.5$ because there exists a basic symmetry in the function $p(z; n, p)$ that enables us to calculate probabilities $P\{Z = z\}$ for $p > 0.5$ from knowledge of the corresponding probabilities for $p < 0.5$. This relationship is:

The probability of z successes in n trials when $p = p_0$ is equal to the probability of $n - z$ successes in n trials when $p = 1 - p_0$; that is,

$$p(z; n, p_0) = p(n - z; n, 1 - p_0). \tag{1.4}$$

This fact can be observed in Exhibit 1.2, where, for example, both $p(3; 10, 0.8)$ and $p(10 - 3; 10, 1 - 0.8) = p(7; 10, 0.2)$ equal 0.0008.

To verify (1.4), note that

$$\binom{n}{n-z} = \frac{n!}{(n-z)!z!} = \frac{n!}{z!(n-z)!} = \binom{n}{z},$$

and thus

$$\begin{aligned} p(z; n, p) &= \binom{n}{z} p^z (1-p)^{n-z} \\ &= \binom{n}{n-z} (1-p)^{n-z} p^z = p(n - z; n, 1 - p), \end{aligned}$$

as asserted.

EXAMPLE 1.5

In Example 1.2 a student took an exam on which there were 5 hard questions and 5 easy questions and the student answered the questions independently. Although the total number of questions answered correctly does not have a binomial distribution, it was noted that if all of the hard questions had the same probability p of being answered correctly, the number Z of correct answers on these questions has a binomial distribution. Similarly, if all of the easy questions had the same probability p^* of being answered correctly, the number Z^* of correct answers on the easy questions has a binomial distribution. Suppose that the student has probability $p = 0.2$ of answering any hard question and probability $p^* = 0.9$ of answering any easy question. What is the probability, $P\{Z = 2\}$, that the student gets 2 hard questions right? What is the probability, $P\{Z^* = 4\}$, that the student answers 4 easy questions correctly?

Because there are $n = 5$ hard questions, Z has a binomial distribution with parameters $n = 5$ and $p = 0.2$. Using Exhibit B.2 of Appendix B, we find that

$$P\{Z = 2\} = p(2; 5, 0.2) = 0.20480.$$

There are also $n = 5$ easy questions and thus Z^* has a binomial distribution with parameters $n = 5$ and $p^* = 0.9$. Exhibit B.2 does not contain a table for $p = 0.9$, but from (1.4) and Exhibit B.2,

$$\begin{aligned} P\{Z^* = 4\} &= p(4; 5, 0.9) = p(5 - 4; 5, 1 - 0.9) \\ &= p(1; 5, 0.1) = 0.32805. \end{aligned}$$ ◆

Exhibit B.2 gives only a limited tabulation of binomial probabilities. More detailed tabulations can be found in National Bureau of Standards (1950), where $n = 2(1)49$, $p = 0.01(0.01)0.50$, and in Romig (1953), where $n = 50(5)100$, $p = 0.01(0.01)0.50$.

Binomial Random Variables as Sums of Independent Bernoulli Variables

We have seen that a binomially distributed random variable Z arises from counting the number of successes in n independent Bernoulli trials, in which on each trial the probability of observing a success is p. On the ith such trial, we can define a Bernoulli random variable X_i, where $X_i = 1$ if the ith trial yields a success, and $X_i = 0$ if the ith trial yields a failure, $i = 1, 2, \ldots, n$. Note that each such X_i has probability model (1.1). Then $\sum_{i=1}^{n} X_i$ is the number of successes in the n trials, as can be seen by noting that the sum $\sum_{i=1}^{n} X_i$ is increased by 1 whenever a trial is a success, but does not change when there is a failure. Consequently, because Z is also the number of successes in these n trials,

$$Z = \sum_{i=1}^{n} X_i. \tag{1.5}$$

We conclude that

> Every binomially distributed random variable Z can be described as being the sum (1.5) of n independent random variables $X_1, \ldots, X_n$, where each X_i has the same Bernoulli distribution (1.1) with parameter p.

Recall that a Bernoulli variable X_i has mean $\mu_{Xi} = p$ and variance $\sigma_{Xi}^2 = p(1 - p)$. It thus follows from (1.5) and the results of Chapter 7 that

$$\begin{aligned} \mu_Z &= \mu_{\sum_{i=1}^{n} X_i} = np, \\ \sigma_Z^2 &= \sigma_{\sum_{i=1}^{n} X_i}^2 = np(1 - p). \end{aligned}$$

We conclude that

> If Z has a binomial distribution with parameters n and p, the mean and variance of Z are
>
> $$\mu_Z = np, \qquad \sigma_Z^2 = np(1 - p),$$
>
> respectively.

EXAMPLE 1.6

A manufacturing process produces parts independently. Each part has the same probability $p = 0.05$ of being damaged. If $n = 800$ parts are produced, what are the mean and variance of the number Z of parts produced that are damaged?

Here Z has a binomial distribution with parameters $n = 800$ and $p = 0.05$. Thus the mean and variance of Z are

$$\mu_Z = np = (800)(0.05) = 40,$$
$$\sigma_Z^2 = np(1 - p) = (800)(0.05)(0.95) = 38,$$

respectively. ◆

No convenient formulas, such as those for the mean and variance, exist for the quantiles of Z (e.g., the median). These can be found from tables of the binomial probabilities $p(z; n, p)$ by using the methods described in Chapter 6. Using (1.3), it can be shown that the mode of Z is the greatest integer less than or equal to $(n + 1)p$; actually, when $(n + 1)p$ is an integer there are two modes: $(n + 1)p$ and $(n + 1)p - 1$. For example, when $n = 10$, $p = 0.5$, the mode of Z is the greatest integer less than or equal to $(10 + 1)(0.5) = 5.5$; that is, the mode of Z is 5. When $n = 9$ and $p = 0.5$, there are two modes: $(n + 1)p = (9 + 1)(0.5) = 5$ and $(n + 1)p - 1 = 4$ (see Exhibit B.2 of Appendix B).

The Practical Significance of the Binomial Distribution

As noted in Example 1.4, the binomial distribution can arise when a simple random sample is selected *with replacement*, and for each unit selected it is determined whether or not the unit possesses a specified property. For example, the unit may be a person and the property may be a "yes" vote on a certain bond issue. If the unit is a machine part, this property may be that the part is defective; if the unit is a leaf, the property may be whether the leaf has worm damage; and so on. Suppose that the proportion of units in the population possessing the property is p. Then, if Z denotes the number of units in the sample (of size n) that possess the given property, Z has a binomial distribution with parameters n and p.

The binomial distribution also arises in contexts other than in random sampling experiments. Examples 1.1, 1.5, and 1.6 are illustrations of the use of the binomial distribution in nonsampling contexts. Still another example is the following:

EXAMPLE 1.7

Genetics

The inheritance of biological characteristics depends on genes, which appear in pairs. In the simplest genetic model, each gene of a pair may assume one of two forms: a or A. Consequently, three combinations, called *genotypes*, may be formed: aa, Aa, AA. (Note that aA and Aa are indistinguishable.)

Fisher and Mather (1936) describe an experiment (called a "backcross") in which the gene controlling straight (W) or wavy (w) hair in mice was segregated. The female parents were all wavy-haired of genotype (ww) and the male parents were all straight-haired of genotype (wW). If the Mendelian laws of inheritance are true, each mouse in the offspring of these parents has a probability 0.5 of being straight-haired (wW) and a probability of 0.5 of being wavy-haired (ww). Assume that the hair character of one offspring is independent of the hair character of any other offspring. Thus, in litters of 8 mice, the theoretical distribution of Z, the number of straight-haired offspring, is the binomial distribution with $n = 8$ and $p = 0.5$. This distribution has the following probability mass function:

z	0	1	2	3	4	5	6	7	8
$p(z)$	0.0039	0.0312	0.1094	0.2188	0.2734	0.2188	0.1094	0.0312	0.0039

In the genetic experiment described by Fisher and Mather, 32 such litters of 8 mice were observed. These litters represent $M = 32$ trials of the random experiment in which Z is observed. If the genetic theory given above truly describes these litters, and if the number $M = 32$ of trials is large enough for the stability of relative frequencies to start to hold, we would expect that

$$\text{r.f.}\{z \text{ straight-haired mice in litter}\} \simeq p(z)$$

for $z = 0, 1, 2, \ldots, 8$. Equivalently, we would anticipate that the frequency of litters in which z straight-haired mice are observed would be equal to $Mp(z) = 32p(z)$. The numbers $32p(z)$ are the **expected frequencies,** or **theoretical frequencies,** of litters with z straight-haired mice among the 32 litters observed. A comparison of the theoretical frequencies with the frequencies actually observed is given in Exhibit 1.3. Note that there are differences between the observed and theoretical frequencies. Whether these differences arise due to failure of the theory to describe the situation ("lack of fit" of the model) or merely due to the fact that the relative frequencies have not completely stabilized ("chance variation") is a question of statistical inference. One way of answering this question is discussed in Chapter 14 under the topic "goodness-of-fit tests." ◆

EXHIBIT 1.3. *Comparison of Observed and Theoretical Frequencies for the Number of Straight-Haired Mice in a Litter of 8 Mice*

Number of Straight-haired Mice in a Litter of 8 Mice	*Observed Frequency*	*Theoretical Frequency*
0	0	0.1
1	1	1.0
2	2	3.5
3	4	7.0
4	12	8.7
5	6	7.0
6	5	3.5
7	2	1.0
8	0	0.1
	32	31.9

2 The Hypergeometric Distribution

In Example 1.4 of Section 1 it was noted that when random samples *without replacement* are drawn, and the number Z of units selected that have a certain property is counted, the distribution of Z is not the binomial distribution because the draws are not independent. Instead, Z has the hypergeometric distribution. Computation of probabilities for the values z of Z has already been discussed in Chapter 3 as part of our discussion of random sampling.

EXAMPLE 2.1

In Example 1.4 there were $N = 100$ fish in a pond of which 40 were of legal size. A sample of $n = 3$ fish we caught; because the fish were not returned to the pond after being caught, sampling was without replacement. The number Z of fish caught that were of legal size was of interest. (Here "being of legal size" is a "success.") What is the probability that $Z = 2$ fish of legal size are caught?

The steps used to find this probability are similar to those used to find probabilities for the binomial distribution. Thus the combined outcomes SSS, SSF, SFS, FSS, SFF, FSF, FFS, FFF of the three trials are listed and their probabilities calculated. The outcomes corresponding to $Z = 2$ are then identified (these are SSF, SFS, FSS) and their probabilities are summed to find the probability $P\{Z = 2\}$. In the binomial case, the outcomes SSF, SFS, FSS had equal probabilities; this is also the case here because

$$P(SSF) = P(SFS) = P(FSS) = \frac{(40)(39)(60)}{(100)(99)(98)} = 0.0965.$$

Consequently,

$$P\{Z = 2\} = (3)(0.0965) = 0.2895.$$

That is, as was the case for the binomial distribution probabilities, we multiply the common probability of each outcome yielding $Z = 2$ by the number of such outcomes in order to find $P\{Z = 2\}$. What is different here is that the draws (trials) are dependent (without replacement), so that the common probability of each outcome (SSF, SFS, FSS) is not the same as when the draws were independent (with replacement). ◆

Derivation of the Hypergeometric Distribution

Assume that we draw a random sample of size n without replacement from a population of N units of which a proportion p are successes and a proportion $1 - p$ are failures. (Note that Np must be an integer.) Let Z denote the number of successes that appear in the sample. We wish to find $p(z) = P\{Z = z\}$, $z = 0, 1, 2, \cdots, n$. To simplify notation, denote the number of successes in the population by M; thus $M = Np$, $N - M = N(1 - p)$.

If there are z successes in the n trials, there must be $n - z$ failures. If $z > M$, there are insufficiently many successes in the population to allow for z successes in the sample; similarly, $n - z > N - M$ means that there are not enough failures in the population to allow $n - z$ failures in the sample. Thus

$$P\{Z = z\} = 0 \qquad \text{if } z > M \quad \text{or} \quad n - z > N - M.$$

As illustrated in Example 2.1, to find $P\{Z = z\}$ for $z \leq M$, $n - z \leq N - M$, one needs to multiply:

(a) The probability that any one outcome has z successes and $n - z$ failures.
(b) The number of distinguishable ways in which $zS's$ and $n - z\ F's$ can be arranged.

As noted when deriving the binomial distribution, the number needed in (b) above is the binomial coefficient $\binom{n}{z}$. The probability of the outcome in which z successes followed by $n - z$ failures occur is

$$P\{SS \cdots SFF \cdots F\} = \frac{[M_{(k)}][(N - M)_{(n-k)}]}{N_{(n)}},$$

because there are $N_{(n)}$ ways to draw an ordered sample of size n from a population of N units, $M_{(k)}$ ways to draw k successes from the M successes in the population, and $(N - M)_{(n-k)}$ ways to draw $n - k$ failures from the $N - M$ failures in the population. Thus

$$\begin{aligned} P\{Z = z\} &= \binom{n}{z} \frac{[M_{(z)}][(N - M)_{(n-z)}]}{N_{(n)}} \\ &= \frac{n!}{z!\,(n - z)!} \frac{M!}{(M - z)!} \frac{(N - M)!}{(N - M - n + z)!} \frac{(N - n)!}{N!} \\ &= \frac{\binom{M}{z}\binom{N - M}{n - z}}{\binom{N}{n}}, \end{aligned}$$

where the fact that

$$K_{(r)} = K(K - 1) \cdots (K - r + 1) = \frac{K!}{(K - r)!}$$

has been used. Recalling that $M = Np$, there are thus two mathematically equivalent ways to express the probability mass function $p(z)$ of Z for non-negative integer values of z:

$$\text{(2.1)} \quad p(z) = \binom{n}{z} \frac{[Np]_{(z)}[N(1 - p)]_{(n-z)}}{N_{(n)}}, \qquad z \leq Np, \quad n - z \leq N(1 - p),$$

$$\text{(2.2)} \quad p(z) = \frac{\binom{Np}{z}\binom{N(1 - p)}{n - z}}{\binom{N}{n}}, \qquad z \leq Np, \quad n - z \leq N(1 - p).$$

Of course, $p(z) = P\{Z = z\} = 0$ when $z > Np$, $n - z > N(1 - p)$, $z < 0$, $z > n$ or z is not an integer.

Notice that the probability mass function $p(z) = P\{Z = z\}$ depends on three quantities; the population size N, the proportion p of successes in the population, and the sample size n. These quantities, N, p, n, are the *parameters of the hypergeometric distribution.*

Note. Instead of p, the number M of successes in the population could serve as a parameter of the hypergeometric distribution. If we are given values of N and p, then $M = Np$ is known; if we are given values of N and M, then $p = M/N$ is known. Consequently, either N, p, n or N, M, n, can be specified as parameters of the hypergeometric distribution.

Calculation of Hypergeometric Probabilities

The hypergeometric distribution depends on the population size N as well as on the parameters n, p it has in common with the binomial distribution. Consequently, any tabulation of hypergeometric probabilities requires considerably more space than does tabulation of binomial probabilities. We have thus not given a tabulation of hypergeometric probabilities in this book. One such tabulation for $N = 2(1)50(10)100$, and various values of n and p, is given by Lieberman and Owen (1961).

To compute individual probabilities $p(z) = P\{Z = z\}$, equation (2.1) is usually more useful than equation (2.2). However, equation (2.2) is useful when tables of binomial coefficients are available.

EXAMPLE 2.2

An auto parts store has $N = 200$ rebuilt starters in stock, of which $M = 4$ are improperly assembled (defective). In response to a purchase order, they randomly select (without replacement) $n = 3$ starters from their stock. Let Z be the number of defective starters chosen. Then Z has the hypergeometric distribution with parameters $N = 200$, $p = 4/200 = 0.02$, and $n = 3$.

What is the probability, $p(2) = P\{Z = 2\}$, that the customer is given two defective starters? Using (2.1) to calculate this probability,

$$p(2) = \binom{3}{2}\frac{(4)_{(2)}(200 - 4)_{(3-2)}}{200_{(3)}} = (3)\frac{[(4)(3)][196]}{(200)(199)(198)} = 0.0009.$$

If (2.2) were used to compute $p(2)$,

$$p(2) = \frac{\binom{4}{2}\binom{200 - 4}{3 - 2}}{\binom{200}{3}} = \frac{(6)(196)}{1{,}313{,}400} = 0.0009.$$

◆

For recursive calculation of probabilities, the formula

$$p(z) = \left\{\frac{[Np - z + 1][n - z + 1]}{[N(1 - p) - n + z][z]}\right\} p(z - 1) \tag{2.3}$$

for $z \leq Np$ and $n - z < N(1 - p)$ can be helpful. For example, in Example 2.2, we can find $p(3)$ by

$$p(3) = \frac{(4 - 3 + 1)(3 - 3 + 1)}{(196 - 3 + 3)(3)} p(2) = \frac{(2)(1)}{(196)(3)} (0.0009) \cong 0.0000.$$

The Mean and Variance of the Hypergeometric Distribution

If Z has a hypergeometric distribution with parameters N, p, n, then Z arises from counting the number of successes in n draws (trials) without replacement from a population in which there are Np successes. We can let $X_i = 1$ if the ith draw yields a success, and $X_i = 0$ if the ith draw yields a failure, $i = 1, 2, \ldots, n$. Then each X_i is a Bernoulli variable, and the argument that led to (1.5) shows that

$$Z = \sum_{i=1}^{n} X_i.$$

However, the Bernoulli variables X_i are not independent because the results of previous draws ($j < i$) change the conditional probability that $X_i = 1$; that is, the draws are dependent.

Even though the Bernoulli variables X_i are not independent, it can be shown that each X_i has the same unconditional probability p of success, so that $P\{X_i = 1\} = p$, $P\{X_i = 0\} = 1 - p$, $i = 1, 2, \ldots, n$. Clearly this is true for X_1 because there are Np ways out of N to choose a success on the first draw. For X_2,

$$\begin{aligned}
P(X_2 = 1) &= P\{X_1 = 1 \text{ and } X_2 = 1\} + P\{X_1 = 0 \text{ and } X_2 = 1\} \\
&= \frac{[Np][Np - 1]}{[N][N - 1]} + \frac{[N(1 - p)][Np]}{[N][N - 1]} \\
&= \frac{Np}{N}\left(\frac{Np - 1 + N(1 - p)}{N - 1}\right) = \frac{Np}{N}(1) = p,
\end{aligned}$$

and similar arguments show that $P\{X_i = 1\} = p$, $i = 3, \ldots, n$. It thus follows that each X_i has mean p, so that by the results of Chapter 7,

(2.4) $$\mu_Z = \mu_{\sum_{i=1}^{n} X_i} = np,$$

which is the same formula obtained for the mean of a binomial distribution.

Although the variances of the X_i are all equal to $p(1 - p)$, the result stated in Chapter 7 about the variance of $Z = \sum_{i=1}^{n} X_i$ fails to hold because X_1, $X_2, \ldots, X_n$ are not independent. By using more general formulas for the variance of a sum that take account of dependence (covariance) between each pair, X_i and X_j $i \neq j$, of variables, it can be shown that

(2.5) $$\sigma_Z^2 = \left(\frac{N - n}{N - 1}\right) np(1 - p).$$

> If Z has a hypergeometric distribution with parameters N, n and p, then the mean and variance of Z are
>
> $$\mu_Z = np, \qquad \sigma_Z^2 = \left(\frac{N - N}{N - 1}\right) np(1 - p),$$
>
> respectively.

The mean of the hypergeometric distribution agrees with the mean of the binomial distribution when both distributions have identical values for the parameters n and p. On the other hand, the variances of the hypergeometric and binomial distributions differ. The ratio of the variance of the hypergeometric distribution to the variance $np(1 - p)$ of the binomial distribution is equal to $(N - n)/(n - 1)$. This number is sometimes called the **finite population factor.** It more correctly could be called the **without replacement factor** because the difference between the two variances is due to the difference between independent trials (binomial distribution) and the dependent trials (hypergeometric distribution) resulting from sampling without replacement.

As the ratio, n/N, between the sample size n and the population size N becomes small,

$$\frac{N - 1}{N - n} = \frac{1 - 1/N}{1 - n/N} \rightarrow 1.$$

Consequently, when n/N is small there is little difference in location and spread between the hypergeometric and the binomial distributions having the same values of the parameters n and p. This suggests that the probabilities assigned to events by these two distributions are approximately equal.

Approximation of the Hypergeometric Distribution by the Binomial Distribution

Suppose that Z has a hypergeometric distribution with parameters N, n, and p. We show that when n and p are fixed, and N is large enough, then

$$p(z) = P\{Z = z\} \approx \binom{n}{z} p^z(1-p)^{n-z}. \tag{2.6}$$

From Equation (2.1),

$$\begin{aligned} p(z) &= \binom{n}{z} \frac{[Np]_{(z)}[N(1-p)]_{(n-z)}}{N_{(n)}} \\ &= \binom{n}{z} \frac{[Np]_{(z)}}{N_{(z)}} \frac{[N(1-p)]_{(n-z)}}{[N-z]_{(n-z)}} \end{aligned}$$

because

$$\begin{aligned} N_{(n)} &= [N(N-1)\cdots(N-z+1)][(N-z)\cdots(N-z-(n-z)+1)] \\ &= N_{(z)}[(N-z)]_{(n-z)}. \end{aligned}$$

As $N \to \infty$,

$$\begin{aligned} \frac{(Np)_{(z)}}{N_{(z)}} &= \left[\frac{Np}{N}\right]\left[\frac{Np-1}{N-1}\right]\cdots\left[\frac{Np-z+1}{N-z+1}\right] \\ &= p\left(p - \frac{1-p}{N-1}\right)\cdots\left(p - \frac{(z-1)(1-p)}{N-z+1}\right) \\ &\to (p)(p)\cdots(p) = p^z \end{aligned}$$

and

$$\begin{aligned} \frac{[N(1-p)]_{(n-z)}}{[N-z]_{(n-z)}} &= \left[\frac{N(1-p)}{N-z}\right]\left[\frac{N(1-p)-1}{N-z-1}\right] \\ &\quad \cdots \left[\frac{N(1-p)-n+z+1}{N-n+1}\right] \\ &= \left[(1-p) + \frac{z(1-p)}{N-z}\right]\left[(1-p) + \frac{z(1-p)-p}{N-z-1}\right] \\ &\quad \cdots \left[(1-p) + \frac{z-p(n-1)}{N-n+1}\right] \\ &\to (1-p)(1-p)\cdots(1-p) = (1-p)^{n-z}. \end{aligned}$$

Consequently, as $N \to \infty$ for fixed n, p, and z,

$$p(z) \to \binom{n}{z} p^z(1-p)^{n-z},$$

as asserted. An inspection of this argument shows that N does not have to be very large with respect to n for the approximation (2.6) to be reasonably accurate.

For example, suppose that $N = 100$, $p = 0.1$, $n = 10$, so that the finite population factor is $(N - n)/(N - 1) = 90/99 = 0.909$. Then a comparison of the exact hypergeometric probabilities (denoted Hyp below) with the probabilities given by the approximation using the binomial distribution (denoted Bin below) with $n = 10$, $p = 0.1$, yields

z	0	1	2	3	4	5	6	7	8	9	10
Hyp	0.330	0.408	0.202	0.052	0.008	0.001	0.000	0.000	0.000	0.000	0.000
Bin	0.349	0.387	0.194	0.057	0.011	0.002	0.000	0.000	0.000	0.000	0.000

As a general rule of thumb, use of the binomial probabilities to approximate the exact hypergeometric probabilities yields results accurate to at least one decimal place when

$$\frac{n}{N} \leq 0.1.$$

In the example above, with $N = 100$, $p = 0.1$, $n = 10$, note that $n/N = 0.1$ exactly, and the largest difference between an exact hypergeometric probability and the binomial probability that approximates it is 0.019.

Uses of the Hypergeometric Distribution

The hypergeometric distribution is important in the analysis of opinion surveys. In such surveys, n individuals are randomly drawn (without replacement) from a population of persons (voters, shoppers, TV viewers) and asked to respond to one or more questions. For one such question, the proportion p of all persons in the population who would choose a particular answer to this question may be of interest. The relative frequency, $\hat{p} = Z/n$, of people surveyed who choose this answer serves to estimate p and may be used to challenge assertions made about the value of p. In evaluating $\hat{p}$ as an estimate of p, or in challenging assertions made about the values of p, the likelihood of observing various values of $\hat{p}$ needs to be evaluated—that is, the distribution of $\hat{p}$, is needed. This distribution can be obtained from the distribution of Z, which has a hypergeometric distribution with parameters N, the size of the population from which people are sampled, n, and p. For convenience, however, and because it is typically the case that the size n of the sample is only a small fraction of the population size N, the exact hypergeometric distribution of Z is usually approximated by the binomial distribution with parameters n and p.

EXAMPLE 2.3

A random sample of $n = 10$ high school seniors is drawn without replacement from among a class of $N = 300$ seniors. Each student sampled is asked if he or she has ever cheated on an examination. Recent surveys of high school seniors have led to the assertion that 40% of all seniors have cheated on at least one examination. In the present survey, only 2 of the 10 seniors interviewed stated that they had cheated on an examination, so that $Z = 2$ and $\hat{p} = 0.2$. How probable is this result if $p = 0.4$?

From Equation (2.1) with $N = 300$, $Np = 300(0.4) = 120$, $N(1 - p) = 300(0.6) = 180$, and $n = 10$.

$$\begin{aligned} P\{\hat{p} = 0.2\} &= P\{Z = 2\} \\ &= \binom{10}{2} \frac{(120)_{(2)}(180)_{(8)}}{(300)_{(10)}} \\ &= (45)\frac{(14{,}280)(9.411631) \times 10^{17})}{5.074224 \times 10^{24}} = 0.1192, \end{aligned}$$

so that the result $\hat{p} = 0.2$ is not improbable assuming that $p = 0.4$. Note that the binomial distribution approximation (with $n = 10$, $p = 0.4$) to this probability is, from Example 1.2, equal to 0.1209. The closeness of this approximation could be anticipated because $n/N = 10/300 = 0.03 < 0.10$. ◆

The hypergeometric distribution is also important in quality control. From an inventory of N items, a random sample of n items is sampled without replacement and the number Z of defective items is noted. This information, together with use of the hypergeometric distribution (2.1) of Z, permits statistical inference to be made concerning the proportion p of defective items in the inventory. Example 2.2 provides an illustration of the probability calculations involved. Again, when the sample size n is small relative to the size N of the inventory, the binomial approximation to the hypergeometric distribution is frequently used.

The hypergeometric distribution also arises in connection with studies of gambling and lottery devices.

EXAMPLE 2.4

Card Playing

Pearson (1924) reported on the records of 25,000 actual deals of the card game whist. His concern was in the determination of whether the shuffling was "perfect." This was done by comparing actual and theoretical results. Exhibit 2.1 gives observed and theoretical frequencies for the event that Z trump cards appear in the first hand of a deal of whist. The data are taken from a sample of 3400 deals of the cards. The theoretical observed frequencies are obtained by multiplying the theoretical probability $P\{Z = z\}$ by 3400. ◆

EXHIBIT 2.1. *Comparison of Observed and Theoretical Frequencies for the Event That Z Trump Cards Appear in the First Hand of a Deal of the Card Game Whist*

Number z of Trump Cards in the Hand	*Observed Frequency*	*Theoretical Frequency Computed from the Hypergeometric Distribution* ($N = 52$, $n = 13$, $p = \frac{1}{4}$)
0	35	43.4
1	290	272.3
2	697	700.2
3	937	973.4
4	851	811.2
5	444	424.0
6	115	141.4
7	21	29.9
8	10	3.7
≥9	0	0.5
	3400	3400.0

EXAMPLE 2.5

The state of California has a state lottery game called Lotto. To play, you select 6 numbers from among the numbers 1, 2, . . . , 49. Then 6 balls are chosen at random from an urn containing balls numbered from 1 to 49. All people whose choice of numbers exactly matches the numbers on the balls drawn share the grand prize. There are various devices sold to help people pick their 6 numbers. However, it appears that certain numbers on the border of one such device tend to be favored. Suppose that there are $M = 25$ of these numbers out of the total population of $N = 49$ numbers, and the device selects $n = 6$ numbers. Thus if all numbers are equally likely to be chosen (the device does not favor any subcollection of numbers), the number Z of numbers on the border of the device chosen by the device should have a hypergeometric distribution with $N = 49$, $M = 25$, $n = 6$. One hundred trials were run using the device, with the value of Z being recorded on each trial. This yielded the following table (Exhibit 2.2) of relative frequencies, r.f.(z), and theoretical probabilities, $p(z)$, for the values z of Z [see Diaconis (1988)].

EXHIBIT 2.2. *Empirical Relative Frequencies and Hypergeometric Probabilities for Z*

z	0	1	2	3	4	5	6
r.f.(z)	0.010	0.040	0.140	0.350	0.300	0.130	0.030
$p(z)$	0.013	0.091	0.250	0.333	0.228	0.016	0.010

Note that $p(z)$ tends to be greater than r.f.(z) for small z-values and less than r.f.(z) for large z-values, thereby supporting the contention that numbers on the border of the device are selected more often than would be the case if the device truly did not favor some numbers over others. ◆

3 The Poisson Distribution

An interesting fact emerges from the study of accident data. Suppose, for example, that X denotes the number of people injured when getting out of an automobile in a given year. For any particular person, the probability of such an accident in the given year is quite small; in fact, it is almost zero. Data reveal, nevertheless, that some such accidents occur every year, and that the yearly number, X, of such accidents varies in an apparently random fashion. Over many years, the relative frequencies of the events $\{X = x\}$, for x a nonnegative integer, exhibit stability. This type of statistically regular behavior has been described by the **Poisson distribution** (named after the French mathematician Simeon D. Poisson, 1781–1840). This discrete distribution has a probability mass function of the form

$$(3.1) \qquad p(x) = P\{X = x\} = \frac{\lambda^x e^{-\lambda}}{x!}, \qquad x = 0, 1, 2, \ldots,$$

where the constant λ can be any positive number. The effect of the parameter λ on the behavior of the probability mass function $p(x)$ can be observed from Exhibit 3.1.

Calculation of Probabilities

Note that for $x = 1, 2, \ldots,$

$$(3.2) \qquad p(x) = \frac{\lambda^x e^{-\lambda}}{x!} = \frac{\lambda}{x}\left[\frac{\lambda^{x-1}e^{-\lambda}}{(x-1)!}\right] = \frac{\lambda}{x}\,p(x-1).$$

EXHIBIT 3.1. *Values of p(x) for the Poisson Distribution when λ = 0.5, 1, 2, 3, 5*

	λ				
x	0.5	1	2	3	5
0	0.6065	0.3679	0.1353	0.0498	0.0067
1	0.3033	0.3679	0.2707	0.1494	0.0337
2	0.0758	0.1839	0.2707	0.2240	0.0842
3	0.0126	0.0613	0.1804	0.2240	0.1404
4	0.0015	0.0153	0.0902	0.1680	0.1755
5	0.0002	0.0031	0.0361	0.1008	0.1755
6		0.0005	0.0120	0.0504	0.1462
7		0.0001	0.0034	0.0216	0.1044
8			0.0009	0.0081	0.0653
9			0.0002	0.0027	0.0363
10				0.0008	0.0181
11				0.0002	0.0082
12				0.0001	0.0034
13					0.0013
14					0.0005
15					0.0002
16					0.0000

The resulting recurrence relation

$$p(x) = \frac{\lambda}{x} p(x - 1), \tag{3.3}$$

together with the initial value $p(0) = e^{-\lambda}$, gives an easy way to calculate Poisson probabilities for any value of the parameter λ. For example, when $\lambda = 1.0$,

$$p(0) = e^{-1} = 0.36788, \qquad p(1) = \tfrac{1}{1}p(0) = 0.36788,$$
$$p(2) = \tfrac{1}{2}p(1) = 0.18394, \qquad p(3) = \tfrac{1}{3}p(2) = 0.06131,$$

and so on. (Compare these results with Exhibit 3.1.)

Of course, it is more convenient to have tables available. Exhibit B.3 in Appendix B gives values of the probability mass function $p(x)$ of the Poisson distribution for $\lambda = 0.10(0.1)10(1)20$. For more detailed tables with $\lambda = 0.001(0.001)1.000(0.01)10.00$, see Kitagawa (1952), and for $\lambda = 0.001(0.001)0.010(0.010)0.30(0.1)15.0(1)100$, see Molina (1942).

Mean, Variance, Quantiles, and Mode of the Poisson Distribution

If X has a Poisson distribution with parameter λ, then

$$\mu_X = \lambda, \qquad \sigma_X^2 = \lambda. \tag{3.4}$$

That is, **the mean and variance of X are both equal to** λ. To verify (3.4), use the recurrence relation (3.3),

$$\mu_X = \sum_{x=0}^{\infty} xp(x) = \sum_{x=1}^{\infty} xp(x) = \sum_{x=1}^{\infty} \lambda p(x-1)$$
$$= \lambda \sum_{i=0}^{\infty} p(i) = \lambda,$$

where we changed index of summation from x to $i = x - 1$ and used the fact that $p(x)$ is a probability mass function. A similar argument (see Exercise T.4) can be used to verify that $\sigma_X^2 = \lambda$.

The recurrence relationship (3.3) also reveals that $p(x) > p(x - 1)$ when $x < \lambda$, and that when $x > \lambda$, $p(x) < p(x - 1)$. Hence $p(x)$ is largest when x is any integer between (and including) $\lambda - 1$ and λ. That is, the mode of X is the integer between $\lambda - 1$ and λ when λ is not an integer, whereas X has two modes, λ and $\lambda - 1$, when λ is an integer. For example, if $\lambda = 1.5$, then Mode(X) = 1, whereas when $\lambda = 2.0$, X has modes at $x = 1$ and $x = 2$.

The quantiles, $Q_X(p)$, of a Poisson-distributed random variable X are most easily found by using the probabilities $p(x)$, $x = 0, 1, 2, \ldots$ and the methods of Chapter 6. For example, when $\lambda = 1$, the median $Q_X(0.50)$ can be found by using Exhibit 3.1 (or Exhibit B.3) and calculating

$$F(0) = p(0) = 0.3679,$$
$$F(1) = p(0) + p(1) = 0.7358.$$

We see that the cumulative distribution function $F(x)$ of X jumps through $p = 0.50$ at $x = 1$. Thus $Q_X(0.50) = 1$.

EXAMPLE 3.1

The yearly number X of accidents at a typical traffic intersection in a large city has a Poisson distribution, and the mean number of accidents per year is $\mu_X = 2.2$. The traffic division of the city keeps records of the yearly number of accidents at each intersection in the city, and wishes to be alerted when the number of accidents is unusually large at any intersection. They decide to target an intersection for safety improvement when the number X of accidents at that intersection is 4 or greater. What is the probability, $P\{X \geq 4\}$, that a typical intersection will have 4 or more accidents in a year?

Because X has a Poisson distribution and the mean of X is 2.2, the parameter λ of the Poisson distribution must, by (3.4), be equal to 2.2. Using Exhibit B.3 of Appendix B with $\lambda = 2.2$ yields

$$\begin{aligned} P\{X \geq 4\} &= p(4) + p(5) + \cdots \\ &= 0.1082 + 0.0476 + 0.0174 + 0.0055 \\ &\quad + 0.0015 + 0.0004 + 0.0001 + 0 \\ &= 0.1807. \end{aligned}$$

The traffic division finds that this probability is too high and would cause them to devote too much attention to intersections that are not really a problem. As an alternative, they decide to use the 99th percentile, $Q_X(0.99)$, of X as a critical value; that is, they will target any intersection for which the yearly number of accidents is $Q_X(0.99)$ or greater. Because the probability that $X \geq Q_X(0.99)$ is no more than $1 - 0.99 = 0.01$, this means that the chance of targeting an intersection that does not need attention will not exceed 0.01. Using Exhibit B.2 with $\lambda = 2.2$ gives

$$F(6) = \sum_{x=0}^{6} p(x) = 0.9751, \qquad F(7) = \sum_{x=0}^{7} p(x) = 0.9925.$$

Because the cumulative distribution function, $F(x)$, of X jumps through 0.99 for $x = 7$, $Q_X(0.99) = 7$, and the traffic division will target an intersection for attention when the number X of accidents at that intersection is 7 or greater. ◆

Applications of the Poisson Distribution

The wide variety of random phenomena that give rise to random variables X having a Poisson distribution is truly astonishing. Some examples for which the Poisson distribution provides a reasonable model are given in this section.

In many biological examples a surface is subdivided into equal sections, from which counts of specimens are made. This occurs, for example, in studies of blood samples or yeast cultures.

EXAMPLE 3.2

Zoology

A horizontal quarry surface was divided into 30 squares about 1 meter on a side. In each square the number X of specimens of the extinct mammal *Litolestes notissimus* was counted. The results are shown in Exhibit 3.2.

EXHIBIT 3.2. *Comparison of Observed and Theoretical Frequencies of the Event $\{X = k\}$ in 30 Trials (Squares) of the Random Experiment in Which the Number X of Specimens of L. notissimus Is Counted*

Number x of Specimens Per Square	*Observed Frequency*	*Theoretical Frequency* ($\lambda = 0.73$)
0	16	14.4
1	9	10.5
2	3	3.9
3	1	0.9
4	1	0.3
≥ 5	0	0.0

The theoretical frequencies were obtained by multiplying each of the probabilities $P\{X = x\} = p(x)$, $x = 0, 1, 2, 3, 4$, for a Poisson distribution with parameter $\lambda = 0.73$, and also $P\{X \geq 5\} = 1 - p(0) - p(1) - \cdots - p(4)$ by the number 30 of trials (squares). Judging by Exhibit 3.2, a Poisson distribution with $\lambda = 0.73$ agrees closely with the observed frequencies. If the number N of squares observed had been larger than 30, we would expect the fit to have been better. ♦

A similar approach is used in studies of the geographical distribution of cancer cases, except that here trials are geographical regions with similar population sizes. For example, Gardner (1989) studied the incidence of childhood cancer in the United Kingdom. A Poisson distribution for the number of cancer cases during 1968–1984 with $\lambda = 1.53$ was fit to 634 areas of Scotland. This distribution then served as a standard against which the incidences of childhood cancer near nuclear installations could be compared.

The Poisson distribution has been applied in virtually every scientific and technical field. Industrial engineers have applied this distribution to model variations in the number of flaws in capacitors, in the number of defects per linear unit of wire and of rope, and in the number of strands in a cross section of thread. Biological applications have been made in modeling the variability of the number of beetle larvae, of the number of fish caught in a day, of the number of photons reaching the retina of the eye, and of counts of bacteria. In dentistry and medicine, the Poisson distribution has been applied to the number of defective teeth per individual and to the number of victims suffering or recovering from various specific diseases. Of course, because of its success in modeling fluctuations in the number of accidents, deaths, fires, and other risks, the Poisson distribution is used extensively in actuarial science (insurance). In the social sciences, the Poisson distribution has been applied to the number of vacancies in the Supreme Court and to the number of labor strikes; this distribution has also been used to model the number of outbreaks of war (see Exercise 25). In education and psychology, the Poisson distribution serves as a basis for theories of learning and recall. Applications of the Poisson distribution have also been made to the number of words misread in a book, to the number of misprints, and to the frequency of earthquakes.

One final application of the Poisson distribution deserves to be mentioned, both because it provides a second illustration of fitting the Poisson distribution to data and because it was influential in promoting the use of the Poisson distribution to describe the variation of important measurements in physics and chemistry.

EXAMPLE 3.3

Radioactive Emissions

Rutherford and Geiger (1910) observed the collisions of alpha particles emitted from a small bar of polonium with a small screen placed at a short distance from the bar. The number of such collisions in each of 2608 eight-minute intervals was recorded; the distance between the bar and screen was gradually decreased so as to compensate for the decay of the radioactive substance. A Poisson distribution with $\lambda = 3.870$ was fitted to the data (see Exhibit 3.3). [See also Rutherford, Chadwick, and Ellis (1930).] ◆

The Poisson Distribution as an Approximation to the Binomial Distribution

Random variables possessing a Poisson distribution can be conceptualized as having arisen from a random experiment consisting of a large number of trials. On each trial a certain rare occurrence may or may not take place. Because the occurrence is rare, the probability of its taking place is small. However, the rarity of the occurrence is offset by the fact that a large number of trials are performed. The resultant of these two effects is the Poisson distribution (3.1). This fact can be demonstrated by obtaining the Poisson distribution as an approximation to the binomial distribution when the two effects are present.

The derivation is as follows. Under a binomial distribution with parameters n and p, $P\{X = x\} = \binom{n}{x} p^x(1 - p)^{n-x}$. Let $p = \lambda/n$, so that a large sample size n will be offset by the diminution of p to produce a constant

EXHIBIT 3.3. *Observed Frequencies of x Alpha Particles in 2608 Trials of the Random Experiment in Which Alpha Particles Emitted from a Small Bar of Polonium Are Counted over an 8-Minute Interval of Time*

Number x of Alpha Particles Observed During the Interval	*Observed Frequency*	*Theoretical Frequency* ($\lambda = 3.3870$)
0	57	54.77
1	203	211.25
2	383	406.85
3	525	524.21
4	532	508.56
5	408	393.81
6	273	252.98
7	139	140.83
8	45	67.81
9	27	28.69
10	10	10.43
11	4	5.22
≥ 12	2	2.59
	2608	2608.00

mean number of successes $\mu_X = np = n(\lambda/n) = \lambda$ for all values of n. Then as $n \to \infty$,

$$\begin{aligned}
P\{X = x\} &= \binom{n}{x}\left(\frac{\lambda}{n}\right)^x\left(1 - \frac{\lambda}{n}\right)^{n-x} \\
&= \frac{\lambda^x}{x!}\left(1 - \frac{\lambda}{n}\right)^n \frac{n!}{(n-x)!\,n^x}\left(1 - \frac{\lambda}{n}\right)^{-x} \\
&= \frac{\lambda^x}{x!}\left(1 - \frac{\lambda}{n}\right)^n \left[\left(\frac{n}{n}\right)\left(\frac{n-1}{n}\right)\cdots\left(\frac{n-x+1}{n}\right)\right]\left(1 - \frac{\lambda}{n}\right)^{-x} \\
&= \frac{\lambda^x}{x!}\left(1 - \frac{\lambda}{n}\right)^n \left[1\left(1 - \frac{1}{n}\right)\left(1 - \frac{2}{n}\right)\cdots\left(1 - \frac{x-1}{n}\right)\right]\left(1 - \frac{\lambda}{n}\right)^{-x} \\
&\cong \frac{\lambda^x}{x!}e^{-\lambda},
\end{aligned}$$

because $(1 - (\lambda/n))^n$ approaches $e^{-\lambda}$ and both $(1 - (\lambda/n))^{-x}$ and the term in square brackets approach 1. Comparing this result with equation (3.1) verifies the assertion that the binomial probability that a rare occurrence will take place exactly x times is closely approximated by the probability (3.1).

For this reason binomial probabilities for small values of p and large values of n are often approximated using the Poisson distribution, with $\lambda = np$.

For example, suppose that $n = 10$ and $p = 0.10$. The comparison between the exact probability mass function for the binomial distribution and the Poisson approximation with $\lambda = np = 1$ is

x	0	1	2	3	4	5	6	7 or more
Binomial	0.349	0.387	0.194	0.057	0.011	0.002	0.000	0.000
Poisson	0.3679	0.3679	0.1839	0.0613	0.0153	0.0031	0.0005	0.0001

If n becomes larger, say 20, and p smaller, say $p = 0.05$, it is still the case that $\lambda = np = 1$, and we obtain the comparison

x	0	1	2	3	4	5	6	7 or more
Binomial	0.3585	0.3774	0.1887	0.0596	0.0133	0.0022	0.0003	0.0000
Poisson	0.3679	0.3679	0.1839	0.0613	0.0153	0.0031	0.0005	0.0001

As can be seen, the agreement between the binomial and Poisson probabilities is good even for n as small as 10. However, it is even closer when $n = 20$. Finally, if $n = 100$ and $p = 0.01$, the comparison becomes

x	0	1	2	3	4	5	6	7 or more
Binomial	0.3660	0.3697	0.1849	0.0610	0.0149	0.0029	0.0005	0.0001
Poisson	0.3679	0.3679	0.1839	0.0613	0.0153	0.0031	0.0005	0.0001

In general, the Poisson distribution with parameter $\lambda = np$ provides a good approximation to the binomial distribution with parameters n and p in cases when n is large and p is small, and when $\lambda = np$ is of moderate size (say, $\lambda \leq 20$).

EXAMPLE 3.4

A bus company tries to keep a supply of new batteries available for replacement use in cold weather. Their records indicate that the probability that a battery will fail to function on a given day is quite small, $p = 0.03$. They have, however, $n = 90$ buses on the streets and currently have only five new batteries in stock. What is the probability that they will not have a sufficient number of batteries to keep all of their buses running?

Let X denote the number of buses whose batteries need replacement. Assuming that batteries fail or do not fail independently of one another, the exact distribution of X is the binomial distribution with parameters $n = 90$ and $p = 0.03$. Because the bus company will not have enough replacement batteries available to keep all buses running if $X > 5$, we need to find $P\{X > 5\}$. Unfortunately, tables of the binomial distribution are not available to help compute this probability. Noting that $n = 90$ is large and $p = 0.03$ is small, approximation of the binomial distribution by the Poisson distribution with $\lambda = np = 90(0.03) = 2.7$ is possible. Using Exhibit B.3 of Appendix B yields

$$\begin{aligned} P\{X > 5\} & \\ &= p(6) + p(7) + \cdots \\ &\cong 0.0362 + 0.0139 + 0.0047 + 0.0014 + 0.0004 + 0.0001 + 0 \\ &= 0.0567. \end{aligned}$$

◆

4 The Geometric Distribution

In a fixed number n of independent Bernoulli trials, each having the same probability p of success, the number Z of successes observed has a binomial distribution. Suppose, however, that instead of fixing the number of trials, it is the number of successes desired that is specified and trials continue to be made until this desired number of successes is observed. For example, suppose that trials continue to be taken until the *first* success is observed, and the number X of trials required to obtain the first success is recorded. In this case, X has the **geometric distribution** with parameter p.

EXAMPLE 4.1

Medicine

A blood bank needs type B negative blood, and must continue purchasing blood from people (without knowing their blood type until it is tested) until a person with type B negative blood is found. If blood types are independent, the random number X of purchases made until a person with type B negative blood appears has a geometric distribution with parameter p equal to the proportion of people in the population who have type B negative blood. ◆

EXAMPLE 4.2

Gambling

The roulette wheel at Monte Carlo has 37 spaces, of which 18 are "red." A gambler continues to bet the same amount on "red" each time until "red" appears. Assuming that the spins of the roulette wheel are made independently (and that the wheel is fair), the number X of bets made by the gambler (including the winning bet that terminates wagering) has a geometric distribution with parameter $p = \frac{18}{37}$. ◆

In the example above, the random variable X, which is the number of trials required to obtain the first success, is related to the cost of running the experiment. If the length of time between trials is uniform, the number X of trials that must be made until a success is observed can also be equated with the *length of time* required until a success occurs. For example, if in Example 4.1 the blood bank makes one blood purchase an hour, then X is the number of hours they must wait until they purchase their first sample of type B negative blood. Thus X frequently refers to a *waiting time* until a certain event occurs.

Derivation of the Geometric Distribution

Assume that independent Bernoulli trials with common probability of success p are observed one at a time until a success is obtained, and let X be the number of trials observed (including the trial resulting in a success). The possible values of X are $x = 1, 2, 3, \ldots$. Because a success can only occur on the last trial taken (or the trials would have been stopped earlier), it follows that in order for X to equal the positive integer x, there must have been $(x - 1)$ failures followed by a success. Thus $X = 1$ if the first trial was a success, and $P\{X = 1\} = P(S) = p$. For $X = 2$, a failure followed by a success must have occurred, and $P\{X = 2\} = P(FS) = P(F)P(S) = (1 - p)p$. Similarly,

$$P\{X = 3\} = P(FFS) = P(F)P(F)P(S) = (1 - p)^2p,$$
$$P\{X = 4\} = P(FFFS) = P(F)P(F)P(F)P(S) = (1 - p)^3p,$$

and for a general positive integer x,

$$P\{X = x\} = P\{FF \cdots FS\} = (1 - p)^{x-1}p,$$

where in each case the fact that the trials are independent with constant probability of success has been used. Thus the probability mass function of X, which is called the geometric distribution with parameter p, is

$$(4.1) \qquad p(x) = \begin{cases} (1 - p)^{x-1}p, & x = 1, 2, \ldots, \\ 0, & x \text{ not a positive integer.} \end{cases}$$

Note that the probabilities $p(x) = P\{X = x\}$ decrease rapidly to 0 as x increases from 1 to infinity. Hence the mode of the geometric distribution always occurs at $x = 1$, regardless of the value of p.

A list of powers $(1-p)^0, (1-p)^1, (1-p)^2, \ldots,$ of increasing degree is called a **geometric series.** The probability mass function $p(x)$ in (4.1) is a constant (p) times such a series. It is this fact that gives the geometric distribution its name.

It should here be mentioned that the terminology "geometric distribution" is sometimes used to refer to the distribution of the number Y of *failures* observed before the first success is obtained, rather than to the distribution of the number X of trials. Note that $Y = X - 1$, because X counts the one successful trial as well as all the unsuccessful ones. Thus

$$p_Y(y) = p_X(y+1) = (1-p)^{y+1-1}p = (1-p)^y p, \qquad y = 0, 1, 2, \ldots,$$

is the probability mass function of Y, and it is this probability mass function that is sometimes said to define the geometric distribution with parameter p. As long as one can clearly distinguish between the number of trials, X, and number of failures, Y, no confusion need result from this lack of uniqueness of terminology.

Probability Calculations for the Geometric Distribution

Unlike the other discrete distributions considered in this chapter, the geometric distribution has a cumulative distribution function $F(x)$ with a simple functional form. For any positive integer x,

$$F(x) = 1 - (1-p)^x, \qquad x = 1, 2, \ldots. \tag{4.2}$$

Also, $F(x) = 0$ for $x < 1$, and $F(x) = F(k)$ when $k \le x < k+1$, $k = 1, 2, \ldots$. This makes the calculation of probabilities of events such as $\{X > a\}$, $\{X \le b\}$, $\{a < X \le b\}$ rather easy.

EXAMPLE 4.3

If X has a geometric distribution with parameter $p = 0.2$, then

$$P\{X > 4\} = 1 - P\{X \le 4\} = 1 - F(4).$$

Because

$$F(4) = 1 - (1 - 0.2)^4 = 1 - 0.4096 = 0.5904,$$

it follows that $P\{X > 4\} = 1 - 0.5904 = 0.4096$. Similarly,

$$\begin{aligned} P\{3 < X \le 5\} &= F(5) - F(3) \\ &= [1 - (1-0.2)^5] - [1 - (1-0.2)^3] \\ &= 0.6723 - 0.4880 = 0.1843, \end{aligned}$$

whereas

$$\begin{aligned} P\{3 \le X \le 5\} &= P\{3 < X \le 5\} + P\{X = 3\} \\ &= 0.1843 + (1 - 0.2)^{3-1}(0.2) \\ &= 0.3123. \end{aligned}$$

◆

Individual probabilities $p(x) = P\{X = x\}$ are also easy to evaluate for all but very large values of x. This fact, and the existence of a convenient formula (4.2) for the cumulative distribution function $F(x)$, makes it unnecessary to table the probability mass function $p(x)$ of the geometric distribution.

To verify equation (4.2), note first that X exceeds x if and only if the first x Bernoulli trials all result in failures F. For example,

$$P\{X > 4\} = P\{FFFF\} = P(F)P(F)P(F)P(F) = (1 - p)^4,$$

and in general for any positive integer x

$$P\{X > x\} = P\{FF \cdots F\} = (1 - p)^x.$$

Because $P\{X > x\} = 1 - F(x)$, it follows that $F(x) = 1 - P\{X > x\} = 1 - (1 - p)^x$, $x = 1, 2, \ldots$, verifying (4.2).

Using (4.2), note that

$$\sum_{x=1}^{\infty} p(x) = \lim_{x \to \infty} F(x) = 1 - \lim_{x \to \infty} (1 - p)^x = 1 - 0 = 1.$$

This result, plus the fact that $p(x) \ge 0$ for all numbers x, verifies that the function (4.1) is a probability mass function.

Descriptive Indices of the Geometric Distribution

We have already remarked that the mode of a geometrically distributed random variable X is always 1, regardless of the value of p. The quantiles $Q_X(p)$ are easily obtained by using the cumulative distribution function $F(x)$. For example, if $p = 0.3$,

$$F(x) = \begin{cases} 1 - (1 - 0.3)^1 = 0.300, & \text{if } x = 1, \\ 1 - (1 - 0.3)^2 = 0.510, & \text{if } x = 2, \\ 1 - (1 - 0.3)^3 = 0.657, & \text{if } x = 3, \end{cases}$$

and so on. Thus $Q_X(0.25) = 1$, because $F(x)$ jumps through 0.25 at $x = 1$. Similarly, $Q_X(0.50) = 2$, and $Q_X(0.60) = 3$.

When X has a geometric distribution with parameter p, the mean μ_X and variance σ_X^2 of X are given by

$$\mu_X = \frac{1}{p}, \qquad \sigma_X^2 = \frac{1-p}{p^2}. \tag{4.3}$$

Note that μ_X increases as p decreases. This is in accord with intuition—the rarer an event, the more trials needed before that event is first observed.

EXAMPLE 4.2 (continued)

In Example 4.2 it was noted that if a gambler continues to bet on "red" in independent spins of a roulette wheel at Monte Carlo until "red" comes up for the first time, the number, X, of bets that the gambler makes has a geometric distribution with parameter $p = \frac{18}{37}$. Thus the mean number of bets that the gambler makes is

$$\mu_X = \frac{1}{p} = \frac{1}{\frac{18}{37}} = 2.06,$$

and the variance of the number of bets made is

$$\sigma_X^2 = \frac{1-p}{p^2} = \frac{1-\frac{18}{37}}{(\frac{18}{37})^2} = 2.17.$$

◆

The results (4.3) can be obtained from the following equalities:

$$\sum_{k=0}^{\infty} r^k = \frac{1}{1-r},$$

$$\sum_{k=0}^{\infty} kr^{k-1} = \frac{1}{(1-r)^2}, \qquad \sum_{k=0}^{\infty} k(k-1)r^{k-2} = \frac{2}{(1-r)^3}, \qquad 0 < r < 1 \tag{4.4}$$

(see Exercise T.9). Using the second equality in (4.4) with $r = 1 - p$ yields

$$\mu_X = \sum_{x=1}^{\infty} xp(x) = \sum_{x=1}^{\infty} x(1-p)^{x-1}p = p\sum_{x=0}^{\infty} x(1-p)^{x-1} = p\left(\frac{1}{p^2}\right) = \frac{1}{p}.$$

Using the third equality in (4.4), similar steps show that

$$E[X(X-1)] = \sum_{x=1}^{\infty} x(x-1)(1-p)^{x-1}p = \frac{2(1-p)}{p^2},$$

from which it follows that

$$\begin{aligned}\sigma_X^2 &= E[X^2] - [E(X)]^2 = E[X(X-1) + X] - [E(X)]^2 \\ &= E[X(X-1)] + \mu_X - \mu_X^2 \\ &= \frac{2(1-p)}{p^2} + \frac{1}{p} - \left(\frac{1}{p}\right)^2 = \frac{1-p}{p^2}.\end{aligned}$$

Applications of the Geometric Distribution

It has already been remarked that the geometric distribution is often used to model waiting times (the number of trials or periods until a certain event occurs). However, this distribution also has other uses.

Example 4.4

Fingerprints

Fingerprints are widely accepted as a method for identifying people. Partial fingerprints, however, are more commonly found in criminal investigations. In this book *Finger Prints*, Galton (1892) identified 10 characteristics that serve as a means of classifying and identifying a person's fingerprint. Not all of the characteristics that serve to identify a fingerprint appear on a portion of that fingerprint, so that the distribution of these characteristics over the surface of the fingerprint is of interest. One can break a fingerprint into a grid of cells and for each cell count the number Y of occurrences of the Galton characteristics. Because it is possible for $Y = 0$ occurrences of Galton characteristics to occur in a cell, the geometric distribution (4.1) defined for the number of *trials* is not a reasonable model for the variation of Y. There is, however, a theoretical justification for using the geometric distribution for the number of *failures* as a model for the variation of Y [see Sclove (1981) and also Example 5.7 of Chapter 11]. Our notation is intended to call attention to the fact that the probability mass function being fitted to the data has the form

$$p(y) = (1-p)^y p, \qquad y = 0, 1, 2, \ldots. \tag{4.5}$$

Sclove (1981) collected data on 598 cells (taken from 39 fingerprints by a sophisticated scheme needed to make observations on cells approximately independent). The observed frequencies for the observed values $y = 0, 1, 2, \ldots$ of Y are shown in Exhibit 4.1 along with theoretical frequencies calculated using the "geometric distribution" (4.5) with $p = 0.737$:

$$\text{theoretical frequency for } \{Y = y\} = 598(1 - 0.737)^y\,(0.737), \qquad y = 0, 1, 2, 3, 4.$$

The agreement between the observed and theoretical frequencies is quite good. ♦

EXHIBIT 4.1 *Observed and Theoretical Frequencies for the Number Y of Occurrences of Galton Characteristics in 598 Fingerprint Cells*

Number y of Occurrences	*Observed Frequency*	*Theoretical Frequency*
0	441	440.73
1	125	115.91
2	25	30.48
3	5	8.02
4	2	2.11
≥ 5	0	0.25
	598	598.00

The System of Doubling the Bet*

The geometric distribution, being of a comparatively simple form, enables us to make some interesting calculations. In particular, the following example provides an illustration of the fact that the mean of a discrete random variable can be infinitely large.

EXAMPLE 4.5

Doubling the Bet

If a gambler has a probability $p = \frac{1}{2}$ of winning on any play of a given game of chance, then from (4.1) the probability of winning the game for the first time at the xth play is $(\frac{1}{2})^x$. Suppose that the gambler follows a betting system of *doubling* his bets until he wins, and that his initial bet is 1 dollar. Then if he loses on the first play of the game, his second bet is 2 dollars; if he loses again, his third bet is 4 dollars, the fourth bet 8 dollars, and so on. This system is advocated by many people as a sure way to win since all losses are recaptured and a dollar is won as soon as the gambler wins for the first time.

Let Z denote the amount the gambler needs in order to be able to continue until he wins for the first time. That is, if he were to win on the first toss, he would only need 1 dollar (to make the first bet). If he does not win until the fifth trial, he needs $1 + 2 + 4 + 8 + 16 = 2^5 - 1$ dollars. In general, if he does not win until the kth play, he needs $2^k - 1$ dollars. Thus Z is a random variable that takes on a value $2^k - 1$ with probability $(\frac{1}{2})^k$ for $k = 1, 2, 3, \ldots$. The probability mass function of Z is given as follows:

z	1	3	7	15	. . .	$2^k - 1$	. . .
$p(z)$	$\frac{1}{2}$	$\frac{1}{4}$	$\frac{1}{8}$	$\frac{1}{16}$	. . .	$\frac{1}{2^k}$	. . .

The expected amount (mean amount) of money needed to sustain the betting system is

$$\mu_Z = \sum_{k=1}^{\infty} (2^k - 1)\left(\frac{1}{2^k}\right) = \frac{1}{2} + \frac{3}{4} + \frac{7}{8} + \frac{15}{16} + \cdots .$$

We see that the size of the terms in the sum above is increasing; consequently, $\mu_Z = \infty$. Thus, *on the average, no finite amount of money is sufficient to sustain this betting system.* ♦

5. *Negative Binomial Distribution*

The geometric distribution is the distribution of the number of independent Bernoulli trials with common probability p of success required to obtain a single success. A natural generalization of this distribution is the distribution of the random variable X that equals the number of trials required until (and including) the trial on which the rth success occurs, $r = 1, 2, \ldots$. The random variable X is said to have the **negative binomial distribution** with parameters r and p.

EXAMPLE 5.1

Fishery Management

To estimate the relative number of fish of a certain species in a given pond, fish are caught one at a time, and for every fish caught, the species of that fish is noted before the fish is returned to the pond. When r fish of the species of interest have been caught, the total number X of fish caught is recorded. Because fish are returned to the pond after being caught, sampling is with replacement, and thus the catches act as independent Bernoulli trials. If there are N fish in the pond, of which M belong to a particular species of interest, the probability that a fish caught is of this species is $p = M/N$. The random variable X thus has a negative binomial distribution with parameters r and $p = M/N$. ♦

EXAMPLE 5.2

Politics

A political canvasser requires $r = 100$ signatures for a petition. The probability that a person will agree to sign the petition is $\frac{1}{10}$, and each person approached acts independently of any other person. Let X denote the number of persons that must be approached before the 100 needed signatures have been gathered. The random variable X has a negative binomial distribution with parameters $r = 100$ and $p = \frac{1}{10}$. ♦

Random variables having a negative binomial distribution are often amenable to a waiting-time interpretation. Suppose that a process is observed at equally spaced intervals of time. During the period between times of observations, some random event may or may not occur. If the event does occur, a success is recorded at the end of the time period; otherwise, a failure is recorded. If observation of the process ends once r successes have been recorded, the number X of time intervals that have elapsed (which is also the number of trials made) gives the total time that the process has been under observation.

EXAMPLE 5.3

A signal light on a reef near a shipping lane is supposed to be lit continually, and is inspected daily. Between the times of inspection, the light may burn out; in this case the light is replaced at the next inspection. The probability that a light burns out during a 1-day interval is 0.01, and whether a light burns out during one daily period is independent of whether lights burn out during all other daily periods. The inspector keeps track of the number of times the light must be replaced, and when the light burns out for the tenth time, places an order for 10 new replacement lights. The number of days X between orders of light bulb replacements thus has a negative binomial distribution with parameters $r = 10$ and $p = 0.01$. ♦

Derivation of the Negative Binomial Distribution

Assume that independent Bernoulli trials with common probability p of success are observed one at a time until the rth success is observed. The random variable X is the required number of trials (including the success that terminates the trials). Because r successes must be observed, the possible values of X are $r, r + 1, r + 2, \ldots$. Note that the last trial observed must be a success, and that for X to equal x, in the $x - 1$ preceding trials, $r - 1$ successes must have been obtained. Further, the last trial and the previous $x - 1$ trials are independent. Let A_x denote the event that the first $x - 1$ trials had $r - 1$ successes. The probability of this event is the probability of observing $r - 1$ successes in $n = x - 1$ trials and can be found using the binomial distribution:

$$\begin{aligned} P(A_x) &= \binom{x-1}{r-1} p^{r-1}(1-p)^{(x-1)-(r-1)} \\ &= \binom{x-1}{r-1} p^{r-1}(1-p)^{x-r}. \end{aligned}$$

The probability of a success on the last (xth) trial is p. Thus, by independence,

$$\begin{aligned} P\{X = x\} &= P\{A_x \textit{ and } \text{success on last trial}\} \\ &= P(A_x)P(S) = \left[\binom{x-1}{r-1} p^{r-1}(1-p)^{x-r}\right]p. \end{aligned}$$

Consequently, the probability mass function of X is

$$(5.1) \qquad p(x) = \begin{cases} \dbinom{x-1}{r-1}(1-p)^{x-r}p^r, & \text{if } x = r, r+1, \ldots, \\ 0, & \text{otherwise.} \end{cases}$$

The probability mass function (5.1) defines the **negative binomial distribution** with parameters r and p. This distribution is sometimes called the **Pascal distribution,** after the French mathematician, Blaise Pascal.

There are situations where one is interested in the number of **failures** $Y = X - r$ obtained before the rth success is observed. (Note that Y is the truly random part of X.) One can find $P\{Y = y\}$ from the probabilities for the number X of trials by noting that

$$P\{Y = y\} = P\{X - r = y\} = P\{X = y + r\}.$$

Thus the probability mass function of Y is

$$(5.2) \qquad p_Y(y) = \binom{y+r-1}{r-1}(1-p)^y p^r, \qquad y = 0, 1, 2, \ldots.$$

As in the case of the geometric distribution, the probability mass function (5.2) is sometimes used to define the negative binomial distribution. This is, for example, true of Williamson and Bretherton (1963), who provide extensive tables of the probability mass function (5.2) rather than of (5.1). (Williamson and Bretherton's notation also differs from that used here; their k, n, p are our r, p, y, respectively.) The probability mass function (5.2) can be generalized to allow noninteger values of r; these generalized probability mass functions are also tabulated by Williamson and Bretherton.

Probability Calculations for the Negative Binomial Distribution

Tables of individual probabilities $p(x)$ for the negative binomial distribution can be prepared by using the recursive formula

$$(5.3)\quad p(x) = \left[\frac{(x-1)(1-p)}{x-r}\right]p(x-1), \qquad x = r+1, r+2, \ldots .$$

and the initial value

$$p(r) = p^r.$$

Thus if $r = 2$, $p = 0.4$,

$$p(2) = (0.4)^2 = 0.1600, \qquad p(3) = \left[\frac{(3-1)(1-0.4)}{3-2}\right]p(2) = 0.1920,$$
$$p(4) = \left[\frac{(4-2)(1-0.4)}{4-2}\right]p(3) = 0.1728,$$
$$p(5) = \left[\frac{(5-1)(1-0.4)}{5-2}\right]p(4) = 0.1382,$$

and so on.

Practical applications of the negative binomial distribution, however, more frequently require calculations of sums of individual probabilities. For such purposes, the following relationship between the negative binomial and binomial distributions can be useful.

> If X has a negative binomial distribution with parameters r and p, then for $x = r, r+1, \ldots .$
>
> $$(5.4)\qquad P\{X > x\} = P\{U < r\},$$
>
> where U has a binomial distribution with parameters $n = x$ and p.

This relationship is easily established by noting that X will exceed x if and only if there are fewer than r successes in the first x trials. (If the first x trials contain fewer than r successes, at least one more success is needed and this will require at least one more trial; thus $X \geq x + 1 > x$. On the other hand, the fact that X exceeds x means that an insufficient number of successes were observed in the first x trials.) The number U of successes in the first x trials has a binomial distribution with parameters $n = x$ and p, and this establishes the equality (5.4).

From (5.4) it follows immediately that the cumulative distribution function $F(x)$ of X is given by

$$F(x) = P\{X \leq x\} = 1 - P\{X > x\} \\ = 1 - P\{U < r\} = P\{U \geq r\} \tag{5.5}$$

for $x = r, r + 1, \ldots$, where U has the binomial distribution with parameters $n = x$ and p. Using $F(x)$, one can find probabilities of intervals of x-values. Consequently, most probability calculations for the negative binomial distribution can be done using tables of binomial probabilities.

EXAMPLE 5.4

A basketball promoter needs two star basketball players for a charity basketball game. The promoter has a long list of basketball stars who have played in previous charity games, and will call names on this list until he finds two stars who agree to play. The promoter believes that 20% of the people on his list would play if asked, and assumes that people make their decisions independently of one another. What is the probability that it takes the promoter no more than 8 calls to secure the two basketball stars?

Let X denote the number of calls that the promoter must make. From the information given, X has a negative binomial distribution with parameters $r = 2$ and $p = 0.20$. The question asks for the probability, $P\{X \leq 8\}$, that no more than 8 calls are made. From (5.5), $P\{X \leq 8\} = P\{U \geq 2\}$, where U has a binomial distribution with parameters $n = 8$ and $p = 0.20$. Using Exhibit B.2 of Appendix B, we have

$$P\{X \leq 8\} = P\{U \geq 2\} = 1 - p_U(0) - p_U(1) \\ = 1 - 0.16777 - 0.33554 = 0.49669.$$

What is the probability, $P\{X > 4\}$, that more than 4 calls are made? From (5.4), $P\{X > 4\} = P\{V < 2\}$, where V has a binomial distribution with $n = 4$ and $p = 0.20$. Using Exhibit B.2 gives

$$P\{X > 4\} = P\{V < 2\} = p_V(0) + p_V(1) \\ = 0.40960 + 0.40960 = 0.81920.$$

Note that tables of binomial probabilities for *different* sample sizes n were used to compute $P\{X \leq 8\}$ and $P\{X > 4\}$, even though X has but one distribution. For a calculation such as

$$P\{4 < X \leq 8\} = P\{X \leq 8\} - P\{X \leq 4\} \\ = P\{U \geq 2\} - (1 - P\{V < 2\}) \\ = 0.49669 - (1 - 0.81920) \\ = 0.31589,$$

one would use two tables of binomial probabilities. ◆

Relations Between the Negative Binomial and Geometric Distributions

The geometric distribution is the special case $r = 1$ (waiting for one success) of the negative binomial distribution. A more useful relationship between the geometric and negative binomial distributions is the following:

> A random variable X having a negative binomial distribution with parameters r and p can be written as the sum
>
> (5.6) $$X = \sum_{i=1}^{r} W_i$$
>
> of r independent random variables $W_1, W_2, \ldots, W_r$, each having a geometric distribution with parameter p.

This assertion follows directly from the way the negative binomial distribution was defined. Suppose that independent Bernoulli trials with probability p of success are observed one at a time. Let W_1 be the number of trials until (up to and including) the trial on which the first success is observed. Let W_2 be the number of *additional* trials until the second success is obtained, W_3 be the number of *additional* trials after that until the third success is obtained, and so on. It is easily seen that $W_1, W_2, \ldots$ are *independent* geometrically distributed random variables, each with parameter p, because each W_i is the number of trials until a success is observed. However, if X is the number of trials until (up to and including) the trial on which the rth success is observed, then X is the sum of $W_1, W_2, \ldots, W_r$, as asserted by (5.6), and X also has a negative binomial distribution with parameters r and p.

Descriptive Indices of the Negative Binomial Distribution

It follows directly from (5.6), facts about the mean and variance of a geometric distribution, and results given in Chapter 7 for means and variances of sums that if X has a negative binomial distribution with parameters r and p,

$$\mu_X = \mu_{\sum_{i=1}^{r} W_i} = r\mu_{W_i} = r\left(\frac{1}{p}\right)$$
$$\sigma_X^2 = \sigma^2_{\sum_{i=1}^{r} W_i} = r\sigma^2_{W_1} = \frac{r(1-p)}{p^2}.$$

That is,

> The mean and variance of a negative binomial distribution with parameters r and p are
>
> (5.7) $$\mu_X = \frac{r}{p}, \qquad \sigma_X^2 = \frac{r(1-p)}{p^2},$$
>
> respectively.

EXAMPLE 5.4 (continued)

The mean number of calls that the promoter must make until he finds $r = 2$ basketball stars is

$$\mu_X = \frac{r}{p} = \frac{2}{0.2} = 10.$$

The variance of the number, X of calls to be made by the promoter is

$$\sigma_X^2 = \frac{r(1-p)}{p^2} = \frac{2(1-0.2)}{(0.2)^2} = 40.$$ ◆

The quantiles of X are most easily found by using (5.5) to compute the cumulative distribution function, $F(x)$, of X for various values of x. To illustrate, suppose that X has a negative binomial distribution with parameters $r = 2$ and $p = 0.30$, and that the median of X is desired. Because the mean and median should be close to each other, as an initial guess for the median one might try the integer x closest to the mean of X. Because $\mu_X = (2)/(0.3) = 6.67$, this is $x = 7$. By (5.5), using tables of the binomial distribution with $n = 7$, $p = 0.3$ (Exhibit B.2 of Appendix B):

$$\begin{aligned} F(7) = P\{U \geq 2\} &= 1 - p_U(0) - p_U(1) \\ &= 1 - 0.08235 - 0.24706 \\ &= 0.67059. \end{aligned}$$

Because $F(7) > 0.50$, try $x = 6$. Using the binomial distribution with $n = 6$, $p = 0.3$,

$$\begin{aligned} F(6) = P\{U \geq 2\} &= 1 - 0.11765 - 0.30253 \\ &= 0.57982. \end{aligned}$$

Because $F(6)$ is still bigger than 0.50, try $x = 5$. Using the binomial distribution with $n = 5$, $p = 0.3$,

$$F(5) = 1 - 0.16807 - 0.36015 = 0.47178.$$

Observing that $F(5) < 0.50 < F(6)$, so that $F(x)$ "jumps through" 0.50 at $x = 6$, it follows that

$$\text{Median}(X) = Q_X(0.5) = 6.$$

To find the 75th percentile of X, $Q_X(0.75)$, note that $F(7) = 0.67059 < 0.75$. Hence try $x = 8$. Using the binomial distribution with $n = 8$ and $p = 0.3$,

$$F(8) = 1 - 0.05765 - 0.19765 = 0.74470.$$

This is still too small, so $x = 9$ is tried, and

$$F(9) = 1 - 0.04035 - 0.15565 = 0.80400.$$

Because $F(8) < 0.75 < F(9)$, it follows that $Q_X(0.75) = 9$.

The mode of X can be found using the recursion relation (5.3). For $r = 1$, the mode of X is always 1. For $r > 1$, $p(x) > p(x - 1)$ when $r + 1 \leq x < [r - (1 - p)]/p$ and $p(x) < p(x - 1)$ when $x > [r - (1 - p)]/p$. Let

$$a = \frac{r - (1 - p)}{p} = \mu_X - \frac{1 - p}{p}.$$

If a is not an integer, then Mode(X) is the integer between $a - 1$ and a. If a is an integer, then both $a - 1$ and a are modes of X. For example, when $r = 2$, $p = 0.4$, then $a = 3.5$ and the mode of X is 3. When $r = 2$, $p = 0.5$, then both $a - 1 = 2$ and $a = 3$ are modes of X.

Applications of the Negative Binomial Distribution

The applications of the negative binomial distribution as a distribution of waiting times have already been indicated. The restriction of the parameter r of this distribution to positive integer values tends to restrict its applicability in other contexts. Instead, a generalization of the negative binomial distribution to permit values of r that are not integers can be made, and it is this more general type of distribution that is widely used to fit data arising in biology, ecology, medicine, and other fields. This generalized negative binomial distribution is discussed in the next section.

6 Other Discrete Distributions

Because researchers are interested in wide varieties of phenomena, no short list of distributions can be expected to serve as models for all data. Very often special classes of distributions are developed to deal with a particular study, only later to find that these distributions have wider applicability. In this section, three such special classes of distributions are described briefly.

The Generalized Negative Binomial Distribution

The negative binomial distribution occurs in a wider range of practical situations than just those in which the random variable X arises as a result of repeated Bernoulli trials. In these other applications, however, it has proved useful to use $Y = X - r$ (which is the random part of X in waiting-time problems) and also to generalize the probability mass function (5.2) of Y to permit noninteger values of the parameter r. To motivate this generalization, note that we can write the probability mass function of $Y = X - r$ in the form

$$p_Y(y) = H(y, r)p^r(1 - p)^y, \qquad y = 0, 1, 2, \ldots, \tag{6.1}$$

where

$$H(y, r) = \frac{(y + r - 1)!}{(r - 1)!\, y!}.$$

To generalize (6.1) to permit noninteger values of r as parameter, the factorial operation (function) $u!$ needs to be extended to cases where u is not an integer. Such an extension is provided by the **gamma function** $\Gamma(a)$; this function is discussed in Appendix A. The properties of $\Gamma(a)$ that are needed here are

$$\Gamma(a + 1) = a\Gamma(a), \qquad \Gamma(1) = 1,$$

so that if u is an integer,

$$\Gamma(u + 1) = u!,$$

and the needed generalization of the factorial function has been found. When r is a positive integer and $y = 0, 1, 2, \ldots,$

$$H(y, r) = \frac{(y + r - 1)!}{(r - 1)!\, y!} = \frac{\Gamma(y + r)}{\Gamma(r)\, \Gamma(y + 1)}.$$

Because the gamma function is defined for integer and noninteger r, the expression above defines $H(y, r)$ for all r. Hence

$$(6.2) \qquad p_Y(y) = \begin{cases} \dfrac{\Gamma(y+r)}{\Gamma(r)\Gamma(y+1)} p^r(1-p)^y, & \text{if } y = 0, 1, 2, \ldots, \\ 0, & \text{otherwise.} \end{cases}$$

This mass function defines the **generalized negative binomial distribution** with parameters $r > 0$ and p, $0 \leq p \leq 1$. Remember that the negative binomial distribution being generalized here is, for integer r, the number of failures $Y = X - r$ and not the number of trials, X.

Computation of the probabilities $p_Y(y) = P\{Y = y\}$ in (6.2) is facilitated by the recursion formula

$$(6.3) \quad p_Y(y) = \left[\frac{(y + r - 1)(1 - p)}{y}\right] p_Y(y - 1), \qquad y = 1, 2, \ldots,$$

and the initial value $p_Y(0) = p^r$.

Thus if $r = 1.2$ and $p = 0.6$,

$$\begin{aligned} p_Y(0) &= (0.6)^{1.2} = 0.5420, \\ p_Y(1) &= \frac{(1 - 1 + 1.2)(1 - 0.6)}{1} p_Y(0) = 0.2602, \\ p_Y(2) &= \frac{(2 - 1 + 1.2)(1 - 0.6)}{2} p_Y(1) = 0.1145, \\ p_Y(3) &= \frac{(3 - 1 + 1.2)(1 - 0.6)}{3} p_Y(2) = 0.0489, \end{aligned}$$

and so on.

When r is an integer, the mean and variance of Y can be obtained from the mean and variance of X given in (5.7). That is,

$$(6.4) \qquad \begin{aligned} \mu_Y &= \mu_{X-r} = \mu_X - r = \frac{r}{p} - r = \frac{r(1-p)}{p}, \\ \sigma_Y^2 &= \sigma_{X-r}^2 = \sigma_X^2 = \frac{r(1-p)}{p^2}. \end{aligned}$$

These formulas for the mean and variance continue to hold for noninteger r.

If $a = \mu_Y - [(1 - p)/p]$ is not an integer, the mode of Y is the integer between $a - 1$ and a. If a is an integer, Y has modes at $a - 1$ and a. Thus if $r = 1.2$, $p = 0.6$, then $a = 0.133$ and the mode of Y is 0. If $r = 1.0$, $p = 0.5$, then $a = 1$ and Y has modes at 0 and 1.

When r is a positive integer, the quantiles of Y can be obtained from the quantiles of $X = Y + r$ and tables of the binomial distribution. That is, for example,

$$Q_Y(0.50) = Q_{X-r}(0.50) = Q_X(0.50) - r$$

and

$$Q_Y(0.75) = Q_X(0.75) - r.$$

A method for finding quantiles of X using binomial tables was illustrated in Section 5. When r is not an integer, no similar useful relationship between probabilities of the generalized negative binomial and binomial distributions exists, and the quantiles of Y are most easily obtained from the probabilities $p_Y(y) = P\{Y = y\}$ of Y using the methods of Chapter 6. Thus when $r = 1.2$ and $p = 0.6$,

$$p_Y(0) = 0.5420, \qquad p_Y(0) + p_Y(1) = 0.5420 + 0.2602 = 0.8044,$$

from which it follows that

$$\begin{aligned} \text{Median}(Y) &= Q_Y(0.50) = 0, \\ Q_Y(0.75) &= 1. \end{aligned}$$

The generalized negative binomial distribution has been applied in a variety of contexts. A theoretical derivation of this distribution was given by Greenwood and Yule (1920) in an attempt to justify the close fit of this distribution to the probabilities associated with the number Y of accidents suffered by a randomly chosen woman working on high-explosive shells. The number of accidents suffered by a specific woman is assumed to have a Poisson distribution with a parameter λ; however, different women have different values of λ, and λ is assumed to be itself a random variable (defined over the population of all women) with a certain continuous distribution. The resulting distribution for Y is a generalized negative binomial distribution (Example 5.7, Chapter 11).

Greenwood and Yule used data gathered on 647 women observed over a period of 5 weeks in order to determine the parameters r and p and to demonstrate the fit of the negative binomial distribution of the data. Their data are shown in Exhibit 6.1.

The theoretical frequencies were obtained by multiplying each of the probabilities $p_Y(y)$, $y = 0, 1, 2, \ldots, 5$ and the probability

$$P\{Y \geq 6\} = 1 - \sum_{y=0}^{5} p_Y(y)$$

EXHIBIT 6.1 *Observed Frequency Distribution for the Number of Accidents Suffered by Women Working on 6-Inch Shells for 5 Weeks*

Number y *of Accidents*	*Observed Frequency of* y *Accidents*	*Theoretical Frequency* ($r = 0.96$, $p = 0.67$)
0	447	440.6
1	132	139.8
2	42	45.3
3	21	14.9
4	3	4.5
5	2	1.3
≥6	0	0.6
	647	647.0

by the number, $M = 647$, of trials. Using the initial value $p_Y(0) = (0.067)^{0.96} = 0.681$,

$$p_Y(1) = 0.216, \quad p_Y(2) = 0.070, \quad p_Y(3) = 0.023,$$
$$p_Y(4) = 0.002, \quad p_Y(5) = 0.002,$$

by use of the recursion formula (6.3). From Exhibit 6.1 it appears that the generalized negative binomial distribution with parameters $r = 0.96$, $p = 0.67$ provide a rather good fit to the data. Note that in this case, $r = 0.96$ is very close to 1, so that the geometric distribution (for the number, Y, of failures) also might give a good fit to these data.

Other examples of applications where the generalized negative binomial distribution has been successfully fit to data are to zircon counts in geological rock samples [Griffiths (1960); $r = 0.239$, $p = 0.186$], to the number of purchases of a particular brand over an extended time period by consumers [Chatfield (1970); $r = 0.041$, $p = 0.315$], to sleeping group sizes of Vervet monkeys [Cohen (1971); see also Boswell, Ord, and Patil (1979); $r = 1.744$, $p = 0.341$], and to the number of red mites on apple leaves in North Carolina [Ludwig and Reynolds (1988); $r = 0.9925$, $p = 0.47$]. An application to the scores made by college football teams (where points scores are reported as multiples Y of the number of points, 7, assigned to a touchdown) is given by Pollard (1973).

The Zeta Distribution

Many discrete quantitative phenomena in the behavioral sciences appear to behave according to a "the-higher-the-fewer" rule. For example, the higher the value of x of an executive's salary X, the fewer executives there are who will have that salary; the higher the frequency of x of a word in prose, the fewer are the words which have that frequency; the larger the size x of a claim against an insurance company, the fewer claims of that size there are; and so on. We might try to approximate the probability mass function of a random variable X that obeys such a the-higher-the-fewer rule by a function of the form

$$(6.5) \qquad p(x) = P\{X = x\} = c\,\frac{1}{x^{\alpha+1}}, \qquad x = 1, 2, \ldots,$$

with $p(x)$ equal to 0 otherwise. Here the constant α, $\alpha > 0$, measures the rate of decrease of the probability of the value x with an increase in x. Because

$$(6.6) \qquad 1 = \sum_{x=1}^{\infty} p(x) = c \sum_{x=1}^{\infty} \frac{1}{x^{\alpha+1}},$$

if (6.5) is to be a probability mass function, the constant c is determined by α, and α is the sole parameter of the distribution (6.5). The function

$$(6.7) \qquad \zeta(s) = \frac{1}{1^s} + \frac{1}{2^s} + \frac{1}{3^s} + \cdots = \sum_{k=1}^{\infty} \frac{1}{k^s}, \qquad s > 1,$$

is well known and arises in many mathematical disciplines. It is called the *Riemann zeta function* [after the German mathematician Georg Friedrich Bernhard Riemann (1826–1866)] and is extensively tabulated [see, e.g., Davis (1935)]. From (6.6) and (6.7) we see that

$$(6.8) \qquad p(x) = \begin{cases} \dfrac{1}{\zeta(\alpha+1)}\,\dfrac{1}{x^{\alpha+1}}, & \text{if } x = 1, 2, 3, \ldots, \\ 0, & \text{otherwise.} \end{cases}$$

A probability distribution of the form (6.8) was used by the Italian economist and sociologist Vilfredo Pareto (1848–1923) to describe the distribution of family incomes. In Pareto's studies, the law fairly accurately described income distribution in almost every country that kept income statistics.

Despite Pareto's contribution to the development of probability laws of the form of (6.8), such probability laws are often associated with the name of G. K. Zipf, because it was Zipf who applied such distributions to a variety of contexts [Zipf (1949)] and, through the success of such applications, succeeded in popularizing distributions of this form.

A third name given to the distribution defined by (6.8) is the **zeta distribution,** because of the relationship of this probability mass function to the

Riemann zeta function. If X has a zeta distribution with parameter α, and if $r > \alpha + 1$, then

$$E[X^r] = \sum_{x=1}^{\infty} x^r \frac{1}{\zeta(\alpha + 1)x^{\alpha+1}} = \frac{1}{\zeta(\alpha + 1)} \sum_{x=1}^{\infty} \frac{1}{x^{\alpha-r+1}}$$
$$= \frac{\zeta(\alpha - r + 1)}{\zeta(\alpha + 1)}, \qquad r = 1, 2, \ldots,$$

and consequently,

(6.9)

$$\mu_X = E[X] = \frac{\zeta(\alpha)}{\zeta(\alpha + 1)}, \qquad \sigma_X^2 = \frac{\zeta(\alpha - 1)\zeta(\alpha + 1) - [\zeta(\alpha)]^2}{[\zeta(\alpha + 1)]^2}, \qquad \alpha > 1.$$

In applying (6.9), it should be noted that $\zeta(s) = \infty$ for $s \leq 1$. Thus when X has a zeta distribution with parameter α, the variance, σ_X^2, of X is infinitely large when $\alpha \leq 1$.

The following example of the application of the probability distribution (6.8) illustrates the usefulness of the zeta distribution in describing the variation of random variables observed in many contexts of human behavior.

EXAMPLE 6.1

Simon and Bonini (1958) ranked the top 10 steel producers at that time according to size. Suppose that every ton of steel that could ever be produced by these firms were to be marked with the rank X of the firm that produced that ton of steel. Thus every ton of steel produced by the largest steel producer (U.S. Steel) would be marked with $X = 1$, whereas every ton produced by the tenth largest producer (Wheeling) would be marked with $X = 10$. If a ton of steel is now drawn at random from the total potential production of the 10 firms, the probability that $X = x$ will be equal to the proportion of the total potential production of steel produced by the xth largest steel producer, $x = 1, 2, \ldots, 10$. Simon and Bonini postulated that $p(x) = P\{X = x\}$ should follow the "higher-the-fewer" rule, so that $p(x)$ would have the form (6.8). This theory can be tested by regarding the tons of steel produced by the steel producers in one year (99,400,000 tons) as trials of the random sampling experiment just described, and comparing the observed production (frequency) of steel by the producer ranked x to the theoretical production given by

$$\text{theoretical production of } x\text{th ranked producer} = (99{,}400{,}000)p(x)$$
$$= (99{,}400{,}000)\frac{1}{\zeta(\alpha + 1)x^{\alpha+1}}, \qquad x = 1, 2, \ldots, 10.$$

Simon and Bonini used $\alpha = 1$, leading to the results shown in Exhibit 6.2. As can be seen from the exhibit, the fit of the theoretical model for production of the firms is very good. ◆

The Truncated Poisson Distribution

Suppose that a random variable X which has a Poisson distribution is observed, but because of the nature of the phenomenon a value of $X = 0$ is unobservable. For example, it might be assumed that the number X of people in a potential social group has a Poisson distribution; if $X = 0$, the group never formed, and thus cannot be observed. To take account of this fact, the range of possible values of X is "truncated" to exclude the value 0.

If X is assumed to have a Poisson distribution with parameter λ, the fact that the event $\{X = 0\}$ cannot occur, or, equivalently, that the event $\{X > 0\}$ has occurred, means that we must redefine the probability model *conditional* on this fact. Thus the probability of the event $\{X = x\}$ is 0 if $x = 0$ and equals

$$P\{X = x | X > 0\} = \frac{P\{X = x \text{ and } X > 0\}}{P\{X > 0\}} = \frac{P\{X = x\}}{1 - P\{X = 0\}} = \frac{\lambda^x e^{-\lambda}/x!}{1 - e^{-\lambda}}$$

for $x = 1, 2, \ldots$.

EXHIBIT 6.2 *Comparison of Actual and Theoretical Ingot Production of 10 Leading Steel Producers in the United States*

Producer	*Rank Order*	*Actual Production*	*Theoretical Production (zeta distribution with $\alpha = 1$)*
U.S. Steel	1	38.7	34.3
Bethlehem	1	18.5	17.1
Republic	3	10.3	11.3
Jones & Laughlin	4	6.2	8.5
National	5	6.0	6.8
Youngstown	6	5.5	5.2
Armeo	7	4.9	4.8
Inland	8	4.7	4.2
Colorado Fuel & Iron	9	2.5	3.8
Wheeling	10	2.1	3.4
		99.4	99.4

A random variable X having a probability mass function

$$p(x) = \begin{cases} \dfrac{\lambda^x}{x!\,(e^\lambda - 1)}, & \text{if } x = 1, 2, \ldots, \\ 0, & \text{otherwise,} \end{cases} \tag{6.10}$$

is said to have a **truncated Poisson distribution** with parameter λ. The mean and variance of such a distribution are

$$\mu_X = \frac{\lambda e^\lambda}{e^\lambda - 1}, \qquad \sigma_X^2 = \frac{\lambda e^{2\lambda} - \lambda(\lambda + 1)e^\lambda}{(e^\lambda - 1)^2}. \tag{6.11}$$

EXAMPLE 6.2

Social Groups

James (1953) studied the size X of small groups that form in a variety of social contexts. Exhibits 6.3 to 6.5 show the fit of truncated Poisson distributions with respective parameters $\lambda = 0.892$, $\lambda = 0.889$, and $\lambda = 1.362$ to the distribution of the number of pedestrians grouped

EXHIBIT 6.3 *Size of Groups of Pedestrians in Eugene, Oregon, on a Spring Morning*

Size x of Group	*Observed Frequency*	*Theoretical Frequency (truncated Poisson, $\lambda = 0.892$)*
1	1486	1501
2	694	670
3	195	199
4	37	44
5	10	8
6	1	1
	2423	2423

EXHIBIT 6.4 *Size of Shopping Groups*

Size x of Group	*Observed Frequency*	*Theoretical Frequency (truncated Poisson, $\lambda = 0.889$)*
1	316	316
2	141	141
3	44	42
4	5	9
5	4	2
	510	510

EXHIBIT 6.5 *Size of Playgroups*

Size x of Group	*Observed Frequency*	*Theoretical Frequency (truncated Poisson, $\lambda = 1.362$)*
1	570	599
2	435	408
3	203	185
4	57	63
5	11	17
6	1	4
7	0	1
	1277	1277

together on a spring morning in Eugene, Oregon, the distribution of the size of shopping groups in department stores and public markets, and the distribution of the size of playgroups in the playgrounds of 14 elementary schools.

In each of Exhibits 6.3 to 6.5, the resulting theoretical frequencies are reasonably close to the actual frequencies, indicating that a truncated Poisson distribution provides a reasonable approximation to the actual distribution of the group sizes. ◆

Notes and References

The book by Johnson, Kotz, and Kemp (1992) discusses a variety of discrete distributions that have been used to model data in a number of different disciplines, and provides extensive lists of references.

Glossary and Summary

The various discrete distributions discussed in Sections 1 to 5 of this chapter are summarized below.

Distribution	*Parameters*	*Probability Mass Function*	*Mean*	*Variance*
Bernoulli	$0 \le p \le 1$	$p^z(1-p)^{1-z}$, $z = 0, 1$	p	$p(1-p)$
Binomial	n = positive integer, $0 \le p \le 1$	$\binom{n}{z} p^z(1-p)^{n-z}$, $z = 0, 1, \ldots, n$	np	$np(1-p)$

Distribution	*Parameters*	*Probability Mass Function*	*Mean*	*Variance*
Hypergeometric	N = positive integer, n = positive integer $0 \le p \le 1$, Np = integer, $n \le N$	$\dfrac{\binom{Np}{z}\binom{N(1-p)}{n-z}}{\binom{N}{n}}$, $z = 0, 1, \ldots, n$ $z \le Np$, $n - z \le N(1-p)$	np	$\left(\dfrac{N-n}{N-1}\right) p(1-p)$
Poisson	$\lambda > 0$	$\dfrac{\lambda^x e^{-\lambda}}{x!}$, $x = 0, 1, 2, \ldots$	λ	λ
Geometric (X = number of trials)	$0 \le p \le 1$	$p(1-p)^{x-1}$, $x = 1, 2, \ldots$	$\dfrac{1}{p}$	$\dfrac{1-p}{p^2}$
(Y = number of failures)	$0 \le p \le 1$	$p(1-p)^y$, $y = 0, 1, \ldots$	$\dfrac{1-p}{p}$	$\dfrac{1-p}{p^2}$
Negative binomial (X = number of trials)	r = positive integer, $0 \le p \le 1$	$\binom{x-1}{r-1}(1-p)^{x-r}p^r$, $x = r, r+1, \ldots$	$\dfrac{r}{p}$	$\dfrac{r(1-p)}{p^2}$
(Y = number of failures)	r = positive integer, $0 \le p \le 1$	$\binom{y+r-1}{r-1}(1-p)^y p^r$, $y = 0, 1, 2, \ldots$	$\dfrac{r(1-p)}{p}$	$\dfrac{r(1-p)}{p^2}$

Other Useful Facts

Bernoulli trial: A random experiment with only two possible outcomes, one of which can be called a **success** and the other a **failure.**

Binomial distribution: Arises as the distribution of the number Z of successes observed in n independent Bernoulli trials with the same probability p of success on each trial. If $n = 1$, the binomial distribution is the same as the Bernoulli distribution.

Hypergeometric distribution: Arises as the distribution of the number Z of successes observed when drawing a random sample of n units without replacement from a population of N units, $M = Np$ of which are successes.

Geometric distribution and negative binomial distribution: The geometric distribution arises as the distribution of the number X of independent Bernoulli trials (with common probability p of a success) that must be seen until (and including) the trial at which the first success is observed. The negative binomial distribution arises as the distribution of the number of trials needed until (and including) the trial on which the rth success is observed. (When $r = 1$, a negative binomial distribution is a geometric distribution.)

Sums:

1. The sum of n independent random variables each having a Bernoulli distribution with parameter p has a binomial distribution with parameters n and p.
2. The sum of r independent random variables each having a geometric distribution with parameter p has a negative binomial distribution with parameters r and p.

Approximations:

1. If $n/N \leq 0.1$, probabilities for the hypergeometric distribution with population size N, sample size n and probability of success p can be closely approximated by corresponding probabilities for a binomial distribution with parameters n and p.
2. If n is large enough, probabilities for the binomial distribution with parameters n and $p = \lambda/n$ can be approximated by corresponding probabilities for the Poisson distribution with parameter λ. (Alternatively, the Poisson approximation to the binomial distribution is used when n is large and p is small, with $\lambda = np$ being of moderate size: $\lambda \leq 20$.)

Relation of binomial and negative binomial probabilities: If X has a negative binomial distribution with parameters r and p, then for $x = r + 1, r + 2, \ldots,$

$$P\{X > x\} = P\{U < r\},$$

where U has a binomial distribution with parameters $n = x$ and p. The special case $r = 1$ yields the useful formula

$$\begin{aligned} F(x) &= P\{X \leq x\} = P\{U \geq 1\} = 1 - P\{U = 0\} \\ &= 1 - (1 - p)^x, \qquad x = 1, 2, \ldots, \end{aligned}$$

for the cumulative distribution function of the geometric distribution (X = number of trials).

Exercises

1. In each of the following situations, a random variable Z is described. Determine for each situation whether or not the variable Z has a binomial distribution. If Z does have a binomial distribution, give the values of the parameters n and p. If Z does not have a binomial distribution, indicate which of the requirements for having a binomial distribution is not met.

(a) A university's athletic program is being reviewed by the National Collegiate Athletic Association (NCAA). As part of the investigation, the NCAA randomly selects 8 different athletes who were seniors and had received varsity letters during the previous year. The NCAA counts the number Z of these athletes who did not graduate. In fact, of 80 athletes who were seniors and received varsity letters last year from the university, 30 had failed to graduate.

(b) Members of a fraternity have purchased 40 of 500 tickets sold for a raffle. Five prizes will be awarded by randomly drawing ticket stubs from a jar, returning a winning stub to the jar after each draw. The random variable Z is the number of prizes won by members of the fraternity.

(c) In a class demonstration, a probability teacher independently tosses a penny, a nickle, a dime, and a quarter and counts the number Z of "heads" that appear. (*Note:* Pennies and nickles are made from a single metal alloy, whereas dimes and quarters are not—they are "clad" coins.)

(d) A certain brand of light bulb is advertised as having a *median* lifetime equal to 1000 hours. Ten light bulbs of this brand are allowed to burn independently until they burn out, and the number Z of such bulbs that last more than 1000 hours is recorded. (*Hint:* What is the probability that a random variable is greater than the median of its distribution?)

2. Follow the instructions given in Exercise 1 for the random variables Z described in each of the situations below.

(a) A large building has 15 elevators. Each elevator has probability 0.10 of developing a mechanical problem this month, and whether one elevator develops such a problem is independent of whether or not any other elevator develops a mechanical problem. The variable Z is the number of elevators that develop mechanical problems this month.

(b) Of 500 student lockers at a city high school, 20 contain weapons. The school principal randomly selects 10 different lockers to search; Z is the number of lockers searched that are found to contain weapons.

(c) A TV game show producer is looking for contestants with unusual ("off-beat") personalities for the show. Three such contestants are needed and the producer will interview potential contestants until the 3 needed contestants are found. From experience, the producer knows that 10% of all potential contestants have suitable off-beat personalities and that the personalities of potential contestants are independent of one another. The variable Z is the number of potential contestants that the producer interviews.

(d) Assume that a random individual is equally likely to be born on any given month of the year, and that the birth month of a husband is independent of the birth month of a wife. In a random sample of 20 married couples (which can be assumed to be taken with replacement), Z is the number of couples who were born in the same month of the year.

3. A quiz consists of 12 multiple-choice questions, each of which has five possible responses, only one of which is correct. A student who has not studied takes this quiz and independently quesses the answer to each question. (That is, for each question, each possible response has equal chance of being chosen.) Let Z denote the number of correct answers made on the quiz by this student.

(a) Does Z have a binomial distribution? If so, what are the values of the parameters n and p?

(b) Find the mean, median, and variance of Z.

(c) Suppose that 6 or more correct answers constitutes a passing grade on the quiz. What is the probability that the student passes?

4. Another student takes the quiz described in Exercise 3, but this student has studied and consequently has probability 0.7 of correctly answering a question. The student answers the questions on the quiz independently. Let Z be the number of correct answers made on the quiz by this student. Complete parts (a) to (c) of Exercise 3 for this student.

5. The Internal Revenue Service has a special letter-opening machine that opens and removes the contents of an envelope. If the envelope is fed improperly into the machine, the contents of the envelope may not be removed or may be damaged. In this case we say that the machine has "failed." Assume that the fate of each envelope entering the machine is independent of the fates of all other envelopes handled by the machine.

(a) If the machine has a probability of failure of 0.1, what is the probability of more than 3 failures occurring in a batch of 10 envelopes? What is the mean number of failures?

(b) If the machine has a probability of failure of 0.05, what is the probability of more than 3 failures occurring in a batch of 20 envelopes? What is the mean number of failures?

(c) If the probability of failure of the machine is 0.01 and a batch of 100 envelopes is to be opened, what is the mean number of failures expected? What is the probability that more than 3 failures will occur? (*Hint:* Use an approximation.)

6. In some military courts, 9 judges are appointed. Both the prosecution and the defense attorneys are entitled to a peremptory challenge of any judge, in which case that judge is removed from the case and is not replaced. Suppose that the prosecution attorney has chosen not to exercise the right of peremptory challenge, and that the defense is limited to two challenges. A defendant is declared guilty only if there are a majority of judges who vote "guilty"; otherwise, the defendant is declared innocent. Suppose that each judge has a probability of 0.6 of voting "guilty" and that judges make their decisions independently. What is the probability that the defendant is declared guilty when there are (a) 9 judges; (b) 8 judges; (c) 7 judges? (Note that a majority of a group of n people consists of more than $n/2$ of these people. Thus a majority of 9 is 5 or more, and a majority of 8 is also 5 or more.) If the defense attorney wants to maximize the probability that his client is declared innocent, which of the following is the attorney's best strategy: (i) challenge no judge, (ii) challenge one judge, or (iii) challenge two judges?

7. The following question is, in essence, the question that the Chevalier de Meré asked Pascal (see Chapter 1): Which of the following is larger?
   (i) The probability that a "6" comes up *at least once* in 4 tosses of a balanced six-sided die.
   (ii) The probability that *at least once* in 24 tosses of two balanced dice, both dice simultaneously show a "6."

   (The Chevalier de Meré thought that he had given a theoretical argument that proved that these probabilities are equal, yet his observations suggested otherwise. He asked Pascal to explain the apparent contradiction.)

8. A hospital has noticed a new skin disease among its patients. Approximately 20% of the patients in the hospital catch this disease. The hospital believes that this disease is not contagious and is caught by exposure to a virus. If this is the case, patients will or will not catch the disease independently of one another. Patients are housed in the hospital in nonprivate rooms, which have four beds. All beds in the hospital are currently occupied.
   (a) If the hospital is correct that the disease is not contagious, what is the probability that exactly one patient in a room will catch the disease?
   (b) There are 75 rooms in the hospital. Let X be the number of rooms in which exactly one patient catches the disease. Find the mean and standard deviation of X. (What is the distribution of X?)
   (c) The frequency distribution for the number Z of patients in a room in the hospital who catch the disease is shown in Exhibit E.1. Fit a binomial distribution with $n = 4$ and $p = 0.2$. Does this distribution fit the data? If not, what alternative theory for how the disease is caught is suggested by the data?

EXHIBIT E.1

Number z of Patients in a Room Who Catch the Disease	*Frequency*
0	31
1	2
2	5
3	17
4	20
	75

9. A large number (53,680) of German families with 8 children were contacted, and for each family, the number Z of male children was noted. The results are given in Exhibit E.2. Fit a binomial distribution with parameters $n = 8$, $p = 0.51$ to these data. [*Remark:* These data are taken from a study by Geissler and are given in Fisher (1950). Geissler's data are also discussed in Simpson, A. Roe, and R. C. Lewontin (1960). The value $p = 0.51$ used in fitting these data is the established probability of a male birth for Caucasian parents.]

EXHIBIT E.2 *Results of Observing the Number Z of Male Children in Each of 53,680 German Families*

Number z of Boys in German Families of 8 Children	*Observed Frequency*
0	215
1	1,485
2	5,331
3	10,649
4	14,959
5	11,929
6	6,678
7	2,092
8	342
	53,680

10. An accountant randomly selects 6 invoices without replacement from among 25 large invoices paid by a city highway department and checks whether the companies sending the invoices received payment. In fact, 10 of the 25 invoices have not been paid. Let Z be the number of invoices not paid out of the 6 invoices chosen by the accountant.

(a) Find the probability that $Z = 3$.
(b) Find the probability that $Z \leq 3$.
(c) Determine the values of the mean, median, and variance of Z.

11. In which of the situations described in Exercises 1 and 2 does the random variable Z have a hypergeometric distribution? For each such situation, give the values of the parameters N, p, and n and find the mean and variance of Z.

12. Sixty patients with symptoms of a certain disease visit a clinic. Twelve of these patients actually have the disease. A definitive diagnosis of this disease depends on a blood test. Individual blood tests are expensive, so the hematologist at the clinic uses the following screening device. The blood of n of the patients is combined and tested. If none of the patients has the disease, the composite (group) test is negative and the patients are sent home. If, however, at least one person has the disease, the composite test will be positive and the patients will then be tested individually to determine who has the disease. Assume that the first n persons whose blood is tested are chosen from among the 60 patients by a simple random sample *without replacement*. What is the probability that the composite blood test will be negative if (a) $n = 2$; (b) $n = 4$; (c) $n = 6$; (d) $n = 10$?

13. A radio manufacturer intends to purchase 100 cathodes from a supply house. He expects that some of these cathodes will be defective and is willing to tolerate 4 defective cathodes in the batch of 100. He checks the quality of the batch by drawing a random sample of 3 cathodes without replacement from among the 100 in the batch, and testing these cathodes carefully. Let Z be the number of defective cathodes that he finds.

(a) If there are 4 defective cathodes in the batch of 100 cathodes, determine the exact values of the probability mass function, $p(z)$, of Z.
(b) For the distribution found in part (a), what is the mode of Z? What is the median of Z? What is the 90th percentile of Z? What are the values of the mean and variance of Z?
(c) The manufacturer decides to reject the batch of cathodes if he finds more than 1 defective cathode. What is the exact probability of rejecting a batch of 100 cathodes containing 4 defective cathodes?

14. Consider a group of 10 people consisting of 7 males and 3 females. A committee of n people will be chosen from among these 10 people by means of a simple random sample without replacement. It is desired that the committee size n should be as small as possible but sufficiently large that the probability that the committee will have at least 1 male and at least 1 female is no less than 0.9. How large should the committee size n be?

15. Let X have a hypergeometric distribution with parameters $N = 10$, $n = 8$, and p to be specified below. Let Y have a hypergeometric distribution with parameters $N = 100$, $n = 8$, and the same p. Finally, let Z have a binomial distribution with parameters $n = 8$ and, again, the same p. Construct a table comparing $P\{X = k\}$, $P\{Y = k\}$, and $P\{Z = k\}$ for $k = 0, 1, 2, \ldots, 8$, in each of the following cases: (a) $p = 0.3$; (b) $p = 0.5$.

16. Fifteen people in a community of N people are opposed to a housing development. Eight people attend a public hearing. Assume that these people make up a simple random sample without replacement from among everyone in the community. For which of the values of N below will the binomial approximation to the hypergeometric distribution yield a good approximation to the probability that a majority (5 or more) of those attending the meeting will oppose the development? (a) $N = 40$; (b) $N = 100$; (c) $N = 150$. Calculate the desired probability for cases (a) and (c).

17. A certain jury venire consists of 300 persons of whom 135 are opposed to the death penalty and 165 favor the death penalty. An initial list of 12 is determined by random selection without replacement from the venire, and questioned by the prosecution and defense attorneys involved in a first-degree murder case.

(a) What is the exact probability that 6 of the people in the list of 12 will oppose the death penalty? What is the approximation to this probability obtained by using the binomial distribution?

(b) Use the binomial approximation to the hypergeometric distribution to find the probability that a majority (7 or more) of the people in the list of 12 will oppose the death penalty.

18. The number X of cars abandoned in a week on a particular highway has a Poisson distribution with parameter $\lambda = 2.0$.

(a) Find the probability, $P\{X \geq 4\}$, that 4 or more cars will be abandoned this week on the highway.

(b) It costs the state \$100 per car to tow away and dispose of an abandoned car. What is the mean cost per week to the state to dispose of abandoned cars?

(c) What is the most probable amount of money that the state must pay in a given week to dispose of abandoned cars?

(d) At present the state pays private tow truck operators \$25 a car to tow away abandoned automobiles. Purchasing and running its own tow truck would cost the state \$50 a week. Should the state purchase its own tow truck? Explain briefly.

19. Under normal circumstances, the number X of new cases of food poisoning reported in a random month in a dormitory cafeteria has a Poisson distribution with mean 0.5.

(a) What is the value of the variance of X? What is the value of the median of X?

(b) The cafeteria is required by the state to request a state inspection of its kitchens whenever 2 or more cases of food poisoning occur in a month. What is the probability that the cafeteria will have to request an inspection this month?

(c) The number of new cases of food poisoning in any one month is independent of the numbers of cases of food poisoning in all other months of the year. Let Z be the number of months (out of the 12 months of the year) that the cafeteria must be inspected by the state. Find the probability, $P\{Z > 1\}$, that the cafeteria is inspected more than once in a year. (*Hint:* What is the distribution of Z? What are the values of the parameters of this distribution?)

20. A manufacturer of cloth inspects every bolt of cloth made. In a day's production, the number X of bolts of cloth that have only minor defects has a Poisson distribution with parameter $\lambda = 16$, whereas the number Y of bolts of cloth that have major defects has a Poisson distribution with parameter $\lambda = 1.2$. The random variables X and Y are independent. Bolts with minor defects can be sold as "irregular" at a loss of \$10. Bolts with major defects must be discarded, resulting in a loss of \$250 per bolt. Thus the total loss per day due to defects in production is

$$T = 10X + 250Y.$$

(a) Find the mean and variance of T.

(b) Find the probability, $P\{T = \$300\}$, of a loss of \$300 in a day. [*Hint:* For what pairs (x, y) of values of X and Y is T equal to 300?]

21. The number X of oil tankers arriving at a certain refinery on a given day has a Poisson distribution with mean 2. The refinery can only service up to 3 tankers a day; if more than 3 tankers arrive in a day, the number in excess of 3 must be sent to another refinery. Let Y be the number of tankers that the refinery will service in a given day.

(a) On a given day, what is the probability that the refinery will have to send tankers away to another refinery? That is, find $P\{X > 3\}$.

(b) What is the value of the mean of Y? This is the mean number of tankers serviced daily by the refinery. (Note that the possible values of Y are 0, 1, 2, and 3. What are the probabilities of these values? Should these probabilities sum to 1?)

(c) The number of tankers sent to other refineries in a day is $X - Y$. The refinery makes a profit of \$10,000 for every tanker they service, but must pay a penalty of \$500 to every tanker that they send to another refinery. Find the mean daily profit made by the refinery.

22. Suppose that every acre of ocean bottom has a probability 0.01 of containing a wrecked ship. Let X denote the number of ships that will be found in 80 acres of ocean bottom randomly chosen by a marine salvage company. Assume that the presence or absence of a wrecked ship on any acre of ocean bottom is independent of the presence or absence of wrecked ships on all other acres of ocean bottom, and that at most one wrecked ship can be found in each acre.

(a) What are the name and parameter values of the exact probability distribution of X?

(b) Can we use the Poisson distribution to approximate probabilities for the values of X? If so, what value of the parameter λ should we use?

(c) Find the probability, $P\{X > 1\}$, that more than one wrecked ship will be found by the marine salvage company.

23. An experiment is run to test whether a ''psychic'' can really read minds. The psychic is placed in one room and the subject whose mind will be read in another room. The subject chooses one of 20 different geometric shapes and is told to picture this shape mentally. The psychic is then supposed to read the subject's mind and correctly identify the shape chosen.

(a) The experimenter believes that the ''psychic'' will be unable to read the subject's mind and thus must guess (pick one of the 20 shapes at random). What is the probability that the psychic guesses correctly?

(b) The experiment is independently repeated 90 times, and the number X of correct answers given by the psychic is recorded. If the psychic is guessing, what is the name of the exact distribution of X? What are the values of the mean and variance of X?

(c) The experimenter decides that if the psychic gets 9 or more correct answers in the 90 repetitions of the experiment, this will be evidence that the psychic can read minds. If, in fact, the psychic is only guessing, what is the probability, $P\{X \geq 9\}$, that the experimenter will come to the false conclusion that the psychic can read minds? (Can you use the Poisson distribution to approximate this probability? What value of λ do you use?)

24. The hypergeometric distribution with parameters N, p, and n can be approximated by the binomial distribution with parameters n and p if n/N is small enough. Further, the binomial distribution with parameters n and p can be approximated by the Poisson distribution with parameter $\lambda = np$ if n is large and p is small. Thus it follows that *the hypergeometric distribution can sometimes be approximated by the Poisson distribution*. Use this fact to answer the following problems.

(a) A manufacturer receives a lot of 10,000 silicon chips, of which 200 are defective. To decide whether or not to accept such lots, the manufacturer takes a random sample of 100 chips (without replacement) and counts the number of defective chips in the sample. If the number X of defective chips in the sample is less than 4, the manufacturer will accept the lot. Find the probability that the manufacturer accepts the lot just received.

(b) A new travel agency gives customers the opportunity to win a free plane trip. Each customer picks a card from a drum containing 1000 cards, each of which has a special rub-off section. If the customer rubs off the cover of this section and finds a picture of a plane, the customer wins the free trip. Only 10 cards in the drum have a picture of the plane on them. If 80 customers pick cards from the drum, what is the probability that no more than 5 of these customers win a free trip?

25. In studying the frequency of wars, Richardson (1944) tallied the number X of wars that began in each of the calendar years from A.D. 1500–1931. The 432 observed values of X are summarized in Exhibit E.3. Fit a Poisson distribution with mean 0.70 to these data.

EXHIBIT E.3. *Results of Observing the Number X of Wars in Each of the Calendar Years 1500–1931*

Number x of Outbreaks of War in a Particular Year	*Observed Frequency*
0	223
1	142
2	48
3	15
4	4
5	0
	432

26. A certain genetic disease appears in 5% of all persons. Patients appear at a clinic as if they were randomly sampled *with* replacement from this population. Let X be the number of patients tested until an individual with the disease is discovered.

(a) What is the mean number μ_X of the patients that must be seen?

(b) What is the probability that no more than 10 patients must be seen?

(c) What is the median number of patients that must be seen?

27. If X has a geometric distribution with parameter p, find the median, Med(X), of X when $p = 0.4$ and when $p = 0.8$. What is the mode of X in both these cases?

28. A sociologist studying the publication behavior of researchers asks each of 363 people to indicate the number of journals to which their most recent published paper was submitted before it was finally published. The data are summarized in Exhibit E.4. Fit a geometric distribution with $p = 0.73$ to these data.

EXHIBIT E.4. *Observations of the Number X of Journals to Which a Sociologist's Most Recent Paper Was Submitted Before Publication*

Number x of Journals	*Observed Frequency*
1	261
2	75
3	24
4	3
≥ 5	0
	363

29. A married couple decides to continue having children until they have 2 daughters. Suppose that each child born to this couple has probability 0.50 of being a girl and that the sexes of successive children are probabilistically independent. Let X be the total number of children born to this couple.
(a) Find $P\{X = 6\}$, $P\{X > 3\}$, $P\{X \leq 6\}$.
(b) What is the mean number of children that the couple will have?
(c) What is the most likely number of children that they will have?
(e) What is the mean number of boys that they will have?

30. Suppose, in Example 5.1, that there are $N = 1000$ fish in the pond and $M = 200$ of these are of the desired species. What is the probability that 10 or more fish that are not of the desired species are caught before 3 fish of the desired species are observed?

31. An organization has two open positions but has such stringent requirements for filling these positions that only 10% of all possible applicants are likely to be qualified. Applicants are interviewed one at a time. The organization will stop interviewing as soon as it finds 2 suitable candidates. Let X denote the number of applicants that the organization must interview until it finds 2 suitable candidates.
(a) Find $P\{X = 6\}$, $P\{X \leq 6\}$, $P\{X > 4\}$.
(b) What is the mean number μ_X of applications that must be processed? What is the mean number of unqualified applications that must be processed?
(c) What is the median number, Med(X), of applications that must be processed?

(d) It costs the organization 0.5 labor-hour to process each application. What is the probability that the cost will exceed 10 labor-hours?

32. When an electric light bulb is purchased, the bulb can be tested to determine whether it is operating properly. If it is defective, the bulb is put aside to be returned to the manufacturer, and another bulb is tested. Bulbs are tested until a good one is found. Suppose that we start with a box of 5 bulbs of which 2 are defective. The experiment consists of testing bulbs *without replacement* until 2 good ones are found. Let X denote the number of bulbs tested. Find the probability mass function of X.

33. The data summarized in Exhibit E.5 list the number Y of traffic accidents incurred during the period 1952–1955 by each of 708 public corporation bus drivers. Fit a negative binomial distribution [failures—see (5.2)] with $r = 4$, $p = 0.66$, to these data.

EXHIBIT E.5. *Observations of the Number X of Traffic Accidents Incurred During 1952–1955 by Each of 708 Bus Drivers*

Number x of Traffic Accidents During the Period	*Observed Frequency of x Accidents*
0	117
1	157
2	158
3	115
4	78
5	44
6	21
7	7
8	6
9	1
10	3
11	1
≥12	0
	708

Source: Data from P. Froggatt. Application of discrete distribution theory to the study of non-communicable events in medical epidemiology, in G. P. Patil, ed., *Random Counts in Biomedical and Social Sciences,* Vol. 2 (University Park, Pa.: Pennsylvania State University Press, 1970).

34. Baseball teams A and B meet in the World Series. Assume that every game is statistically independent of the other games between these teams and that the teams are evenly matched. In the World Series, the first team to win 4 games wins the series.

(a) What is the probability that A wins the series in k games? What is the conditional probability that A wins the series given that the series lasts k games? Answer these questions for $k = 4, 5, 6$, and 7.

(b) What is the probability that A wins the World Series?

(c) If A is a better team than B, and has a probability of 0.60 of winning any game, what is the probability that team A wins the World Series?

Theoretical Exercises

T.1. Let

$$p(z; n, p) = \binom{n}{z} p^z (1 - p)^{n-z}, \qquad z = 0, 1, 2, \ldots, n.$$

(a) Show that $zp(z; n, p) = np[p(z - 1; \; n - 1, p)]$, $z = 1, 2, \ldots, n$.

(b) Show that $z(z - 1)p(z; n, p) = (z - 1)\, np\, [p(z - 1; n - 1, p)]$
$= n(n - 1)p^2\, [p(z - 2; n - 2, p]$,
$z = 2, \ldots, n$.

(c) Use the recursion formulas (a) and (b) to show that if Z has the binomial distribution with parameters n and p, then

$$\mu_Z = np, \qquad E[Z(Z - 1)] = n(n - 1)\, p^2, \qquad \sigma_Z^2 = np(1 - p).$$

Hint: $m! = m[(m - 1)!]$ Also, $E[Z^2] = E[Z(Z - 1)] + \mu_Z$.

T.2 The **binomial expansion**

$$(a + b)^n = \sum_{z=0}^{n} \binom{n}{z} a^z b^{n-z}$$

gives the binomial distribution its name. Use this expansion to show that:

(a) The sum of the binomial probabilities $p(z; n, p)$ over $z = 0, 1, \ldots, n$ equals 1.

(b) The moment generating function of Z, where Z has a binomial distribution with parameters n and p, is

$$M(t) = E[e^{tZ}] = (pe^t + (1 - p))^n.$$

T.3. Use moment generating functions to show that if Z_1 and Z_2 are probabilistically independent, binomially distributed random variables with common probability p of success and sample sizes $n = n_1$, $n = n_2$, respectively, then $Z_1 + Z_2$ has a binomial distribution with parameters $n = n_1 + n_2$ and p.

T.4. (a) Show that if $p(x)$ is the probability mass function of the Poisson distribution with parameter λ, then

$$x(x - 1)p(x) = \lambda^2 p(x - 2).$$

(b) Show that if X has the Poisson distribution with parameter λ, then $E[X(X - 1)] = \lambda^2$, and $\sigma_X^2 = \lambda$.

T.5. Show that the moment generating function $M(t) = E(e^{tX})$ of X, where X has the Poisson distribution with parameter λ, is

$$M(t) = e^{-\lambda + \lambda e^t}.$$

[*Hint*: $e^a = \sum_{i=0}^{\infty} (a^i/i!)$, for any $a > 0$.] Use $M(t)$ to find μ_X and σ_X^2.

T.6. Use moment generating functions to show that if X_1, X_2 are independent Poisson variables with parameters λ_1, λ_2, respectively, then $X_1 + X_2$ has a Poisson distribution with parameter $\lambda = \lambda_1 + \lambda_2$.

T.7. If X has a geometric distribution with parameter p, find

$$P(X > k + 1 \mid X > k)$$

for $k = 1, 2, \ldots$.

T.8. Show that if X has a geometric distribution with parameter p, the moment generating function $M(t) = E(e^{tX})$ of X is

$$M(t) = \frac{pe^2}{1 - (1 - p)e^t}, \qquad \text{if } t < -\log(1 - p).$$

[*Hint:* If $0 \le r < 1$, $\sum_{k=1}^{\infty} r^{k-1} = (1 - r)^{-1}$.] Use $M(t)$ to find μ_X and σ_X^2.

T.9. Let s be any positive number. For any $x \neq 1$, and for $k = 1, 2, \ldots$,

$$\begin{aligned}\left(\frac{d}{dx}\right)^k \frac{1}{(1 - x)^s} &= [s(s + 1) \cdots (s + k - 1)] \frac{1}{(1 - x)^{s+k}} \\ &= \frac{\Gamma(s + k)}{\Gamma(s)} \frac{1}{(1 - x)^{s+k}}.\end{aligned}$$

Note that for $-1 \le x < 1$, and for each $k = 1, 2, \ldots$, these derivatives are continuous in x. Thus, by Taylor's theorem [see, e.g., Protter and Morrey (1985) and Taylor (1955)], for each $n \ge 1$ there exists a number ω between x and 0 such that

$$\frac{1}{(1-x)^s} = 1 + \sum_{k=1}^{n} \frac{\Gamma(s+k)}{\Gamma(s)} \frac{x^k}{k!} + \frac{\Gamma(s+n+1)}{\Gamma(s)\,(n+1)!} \frac{x^{n+1}}{(1-\omega)^{s+n+1}}.$$

(a) Show that if $0 < x < 1$, then

$$\lim_{n \to \infty} \frac{\Gamma(s+n+1)}{\Gamma(s)\,(n+1)!} \frac{x^{n+1}}{(1-\omega)^{s+n+1}} = 0.$$

(b) Hence [see, e.g., Protter and Morrey (1985)] conclude that

$$\frac{1}{(1-x)^s} = \sum_{k=0}^{\infty} \frac{\Gamma(s+k)}{\Gamma(s)k!} x^k.$$

This is the *generalized negative binomial expansion* of $(1-x)^{-s}$. In the special case where s is a positive integer and $x = 1 - r$, $0 \le r < 1$, we obtain

$$\frac{1}{r^s} = \sum_{k=0}^{\infty} \frac{(s+k-1)!}{(s-1)!\,k!} (1-r)^k.$$

Special cases ($s = 2, 3$) of this result were used to find the mean and variance of the geometric distribution.

(c) Prove that for $r > 0$, $0 < p < 1$,

$$\sum_{y=0}^{\infty} \frac{\Gamma(y+r)}{\Gamma(r)\Gamma(y+1)} p^r(1-p)^y = 1,$$

which shows that the probabilities of the generalized negative binomial distribution, with parameters r, p, sum to 1.

(d) Show that if Y has a generalized binomial distribution with parameters r, p, then

$$\mu_Y = \frac{r(1-p)}{p}, \qquad \sigma_Y^2 = \frac{r(1-p)}{p^2}.$$

T.10. The *discrete uniform distribution* on the numbers $1, 2, \ldots, K$, is defined by the probability mass function

$$p(x) = \begin{cases} \dfrac{1}{K}, & x = 1, 2, 3, \ldots, K, \\ 0, & \text{otherwise.} \end{cases}$$

The constant K is the parameter of this distribution. Suppose that X has the discrete uniform distribution with parameter K.

(a) Using the well-known summation identities

$$\sum_{i=1}^{K} i = \frac{K(K = 1)}{2}, \qquad \sum_{i=1}^{K} i^2 = \frac{K(K+1)(2K+1)}{6},$$

find the mean and variance of X.

(b) Find the cumulative distribution function, $F(x)$, of x.

(c) Find the median of X. (*Hint:* There are two cases: K odd and K even.)

9

The Normal Distribution

1 Introduction

Of all distributions of continuous random variables, the most widely studied and frequently used is the **normal distribution**. The form of the normal distribution was discovered early in the history of probability as a convenient approximation to binomial probabilities (Section 5); credit for this discovery is usually given to the French mathematician Abraham de Moivre (1667–1754). Both Pierre Simon Laplace (1749–1827) and Carl Friedrich Gauss (1777–1855) proposed this distribution as a "law of errors" to describe the variability of measurement errors in the physical sciences. Empirical support for the law of errors in the context of astronomical observations was given by Bessel (1818). However, acceptance of the normal distribution by astronomers and physicists was based less on its empirical fit to data than on the theoretical support given to it by Laplace's **Central Limit Theorem** (Section 5), and Gauss's (1809) use of this distribution to solve the important statistical problem of combining observations. Because Gauss played such a prominent role in demonstrating the usefulness of the normal distribution, the normal distribution is often also called the **Gaussian distribution.**

Among the first scientists to apply the normal distribution to physiological and behavioral phenomena were Adolfe Quetelet (1796–1874) and Sir Frances Galton (1822–1911). Quetelet argued that the normal distribution could be used to model highly variable measurements on human populations. That is, populations of such measurements could be thought of as having a common shape (that of the bell-shaped normal density) centered about a center of gravity (mean). Quetelet called this center of gravity the "average" or "normal" man (*l'homme moyen*). If populations of measurements had similar shapes, they could be compared by comparing these ideal "average men." In this way, one might be able to ascertain how various causal factors affected human attributes, rather than trying to sort out their effects on an individual-by-individual basis. This important conceptualization of a stable form for human population measurements, based on the normal distribution, gave rise to modern experimental sociology. Although the name "normal distribution" may have arisen from Quetelet's *l'homme moyen*, it is more likely that it came from Wilhelm Lexis's (1837–1914) work on vital statistics. Influenced by Quetelet's work, Lexis fitted normal curves to empirical distributions of age at death; those deaths that fit the curve were called "normal" by Lexis.

Galton was also interested in describing populations. He was excited by Quetelet's description of populations in terms of deviations from an average, as described by the normal curve. Like Quetelet, he used the normal curve to *define* a population; if data did not fit the normal curve, the data were not all from the same population. This suggested that biological populations could be *classified* into more homogeneous groups (taxa), each described by the characteristic normal curve.

The work of Quetelet and Galton was followed in the early twentieth century by numerous studies of biological, psychological, and anthropometric data in which fitting normal distributions to sets of data was an important part of the scientific argument. For example, Pearl (1905) studied variations in brain weight in Swedish males, Latter (1902) fit the normal distribution to sizes of birds' eggs. Macdonnell (1902) compared measurements on criminals and noncriminals to see if criminality could be predicted from any of these measurements, and Greenwood (1904) similarly tried classifying diseased and nondiseased hearts using weights of human viscera. Ebbinghaus (1885) applied Quetelet's approach to errors in memorizing nonsense syllables, while the field of psychometrics was based on the assumption that the variability of human abilities and attributes could be described by the normal curve. (That point of view explains in part the notion of "grading on the curve.")

Statistics, in its modern form, grew out of the early work of Laplace, Gauss, Quetelet, and Galton on population modeling and the combining and comparison of data. Because the normal distribution was central to the approach of these early investigators, it provided the motivation and justification for many of the methods of data analysis developed by statisticians. It may seem that the normal distribution is the only important continuous distribution. It is thus worth mentioning that these early pioneers were well aware that the variations of physical, biological, and behavioral variables were not always fit well by the normal distribution, and that other continuous distributions (Chapter 10) could be useful.

2 The Normal Density and Its Properties

The probability density function $f(x)$ of a normal distribution is defined by the equation

$$(2.1) \qquad f(x) = \frac{1}{\sqrt{2\pi\sigma^2}} \exp\left[-\frac{(x-\mu)^2}{2\sigma^2}\right], \qquad -\infty < x < \infty.$$

The constants μ $(-\infty < \mu < \infty)$, and σ^2 $(\sigma^2 > 0)$ are the parameters of the normal distribution. The graph of $f(x)$, which is a bell-shaped curve, is shown in Exhibit 2.1. Note that this graph is symmetric about the point $x = \mu$ and extends over the entire horizontal axis.

As suggested by our notation, the parameter μ gives the value of the mean μ_X of the normal density (2.1), whereas the parameter σ^2 gives the value of the variance σ_X^2. As a shorthand way of saying that the random variable X has a normal distribution with mean μ and variance σ^2, we write

$$X \sim \mathcal{N}(\mu, \sigma^2).$$

Thus $X \sim \mathcal{N}(80, 16)$ means that the random variable X has a normal distribution with mean $\mu = 80$ and variance $\sigma^2 = 16$. The symbol "$\sim$," which means "is distributed as," is a widely utilized notation in probability and statistics.

Because $f(x)$ is symmetric about $x = \mu$ and achieves its maximum value at $x = \mu$ (See Exhibit 2.1), the median and the mode of the normal distribu-

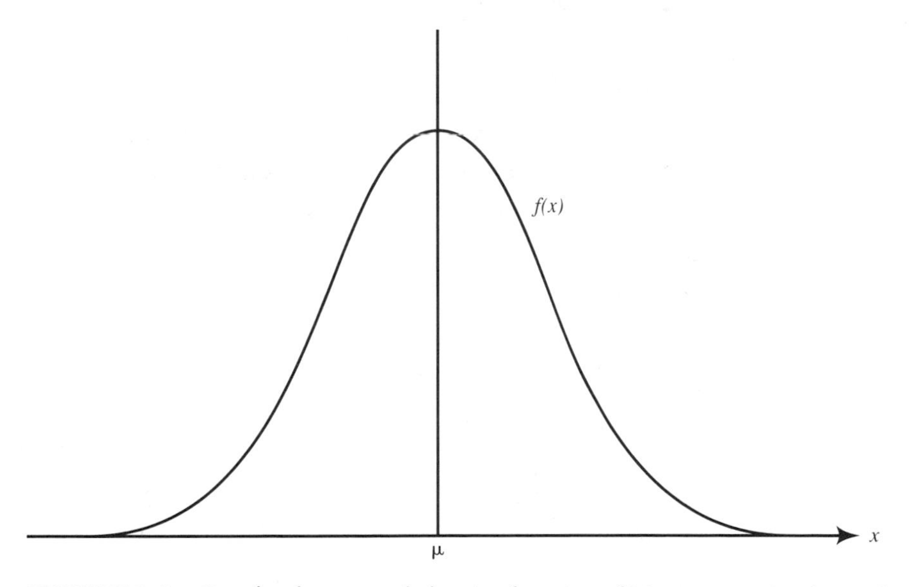

EXHIBIT 2.1 *Graph of a normal density function, f(x), symmetric about the point μ.*

tion both have the value μ. Also because of the symmetry of $f(x)$ about $x = \mu$, all odd-order central moments of the normal distribution equal 0. That is,

$$E[X - \mu]^r = 0 \qquad \text{for} \quad r = 1, 3, 5, 7, \ldots .$$

In particular, $E[X - \mu] = E[X] - \mu = 0$, verifying our assertion that the mean of X equals the constant μ in (2.1). Also, the measure of skewness, $\gamma_1 = E[X - \mu]^3/\sigma^3$, introduced in Chapter 6 equals 0. It is shown below that the even central moments of the distribution (2.1) are given by

$$E[X - \mu]^r = \frac{\sigma^r 2^{r/2}\Gamma(\frac{1}{2}(r + 1))}{\sqrt{\pi}}, \qquad r = 2, 4, \ldots$$

where $\Gamma(a)$ is the gamma function (Appendix A). Because

$$\Gamma\left(\frac{3}{2}\right) = \frac{1}{2}\Gamma\left(\frac{1}{2}\right) = \frac{1}{2}\sqrt{\pi}, \qquad \Gamma\left(\frac{5}{2}\right) = \frac{3}{2}\Gamma\left(\frac{3}{2}\right) = \frac{3}{4}\sqrt{\pi},$$

it follows that the variance $E[X - \mu]^2$ of X is σ^2, as asserted, and that the index of kurtosis (Chapter 6) of the normal distribution is

$$\gamma_2 = \frac{E[X - \mu]^4}{\sigma^4} = 3.$$

This value serves as a standard against which the kurtosis values of other distributions are assessed.

Formal Verifications★

It has been asserted that the function $f(x)$ in (2.1) is a density function. Although it is easily seen that $f(x)$ is always nonnegative, it is not obvious that the area under the curve of $f(x)$ equals 1. Thus we need to show that

$$\int_{-\infty}^{\infty} f(x)\, dx = 1. \tag{2.2}$$

Make the change of variables from x to $z = (x - \mu)/\sigma$ in (2.2). Then

$$\int_{-\infty}^{\infty} \frac{1}{\sqrt{2\pi\sigma^2}} \exp\left[-\frac{1}{2}\frac{(x - \mu)^2}{\sigma^2}\right] dx = \int_{-\infty}^{\infty} \frac{1}{\sqrt{2\pi}} \exp\left(-\frac{1}{2}z^2\right) dz \tag{2.3}$$
$$= \frac{2}{\sqrt{2\pi}} \int_0^{\infty} \exp\left(-\frac{1}{2}z^2\right) dz,$$

where the last equality holds because $\exp(-\frac{1}{2}z^2)$ is an even function of z. Equation (A.7) of Appendix A states that for all $s > 0$,

(2.4) $$\int_0^\infty z^{2s-1} \exp(-\tfrac{1}{2}z^2)\, dz = 2^{s-1}\,\Gamma(s).$$

Hence, applying (2.4) with $s = \frac{1}{2}$ to (2.3), yields

$$\int_{-\infty}^\infty f(x)\, dx = \frac{2}{\sqrt{2\pi}}\left[\frac{\Gamma(\frac{1}{2})}{\sqrt{2}}\right] = \frac{\Gamma(\frac{1}{2})}{\sqrt{\pi}} = 1,$$

because $\Gamma(\frac{1}{2}) = \sqrt{\pi}$ (see Appendix A). Thus (2.2) has been demonstrated.

With the same change of variables used in (2.3),

(2.5) $$E[X - \mu]^r = \int_{-\infty}^\infty (x - \mu)^r f(x)\, dx = \sigma^r \int_{-\infty}^\infty \frac{z^r \exp(-\frac{1}{2}z^2)}{\sqrt{2\pi}}\, dz.$$

When r is an *odd* integer, $z^r \exp(-\frac{1}{2}z^2)$ is an odd function of z, so that

$$E[X - \mu]^r = 0 \qquad \text{for} \quad r = 1, 3, 5, \ldots.$$

When r is an even integer, $z^r \exp\{-\frac{1}{2}z^2\}$ is an even function of z. It follows from (2.4) and (2.5) that

$$E[X - \mu]^r = \frac{\sigma^r 2}{\sqrt{2\pi}} \int_0^\infty z^r \exp\left(-\frac{1}{2}z^2\right) dz = \frac{\sigma^r 2}{\sqrt{2\pi}}\left[\Gamma\left(\frac{1}{2}(r + 1)\right)2^{(r-1)/2}\right]$$
$$= \frac{\sigma^r 2^{r/2}\Gamma(\frac{1}{2}(r + 1))}{\sqrt{\pi}}, \qquad r = 2, 4, \ldots.$$

3 Probability Calculations: The Standard Normal Distribution

If X is any random variable, the transformation

(3.1) $$Z = \frac{X - \mu_X}{\sigma_X},$$

called the **standardization** of X, converts X to a new variable that has mean 0, variance 1, and which takes values on a scale independent of the units of measurement in which X was originally measured. The **standardized random variable** Z expresses the distance of X from its mean μ_X in standard deviation units. Thus $Z = 2.3$ means that the value of X is 2.3 standard deviation units larger than the mean μ_X of its distribution.

EXAMPLE 3.1

Two examinations were given in a certain class. Scores on the first examination had mean $\mu_X = 80$ and a standard deviation $\sigma_X = 4$. The second examination was harder, with $\mu_X = 65$ and $\sigma_X = 5$. A student had scores of 82 on the first examination and 75 on the second examination. Although the student's score on the first examination was larger, the standardized scores of $Z = (82 - 80)/4 = 0.5$ and $Z = (75 - 65)/5 = 2.0$ show that the student's performance relative to the class was superior on the second exam. ◆

The standardization of X to Z changes values on the horizontal axis in Exhibit 2.1, but otherwise does not change the shape of the density. Consequently, Z has a normal distribution. Because Z has mean 0 and variance 1, the distribution of Z is the $\mathcal{N}(0, 1)$ distribution, which is (for obvious reasons) called the **standard normal distribution.** This fact is worth remembering:

> If X has a normal distribution with mean μ and variance σ^2, then the standardization $Z = (X - \mu)/\sigma$ of X has the standard normal distribution. That is, $Z \sim \mathcal{N}(0, 1)$.

Formal verification of this fact is obtained by noting that for any number z,

$$
\begin{aligned}
F_Z(z) &= P\{Z \leq z\} = P\left\{\frac{X - \mu}{\sigma} \leq z\right\} = P\{X \leq \sigma z + \mu\} \\
&= F_X(\sigma z + \mu).
\end{aligned}
\tag{3.2}
$$

Now differentiate both sides of the equality (3.2) with respect to z, using the chain rule on the right-hand side and the fact that $(d/dx)F_X(x) = f(x)$. Thus, from (2.1), the density function of Z is

$$
\begin{aligned}
f_Z(z) = \frac{d}{dz} F_Z(z) &= \frac{d(\sigma z + \mu)}{dz} \left[\frac{d}{dx} F_X(x) \Big|_{x=\sigma z+\mu} \right] \\
&= \sigma f(\sigma z + \mu) \\
&= \sigma \left\{ \frac{1}{\sqrt{2\pi\sigma^2}} \exp\left[-\frac{1}{2} \frac{(\sigma z + \mu - \mu)^2}{\sigma^2} \right] \right\} \\
&= \frac{1}{\sqrt{2\pi}} \exp\left(-\frac{1}{2} z^2 \right).
\end{aligned}
$$

This density function has the form (2.1) of the normal density function with $\mu = 0$, $\sigma^2 = 1$, verifying that $Z \sim \mathcal{N}(0, 1)$.

For normal distribution probability calculations, the great advantage of standardizing X is that the cumulative distribution functions $F_X(x)$ of all normally distributed random variables can be obtained from a single table of the cumulative distribution function $F_Z(z)$ of the standard normal distribution. Indeed, when $X \sim \mathcal{N}(\mu, \sigma^2)$, then

$$\begin{aligned} F_X(x) &= P\{X \leq x\} = P\left\{\frac{X-\mu}{\sigma} \leq \frac{x-\mu}{\sigma}\right\} \\ &= P\left\{Z \leq \frac{x-\mu}{\sigma}\right\} = F_Z\left(\frac{x-\mu}{\sigma}\right). \end{aligned}$$

To summarize:

> If $X \sim \mathcal{N}(\mu, \sigma^2)$, and $Z \sim \mathcal{N}(0, 1)$, then
>
> (3.3) $$F_X(x) = F_Z\left(\frac{x-\mu}{\sigma}\right).$$

Exhibit B.4 of Appendix B gives values of the cumulative distribution function $F_Z(z)$ of a standard normal distribution for $z = 0.00(0.01)4.09$. Because of symmetry (see Exhibit 3.1),

> (3.4) $$F_Z(-z) = P\{Z \leq -z\} = P\{Z \geq z\} = 1 - F_Z(z).$$

For example,

$$F_Z(-1.96) = 1 - F_Z(1.96) = 1 - 0.97500 = 0.02500.$$

Thus only values of $F_Z(z)$ for $z \geq 0$ need to be given in Exhibit B.4.

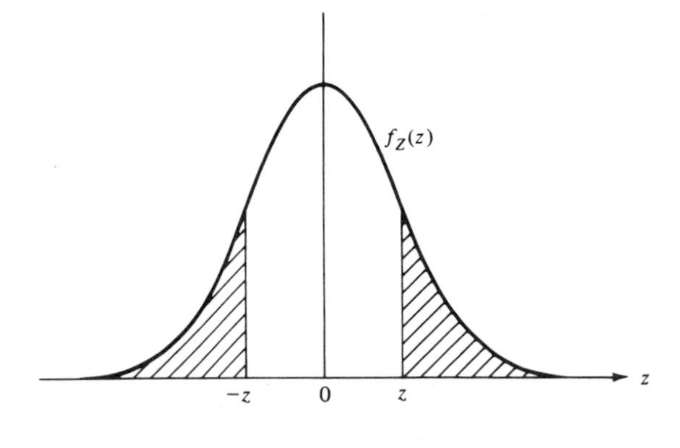

EXHIBIT 3.1 *Graph of the density of the $\mathcal{N}(0, 1)$ distribution. The shaded area $P\{Z \leq -z\}$ equals the shaded area $P\{Z \geq z\}$ under the graph to the right of $Z = z$.*

EXAMPLE 3.2

Two students take a reading comprehension test for which the scores X are regarded as having a normal distribution with mean $\mu = 80$ and variance $\sigma^2 = 16$.[1] One student obtains a score of 85, and the other student obtains a score of 70. Because $X \sim \mathcal{N}(80, 16)$, the quantile rank of the first student among the total population of people taking the test is

$$P\{X \leq 85\} = F_X(85) = F_Z\left(\frac{85 - 80}{\sqrt{16}}\right) = F_Z(1.25) = 0.89435.$$

Using (3.4), the quantile rank of the second student is

$$\begin{aligned} F_X(70) &= F_Z\left(\frac{70 - 80}{\sqrt{16}}\right) = F_Z(-2.5) \\ &= 1 - F_Z(2.5) = 1 - 0.99379 = 0.00621. \end{aligned}$$

That is, the first student has a score exceeding that of 89.435% of the population, whereas the second student's score exceeds that of only 0.62% of the population. ◆

If $X \sim \mathcal{N}(\mu, \sigma)$, then X is a continuous random variable and

$$\begin{aligned} (3.5) \quad P\{a \leq X \leq b\} &= P\{a < X < b\} = P\{a \leq X < b\} = P\{a < X \leq b\} \\ &= F_X(b) - F_X(a) = F_Z\left(\frac{b - \mu}{\sigma}\right) - F_Z\left(\frac{a - \mu}{\sigma}\right). \end{aligned}$$

When $b = \infty$, (3.5) reduces to

$$P\{X \geq a\} = P\{X > a\} = 1 - F_X(a) = 1 - F_Z\left(\frac{a - \mu}{s}\right)$$

and gives the **tail probability** of the distribution of X.

EXAMPLE 3.2 (continued)

The fraction of all scores on the reading comprehension test between 78 and 84 is

$$\begin{aligned} P\{78 \leq X \leq 84\} &= F_Z\left(\frac{84 - 80}{\sqrt{16}}\right) - F_Z\left(\frac{78 - 80}{\sqrt{16}}\right) \\ &= F_Z(1.00) - F_Z(-0.50) \\ &= F_Z(1.00) - [1 - F_Z(0.50)] \\ &= 0.84134 - (1 - 0.69146) = 0.53280. \end{aligned}$$ ◆

[1] The scores are actually discrete (integers), but are commonly treated as being continuous both because this provides a convenient simplification, and because the ability assumed to be measured by the scores is thought to be a continuous variable.

EXAMPLE 3.3

Suppose that skeletal heights (in centimeters), X, of eighteenth-century European males are known to have a $\mathcal{N}(170,100)$ distribution. If the height of the skeleton of a male discovered in a French graveyard is 188 cm, then

$$P\{X \geq 188\} = 1 - F_X(188) = 1 - F_Z\left(\frac{188 - 170}{\sqrt{100}}\right)$$
$$= 1 - 0.96407 = 0.03593.$$

Thus the person whose skeleton was found would have been unusually tall compared to other males at that time. ◆

The graph of the cumulative distribution function $F_X(x)$ of a normally distributed random variable X is shown in Exhibit 3.2. Notice that the graph has an S-shaped form. The graph of the cumulative distribution function $F_X(x)$ of a normally distributed random variable is often called a **normal ogive.**

In computing probabilities for the normal distribution, a rough sketch of the graph of the density function showing the required area under the graph is often helpful. For example, suppose that $X \sim \mathcal{N}(15, 49)$ and that we want to calculate the probability that $X < 8$ or $X > 29$. A sketch such as Exhibit 3.3 tells us that the desired probability is the sum of $F_X(8) = F_Z(-1.0) = 0.15866$ and $1 - F_X(29) = 1 - F_Z(2) = 0.00275$.

Exact Normal Probability Calculations Compared to Bienaymé-Chebychev Bounds

Using (3.4) and (3.5), we can compute the probability that X is within k standard deviations on each side of its mean μ:

$$P\{-k\sigma \leq X - \mu \leq k\sigma\} = P\{\mu - k\sigma \leq X \leq \mu + k\sigma\}$$
$$= F_Z(k) - F_Z(-k) = 2F_Z(k) - 1.$$

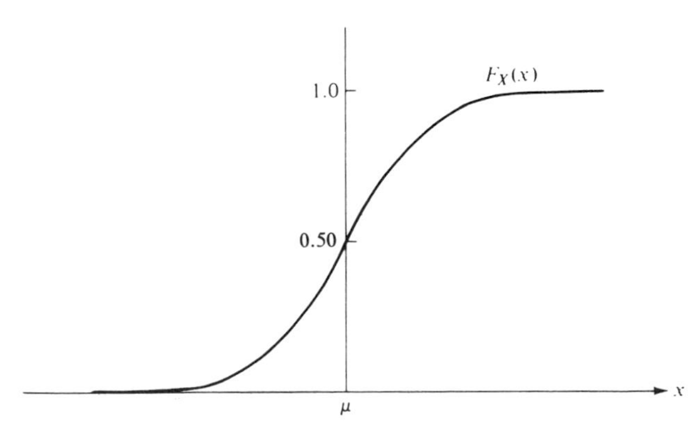

EXHIBIT 3.2 *Graph of the cumulative distribution function of a $\mathcal{N}(\mu, \sigma^2)$ distribution.*

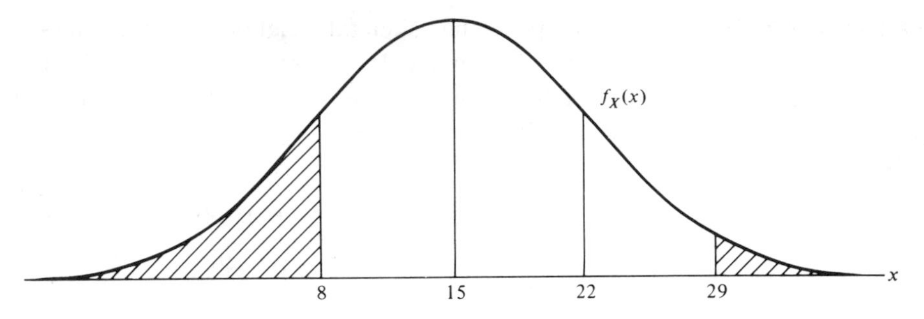

EXHIBIT 3.3 *The sum of the shaded areas gives the probability $P\{X < 8$ or $X > 29\}$ for a $\mathcal{N}(15, 49)$ distribution.*

From Exhibit B.4 of Appendix B, we obtain Exhibit 3.4, which gives values for $P\{-k\sigma \leq X - \mu \leq k\sigma\}$ for $k = 0.5(0.5)3.0$. (A pictorial representation of these probabilities appears in Exhibit 3.5). It is of interest to compare these probabilities with the lower bounds for such probabilities given by the Bienaymé–Chebychev inequality (see Chapter 6).

Quantiles of the Normal Distribution

Because the normal distribution is the distribution of a continuous random variable, to find the pth quantile $Q_X(p)$ we can solve the equation

$$F_X(x) = p$$

for x. However, if $X \sim \mathcal{N}(\mu, \sigma^2)$, then

$$p = F_X(x) = F_Z\left(\frac{x - \mu}{\sigma}\right), \qquad \text{where } Z \sim \mathcal{N}(0,1).$$

EXHIBIT 3.4 *Comparison of the Probability That a Normally Distributed Random Variable Is Within k Standard Deviations of Its Mean with the Bienaymé–Chebychev Bound*

k	$P\{-k\sigma \leq X - \mu \leq k\sigma\}$	*Bienaymé–Chebychev Bound*
0.5	0.38292	0.00000
1.0	0.68268	0.00000
1.5	0.86638	0.55556
2.0	0.95450	0.75000
2.5	0.98758	0.84000
3.0	0.99730	0.88889

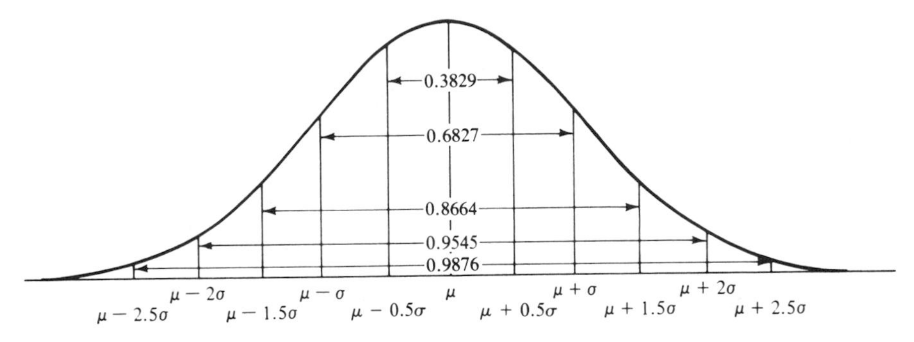

EXHIBIT 3.5 *The areas indicated by the arrows give the probability that a normally distributed random variable is within k standard deviations of its mean μ.*

This suggests finding $Q_X(p)$ in two steps:

1. Find z^* such that $F_Z(z^*) = p$. Note that such a $z^* = Q_Z(p)$.
2. Solve $(x - \mu)/\sigma = z^* = Q_Z(p)$ for x; that is, $x = \sigma z^* + \mu$.

The result of these two steps is the following equation:

$$Q_X(p) = \mu + \sigma Q_Z(p). \tag{3.6}$$

Equation (3.6) tells us how to find $Q_X(p)$ once we find $Q_Z(p)$. To find the pth quantile, $Q_Z(p)$, of $Z \sim \mathcal{N}(0, 1)$, we use linear interpolation in Exhibit B.4 of Appendix B. A brief discussion of linear interpolation is provided in Appendix B.

EXAMPLE 3.4

The yield of hay X in hundredweights per acre is assumed to have a normal distribution with mean $\mu = 28$ and variance $\sigma^2 = 16$. What is the 0.75th quantile (75th percentile) $Q_X(0.75)$ of X?

To find $Q_X(0.75)$, we first need to find $Q_Z(0.75)$. From Exhibit B.4, $F_Z(0.67) = 0.74857$ and $F_Z(0.68) = 0.75175$, so that we know $Q_Z(0.75)$ is between 0.67 and 0.68.

Using linear interpolation (Appendix B) to solve for $Q_Z(0.75)$ gives us

$$\begin{aligned} Q_Z(0.75) &= 0.67 + \frac{0.75000 - F_Z(0.67)}{F_Z(0.68) - F_Z(0.67)}(0.68 - 0.67) \\ &= 0.67 + \frac{0.75000 - 0.74857}{0.75175 - 0.74857}(0.01) = 0.674. \end{aligned}$$

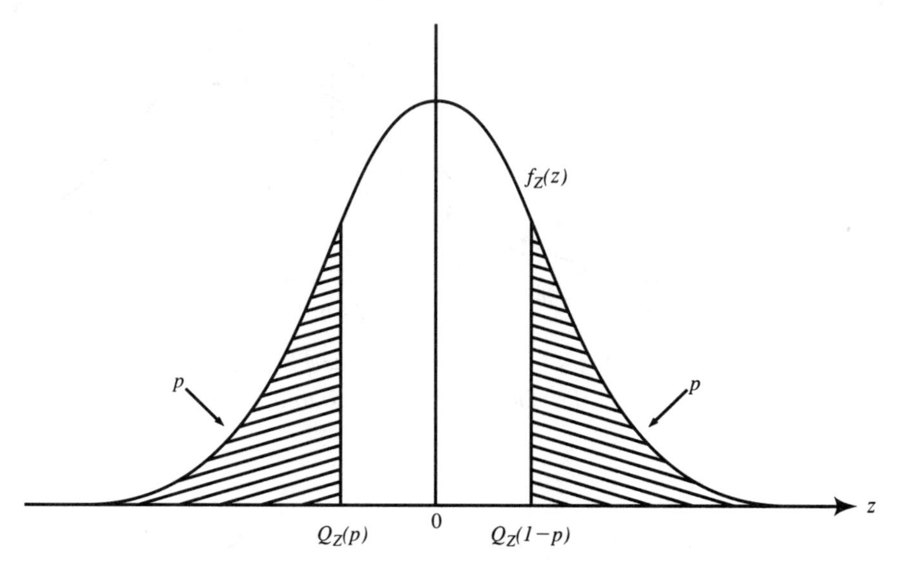

EXHIBIT 3.6 *By symmetry, because the shaded areas are equal, $Q_Z(p)$ is the same distance to the left of 0 as $Q_Z(1 - p)$ is to the right of 0.*

Using equation (3.6) yields

$$Q_X(0.75) = 28 + 4(0.674) = 30.696.$$ ◆

To find the 0.25th quantile, $Q_X(0.25)$ in Example 3.4, we first want to find $Q_Z(0.25)$. The values of $F_Z(z)$ in Exhibit B.4 of Appendix B are never less than 0.50. However, the symmetry of the standard normal density function about 0 (Exhibit 3.6) implies that

$$Q_Z(p) = -Q_Z(1 - p). \tag{3.7}$$

Thus $Q_Z(0.25) = -Q_Z(0.75) = -0.674$, and

$$Q_X(0.25) = \mu + \sigma Q_Z(0.25) = 28 + 4(-0.674) = 25.304.$$

4 Fitting a Normal Distribution

We now consider a classical example of the applicability of the normal distribution.

EXAMPLE 4.1

Intelligence Quotients

In his well-known book *The Intelligence of School Children*, Terman (1919) describes several studies on school children that utilized the Stanford Revision of the Binet–Simon Intelligence Scale. In one of these studies, 112 children (65 boys and 47 girls) attending five kindergarten classes in San Jose and San Mateo, California, were measured on this intelligence scale. The original ungrouped (raw) data are given in Exhibit 4.1; the data are grouped in Exhibit 4.2 and are graphed as a

EXHIBIT 4.1 *Original IQ Data in the Terman Study of 112 Children Attending Kindergarten Classes in San Jose and San Mateo, California*[a]

152	146	142	136	130(2)	129
126(2)	125	124(2)	123	122	121(5)
120	119(2)	118	117(2)	114(6)	113(4)
112(2)	111(2)	110(4)	109(5)	108(2)	107(4)
106(3)	105	103(3)	102(5)	102(2)	100(3)
99	98(3)	97(2)	96(4)	94(2)	93(3)
92	91(2)	90(4)	88	86(2)	85(3)
84	82	81	80(4)	79	77
76	75	72			

[a] If a value appeared more than once in the data, the number of times that it appeared appears in parentheses after the value.

Source Lewis M. Terman, *The Intelligence of School Children* (Boston: Houghton Mifflin Company, 1919). Reprinted with permission of the publisher.

EXHIBIT 4.2 *Observed Frequency Distribution of IQ of 112 Children in Terman Study*

Interval of IQ Values	*Midpoint of Interval*	*Frequency*
60–69	65	1
70–79	75	5
80–89	85	13
90–99	95	22
100–109	105	28
110–119	115	23
120–129	125	14
130–139	135	3
140–149	145	2
150–159	155	1
		112

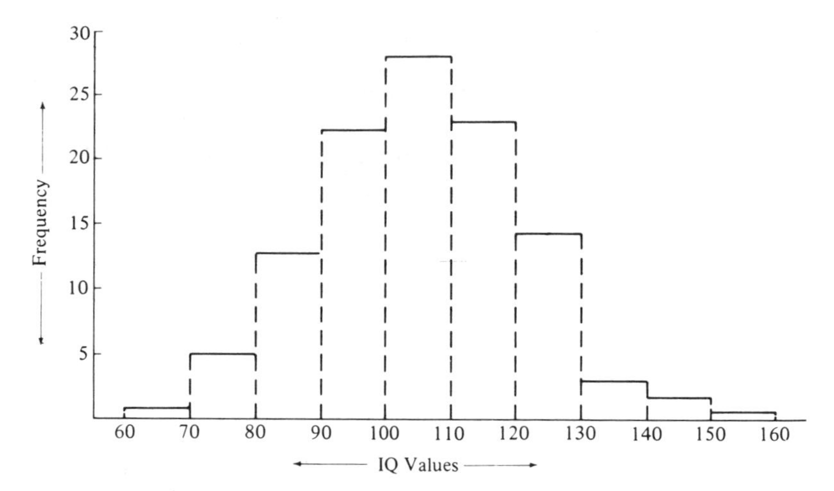

EXHIBIT 4.3 *Histogram corresponding to grouped data in the Terman study.*

frequency histogram in Exhibit 4.3. It appears from Exhibit 4.3 that a normal distribution might serve as an approximation to the actual probability distribution of intelligence scores (IQs).

To fit a normal distribution to this data,[2] we estimate μ by $\hat{\mu}_X = 104.5$ and σ^2 by $\hat{\sigma}_X^2 = 263.66$, which are the mean and variance of the data of Exhibit 4.1 (see Chapter 6, Notes). A $\mathcal{N}(104.5, 263.66)$ distribution is then fitted to the data.

The resulting theoretical frequencies can be compared with the observed frequencies. Except for the need to compute probabilities of intervals of values, rather than individual values, this is the same approach as the one used to check the fit of discrete distributions (Chapter 8).

Before calculating probabilities of the intervals in Exhibit 4.3, we calculate cumulative probabilities $F_X(x)$ at each of the endpoints x of these intervals. Thus

$$F_X(60) = F_Z\left(\frac{60 - 104.5}{\sqrt{263.66}}\right) = F_Z(-2.74) = 1 - F_Z(2.74) = 0.00347,$$

and so on. Our results appear in Exhibit 4.4.

Now, theoretical probabilities of the intervals in Exhibit 4.3 are found using equation (3.5):

[2] See footnote 1.

EXHIBIT 4.4 Theoretical Cumulative Probabilities for the Distribution in Exhibit 4.2

x	$F_X(x)$	x	$F_X(x)$
60	0.00347	120	0.83147
70	0.01659	130	0.94179
80	0.06552	140	0.98574
90	0.18406	150	0.99752
100	0.38974	160	0.99970
110	0.63307		

$$P\{-\infty < X \leq 60\} = F_X(60) = 0.00347,$$
$$P\{60 \leq X \leq 70\} = F_X(70) - F_X(60) = 0.01312,$$
$$\vdots$$
$$P\{150 \leq X \leq 160\} = F_X(160) - F_X(150) = 0.00218$$

and $P\{160 \leq X < \infty\} = 1 - F_X(160) = 0.00030$. Each such theoretical probability then is multiplied by 112 to yield the theoretical frequencies in Exhibit 4.5.

EXHIBIT 4.5 Comparison of Observed and Theoretical Frequency Distributions for the IQs of 112 Children in the Terman Study

Interval of IQ Scores	*Observed Frequency*	*Theoretical Frequency*
Less than 60	0	0.39
60–69	1	1.46
70–79	5	5.48
80–89	13	13.28
90–99	22	23.04
100–109	28	27.25
110–119	23	22.22
120–129	14	12.36
130–139	3	4.92
140–149	2	1.32
150–159	1	0.24
More than 160	0	0.03
	112	111.99

The normal distribution with parameters $\mu = 104.5$ and $\sigma^2 = 263.66$ appears to provide a reasonably close approximation to the distribution of IQ scores. ◆

Remark. It is a more common practice to construct the histogram in Exhibit 4.3 using intervals of the form [59.5, 69.5], [69.5, 79.5], and so on. Although this way of constructing the histogram would require us to compute probabilities of events such as $\{59.5 \leq X \leq 69.5\}$ and $\{X \geq 159.5\}$, the resulting theoretical frequencies would differ only slightly from those given in Exhibit 4.5.

5 The Central Limit Theorem

We have observed that the normal distribution has been used to describe the variation in many quantitative phenomena. However, the normal distribution is also useful in other ways. Suppose that we are interested in estimating the mean μ_X of a random variable X, and that we have made n independent observations $X_1, X_2, \ldots, X_n$ on X; that is, observations are taken from n independent trials of the random experiment that generated the random variable X. We can estimate μ_X by the sample average

$$\overline{X} = \frac{1}{n} \sum_{i=1}^{n} X_i, \tag{5.1}$$

which, because each observation X_i is a random variable, is itself a random variable. We cannot expect $\overline{X}$ always to be equal to μ_X, because every time we take n independent observations on X, a new value of $\overline{X}$ will be obtained. We can, however, measure how closely $\overline{X}$ varies about μ_X by calculating the probability

$$P\{-\varepsilon \leq \overline{X} - \mu_X \leq \varepsilon\} = P\{\mu_X - \varepsilon \leq \overline{X} \leq \mu_X + \varepsilon\} \tag{5.2}$$

that a specified accuracy $\varepsilon > 0$ is obtained. The distribution of $\overline{X}$, and thus the probability (5.2), can be difficult to obtain, especially when the distribution of X is not simple.

In Chapter 7 we saw that the mean $\mu_{\overline{X}}$ of the distribution $\overline{X}$ is always the same as the mean μ_X of X, and that the variance, $\sigma^2_{\overline{X}}$, of $\overline{X}$ always equals (σ^2_X/n). It is a remarkable fact that for a sufficiently large sample size n, the probability (5.2) can be closely approximated by assuming that $\overline{X}$ has a normal distribution with mean μ_X and variance (σ^2_X/n). Consequently,

$$P\{-\varepsilon \leq \overline{X} - \mu_X \leq \varepsilon\} = P\left\{\frac{-\varepsilon}{\sqrt{\sigma_X^2/n}} \leq \frac{\overline{X} - \mu_X}{\sqrt{\sigma_X^2/n}} \leq \frac{\varepsilon}{\sqrt{\sigma_X^2/n}}\right\}$$

$$(5.3) \qquad \simeq P\left\{-\frac{\varepsilon\sqrt{n}}{\sigma_X} \leq Z \leq \frac{\varepsilon\sqrt{n}}{\sigma_X}\right\}$$

$$= 2F_Z\left(\frac{\varepsilon\sqrt{n}}{\sigma_X}\right) - 1,$$

where $Z \sim \mathcal{N}(0, 1)$. The basis of this assertion, and of the approximation (5.3), is the famous Central Limit Theorem.

EXAMPLE 5.1

Independent measurements $X_1, X_2, \ldots, X_n$ are taken of the true distance D in light-years between an observatory and a distant star. These measurements are known to have a common distribution with mean $\mu_X = D$ and variance $\sigma_X^2 = 4$. To estimate D with an accuracy of ± 0.25 light-year, an astronomer might decide to use the sample average of $n = 100$ observations. Using (5.3) with $\varepsilon = 0.25$, $n = 100$, the probability of obtaining the desired accuracy is

$$P\{-0.25 \leq \overline{X} - D \leq 0.25\}$$
$$\cong 2F_Z\left(\frac{\sqrt{100}(0.25)}{\sqrt{4}}\right) - 1 = 2(0.89435) - 1 = 0.78870.$$

When $n = 400$ observations are used, this probability increases to

$$P\{-0.25 \leq \overline{X} - D \leq 0.25\} \cong 2F_Z\left(\frac{\sqrt{400}(0.25)}{\sqrt{4}}\right) - 1 = 0.98758. \quad \blacklozenge$$

A formal statement of the Central Limit Theorem is the following.

Central Limit Theorem. Let $X_1, X_2, \ldots, X_n$ be mutually independent observations on a random variable X having a well-defined mean μ_X and variance σ_X^2. Let

$$Z_n = \frac{\overline{X} - \mu_X}{\sigma_X/\sqrt{n}},$$

and let $F_{Z_n}(z)$ be the cumulative distribution function of the random variable Z_n. Then for all z, $-\infty < z < \infty$,

$$\lim_{n\to\infty} F_{Z_n}(z) = F_Z(z),$$

where $F_Z(z)$ is the cumulative distribution function of $Z \sim \mathcal{N}(0, 1)$.

A proof of the central limit theorem is given in Section 6.

It follows directly from the Central Limit Theorem that for any fixed interval $[a, b]$,

$$(5.4) \quad \lim_{n\to\infty} P\{a \le Z_n \le b\} = \lim_{n\to\infty} [F_{Z_n}(b) - F_{Z_n}(a)] = F_Z(b) - F_Z(a).$$

However, our earlier application (5.3) of the central limit theorem involved an interval whose endpoints varied with the sample size n. Because it can be shown that the limit in (5.4) is achieved uniformly in z, $-\infty < z < \infty$, it then follows that for any sequence of numbers $\{z_n\}$, we have

$$(5.5) \qquad \lim_{n\to\infty} [F_{Z_n}(z_n) - F_Z(z_n)] = 0.$$

That is, for n large enough, $F_{Z_n}(z_n) \simeq F_Z(z_n)$. It follows from (5.5) and the definition of Z_n that for n large enough,

$$(5.6) \qquad P\{\overline{X} \le t\} = P\left\{Z_n \le \frac{t - \mu_X}{\sigma_X/\sqrt{n}}\right\} \cong F_Z\left(\frac{t - \mu_X}{\sigma_X/\sqrt{n}}\right).$$

The result (5.3) now follows directly from (5.6). Equation (5.6) also justifies the informal statement of the central limit theorem used in motivating (5.3):

> For large enough sample size n, the sample average $\overline{X}$ is approximately normally distributed with mean μ_X and variance σ_X^2/n.

EXAMPLE 5.2

An insurance company has insured 50 oil pumping platforms in the North Atlantic against liability for environmental damage. Their underwriter tells them that the profit X, in thousands of dollars, on any one such policy can be modeled as having a continuous distribution with mean $\mu_X = 17.5$ thousand dollars and variance $\sigma_X^2 = 1.62$. If the profits $X_1, X_2, \ldots, X_{50}$ obtained from these policies are mutually independent, what is the probability that their average profit $\overline{X}$ per policy will exceed 18 thousand dollars ($18,000)?

Assuming that $n = 50$ is sufficiently large for the approximation (5.6) provided by the Central Limit Theorem to be close, we act as if $\overline{X}$ has a normal distribution with mean $\mu_{\overline{X}} = \mu_X = 17.5$ and variance $\sigma_{\overline{X}}^2 = \sigma_X^2/n = 1.62/50 = 0.0324$. Thus

$$\begin{aligned} P\{\overline{X} > 18\} &\cong 1 - F_Z\left(\frac{18 - 17.5}{\sqrt{0.0324}}\right) = 1 - F_Z(2.78) \\ &= 1 - 0.99728 = 0.00272. \end{aligned}$$

The insurance company might also want to find the 0.90th quantile (90th percentile) of the distribution of their average profit $\overline{X}$. From linear interpolation in Exhibit B.4 of Appendix B, we obtain

$$Q_Z(0.90) = 1.28 + \frac{0.90000 - 0.89973}{0.90147 - 0.89973}(1.29 - 1.28) = 1.282,$$

and hence

$$\begin{aligned} Q_{\overline{X}}(0.90) &\cong \mu_{\overline{X}} + \sigma_{\overline{X}}\, Q_Z(0.90) \\ &= 17.5 + (\sqrt{0.0324})(1.282) \\ &= 17.731 \text{ thousand dollars.} \end{aligned}$$

◆

The Central Limit Theorem also applies to sums (totals)

$$S_n = \sum_{i=1}^{n} X_i$$

of independent observations $X_1, X_2, \ldots, X_n$ on a random variable X. In Chapter 7 it is shown that

$$\mu_{S_n} = n\mu_X, \qquad \sigma_{S_n}^2 = n\sigma_X^2.$$

Because $S_n = n\overline{X}$ is just a constant multiple of $\overline{X}$, and $\overline{X}$ is approximately normally distributed,

$$\begin{aligned} P\{S_n \le t\} &= P\{n\overline{X} \le t\} = P\left\{\overline{X} \le \frac{t}{n}\right\} \\ &\cong F_Z\left(\frac{t/n - \mu_X}{\sigma_X/\sqrt{n}}\right) \equiv F_Z\left(\frac{t - n\mu_X}{\sqrt{n}\sigma_X}\right) \end{aligned}$$

suggesting that for large enough n, we can treat S_n as if it were (approximately) normally distributed.

> For large enough values of n, the sample total (sum) S_n of mutually independent observations $X_1, X_2, \ldots, X_n$ on a random variable X has approximately a normal distribution with mean $\mu_{S_n} = n\mu_X$ and variance $\sigma_{S_n}^2 = n\sigma_X^2$.

EXAMPLE 5.3

Airlines check baggage without weighing each bag, but because the total weight is important in determining fuel consumption, an airline has made a study of the weight X of such bags. For a randomly chosen suitcase, the weight X (in pounds) is a random variable with mean $\mu_X = 35$ and variance $\sigma_X^2 = 300$. The cargo area has capacity for $n = 100$ pieces of baggage. If the cargo area of a plane is completely filled, what is the probability that the total weight $S_{100} = \Sigma_{i=1}^{100} X_i$ of the bags is no more than 4000 pounds? What is the 99th percentile of the distribution of total baggage weight S_{100} when the hold is filled?

Using the Central Limit Theorem approximation, the distribution of S_{100} is approximately normal with mean and variance

$$\mu_{S_{100}} = (100)\mu_X = 100(35) = 3500,$$
$$\sigma_{S_{100}}^2 = (100)\sigma_X^2 = 100(300) = 30{,}000,$$

respectively. Thus

$$P\{S_{100} \leq 4000\} \simeq F_Z\left(\frac{4000 - 3500}{\sqrt{30{,}000}}\right) = F_Z(2.89) = 0.99807.$$

Also, because

$$Q_Z(0.99) = 2.32 + \left[\frac{0.99000 - 0.98983}{0.99010 - 0.98983}\right](2.33 - 2.32) = 2.326,$$

we obtain

$$Q_{S_{100}}(0.99) \cong 3500 + \sqrt{30{,}000}\, Q_Z(0.99)$$
$$= 3500 + (173.205)(2.326) = 3902.875 \text{ pounds.}$$

◆

Sums and Averages of Normally Distributed Random Variables

There is one case where $\overline{X}$ and S_n have *exact* normal distributions *regardless* of how many independent observations $X_1, X_2, \ldots, X_n$ are taken on X. This is the case when X has a $\mathcal{N}(\mu, \sigma^2)$ distribution.

In general, when $X_1, X_2, \ldots, X_n$ are mutually independent observations on $X \sim \mathcal{N}(\mu, \sigma^2)$, then $\overline{X}$ has an exact $\mathcal{N}(\mu, \sigma^2/n)$ distribution and S_n has an exact $\mathcal{N}(n\mu, n\sigma^2)$ distribution for any sample size $n = 1, 2, \ldots$.

EXAMPLE 5.4 The reaction time X (in seconds) to a certain stimulus is normally distributed with mean $\mu = 0.1$ second and variance $\sigma^2 = 1.69(10^{-4})$ (second)^2. For a sample of $n = 9$ reaction times, we know that $\overline{X}$ is exactly normally distributed with mean $\mu_{\overline{X}} = 0.10$ and variance $\sigma_{\overline{X}}^2 = (1.69)(10^{-4})/9$. Thus

$$P\{\overline{X} > 0.11\} = 1 - F_Z\left(\frac{0.11 - 0.10}{\sqrt{(1.69)(10^{-4})/9}}\right)$$
$$= 1 - F_Z(2.31) = 1 - 0.98956 = 0.01044$$

exactly. ◆

EXAMPLE 5.5 Suppose that cars have (random) first-year maintenance costs X which are normally distributed with mean $\mu_X = \$100$ and variance $\sigma_X^2 = 625$ (dollars)^2. We wish to determine an amount c such that the total cost $S_4 = X_1 + X_2 + X_3 + X_4$ of maintaining four cars will have a probability of only 0.025 of exceeding c.

Because $P\{S_4 > c\} = 1 - P\{S_4 \leq c\}$ must equal 0.025, the number c is the $(1 - 0.025)$th quantile of the distribution of S_4. That is, we need to find $Q_{S_4}(0.975)$. Note that $F_Z(1.96) = 0.97500$ when $Z \sim \mathcal{N}(0, 1)$, so that $Q_Z(0.975) = 1.96$. Because S_4 has exactly a $\mathcal{N}((4)(100), (4)(625)) = \mathcal{N}(400, 2500)$ distribution, we obtain

$$c = Q_{S_4}(0.975) = 400 + \sqrt{2500}(1.96) = \$498.$$ ◆

The Accuracy of the Central Limit Theorem

The informal statements of the Central Limit Theorem for sums and averages include the qualification "for large enough sample size n." How large must n be before we can approximate the distributions of $\overline{X}$ and S_n by an appropriate normal distribution? Because $\overline{X}$ and S_n are determined from the observations $X_1, \ldots, X_n$, and each such observation has the distribution of X, the distribution of X governs what the distributions of $\overline{X}$ and S_n are for any sample size n. Consequently, the distribution of X determines how well the distributions of $\overline{X}$ and S_n are approximated by the normal distribution.

To see the influence of the distribution of X and the sample size n on the shapes of the density functions of $\overline{X}$, in Exhibit 5.1 we have graphed the density function of $\overline{X}$ for sample sizes of $n = 1, 2, 5, 10, 20, 30$ for the following choices of the density function $f(x)$ of X:

$$\text{(a)} \quad f(x) = \begin{cases} 1/2\sqrt{3}, & \text{if } -\sqrt{3} \leq x \leq \sqrt{3}, \\ 0, & \text{otherwise.} \end{cases}$$

$$\text{(b)} \quad f(x) = \begin{cases} e^{-(x+1)}, & \text{if } x > -1, \\ 0, & \text{otherwise.} \end{cases}$$

For both choices of $f(x)$, X has mean 0 and variance 1. Note that for $n = 1$, $\overline{X} = X_1$ has density $f(x)$,so that Exhibit 5.1 displays $f(x)$ when $n = 1$. For larger values of n, the horizontal scales of the graphs in Exhibit 5.1 have been adjusted to compensate for the fact that $\overline{X}$ has variance $n^{-1}\sigma^2 = 1/n$ decreasing in n. (The densities graphed are those of the standardization $Z_n = n^{1/2}\overline{X}$ of $\overline{X}$.) In this way, the shapes of the densities of $\overline{X}$ are more easily compared.

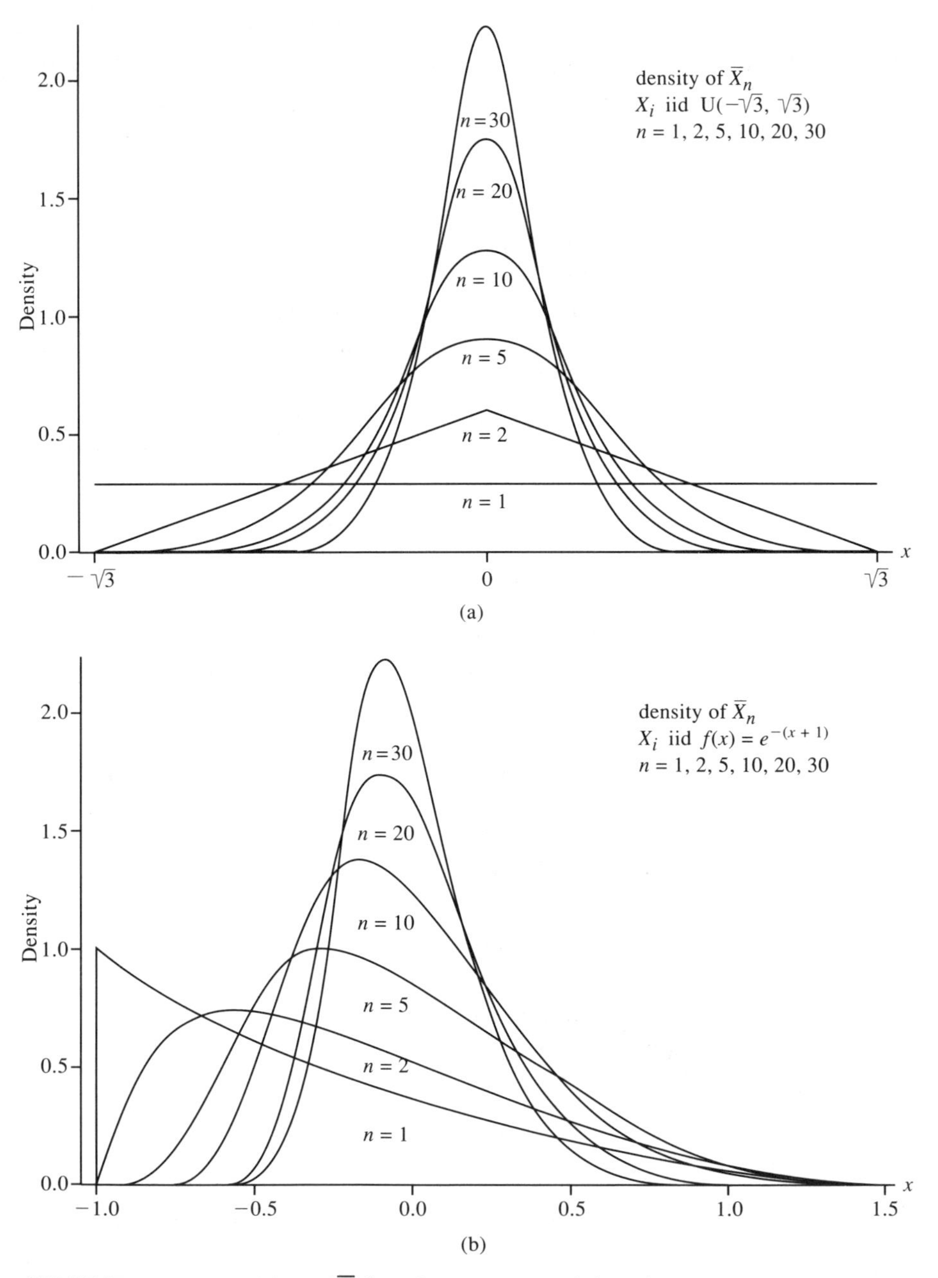

EXHIBIT 5.1 *Densities of $\overline{X}$ for choices (a) and (b) when n = 1, 2, 5, 10, 20, 30.*

From Exhibit 5.1 we see that the density of $\overline{X}$ for choice (a) has a nearly bell-shaped form for $n = 5$ and is indistinguishable from the normal density for $n = 20, 30$. In contrast, the density of $\overline{X}$ for choice (b) continues to display some asymmetry even for $n = 20$, although it is approximately bell-shaped for $n = 30$.

Perhaps the major difference between the densities for X ($n = 1$) for choices (a) and (b) is that the density in case (a) shares with the normal density the property of being symmetric about its mean (although it lacks a bell shape). The density for choice (b), on the other hand, is highly skewed (asymmetric). In this respect, the density for choice (a) is more normal-like than the density for choice (b).

To this comparison, we can add the already noted fact that if X is normally distributed, then so is $\overline{X}$ for all n, even for $n = 1$. Here the distribution of X is normal from the start. It appears from this fact and our discussion of Exhibit 5.1 that "the closer the distribution of X is to a normal bell shape, the smaller the sample size n required to make the distribution of $\overline{X}$ approximately normally (bell) shaped." A second observation arising from Exhibit 5.1 is that "the more asymmetric (*skewed*) the distribution of X is, the larger n must be to make $\overline{X}$ have an approximately normal shape." Because the densities of $\overline{X}$ and $S_n = n\overline{X}$ have similar shapes, these two points also apply to normal approximations to the distribution of S_n. Although these statements are helpful, in that they suggest what properties of the density of X to look for, they fail to give precise guidelines on how large n must be.

Unfortunately, there is no universal rule to state when a sample size n is "large enough" for the Central Limit Theorem to provide a good approximation to the distribution of $\overline{X}$ or S_n. If the common distribution of the observations $X_1, X_2, \ldots, X_n$ is not extremely skewed (asymmetric), and also resembles the normal distribution in the sense of having a single mode (unimodal), experience indicates that a sample size of $n \geq 30$ is usually sufficient to provide at least one-decimal accuracy when computing probabilities of events involving the average $\overline{X}$ or sum S_n.

Approximating Discrete Variables: The Continuity Correction

At the beginning of this chapter it was mentioned that the normal distribution first arose as an approximation to the binomial distribution. Because the binomial distribution is a discrete distribution and the normal distribution is continuous, this requires some explanation. Note that if X is a discrete random variable, then $P\{X = x\} > 0$ for certain values x (called the possible values of X). In contrast, when X is a continuous random variable, we have noted that $P\{X = x\} = 0$ for *all* numbers x. How then can we use a continuous distribution to approximate a discrete distribution? Our experience with

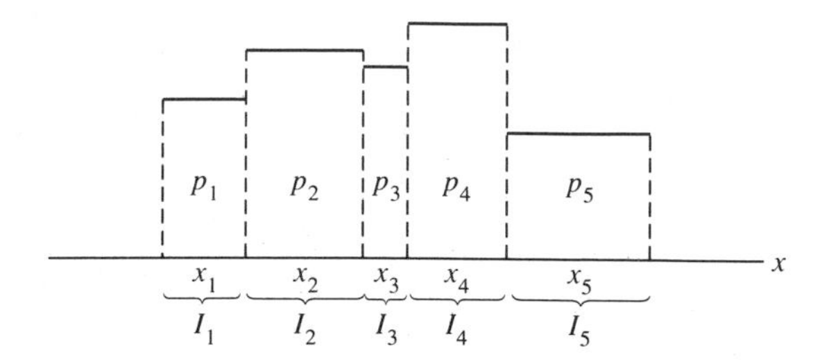

EXHIBIT 5.2 *Histogram constructed for the discrete random variable X.*

grouped data and histograms (Chapter 5; see also the appendix to Chapter 6) suggests a possible approach. Suppose that a discrete random variable X has distinct possible values $x_1, x_2, \ldots, x_m$, ordered as $x_1 < x_2 < \cdots < x_m$. (The number m of such values can be infinite, but here we will discuss only the case where m is finite.) Each such possible value x_i is regarded as the midpoint of the interval

$$(5.7) \quad I_i = \left[x_i - \frac{x_i - x_{i-1}}{2}, \quad x_i + \frac{x_{i+1} - x_i}{2}\right], \qquad i = 2, \ldots, m-1,$$

whose endpoints are midway between the possible values. Regard $p_i = P\{X = x_i\}$ as the relative frequency of the interval whose midpoint is x_i, and construct a rectangle over this interval with area equal to p_i, $i = 2, \ldots, m - 1$. For x_1 construct a rectangle over the interval $I_1 = [x_1 - \frac{1}{2}(x_2 - x_1), x_1 + \frac{1}{2}(x_2 - x_1)]$ of area p_1, whereas for x_m construct a rectangle over the interval $I_m = [x_m - \frac{1}{2}(x_m - x_{m-1}), x_m + \frac{1}{2}(x_m - x_{m-1})]$ of area p_m. The result is a (modified) relative-frequency histogram (Exhibit 5.2). The graph traced by the tops of the rectangles in Exhibit 5.2 is the graph of a density function $\hat{f}(x)$ having the property that

$$(5.8) \qquad P\{X = x_i\} = \int_{x \in I_i} \hat{f}(x)\, dx, \qquad i = 1, 2, \ldots, m.$$

If we wish to approximate the probabilities for the discrete random variable X by using the density function $f(x)$ of a continuous distribution, this suggests that we replace $\hat{f}(x)$ in (5.8) by $f(x)$, yielding the approximation

$$(5.9) \qquad P\{X = x_i\} \cong \int_{x \in I_i} f(x)\, dx, \qquad i = 1, \ldots, m.$$

If we want to find probabilities of events E describing X, we simply add the approximate probabilities (5.9) corresponding to all possible values x_i of X that are included in E. This approach to approximating probabilities for a discrete variable X is called making a **continuity correction.** This is somewhat of a misnomer; what we are actually doing is correcting for the discontinuity of X.

The continuity correction is most commonly applied when the possible values of X are equally spaced; that is: $x_2 = x_1 + \delta$, $x_3 = x_1 + 2\delta, \ldots, x_m = x_1 + (m - 1)\delta$. In this case, (5.8) becomes

$$P\{X = x_i\} \cong \int_{x_i-(1/2)\delta}^{x_i+(1/2)\delta} f(x)\,dx.$$

For example, when the possible values of X are integers (as is the case when X has a binomial distribution), $\delta = 1$ and

$$P\{X = i\} \cong \int_{i-1/2}^{i+1/2} f(x)\,dx.$$

If we want to approximate the probability of an event such as $\{2 \leq X \leq 5\}$, then

$$\begin{aligned} P\{2 \leq X \leq 5\} &= \sum_{i=2}^{5} P\{X = i\} \cong \sum_{i=2}^{5} \int_{i-1/2}^{i+1/2} f(x)\,dx \\ &= \int_{2-1/2}^{5+1/2} f(x)\,dx = \int_{1.5}^{5.5} f(x)\,dx. \end{aligned}$$

On the other hand,

$$P\{2 < X \leq 5\} = \sum_{i=3}^{5} P\{x = i\} \cong \int_{3-1/2}^{5+1/2} f(x)\,dx = \int_{2.5}^{5.5} f(x)\,dx,$$

$$P\{2 \leq X < 5\} = \sum_{i=2}^{4} P\{X = i\} \cong \int_{2-1/2}^{4+1/2} f(x)\,dx = \int_{1.5}^{4.5} f(x)\,dx.$$

A picture of the event desired is often helpful. Thus if we want to find $P\{X > 2\}$, we see from Exhibit 5.3 that the desired event includes the values 3, 4, 5, . . . , and thus

$$P\{X > 2\} \cong \int_{2.5}^{\infty} f(x)\,dx.$$

At this point some examples might be helpful.

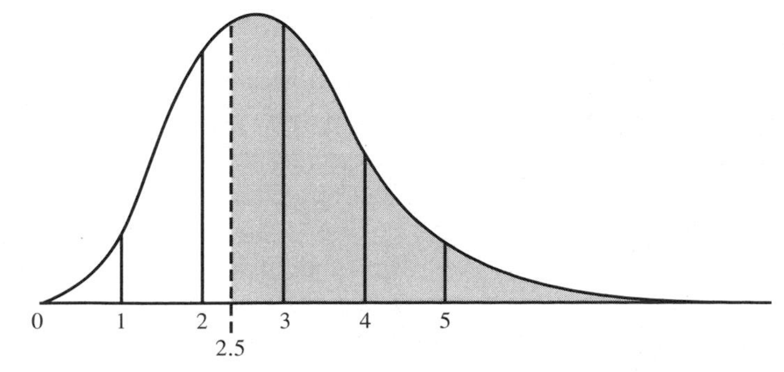

EXHIBIT 5.3 *The event $\{X > 2\}$ includes possible values 3, 4, 5.*

EXAMPLE 5.6

Suppose that X is a discrete random variable with probability mass function

$$p(x) = \begin{cases} \dfrac{x}{15}, & x = 1, 2, 3, 4, 5, \\ 0, & \text{otherwise.} \end{cases}$$

Think of approximating probabilities for X using a density function

$$f(x) = \begin{cases} \dfrac{x}{18}, & 0 \le x \le 6, \\ 0, & \text{otherwise,} \end{cases}$$

which, like $p(x)$, increases linearly with x. Suppose that we want to approximate $P\{X = 3\}$. Because the possible values of X are integers, $\delta = 1$ and

$$P\{X = 3\} \cong \int_{3-1/2}^{3+1/2} f(x)\,dx = \int_{2.5}^{3.5} \frac{1}{18}\,x\,dx = \frac{x^2}{36}\bigg|_{2.5}^{3.5} = \frac{1}{6}.$$

The correct value of $P\{X = 3\} = \frac{3}{15} = \frac{1}{5}$ is close to this approximate value.

Similarly,

$$P\{X > 2\} \cong \int_{2.5}^{\infty} f(x)\,dx = \int_{2.5}^{6} \frac{1}{18}\,x\,dx + \int_{6}^{\infty} 0\,dx = \frac{x^2}{36}\bigg|_{2.5}^{6} = 0.826.$$

versus the correct value of $(3 + 4 + 5)/18 = 0.667$. Here the approximation is not very good, but the purpose of this example was to illustrate the continuity correction, not to advocate use of the density $f(x)$ as an approximation. ♦

EXAMPLE 5.7

A gambler decides to play a card game 100 times and to bet \$10 each time. Assume that the hands are independently played, and that on each play the gambler's probability of winning \$10 is 0.49; the probability of losing \$10 is 0.51. Let X_i $(i = 1, 2, \ldots, 100)$ denote the gambler's winnings on the ith play. Then

$$\begin{aligned} \mu_{X_i} &= \$10(0.49) + (-\$10)(0.51) = -\$0.20, \\ \sigma_{X_i}^2 &= (\$10 + \$0.20)^2(0.49) + (-\$10 + \$0.20)^2(0.51) \\ &= 99.96 \text{ dollars squared,} \end{aligned}$$

The gambler's total winnings are $S_{100} = \sum_{i=1}^{100} X_i$. Note that each X_i is a discrete random variable whose values are spaced \$20 apart.

What is the probability $P\{S_{100} > 0\}$ that the gambler finishes with a profit? Here S_{100} is the sum of the independent random variables X_i. The

Central Limit Theorem suggests that we use the normal distribution with mean and variance

$$\mu_{S_{100}} = 100\mu_X = 100(-0.20) = -20.00,$$
$$\sigma^2_{S_{100}} = 100\sigma^2_X = 100(99.96) = 9996,$$

respectively, to approximate probabilities for S_{100}. Using the continuity correction, and letting $f(x)$ be the $\mathcal{N}(-20, 9996)$ density function, we have

$$\begin{aligned} P\{S_{100} > 0\} &\cong \int_{0-(1/2)(20)}^{\infty} f(x)\,dx = \int_{-10}^{\infty} f(x)\,dx \\ &= 1 - F_Z\left(\frac{-10 - (-20)}{\sqrt{9996}}\right) = 1 - F_Z(0.30) \\ &= 0.38209, \end{aligned}$$

where Z has a $\mathcal{N}(0, 1)$ distribution. Similarly, the approximate probability, $P\{S_{1000} > 0\}$, that the gambler makes a profit if 1000 hands are played (instead of 100) is $1 - F_Z(0.66) = 0.25463$. Calculations of this type show that the probability that the gambler finishes with a profit decreases with the number of plays. ◆

Normal Approximation to the Binomial Distribution

The discussion above tells us how we can use the continuous normal distribution to approximate the discrete binomial distribution but doesn't tell us why we should do this or which normal distribution to use. However, recall from Chapter 8, Section 1, that a binomial random variable X with parameters n and p can be considered to have arisen as a sum $X = \sum_{i=1}^{n} W_i$ of n independent Bernoulli random variables $W_1, \ldots, W_n$. Because we know that each W_i has mean $\mu_W = p$ and variance $\sigma^2_W = p(1 - p)$, the Central Limit Theorem states that X is approximately normally distributed with mean and variance

$$\mu_X = n\mu_W = np, \qquad \sigma^2_X = n\sigma^2_W = np(1 - p),$$

respectively, when n is "sufficiently large."

Exhibit 5.4 shows histograms for the standardization

$$Z_n = \frac{X - np}{\sqrt{np(1 - p)}}$$

of X constructed by the method used to construct Exhibit 5.2. We have displayed three choices for p, namely, $p = 0.2, 0.5, 0.6$ and chosen $n = 1, 5, 15, 30, 50$ for each choice. Notice that in each case, the original Bernoulli variable W corresponds to X for $n = 1$. From Exhibit 5.4 we can see that as n grows large, the histograms become more and more bell-shaped in each

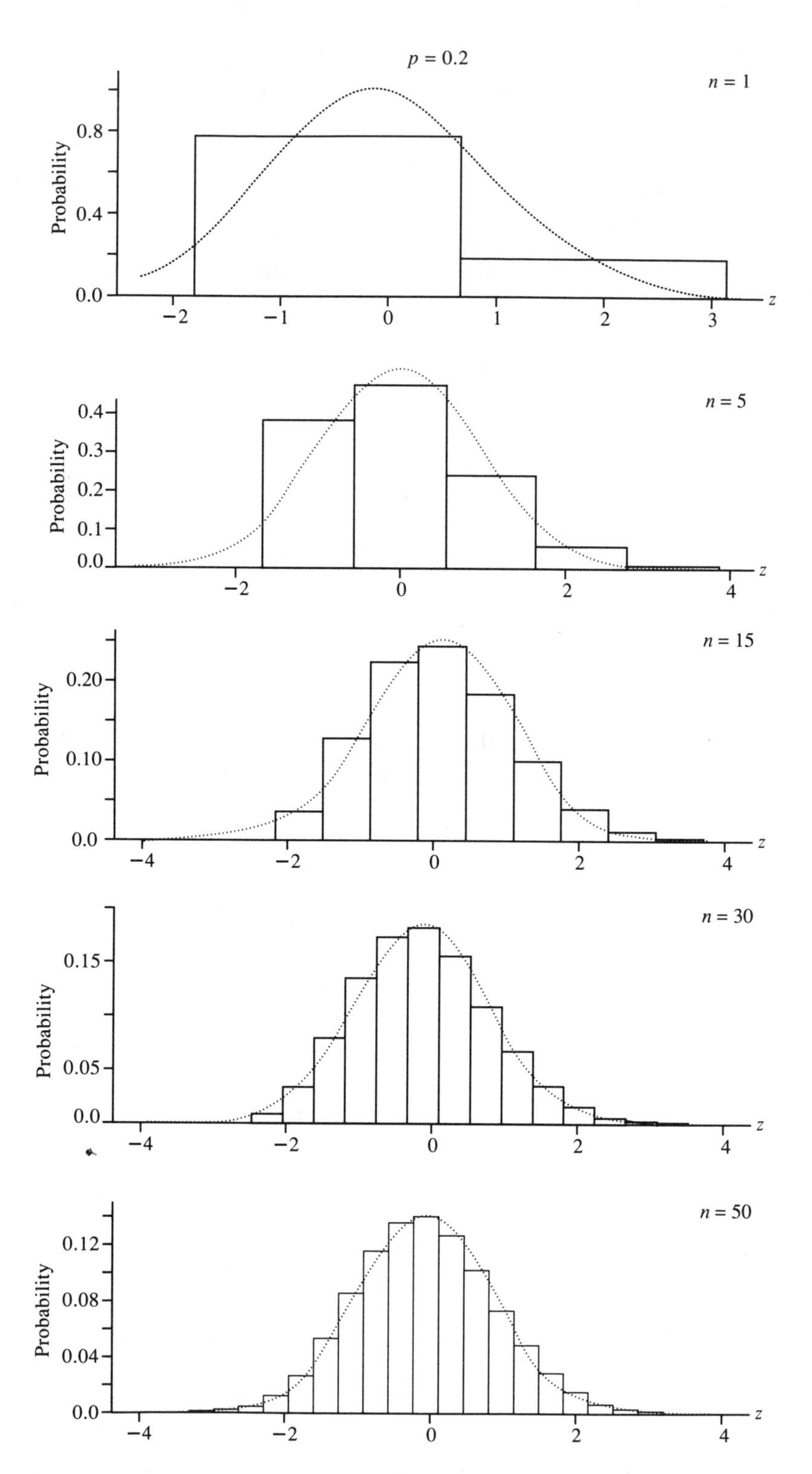

EXHIBIT 5.4 *Histograms for binomial distributions as approximated by the standard normal density.*

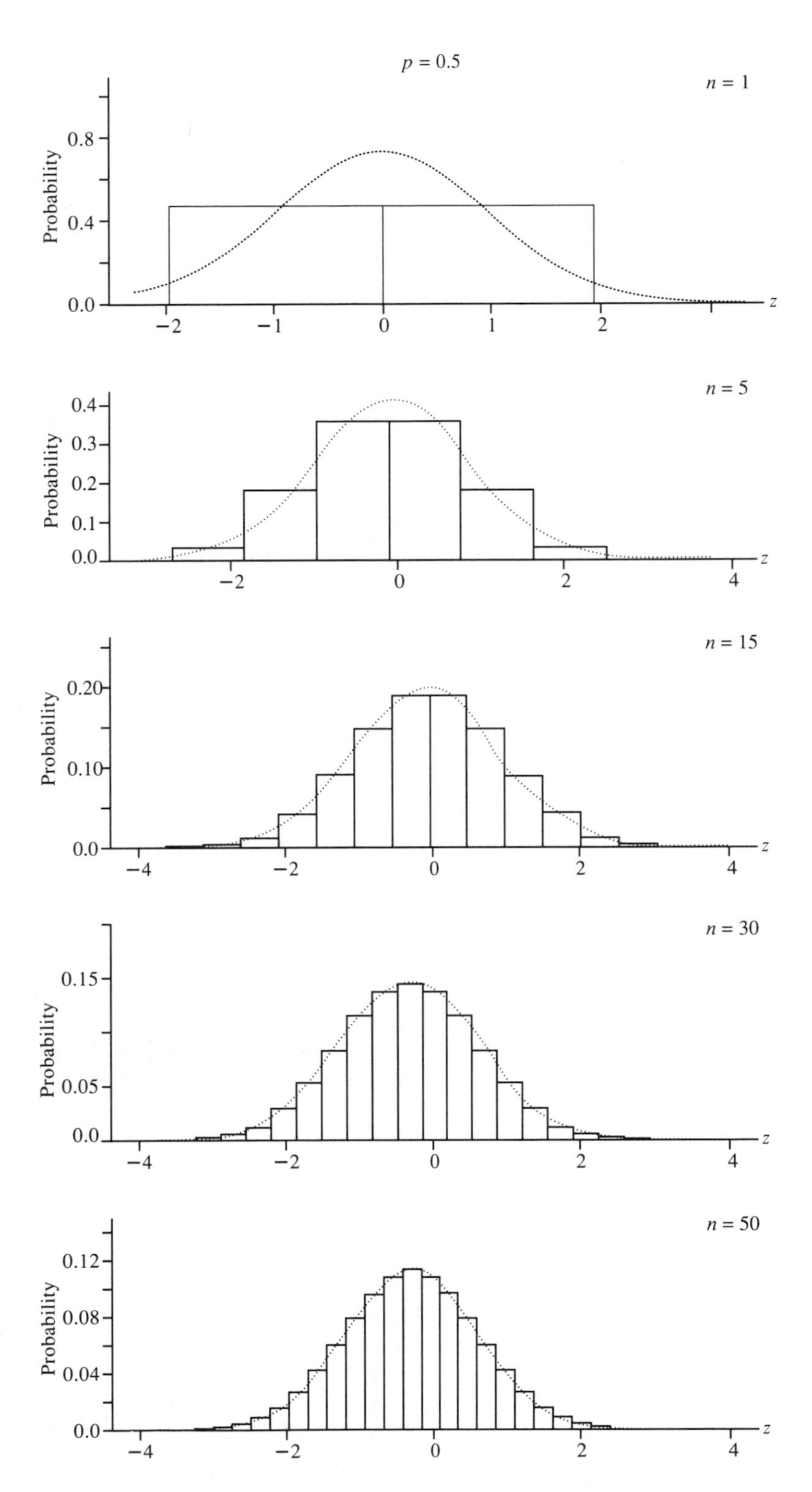

EXHIBIT 5.4 *(continued)*

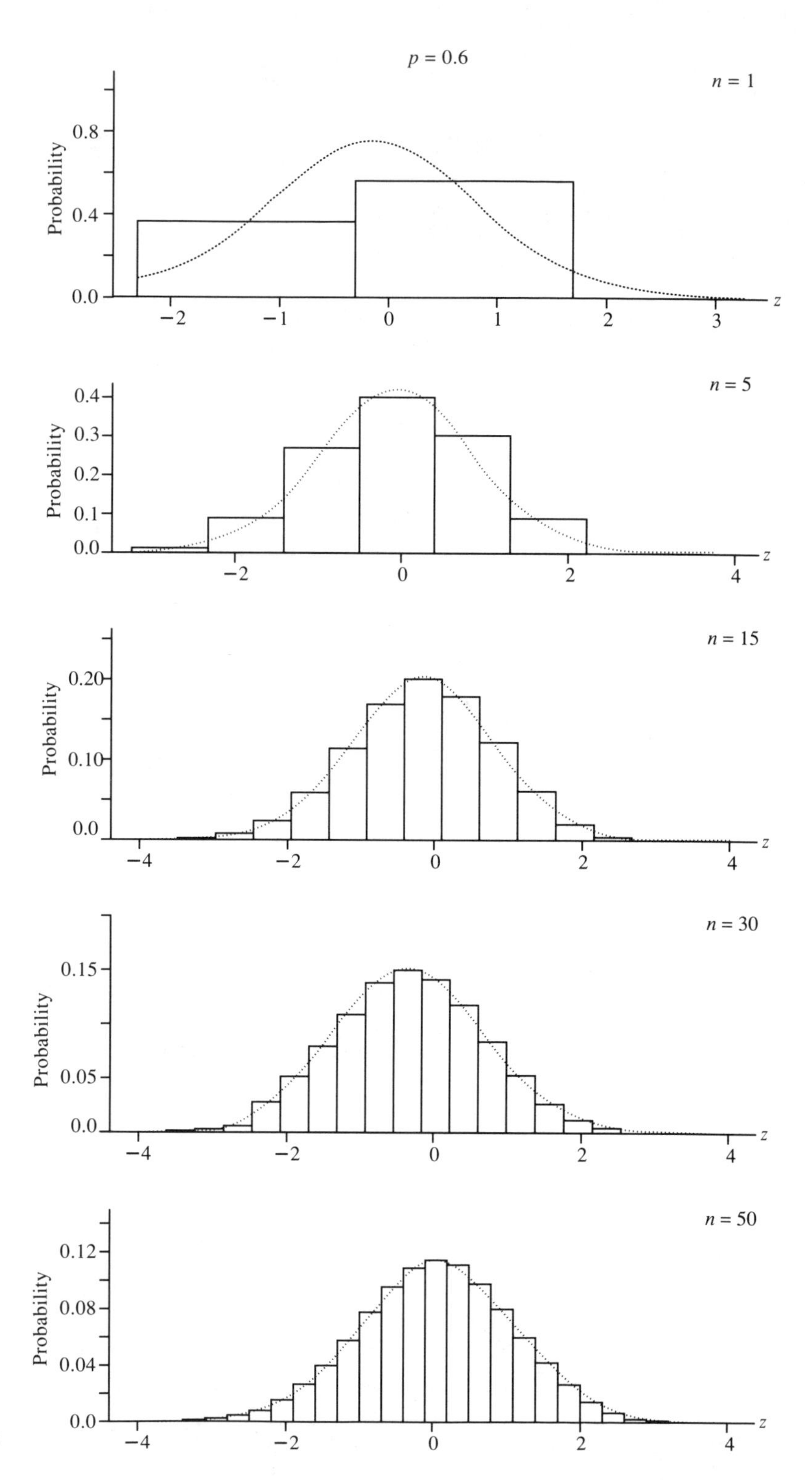

EXHIBIT 5.4 *(continued)*

case, with normality already being a good approximation to the histogram for X in the case $p = 0.5$ when $n = 15$. As seen from the figure for $n = 1$, the case $p = 0.5$ is the most symmetric one. As we might predict from our earlier discussion, convergence to normality as n increases is slowest for the most highly skewed case ($p = 0.2$).

Years of experience with the normal distribution approximation to the binomial distribution has yielded a very simple rule for when this approximation is sufficiently accurate. If the continuity correction is used, the normal approximation will give good approximations when

$$\sigma_X^2 = np(1 - p) \geq 3. \tag{5.11}$$

Note that this rule accounts for the role of asymmetry of the underlying Bernoulli distribution in the approximation. In the symmetric case $p = 0.5$, the quantity $p(1 - p)$ achieves its largest value, and the inequality (5.11) holds for $n \geq 12$. As p moves away from 0.5 (greater asymmetry), $p(1 - p)$ becomes smaller. Thus, in the most asymmetric case, $p = 0.2$, in Exhibit 5.4, $p(1 - p) = 0.16$ and we require that $n \geq 19$. For $n \geq 30$, (5.11) will be satisfied except for very small p or very large p.

Remark. If the inequality (5.11) fails to hold, it is usually better to use the Poisson distribution to approximate the binomial distribution.

If the inequality (5.11) for n is satisfied, we can use the normal distribution with mean np and variance $np(1 - p)$ to approximate the binomial distribution. We accomplish this approximation in two steps:

1. Use the continuity correction to replace the event whose probability is desired by one more suited to finding probabilities by area. Thus $\{X = x\}$ is replaced by $\{x - \frac{1}{2} \leq X \leq x + \frac{1}{2}\}$, $\{X \leq x\}$ is replaced by $\{X \leq x + \frac{1}{2}\}$, and so on. Remember that if an integer x is included in the region whose probability you want, so must be all values in the interval $[x - \frac{1}{2}, x + \frac{1}{2}]$. If you don't want x, no value in $[x - \frac{1}{2}, x + \frac{1}{2}]$ should be included.
2. Now pretend that $X \sim \mathcal{N}(np, np(1 - p))$, and standardize X so that Exhibit B.4 of Appendix B can be used to find the desired probability.

EXAMPLE 5.8

Suppose that X has a binomial distribution with parameters $n = 100$, $p = 0.2$. Find $P\{14 \leq X \leq 26\}$.

Here $\sigma_X^2 = 100(0.2)(0.8) = 16 \geq 3$, so that the approximation of the binomial distribution by the normal distribution can be used. Further, $\mu_X = 100(0.2) = 20$. Because

$$P\{14 \leq X \leq 25\} = \sum_{x=14}^{25} P\{X = x\},$$

we want to sum the probabilities of $x = 14, 15, \ldots, 25$. Thus, in our normal approximation, we need the probability of the interval $[14 - \frac{1}{2}, 25 + \frac{1}{2}]$ for the $\mathcal{N}(20, 16)$ distribution:

$$\begin{aligned} P\{14 \leq X \leq 25\} &\cong F_Z\left(\frac{25 + \frac{1}{2} - 20}{\sqrt{16}}\right) - F_Z\left(\frac{14 - \frac{1}{2} - 20}{\sqrt{16}}\right) \\ &= F_Z(1.375) - F_Z(-1.625) \\ &= 0.91543 - (1 - 0.94792) = 0.86335, \end{aligned}$$

using linear interpolation in Exhibit B.4 of Appendix B. If, instead, we had been asked for $P\{14 < X \leq 25\}$, we would not want $P\{X = 14\}$ in our sum, and thus would delete the interval $[14 - \frac{1}{2}, 14 + \frac{1}{2}]$ from the interval whose probability we compute using the normal approximation. That is,

$$\begin{aligned} P\{14 < X \leq 25\} &\cong F_Z\left(\frac{25 + \frac{1}{2} - 20}{\sqrt{16}}\right) - F_Z\left(\frac{14 + \frac{1}{2} - 20}{\sqrt{16}}\right) \\ &= F_Z(1.375) - F_Z(-1.375) \\ &= 2F_Z(1.375) - 1 = 0.83086. \end{aligned}$$

◆

EXAMPLE 5.9

A genetic theory asserts that 25% of the offspring of a crossbreeding experiment will have the characteristic C. A total of $n = 50$ offspring are observed and only six are found to have characteristic C. If the genetic theory is true, the number X of offspring observed having characteristic C has a binomial distribution with parameters $n = 50$ and $p = 0.75$. Note that $\mu_X = 50(0.25) = 17.5$ and $\sigma_X^2 = np(1 - p) = 50(0.25)(0.75) = 9.375 \geq 3$. Using the normal approximation to the binomial distribution with continuity correction, the probability of observing six or fewer offspring with characteristic C is approximately

$$\begin{aligned} P\{X \leq 6\} &\cong F_Z\left(\frac{6 + \frac{1}{2} - 12.5}{\sqrt{9.375}}\right) \\ &= F_Z(-1.96) = 1 - F_Z(1.96) = 0.02500. \end{aligned}$$

Thus we have observed an outcome belonging to an event that is very unlikely if the genetic theory is correct.

To approximate the probability of observing *exactly* 10 offspring having the characteristic C, note that

$$\begin{aligned} P\{X = 10\} &\cong F_Z\left(\frac{10 + \frac{1}{2} - (50)(0.75)}{\sqrt{(50)(0.25)(0.75)}}\right) - F_Z\left(\frac{10 - \frac{1}{2} - (50)(0.25)}{\sqrt{(50)(0.25)(0.75)}}\right) \\ &= F_Z(-0.65) - F_Z(-0.98) = 0.25785 - 0.16354 \\ &= 0.09431. \end{aligned}$$

◆

A Justification of the Normal Distribution by the Central Limit Theorem

The Central Limit Theorem applied to sums of independent random variables helps to explain why so many quantitative phenomena have (at least approximately) a normal distribution. For example, physical scientists of the nineteenth century thought of errors of measurement as resulting from the sum of small independent adjustments, due to a variety of causes, that occurred during the act of measurement. Actually, they originally thought of these adjustments are being independent Bernoulli variables, with possible values a and $-a$ rather than 1 and 0 ($a > 0$ a small number). DeMoivre's demonstration that the distribution of sums of such variables could be approximated by a normal distribution then justified use of the normal distribution as a law of errors. Laplace's Central Limit Theorem then showed that the restriction to Bernoulli adjustments was unnecessary, but still required the adjustments to have identical distributions. Even this requirement can be weakened, so that there are now Central limit Theorems for sums of nonidentically distributed and even nonindependent (dependent) random variables. These results do require restrictions, however; no one variable or group of variables in the sum can have dramatically higher variability (or magnitude) than the remaining variables, and only weak dependence is permitted between most pairs of variables in the sums. Nevertheless, these results imply that if a random variable X can be conceived of as resulting from a sum of a large number of (nearly) independent adjustments (causes, growths) of comparable magnitudes, then X can be expected to have a distribution of approximately normal (bell-shaped) form.

Such a conceptualization clearly applies to many quantitative phenomena observed in the physical, biological, behavioral, and social sciences. For example, the size X of crystals can be thought of as resulting from accumulations of molecules, the height X of a human skeleton arises from a large number of growth spurts during the growing years, and the score X of a person on an ability test is the sum of that person's scores on the individual items of the test. Although such conceptualizations are strongly suggestive, ultimately the most convincing proof that X is normally distributed rests on empirical verification—fitting normal distributions to observed data on X. Quetelet's concept of the average man as an idealization of the mean of a normal distribution took hold of the imagination of behavioral and biological scientists partly because he was able to demonstrate the fit of the normal distribution to observed data on human populations (Stigler, 1986). We have already noted how biologists and behavioral scientists, such as Galton and Ebbinghaus, followed Quetelet's example, thereby building up a rich storehouse of information on the nature of the variability of biological and behavioral measurements. Modern statisticians, using more sophisticated methods of fitting, have demonstrated that the accuracy of these fits to normal distributions may have been overstated. In particular, the law of errors does not adequately describe the errors of measurement made in some classic nineteenth- and twentieth-century physical experiments (Stigler, 1977), and such anthropometric measures as systolic blood pressure and serum cholesterol concentration are perhaps better fit by other continuous distributions (Chapter 10). Even so, it is remarkable how widely the normal distribution applies as an approximate model for the variation of quantitative phenomena.

6. The Moment Generating Function and a Proof of the Central Limit Theorem★

In this section the moment generating function of the normal distribution is obtained, and it is shown that sums and averages of independent normally distributed random variables are themselves normally distributed. The section ends with a proof of the Central Limit Theorem.

The Moment Generating Function of the Normal Distribution

Suppose that $X \sim \mathcal{N}(\mu, \sigma^2)$, so that (2.1) gives the probability density function of X. Thus the moment generating function of X is, for any real number t, given by

$$
\begin{aligned}
M(t) = E(e^{tX}) &= \int_{-\infty}^{\infty} e^{tx} \frac{e^{-(1/2)(x-\mu)^2/\sigma^2}}{\sqrt{2\pi\sigma^2}} \\
&= \int_{-\infty}^{\infty} \frac{\exp\{tx - \frac{1}{2}(x - \mu)^2/\sigma^2\}}{\sqrt{2\pi\sigma^2}} \\
&= e^{\mu t + (1/2)\sigma^2 t^2} \int_{-\infty}^{\infty} \frac{e^{-(1/2)Q(x)}}{\sqrt{2\pi\sigma^2}}\, dx,
\end{aligned}
\tag{6.1}
$$

where

$$
Q(x) = \frac{[x - (\sigma^2 t + \mu)]^2}{\sigma^2}.
$$

We recognize $(2\pi\sigma^2)^{-1/2} \exp[-\frac{1}{2}Q(x)]$ as being the density function of a normal distribution with mean $(\sigma^2 t + \mu)$ and variance σ^2. Thus the integral of this density over $-\infty < x < \infty$ equals 1. It follows from (6.1) that

$$
M(t) = \exp(\mu t + \tfrac{1}{2}\sigma^2 t^2). \tag{6.2}
$$

Recall from Chapter 7, Section 3, that if $X_1, X_2, \ldots, X_n$ are mutually independent random variables with respective moment generating functions

$M_{X_1}(t), M_{X_2}(t), \ldots, M_{X_n}(t)$, the moment generating function of any weighted sum $Y = \sum_{i=1}^n a_i X_i$ of these variables is given by the product

$$M_Y(t) = M_{X_1}(a_1 t) M_{X_2}(a_2 t) \cdots M_{X_n}(a_n t). \tag{6.3}$$

If X_i has a normal distribution with mean μ_i and variance σ_i^2, $i = 1, 2, \ldots, n$, then for any real number t,

$$\begin{aligned} M_Y(t) &= \prod_{i=1}^n \exp[\mu_i(a_i t) + \tfrac{1}{2}\sigma_i^2(a_i t)^2] \\ &= \exp\left[\left(\sum_{i=1}^n a_i\mu_i\right)t + \tfrac{1}{2}\left(\sum_{i=1}^n a_i^2\sigma_i^2\right)t^2\right], \end{aligned}$$

which is the moment generating function of a normal distribution with mean $\mu_Y = \sum_{i=1}^n a_i\mu_i$ and variance $\sigma_Y^2 = \sum_{i=1}^n a_i^2\sigma_i^2$. By the uniqueness property of moment generating functions, it follows that $Y \sim \mathcal{N}(\sum_{i=1}^n a_i\mu_i, \sum_{i=1}^n a_i^2\sigma_i^2)$. We have thus shown that

> Any weighted sum $\sum_{i=1}^n a_i X_i$ of independent normally distributed random variables $X_1, \ldots, X_n$ is normally distributed:
>
> $$\sum_{i=1}^n a_i X_i \sim \mathcal{N}(\sum_{i=1}^n a_i\mu_i, \sum_{i=1}^n a_i^2\sigma_i^2).$$

Note that the sum $S_n = \sum_{i=1}^n X_i$ and average $\overline{X} = \frac{1}{n}\sum_{i=1}^n X_i$ of $X_1, X_2, \ldots, X_n$ are both weighted sums. Consequently, if $X_1, X_2, \ldots, X_n$ are mutually independent normally distributed random variables, then both S_n and $\overline{X}$ are normally distributed:

$$S_n \sim \mathcal{N}\left(\sum_{i=1}^n \mu_i, \sum_{i=1}^n \sigma_i^2\right), \qquad \overline{X} \sim \mathcal{N}\left(\frac{1}{n}\sum_{i=1}^n \mu_i, \frac{1}{n^2}\sum_{i=1}^n \sigma_i^2\right). \tag{6.4}$$

If $X_1, X_2, \ldots, X_n$ are mutually independent observations of a normally distributed random variable X having mean μ and variance σ^2, then from (6.4)

$$S_n \sim \mathcal{N}(n\mu, n\sigma^2), \qquad \overline{X} \sim \mathcal{N}\left(\mu, \frac{1}{n}\sigma^2\right). \tag{6.5}$$

This verifies the assertions made in Section 5 concerning sums and averages of independent normally distributed variables.

Proof of the Central Limit Theorem

The Central Limit Theorem indirectly concerns the average $\overline{X} = 1/n \sum_{i=1}^{n} X_i$ of mutually independent observations of a random variable X having mean μ and (finite) variance σ^2. However, the assertion of the theorem refers specifically to the standardization

$$Z_n = \frac{\overline{X} - \mu_{\overline{X}}}{\sigma_{\overline{X}}} = \frac{\overline{X} - \mu}{\sigma/\sqrt{n}}$$

of $\overline{X}$. If $F_{Z_n}(z)$ is the cumulative distribution function of Z_n and $F_Z(z)$ is the cumulative distribution function of a standard normal variable Z, it is asserted that for all real numbers z,

$$\text{(6.6)} \qquad \lim_{n\to\infty} F_{Z_n}(z) = F_Z(z).$$

Note that

$$\text{(6.7)} \qquad Z_n = \frac{\overline{X} - \mu}{\sigma/\sqrt{n}} = \frac{n(\overline{X} - \mu)}{n(\sigma/\sqrt{n})} = \frac{S_n - n\mu}{\sqrt{n\sigma^2}}$$

is also the standardization of the sum $S_n = \sum_{i=1}^{n} X_i$. Thus the Central Limit Theorem also concerns the sum S_n.

We now prove (6.6) using the method of moment generating functions—in particular the preservation of limits property (Chapter 7, Section 3) of moment generating functions will be used. To do so, we will need to assume that the random variable X on which the observations $X_1, X_2, \ldots, X_n$ are taken has a moment generating function $M(t)$ defined on an open interval containing $t = 0$. Because not all random variables have moment generating functions, this is an extra assumption. We have adopted this assumption to keep our mathematical argument simple. Existence of a moment generating function implies that all central moments $E(X - \mu)^r$, $r = 2, 3, \ldots$, of X are well defined and finite, whereas the statement of the Central Limit Theorem required only that the second central moment, $\sigma^2 = E(X - \mu)^2$, be finite. Note that

$$\text{(6.8)} \qquad Z_n = \frac{S_n - n\mu}{\sqrt{n\sigma^2}} = \frac{\sum_{i=1}^{n}(X_i - \mu)}{\sqrt{n}(\sigma)} = \frac{1}{\sqrt{n}} \sum_{i=1}^{n} \left(\frac{X_i - \mu}{\sigma}\right) = \frac{1}{\sqrt{n}} \sum_{i=1}^{n} Y_i,$$

where $Y_i = (X_i - \mu)/\sigma$ is the standardization of X_i; $i = 1, 2, \ldots, n$. Let $Y = (X - \mu)/\sigma$ be the corresponding standardization of X. Thus $Y_1, Y_2, \ldots, Y_n$ are mutually independent (because $X_1, X_2, \ldots, X_n$ are independent) observations of Y. Note that Y has mean $\mu_Y = 0$ and variance $\sigma_Y^2 = 1$. By Property

3 of moment generating functions (Chapter 7, Section 3), the moment generating function of Y (and each of $Y_1, Y_2, \ldots, Y_n$) is

$$M_Y(t) = e^{-(\mu/\sigma)t} M\left(\frac{t}{\sigma}\right).$$

Thus the moment generating function of $Z_n = \sum_{i=1}^n Y_i/\sqrt{n}$ is

$$\begin{aligned} M_{Z_n}(t) &= M_{Y_1}\left(\frac{t}{\sqrt{n}}\right) M_{Y_2}\left(\frac{t}{\sqrt{n}}\right) \cdots M_{Y_n}\left(\frac{t}{\sqrt{n}}\right) \\ &= \left[M_Y\left(\frac{t}{\sqrt{n}}\right)\right]^n. \end{aligned}$$

Let $M_Y^{[k]}(t) = d^k M_Y(t)/(dt)^k$ denote the kth derivative of $M_Y(t)$. Because Y has mean 0 and variance 1, Property 2 of moment generating functions (Chapter 7) implies that

$$\begin{aligned} &M_Y^{[1]}(0) = E[Y] = 0, \qquad M_Y^{[2]}(0) = E[Y^2] = \sigma_Y^2 = 1, \\ &M^{[3]}(0) = E[Y^3] = E\left[\frac{X-\mu}{\sigma}\right]^3 = \frac{E[X-\mu]^3}{\sigma^3} < \infty, \end{aligned} \tag{6.10}$$

For large values of n, $t/\sqrt{n}$ is close to zero. This fact suggests expanding $M_Y(t/\sqrt{n})$ in a Taylor's series about $t/\sqrt{n} = 0$. Using (6.10), we obtain

$$\begin{aligned} M_Y\left(\frac{t}{\sqrt{n}}\right) &= M_Y(0) = M_Y^{[1]}(0)\frac{t}{\sqrt{n}} + M_Y^{[2]}(0)\frac{(t/\sqrt{n})^2}{2!} + R_n \\ &= 1 + (0)\frac{t}{\sqrt{n}} + (1)\frac{t^2}{2n} + R_n = 1 + \frac{t^2 + 2nR_n}{2n}, \end{aligned} \tag{6.11}$$

where

$$R_n = \frac{(\tilde{t})^3}{3!\,(\sqrt{n})^3} M_Y^{[3]}\left(\frac{\tilde{t}}{\sqrt{n}}\right) = \frac{(\tilde{t})^3}{6n^{3/2}} M_Y^{[3]}\left(\frac{\tilde{t}}{\sqrt{n}}\right)$$

and $\tilde{t}$ is a number between 0 and t. Because $M_Y^{[3]}(t)$ is diffentiable with respect to t at $t = 0$, and thus is continuous in t at $t = 0$, it can be shown that

$$\lim_{n\to\infty} M_Y^{[3]}\left(\frac{\tilde{t}}{\sqrt{n}}\right) = M_Y^{[3]}(0) = E[Y^3] < \infty,$$

and thus that

$$\lim_{n\to\infty} nR_n = \lim_{n\to\infty} \frac{n\tilde{t}^3}{6n^{3/2}} M^{[3]}\left(\frac{\tilde{t}}{\sqrt{n}}\right) = 0. \tag{6.12}$$

It follows from (6.9), (6.11), and (6.12) that

$$\begin{aligned}\lim_{n\to\infty} M_{Z_n}(t) &= \lim_{n\to\infty}\left[M_Y\left(\frac{t}{\sqrt{n}}\right)\right]^n \\ &= \lim_{n\to\infty}\left(1+\frac{t^2+nR_n}{2n}\right)^n = e^{(1/2)t^2} = M_Z(t),\end{aligned}$$

where $Z \sim \mathcal{N}(0, 1)$ and t is any number in the interval about $t = 0$ for which $M_X(t)$ is defined. By the preservation of limits property of moment generating functions, it follows that (6.6) is true. This completes the proof of the Central Limit Theorem.

A proof of the Central Limit Theorem without our extra assumption that $M_X(t)$ exists can be given using steps similar to those used above, but with characteristic functions (Chapter 7) used in place of moment generating functions. Such a proof still requires an extra assumption—namely that the third central moment, $E(X - \mu)^3$, is well defined and finite. Removing this last condition, so that only existence and finiteness of the variance σ^2 of X is required, requires more difficult analytic arguments.

Glossary and Summary

Normal distribution: A continuous distribution whose density has the characteristic bell-shaped curve corresponding to the graph of the function $f(x) = (2\pi\sigma^2)^{-1/2} \exp[-(x - \mu)^2/2\sigma^2]$.

Gaussian distribution: Another name for the normal distribution.

Central Limit Theorem: The result which asserts that sums or averages of a sufficiently large number of independent, identically distributed random variables have distributions that are approximately normal.

Standardization: The transformation of a random variable X with mean μ and variance σ^2 to a new variable $Z = (X - \mu)/\sigma$ having mean 0 and variance 1. Regardless of what units of measurement are used for X, the variable Z is dimensionless, thus allowing comparisons between measurements in different scales or between observations from different populations. The value of Z expresses the deviation of X from its mean as a multiple of the standard deviation of X.

Standard normal distribution: A normal distribution with mean 0 and variance 1 that results from standardizing a normally distributed random variable.

Normal ogive: The S-shaped graph of the cumulative distribution function of a normal distribution.

Continuity correction: A correction made when approximating the probabilities of a discrete distribution by the probabilities of a continuous distribution. The probability of any possible value x of the discrete variable is found by the area under the curve of the density of the continuous distribution over an interval whose midpoint is x and whose endpoints are midway between x and the two possible values of the discrete variable that are adjacent (on either side) to x.

Facts About the Normal Distribution

1. $X \sim \mathcal{N}(\mu, \sigma^2)$ means that the random variable X has a normal distribution with mean μ, variance σ^2, and density

$$f(x) = \frac{1}{\sqrt{2\pi\sigma^2}} \exp\left[-\frac{(x-\mu)^2}{2\sigma^2}\right], \qquad -\infty < x < \infty.$$

In this case the median and mode of X are equal to the mean μ. Also,

$$E[X-\mu]^r = \begin{cases} 0, & r = 1, 3, 5, \ldots \\ \dfrac{\sigma^r 2^{r/2}\Gamma(\frac{1}{2}(r+1))}{\sqrt{\pi}}, & r = 2, 4, 6, \ldots, \end{cases}$$

so that the skewness γ_1 of the normal distribution is 0 and the kurtosis γ_2 is 3.

2. If $X \sim \mathcal{N}(\mu, \sigma^2)$, then $Z = (X - \mu)/\sigma \sim \mathcal{N}(0, 1)$, and

$$P\{X \le x\} = F_X(x) = F_Z\left(\frac{x-\mu}{\sigma}\right), \qquad Q_X(p) = \mu + \sigma Q_Z(p).$$

3. If $X \sim \mathcal{N}(\mu, \sigma^2)$, the moment generating function of X is

$$M_X(t) = \exp(t\mu + \tfrac{1}{2}t^2\sigma^2), \qquad -\infty < t < \infty.$$

4. For the standard normal distribution, $Z \sim \mathcal{N}(0, 1)$: $f_Z(z) = f_Z(-z)$, $F_Z(-z) = 1 - F_Z(z)$, $Q_Z(p) = -Q_Z(1-p)$, $0 \le p \le 1$, where $f_Z(z)$ is the density function and $F_Z(z)$ is the cumulative distribution function of Z.
5. If $X_1, X_2, \ldots, X_n$ are independent random variables, with $X_i \sim \mathcal{N}(\mu_i, \sigma_i^2)$, $i = 1, 2, \ldots, n$, and if $a_1, a_2, \ldots, a_n$ are constants, then

$$\sum_{i=1}^{n} a_i X_i \sim \mathcal{N}\left(\sum_{i=1}^{n} a_i\mu_i, \sum_{i=1}^{n} a_i^2\sigma_i^2\right).$$

Thus if $X_1, X_2, \ldots, X_n$, are independent $\mathcal{N}(\mu, \sigma^2)$ random variables,

$$S_n = \sum_{i=1}^{n} X_i \sim \mathcal{N}(n\mu, n\sigma^2),$$

$$\overline{X} = \frac{1}{n}\sum_{i=1}^{n} X_i \sim \mathcal{N}\left(\mu, \frac{\sigma^2}{n}\right).$$

6. **Central Limit Theorem.** If $X_1, X_2, \ldots$ is a sequence of independent identically distributed random variables with common mean μ and common variance $\sigma^2 < \infty$, and

$$\overline{X}_n = \frac{1}{n}\sum_{i=1}^{n} X_i, \qquad S_n = \sum_{i=1}^{n} X_i, \qquad Z_n = \frac{\overline{X} - \mu}{\sigma/\sqrt{n}} = \frac{S_n - n\mu}{(\sqrt{n})\sigma}$$

then

$$\lim_{n\to\infty} F_{Z_n}(z) = F_Z(z)$$

for all real numbers z, where $F_{Z_n}(z)$ is the cumulative distribution of Z_n and $F_Z(z)$ is the cumulative distribution function of $Z \sim \mathcal{N}(0, 1)$. Further, if $z_1, z_2, \ldots$ is any sequence of numbers such that $\lim_{n\to\infty} z_n = z$, then

$$\lim_{n\to\infty} F_{Z_n}(z_n) = F_Z(z)$$

Informally, for large enough n, $\overline{X}_n$ has approximately a $\mathcal{N}(\mu, \sigma^2/n)$ distribution and S_n has approximately a $\mathcal{N}(n\mu, n\sigma^2)$ distribution.

The smallest sample size n necessary for $\overline{X}_n$ or S_n to be approximately normally distributed depends on the shape of the common distribution of $X_1, X_2, \ldots$. Loosely speaking, the more this shape resembles the shape of the normal distribution (symmetric, single peak, bell-shaped), the smaller is the needed sample size. For distributions that are not extremely skewed and have a single peak, a sample size of $n \geq 30$ usually suffices.

7. Probabilities for the binomial distribution can be approximated, using a continuity correction, by probabilities of the normal distribution with mean $\mu = np$ and variance $\sigma^2 = np(1 - p)$ provided that $np(1 - p) \geq 3$.

Exercises

1. If $Z \sim \mathcal{N}(0, 1)$, $A = \{-0.3 \le Z \le 0.7\}$, and $B = \{0.1 \le Z \le 3.0\}$, find:
   (a) $P(A)$. (b) $P(B)$. (c) $P(A \cup B)$.
   (d) $P(A \cap B)$. (e) $P(A \mid B)$. (f) $P(B \mid A)$.
2. If $X \sim \mathcal{N}(3, 4)$, find:
   (a) $P\{X > 5\}$. (b) $P\{X \ge 5\}$.
   (c) $P\{X \le 5\}$. (d) $P\{3 < X \le 5\}$.
   (e) $P\{2 \le X \le 5\}$. (f) $P\{2 < X \le 5\}$.
   (g) $P\{X < 2 \text{ or } X \ge 5\}$. (h) $P\{X \le 3 \text{ or } X > 5\}$.
   (i) $P\{X > 3 \text{ or } X \le 5\}$.
3. Two commonly used scales yield values $X \sim \mathcal{N}(100, 324)$ and $Y \sim \mathcal{N}(50,100)$. If both scales measure the same phenomenon, what Y-score is equivalent to an X-score of 127?
4. If $X \sim \mathcal{N}(100, 225)$, find (a) the 0.25th quantile $Q_X(0.25)$; (b) the 0.95th quantile $Q_X(0.95)$; (c) the median, Med (X); (d) the interval $[a, b]$ of length $b - a = 10$ having the highest probability $P\{a \le X \le b\}$.
5. The scores X of people on a personality test are normally distributed, but the mean μ and variance σ^2 of the scores on the test are supposedly confidential. If we know that $Q_X(0.90) = 87$ and $Q_X(0.95) = 96$, find μ and σ^2.
6. Men's shirts are classified according to size as S, M, L, XL, corresponding, respectively, to neck circumferences of under 15 inches (for S), between 15 and 16 inches (for M), between 16 and 17 inches (for L), and over 17 inches (for XL). Suppose that neck circumferences X for adult males have a $\mathcal{N}(15.75, 0.49)$ distribution.
   (a) How many shirts should be manufactured in each category of shirt size if 1000 shirts are to be manufactured?
   (b) If you wanted to define categories S, M, L, XL so that each category contained 25% of the total population of adult males, what neck sizes would you assign to each of these categories?
7. The height X in inches of a randomly chosen male is a random variable having a normal distribution with mean $\mu_X = 70$ and variance $\sigma_X^2 = 4$.
   (a) What is the probability that a randomly chosen male will either be smaller than 68 inches *or* taller than 75 inches?
   (b) What is the 75th percentile $Q_X(0.75)$ of the distribution of X?
   (c) If we take a random sample of 10 males *with* replacement, what is the probability that exactly 5 of these males will be 70 inches tall or taller?
8. The concentration X of a pesticide in uncontaminated cows has a normal distribution with mean $\mu = 1$ and variance $\sigma^2 = 0.25$. The U.S. Department of Agriculture has a meat inspection program. If the ob-

served concentration X of pesticide exceeds a certain critical constant c, meat is declared unsafe.

(a) If $c = 1.3$, find $P\{X > 1.3\}$.

(b) Find the critical constant c so that if a cow is uncontaminated, then $P\{X > c\} = 0.01$.

9. The maximum height X (in feet) that the tide reaches at a given coastal town on a random day is a random variable having a normal distribution with mean $\mu = 17$ and variance $\sigma^2 = 25$. The town has a dike of height $H = 30$ feet to prevent flooding.

(a) Find the probability $P\{X > 30\}$ of a flood.

(b) Assume that the maximum tide heights $X_1, X_2, \ldots$ on days 1, 2, . . . following a flood are statistically independent of one another, and all have the same normal $\mathcal{N}(17, 25)$ distribution. Let Y be the number of days until the next flood. Name the distribution of Y and give the values of its mean and variance. (*Reminder*: A flood occurs on day i if $X_i > 30$.)

10. Simple flashlights generally contain two or three batteries. Suppose that the life of each battery is normally distributed with a mean of 30 hours and a variance of 25 (hours)^2. The flashlight will cease to function if one or more of its batteries fails. Assuming that the lifetimes of batteries in a flashlight are probabilistically independent, find the probability that a flashlight will operate for at least 40 hours if the flashlight has (a) 2 batteries; (b) 3 batteries. If you take a 2-battery flashlight and a 3-battery flashlight on a trip, what is the probability that at least one of these flashlights will last throughout a trip of 40 hours?

11. A genetic theory asserts that 40% of the offspring of a crossbreeding experiment will have the characteristic C. Suppose that 24 offspring of such a crossbreeding are observed. Let X be the number of observed offspring having the characteristic C.

(a) Name the exact distribution of X and give the value(s) of the parameter(s) of this distribution.

(b) Use an approximation to the exact distribution of X to find the probability, $P\{X \geq 12\}$, that at least 12 of the observed offspring have the characteristic C.

(c) Suppose that we independently observe offspring of the crossbreeding experiment one at a time, continuing until we have found 12 offspring with the characteristic C. What is the expected (mean) number of offspring that we will observe?

12. Test scores X on a reading comprehension test are known to have a normal distribution with mean $\mu_X = 70$ and variance $\sigma_X^2 = 16$.

(a) What proportion of all people who take this test have scores between 65 and 75? That is, find $P\{65 \leq X \leq 75\}$.

(b) Find $Q_X(0.90)$.

(c) Suppose that 9 persons independently take this reading comprehension test. What is the probability that no more than 3 of these obtain scores greater than the 90th percentile, $Q_X(0.90)$?

13. The serum cholesterol level X in 14-year-old boys has approximately a normal distribution with mean $\mu = 170$ and variance $\sigma^2 = 900$.

(a) Find the probability, $P\{160 \leq X \leq 200\}$, that a randomly chosen boy has serum cholesterol level between 160 and 200.

(b) Given that a boy's serum cholesterol level is known to exceed 170, what is the conditional probability, $P\{X \leq 200 \mid X > 170\}$, that his serum cholesterol level is no greater than 200?

(c) Unusually high serum cholesterol levels are a cause for concern. Find the level c such that only 0.5% of the population of 14-year-old boys have serum cholesterol levels higher than c. That is, find c to satisfy $P\{X > c\} = 0.005$.

14. An actuary fits the Poisson distribution to accident data, using $n = 900$ observations $X_1, X_2, \ldots X_{900}$, where each X_i is an independent observation of the number X of accidents in a given time period. The parameter λ of the Poisson distribution is estimated by $\overline{X}$, and we want our estimate of λ not to be in error by more than ± 0.1. Using the Central Limit Theorem, find the probability $P\{-0.1 \leq \overline{X} - \lambda \leq 0.1\}$ that the desired accuracy is achieved if the value of λ is actually equal to 9.

15. Let X be the number of children of elementary school age in a randomly chosen family. Suppose that the probability model for X is given by

X	0	1	2	3
$P\{X = x\}$	0.35	0.45	0.15	0.05

(a) Find μ_X and σ_X^2.

(b) A school district expects 200 families to move into this district. This means that additional classroom space will be needed for $Y = \sum_{i=1}^{200} X_i$ new children. The district believes that they have room for 150 extra children. Use the Central Limit Theorem with continuity correction to find the probability $P\{Y > 150\}$ that there will be insufficient room to accommodate all of the new children.

16. The probability of obtaining a male infant in a random birth is $p = 0.51$. Use the Central Limit Theorem, both with and without the continuity correction, to answer the following.

(a) In $n = 100$ probabilistically independent births, what is the approximate probability of obtaining 50 or fewer male births?

(b) In $n = 400$ births, what is the probability of obtaining 200 or fewer male births?

(c) In $n = 1000$ births, what is the probability of obtaining 500 or fewer male births?

Reach conclusions about (i) the importance of the continuity correction as n increases, and (ii) the probability of 50% or fewer male births as n increases.

17. Let Y have a binomial distribution with $p = 0.80$ and $n = 100$. Use the Central Limit Theorem with the continuity correction to find:
(a) $P\{Y \leq 80\}$. (b) $P\{60 < Y \leq 90\}$.
(c) $P\{70 \leq Y\}$. (d) $P\{Y = 80\}$.

18. Students have a choice of final exams:
(i) A 100-question true–false test (passing grade: 55 correct or higher).
(ii) A 100-question multiple-choice test with 5 possible choices on each question (passing grade: 24 correct or higher).
Use the Central Limit Theorem to determine which test the student has the greatest probability of passing if the student must guess the answers. (Do you need to make a continuity correction?)

19. A fleet owner has 100 automobiles in need of maintenance. If the cost for a single vehicle has the $\mathcal{N}(50,225)$ distribution and the costs for distinct cars are independent, what is the probability that the total cost for all 100 cars is no more than $5500?

20. A cable TV franchise offers a month's free viewing as an incentive to customers to subscribe to HBO. The franchise earns $60 for each subscription and estimates that 60% will subscribe. If 500 offers are made, use the normal approximation to estimate the probability that the revenue from these offers is greater than $18,000.

21. A video game arcade estimates the number of quarters taken in each week by weighing them. The number X of quarters per pound has mean $\mu = 80$, and the standard deviation of X is 0.25. In a given week the arcade took in 900 pounds of quarters.
(a) What is the expected total number of quarters?
(b) Find A so that the probability that there are no fewer than A quarters in 900 pounds is at most 0.001. That is, find A so that $P\{\Sigma_1^{900} X_i \geq A\} \leq 0.001$.

22. Let $X_1, X_2, \ldots, X_{100}$ be independent, normally distributed random variables with mean 6 and variance 100. [That is, each $X_i \sim \mathcal{N}(6,100)$.]
(a) Find the probability that $X_1 > 9$.
(b) Find the probability that the average, $\overline{X} = \frac{1}{100} \Sigma_{i=1}^{100} X_i$, of the X_i's exceeds 9. That is, find $P\{\overline{X} > 9\}$.
(c) Find $E(\Sigma_{i=1}^{100} X_i^2)$, the mean of the sum of the squares of the X_i's.

23. The Internal Revenue Service uses a letter-opening machine to remove checks from envelopes. This machine has a probability $p = 0.03$ of failure for any envelope. On a certain day, $n = 500$ envelopes are independently treated by the machine. Let Y be the number of envelopes for which the machine fails to remove a check.
(a) What is the exact distribution of Y?
(b) Use the Poisson distribution to approximate $P\{Y \leq 5\}$.

(c) Use the Central Limit Theorem with continuity correction to approximate $P\{Y \leq 5\}$.

24. A shipment of 30,000 cans of peaches contains 300 damaged cans. A random sample of $n = 100$ of these cans is randomly selected *without* replacement for a quality control inspection. Let Y be the number of damaged cans observed in the sample.

(a) What is the exact distribution of Y?

(b) If the cans had been sampled *with* replacement, what would be the exact distribution of Y?

(c) Use the Central Limit Theorem with continuity correction to approximate $P\{Y \geq 3\}$.

(d) Use the Poisson distribution to approximate $P\{Y \geq 3\}$.

25. The College Entrance Examination Board Advanced Placement Examination in English was administered on May 18, 1964, to 11,329 secondary school students seeking advanced placement in college. The test consists of three parts: (i) analysis of a poem, (ii) literature, and (iii) composition. A detailed distributional analysis of 370 of the composition scores appears in Exhibit E.1. Fit a theoretical normal distribution with mean $\mu_X = 43$ and variance $\sigma_X^2 = 100$ to the data by calculating theoretical frequences for the intervals of scores given. (*Remark:* It would be currently more acceptable to use intervals of the form [7.5, 11.5], [11.5, 15.5], and so on, rather than the kinds of intervals used in Exhibit E.1.)

EXHIBIT E.1: *Distribution of Scores X on the Composition Portion of the College Entrance Examination Board Advanced Placement Examination in English*

Interval of Scores	*Observed Frequency*	*Interval of Scores*	*Observed Frequency*
68–71	2	32–35	34
64–67	6	28–31	28
60–63	13	24–27	17
56–59	21	20–23	3
52–55	35	16–19	1
48–51	41	12–15	1
44–47	58	8–11	1
40–43	63		370
36–39	49		

26. In data collected at the Lick Observatory on 80 bright stars in a celestial area, the radial velocity X is measured. The data [reported by Trumpler and Weaver (1953)] are shown in Exhibit E.2. Fit a theoretical normal distribution with mean $\mu_X = -21.0$ and variance $\sigma_X^2 = 289$ to these data. (See the remark after Exercise 25.)

EXHIBIT E.2: *Frequency Distribution of Radial Velocity X for Bright Stars*

Interval of Velocities	*Midpoint of Interval*	*Observed Frequency*
−80 to −70	−75	1
−70 to −60	−65	2
−60 to −50	−55	2
−50 to −40	−45	2
−40 to −30	−35	8
−30 to −20	−25	24
−20 to −10	−15	26
−10 to 0	−5	11
0 to 10	5	2
10 to 20	15	1
20 to 30	25	1
		80

Theoretical Exercises

T.1. Let $X \sim \mathcal{N}(\mu, \sigma^2)$. Show that $Y = aX + b \sim \mathcal{N}(a\mu + b, a^2\sigma^2)$ by finding the density function of $f_Y(y)$ of y. Show that if $Z \sim \mathcal{N}(0, 1)$, then $\sigma Z + \mu \sim \mathcal{N}(\mu, \sigma^2)$.

T.2. Let $X \sim \mathcal{N}(0, \sigma^2)$. If (a) $Y = |X|$, and (b) $Y = X^2$, find the cumulative distribution function of $F_Y(x)$ and then the probability density function of $f_Y(x)$ of Y in each case.

T.3. Suppose that X_1, X_2 are independent random variables, each having a normal distribution with mean $\mu \neq 0$ and variance σ^2. Among all linear combinations $a_1X_1 + a_2X_2$ of X_1, X_2, that have mean equal to μ, find the one that maximizes $P\{-\varepsilon \leq a_1X_1 + a_2X_2 - \mu \leq \varepsilon\}$ for each $\varepsilon > 0$. (*Hint:* Argue that the linear combination with the smallest variance maximizes the probability. Then find a_1, a_2 such that $a_1X_1 + a_2X_2$ has mean μ and smallest variance.)

T.4. If $X \sim \mathcal{N}(\mu, \sigma^2)$ and a is a positive constant, show that $g(t) = P\{t - a \leq X \leq t + a\}$ is decreasing in t for $t < \mu$ and increasing in t for $t > \mu$. Hence conclude that $g(t)$ is minimized when $t = \mu$ [see also Exercise 4(d)].

T.5. If $X \sim \mathcal{N}(0, \sigma^2)$, find the moment generating function of $Y = X^2$. Use this result to find $E[X^{2r}]$, $r = 1, 2, 3$.

T.6.★ Let $X \sim \mathcal{N}(\mu, \sigma^2)$. The central moments $E[X - \mu]^r$ for $r = 0, 1, \ldots$, were found in Section 2. Show that $E[X - \mu]^r$ does not exist for $r = -1, -2, \ldots$.

T.7. The hypergeometric distribution with parameters N = population size, n = sample size, p = probability of success is approximated by the binomial distribution when p is small relative to N. The binomial distribution with parameters n and p is approximated by the normal distribution (with continuity correction) when n is large enough that $np(1 - p) \geq 3$. Consequently, when both N and n are large, the hypergeometric distribution can be approximated by the normal distribution.

(a) Under what conditions on N, n, p would you suggest using the normal approximation to the hypergeometric distribution? What mean and variance should the approximating normal distribution have? Do you need a continuity correction?

(b) If X has the hypergeometric distribution with $N = 5000$, $n = 100$, $p = 0.2$, would the normal distribution approximation be a good one? Explain. If you think the normal approximation can be used, use this approximation to find $P\{16 < X \leq 25\}$.

T.8.★ Argue that the negative binomial distribution with parameters r and p can be approximated by the normal distribution when r is sufficiently large. (*Hint:* Use the relation between probabilities for negative binomial and binomial distributions given in Chapter 8, Section 5.) What requirements need to be placed on r and p? What mean and variance should be used for the approximating normal distribution? Is a continuity correction needed?

T.9.★ Let X have a Poisson distribution with parameter λ. Show that the moment generating function of $(X - \lambda)/\sqrt{\lambda}$ converges as $\lambda \to \infty$ to the moment generating function of a standard normal distribution. How would you use this result to approximate probabilities for a Poisson-distributed variable X when λ is large? Illustrate by approximating $P\{X \leq 25\}$ when $\lambda = 30$.

10

Special Distributions: Continuous Case

In Chapter 8 we presented a variety of discrete distributions that often arise in applications. In the present chapter we discuss continuous distributions other than the normal distribution that are frequently used: namely, the exponential, gamma (and chi-squared), Weibull, uniform, beta, lognormal, and Cauchy distributions. We continue the practice, begun in Chapters 8 and 9, of omitting identifying subscripts on density functions and cumulative distribution functions when only one random variable is under discussion.

1 The Exponential Distribution

A light bulb is turned on and left to burn until it burns out. How long will it last? An apprentice is given a learning task to master. How long will it take to complete the task? An alpha particle is emitted from a sample of radioactive material. How long will it be before the next alpha particle is emitted?

Many scientific experiments involve the measurement of the duration of time, X, between an initial point of time and the occurrence of some phenomenon of interest. For example, psychologists have studied the length of time X between successive presses of a bar by a rat placed in a Skinner box [see Mueller (1950)]. Engineers have investigated the period of time X between successive failures of components in complex machines. Telephone companies are interested in the duration X of a telephone call. Psychologists and biophysicists have been interested in the time X between successive electrical impulses in the spinal cords of various mammals.

If we make repeated measurements of X in such experiments, the histogram obtained often has the form shown in Exhibit 1.1. A probability density

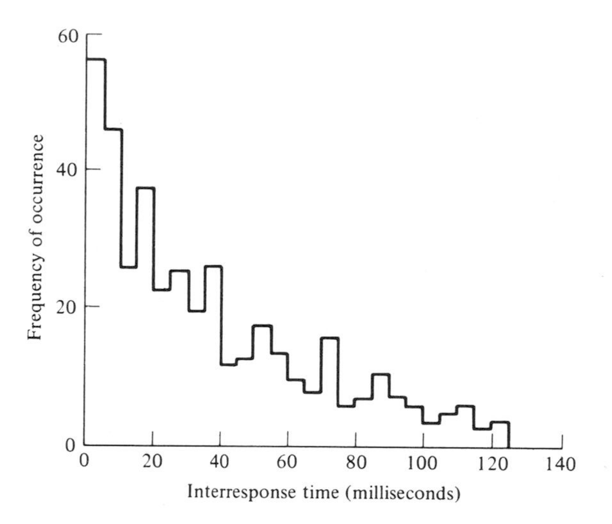

EXHIBIT 1.1 *Frequency distribution of interresponse times in the spontaneous activity of a single spinal interneuron. The distribution consists of 391 intervals recorded by a micropipet in the spinal cord of a cat. [Reprinted with permission from W. J. McGill (1963) in Luce, Bush, and Galanter,* Handbook of Mathematical Psychology. *New York: John Wiley & Sons, Inc.]*

function whose graph resembles this histogram is the exponential density function:

$$f(x) = \begin{cases} \theta e^{-\theta x}, & \text{if } x \geq 0, \\ 0, & \text{if } x < 0, \end{cases} \tag{1.1}$$

where $\theta > 0$. Any random variable X having the density function (1.1) is said to have an **exponential distribution with parameter θ**. Exhibit 1.2 provides graphs of the exponential density function for $\theta = 0.1$, 0.5, 1.0, and 2.0. Notice that the graph of an exponential density function $f(x)$ falls off rapidly as x increases from 0 to infinity, with the steepness of the fall greatest for large values of the parameter θ.

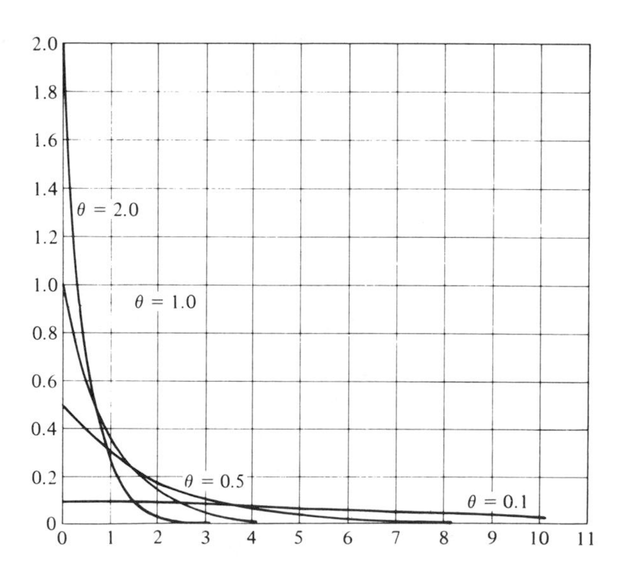

EXHIBIT 1.2 *Graphs of exponential density functions for $\theta = 0.1$, 0.5, 1.0, and 2.0.*

Probability Calculations

The cumulative distribution function $F(x)$ of the exponential distribution with parameter θ is given by the expression

$$F(x) = \begin{cases} 0, & \text{if } x < 0, \\ 1 - e^{-\theta x}, & \text{if } x \geq 0. \end{cases} \tag{1.2}$$

To see this, note that because the density $f(x)$ is 0 for all values $x < 0$, it follows that $F(x) = 0$ when $x < 0$. When $x \geq 0$,

$$F(x) = \int_0^x f(t)\, dt = \int_0^x \theta e^{-\theta t}\, dt = -e^{-\theta t}\Big|_0^x = -e^{-\theta x} + 1.$$

Recall from Chapter 5, "Notes and References," that

$$\overline{F}(x) = 1 - F(x) = P\{X > x\}$$

is called the survival function, or tail probability function of the distribution of X. From (1.2),

$$\overline{F}(x) = \begin{cases} 1 & \text{if } x < 0, \\ e^{-\theta x}, & \text{if } x \geq 0. \end{cases}$$

Because $\overline{F}(x)$ has a convenient form and

$$P\{a < X \leq b\} = F(b) - F(a) = (1 - F(a)) - (1 - F(b)) = \overline{F}(a) - \overline{F}(b),$$

it is easier to use $\overline{F}(x)$ than $F(x)$ to obtain probabilities for the exponential distribution. Thus, for $0 \leq a < b$,

$$\begin{aligned} P\{a \leq X \leq b\} &= P\{a \leq X < b\} = P\{a < X < b\} = P\{a < X \leq b\} \\ &= \overline{F}(a) - \overline{F}(b) = e^{-\theta a} - e^{-\theta b}, \end{aligned}$$

because X is a continuous random variable. To obtain these probabilities one needs to be able to compute e^{-t}. Because most scientific calculators can perform this calculation, tables of e^{-t} have not been included in this book. Note, however, that the probability of observing the value 0 for a discrete random variable Y having the Poisson distribution with parameter $\lambda = t$ is equal to e^{-t}. Thus Exhibit B.3 of Appendix B can be used to find values of e^{-t}. For example, $e^{-0.6} = P\{Y = 0\}$ for Y having a Poisson distribution with parameter $\lambda = 0.6$. Using Exhibit B.3, we find that $e^{-0.6} = 0.5488$.

EXAMPLE 1.1

Mueller (1950) describes an experiment in which white rats were periodically conditioned. The duration of time X in seconds between presses of a rat on a bar has an exponential distribution with parameter $\theta = 0.20$. Consequently, the probability that it takes a rat between 1 and 3 seconds to press the bar is

$$P\{1 \le X \le 3\} = e^{-(0.20)(1)} - e^{-(0.20)(3)} = 0.8187 - 0.5488 = 0.2699.$$

Similarly,

$$P\{X > 3\} = e^{-(0.20)(3)} = 0.5488. \quad \blacklozenge$$

Descriptive Indices

The quantiles of the exponential distribution can be obtained from the cumulative distribution function $F(x)$. To find $Q_X(p)$ for a given value of p, we solve

$$F(x) = 1 - e^{-\theta x} = p$$

for x. The solution is

$$Q_X(p) = \frac{-\log(1 - p)}{\theta} = \frac{\log[1/(1 - p)]}{\theta}. \tag{1.3}$$

For example, if $\theta = 0.1$, the median of X is

$$\text{Med}(X) = Q_X(0.5) = \frac{\log(1/0.5)}{0.1} = \frac{\log 2}{0.1} = 6.93.$$

Note from Exhibit 1.2 that the mode of X is always 0, regardless of the value of θ.

When X has an exponential distribution with parameter θ, the mean and variance of X are

$$\mu_X = \frac{1}{\theta}, \qquad \sigma_X^2 = \frac{1}{\theta^2}, \tag{1.4}$$

respectively. Instead of specifying an exponential distribution by stating the value of θ, in many cases the value of the mean μ_X is specified. Note that

$$\theta = \frac{1}{\mu_X}.$$

Thus if we are told that X has an exponential distribution with mean $\mu_X = 100$, we know that $\theta = 1/100 = 0.01$. [Because $\sigma_X^2 = \mu_X^2$, as can be seen from (1.4), we also know that $\sigma_X^2 = (100)^2 = 10{,}000$.]

To verify the formulas in (1.4), note that for any $t > 0$,

$$E[X^t] = \int_0^\infty x^t \theta e^{-\theta x}\, dx = \frac{1}{\theta^t} \int_0^\infty v^t e^{-v}\, dv,$$

where we have changed the variable of integration from x to $v = \theta x$. From Appendix A,

$$\int_0^\infty v^t e^{-v}\, dv = \Gamma(t + 1),$$

where $\Gamma(z)$ is the gamma function. Thus

$$E[X^t] = \frac{\Gamma(t + 1)}{\theta^t}, \qquad t > 0.$$

Recalling that $\Gamma(k + 1) = k!$ for $k = 1, 2, \ldots$, we conclude that $\mu_X = E[X] = 1/\theta$ and that

$$\sigma_X^2 = E[X^2] - (E[X])^2 = \frac{2!}{\theta^2} - \left(\frac{1}{\theta}\right)^2 = \frac{1}{\theta^2},$$

as asserted in (1.4).

Fitting the Exponential Distribution to Data

In a study of the performance of the air-conditioning systems of its jet passenger airplanes, an airplane manufacturer recorded for each plane the durations X of time in hours between successive failures of its air-conditioning system [Proschan (1963)]. The 30 observations obtained from one plane are listed in Exhibit 1.3.

The sample average of these data is 59.6 hours, so that we might try to fit an exponential distribution with mean $\mu_X = 59.6$ (or parameter $\theta = 1/59.6 = 0.0168$) to these data. Exhibit 1.4 shows a modified relative fre-

EXHIBIT 1.3 *Durations of Time in Hours Between Successive Failures of the Air-Conditioning System of a Jet Airplane*

23	261	87	7	120	14	62	47	225	71	246	21	42	20	5
12	120	11	3	14	71	11	14	11	16	90	1	16	52	95

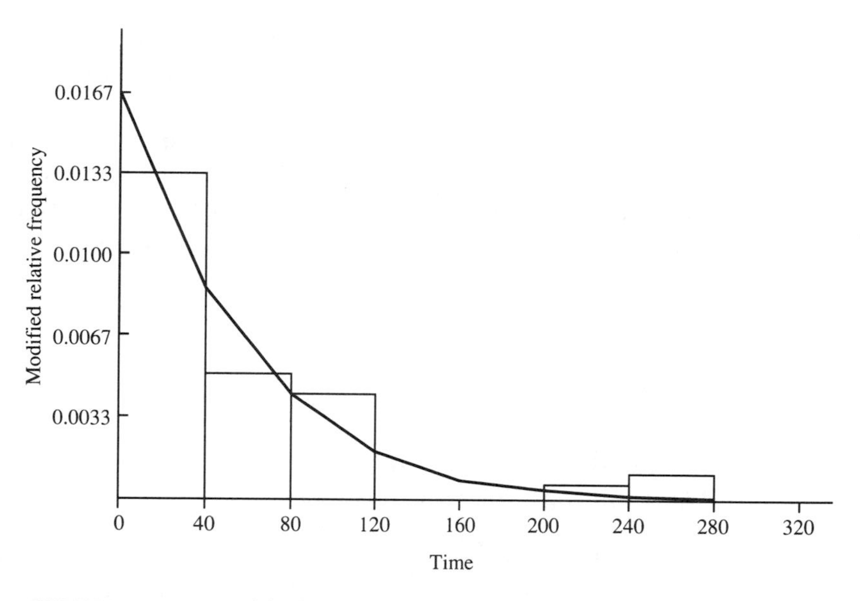

EXHIBIT 1.4 *Modified relative frequency histogram and fitted expontential density ($\theta = 0.0168$) for the data of Exhibit 1.3.*

quency histogram of the data from Exhibit 1.3, together with a graph of the density (1.1) with $\theta = 0.0168$. There appears to be good fit of the density to the histogram except for large values of X.

For lifetime or duration data, modelers prefer to graphically compare the empirical survival function

$$\overline{F}_n(x) = \frac{\text{number of observations} > x}{n}$$

to the theoretical survival function $\overline{F}(x)$. Because the logarithm of $\overline{F}(x)$ is a linear function of x,

$$-\log \overline{F}(x) = -\log(e^{-\theta x}) = \theta x,$$

it can be more revealing to plot $y = -\log \overline{F}(x)$ against the observed x-values. If the resulting plot fits closely to a straight line $y = \theta x$ through the origin, this supports the hypothesis that X has an exponential distribution. Further, the slope of such a straight line directly provides an estimate of θ. Unfortunately, $\overline{F}(x)$ is equal to 0 for the largest observed value of X, and a computer cannot plot $-\log \overline{F}(x)$ for this x-value. Instead,

$$y^* = -\log \left[\frac{n\overline{F}_n(x) + \frac{1}{2}}{n + 1}\right]$$

is plotted against x. This adjustment does not matter if the number n of observations of X is moderately large. Exhibit 1.5 displays such a plot for the data of Exhibit 1.3.

From Exhibit 1.5 we see that the plot of y^* versus x is nearly linear, except for two large values ($x = 225$, $x = 246$) of x. This confirms, in greater detail, our earlier conclusion that the exponential distribution can serve as an approximate model for the duration, X, of performance of the plane's air-conditioning system. The airplane manufacturer, however, might be interested in determining why two of the three largest observations of X deviated from this model. The data may be indicating that once an air-conditioning system survives a "burn-in" period, it is more resistant to failure.

Lack of Memory Property of the Exponential Distribution

Suppose that the random variable X measures the duration of time until the occurrence of a given phenomenon and that it is known that X has an exponential distribution with parameter θ. At the present time, x^*, the phenomenon of interest, has not yet occurred. Thus we know that $X > x^*$. Let $Y = X - x^*$ equal the additional time that we must wait for the phenomenon to occur. What is the *conditional distribution* of Y given that we have already waited x^* units of time?

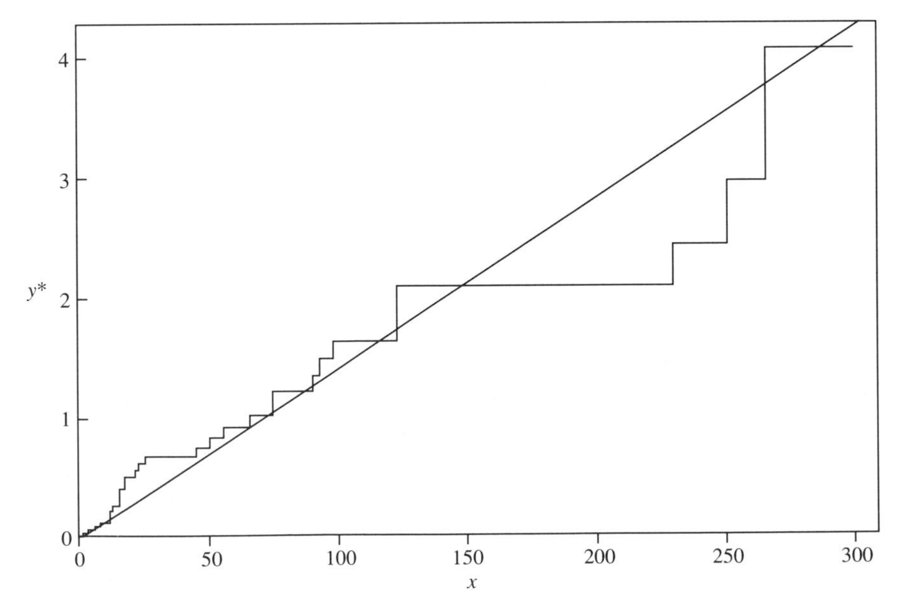

EXHIBIT 1.5 *Plot of y^* versus x.*

The answer may be surprising. Given that we have already waited x^* units of time, the length $Y = X - x^*$ of additional time that we must wait has the exponential distribution with parameter θ. That is, given that $X > x^*$, Y has the same distribution, conditionally, as the original waiting time X had unconditionally. It is as if we start afresh, forgetting the fact that we have waited x^* units of time already. In this sense *the exponential distribution has no memory of the past.*

If the length X of a telephone conversation has an exponential distribution, the foregoing property implies that if you have been waiting at a phone booth for 10 minutes for a conversation to end, the remaining time that you have to wait has the *same* distribution as if the conversation had just started. Consequently, if one phone booth has just been occupied as you arrive and another phone booth has been occupied for some time, it does not matter which phone booth you decide to wait for, because the probability distribution of the length of time you must wait will be the same for either phone booth.

To verify the foregoing assertion, note that the conditional survival function of $Y = X - x^*$ given that the event $A = \{X > x^*\}$ has occurred is

$$\begin{aligned}\bar{F}(y|A) = P\{Y > y|A\} &= \frac{P\{X - x^* > y \text{ and } X > x^*\}}{P\{X > x^*\}} \\ &= \frac{P\{X > x^* + y\}}{P\{X > x^*\}} = \frac{e^{-\theta(x^*+y)}}{e^{-\theta x^*}} = e^{-\theta y},\end{aligned}$$

where we have used the fact that the event $\{X > x^* + y\}$ includes the event $\{X > x^*\}$ for $y \geq 0$. Consequently,

$$\bar{F}(y|A) = \begin{cases} 1, & \text{if } y < 0, \\ e^{-\theta y}, & \text{if } y \geq 0, \end{cases} \tag{1.5}$$

which is the survival function of an exponential random variable with parameter θ.

A Characterization of the Exponential Distribution*

Among all distributions of nonnegative continuous variables, only the exponential distributions "have no memory." To see this, let

$$\bar{F}(s) = P\{X > s\}, \qquad s > 0.$$

The memoryless property asserts that

$$P\{X > s + t|X > s\} = P\{X > t\}, \qquad s > 0, \quad t > 0. \tag{1.6}$$

But

$$(1.7)\quad P\{X > s + t | X > s\} = \frac{P(\{X > s + t\} \cap \{X > s\})}{P\{X > s\}} = \frac{P\{X > s + t\}}{P\{X > s\}}.$$

Thus (1.6) and (1.7) together yield the famous functional equation

$$(1.8)\qquad \bar{F}(t + s) = \bar{F}(s)\bar{F}(t), \qquad s > 0, \quad t > 0,$$

which is known as the *Cauchy equation* [see, e.g., Aczél (1966)]. Under mild regularity conditions on the function $\bar{F}$, the only solution of (1.8) is

$$\bar{F}(x) = e^{-cx},$$

where c is any constant. However, $F(x) = 1 - \bar{F}(x)$ is a cumulative distribution function, and thus $\lim_{x\to\infty} F(x) = 1$. This forces c to be positive, and we recognize $\bar{F}(x)$ as being the survival function of the exponential distribution with parameter $\theta = c$.

A related characterization of the exponential distribution as the distribution of a nonnegative continuous random variable having constant hazard function is given in the "Notes and References" of Chapter 5.

2 The Gamma Distribution

Visualize a system (such as a radio, an assembly line, an airplane, etc.) in which the proper functioning of a component is essential for the functioning of the system as a whole. To increase the reliability of the system, the system may be designed to carry an original and $(r - 1)$ spare components. When the original component fails, one of the $(r - 1)$ spare components is activated to take its place. If this component fails, one of the $(r - 2)$ remaining spare components takes over. This process continues until the original component and all of the $(r - 1)$ spares have failed, at which point the entire system suffers failure. Assuming that all components of the system other than the essential component have infinite lifetimes, the lifetime (time until failure) of the entire system is the sum $Y = \sum_{i=1}^{r} X_i$ of the lifetimes $X_1, X_2, \ldots, X_r$ of the r duplicates of the essential component. If the lifetimes $X_1, X_2, \ldots, X_r$ are probabilistically independent and each has the same exponential distribution with parameter θ, the lifetime $Y = \sum_{i=1}^{r} X_i$ of the system has the **gamma distribution** with density function

$$(2.1)\qquad f(y) = \begin{cases} \dfrac{\theta^r y^{r-1} e^{-\theta y}}{\Gamma(r)}, & \text{if } y \geq 0, \\ 0, & \text{if } y < 0, \end{cases}$$

where θ and r are positive numbers and $\Gamma(r)$ is the gamma function. The constants θ and r in (2.1) are the parameters of the gamma distribution. The gamma function $\Gamma(r)$ in the formula (2.1) for the gamma distribution gives this distribution its name. Although the parameter r of the gamma distribution is an integer in the example of the system having spare components, this need not always be the case. In many applications of the gamma distribution, r is not an integer.

Example 2.1 **Radioactivity**

Slack and Krumbein (1955) describe an experiment in which a representative measure Y of radioactivity (alphas per minute) within a sample of Pennsylvania black shale is obtained. A gamma distribution with parameters $\theta = 0.9083$ and $r = 2.493$ is found to provide a reasonably close fit to the observed frequency distribution of Y. ◆

Probability Calculations: The Standard Gamma Density

When Y has a gamma distribution with parameters θ and r, probabilities can be obtained from the cumulative distribution function $F(y)$. Except in special cases, $F(y)$ does not have a simple functional form, and thus must be tabulated. Since the distribution of Y has two parameters θ and r, the problem of tabulation appears formidable. Fortunately, as in the case of the normal distribution, some standardization is possible.

Note that if $V = \theta Y$, then

$$(2.2)\quad F_V(v) = P\{V \leq v\} = P\{\theta Y \leq v\} = P\{Y \leq v/\theta\} = \int_0^{v/\theta} \frac{\theta^r y^{r-1} e^{-\theta y}}{\Gamma(r)}\, dy.$$

Change the variable of integration in (2.2) from y to $z = \theta y$. Then

$$F_V(v) = \int_0^v \frac{z^{r-1} e^{-z}}{\Gamma(r)}\, dz,$$

from which the probability density of V is seen to be

$$(2.3)\qquad f(v) = \begin{cases} \dfrac{v^{r-1} e^{-v}}{\Gamma(r)}, & \text{if } v \geq 0, \\ 0, & \text{if } v < 0. \end{cases}$$

We recognize this density as being the special case of the gamma density (2.1) with $\theta = 1$. A random variable V with the density function (2.3) is said to have the **standard gamma distribution with parameter r.** We have shown that if Y has a gamma distribution with parameters θ and r, then $V = \theta Y$ has the standard gamma density with parameter r. Consequently,

$$F_Y(y) = P\{Y \le y\} = P\{\theta Y \le \theta y\} = P\{V \le \theta y\} = F_V(\theta y). \tag{2.4}$$

> If Y has a gamma distribution with parameters θ and r and V has the standard gamma distribution with parameter r, then
>
> $$F_Y(y) = F_V(\theta y).$$

Therefore, values of the cumulative distribution function $F_Y(y)$ of a gamma distribution with parameters θ and r can be obtained from tables of the cumulative distribution function $F_V(v)$ of the standard gamma distribution with parameter r.

Graphs of the standardized gamma density function are given in Exhibit 2.1(a) for $r = 0.5$, 2, 4, and 5. Note that when $r \le 1$, the graph of the

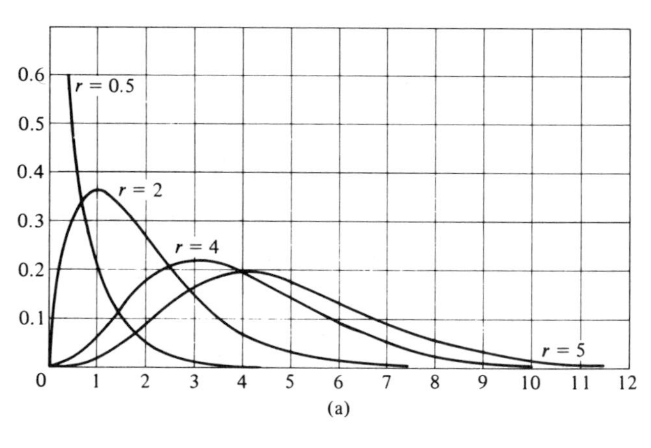

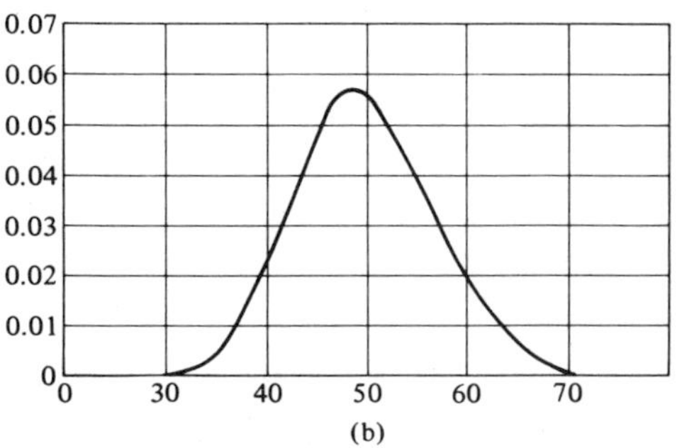

EXHIBIT 2.1 *Graphs of the standardized gamma density function for (a) $r = 0.5, 2, 4, 5$; (b) $r = 50$.*

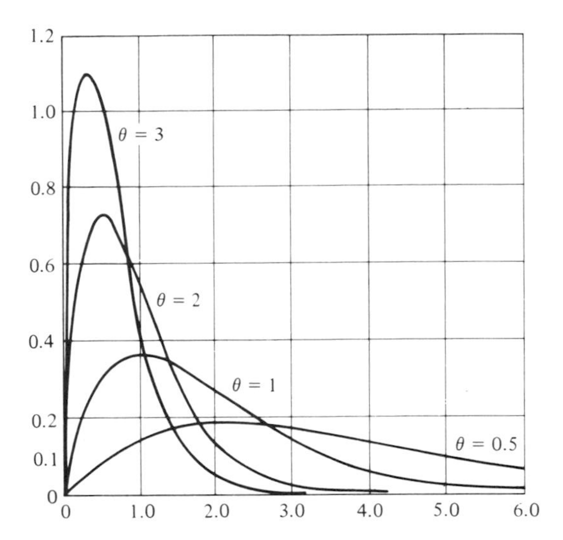

EXHIBIT 2.2 *Graphs of the gamma density function for $r = 2$, $\theta = 0.5, 1, 2, 3$.*

corresponding standardized gamma density function drops off steeply, just as the case of the exponential density function. Indeed, when $r = 1$ the density function of the standardized gamma function is identical to the density function of the exponential distribution with parameter $\theta = 1$. When $r > 1$, the graph of the standardized gamma density function takes on a "humped" form. The density function has a long right-hand "tail," but this skewness becomes less pronounced as r increases. For r large enough (say, $r \geq 50$), the standard gamma distribution closely resembles a normal distribution with mean and variance both approximately equal to r [see Exhibit 2.1(b)].

Exhibit 2.1 shows the influence of the parameter r on the shape of the gamma density. The parameter θ in the (nonstandardized) gamma density (2.1) influences the shape of the density in the same manner as θ influences the shape of the exponential density (1.1); that is, as θ increases, the density is "squeezed" in such a manner that the right-hand tail of the density function moves closer to the vertical axis (see Exhibit 2.2).

The function

$$(2.5) \qquad I_r(t) = \int_0^t \frac{z^{r-1}e^{-z}}{\Gamma(r)}\, dz, \qquad t \geq 0,$$

is extensively tabulated and is known as the **incomplete gamma function.** Values of $I_r(t)$ for $r = 0.5(0.5)9.5$, $t = 0.2(0.2)8.0(1.0)15.0$, are given in Exhibit B.5 of Appendix B. If we adopt the convention that $I_r(t) = 0$ for $t < 0$, then $I_r(t)$ gives the cumulative distribution function of the standard gamma distribution with parameter r. Equations (2.4) and (2.5) yield the conclusion:

> If Y has a gamma distribution with parameters θ and r, then
>
> (2.6) $$F_Y(y) = I_r(\theta y).$$

Consequently,

$$\overline{F}(y) = P\{Y > y\} = 1 - F(y) = 1 - I_r(\theta y)$$

and

$$P\{a \le Y \le b\} = F(b) - F(a) = I_r(\theta b) - I_r(\theta a).$$

Remark. An important property of the incomplete gamma function (2.5) is the recursion formula,

(2.7) $$I_m(t) - I_{m+1}(t) = \frac{t^m e^{-t}}{\Gamma(m+1)}.$$

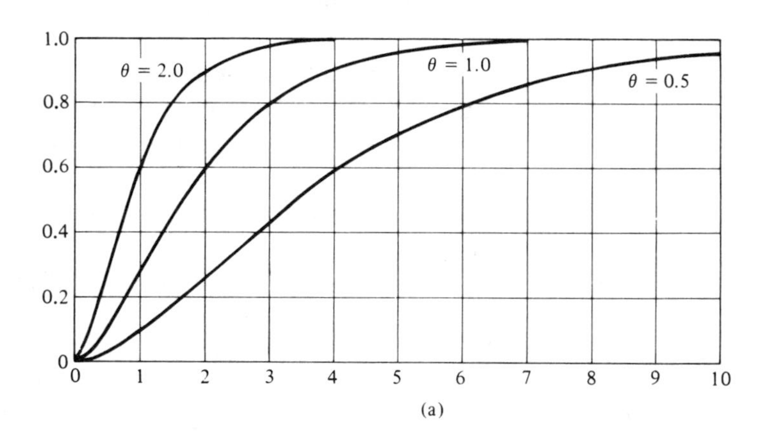

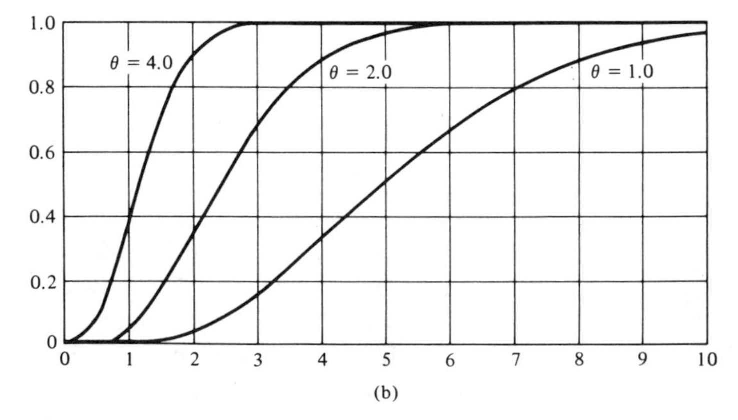

EXHIBIT 2.3 *Graphs of the cumulative distribution functions for gamma distributions with parameters (a) $r = 2$, $\theta = 0.5$, 1.0, 2.0; (b) $r = 5$, $\theta = 1.0$, 2.0, 4.0.*

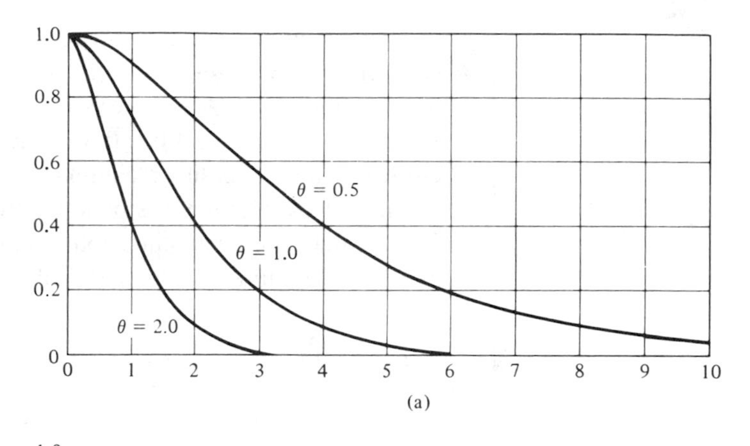

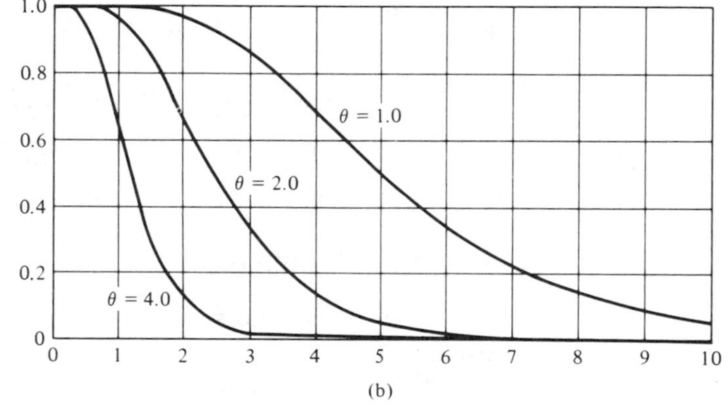

EXHIBIT 2.4 *Graphs of the survival functions for gamma distributions with parameters (a) $r = 2$, $\theta = 0.5, 1.0, 2.0$; (b) $r = 5$, $\theta = 1.0, 2.0, 4.0$.*

To verify this formula, integrate by parts in (2.5):

$$\begin{aligned} I_{m+1}(t) &= \int_0^t \frac{z^m e^{-z}}{\Gamma(m+1)}\,dz = -\frac{z^m e^{-z}}{\Gamma(m+1)}\bigg|_0^t + m\int_0^t \frac{z^{m-1}e^{-z}}{\Gamma(m+1)}\,dz \\ &= \frac{-t^m e^{-t}}{\Gamma(m+1)} + \int_0^t \frac{z^{m-1}e^{-z}}{\Gamma(m)}\,dz \\ &= \frac{-t^m e^{-t}}{\Gamma(m+1)} + I_m(t). \end{aligned}$$

Equation (2.6) and Exhibit B.5 of Appendix B allow us to graph either the cumulative distributions functions $F(y)$ or the survival functions $\overline{F}(y)$ of random variables Y having a gamma distribution. Graphs of the cumulative distribution functions and survival functions for random variables having gamma distributions with various pairs of parameters (θ, r) are displayed in Exhibits 2.3 and 2.4, respectively.

EXAMPLE 2.2

The time Y in minutes that it takes a child to complete a complex learning task is measured. One theory of learning asserts that a complex task is solved by independently and sequentially solving r simple tasks, each of approximately the same difficulty. If the time X_i taken to solve the ith simple task is exponentially distributed with parameter θ for $i = 1, 2, \ldots, r$, the total length of time $Y = \sum_{i=1}^{r} X_i$ required to finish the entire learning task has a gamma distribution with parameters θ and r. From past experience it is believed that $\theta = 0.40 = (2.50)^{-1}$ and $r = 5$. To calculate the probability that a child cannot finish the learning task in, say, 30 minutes, Exhibit B.5 of Appendix B yields

$$P\{Y > 30\} = 1 - I_5\left(\frac{30}{2.5}\right) = 1 - I_5(12) = 1 - 0.99240 = 0.00760.$$

Similarly,

$$\begin{aligned} P\{15 \le Y \le 20\} &= I_5\left(\frac{20}{2.5}\right) - I_5\left(\frac{15}{2.5}\right) = I_5(8) - I_5(6) \\ &= 0.90037 - 0.71494 = 0.18543. \end{aligned}$$ ◆

Descriptive Indices

When V has the standard gamma distribution with parameter r, then using (2.3),

$$\begin{aligned} E(V^k) &= \int_0^\infty v^k f_V(v)\, dv = \int_0^\infty \frac{v^k v^{r-1} e^{-v}}{\Gamma(r)}\, dv \\ &= \frac{1}{\Gamma(r)}\left[\int_0^\infty v^{k+r-1} e^{-v}\, dv\right] = \frac{\Gamma(k+r)}{\Gamma(r)}. \end{aligned}$$

Thus if Y has a gamma distribution with parameters θ and r,

$$E(Y^k) = E(\theta^{-k}(\theta Y)^k) = \theta^{-k} E(V^k) = \frac{\Gamma(k+r)}{\theta^k \Gamma(r)} \tag{2.8}$$

by the linearity property of expected values and the fact that $V = \theta Y$ has a standard gamma distribution with parameter r.

The special cases $k = 1, 2$, of (2.8) yield

$$\begin{aligned} \mu_Y &= E(Y) = \frac{\Gamma(1+r)}{\theta\Gamma(r)} = \frac{r\Gamma(r)}{\theta\Gamma(r)} = \frac{r}{\theta}, \\ \sigma_Y^2 &= E(Y^2) - (\mu_Y)^2 = \frac{\Gamma(2+r)}{\theta^2\Gamma(r)} - \left(\frac{r}{\theta}\right)^2 = \frac{(r+1)r\Gamma(r)}{\theta^2\Gamma(r)} - \frac{r^2}{\theta^2} \\ &= \frac{r(r+1) - r^2}{\theta^2} = \frac{r}{\theta^2}. \end{aligned}$$

These results can also be obtained for integer r by recalling that in this case Y has the distribution of the sum $\Sigma_{i=1}^{r} X_i$ of r independent exponentially distributed random variables with mean $1/\theta$ and variance $1/\theta^2$.

> If Y has a gamma distribution with parameters θ and r, then
>
> $$\mu_Y = \frac{r}{\theta}, \qquad \sigma_Y^2 = \frac{r}{\theta^2}.$$

The quantiles of Y can be found from (2.6) and interpolation in Exhibit B.5 of Appendix B. To find $Q_Y(p)$, we must solve

$$p = F(y) = I_r(\theta y)$$

for y. We do so as follows:

1. Find v^* such that $I_r(v^*) = p$. Note that $v^* = Q_V(p)$, where V has a standard gamma distribution with parameter r.
2. Solve $\theta y = v^*$ for y.

> If Y has a gamma distribution with parameters θ and r, then
>
> $$Q_Y(p) = \frac{Q_V(p)}{\theta},$$
>
> where V has the standard gamma distribution with parameter r.

EXAMPLE 2.3

Let Y have a gamma distribution with parameters $\theta = 0.10$ and $r = 3$. We wish to find the median, $Q_Y(0.50)$, of Y. Thus we look in Exhibit B.5 of Appendix B, in the column for $r = 3$ and try to find t such that $I_3(t) = 0.50$. We find that $I_3(2.6) = 0.48157$, $I_3(2.8) = 0.53055$. Using linear interpolation yields

$$Q_V 0.50 = 2.6 + \frac{0.50000 - 0.48157}{0.53055 - 0.48157}(2.8 - 2.6) = 2.608.$$

Thus

$$Q_Y(0.50) = \frac{2.608}{0.10} = 26.08.$$

◆

The mode of a gamma distribution with parameters θ and r is

$$\text{Mode}(Y) = \begin{cases} 0, & \text{if } r \leq 1, \\ \dfrac{r-1}{\theta}, & \text{if } r > 1. \end{cases}$$

Sums of Gamma Variables

Suppose that $Y_1, Y_2, \ldots, Y_n$ are mutually independent observations from a gamma distribution with parameters θ and r. What are the distributions of the sum $S_n = \sum_{i=1}^{n} Y_i$ and average $\overline{Y} = \frac{1}{n} S_n$ of $Y_1, Y_2, \ldots, Y_n$?

The answer to this question is that both S_n and $\overline{Y}$ have gamma distributions. The distribution of S_n is the gamma distribution with parameters θ and nr, whereas $\overline{Y} = n^{-1}S_n$ has a gamma distribution with parameters θn and nr.

More generally, if Y_i has a gamma distribution with parameters θ and r_i, $i = 1, 2, \ldots, n$, and $Y_1, \ldots, Y_n$ are independent, then for any positive constant c it can be shown that $U = c^{-1} \sum_{i=1}^{n} Y_i$ has a gamma distribution with parameters $c\theta$ and $\sum_{i=1}^{n} r_i$. When the r_i's are positive integers, this result can be shown by using the motivation of how the gamma distribution arises in practice given at the start of this section. That is, we can think of each Y_i as being the sum $\sum_{j=1}^{r_i} X_{ij}$ of independent exponentially distributed random variables X_{ij}, where the exponential distributions all have parameter θ. Because the Y_i's are given to be independent, we can assume that the X_{ij}'s are mutually independent, $1 \leq j \leq r_i$, $1 \leq i \leq n$. Consequently,

$$\sum_{i=1}^{n} Y_i = \sum_{i=1}^{n} \sum_{j=1}^{r_i} X_{ij}$$

is the sum of $\sum_{i=1}^{n} r_i$ independent exponentially distributed random variables, each with parameter θ, and thus $\sum_{i=1}^{n} Y_i$ has a gamma distribution with parameters θ and $r = \sum_{i=1}^{n} r_i$. The cumulative distribution function $F_U(u)$ of $U = c^{-1} \sum_{i=1}^{n} Y_i$ is

$$\begin{aligned} F_U(u) = P\{U \leq u\} &= P\left\{c^{-1}\sum_{i=1}^{n} Y_i \leq u\right\} \\ &= P\left\{\sum_{i=1}^{n} Y_i \leq cu\right\} = I_r(\theta cu). \end{aligned}$$

Consequently, by (2.6), $F_U(u)$ is the cumulative distribution function of a gamma distribution with parameters $c\theta$ and r.

If some of the parameters r_i are not integers, we cannot use this heuristic argument, but the result is still true (see Exercise T.2).

The Chi-Squared Distribution

If Z has a standard normal distribution, then $Y = Z^2$ has cumulative distribution function

$$
\begin{aligned}
F_Y(y) &= P\{Y \le y\} = P\{Z^2 \le y\} \\
&= P\{-\sqrt{y} \le Z \le \sqrt{y}\} \\
&= \int_{-\sqrt{y}}^{\sqrt{y}} \frac{1}{\sqrt{2\pi}} e^{-(1/2)z^2}\, dz \\
&= 2\int_0^{\sqrt{y}} \frac{1}{\sqrt{2\pi}} e^{-(1/2)z^2}\, dz, \qquad y \ge 0,
\end{aligned} \tag{2.9}
$$

where we have used the fact that $\exp(-\frac{1}{2}z^2)$ is an even function of z. For $y < 0$, $F_Y(y) = P\{Z^2 \le y\}$ is clearly equal to 0. Now make the change of variables $t = z^2$ in (2.9) to obtain

$$
\begin{aligned}
F_Y(y) &= 2\int_0^{\sqrt{y}} \frac{1}{\sqrt{2\pi}} e^{-1/2z^2}\, dz = \frac{2}{\sqrt{2\pi}} \int_0^y \frac{1}{2t^{1/2}} e^{-(1/2)t}\, dt \\
&= \int_0^y \frac{(\frac{1}{2})^{1/2} t^{1/2-1}\, e^{-(1/2)t}\, dt}{\Gamma(\frac{1}{2})}
\end{aligned} \tag{2.10}
$$

because $\Gamma(\frac{1}{2}) = \sqrt{\pi}$. We recognize the integrand in the last integral in (2.10) as being the probability density function (2.1) of a gamma distribution with parameters $\theta = \frac{1}{2}$, $r = \frac{1}{2}$. Consequently, we have shown that the square Z^2 of a standard normal random variable Z has a gamma distribution with parameters $\theta = \frac{1}{2}$, $r = \frac{1}{2}$.

Because of the special role that sums of squares of n independent (standard) normal random variables Z_i play in statistics, the distribution of the sum of squares $\sum_{i=1}^n Z_i^2$ has a special name: **chi-squared distribution with n "degrees of freedom."** The special notation χ_n^2 represents this distribution. By $U \sim \chi_n^2$, we mean that the variable U has a chi-squared distribution with n degrees of freedom. The degrees of freedom n of the χ_n^2 distribution is the only parameter of that distribution.

We have shown that if Z is a standard normal random variable, Z^2 has a gamma distribution with parameters $\theta = \frac{1}{2}$, $r = \frac{1}{2}$. By definition, $Z^2 \sim \chi_1^2$. Thus the χ_1^2 distribution is the same as the gamma distribution with parameters $\theta = \frac{1}{2}$, $r = \frac{1}{2}$. Because the sum of independent gamma distributed variables Y_i, each with parameters $\theta = \frac{1}{2}$, $r_i = \frac{1}{2}$, has a gamma distribution with parameters $\theta = \frac{1}{2}$, $r = \sum_{i=1}^n r_i = \frac{1}{2}n$, it follows that the sum of squares $\sum_{i=1}^n Z_i^2$ of n independent standard normal random variables has a gamma distribution with parameters $\theta = \frac{1}{2}$, $r = \frac{1}{2}n$. Because the distribution of $\sum_{i=1}^n Z_i^2$ is defined to be the χ_n^2 distribution, it follows that the χ_n^2 distribution is the same as the gamma distribution with parameters $\theta = \frac{1}{2}$, $r = \frac{1}{2}n$.

Consequently, we can find the cumulative distribution function, probabilities, and descriptive indices of chi-squared distributions from the corresponding results given in this section for gamma distributions.

If $Y \sim \chi^2_n$, then

$$F_Y(y) = P\{Y \le y\} = I_{n/2}(y/2),$$
$$\mu_Y = \frac{1}{2}\frac{n}{1/2} = n, \qquad \sigma^2_Y = \frac{n/2}{(1/2)^2} = 2n,$$
$$Q_Y(p) = \frac{Q_V(p)}{1/2} = 2Q_V(p),$$

where $Q_V(p)$ is the pth quantile of a standard gamma distribution with parameter $r = n/2$.

Fitting the Gamma Distribution to Data

To illustrate fitting the gamma distribution to data, consider the following example.

EXAMPLE 2.4

Although most measurements on human beings have approximate normal distributions, systolic blood pressures appear to be an exception. Exhibit 2.5 presents a grouped frequency table for systolic blood pressures Y (in millimeters of mercury) in a random sample of 1000 men aged 45 to 59. [The relative frequencies are equal to those given in Roberts (1975)]. A modified relative frequency histogram of these data appears in Exhibit 2.6. Note that the distribution (histogram) is skewed to the right, so that the normal distribution is not an appropriate model.

One might consider fitting a gamma distribution to these data, but whereas the gamma density is positive at all values of y greater than zero, the histogram of Exhibit 2.6 lies above the y-axis only for values $y \ge 90$. Thus we would like to shift the gamma density to the right on the y-axis. To do this we can assume that $Y - Y_0$ has a gamma distribution for some constant Y_0. This constant Y_0 might be regarded as a lower threshold for systolic blood pressure, below which the health of the person is seriously impaired. For present purposes, we choose $Y_0 = 80$ and fit a gamma distribution with parameters $r = 9$, $\theta = 1/6.5 = 0.1538$ to $Y - 80$. This implies a density function for Y of the form

$$f(y) = \frac{(0.1538)^9(y - 80)^{9-1}\exp[-(0.1538)(Y - 80)]}{\Gamma(9)}$$

EXHIBIT 2.5 *Grouped Frequency Table of Systolic Blood Pressures Y from 1000 Male Subjects Aged 45–59*

Interval	*Observed Frequency*	*Theoretical Frequency ($Y - 80$ has Gamma Distribution $r = 9$, $\theta = 1/6.5 = 0.1538$)*
80–90	0	0.03
90–100	4	4.44
100–110	37	41.38
110–120	105	123.20
120–130	214	195.54
130–140	220	209.75
140–150	180	172.54
150–160	94	117.24
160–170	65	68.98
170–180	35	36.29
180–190	23	17.45
190–200	20	7.80
≥ 200	3	5.36

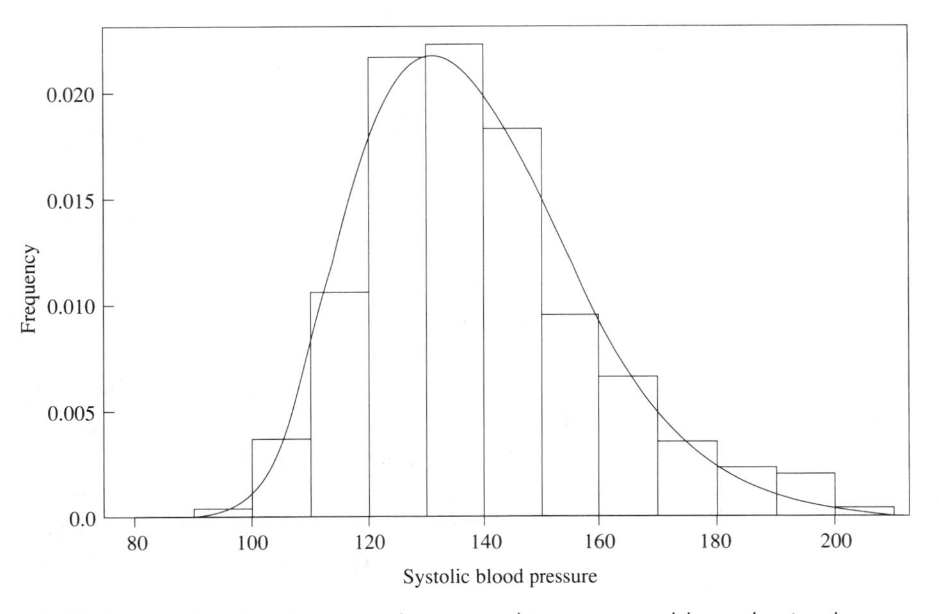

EXHIBIT 2.6 *Modified relative frequency histogram and hypothesized shifted gamma density for the data in Exhibit 2.5.*

for $y \geq 80$ and $f(y) = 0$ for $y < 80$. Such a density function is a *shifted gamma density* and is sometimes called a *three-parameter gamma density*. (The third parameter is Y_0.) The density $f(y)$ is plotted in Exhibit 2.6 for comparison with the modified relative frequency histogram of Y. It appears to fit the histogram reasonably well.

To evaluate further the fit of this shifted gamma distribution, theoretical frequencies are given in Exhibit 2.5 for each of the grouping intervals for Y. Calculation of such theoretical frequencies is illustrated for the interval $\{160 \leq Y \leq 170\}$ in the following:

$$\begin{aligned}\text{theoretical frequency of } \{160 \leq Y \leq 170\} \\ &= (1000)P\{160 - 80 \leq Y - 80 \leq 170 - 80\} \\ &= (1000)\left[I_9\left(\frac{90}{6.5}\right) - I_9\left(\frac{80}{6.5}\right)\right] \\ &= 68.98.\end{aligned}$$

Although the values of the incomplete gamma function $I_9(t)$ could have been approximated using interpolation in Exhibit B.5 of Appendix B, a more accurate computer program was instead utilized to obtain the theoretical frequencies. A comparison of the observed and theoretical frequencies in Exhibit 2.5 again indicates a reasonably good fit, except perhaps for large values of Y. ◆

A Connection Among the Exponential, Gamma, and Poisson Distributions

The exponential, gamma, and Poisson distributions are related in an interesting way. Suppose that we have a batch of light bulbs whose lifetimes are each exponentially distributed with parameter θ. We start at a given time (say, $t = 0$) and burn one bulb until extinction. We replace that bulb instantly with another bulb, wait until that bulb burns out, replace it with another, wait until this third bulb burns out, and so on. At a given time t, we stop this process and count the number L of light bulbs that we have burned out. It then can be shown that L is a random variable having the Poisson distribution with parameter $\lambda = \theta t$.

To see this, we make use of the gamma distribution. Let us find the probability of the event $\{L \geq k\}$ that at least k bulbs have been burned out when we stop the experiment at time t. Let $X_1, X_2, \ldots, X_k$ be the lifetimes of the first k bulbs that we (might) have burned. Then, $X_1, X_2, \ldots X_k$ are, by assumption, probabilistically independent of one another, and each lifetime X_i has an exponential distribution with parameter θ. We are able to burn at least these k bulbs if $X_1 + X_2 + \cdots + X_k \leq t$, because in that case the k bulbs will have all burned out (one after the other) before our time is up. Recall that $Y = \sum_{i=1}^{k} X_i$ has a gamma distribution with parameters θ and k.

Hence, using (2.7) repeatedly yields

$$P\{L \geq k\} = P\{Y \leq t\} = I_k(\theta t) = -\sum_{i=1}^{k-1} \frac{(\theta t)^i e^{-\theta t}}{i!} + I_1(\theta t)$$
$$= 1 - \sum_{i=1}^{k-1} \frac{(\theta t)^i e^{-\theta t}}{i!},$$

because $I_1(\theta t) = \int_0^{\theta t} e^{-v} dv = 1 - e^{-\theta t}$. However,

$$P\{L = k\} = P\{L \geq k\} - P\{L \geq k + 1\}$$
$$= \left[1 - \sum_{i=1}^{k-1} \frac{(\theta t)^i e^{-\theta t}}{i!}\right] - \left[1 - \sum_{i=1}^{k} \frac{(\theta t)^i e^{-\theta t}}{i!}\right] = \frac{(\theta t)^k e^{-\theta t}}{k!},$$

so that the probability mass function $p_L(k) = P\{L = k\}$ of the number L of bulbs burned out is that of a Poisson distribution with parameter $\lambda = \theta t$.

We have shown by this example that L has a Poisson distribution with parameter $\lambda = \theta t$. In the process, however, we have also obtained the following useful method for computing probabilities for the gamma distribution:

> If Y has a gamma distribution with parameters θ and r, r a positive integer, then
>
> $$F(y) = P\{Y \leq y\} = P\{X > r - 1\},$$
>
> where X has the Poisson distribution with parameter $\lambda = \theta y$. Also,
>
> $$\overline{F}(y) = P\{Y > y\} = P\{X \leq r - 1\}.$$

EXAMPLE 2.5

Suppose that Y has a gamma distribution with parameters $\theta = 0.5$ and $r = 5$. To find the probability that Y exceeds its mean $\mu_Y = r/\theta = 5/0.5 = 10$, we compute

$$P\{Y > 10\} = \overline{F}(10) = P\{X \leq 5 - 1 = 4\},$$

where X has the Poisson distribution with parameter $\lambda = (0.5)(10) = 5$. Using Exhibit B.3, Appendix B, gives

$$P\{X \leq 4\} = 0.0067 + 0.0337 + 0.0842 + 0.1404 + 0.1755$$
$$= 0.4405.$$

Also, the probability that Y is between 8.2 and 12.0 is given by

$$\begin{aligned} P\{8.2 \le Y \le 12.0\} &= \bar{F}(8.2) - \bar{F}(12.0) \\ &= P\{X \le 4\} - P\{X^* \le 4\}, \end{aligned}$$

where X has the Poisson distribution with parameter $\lambda = (0.5)(8.2) = 4.1$ and X^* has the Poisson distribution with parameter $\lambda = (0.5)(12.0) = 6$. Thus, from Exhibit B.3,

$$\begin{aligned} \Gamma\{8.2 \le Y \le 12.0\} &= (0.0166 + 0.0679 + \cdots + 0.1951) \\ &\quad - (0.0025 + 0.0149 + \cdots + 0.1339) \\ &= 0.5993 - 0.2718 = 0.3275. \end{aligned}$$

◆

3 The Weibull Distribution

Important examples of nonnegative random variables occurring in applications are lifetimes, waiting times, learning times, durations of epidemics, and traveling times. Nontemporal examples of nonnegative random variables include material strengths, particle dimensions, radioactive intensities, rainfall amounts, and costs of industrial accidents. Although exponential or gamma distributions provide reasonable fits to the frequency distributions of some of these random variables, they may not fit as closely as desired. Hence other classes of distributions have been introduced to explain the variability of some of these phenomena. One such class of distributions are the **Weibull distributions**, named after the Swedish physicist Waloddi Weibull.

If a random variable X has a Weibull distribution, the probability density function $f(x)$ of X has the form

$$(3.1) \qquad f(x) = \begin{cases} \dfrac{\beta}{\alpha}\left(\dfrac{x-\nu}{\alpha}\right)^{\beta-1} \exp\left[-\left(\dfrac{x-\nu}{\alpha}\right)^{\beta}\right], & \text{if } x \ge \nu. \\ 0, & \text{if } x < \nu. \end{cases}$$

The three constants $\beta > 0$, $\alpha > 0$, and $\nu \ge 0$ are the parameters of the Weibull distribution. Note that the parameter ν tells us the smallest possible value of the random variable X. The parameter β determines the shape of the density (3.1), whereas $1/\alpha$ plays the same role that the parameter θ did for the family of gamma densities. The roles of β and α are illustrated in Exhibits 3.1 and 3.2; in Exhibit 3.1, $f(x)$ is graphed for $\beta = 2.0$, $\nu = 0.0$, $\alpha = 0.5$, 1.0, 2.0, and in Exhibit 3.2 $f(x)$ is graphed for $\beta = 10.0$, $\nu = 0.0$, $\alpha = 0.5$, 1.0, 2.0. Finally, it is shown that ν is a measure of location for the density (3.1) by graphing $f(x)$ for $\beta = 0.5$, $\alpha = 1.0$, $\nu = 0.0$, 0.5 in Exhibit 3.3(a) and for $\beta = 2.0$, $\alpha = 1.0$, $\nu = 0.0$, 0.5 in Exhibit 3.3(b). From (3.1) we see that the

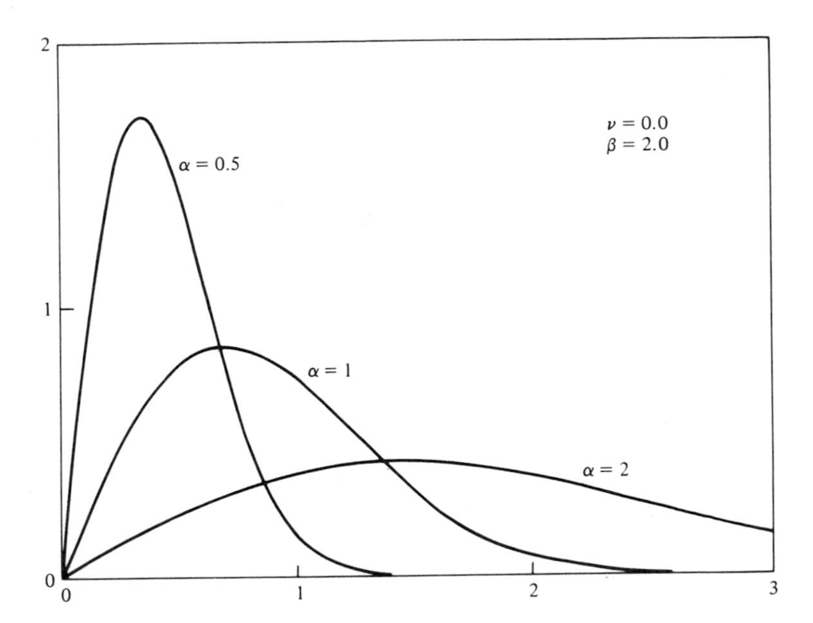

EXHIBIT 3.1 *Graphs of the Weibull density function for $\nu = 0.0$, $\beta = 2.0$, $\alpha = 0.5$, 1.0, 2.0.*

exponential distribution with parameter θ is the special case $\beta = 1$, $\alpha = 1/\theta$, $\nu = 0$ of the Weibull distribution.

Probability Calculations

The cumulative distribution function of the Weibull distribution is

$$F(x) = \begin{cases} 1 - \exp\left[-\left(\dfrac{x-\nu}{\alpha}\right)^{\beta}\right], & \text{if } x \geq \nu. \\ 0, & \text{if } x < \nu. \end{cases} \tag{3.2}$$

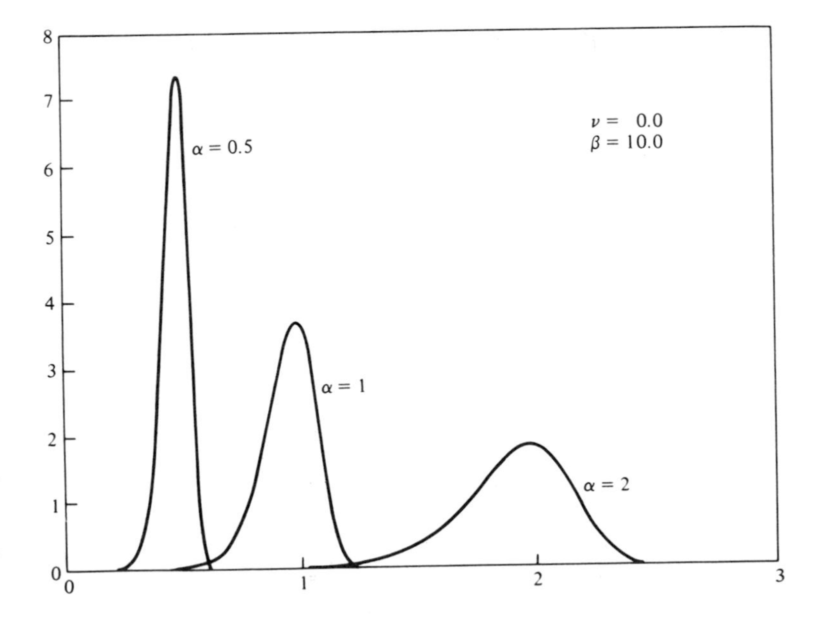

EXHIBIT 3.2 *Graphs of the Weibull density function for $\nu = 0.0$, $\beta = 10.0$, $\alpha = 0.5$, 1.0, 2.0.*

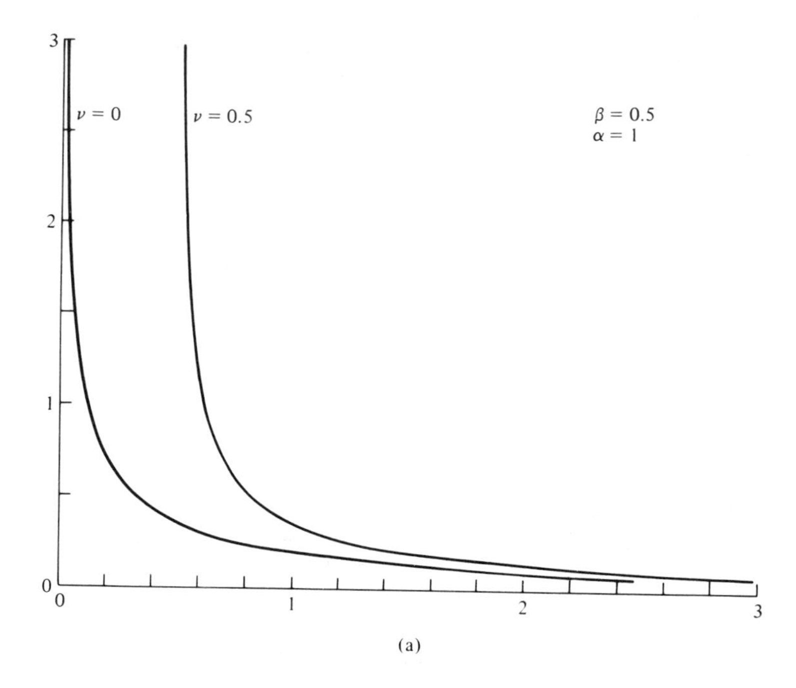

(a)

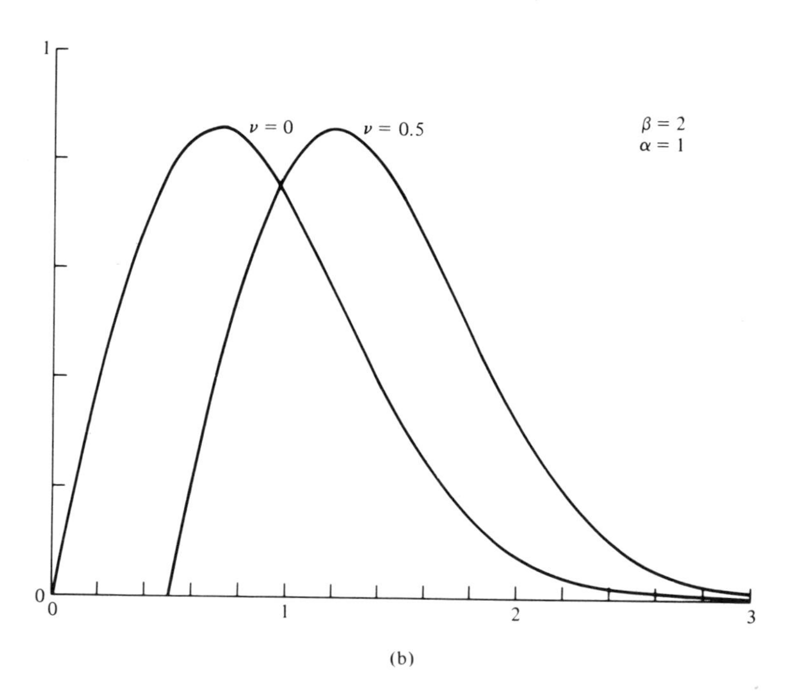

(b)

EXHIBIT 3.3 *Graphs of the Weibull density functions for (a) $\beta = 0.5$, $\alpha = 1.0$, $\nu = 0.0$, 0.5; (b) $\beta = 2.0$, $\alpha = 1.0$, $\nu = 0.0$, 0.5.*

To verify (3.2), note first that because the density $f(x)$ is 0 for $x < \nu$, it must likewise be the case that $F(x) = 0$ when $x < \nu$. For $x \geq \nu$, note that by the chain rule,

$$\frac{d}{dt}\left[\exp\left[-\left(\frac{t-\nu}{\alpha}\right)^{\beta}\right]\right] = -\frac{\beta}{\alpha}\left(\frac{t-\nu}{\alpha}\right)^{\beta-1}\exp\left[-\left(\frac{t-\nu}{\alpha}\right)^{\beta}\right].$$

Consequently,

$$\begin{aligned} F(x) &= \int_{\nu}^{x}\frac{\beta}{\alpha}\left(\frac{t-\nu}{\alpha}\right)^{\beta-1}\exp\left[-\left(\frac{t-\nu}{\alpha}\right)^{\beta}\right]dt = \left[-\exp\left\{-\left(\frac{t-\nu}{\alpha}\right)^{\beta}\right\}\right]_{\nu}^{x} \\ &= -\exp\left[-\left(\frac{x-\nu}{\alpha}\right)^{\beta}\right] + 1, \end{aligned}$$

as asserted in (3.2).

Because the survival function of the Weibull distribution has the simpler form

$$\bar{F}(x) = 1 - F(x) = \begin{cases} \exp\left[-\left(\dfrac{x-\nu}{\alpha}\right)^{\beta}\right], & x \geq \nu, \\ 1, & x < \nu, \end{cases} \tag{3.3}$$

it is usually more convenient to use the survival function, rather than the cumulative distribution function, to compute probabilities. If X has a Weibull distribution with parameters β, α, and ν, then because X is a continuous random variable,

$$\begin{aligned} P\{a \leq X \leq b\} &= P\{a < X \leq b\} = P\{a \leq X < b\} = P\{\alpha < X < b\} \\ &= \bar{F}(a) - \bar{F}(b) \\ &= \exp\left[-\left(\frac{a-\nu}{\nu}\right)^{\beta}\right] - \exp\left[-\left(\frac{b-\nu}{\alpha}\right)^{\beta}\right]. \end{aligned}$$

Relations Between the Weibull and Exponential Distributions

There are two interesting relationships between the exponential and Weibull distributions:

1. If X has a Weibull distribution with parameters β, α, ν, then $Y = [(X - \nu)/\alpha]^{\beta}$ has an exponential distribution with parameter 1.
2. If Y has an exponential distribution with parameter $\theta = 1$, then $X = \alpha Y^{1/\beta} + \nu$ has a Weibull distribution with parameters β, α, ν.

In support of assertion 1, note that if X has an Weibull distribution with parameters β, α, ν, then if $Y = [(X - \nu)/\alpha]^{\beta}$ we have, by (3.2),

$$\begin{aligned} F_Y(y) &= P\{Y \le y\} = P\left\{\left(\frac{X - \nu}{\alpha}\right)^{\beta} \le y\right\} \\ &= P\{X \le \alpha y^{1/\beta} + \nu\} = F_X(\alpha y^{1/\beta} + \nu) \\ &= 1 - \exp\left[-\left(\frac{\alpha y^{1/\beta} + \nu - \nu}{\alpha}\right)^{\beta}\right] \\ &= 1 - \exp(-y) \end{aligned}$$

for $y \ge 0$. (The transformation from X to Y is monotone increasing because $\beta, \alpha > 0$.) Note also that because $P\{X < \nu\} = 0$, $Y = [(X - \nu)/\alpha]^{\beta}$ can never be negative; thus $F_Y(y) = 0$ for $y < 0$. Hence

$$F_Y(y) = \begin{cases} 1 - \exp(-y), & \text{if } y \ge 0, \\ 0, & \text{if } y < 0, \end{cases} \tag{3.4}$$

which is the cumulative distribution function (1.2) of the exponential distribution with parameter $\theta = 1$. Because the cumulative distribution function of a random variable determines its distribution, assertion 1 has been verified.

On the other hand, if Y has the exponential distribution with parameter $\theta = 1$, then Y has the cumulative distribution function (3.4). Note that $X = \alpha Y^{1/\beta} + \nu$ always exceeds ν, because $P\{Y \ge 0\} = 1$. Thus the cumulative distribution function $F(x)$ of X equals 0 when $x < \nu$. For $x \ge \nu$,

$$\begin{aligned} F(x) &= P\{X \le x\} = P\{\alpha Y^{1/\beta} + \nu \le x\} \\ &= P\left\{Y \le \left(\frac{x - \nu}{\alpha}\right)^{\beta}\right\} = F_Y\left(\left(\frac{x - \upsilon}{\alpha}\right)^{\beta}\right) \\ &= 1 - \exp\left[-\left(\frac{x - \nu}{\alpha}\right)^{\beta}\right]. \end{aligned}$$

Thus $F(x)$ has the form (3.2), and we have shown that X has a Weibull distribution with parameters β, α, ν.

Remark. Both the gamma family of distributions and the Weibull family of distributions generalize the exponential family of distributions in the sense that an exponentially distributed random variable X with parameter θ can also be said to have a gamma distribution with parameters θ and $r = 1$, and a Weibull distribution with parameters $\alpha = 1/\theta$, $\beta = 1$, and $\nu = 0$. However, there are gamma distributions that are not also Weibull distributions (e.g., any gamma distribution with parameter $r \neq 1$) and Weibull distributions that are not gamma distributions (e.g., any Weibull distributions with $\beta \neq 1$ or $\nu \neq 0$).

Descriptive Indices of the Weibull Distribution

We have seen that if X has a Weibull distribution with parameters β, α, and ν, and Y has an exponential distribution with parameter $\theta = 1$, then X and $\alpha Y^{1/\beta} + \nu$ have the same distribution, and hence have the same mean, variance, and other moments. Thus,

$$\mu_X = E(\alpha Y^{1/\beta} + \nu) = \alpha E(Y^{1/\beta}) + \nu,$$
$$E(X^2) = E(\alpha Y^{1/\beta} + \nu)^2 = \alpha^2 E(Y^{2/\beta}) + 2\nu\alpha E(Y^{1/\beta}) + \nu^2.$$

Recall that if Y has an exponential distribution with parameter $\theta = 1$ and s is a positive constant, then

$$E(Y^s) = \Gamma(s + 1).$$

Consequently,

$$\mu_X = \alpha E(Y^{1/\beta}) + \nu = \alpha\Gamma\left(1 + \frac{1}{\beta}\right) + \nu,$$

and

$$\begin{aligned}\sigma_X^2 &= E(X^2) - \mu_X^2 \\ &= \alpha^2\Gamma\left(1 + \frac{2}{\beta}\right) + 2\nu\alpha\Gamma\left(1 + \frac{1}{\beta}\right) + \nu^2 - \left[\alpha\Gamma\left(1 + \frac{1}{\beta}\right) + \nu\right]^2 \\ &= \alpha^2\left[\Gamma\left(1 + \frac{2}{\beta}\right) - \left(\Gamma\left(1 + \frac{1}{\beta}\right)\right)^2\right].\end{aligned}$$

Summarizing, we have:

If X has a Weibull distribution with parameters β, α, and ν, then

$$\mu_X = \alpha\Gamma\left(1 + \frac{1}{\beta}\right) + \nu, \qquad \sigma_X^2 = \alpha^2\left[\Gamma\left(1 + \frac{2}{\beta}\right) - \left(\Gamma\left(1 + \frac{1}{\beta}\right)\right)^2\right].$$

Note that the variance σ_X^2 of a random variable X having a Weibull distribution with parameters β, α, and ν depends on β and α, but not on ν.

This should not be surprising because we have already seen (Exhibit 3.3) that ν acts as a measure of location and does not affect the shape of the graph of the density function $f(x)$.

The pth quantile, $Q_X(p)$, of the Weibull distribution with parameters β, α, ν is found by solving

$$F(x) = 1 - \exp\left[-\left(\frac{x - \nu}{\alpha}\right)^{\beta}\right] = p$$

for x. Thus

$$Q_X(p) = \alpha\left[\log\left(\frac{1}{1 - p}\right)\right]^{1/\beta} = \nu. \tag{3.5}$$

For $\beta \leq 1$, the mode of the Weibull distribution is ν, regardless of the value of the parameter α. For $\beta > 1$,

$$\text{Mode}(X) = \alpha\left(\frac{\beta - 1}{\beta}\right)^{1/\beta} + \nu.$$

A Theoretical Explanation

The experience of many investigators has shown that Weibull distributions provide good probability models for describing ''length of life'' and other endurance data. One explanation for this success is related to the following ''weakest link'' interpretation of endurance. Suppose that a system is put under stress. Think of the system as being composed of a large number of separate parts, each of which has its own random endurance time (lifetime). Assume that the lifetimes of these parts are mutually independent. If any one of these parts fails under stress, the entire system experiences failure. For example, the system could be a metal chain composed of a large number of links. If any link breaks, so does the chain. Thus the lifetime of the object is equal to the minimum lifetime of any of its parts. If the lifetime of any system has this property, it can be shown that a Weibull distribution provides a close approximation to the distribution of the system lifetime.

Fitting the Weibull Distribution to Data

In a paper published in 1951, Weibull provided three examples in which a distribution of the form (3.1) was fit to data. Since that time, there have been numerous other applications of the Weibull distribution.

EXAMPLE 3.1

Size Distribution of Particles

The term "fly ash" refers to fine solid particles of noncombustible ash that are carried out of a bed of solid fuel by the draft of combustion. Exhibit 3.4 shows a frequency distribution of the size (particle diameter) of fly ash, based on $n = 211$ observations, and the corresponding theoretical frequencies computed under the assumption that X has a Weibull distribution with parameters $\beta = 2.331$, $\alpha = 6.41$, $\nu = 1.50$.

As an example of the computations that yielded the theoretical frequencies in Exhibit 3.4, the entry for the interval $1.5 \leq x \leq 2.5$ was computed by multiplying the number of trials, 211, by the probability

$$P\{1.5 \leq X \leq 2.5\} = \left\{\exp\left[-\left(\frac{1.5 - 1.5}{6.41}\right)^{2.331}\right] - \exp\left[\left(\frac{2.5 - 1.5}{6.41}\right)^{2.331}\right]\right\}$$
$$= 1 - 0.9869 = 0.0131.$$ ♦

We can also use the survival function (3.3) to evaluate the fit of the Weibull distribution to data. In Exhibit 3.5 we compare the theoretical survival function and sample survival function based on the data in Example 3.1. This comparison suggests that the variability of the random variable in this example is adequately described by a Weibull distribution.

EXHIBIT 3.4 *Observed Frequency Distribution of the Size X of Fly Ash (in 20-Micron Units)*

Interval of Sizes	*Observed Frequency*	*Theoretical Frequency ($\beta = 2.331$, $\alpha = 6.41$, $\nu = 1.50$)*
1.5–2.5	3	2.76
2.5–3.5	11	10.76
3.5–4.5	20	19.52
4.5–5.5	22	26.73
5.5–6.5	29	30.74
6.5–7.5	41	30.95
7.5–8.5	24	27.73
8.5–9.5	25	22.32
9.5–10.5	13	16.23
10.5–11.5	9	10.68
11.5–12.5	5	6.33
12.5–13.5	6	3.42
>13.5	3	2.83
	211	211

Source: Data from W. Weibull, A statistical distribution function of wide applicability. *Journal of Applied Mechanics,* vol. 18, pp. 293–97 (1951).

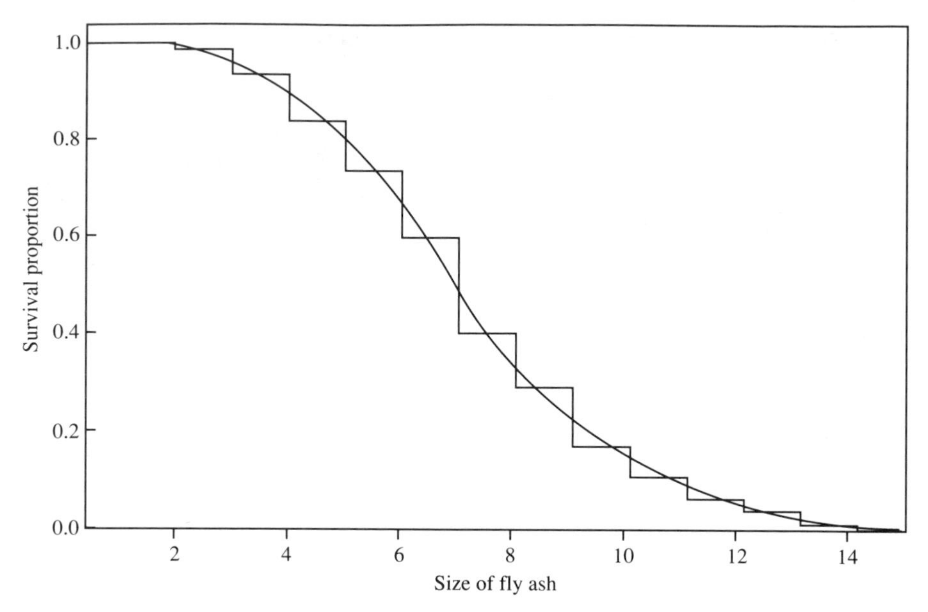

EXHIBIT 3.5 *Comparison of the observed and theoretical survival functions for the data of Example 3.1. The theoretical survival function is based on the Weibull distribution with parameters $\beta = 2.331$, $\alpha = 6.41$, $\nu = 1.5$.*

Note. In this example, only data that had already been grouped were available, so that the sample survival function $\overline{F}_n(x)$ could only be evaluated and grouped at the endpoints of the grouping intervals. A more detailed and revealing graphical comparison could be done if the individual data values were available. Note that for $x > \nu$,

$$\begin{aligned}\log(-\log \overline{F}(x)) &= \log(-\log\{\exp - [(x - \nu)/\alpha]^{\beta}\} \\ &= \beta \log(x - \nu) - \beta \log \alpha.\end{aligned}$$

If a value for ν can be specified, this suggests plotting values of $y = \log(-\log \overline{F}_n(x))$ versus $z = \log(x - \nu)$ for each of the observed x-values. As in the case of the exponential distribution, the problem of taking the logarithm of 0 prompts us to make the adjustment $y^* = \log\{-\log[(n\overline{F}_n(x) + \frac{1}{2})/(n + 1)]\}$, and to graph y^* versus z instead. If X has a Weibull distribution, the resulting plot will be close to a straight line with slope β and intercept equal to $-\beta \log \alpha$. Rather than taking the effort to make the calculations of y^* and z, modelers make use of specially scaled Weibull paper to plot $\overline{F}_n(x)$ versus x.

4 The Uniform Distribution

A continuous random variable X has a **uniform distribution** over the interval from a to b, denoted by $X \sim U[a, b]$, if intervals of possible values that have equal length have equal probability of containing an observed value of X. The probability density function $f(x)$ of a uniformly distributed random variable X on a to b is given by the equation

$$f(x) = \begin{cases} \dfrac{1}{b-a}, & \text{if } a \le x \le b, \\ 0, & \text{if } x < a \text{ or } x > b. \end{cases} \tag{4.1}$$

The interval endpoints a and b, $-\infty < a < b < \infty$, are the parameters of the uniform distribution. Note that a is the smallest possible value of X and that b is the largest possible value of X. The density (4.1) is uniform in value over the interval $[a, b]$.

Graphs of the density functions $f(x)$ for a $U[0, 1]$ and a $U[1, 3]$ distribution are shown in Exhibit 4.1. From Exhibit 4.1, we can see why the uniform distribution is also called the **rectangular distribution.**

Probability Calculations

Probability calculations for the uniform distribution are quite easy to perform. Suppose, for example, that $X \sim U[a, b]$; then

$$F(x) = \begin{cases} 0, & \text{if } x < a, \\ \dfrac{x-a}{b-a}, & \text{if } a \le x \le b, \\ 1, & \text{if } x > b. \end{cases} \tag{4.2}$$

The graph of $F(x)$ for $a = 1$, $b = 3$, is shown in Exhibit 4.2.

EXAMPLE 4.1

A person who wants to catch a 6:00 P.M. train takes a cab at 5:43 P.M. from an intersection that under ideal driving conditions is 10 minutes from the train station, but which can be as much as 50 minutes driving time away from the station if traffic is heavy. Assume that $X \sim U[10, 50]$, where X is the driving time from the intersection to the station, and assume that upon reaching the station, the person needs 2 minutes to board the train. Thus if the person is to catch the train, X must be no greater than 15 minutes. The probability that the person catches the train is thus

$$P\{X \le 15\} = F(15) = \frac{15-10}{50-10} = \frac{1}{8} = 0.125.$$

◆

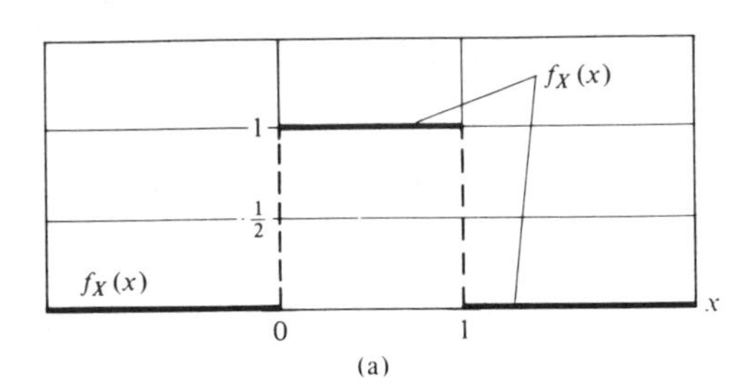

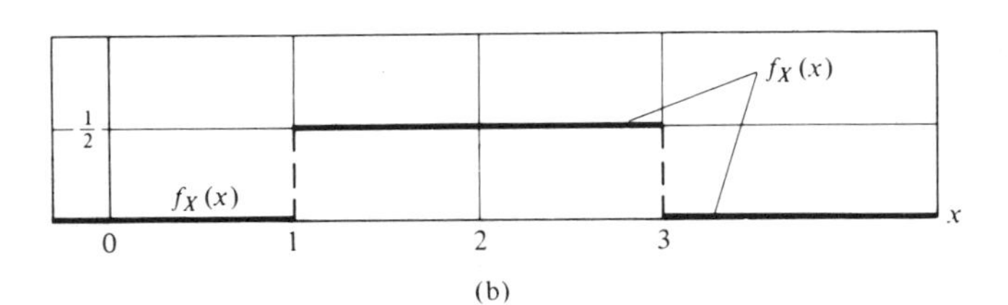

EXHIBIT 4.1 *Graphs of the uniform probability density function $f(x)$ for (a) $a = 0$, $b = 1$; (b) $a = 1$, $b = 3$.*

EXAMPLE 4.2

A biologist takes homing pigeons from their cote and keeps them for several days in a dark room. If the pigeons orient themselves by light, they may be confused when they are brought back into the open air. The assumption is that once released, a confused pigeon will pick a direction in which to fly according to a uniform distribution. More precisely, if the bird is released at a point A (see Exhibit 4.3), its line of flight will make an angle of X degrees with the fixed line of orientation L, where $X \sim U[0, 360]$. It has been observed that correct lines of flight, taken by pigeons who were not confused, vary from an angle of 210 degrees from line L to an angle of 220 degrees from line L. Under the assumption of a uniform distribution, the probability that a confused pigeon chooses a correct line of flight is

$$P\{210 \le X \le 220\} = F(220) - F(210) = \frac{220 - 0}{360 - 0} - \frac{210 - 0}{360 - 0} = \frac{1}{36}.$$

Thus, under the assumption of a uniform distribution, it is rather improbable that a confused pigeon will choose a correct line of flight. ♦

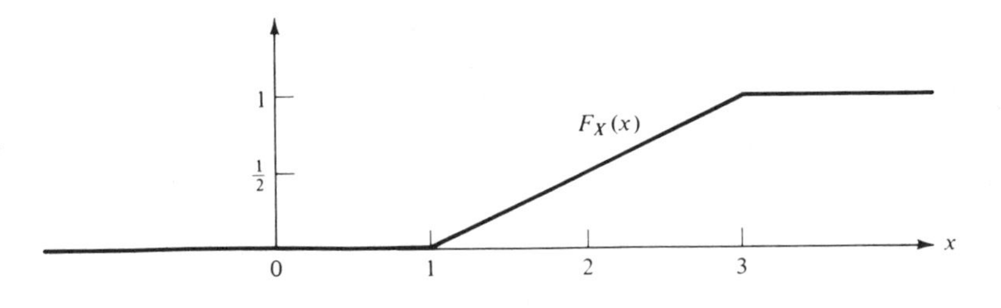

EXHIBIT 4.2 *Graph of the cumulative distribution function $F(x)$ of the uniform distribution with parameters $a = 1$, $b = 3$.*

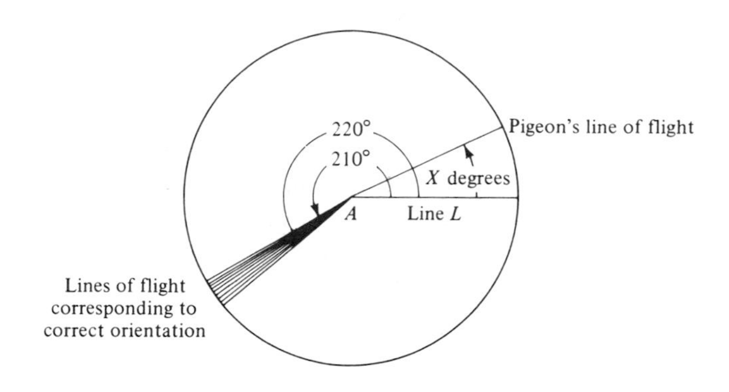

EXHIBIT 4.3 *Diagram illustrating the experiment described in Example 4.2. The pigeon is released at point A. Its line of flight makes an angle of X degrees with the fixed line L. Lines of flight making angles of 210–220 degrees correspond to orientations ordinarily taken by pigeons returning home.*

Descriptive Indices of the Uniform Distribution

If $X \sim U[a, b]$, then for $k \geq 0$,

$$E(X^k) = \int_a^b x^k\left(\frac{1}{b-a}\right) dx = \left(\frac{1}{b-a}\right) \frac{x^{k+1}}{k+1}\Bigg|_a^b = \frac{b^{k+1} - a^{k+1}}{(k+1)(b-a)},$$

from which

$$\text{(4.3)} \qquad \begin{aligned} \mu_X &= E(X) = \frac{b^2 - a^2}{2(b-a)} = \frac{b+a}{2}, \\ \sigma_X^2 &= E(X^2) - (\mu_X)^2 = \frac{b^3 - a^3}{3(b-a)} - \left(\frac{b+a}{2}\right)^2 = \frac{(b-a)^2}{12}. \end{aligned}$$

Also, the pth quantile $Q_X(p)$ is the solution, for x, of the equation

$$F(x) = \frac{x-a}{b-a} = p,$$

so that

$$\text{(4.4)} \qquad Q_X(p) = p(b-a) + a = pb + (1-p)a.$$

Finally, all values of x between a and b are modes of the $U[a, b]$ distribution (see Exhibit 4.1).

Applications

Rounding Errors. In measurement theory, the uniform distribution with $a = 0$, $b = (0.5)10^{-k}$ is often used to represent the distribution of "rounding-off" errors in values tabulated to the nearest k decimal places. That is, the unsigned difference, $X = |Y - Y_R|$, between the actual observed value Y of a random variable and the rounded-off value Y_R is assumed to have a uniform distribution with parameters $a = 0$ and $b = (0.5)10^{-k}$.

Computer Simulations. The $U[0, 1]$ distribution is widely used in what are called **Monte Carlo simulation methods.** Computer programs called **random number generators** provide a succession of values that resemble, in their variation, observations obtained from the $U[0, 1]$ distribution. Using the output of such a program, we can generate observations made upon a random variable Y having *any* continuous cumulative distribution function $F_Y(y)$ as follows:

1. Use the random number generator to generate a value x of $X \sim U[0, 1]$.
2. Solve the equation $F_Y(y) = x$ for y. This solution y^* is then the observed value of the simulated random variable Y. [If the solution is not unique, we let y^* be the smallest solution of $F_Y(y) = x$.]

Analytic and graphical methods for solving the equation $F_Y(y) = x$ were discussed in Chapter 6, Section 3. By repeating steps 1 and 2 over and over, any desired number of observations of Y can be simulated.

EXAMPLE 4.3

Simulation of an Exponentially Distributed Random Variable

Suppose that we desire to simulate a random variable Y having an exponential distribution with parameter $\theta = 1$. The cumulative distribution function of Y is $F_Y(y) = 1 - e^{-y}$. Using a random number generator, we generate a value x of $X \sim U[0, 1]$, and solve

$$F_Y(y) = 1 - e^{-y} = x$$

for y. Thus $y^* = -\log(1 - x)$. For example, if we generate $x = 0.93$ for X, then $y^* = -\log(0.07) = 2.65926$. ♦

5 The Beta Distribution

There are many continuous quantitative phenomena that take on values bounded above and below by known numbers a and b. We have given examples of such phenomena in Section 4. Other examples include (1) the distance from one end of a steel bar of known length to the point at which failure occurs when the bar is placed under stress, (2) the proportion of total farm acreage spoiled by a certain kind of fungus, (3) the percentage of defective items in a given shipment of items, (4) the ratio of the length of the femur to the total length of the leg of a given person, and (5) the fraction of people in a given population who are able to answer a given question correctly at a given time. Of these last examples, notice that examples 2, 4, and 5 all involve random variables that are fractions; thus these variables have possible values between $a = 0$ and $b = 1$. The random variable described in example 3 is a percentage and has possible values between $a = 0$ and $b = 100$. Finally, the random variables mentioned in example 1 have possible values between $a = 0$ and $b =$ length of the steel bar.

A class of distributions that includes the uniform distribution and is rich enough to provide models for most random variables having a restricted range of possible values is the class of *beta distributions*. Because the class of beta distributions is defined only for random variables Y in $[0, 1]$, if X is in the interval $[a, b]$, we first find a distribution to fit the transformed random variable

$$Y = \frac{X - a}{b - a},$$

which has possible values between 0 and 1. Once we have found a beta distribution that describes the variability of Y, we transform back to obtain the distribution of X.

EXAMPLE 5.1

Let X be the distance (in inches), from the end of a 12-inch steel bar to the point at which the bar breaks under stress. Then X is a random variable with possible values between $a = 0$ and $b = 12$. To find a distribution for X, we first find a beta distribution that describes the variation of $Y = (X - 0)/(12 - 0) = X/12$. ◆

EXAMPLE 5.2

Let X denote the proportion of a geographical region that receives rain when a certain type of cloud formation is observed over that region. Because X is a proportion, the possible values of X are between $a = 0$ and $b = 1$. In this case, $Y = X$, so that no transformation is needed. ◆

If X has a beta distribution, then the probability density function $f(x)$ of X has the form

$$(5.1) \qquad f(x) = \begin{cases} \dfrac{x^{r-1}(1 - x)^{s-1}}{B(r, s)}, & \text{if } 0 \leq x \leq 1, \\ 0, & \text{if } x < 0 \text{ or } x > 1. \end{cases}$$

Here r and s are positive numbers and are the parameters of the beta distribution. The constant $B(r, s)$, called the *beta function* (see Appendix A), is defined by

$$B(r, s) = \frac{\Gamma(r)\Gamma(s)}{\Gamma(r+s)}. \tag{5.2}$$

To illustrate the great diversity of shapes taken on by the graph of the probability density function $f(x)$ of the beta distribution, Exhibit 5.1 gives graphs of $f(s)$ for various choices of r and s. The case $r = s = 1$, not shown in Exhibit 5.1, yields the $U[0, 1]$ distribution. For $r < 1$ or $s < 1$, the density $f(x)$ is infinitely large at $x = 0$ or $x = 1$, respectively. When $r < 1$ and $s \geq 1$, the graph of $f(x)$ is decreasing in x and concave, while it is increasing and concave when $r \geq 1$ and $s < 1$. When r and s are both less than 1, the graph is "U-shaped," while when $r > 1$ and $s > 1$, the graph has a single hump. Finally, when $r = s$, the graph of $f(x)$ is symmetric about $x = \frac{1}{2}$; otherwise, it is skewed.

Probability Calculations for the Beta Distribution

If X has a beta distribution with parameters r and s, the cumulative distribution function $F(x)$ of X is given by

$$F(x) = \begin{cases} 0, & \text{if } x < 0, \\ I_x(r, s), & \text{if } 0 \leq x < 1, \\ 1, & \text{if } x \geq 1, \end{cases} \tag{5.3}$$

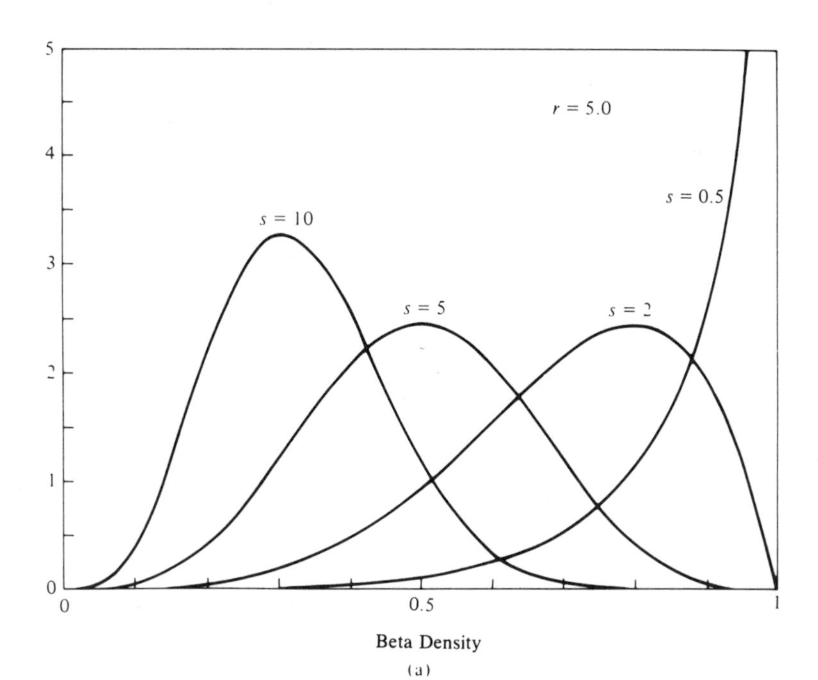

EXHIBIT 5.1 *Graphs of the beta density function for (a) r = 5.0, s = 0.5, 2.0, 5.0, 10.0; (b) r = 2.0, s = 0.5, 1.0, 2.0, 10.0; (c) r = 0.5, s = 0.5, 2.0, 10.0.*

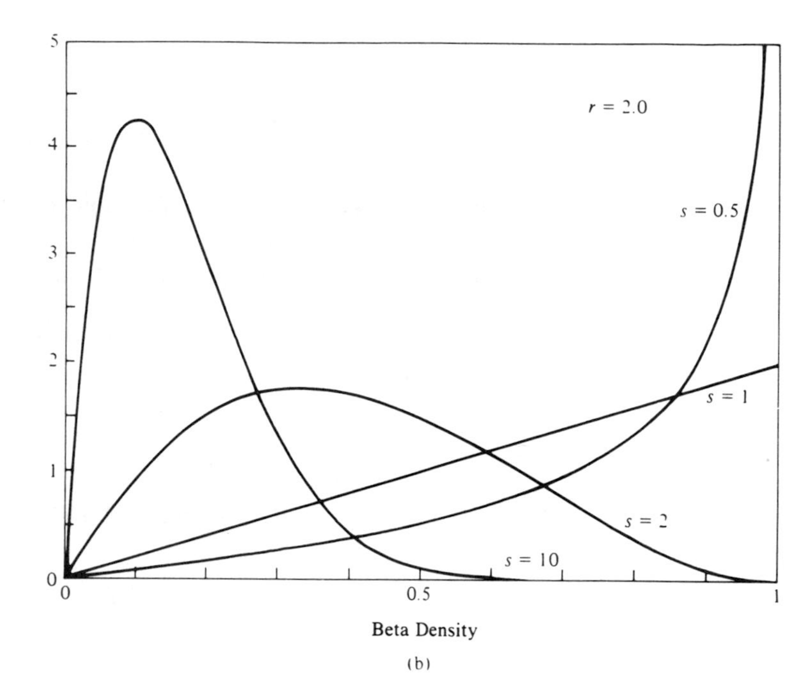

(b)

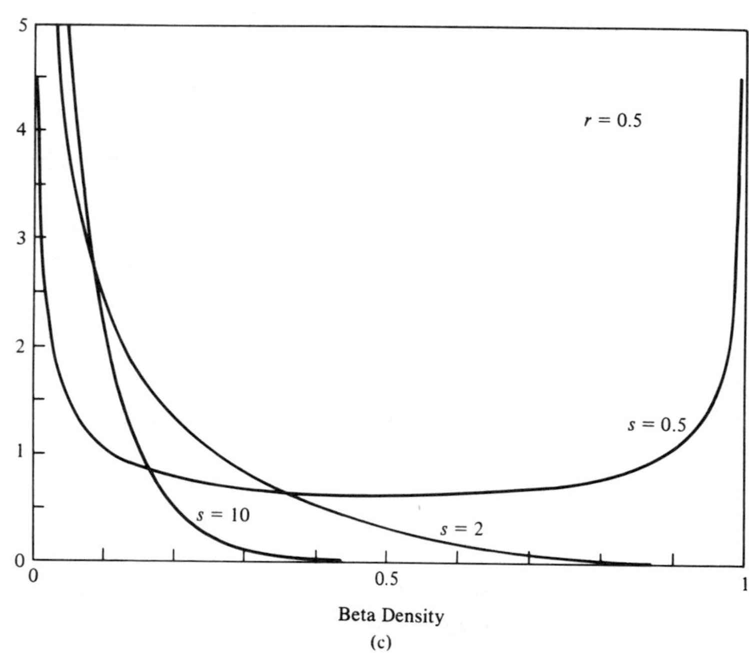

(c)

EXHIBIT 5.1 *(continued)*

where

$$I_x(r, s) = \int_0^x \frac{t^{r-1}(1-t)^{s-1}}{B(r, s)}\, dt \tag{5.4}$$

is known as the **incomplete beta function.** Except in special cases (e.g., $s = 1$), this function cannot be expressed in terms of a simple formula involving r, s, and x and thus must be tabled. Because we need a separate table for every pair (r, s) of parameter values, a complete book of tables is required to give $F(x)$ for all pairs (r, s) that might be met in practice. An extensive set of tables for the incomplete beta function is *Tables of the Incomplete Beta Function* (1934), edited by Karl Pearson. (In that book r, q is used rather than our r, s.)

In certain special cases, it is possible to write explicit formulas for $I_x(r, s)$. For example, when $r > 0$ and $s = 1$, and $0 \le x \le 1$,

$$I_x(r, 1) = \int_0^x \frac{t^{r-1}\, dt}{B(r, 1)} = \frac{t^r}{rB(r, 1)}\bigg|_0^x = x^r,$$

because $B(r, 1) = \Gamma(r)\Gamma(1)/\Gamma(r+1) = 1/r$. Similarly, when $r = 1$ and $s > 0$,

$$I_x(1, s) = 1 - (1-x)^s.$$

When $r > 0$, $s \ge 1$, integration by parts in (5.4) yields the recursion relation

$$\begin{aligned} I_x(r, s) &= \int_0^x \frac{t^{r-1}(1-t)^{s-1}}{B(r, s)}\, dt = \frac{t^r(1-t)^{s-1}}{rB(r, s)}\bigg|_0^x + \int_0^x \frac{t^r(1-t)^{s-2}\, dt}{[rB(r, s)/(s-1)]} \\ &= \frac{\Gamma(r+s)x^r(1-x)^{s-1}}{\Gamma(r+1)\Gamma(s)} + I_x(r+1, s-1), \end{aligned} \tag{5.5}$$

using the readily verified fact that

$$\frac{rB(r, s)}{s-1} = B(r+1, s-1).$$

Now when s is a positive integer, and $r \ge 0$, (5.5) can be used repeatedly to derive the result

$$I_x(r, s) = \sum_{i=0}^{s-1} \left[\frac{\Gamma(r+s)}{\Gamma(r+i+1)\Gamma(s-i)}\right] x^{r+i}(1-x)^{s-i-1}. \tag{5.6}$$

Finally, when both r and s are positive integers, then $r + s$ is a positive integer, $r + i$ is an integer, and

$$\frac{\Gamma(r + s)}{\Gamma(r + i + 1)\Gamma(s - i)} = \frac{(r + s - 1)!}{(r + i)!\,(s - i - 1)!} = \binom{r + s - 1}{r + i}.$$

Substituting into (5.6) and changing the index of summation from i to $j = r + i$, we obtain the following useful result: If r and s are both positive integers, then for $0 \leq x \leq 1$,

$$I_x(r, s) = \sum_{j=r}^{r+s-1} \binom{r + s - 1}{j} x^j(1 - x)^{r+s-1-j}. \tag{5.7}$$

We recognize the sum on the right side of (5.7) as being the probability that a random variable U with a binomial distribution with parameters $n = r + s - 1$ and $p = x$ is greater than or equal to r.

Relationship Between Beta and Binomial Distributions. If X has a beta distribution with parameters r, s which are both positive integers, then for $0 \leq x \leq 1$,

$$F(x) = P\{X \leq x\} = P\{U \geq r\} = 1 - F_U(r - 1), \tag{5.8}$$

where U has a binomial distribution with parameters $n = r + s - 1$ and $p = x$.

EXAMPLE 5.3

Let X be the fraction of saturation of dissolved oxygen in a river at a given point, and suppose that the distribution of X in pure water is a beta distribution with parameters $r = 3$, $s = 2$. A water quality monitoring system declares the river water to be impure when the observed value of X goes below a certain threshold level x^*. This system should not falsely declare the river water to be impure too often (a "false alarm") when the water actually is pure. If $x^* = 0.40$, then from (5.8) and Exhibit B.2 of Appendix B with $n = 3 + 2 - 1 = 4$ and $p = 0.40$,

$$\begin{aligned} P\{\text{false alarm}\} &= P\{X \leq 0.40\} = F(0.40) = P\{U \geq 3\} \\ &= 0.15360 + 0.02560 = 0.17920. \end{aligned}$$

If this probability of false alarm is too large, the threshold can be lowered. For the threshold level $x^* = 0.20$, Exhibit B.2 with $n = 4$ and $p = 0.20$ is used to find the probability of a false alarm,

$$P\{X \leq 0.20\} = F(0.20) = P\{U \geq 3\} = 0.02560 + 0.00160 = 0.02720,$$

which is satisfactorily low. ◆

Using the cumulative distribution function $F(x)$ and either (5.8) or tables of $I_x(r, s)$, probabilities of events such as $\{a \leq X \leq b\}$ and $\{X > a\}$ can be obtained.

EXAMPLE 5.4

A part of sleep called *REM sleep* is highly associated with dreaming. Suppose that the fraction, X, of the total sleeping time of normal sleepers spent in REM sleep has a beta distribution with parameters $r = 12$, $s = 48$. A "typical range of values" for X is defined to be an interval $[x_1, x_2]$ such that $P\{x_1 \leq X \leq x_2\} = 0.95$. It is suggested that $x_1 = 0.10$ and $x_2 = 0.30$ meet this requirement. Although $r = 12$ and $s = 48$ are both integers, so that (5.8) applies, $r + s - 1 = 59$ is too large for us to use Exhibit B.2 of Appendix B. Exhibit 5.2 is a condensed table of values of $I_x(12, 48)$, for $x = 0.00(0.02)0.48$. Note that the first decimal place of x indexes the row and the last decimal place the column. That is, the value of $I_{0.12}(12, 48)$ is found at the intersection of the row marked 0.1 and the column marked 0.02; $I_{0.12}(12, 48) = 0.04552$. Using this table and (5.3),

$$\begin{aligned} P\{0.10 \leq X \leq 0.30\} &= F(0.30) - F(0.10) \\ &= I_{0.30}(12, 48) - I_{0.10}(12, 48) \\ &= 0.96507 - 0.01280 = 0.95227. \end{aligned}$$ ◆

One final, and useful result should be noted:

> If X has a beta distribution with parameters r and s, then $1 - X$ has a beta distribution with parameters s and r.

Indeed, if $V = 1 - X$, then it follows from (5.4) that

$$\begin{aligned} F_V(v) &= P\{V \leq v\} = P\{1 - X \leq v\} = P\{X \geq 1 - v\} \\ &= \int_{1-v}^{1} \frac{x^{r-1}(1 - x)^{s-1}}{B(r, s)}\, dx = \int_0^v \frac{t^{s-1}(1 - t)^{r-1}}{B(s, r)}\, dt = I_v(s, r), \end{aligned}$$

EXHIBIT 5.2 *Values of the Incomplete Beta Function $I_x(12, 48)$*

x	0.00	0.02	0.04	0.06	0.08
0.0	0.00000	0.00000	0.00000	0.00017	0.00219
0.1	0.01280	0.04552	0.11496	0.22646	0.37037
0.2	0.52582	0.67041	0.78857	0.87468	0.93127
0.3	0.96507	0.98353	0.99279	0.99707	0.99889
0.4	0.99961	0.99987	0.99996	0.99999	1.00000

where we have changed variable of integration from x to $t = 1 - x$ and used the fact that

$$B(r, s) = \frac{\Gamma(r)\Gamma(s)}{\Gamma(r + s)} = B(s, r).$$

Because $F_V(v) = I_v(s, r)$ is, by (5.3), the cumulative distribution function of a beta distribution with parameters s and r, verification of the assertion is complete.

Our argument has also shown that

$$I_x(r, s) = 1 - I_{1-x}(s, r) \tag{5.9}$$

because

$$\begin{aligned} I_x(r, s) &= P\{X \le x\} = P\{1 - X \ge 1 - x\} = 1 - F_{1-X}(1 - x) \\ &= 1 - I_{1-x}(s, r). \end{aligned}$$

The result (5.9) is very useful for tabling the incomplete beta function, because it means that we do not have to tabulate $I_x(r, s)$ for values of x in the interval (0.50, 1].

As an example of the use of (5.9), suppose that X has a beta distribution with parameters $r = 48$ and $s = 12$, and that we wish to calculate $P\{0.58 \le X \le 0.72\}$. Because Exhibit 5.2 gives values of the cumulative distribution function of a beta distribution with parameters $r = 12$ and $s = 48$, and $1 - X$ has such a distribution,

$$\begin{aligned} P\{0.58 \le X \le 0.72\} &= P\{1 - 0.72 \le 1 - X \le 1 - 0.58\} \\ &= F_{1-X}(0.42) - F_{1-X}(0.28) \\ &= 0.99987 - 0.93127 = 0.06860. \end{aligned}$$

Descriptive Indices of the Beta Distribution

It is shown in Appendix A that when $a \ge 0$, $b \ge 0$,

$$B(a, b) = \int_0^1 u^{a-1}(1 - u)^{b-1}\, du.$$

Thus when the random variable X has a beta distribution with parameters r and s,

$$E(X^k) = \frac{1}{B(r, s)} \int_0^1 x^{r+k-1}(1 - x)^{s-1}\, dx = \frac{B(r + k, s)}{B(r, s)} = \frac{\Gamma(r + k)\Gamma(r + s)}{\Gamma(r)\Gamma(r + s + k)},$$

for $k \geq 0$. Applying this result when $k = 0$ shows that $\int_0^1 f(x)\,dx = 1$. Because $f(x) \geq 0$ for all x, $f(x)$ is indeed a probability density function. Further,

$$\mu_X = E(X) = \frac{\Gamma(r+1)\Gamma(r+s)}{\Gamma(r)\Gamma(r+s+1)} = \frac{[r\Gamma(r)]\Gamma(r+s)}{\Gamma(r)[(r+s)\Gamma(r+s)]} = \frac{r}{r+s}$$

and

$$\begin{aligned}\sigma_X^2 = E(X^2) - (\mu_X)^2 &= \frac{\Gamma(r+2)\Gamma(r+s)}{\Gamma(r)\Gamma(r+s+2)} - \left(\frac{r}{r+s}\right)^2 \\ &= \frac{[(r+1)(r)\Gamma(r)]\Gamma(r+s)}{\Gamma(r)[(r+s+1)(r+s)\Gamma(r+s)]} - \left(\frac{r}{r+s}\right)^2 \\ &= \frac{(r+1)r}{(r+s+1)(r+s)} - \left(\frac{r}{r+s}\right)^2 = \frac{rs}{(r+s)^2(r+s+1)}.\end{aligned}$$

> If X has a beta distribution with parameters r and s, then
>
> $$\mu_X = \frac{r}{r+s}, \qquad \sigma_X^2 = \frac{rs}{(r+s)^2(r+s+1)}.$$

The quantiles $Q_X(p)$ of a beta distribution must be found by interpolation in tables of $F(x) = I_x(r, s)$. For example, if X has a beta distribution with parameters $r = 12$, $s = 48$, then from Exhibit 5.2 the 0.90th quantile, $Q_X(0.90)$, of X is

$$Q_X(0.90) = 0.26 + \frac{0.90000 - 0.87468}{0.93127 - 0.87468}(0.28 - 0.26) = 0.269$$

because $F(0.26) = I_{0.26}(12, 48) = 0.87468$, and $F(0.28) = I_{0.28}(12, 48) = 0.93127$. As can be seen from Exhibit 5.1, the beta distribution can have one mode or two widely separated modes (at $x = 0$ and $x = 1$).

Fitting the Beta Distribution to Data

EXAMPLE 5.5

Let X be the proportion of change in wholesale prices of commodities from one year to the next when the prices fall. That is, if the first year price is d_1 and the second year price is $d_2 \leq d_1$, then $X = (d_1 - d_2)/d_1$. If prices rise (i.e., $d_1 < d_2$), no value of X is recorded. A total of 2314 cases of falling commodity prices were observed. These data are summarized in the form of a frequency distribution in Exhibit 5.3.

EXHIBIT 5.3 *Frequency Distribution of the Proportion X of Change in Falling Wholesale Prices of Commodities from One Year to the Next*

Interval	*Observed Frequency*	*Theoretical Frequency (beta distribution, $r = 1$, $s = 10.63$)*
$0.00 \leq x < 0.04$	780	814.64
$0.04 \leq x < 0.08$	567	545.63
$0.08 \leq x < 0.12$	373	359.15
$0.12 \leq x < 0.16$	227	231.96
$0.16 \leq x < 0.20$	147	146.74
$0.20 \leq x < 0.24$	84	90.73
$0.24 \leq x < 0.28$	49	54.71
$0.28 \leq x < 0.32$	43	32.07
$0.32 \leq x < 0.36$	17	18.22
$0.36 \leq x < 0.40$	12	10.00
$0.40 \leq x < 0.44$	9	5.27
$0.44 \leq x < 0.48$	3	2.66
$0.44 \leq x < 0.52$	2	1.27
$0.52 \leq x < 0.56$	1	0.57
$x \geq 0.56$	0	0.31
	2314	2313.93

In Exhibit 5.3, theoretical frequencies for a beta distribution with parameters $r = 1$, $s = 10.63$, are calculated for comparison to the observed frequencies. Probability calculations for this beta distribution are quite simple because, as noted previously, the cumulative distribution function of X is

$$F(x) = I_x(1, 10.63) = 1 - (1 - x)^{10.63}.$$

Thus, for example,

$$\begin{aligned} P\{0.12 \leq X < 0.16\} &= I_{0.16}(1, 10.63) - I_{0.12}(1, 10.63) \\ &= (1 - 0.12)^{10.63} - (1 - 0.16)^{10.63} \\ &= 0.25695 - 0.15671 = 0.10024, \end{aligned}$$

so that the theoretical frequency of the interval $[0.12, 0.16)$ is $(2314)(0.10024) = 231.96$.

In Exhibit 5.4, a modified relative frequency histogram based on the data in Exhibit 5.3 is graphed along with the probability density function of the beta distribution with parameters $r = 1$ and $s = 10.63$. From Exhibits 5.3 and 5.4, it appears that the beta distribution provides a reasonably close fit to the observed variation of X. ◆

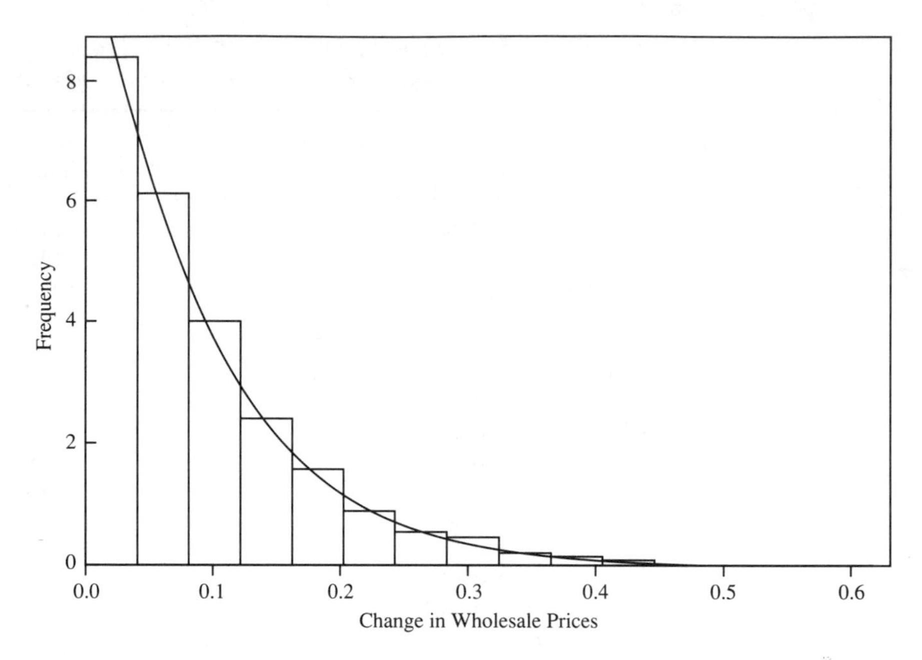

EXHIBIT 5.4 *Modified relative frequency histogram of data in Exhibit 5.3 compared to probability density function of a beta distribution with $r = 1$, $s = 10.63$.*

6 The Lognormal and Cauchy Distributions

Even though the normal, exponential, gamma, Weibull, uniform, and beta distributions are frequently used to model the variability of continuous random phenomena, many other distributions have proved to be useful in certain contexts. A survey of continuous distributions [Johnson and Kotz (1970)] lists over 50 different families of such distributions. Thus no short list of distributions can be expected to serve as models for all quantitative random phenomena.

In this section we describe briefly two additional families of continuous distributions, the lognormal distributions and the Cauchy distributions, both because of their importance as models of the variation of quantitative random phenomena, and also because these distributions illustrate facts and concepts about continuous probability distributions that are worthy of attention.

The Lognormal Distribution

The applicability of the normal distribution to many quantitative random phenomena can be justified theoretically by assuming that such quantitative random phenomena arise from the summation of many probabilistically independent and identically distributed random causes. The **lognormal distribution** arises from the product of many independent and identically distributed random variables. A random variable X has approximately a lognormal distribution if we can conceive of X as being equal to the product $\Pi_1^n W_i$, where $W_1, W_2, \ldots, W_n$ are probabilistically independent realizations of the same nonnegative random variable W. The name "lognormal" given to such a distribution comes from noting that

$$(6.1) \qquad Y = \log X = \log \prod_{i=1}^{n} W_i = \sum_{i=1}^{n} \log W_i.$$

Thus $Y = \log X$ is the sum of a large number n of independent, identically distributed random variables, and hence by the Central Limit Theorem, $Y = \log X$ is (approximately) normally distributed. A nonnegative random variable X has a lognormal distribution whenever $Y = \log X$ has a normal distribution.

Suppose that the random variable X has a lognormal distribution and that $Y = \log X$ is normally distributed with mean $\mu_Y = \xi$ and variance $\sigma_Y^2 = \delta^2$. Then $X = e^Y$, so that the cumulative distribution function of X for $x \geq 0$ is

$$(6.2) \quad F_X(x) = P\{X \leq x\} = P\{e^Y \leq x\} = P\{Y \leq \log x\} = F_Z\left(\frac{\log x - \xi}{\delta}\right),$$

where $F_Z(z)$ is the cumulative distribution function of a standard normal variable. Differentiating both sides of the equality

$$F_X(x) = F_Z\left(\frac{\log x - \xi}{\delta}\right), \qquad x \geq 0,$$

using the chain rule, we find that

$$f_X(x) = \frac{d}{dx} F(x) = \frac{d}{dx}\left(\frac{\log x - \xi}{\delta}\right) f_Z\left(\frac{\log x - \xi}{\delta}\right)$$

$$= \left(\frac{1}{x\delta}\right) \frac{\exp\left[-\frac{1}{2}\frac{(\log x - \xi)^2}{\delta^2}\right]}{\sqrt{2\pi}}, \qquad x \geq 0.$$

Here we have used the fact that (Chapter 9)

$$\frac{d}{dz} F_Z(z) = f_Z(z) = \frac{\exp(-\frac{1}{2}z^2)}{\sqrt{2\pi}}.$$

Thus the probability density function of X is

$$(6.3)\qquad f_X(x) = \begin{cases} \dfrac{\exp[-(1/2\delta^2)(\log x - \xi)^2]}{x\sqrt{2\pi\delta^2}}, & \text{if } x \geq 0, \\ 0, & \text{if } x < 0. \end{cases}$$

Note that because X is a nonnegative variable, $f_X(x) = 0$ for $x < 0$.

Because the density $f_X(x)$ depends on the constants ξ and $\delta^2 > 0$, as well as on x, these constants ξ, δ^2 are the parameters of the lognormal distribution.

In contrast to the graph of the density function of the normal distribution, the graph of the probability density function of the lognormal distribution is not a symmetric, bell-shaped curve. Rather, the lognormal distribution is positively skewed. In Exhibit 6.1, graphs of the probability density function (6.3) of the lognormal distribution are given for various values of ξ and δ^2.

Computing Probabilities for the Lognormal Distribution

As Equation (6.2) indicates, probabilities for the lognormal distribution can be obtained from Exhibit B.4 of Appendix B, which gives the cumulative distribution function of a standard normal distribution. That is, for $x > 0$,

$$F(x) = P\{X \leq x\} = F_Z\left(\frac{\log x - \xi}{\delta}\right),$$

where $F_Z(z)$ is the cumulative function of a standard normal random variable Z. Similarly, the survival function of the lognormal distribution is for $x > 0$,

$$\begin{aligned} \bar{F}_X(x) &= P\{X > x\} = 1 - P\{X \leq x\} = 1 - F_Z\left(\frac{\log x - \xi}{\delta}\right) \\ &= F_Z\left(\frac{\xi - \log x}{\delta}\right). \end{aligned}$$

Because negative values of X are not observed when X has a lognormal distribution, $F_X(x) = 0$ and $\bar{F}_X(x) = 1$ for $x < 0$. Finally, noting that the lognormal distribution is a continuous distribution,

$$\begin{aligned} P\{a \leq X \leq b\} &= P\{a < X \leq b\} = P\{a \leq X < b\} = P\{a < X < b\} \\ &= F_Z\left(\frac{\log b - \xi}{\delta}\right) - F_Z\left(\frac{\log a - \xi}{\delta}\right) \end{aligned}$$

for $0 < a \leq b$.

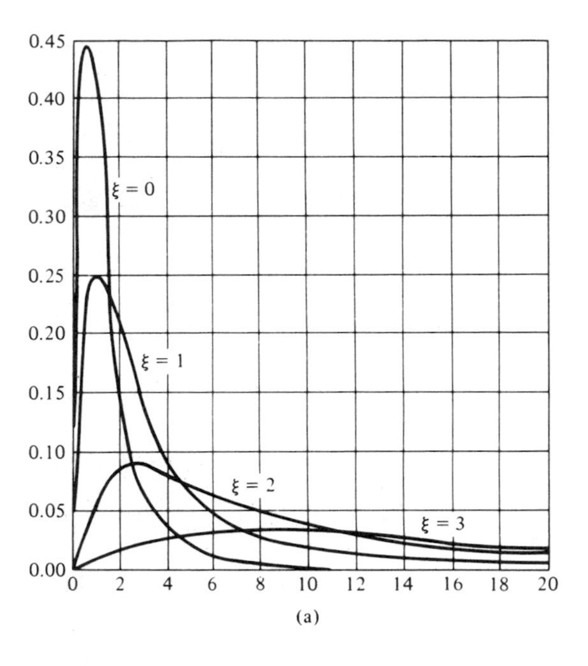

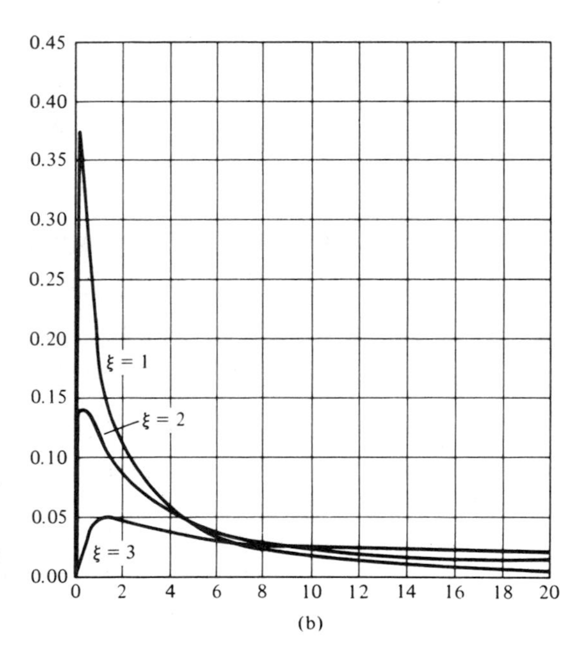

EXHIBIT 6.1 *Graphs of the probability density function of the lognormal distribution for (a) $\delta^2 = 1$, $\xi = 0, 1, 2, 3$; (b) $\delta^2 = 3$, $\xi = 1, 2, 3$.*

Mean and Variance of the Lognormal Distribution

If X has a lognormal distribution with parameters ξ and δ^2, then changing variables from x to $y = \log x - \xi$ (so that $x = e^{\xi}e^{y}$ and $-\infty < y < \infty$), we have

$$\begin{aligned} E(X^r) &= \int_0^{\infty} x^r f(x)\,dx = \int_0^{\infty} \frac{x^{r-1}}{\sqrt{2\pi\delta^2}} \exp\left[-\frac{1}{2\delta^2}(\log x - \xi)^2\right] dx \\ &= \int_{-\infty}^{\infty} \frac{1}{\sqrt{2\pi\delta^2}} \exp\left(r\xi + ry - \frac{y^2}{2\delta^2}\right) dy \\ &= \exp(r\xi + \tfrac{1}{2}\delta^2 r^2) \int_{-\infty}^{\infty} \frac{1}{\sqrt{2\pi\delta^2}} \exp\left[-\frac{1}{2\delta^2}(y^2 - 2\delta^2 ry + \delta^4 r^2)\right] dy \\ &= \exp(r\xi + \tfrac{1}{2}\delta^2 r^2) \int_{-\infty}^{\infty} \frac{1}{\sqrt{2\pi\delta^2}} \exp\left[-\frac{1}{2\delta^2}(y - \delta^2 r^2)^2\right] dy \\ &= \exp(r\xi + \tfrac{1}{2}\delta^2 r^2) \end{aligned}$$

for $r \geq 0$. We have thus shown that

$$E(X^r) = \exp(r\xi + \tfrac{1}{2}\delta^2 r^2), \qquad \text{for} \quad r \geq 0.$$

In particular,

$$\mu_X = E(X) = e^{\xi + (1/2)\delta^2}, \qquad \sigma_X^2 = e^{2\xi + \delta^2}(e^{\delta^2} - 1). \tag{6.4}$$

Uses of the Lognormal Distribution

The lognormal distribution is used and known under a variety of alternative names. Because Galton (1879) and McAlister (1879) were perhaps the first scientists to use the lognormal distribution, the lognormal distribution is sometimes known as the Galton–McAlister distribution. In economics, the lognormal distribution is often applied to production data, and in such contexts it is called the Cobb–Douglas distribution [e.g., see Dhrymes (1962)]. In psychophysical studies, the lognormal distribution has been mentioned by Fechner (1897), while Gaddum (1945) and Bliss (1934) have applied this distribution to the study of critical doses of drugs in human beings and animals. Among other nonnegative quantitative random phenomena whose variation has been modeled by the lognormal distribution are

1. Particle sizes in naturally occurring aggregates [Hatch and Choute (1929); Krumbein (1936); Herdan (1960)].
2. Lengths of words [Herdan (1958)], and sentences [Williams (1940)].
3. Concentrations of the chemical elements in geological materials [Ahrens (1954a,b; 1957); Chayes (1954)].

4. Lifetimes of mechanical and electrical systems [Epstein (1947, 1948)] and other survival data [Feinlieb (1960); Goldthwaite (1961); Adams (1962)].
5. Abundance of species of animals [Grundy (1951)].
6. Incubation periods of infectious diseases [Sartwell (1950)].

From this list it is apparent that the lognormal distributions are important competitors to the exponential, gamma, or Weibull distributions as models for nonnegative phenomena. We have seen that important properties of the lognormally distributed random variable X are obtained by transforming this variable to a new variable $Y = \log X$. This illustrates the value of transformations of variables to simplify probability calculations.

The Cauchy Distribution

Mathematical difficulties that arise in connection with models of physical or biological phenomena are often regarded as theoretical curiosities of little relevance to experimental practice. Even though there may often be some truth in this belief, it is not necessarily always correct.

Within the context of probability theory, the **Cauchy distribution** [named after the French mathematician Augustin Cauchy (1789–1857)] exhibits certain nonintuitive behavior, behavior that appears inconsistent with certain consequences of the frequency interpretation of probability theory.

The probability density function $f(x)$ of a random variable X having a Cauchy distribution has the form

$$f(x) = \frac{1}{\pi} \frac{1}{1 + (x - \theta)^2} \tag{6.5}$$

for all real numbers x. Graphs of this density function for $\theta = -1, 0, 1$ appear in Exhibit 6.2. The graph of a Cauchy probability density function, as that of the normal probability density function, is symmetric about a central value (in this case, θ) that is both the unique median and the unique mode of the distribution. Thus the parameter θ is a measure of location for the Cauchy distribution.

Apparently, then, the Cauchy distribution shares many properties with the normal distribution. In Exhibit 6.3 the density function of a $\mathcal{N}(0, 1)$ and $\mathcal{N}(0, 2)$ distribution, and a Cauchy distribution with parameter $\theta = 0$, are graphed together. Notice that the Cauchy distribution is flatter and more "spread out" than the $\mathcal{N}(0, 1)$ distribution, but not as flat as the $\mathcal{N}(0, 2)$ distribution. Graphically, it would appear that the Cauchy distribution is mathematically well behaved.

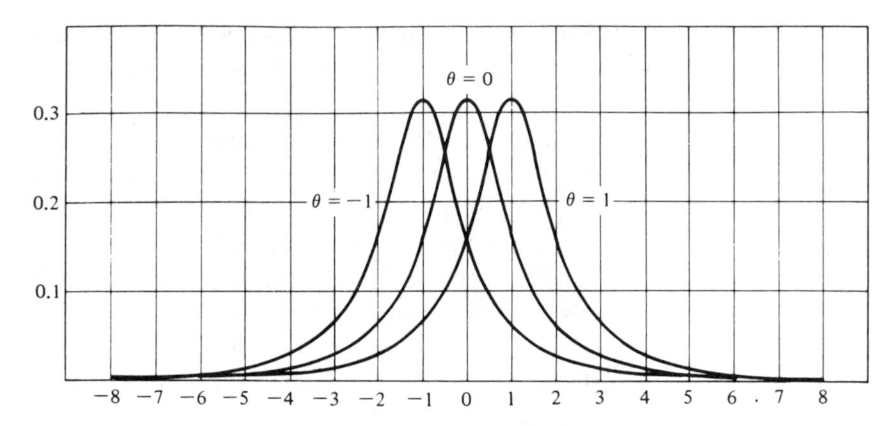

EXHIBIT 6.2 *Graphs of the Cauchy density function for $\theta = -1, 0, 1$.*

If X has a Cauchy distribution, however, its mean μ_X is not well defined. To see this, note that if μ_X is well defined, then

$$\mu_X = \int_{-\infty}^{\infty} xf(x)\,dx = \int_{-\infty}^{\infty} (x - \theta)f(x)\,dx + \theta \int_{-\infty}^{\infty} f(x)\,dx$$
$$= \int_{-\infty}^{\infty} (x - \theta)f(x)\,dx + \theta.$$

The integral of $(x - \theta)f(x)$ over $-\infty < x < \infty$ is not well defined, as we now demonstrate. If we make the change of variables from x to $v = x - \theta$, then

$$\text{(6.6)} \quad \int_{-\infty}^{\infty} (x - \theta)f(x)\,dx = \frac{1}{\pi}\int_{-\infty}^{\infty} \frac{x - \theta}{1 + (x - \theta)^2}\,dx = \frac{1}{\pi}\int_{-\infty}^{\infty} \frac{v}{1 + v^2}\,dv$$
$$= \frac{1}{\pi}\left(\int_{-\infty}^{0} \frac{v}{1 + v^2}\,dv + \int_{0}^{\infty} \frac{v}{1 + v^2}\,dv\right).$$

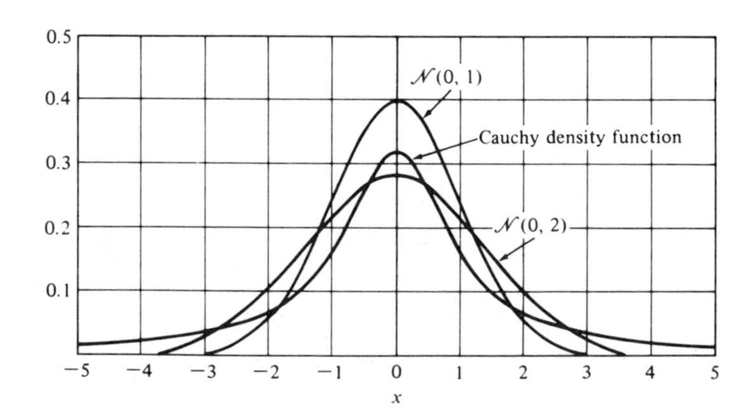

EXHIBIT 6.3 *Comparison of the $\mathcal{N}(0,1)$ and $\mathcal{N}(0,2)$ density functions with the Cauchy density function with $\theta = 0$.*

Because $(d/dv)\log(1 + v^2) = 2v/(1 + v^2)$,

$$\int_0^\infty \frac{v}{1 + v^2}\, dv = \frac{1}{2} \log(1 + v^2)\Big|_0^\infty = \infty.$$

Noting that $v/(1 + v^2)$ is an odd function of v, we have

$$\int_0^\infty \frac{v}{1 + v^2}\, dv = - \int_{-\infty}^0 \frac{v}{1 + v^2}\, dv,$$

and thus the integral in (6.6) is equal to the difference of two infinitely large numbers, and consequently is not well defined.

The fact that the mean of a Cauchy distribution is not well defined may be viewed as a mathematical peculiarity irrelevant to practice. However, suppose that we take independent observations $X_1, X_2, \ldots, X_n$ of a random variable X which has a Cauchy distribution with parameter θ and compute the sample average $\overline{X} = \sum_1^n X_i/n$. The frequency interpretation of probability leads us to believe that the variability of $\overline{X}$ about some central value, say θ, decreases as the number of observations increases, and that for very large n, $\overline{X} \simeq \theta$. We have seen how this fact, called the Law of Large Numbers, could be of use in experimental practice. However, for the Cauchy distribution, it is shown in Chapter 13 that *the sample average* $\overline{X}$ *has the same distribution as* X (i.e., the Cauchy distribution) *no matter how many observations of* X *are taken*, and thus the variability of $\overline{X}$ about θ remains the same even if we take an infinitely large number n of observations. This apparent contradiction to the law of large numbers is explained by the fact that the Cauchy distribution does not have a well-defined mean. A sufficient condition for the law of large numbers to be valid for a given probability distribution is that the mean of this distribution be well defined and finite.

It is tempting to dismiss the Cauchy distribution as a mathematical invention having ''no practical importance.'' However, this distribution is useful in certain scientific contexts: in mechanics and electrical theory, in psychophysics [Urban (1909)], in physical anthropology [Fieller (1932)], and in measurement and calibration problems in many fields of science. Perhaps the most common occurrence of the Cauchy distribution arises from the use of ratios. If independent random variables U and V each have a standard normal distribution, then $X = U/V$ has a Cauchy distribution.

Glossary and Summary

The various continuous distributions discussed in Sections 1 through 6 of this chapter are summarized below. The first summary gives the functional forms of the density, cumulative distribution function, and survival function of each type of distribution, together with the appropriate ranges for the argument x of these functions. Note that in this summary, $I_r(t)$ is the incomplete gamma function tabled in Exhibit B.5 of Appendix B, $I_x(r, s)$ is the incomplete beta function, and $F_Z(z)$ is the cumulative distribution function of the standard normal distribution tabled in Exhibit B.4 of Appendix B. Also, $\Gamma(r)$ is the gamma function and $B(r, s)$ is the beta function (see Appendix A).

Distribution and Parameters	*Probability Density Function*	*Cumulative Distribution Function*	*Survival Function*	*Range*
Exponential ($\theta > 0$)	$\theta e^{-\theta x}$	$1 - e^{-\theta x}$	$e^{-\theta x}$	$x > 0$
Gamma ($r > 0,\ \theta > 0$)	$\dfrac{\theta^r x^{r-1} e^{-\theta x}}{\Gamma(r)}$	$I_r(\theta x)$	$1 - I_r(\theta x)$	$x > 0$
Weibull ($-\infty < \nu < \infty,$ $\alpha > 0,\ \beta > 0$)	$\dfrac{\beta}{\alpha}\left(\dfrac{x-\nu}{\alpha}\right)^{\beta-1} \times \exp\left[-\left(\dfrac{x-\nu}{\alpha}\right)^{\beta}\right]$	$1 - \exp\left[-\left(\dfrac{x-\nu}{\alpha}\right)^{\beta}\right]$	$\exp\left[-\left(\dfrac{x-\nu}{\alpha}\right)^{\beta}\right]$	$x > \nu$
Uniform ($-\infty < a < b < \infty$)	$\dfrac{1}{b-a}$	$\dfrac{x-a}{b-a}$	$\dfrac{b-x}{b-a}$	$a \le x \le b$
Beta ($r, s > 0$)	$\dfrac{x^{r-1}(1-x)^{s-1}}{B(r, s)}$	$I_x(r, s)$	$1 - I_x(r, s) = I_{1-x}(s, r)$	$0 \le x \le 1$
Lognormal ($-\infty < \xi < \infty,$ $\delta^2 > 0$)	$\dfrac{\exp\left[-\dfrac{(\log x - \xi)^2}{2\delta^2}\right]}{\sqrt{2\pi x^2\delta^2}}$	$F_Z\left(\dfrac{\log x - \xi}{\delta}\right)$	$F_Z\left(\dfrac{\xi - \log x}{\delta}\right)$	$x > 0$
Cauchy ($-\infty < \theta < \infty$)	$\dfrac{1}{\pi[1 + (x-\theta)^2]}$	$\dfrac{1}{2} + \dfrac{\tan^{-1}(x-\theta)}{\pi}$	$\dfrac{1}{2} - \dfrac{\tan^{-1}(x-\theta)}{\pi}$	$-\infty < x < \infty$

The second summary gives the mean, variance, and pth quantile for each type of distribution.

Distribution	*Mean*	*Variance*	*pth Quantile*
Exponential ($\theta > 0$)	$\frac{1}{\theta}$	$\frac{1}{\theta^2}$	$\frac{\log[1/(1-p)]}{\theta}$
Gamma ($r > 0, \theta > 0$)	$\frac{r}{\theta}$	$\frac{r}{\theta^2}$	Requires interpolation in tables
Weibull ($-\infty < \nu < \infty$, $\alpha > 0, \beta > 0$)	$\nu + \alpha\Gamma\left(1 + \frac{1}{\beta}\right)$	$\alpha^2\left[\Gamma\left(1 + \frac{2}{\beta}\right) - \Gamma^2\left(1 + \frac{1}{\beta}\right)\right]$	$\nu + \alpha\left[\log\left(\frac{1}{1-p}\right)\right]^{1/\beta}$
Uniform ($-\infty < a,b < \infty$)	$\frac{a+b}{2}$	$\frac{(b-a)^2}{12}$	$a + p(b - a)$
Beta ($r > 0, s > 0$)	$\frac{r}{r+s}$	$\frac{rs}{(r+s)^2(r+s+1)}$	Requires interpolation in tables
Lognormal ($-\infty < \xi < \infty$, $\delta^2 > 0$)	$\exp\left(\xi + \frac{1}{2}\delta^2\right)$	$\exp(2\xi + \delta^2) \times [\exp(\delta^2) - 1]$	$\exp[\xi + \delta^2 Q_Z(p)]$
Cauchy ($-\infty < \theta < \infty$)	Not well defined	∞	$\theta + \tan[\pi(p - \frac{1}{2})]$

Note: $Q_Z(p)$ is the 100pth percentile of the standard normal distribution.

Note. The *chi-squared distribution with n degrees of freedom* is a special case of the gamma distribution with parameters $r = n/2$ and $\theta = \frac{1}{2}$. It arises in statistics as the distribution of the sum of squares, $\sum_{i=1}^{n} Z_i^2$, of independent standard normal random variables $Z_1, \ldots, Z_n$. The notation $X \sim \chi_n^2$ means that X has the chi-squared distribution with n degrees of freedom.

Some Other Useful Facts

1. The exponential distribution "has no memory." That is, if X has the exponential distribution with parameter θ, and it is known that $X > a$, then conditional on this fact, $X - a$ has an exponential distribution with parameter θ. Consequently, $P\{X - a > t \mid X > a\} = P\{X > t\}$. If a distribution of a continuous nonnegative random variable "has no memory," this distribution must be an exponential distribution.
2. The exponential distribution with parameter θ is a gamma distribution with parameters $r = 1$ and θ, and is also a Weibull distribution with parameters $\nu = 0$, $\alpha = 1/\theta$ and $\beta = 1$. Two other relations between the exponential and Weibull distributions are:
   (a) If X has a Weibull distribution with parameters ν, α, and β, then $[(X - \nu)/\alpha]^\beta$ has an exponential distribution with parameter $\theta = 1$.

(b) If X has an exponential distribution with parameter $\theta = 1$, then $\alpha X^{1/\beta} + \nu$ has a Weibull distribution with parameters ν, α, and β.

3. The distribution of a sum $\sum_{i=1}^{r} X_i$ of r independent exponentially distributed random variables X_i with common parameter θ is a gamma distribution with parameters r and θ. The distribution of the sum $\sum_{i=1}^{n} X_i$ of n independent gamma distributed random variables, where X_i has parameters r_i and θ, $i = 1, \ldots, n$, is a gamma distribution with parameters $r = \sum_{i=1}^{n} r_i$ and θ.
4. If X has a gamma distribution with parameters r, a positive integer, and θ, then for $x > 0$,

$$P\{X \leq x\} = P\{L > r - 1\},$$

   where L has a Poisson distribution with parameter $\lambda = \theta x$.
5. If X has a beta distribution with parameters r, s both positive integers, then

$$P\{X \leq x\} = P\{U \geq r\},$$

   where U has a binomial distribution with parameters $n = r + s - 1$ and $p = x$.
6. If X has a beta distribution with parameters r and s, then $1 - X$ has a beta distribution with parameters s and r.
7. The lognormal distribution derives its name from the fact that if X has a lognormal distribution with parameters ξ and δ^2, then $\log X$ has a normal distribution with mean ξ and variance δ^2. Conversely, if $Y \sim \mathcal{N}(\mu, \sigma^2)$, then $\exp(Y)$ has a lognormal distribution with parameters $\xi = \mu$ and $\delta^2 = \sigma^2$.
8. If $X_1, X_2, \ldots, X_n$ are independent Cauchy random variables with common parameter θ, then $\overline{X} = (1/n) \sum_{i=1}^{n} X_i$ also has a Cauchy distribution with parameter θ. Consequently, in this case, the Law of Large Numbers fails to hold; that is, the distribution of $\overline{X}$ does not become more concentrated about a central value as $n \to \infty$.
9. If $X_1, X_2, \ldots, X_m$ are independent chi-squared random variables with $X_i \sim \chi^2_{n_i}$, $i = 1, 2, \ldots, m$, then $\sum_{i=1}^{m} X_i \sim \chi^2_n$ where $n = \sum_{i=1}^{m} n_i$.

Simulation: One way to simulate an observation of a random variable X having a continuous cumulative distribution function $F(x)$ is first to use a random number generator to obtain an observation U uniformly distributed on the interval $[0, 1]$, and then to solve $F(x) = U$ for x. The resulting solution is the simulated value of X. This is a special case of methods used in Monte Carlo simulation studies.

Exercises

1. Rasch (1960) considers a stochastic model for reading time in which the time X in minutes required for a person to read a given passage is a random variable following an exponential distribution with parameter θ. Experimental evidence indicates that for a certain reading exercise the value of θ is 2.0.
   (a) Find the probability of the event that it takes a person more than 1 minute to read the reading exercise.
   (b) Find the median and 0.90th quantile of the distribution of the time X required to read the reading exercise.
2. The distribution of the duration of pauses (and the duration of vocalizations) that occur in a monologue has an exponential distribution [Jaffe and Feldstein (1970)]. If the mean duration of pauses is 0.70 second, what is the variance of the duration of pauses? What is the 0.90th quantile?
3. The weight X of titanium ore (in tons) produced on a random day has an exponential distribution with mean $\mu_X = 2$ tons.
   (a) Find the probability $P\{1 \leq X \leq 3\}$ that between 1 and 3 tons of titanium ore will be produced in a day.
   (b) What is the median amount of titanium ore produced in a day?
   (c) If it is known that 2 tons of ore have already been produced today, what is the conditional probability that more than 4 tons will have been produced by the end of the day?
   (d) Let X_1, X_2, X_3, X_4, X_5 be the amounts of titanium ore produced in 5 random days. Assume these amounts are mutually probabilistically independent. What is the probability that in exactly 4 of the 5 days the amounts of ore produced exceeded 2 tons?
4. The lifetime T of a phonograph needle has an exponential distribution with mean $\mu = 200$ hours.
   (a) What is the probability, $P\{T > 400\}$, that the needle will last for more than 400 hours?
   (b) A needle has been played for 400 hours and still functions. Note that the needle has already lasted beyond its expected (mean) lifetime. Should you buy a new needle? Explain.
5. The length of time X, in hours, required for a truck to complete a round trip is exponentially distributed with parameter $\theta = \frac{1}{48}$. Given that a truck has already been gone for 48 hours, what is the conditional probability that this truck will arrive within the next hour? What is the mean additional length of time required until the truck completes its route, given that the truck has already been gone for 48 hours?
6. The posttreatment survival time X, in years, of a patient given treatment for cancer is often assumed to have an exponential distribution. A

new treatment has been proposed for a particularly deadly kind of cancer and will be tested in a clinical study. Because the study can only be run for a fixed length T of time, it is possible that a patient given the treatment will live past the endpoint of the study (i.e., X may exceed T). In this case, suppose that the lifetime recorded for that patient is arbitrarily set equal to $T + \frac{1}{2}$. Thus the survival time Y recorded for a patient is

$$Y = \begin{cases} X, & \text{if } X \leq T, \\ T + \frac{1}{2}, & \text{if } X > T. \end{cases}$$

Suppose that X has an exponential distribution with mean $\mu_X = 2$ years and that $T = 3$ years.

(a) Find the probabilities of the following events: (i) $P\{Y = 3.5\}$; (ii) $P\{Y = 2.0\}$; (iii) $P\{Y \leq 1\}$; (iv) $P\{Y > 2\}$.

(b) Find the survival function $\bar{F}_Y(y) = P\{Y > y\}$ of Y.

(c) Find the median, $Q_y(0.50)$, and 90th percentile, $Q_y(0.90)$, of Y.

(d) Find the mean, μ_Y, of Y. (The answer is *not* 2.)

7. The magnitude M of an earthquake, as measured on the Richter scale, is a random variable. As a consequence of studies of earthquakes with large magnitudes (say, magnitudes exceeding 3.25), it has been suggested that the "excess" $X = M - 3.25$ follows an exponential distribution. Over the period January 1934 to May 1943, the magnitudes of earthquakes were recorded. The resulting data are summarized by Gutenberg and Richter (1944). From these data the frequency distribution in Exhibit E.1 has been constructed.

EXHIBIT E.1. *Observed Frequency Distribution of Earthquakes with Large Magnitudes, January 1934 to May 1943*

Midpoint, m, of Interval of Magnitudes	*Midpoint, x, of Interval of Excesses*	*Observed Frequency of Earthquakes Whose "Excess" $X = M - 3.25$ Is between $x - 0.25$ and $x + 0.25$*
3.5	0.25	579
4.0	0.75	311
4.5	1.25	108
5.0	1.75	32
5.5	2.25	13
6.0	2.75	5
6.5	3.25	2
		1050

(a) Determine theoretical frequencies for the intervals in Exhibit E.1 assuming that X has an exponential distribution with parameter $\theta = 1.70$. You will need to add the interval $[3.50, \infty)$ to your table. The observed frequency of this interval is 0. What is the theoretical frequency of this interval?

(b) Assuming that $X = M - 3.25$ has an exponential distribution with parameter $\theta = 1.70$, what is the probability that the magnitude M of an earthquake with large magnitude will exceed 6 on the Richter scale?

8. Suppose that X and Y are probabilistically independent random variables, and that both X and Y have an exponential distribution with parameter $\theta = 0.10$.

(a) Find $P\{X > 5.0 \text{ and } Y > 5.0\}$.

(b) Find a general formula for $P\{X > t \text{ and } Y > t\}$.

(c) Let $W = \min\{X, Y\}$. Find the probability density function of W, and find μ_W.

9. The gamma distribution has been used in a medical application by Masuyama and Kuroiwa (1952). They considered the sedimentation rate X at various stages during normal pregnancy. In the 30th week after commencement of pregnancy, their data were fit approximately by a gamma distribution with $r = 5.07$ and $\theta = 1/9.98$. Instead, assume that X has a gamma distribution with parameters $r = 5$ and $\theta = 0.1$. Find Med(X), the median sedimentation rate in the 30th week after commencement of pregnancy. Compute $P\{X > 50\}$, $P\{10 \leq X \leq 90\}$, the mean μ_X, and the variance σ_X^2 of X.

10. Show that for all values of the parameters θ, r of a gamma distribution, the mode is less than the mean.

11. Suppose that a student is asked to read 5 similar reading exercises one after the other. Also assume that the student's reading times for these exercises are probabilistically independent random variables, each having an exponential distribution with parameter $\theta = 2$. Let Y be the total time required for the student to read all 5 exercises.

(a) Find μ_Y and σ_Y^2.

(b) Find the median and (0.90)th quantile $Q_Y(0.90)$ of the distribution of Y.

(c) Find $P\{Y \leq 2.5\}$, $P\{Y > 5\}$, and $P\{1 \leq Y < 5\}$.

12. There are 4 customers in line at a bank teller's window. The length of time X in minutes needed to service a random customer has an exponential distribution with mean $\mu_X = 5$ minutes. As each customer finishes, the teller instantly starts the next customer. The times X_1, X_2, X_3, X_4 needed to service the 4 customers are mutually independent.

(a) What is the mean, μ_T of the total amount of time $T = \sum_{i=1}^{4} X_i$ needed to service the 4 customers? What is the variance, σ_T^2, of T?

(b) The bank will close in 15 minutes. What is the probability that the teller will finish servicing all 4 customers before the bank closes?

(c) What is the probability that the teller finishes servicing no more than 3 of the 4 customers before the bank closes? (Remember that the bank closes in 15 minutes.) (*Hint:* Let Y be the number of customers that are served by the teller. What is the name of the distribution of Y?)

13. Use the relationship between the probabilities of a gamma distribution and those of the Poisson distribution to find the following probabilities for Y having a gamma distribution with parameters $r = 3$, $\theta = 0.10$: (a) $P\{Y \leq 26\}$; (b) $P\{Y > 40\}$; (c) $P\{10 \leq Y \leq 40\}$.

14. The gamma distribution has been used as a model for the lifetimes of metals subjected to stress. In one experiment, 101 rectangular strips of aluminum of standardized dimensions were submitted to repeated alternating stresses (at a frequency of 18 cycles per second). The lifetimes of these strips of aluminum (expressed in thousands of cycles) are summarized in Exhibit E.2. Using grouping intervals of length 180 starting at 370 and ending at 2530, construct a table of observed frequencies and theoretical frequencies for the data in Exhibit E.2 assuming that the lifetimes have a gamma distribution with parameters $\theta = 0.01$, $r = 13$.

15. Let X have an exponential distribution with parameter $\theta = 1$ and let Y

EXHIBIT E.2. *Lifetimes Under Periodic Loading, Maximum Stress 21,000 psi 18 Cycles per Second, 6061-T6 Aluminum Coupon Cut Parallel to the Direction of Rolling*[a]

370	988	1120	1269	1420	1530	1730	1893
706	990	1134	1270	1420	1540	1750	1895
716	1000	1140	1290	1450	1560	1750	1910
746	1010	1199	1293	1452	1567	1763	1923
785	1016	1200	1300	1475	1578	1768	1940
797	1018	1200	1310	1478	1594	1781	1945
844	1020	1203	1313	1481	1602	1782	2023
855	1055	1222	1315	1485	1604	1792	2100
858	1085	1235	1330	1502	1608	1820	2130
886	1102	1238	1355	1505	1630	1868	2215
886	1102	1252	1390	1513	1642	1881	2268
930	1108	1258	1416	1522	1674	1890	2440
960	1115	1262	1419	1522			

Source: Z. W. Birnbaum and S. C. Saunders, A statistical model for the life-length of materials, *Journal of the American Statistical Association,* vol. 53, pp. 151–60 (1958). Reprinted by permission.

[a] Observations listed in increasing order.

have a gamma distribution with parameters $r = 2$ and $\theta = 2$. Note that $\mu_X = \mu_Y$.

(a) Compare σ_X^2 with σ_Y^2.

(b) Compare $P\{X - \mu_X > 2\sigma_X\}$ with $P\{Y - \mu_Y > 2\sigma_Y\}$.

16. The errors, X, of measurement made when measuring a physical constant by a certain process are assumed to have a normal distribution with mean 0 and variance σ^2. If $n = 5$ independent measurements X_1, X_2, X_3, X_4, X_5 are made of the constant, the average of the squares of these errors, $Y = \sum_{i=1}^5 X_i^2/5$, is used to estimate σ^2. Note that the standardization $Z_i = (X_i - 0)/\sigma$ of X_i has a standard normal distribution, $Z_i \sim \mathcal{N}(0, 1)$, $i = 1, 2, \ldots, 5$, and that $Y = \sigma^2(\sum_{i=1}^5 Z_i^2/5)$.

(a) Show that $E[Y] = \sigma^2$.

(b) Find the probability $P\{Y \le \sigma^2\}$ that Y is less than σ^2. (*Hint:* What is the distribution of $\sum_{i=1}^5 Z_i^2$?)

(c) Find the median, $Q_Y(0.50)$, of Y.

17. If $X \sim \chi_{10}^2$, find:

(a) $P\{X \le 13\}$, $P\{X > 9\}$, $P\{8 \le X \le 12\}$.

(b) The median of X.

(c) $Q_X\,(0.95)$.

18. (a) Explain why the Central Limit Theorem can be used to find approximate probabilities for the gamma distribution with parameters r and θ when r is a large integer (say, $r \ge 30$).

(b) Use the Central Limit Theorem to find $P\{X \le 40\}$ when X has a gamma distribution with parameters $\theta = 1$, $r = 30$.

19. It is known that the duration X, in days, of an epidemic of a certain infectious disease has a Weibull distribution with parameters $\beta = 2$, $\alpha = 10$, and $\nu = 2$.

(a) Find $P\{X > 14\}$ and $P\{4 \le X \le 7\}$.

(b) Find μ_X and σ_X^2.

(c) Find Med(X).

(d) Find the mode of X.

(e) If an epidemic of the disease has just been reported, predict the length of time (in days) that the epidemic will last.

20. The Weibull distribution has been used by Indow (1971) as a model for determining the effect of advertising. Let X be the length of time (in days) after the end of an advertising campaign that a person is able to remember the name of the product being advertised. In a particular experiment a brand of chocolate was used, and Indow fit a Weibull distribution with $\beta = 0.98$, $\alpha = 7360$, and $\nu = 1.0$ to the variation of X.

(a) What is the median, Med(X), of X? What is μ_X?

(b) What proportion of all persons could be expected to remember the advertised brand of chocolate 1 week (7 days) after the advertising campaign ended?

21. Medley, Anderson, Cox, and Billard (1987) modeled the incubation time X in years between infection with the HIV virus and the onset of full-blown AIDS. For adult cases in the age range 5 to 59 years, they found that X has a Weibull distribution with parameters $\nu = 0$, $\beta = 2.396$, and $\alpha = 9.2851$.

(a) Find the probability that the incubation time X of a randomly chosen AIDS patient is greater than (i) 10 years; (ii) 20 years.

(b) Find the median incubation time $Q_X(0.50)$ for AIDS.

(c) Find the mean incubation time, μ_X, for AIDS. (You will need to use the table of the gamma function given in Appendix A.)

22.★ If X has a Weibull distribution with parameters $\beta = 3$, $\alpha = 5$, $\nu = 1$, find the distribution of:

(a) $Y = X - 1$. (b) $Z = (X - 1)/5$. (c) $V = (X - 1)^2/25$.

23. A bus travels between two cities, A and B, which are 100 miles apart. If the bus has a breakdown, the distance X of the point of the breakdown from city A has a $U[0, 100]$ distribution.

(a) There are service garages in city A, city B, and midway between cities A and B. If a breakdown occurs, a tow truck is sent from the garage closest to the point of breakdown. What is the probability that the tow truck has to travel more than 10 miles to reach the bus?

(b) Would it be more "efficient" if the three service garages were placed 25, 50, and 75 miles from city A?

24. Buses run on the half-hour, and on any given day a commuter has a time X of arrival at a bus stop that is uniformly distributed between 8:15 and 8:45 A.M.

(a) What is the probability that the commuter will have to wait for a bus (i) more than 15 minutes; (ii) more than 20 minutes?

(b) What is the mean length of time that the commuter must wait for a bus?

25. A random number generator has been used to obtain an observation U from a uniform distribution on $[0, 1]$. Suppose that $U = 0.45$ has been obtained. Use this value of U to simulate:

(a) X having a normal distribution with mean 2 and variance 4.

(b) X having an exponential distribution with parameter $\theta = 0.25$.

(c) X having a gamma distribution with parameters $r = 2$ and $\theta = 0.25$.

(d) X having a Weibull distribution with parameters $\nu = 0$, $\alpha = 2$, $\beta = 3$.

(e) X having a log-normal distribution with parameters $\xi = 0$, $\delta^2 = 1$.

26. Find the moment generating function $M(t)$ of the uniform distribution on $[a, b]$. Use $M(t)$ to determine the mean and variance of the $U[a, b]$ distribution.

27. If $X \sim U[a, b]$, find $\mu_X^{(3)} = E(X - \mu_X)^3$ and $\mu_X^{(4)} = E(X - \mu_X)^4$. Comment on the skewness and kurtosis of the uniform distribution.

28. Let X and Y be independent $U[1, 2]$ random variables.

(a) Find $P\{X > 1.5 \text{ and } Y > 1.5\}$, $P\{X \le 1.5 \text{ and } Y \le 1.5\}$, and $P\{X \le 1.5 \text{ and } Y > 1.5\}$.

(b) Let $U = \min\{X, Y\}$. For $1 \le x \le 2$, show that $P\{X > x \text{ and } Y > x\} = P\{U > X\} = \overline{F}_U(x)$, and find $\overline{F}_U(x)$ as a function of x.

(c) Let $V = U - 1$. Use the result of part (b) to show that V has a beta distribution, and find the parameters r, s of this distribution.

29. The monthly fire incidence (monthly total ÷ yearly total) in buildings in England and Wales in 1961 is given in Exhibit E.3. Let X equal the exact time in days (after 12:00 A.M. on the morning of January 1, 1961) that a given fire occurs in a building in England and Wales. Assume that $X \sim U[0, 365]$ and compute theoretical incidence figures to compare with the observed incidences in Exhibit E.3. (For example, the theoretical incidence for February would be $\frac{28}{365} = 0.0767$.) Does it appear that the $U[0, 365]$ distribution provides a good model for the variation of X?

EXHIBIT E.3. *Monthly Fire Incidences in Buildings in England and Wales in 1961*

Month	*Incidence of Fires*
January	0.0936
February	0.0739
March	0.0917
April	0.0704
May	0.0861
June	0.0809
July	0.0771
August	0.0699
September	0.0705
October	0.0842
November	0.0942
December	0.1075

30. Suppose that X has a beta distribution with parameters $r = 4$ and $s = 3$.

(a) Find $P\{X \le 0.5\}$.
(b) Find $P\{X > 0.7\}$.
(c) Find $P\{0.4 \le X \le 0.6\}$.
(d) Find μ_X and σ_X^2.
(e) Find the median of X.

[*Hint:* For parts (a), (b), (c), and (e), use the relationship between the probabilities of the beta and the binomial distributions.]

31. Ore samples from the Transvaal vein of level 10 of the Frisco mine were assayed for metal content. For each of 1000 ore samples, the ratios X of the weight of copper (Cu) to the sum of the weights of copper (Cu) and lead (Pb) were measured. Exhibit E.4 gives a frequency distribution for X.

(a) Using the values $r = 1.13$, $s = 3.00$ and the formulas for calculating probabilities for the beta distribution, find theoretical frequencies for the intervals [0.00, 0.04], and [0.04, 0.08] and compare these to the observed frequencies in Exhibit E.4.

(b) Find theoretical frequencies for all of the intervals in Exhibit E.4 using the beta distribution with $r = 1.00$, $s = 3.00$. Is the fit any better than in part (a)? [Compare only for the intervals whose theoretical frequencies you calculated in part (a).]

32. Let X have a beta distribution with parameters r and s. Find the probability density function of $Y = (b - a)X + a$, where $-\infty < a < b < \infty$. Note that Y takes on values in $[a, b]$. What are the mean, μ_Y, and variance σ_Y^2 of Y?

33. Rogers (1977) assumed that losses X, in thousands of British pounds, from large fires have a lognormal distribution. For fire losses in the British textile industry (among multistory buildings lacking an automatic sprinkler system), he found that the mean loss per fire was 25,200 pounds. Thus $\mu_X = 25.2$. Suppose that $\sigma_X^2 = 100$.

(a) Solve the equations

$$\mu_X = e^{\xi+\frac{1}{2}\delta^2} = 25.2, \qquad \sigma_X^2 = (e^{\xi+\frac{1}{2}\delta^2})^2(e^{\delta^2} - 1) = 100,$$

for ξ and δ^2.

EXHIBIT E.4. *Frequency Distribution of the Ratio Cu/(Cu + Pb) in 1000 Samples of Ore*

Midpoint, x, of Interval of Values	*Observed Frequency of Interval [x − 0.02, x + 0.02]*	*Midpoint, x, of Interval of Values*	*Observed Frequency of Interval [x − 0.02, x + 0.02]*
0.02	209	0.54	5
0.06	264	0.58	4
0.10	170	0.62	8
0.14	105	0.66	1
0.18	58	0.70	5
0.22	53	0.74	4
0.26	33	0.78	5
0.30	22	0.82	2
0.34	12	0.86	1
0.38	11	0.90	1
0.42	11	0.94	0
0.46	10	0.98	2
0.50	4		$N = 1000$

Source: Based on G. S. Koch, Jr., and R. F. Link, *Statistical Analysis of Geological Data,* Vol. 2 (New York: John Wiley & Sons, Inc., 1971). Reprinted by permission.

(b) Find the probability, $P\{X \geq 50\}$, that a large fire in the British textile industry results in a loss of more than 50,000 pounds.
(c) What is the median loss, $Q_X(0.50)$, in thousands of British pounds, due to large fires in the British textile industry?

Theoretical Exercises

T.1. Show that the moment generating function

$$M(t) = E(e^{tX})$$

of an exponential distribution with parameter θ is $M(t) = \theta/(\theta + t)$ for $t > -\theta$. Use this result to find the mean and the variance of the exponential distribution.

T.2. Note that for any $r > 0$, $c > 0$,

$$\int_0^\infty \frac{c^r x^{r-1} e^{-cx}}{\Gamma(r)} = 1,$$

because this integral gives the total probability of a gamma distribution with parameters $r > 0$, $c > 0$. Hence

$$\int_0^\infty x^{r-1} e^{-cx}\, dx = \frac{\Gamma(r)}{c^r}.$$

(a) Use this result to show that the moment generating function of the gamma distribution with parameters $r > 0$, $\theta > 0$ is

$$M(t) = \left(\frac{\theta}{\theta + t}\right)^r \quad \text{for} \quad t > -\theta.$$

(b) Using part (a) and the result of Exercise T.1, prove the result stated in the text that the sum $Y = \Sigma_{i=1}^{r} X_i$ of r mutually independent, exponentially distributed random variables $X_1, \ldots, X_r$, each with the same parameter θ, has the gamma distribution with parameters r and θ. (See Chapter 7. You may want to use induction on r.)
(c) Suppose that X_i has a gamma distribution with parameters θ and r_i, $i = 1, \ldots, n$, and that $X_1, \ldots, X_n$ are mutually independent. Prove the result stated in the text that $\Sigma_{i=1}^{n} X_i$ has a gamma distribution with parameters $r = \Sigma_{i=1}^{n} r_i$ and θ.

T.3. Verify the assertions made in the "Glossary and Summary" about the cumulative distribution function, survival function and pth quantile of a Cauchy distribution with parameter θ. That is, show that if X has a Cauchy distribution with parameter θ, then

$$F_X(x) = P\{X \leq x\} = \frac{1}{2} + \frac{\tan^{-1}(x - \theta)}{\pi},$$

$$\overline{F}_X(x) = P\{X > x\} = \frac{1}{2} - \frac{\tan^{-1}(x - \theta)}{\pi},$$

$$Q_X(p) = \tan[\pi(p - \tfrac{1}{2})].$$

T.4.★ Show that the moments $E[X^k]$ of the lognormal distribution with parameters ξ and δ^2 exist for $k = 1, 2, \ldots$. Nevertheless, show that the moment generating function $M(t)$ of this distribution does not exist for $t > 0$.

T.5. Find a general formula giving the mode (or modes) of a beta distribution with parameters r and s.

T.6. There is no nice functional form for the moment generating function $M(t)$ of the beta distribution. However, the moment generating function $M^*(t)$ of $\log(X)$, where X has a beta distribution with parameters r and s, is easily obtained. Find a formula for $M^*(t)$.

T.7. If X_1 has a lognormal distribution with parameters ξ_1, δ_1^2, X_2 has a lognormal distribution with parameters ξ_2, δ_2^2, and X_1, X_2 are probabilistically independent, show that $Y = X_1X_2$ has a lognormal distribution with parameters $\xi_1 + \xi_2$, $\delta_1^2 + \delta_2^2$.
(*Hint:* Note that $\log Y = \log X_1 + \log X_2$.)

11

Bivariate Distributions

1 Introduction

Many important problems in science and technology concern questions of how the values of one measurable quantity can be used to predict, control, or relate to another quantity. For example, physicists predict the motion of a body from knowledge of the forces acting on that body, aeronautical engineers use information obtained from automatic sensors to control the position of a space satellite, and biologists relate chemical concentrations in the environment to the frequency of cancerous tumors in animals. Similarly, sociologists relate group size to problem-solving productivity, psychologists associate personality test scores with anxiety level, and educators predict college achievement from high school grades. Indeed, virtually all scientific and technological disciplines require, as part of their development, knowledge of the interrelationships among the variables that compose their empirical framework.

Mechanistic models for natural phenomena express relationships between variables in terms of exact mathematical formulas. Probability models, on the other hand, express such relationships either in terms of probabilities of joint occurrence, or in terms of "average" or "typical" relationships. When probability models describe the joint variability of more than one random variable, the resulting model is said to be the **joint** (or **multivariate**) **distribution** of these variables. For simplicity, we concentrate in this chapter on the joint distribution of two random variables (a **bivariate distribution**), and limit our consideration to cases where either both variables are discrete (discrete bivariate distributions), or where both variables are continuous (continuous bivariate distributions). The most widely used continuous bivariate distribution, the bivariate normal distribution, is discussed in some detail in Chapter 12 because of its importance as a background for classical statistical methodology. At the end of this chapter we briefly discuss how concepts for bivariate distributions can be generalized to joint distributions of three or more random variables.

2 Determining Bivariate Distributions

By the **bivariate distribution** of two random variables X and Y, we mean a probability model for the relative frequencies with which values of X occur jointly with values of Y. In analogy with distributions for one random variable (**univariate distributions**), the bivariate distribution of X and Y can always be determined from the **joint cumulative distribution function**

$$F(x, y) = P\{X \leq x \text{ and } Y \leq y\}. \tag{2.1}$$

However, for applications it is usually more convenient to work with functions analogous to the univariate mass functions and density functions discussed in Chapter 5. Thus when X and Y are both discrete random variables, their joint distribution is determined by the **joint probability mass function:**

$$p(x, y) = P\{X = x \text{ and } Y = y\}. \tag{2.2}$$

When X and Y are both continuous random variables with a joint cumulative distribution function $F(x, y)$ satisfying

$$\frac{\partial^2}{\partial x\, \partial y} F(s, y) = \frac{\partial^2}{\partial y\, \partial x} F(x, y),$$

then the joint distribution of X and Y is determined by the **joint probability density function:**

$$f(x, y) = \frac{\partial^2}{\partial x\, \partial y} F(x, y) = \frac{\partial^2}{\partial y\, \partial x} F(x, y). \tag{2.3}$$

Joint probability mass functions and joint probability density functions have characterizations similar to those of probability mass functions and probability density functions of a single random variable. That is:

A function $p(x, y)$ is the joint probability mass function of a pair of discrete random variables X, Y if and only if

(2.4)
(i) $p(x, y) \geq 0$ for all (x, y) pairs, and is positive for only a countable number of such pairs.
(ii) $\Sigma_x \Sigma_y p(x, y) = \Sigma_y \Sigma_x p(x, y) = 1$.

A function $f(x, y)$ is the joint probability density function of continuous random variables X, Y if and only if

(2.4′)
(i′) $f(x, y) \geq 0$ for all (x, y) pairs.
(ii′) $\int_{-\infty}^{\infty} \int_{-\infty}^{\infty} f(x, y)\, dx\, dy = \int_{-\infty}^{\infty} \int_{-\infty}^{\infty} f(x, y)\, dy\, dx = 1$.

The reasons for the requirements (2.4) and (2.4′) are that probabilities must be nonnegative and that the probability of the entire sample space must be 1.

An example of a joint probability mass function is

$$p(x, y) = \begin{cases} \dfrac{x + y}{54}, & x = 1, 2, 3, \quad y = 1, 2, 3, 4, \\ 0, & \text{otherwise.} \end{cases} \tag{2.5}$$

Note that $p(x, y) \geq 0$ for all (x, y) pairs and that

$$\sum_{x=1}^{3} \sum_{y=1}^{4} p(x, y) = \sum_{x=1}^{3} \sum_{y=1}^{4} \left(\frac{x}{54} + \frac{y}{54}\right) = \sum_{x=1}^{3} \left(4\frac{x}{54} + \frac{10}{54}\right)$$
$$= \frac{24}{54} + \frac{3(10)}{54} = 1.$$

An example of a joint probability density function is

$$f(x, y) = \begin{cases} \exp[-x(y + 1)^2], & x > 0, y > 0, \\ 0, & \text{otherwise.} \end{cases} \tag{2.6}$$

Because e^a is always nonnegative for all values of a, $f(x, y) \geq 0$ all (x, y). Further,

$$\int_{-\infty}^{\infty}\int_{-\infty}^{\infty} f(x, y)\, dx\, dy = \int_0^{\infty}\left(\int_0^{\infty} e^{-x(y+1)^2}\, dx\right)dy$$
$$= \int_0^{\infty}\left(\left[\frac{-e^{-x(y+1)^2}}{(y+1)^2}\right]_{x=0}^{x=\infty}\right)dy$$
$$= \int_0^{\infty}\frac{1}{(1+y)^2}\, dy = -\frac{1}{1+y}\Big|_0^{\infty} = 1.$$

Contingency Tables

When X and Y are discrete variables having a finite number of possible values, it can be helpful to table their probabilities $p(x, y)$ in the form of a **contingency table** such as Exhibit 2.1. The contingency table obtained from the joint probability mass function (2.5) is the following:

	y			
x	1	2	3	4
1	2/54	3/54	4/54	5/54
2	3/54	4/54	5/54	6/54
3	4/54	5/54	6/54	7/54

For example, the entry in the row for $x = 1$ and the column for $y = 2$ is $p(1, 2) = 3/54$. Because the sum of $p(x, y)$ over all (x, y) values must be equal to 1, *the sum of the entries in a contingency table is always equal to* 1.

As an alternative to a contingency table, the joint probability mass function $p(x, y)$ can be graphed as in Exhibit 2.2 The possible pairs (x, y) of values are represented as points in the plane, and from these points lines are drawn perpendicular to the plane with height equal to $p(x, y)$.

EXHIBIT 2.1 *Contingency Table for Two Discrete Random Variables X and Y, Where X Has Three Possible Values, x_1, x_2, x_3, and Y Has Four Possible Values, y_1, y_2, y_3, y_4*

	y			
x	y_1	y_2	y_3	y_4
x_1	$p(x_1, y_1)$	$p(x_1, y_2)$	$p(x_1, y_3)$	$p(x_1, y_4)$
x_1	$p(x_2, y_1)$	$p(x_2, y_2)$	$p(x_2, y_3)$	$p(x_2, y_4)$
x_3	$p(x_3, y_1)$	$p(x_3, y_2)$	$p(x_3, y_3)$	$p(x_3, y_4)$

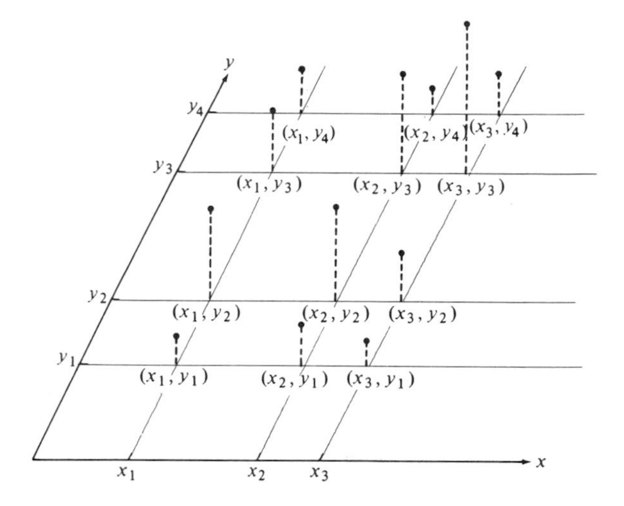

EXHIBIT 2.2 *Graph of a joint probability mass function p(x, y).*

EXAMPLE 2.1

Two professional wine tasters are asked to rate a wine using a rating scale of 1, 2, 3, 4, or 5. The wine tasted is chosen randomly from among table wines marketed nationally. Let wine taster I's rating be denoted X and wine taster II's rating be denoted Y. A possible contingency table for this experiment is given in Exhibit 2.3. The row and column sums of contingency tables are known as **marginal totals** or **marginal probabilities**. For example, the marginal total for row 1 in Exhibit 2.3 gives the probability that $X = 1$. A graph of the joint probability mass function $p(x, y)$ defined by Exhibit 2.3 appears in Exhibit 2.4. ◆

EXHIBIT 2.3 *Bivariate Probability Mass Function for Two Wine Tasters Exhibited in the Form of a Contingency Table*

Rating x of Wine Taster I	*Rating y of Wine Taster II*					
	1	2	3	4	5	Total
1	0.03	0.02	0.01	0.00	0.00	0.06
2	0.02	0.08	0.05	0.02	0.01	0.18
3	0.02	0.06	0.22	0.05	0.01	0.36
4	0.01	0.02	0.06	0.18	0.02	0.29
5	0.00	0.01	0.01	0.03	0.06	0.11
Total	0.08	0.19	0.35	0.28	0.10	1.00

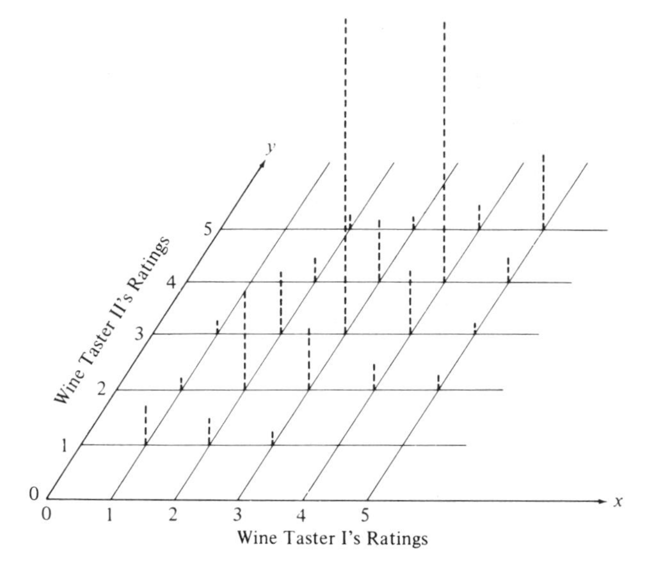

EXHIBIT 2.4 *Graph of the bivariate mass function p(x, y) corresponding to the probability model for joint ratings of wine tasters I and II associated with Exhibit 2.3.*

Probability Calculations

In obtaining probabilities for jointly distributed random variables X and Y, it is helpful to think of the pair (X, Y) as the coordinates of a random point in the plane and to describe any event of interest concerning X and Y in terms of a region R in the plane. For example, the event "Y equals X" can be written as $\{(X, Y) \text{ in } R\}$, where $R = \{(x, y) : y = x\}$ is the collection of points in the plane falling on the line $y = x$. Similarly, the event "Y exceeds X" can be written as $\{(X, Y) \text{ in } R\}$, where $R = \{(x, y) : y > x\}$ is the half-plane lying strictly above and to the left of the line $y = x$.

If R is any region in the plane and X and Y are discrete random variables with joint probability mass function $p(x, y)$, then

$$P\{(X, Y) \text{ in } R\} = \sum_{(x,y) \text{ in } R} p(x, y) \tag{2.7}$$

(see Exhibit 2.5). We often write $P(R)$ as an abbreviated notation for $P\{(X, Y) \text{ in } R\}$.

EXAMPLE 2.1 *(continued)*

The contingency table, Exhibit 2.3, for the ratings X, Y of the wine tasters shows that the highest probabilities are given to pairs (x, y), where $x = y$, indicating agreement among the wine tasters. To compute the probability $P\{X = Y\}$ that the tasters give identical ratings to a

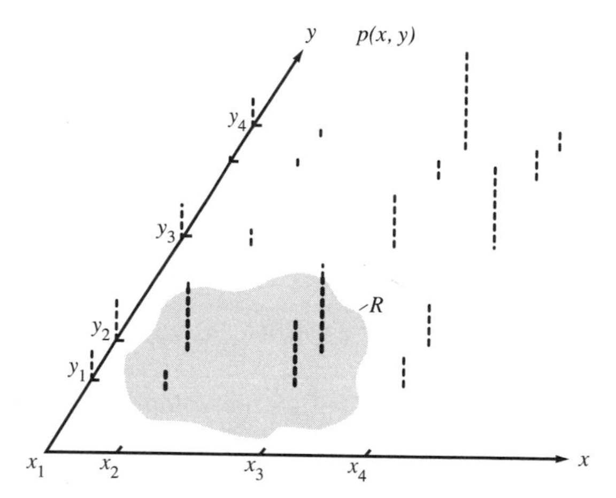

EXHIBIT 2.5 A region R in the plane. The probability that the random point (x, y) falls in R is the sum of the heights p(x, y) of the dotted lines that correspond to points (x, y) in R.

randomly chosen wine, let $R = \{(x, y) : x = y\}$, so that

$$
\begin{aligned}
P\{X = Y\} = P(R) &= \sum_{x=y} p(x, y) \\
&= p(1, 1) + p(2, 2) + p(3, 3) + p(4, 4) + p(5, 5) \\
&= 0.03 + 0.08 + 0.22 + 0.18 + 0.06 = 0.57.
\end{aligned}
$$
◆

When X and Y are continuous random variables with joint probability density function $f(x, y)$, then

$$
(2.8) \qquad P(R) = P\{(X, Y) \text{ in } R\} = \int_R\int f(x, y)\, dx\, dy,
$$

provided that the double integral in (2.8) is well defined. This double integral gives the volume over the region R between the (x, y)-plane and the surface determined by the graph of $f(x, y)$. Exhibits 2.6 to 2.8 give examples of such volumes when the bivariate density function $f(x, y)$ has the form

$$
(2.9) \qquad f(x, y) = \begin{cases} 1, & \text{if } 0 \le x \le 1, \quad 0 \le y \le 1, \\ 0, & \text{otherwise.} \end{cases}
$$

This density defines the **bivariate uniform distribution over the unit square.**

EXAMPLE 2.2

Suppose that X and Y are continuous random variables with the joint probability density function (2.9). Let $R = \{(x, y) : 0 \le x \le \frac{1}{4}$ and $\frac{1}{2} \le y \le \frac{3}{4}\}$. In this case,

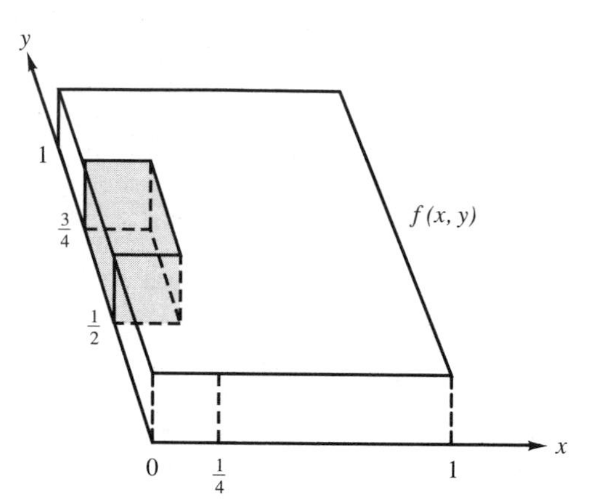

EXHIBIT 2.6 *The probability $P\{0 \le X \le \frac{1}{4} \text{ and } \frac{1}{2} \le Y \le \frac{3}{4}\}$ equals the volume of the shaded rectangle.*

$$\begin{aligned} P\{0 \le X \le \tfrac{1}{4} \text{ and } \tfrac{1}{2} \le Y \le \tfrac{3}{4}\} &= P\{(X, Y) \text{ in } R\} \\ &= \int_{1/2}^{3/4} \int_0^{1/4} f(x, y)\, dx\, dy \\ &= \int_{1/2}^{3/4} \int_0^{1/4} 1\, dx\, dy = \tfrac{1}{16} \end{aligned}$$

is the volume illustrated in Exhibit 2.6.

Of course, the region R can be of any shape (as in Exhibit 2.7). As another example, if R is a circle (see Exhibit 2.8) centered at the point $(\frac{1}{2}, \frac{1}{2})$ and with radius $\frac{1}{2}$, then

$$\begin{aligned} P\{(X - \tfrac{1}{2})^2 + (Y - \tfrac{1}{2})^2 \le \tfrac{1}{4}\} &= \iint_{(x-1/2)^2+(y-1/2)^2 \le 1/4} 1\, dx\, dy \\ &= \int_0^1 \left(\int_{(1/2)-\sqrt{y(1-y)}}^{(1/2)+\sqrt{y(1-y)}} 1\, dx \right) dy \\ &= \int_0^1 2y^{1/2}(1 - y)^{1/2}\, dy = 2B(\tfrac{3}{2}, \tfrac{3}{2}) \\ &= \frac{2[\Gamma(\frac{3}{2})]^2}{\Gamma(3)} = \frac{2(\frac{1}{2}\sqrt{\pi})^2}{2!} = \frac{\pi}{4} = 0.78539, \end{aligned}$$

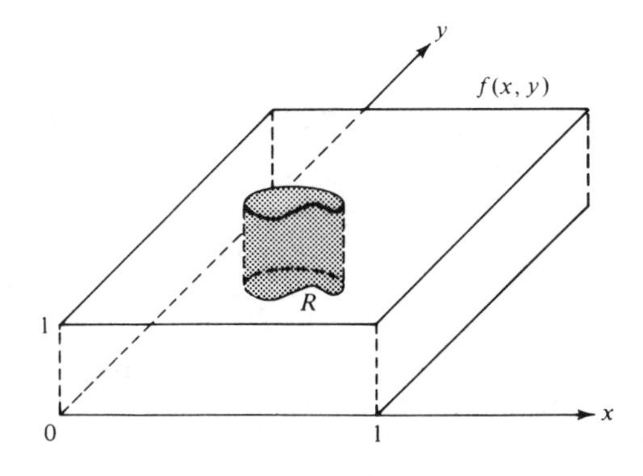

EXHIBIT 2.7 *The shaded volume equals the probability $P\{(X, Y)$ in $R\}$.*

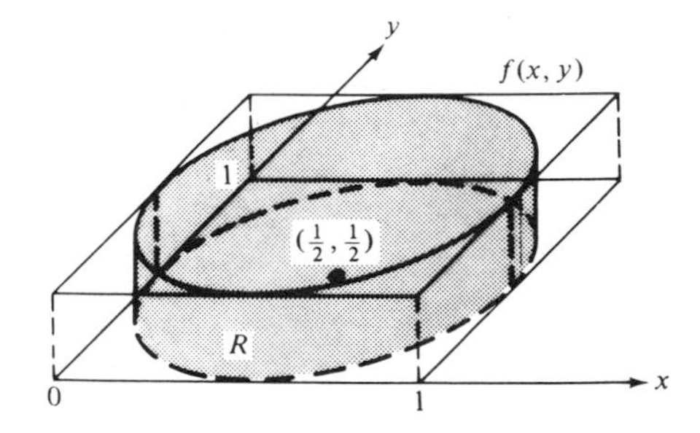

EXHIBIT 2.8 *The probability of the event $\{(X - \frac{1}{2})^2 + (Y - \frac{1}{2})^2 \le \frac{1}{4}\}$ equals the volume of the shaded cylinder whose base is the circle R.*

where use has been made of facts about the beta and gamma functions (Appendix A). Alternatively, this probability is the volume of a cylinder with height 1 and base $\pi(\frac{1}{2})^2$, as shown in Exhibit 2.8. ◆

For two continuous variables X and Y with a joint probability density function $f(x, y)$, the interpretation of $P\{(X, Y) \text{ in } R\}$ in terms of volume over a region R under the graph of $f(x, y)$ yields the following important consequence: *Whenever the area of R is 0, then the probability that (X, Y) falls in R is equal to 0.* For example, for any two numbers x_0 and y_0,

$$P\{X = x_0 \text{ and } Y = y_0\} = P(\{(x, y) : (x, y) = (x_0, y_0)\}) = 0,$$

because the area of a point is zero. More interestingly, because a line also has zero area,

$$P\{X = Y\} = P\{X - Y = 0\} = P(\{(x, y) : x - y = 0\}) = 0,$$

in contrast to the nonzero probability that the discrete variables X and Y in Example 2.1 are equal. In general, *when X and Y are continuous random variables having a joint probability density function, the probability that (X, Y) lies along any particular straight line is zero.*

Marginal Distributions

Not only does the joint distribution of the random variables X and Y contain all the relevant probabilistic information about the *joint* variation of X and Y; it also contains all information about the variations of X and of Y considered *individually*. For example, from the joint probability mass function $p(x, y)$ of two discrete random variables X and Y, we can obtain the individual probability mass functions $p_X(x)$ and $p_Y(y)$ of X and Y, respectively, by means of the formulas

$$(2.10) \qquad p_X(x) = \sum_y p(x, y), \qquad p_Y(y) = \sum_x p(x, y).$$

To verify (2.10), define the regions R_1 and R_2 by

$$R_1 = \{(x, y) : x = x_0,\ y \text{ has any value}\},$$
$$R_2 = \{(x, y) : y = y_0,\ x \text{ has any value}\}.$$

Then

$$p_X(x_0) = P\{X = x_0\} = P(R_1) = \sum_{(x,y) \text{ in } R_1} p(x, y) = \sum_y p(x_0, y),$$

$$p_Y(y_0) = P\{Y = y_0\} = P(R_2) = \sum_{(x,y) \text{ in } R_2} p(x, y) = \sum_x p(x, y_0).$$

As we have noted earlier, $p_X(x) = \Sigma_y\, (p(x, y)$ and $p_Y(y) = \Sigma_x\, p(x, y)$ can be obtained from the row sums and column sums, respectively, of a contingency table (such as Exhibit 2.3). For this reason, $p_X(x)$ and $p_Y(y)$ are referred to as **marginal probability mass functions** of the joint probability mass function $p(x, y)$.

EXAMPLE 2.1 *(contined)*

The marginal probability mass functions $p_X(x)$ and $p_Y(y)$ of the ratings X and Y of wine tasters I and II, respectively, can be obtained from the row and column totals of Exhibit 2.3, and are summarized in Exhibit 2.9. Note that the marginal distributions of X and Y are nearly the same. This result might be anticipated from the fact that both wine tasters seem to be using similar standards to rate the wines. ◆

EXHIBIT 2.9 *Marginal Probability Mass Functions of X and Y Based on the Joint Probability Distribution Given in Exhibit 2.3*

Rating, t	*Row Sum,* $p_X(t)$	*Column Sum,* $p_Y(t)$
1	0.06	0.08
2	0.18	0.19
3	0.36	0.35
4	0.29	0.28
5	0.11	0.10

When X and Y are continuous random variables with joint probability density function $f(x, y)$, then the **marginal density functions** $f_X(x)$ and $f_Y(y)$ of $f(x, y)$, which are the individual probability density functions of X and Y, respectively, are obtained by means of the formulas

$$(2.11) \qquad f_X(x) = \int_{-\infty}^{\infty} f(x, y)\, dy, \qquad f_Y(y) = \int_{-\infty}^{\infty} f(x, y)\, dx.$$

To verify (2.11) in the case of $f_X(x)$, we first obtain the marginal cumulative distribution function, $F_X(x_0)$, of X. To do so, we use (2.8) and the region

$$R_1 = \{(x, y) : x \leq x_0, \text{ any value}\}.$$

Thus

$$F_X(x_0) = P\{X \leq x_0\} = P(R_1) = \int_{-\infty}^{x_0} \left[\int_{-\infty}^{\infty} f(x, y)\, dy\right] dx.$$

Differentiating $F_X(x_0)$ with respect to x_0 and making use of the fundamental theorem of calculus establishes (2.11) for $f_X(x)$. The formula in (2.11) for $f_Y(y)$ is obtained in a similar fashion.

EXAMPLE 2.2 *(continued)*

For the bivariate probability density function $f(x, y)$ which equals 1 if $0 \leq x \leq 1$, $0 \leq y \leq 1$, and equals 0 otherwise, let us determine the marginal density function $f_X(x)$ of x. Note first that if $x < 0$ or $x > 1$, then $f(x, y) = 0$ for all y and $f_X(x) = \int_{-\infty}^{\infty} f(x, y)\, dy = 0$. When $0 \leq x \leq 1$, $f(x, y)$ is positive only for $0 \leq y \leq 1$. Thus

$$f_X(x) = \int_{-\infty}^{\infty} f(x, y)\, dy = \int_0^1 1\, dy = 1, \qquad 0 \leq x \leq 1.$$

We conclude that

$$f_X(x) = \begin{cases} 1, & \text{if } 0 \leq x \leq 1, \\ 0, & \text{if } x < 0 \text{ or } x > 1, \end{cases}$$

which is the probability density function of the uniform distribution on [0, 1]. In a similar manner Y also has a marginal uniform distribution on [0, 1]. ◆

EXAMPLE 2.3

Consider the bivariate probability density function

(2.12)

$$f(x, y) = \begin{cases} 360xy^2(1 - x - y), & \text{if } 0 \leq x, \quad y \leq 1, \quad 0 \leq x + y \leq 1, \\ 0, & \text{otherwise,} \end{cases}$$

which might model the joint distribution of the proportions X, Y of two types of metal in ore samples from a mine. To obtain the marginal density, $f_Y(y)$, of the proportion Y of the second type of metal, notice first from (2.12) that when $y < 0$ or $y > 1$, $f(x, y) = 0$ for all x. Consequently, $f_Y(y) = \int_{-\infty}^{\infty} f(x, y)\,dx = 0$ when $y < 0$ or $y > 1$. When $0 \le y \le 1$, then $f(x, y)$ is positive only for $0 \le x \le 1 - y$, and

$$f_Y(y) = \int_{-\infty}^{\infty} f(x, y)\,dx = 360y^2 \int_0^{1-y} [x(1 - y) - x^2]\,dx$$
$$= 360y^2\left[(1 - y)\left(\left(\frac{x^2}{2}\Big|_0^{1-y}\right) - \frac{x^3}{3}\Big|_0^{1-y}\right)\right]$$
$$= 60y^2(1 - y)^3.$$

We conclude that

$$f_Y(y) = \begin{cases} 60y^2(1 - y)^3, & 0 \le y \le 1, \\ 0, & y < 0 \text{ or } y > 1, \end{cases}$$

and recognize this density as that of the beta distribution with parameters $r = 3$, $s = 4$ (Section 5, Chapter 10; note that $1/B\,(3, 4) = \Gamma(7)/[\Gamma(3)\Gamma(4)] = 60$). In a similar fashion, it can be shown that the marginal density function $f_X(x)$ of the proportion X of the first type of metal is that of the beta distribution with parameters $r = 2$, $s = 5$:

$$f_X(x) = \begin{cases} 30x(1 - x)^4, & 0 \le x \le 1, \\ 0, & x < 0 \text{ or } x > 1. \end{cases}$$ ◆

It should be emphasized that the marginal distributions obtained from the joint distribution of any pair (X, Y) of jointly distributed random variables must be the same as the distributions that we would have obtained if we had considered the variables X and Y individually.

EXHIBIT 2.10 *Two Joint Discrete Probability Distributions with the Same Marginal Distributions*

(a)

	y		
x	1	0	Total
1	0.40	0.20	0.60
0	0.30	0.10	0.40
Total	0.70	0.30	1.00

(b)

	y		
x	1	0	Total
1	0.42	0.18	0.60
0	0.28	0.12	0.40
Total	0.70	0.30	1.00

Although the individual distributions of X and Y can be obtained in the form of marginal distributions from knowledge of the joint distribution of X and Y, the converse is not true. That is, *the marginal distributions of two jointly distributed random variables do not determine the joint distribution of these random variables*. To verify this fact, we need only show that two *different* joint distributions can have the *same* marginal distribution for X and the *same* marginal distribution for Y. Exhibit 2.10 shows two different contingency tables that have the same marginal distributions.

EXAMPLE 2.4

Suppose that X and Y have the joint probability density function

(2.13)
$$f(x, y) = \begin{cases} 1800x(1 - x)^4y^2(1 - y)^3, & \text{if } 0 \le x \le 1, \quad 0 \le y \le 1, \\ 0, & \text{otherwise.} \end{cases}$$

For $0 \le x \le 1$,

$$f_X(x) = \int_{-\infty}^{\infty} f(x, y)\, dy = 1800x(1 - x)^4\left[\int_0^1 y^2(1 - y)^3\, dy\right].$$

From Appendix A,

$$\int_0^1 y^2(1 - y)^3\, dy = B(3, 4) = \frac{\Gamma(3)\Gamma(4)}{\Gamma(3 + 4)} = \frac{2!\, 3!}{6!} = \frac{1}{60}.$$

Thus

$$f_X(x) = \begin{cases} 30x(1 - x)^4, & \text{if } 0 \le x \le 1 \\ 0, & \text{if } x < 0 \text{ or } x > 1, \end{cases}$$

which is the same marginal density for X obtained in Example 2.3. Similarly, the marginal density, $f_Y(y)$, for Y obtained from the joint probability density function (2.13) can be shown to be the same as that obtained from the joint probability density function (2.12). However, the joint density functions (2.12) and (2.13) define different distributions, as can be seen from the fact that (2.12) assigns zero density to points (x, y) for which $0 \le 1 - x < y < 1$, whereas (2.13) assigns positive density to these points. ◆

For each definition or result presented in this section, we have given separate treatments for the case of discrete bivariate distributions and for the case of continuous bivariate distributions. There is, however, a similarity between these two cases, in which:

1. The discrete probability mass functions $p(x, y)$, $p_X(x)$, $p_Y(y)$ correspond to the probability density functions $f(x, y)$, $f_X(x)$, $f_Y(y)$, respectively.
2. Sums of joint or marginal probability mass functions correspond to (and have properties similar to) integrals of joint or marginal probability density functions [compare equations (2.9) and (2.10)].

Consequently, it is both unnecessary and repetitious to continue to give detailed presentations or arguments for both discrete and continuous cases. Verification of a result for one case will suggest how to verify the corresponding result for the other case, making use of the similarities noted above.

3 Descriptive Indices of Bivariate Distributions

For univariate distributions, the mean and variance are frequently used as descriptive indices of location and of dispersion, respectively. The measure of location typically used for bivariate distributions of random variables X and Y is the pair (μ_X, μ_Y), where μ_X is the mean of the marginal distribution of X and μ_Y is the mean of the marginal distribution of Y. The pair (μ_X, μ_Y) is called the **joint mean** or **mean vector** of the joint distribution of X and Y.

Although μ_X and μ_Y can be computed from the respective marginal distributions of X and Y, these quantities can also be computed directly from the joint distribution of X and Y. To do so, it is helpful to define the bivariate generalization of the notion of the expected value of a function.

Expected Value of a Function of Two Jointly Distributed Random Variables

Let $h(x, y)$ be any function of x and y. Then if the random variables X and Y have a joint distribution, the **expected value of $h(X, Y)$** is defined to be:

$$(3.1) \quad E[h(X, Y)] = \begin{cases} \sum_x \sum_y h(x, y)p(x, y) = \sum_y \sum_x h(x, y)p(x, y), & \textbf{discrete case,} \\ \int_{-\infty}^{\infty} \int_{-\infty}^{\infty} h(x, y)f(x, y)\,dx\,dy, \\ \quad = \int_{-\infty}^{\infty} \int_{-\infty}^{\infty} h(x, y)f(x, y)\,dy\,dx, & \textbf{continuous case,} \end{cases}$$

provided that one of the double sums or integrals in (3.1) is well defined.

Note. A double sum $\sum_x \sum_y S(x, y)$ is well defined if the corresponding double sum $\sum_x \sum_y |S(x, y)|$ of unsigned values of $S(x, y)$ yields a finite value. Similarly, a double integral such as $\int_{-\infty}^{\infty} \int_{-\infty}^{\infty} S(x, y)\, dy\, dx$ is well defined if the corresponding double integral $\int_{-\infty}^{\infty} \int_{-\infty}^{\infty} |S(x, y)|\, dy\, dx$ is finite. If one of the double sums in (3.1) is well defined, so is the other, and (as shown) their values are equal. A similar fact is true for double integrals in (3.1). This implies that we can find $E[h(X, Y)]$ either by summing (integrating) first over values of x and then over values of y, or by summing (integrating) first over y and then over x.

EXAMPLE 3.1

The bivariate density function

$$f(x, y) = \begin{cases} 360xy^2(1 - x - y), & 0 \le x, \quad y \le 1, \quad 0 \le x + y \le 1, \\ 0, & \text{otherwise,} \end{cases}$$

was used in Example 2.3 as a model for the variation of the (continuous) proportions X, Y of two types of metal. The function $h(X, Y) = X/Y$ gives the ratio of the two proportions. The expected value of this function is

$$\begin{aligned} E[X/Y] &= \int_{-\infty}^{\infty} \int_{-\infty}^{\infty} \frac{x}{y} f(x, y)\, dy\, dx \\ &= \int_0^1 \left[\int_0^{1-x} \frac{x}{y} [360xy^2(1 - x - y)]\, dy \right] dx \end{aligned}$$

because $f(x, y)$ is positive only when $0 \le y \le 1 - x$ and $0 \le x \le 1$. The inner integral is

$$\begin{aligned} \int_0^{1-x} 360x^2y(1 - x - y)dy &= 360x^2\left[(1 - x)\frac{y^2}{2} - \frac{y^3}{3}\right]_0^{1-x} \\ &= 60x^2(1 - x)^3. \end{aligned}$$

Thus, using facts given in Appendix A about the beta function, we have

$$E[X/Y] = \int_0^1 60x^2(1 - x)^3\, dx = 60B(3, 4) = 1. \qquad \blacklozenge$$

The following properties of expected values are obtained directly from corresponding properties of integrals and sums.

For any two functions $h_1(x, y)$ and $h_2(x, y)$ having expected values, and any constants a and b,

$$E[ah_1(X, Y) + bh_2(X, Y)] = aE[h_1(X, Y)] + bE[h_2(X, Y)]. \tag{3.2}$$

If $h(x, y) \geq 0$ for all x and y, then

$$E[h(X, Y)] \geq 0. \tag{3.3}$$

Equation (3.2) generalizes the linearity property of expected values proved for the expected value of a function of a single random variable in Chapter 6. Similarly, equation (3.3) generalizes the preservation of inequalities property of expected values.

When $h(x, y)$ is a function solely of one variable (say x), the bivariate generalization of the definition of an expected value given by (3.1) yields the same result as the definition of the expected value relative to the marginal distribution (of X). That is, if $h(x, y) = g(x)$, then

$$E[g(X)] = \begin{cases} \Sigma_x \Sigma_y \, g(x)p(x, y) = \Sigma_x \, g(x)p_X(x), & \textbf{discrete case,} \\ \int_{-\infty}^{\infty} \int_{-\infty}^{\infty} g(x)f(x, y) \, dy \, dx = \int_{-\infty}^{\infty} g(x)f_X(x) \, dx, & \textbf{continuous case.} \end{cases} \tag{3.4}$$

The verification of (3.4) in the discrete case is

$$E[g(X)] = \sum_x \sum_y g(x)p(x, y) = \sum_x g(x)\left[\sum_y p(x, y)\right] = \sum_x g(x) \, p_X(x),$$

where the last equality is a consequence of equation (2.9). It follows from this fact that μ_X, σ_X^2, μ_Y and σ_Y^2 can either be computed as expected values relative to the appropriate marginal distribution, or as expected values

$$\mu_X = E[X], \qquad \mu_Y = E[Y], \qquad \sigma_X^2 = E(X - \mu_X)^2, \qquad \sigma_Y^2 = E(Y - \mu_Y)^2,$$

relative to the joint distribution of X and Y. Using the latter characterization of μ_X and μ_Y, and (3.2) with $h_1(x, y) = x$, $h_2(x, y) = y$, we obtain

$$\mu_{aX+bY} = E[aX + bY] = a\mu_X + b\mu_Y. \tag{3.5}$$

Examples of the use of equation (3.5) have already been given in Chapter 7.

The Covariance

In the search for measures of dispersion for bivariate distributions, the variances σ_X^2, σ_Y^2 of X and Y come naturally to our attention. However, these indices are determined only by the marginal distributions of X and Y, respectively. Because the marginal distributions of X and Y do not determine the joint distribution, it is clear that σ_X^2 and σ_Y^2, by themselves cannot tell us how X and Y vary jointly. That is, we lack a measure of the **interrelationship** or **covariation** of X and Y.

Note that σ_X^2 measures dispersion along the x-direction in the plane (Exhibit 3.1) and that σ_Y^2 measures dispersion along the y-direction in the plane (Exhibit 3.2). Correspondingly, σ_{aX+bY}^2 should measure dispersion along the $(ax + by)$-direction in the plane. A knowledge of σ_{aX+bY}^2 for all values of a and b would thus yield information about the joint variability of X and Y in all directions. However, using (3.2) repeatedly gives

$$(3.6)\ \sigma_{aX+bY}^2 = E(aX + bY - \mu_{aX+bY})^2 = E(aX + bY - a\mu_X - b\mu_Y)^2$$
$$= E[a^2(X - \mu_X)^2 + 2ab(X - \mu_X)(Y - \mu_Y) + b^2(Y - \mu_Y)^2]$$
$$= a^2\sigma_X^2 + 2abE[(X - \mu_X)(Y - \mu_Y)] + b^2\sigma_Y^2.$$

Because σ_X^2 and σ_Y^2 only measure how X and Y vary individually, it follows from (3.6) that the index

$$(3.7)\qquad \sigma_{XY} = E[(X - \mu_X)(Y - \mu_Y)],$$

called the **covariance** of X and Y, conveys information about how X and Y covary jointly.

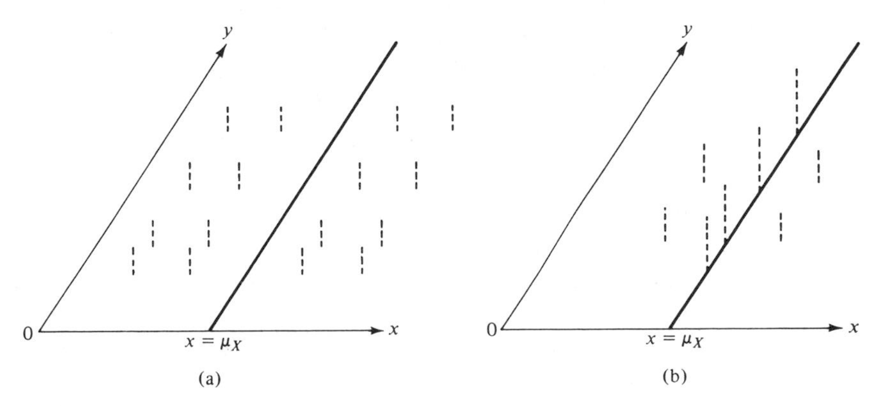

EXHIBIT 3.1 *The variance σ_X^2 shows the dispersion of the joint probability mass function $p(x, y)$ around the line $x = \mu_X$. The dispersion in (a) is larger than the dispersion in (b).*

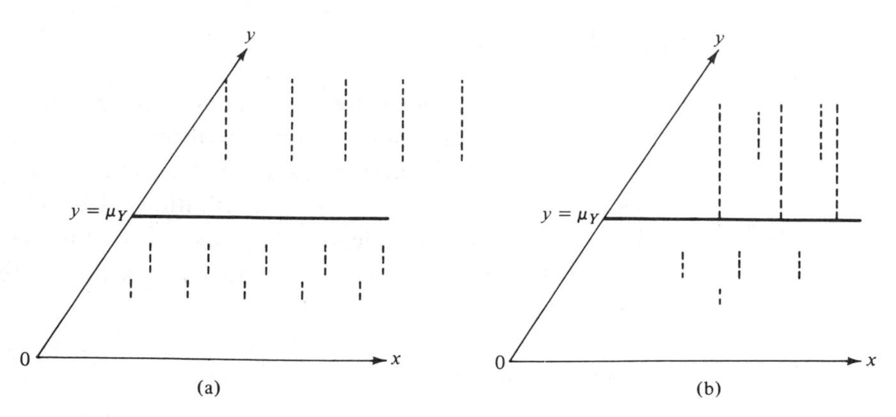

EXHIBIT 3.2 *The variance σ_Y^2 shows the dispersion of the joint probability mass function p(x, y) around the line $y = \mu_Y$. The dispersion in (a) is larger than the dispersion in (b).*

In place of the defining formula (3.7) for the covariance of σ_{XY}, it is usually more convenient to use the alternative computational formula:

$$\text{(3.8)} \qquad \sigma_{XY} = E[XY] - E[X]E[Y] = E[XY] - \mu_X\mu_Y$$

to compute σ_{XY}. To verify (3.8), note that by applying repeatedly the linearity property (3.2) of expected values,

$$\begin{aligned}\sigma_{XY} &= E(X - \mu_X)(Y - \mu_Y) = E[XY] - \mu_X E[Y] - \mu_Y E[X] + \mu_X\mu_Y \\ &= E[XY] - \mu_X\mu_Y.\end{aligned}$$

The result (3.8) is very similar to the alternative computational formula for the variance σ_X^2 given in Chapter 6. This is not surprising, for the variance σ_X^2 is simply the covariance of X with itself. That is,

$$\sigma_{XX} = E[(X - \mu_X)(X - \mu_X)] = E(X - \mu_X)^2 = \sigma_X^2.$$

Similarly, $\sigma_{YY} = \sigma_Y^2$.

The covariance σ_{XY} can be positive or negative. It is positive when large values of X correspond to large values of Y and small values of X correspond to small values of Y. It is negative when X and Y move in opposite directions (see Exhibit 3.3). When X and Y are both on the same sides of their respective means, $(X - \mu_X)(Y - \mu_Y)$ is positive. When X and Y fall on opposite sides of their means, $(X - \mu_X)(Y - \mu_Y)$ is negative. The covariance σ_{XY} is thus positive if (X, Y) falls most often in regions where X and Y tend to be both large or both small, and negative when (X, Y) falls most often in regions

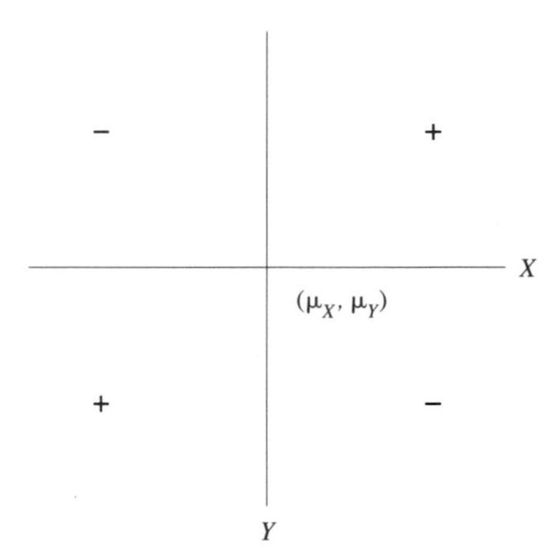

EXHIBIT 3.3 *Regions where $(X - \mu_X)(Y - \mu_Y)$ contribute positive and negative mass to the covariance.*

where Y is large when X is small, or vice versa. The covariance σ_{XY} is zero if contributions from all four regions shown in Exhibit 3.3 are in balance.

Consequently, if X and Y denote a student's high school and college grade-point averages, respectively, we expect the covariance of X and Y to be positive. If X denotes the amount of minerals in a water supply and Y is a measure of pipe corrosion, we also expect a positive covariance. On the other hand, if X denotes the amount of alcohol in the blood and Y is a measure of motor coordination, the covariance between X and Y should be negative.

The following examples illustrate the computation of σ_{XY}.

EXAMPLE 3.2

A study made of the relationship between the number X of visits made by representatives of a computer manufacturer to potential customers and the number Y of large computer systems sold to these customers yielded the following joint distribution, summarized in a contingency table:

	y			
x	0	1	2	$p_X(x)$
1	0.20	0.10	0.01	0.31
2	0.15	0.30	0.06	0.51
3	0.03	0.05	0.10	0.18
$p_Y(y)$	0.38	0.45	0.17	

From the marginal probability mass functions $p_X(x)$ and $p_Y(y)$, the means of X and Y are found to be

$$\mu_X = 1.87, \qquad \mu_Y = 0.79.$$

Also,

$$\begin{aligned} E[XY] &= \sum_{x=1}^{3}\sum_{y=1}^{3} xyp(x, y) \\ &= (1)(0)(0.20) + (1)(1)(0.10) + \cdots \\ &\quad + (3)(1)(0.05) + (3)(2)(0.10) \\ &= 1.71, \end{aligned}$$

so that using (3.8),

$$\sigma_{XY} = E[XY] - \mu_X\mu_Y = 1.71 - (1.87)(0.79) = 0.2327.$$

The positive value of the covariance σ_{XY} indicates a positive association between the attention paid to a customer (measured by the number of X of visits) and the number Y of sales made to that customer. ◆

EXAMPLE 3.1 (continued)

In this example, X and Y are the proportions of two metals in ore samples from a mine, and have the joint probability density function

$$f(x, y) = \begin{cases} 360xy^2(1 - x - y), & \text{if } 0 \le x, y \le 1, 0 \le x + y \le 1, \\ 0, & \text{otherwise,} \end{cases}$$

In Example 2.3 it was shown that the marginal distribution of X is a beta distribution with parameters $r = 2$, $s = 5$; also, the marginal distribution of Y is a beta distribution with parameters $r = 3$, $s = 4$. It follows from Chapter 10, Section 5 that the means of X Y are $\mu_X = 2/(2 + 5) = \frac{2}{7}$ and $\mu_Y = 3/(3 + 4) = \frac{3}{7}$, respectively. Now

$$\begin{aligned} E[XY] &= \int_{-\infty}^{\infty}\int_{-\infty}^{\infty} xyf(x, y)\,dx\,dy = \int_0^1\int_0^{1-x} xy[360xy^2(1 - x - y)]\,dy\,dx \\ &= 360\int_0^1 x^2\left[\int_0^{1-x} y^3(1 - x - y)\,dy\right] dx \\ &= 360\int_0^1 x^2\left[(1 - x)\left(\frac{y^4}{4}\bigg|_0^{1-x}\right) - \left(\frac{y^5}{5}\bigg|_0^{1-x}\right)\right] \\ &= 18\int_0^1 x^2(1 - x)^5\,dx. \end{aligned}$$

From Appendix A, $\int_0^1 x^2(1 - x)^5\,dx = B(3, 6) = 1/168$. Thus $E[XY] = 18/168$ and

$$\sigma_{XY} = E[XY] - \mu_X\mu_Y = \frac{18}{168} - \left(\frac{2}{7}\right)\left(\frac{3}{7}\right) = -0.015.$$

It is not surprising that σ_{XY} is negative. Because the sum $X + Y$ of the proportions X, Y cannot exceed 1, large values of X force the values of Y to be small, and vice versa. Because X and Y tend to vary in opposite directions, we are not surprised to find that the covariance σ_{XY} between X and Y is negative. ♦

The Correlation Coefficient

As a measure of how closely X and Y covary, the covariance possesses an important defect: its value depends on the scales of measurement used for X and Y. A change of scale of measurement for X replaces X by the linear transformation $aX + b$ of X; similarly, a change of scale for Y replaces Y by $cY + d$. After such changes of scale, the covariance measured is that between $aX + b$ and $cY + d$, which is

$$\begin{aligned}\sigma_{aX+b,cY+d} &= E(aX + b - \mu_{aX+b})(cY + d - \mu_{cY+d}) \\ &= E(aX + b - a\mu_X - b)(cY + d - c\mu_Y - d) \\ &= acE[(X - \mu_X)(Y - \mu_Y)] = ac\sigma_{XY}.\end{aligned}$$

That is,

(3.9) $$\sigma_{aX+b,cY+d} = ac\sigma_{XY}.$$

Thus if X is the length in inches of a randomly selected rod and Y is a subject's estimate of that length (in inches), we might find that $\sigma_{XY} = 1.44$ and conclude that the estimates of length have a substantial positive association with the true lengths. Switching to measuring in feet, however, produces measurements $X^* = X/12$ and $Y^* = Y/12$ for which the covariance

$$\begin{aligned}\sigma_{X^*Y^*} &= \sigma_{X/12,Y/12} = (\tfrac{1}{12})(\tfrac{1}{12})\sigma_{XY} \\ &= \tfrac{1}{144}(1.44) = 0.01\end{aligned}$$

is discouragingly low. The nature of the observations, however, has not changed, so that this change in our measure of covariation must be illusory, and reflects the dependence of the covariance index on the scales of measurement for X and Y.

As a possibly scale-free index of independence between X and Y, consider

(3.10) $$\rho_{XY} = \frac{\sigma_{XY}}{\sigma_X\sigma_Y}.$$

Because the standard deviations σ_X, σ_Y of X and Y are positive, ρ_{XY} has the same sign as σ_{XY} and is equal to 0 if and only if $\sigma_{XY} = 0$. Thus ρ_{XY} conveys information about the existence and nature of any dependence between X and Y similar to that provided by σ_{XY}. On the other hand,

$$\rho_{aX+b,cY+d} = \frac{\sigma_{aX+b,cY+d}}{\alpha_{aX+b}\sigma_{cY+d}} = \frac{ac\sigma_{XY}}{|a|\sigma_X|c|\sigma_Y} = (\text{sign of } ac)\,\frac{\sigma_{XY}}{\sigma_X\sigma_Y}.$$

Consequently,

(3.11)	$\rho_{aX+b,cY+d} = \rho_{XY}, \quad \text{all } ac > 0,\ b,\ d,$

so that as long as a change in scale does not reverse the order of the values of X relative to Y (i.e., $ac > 0$), the index ρ_{XY} is independent of any changes of scale. Said another way, ρ_{XY} is independent of the units of measurement used to express X and Y. The index ρ_{XY} is called the **correlation coefficient** between X and Y.

Note that from (3.10) it follows that

(3.12)	$\sigma_{XY} = \sigma_X\sigma_Y\rho_{XY},$

and from (3.6), (3.7), and (3.12) we have

$$\text{(3.13)} \qquad \begin{aligned} \sigma^2_{aX+bY} &= a^2\sigma^2_X + b^2\sigma^2_Y + 2ab\sigma_{XY} \\ &= a^2\sigma^2_X + b^2\sigma^2_Y + 2ab\sigma_X\sigma_Y\rho_{XY}. \end{aligned}$$

We can use (3.13) to show that the correlation coefficient ρ_{XY} assumes values between -1 and 1. To demonstrate this fact, apply (3.13) to

$$Z = \left(\frac{1}{\sigma_Y}\right)Y + \left(-\frac{\rho_{XY}}{\sigma_X}\right)X.$$

Noting that $a = -\rho_{XY}/\sigma_X$, $b = 1/\sigma_Y$ in (3.13),

$$\begin{aligned} \sigma^2_Z &= \left(-\frac{\rho_{XY}}{\sigma_X}\right)^2\sigma^2_X + \left(\frac{1}{\sigma_Y}\right)^2\sigma^2_Y + 2\left(-\frac{\rho_{XY}}{\sigma_X}\right)\left(\frac{1}{\sigma_Y}\right)\sigma_X\sigma_Y\rho_{XY} \\ &= 1 - \rho^2_{XY}. \end{aligned}$$

Because the variance, σ_Z^2, must be nonnegative, it follows that $1 - \rho_{XY}^2 \geq 0$, or

(3.14)	$-1 \leq \rho_{XY} \leq 1.$

Note that if $\rho_{XY} = \pm 1$, then $\sigma_Z^2 = 0$ and Z does not vary from its mean μ_Z. That is, $P\{Z = \mu_Z\} = 1$ or, equivalently,

$$Y = \left(\frac{\sigma_Y}{\sigma_X}\rho_{XY}\right)X + \mu_Z = \left(\frac{\sigma_Y}{\sigma_X}\rho_{XY}\right)X + \left(\mu_Y - \frac{\sigma_Y}{\sigma_X}\rho_{XY}\mu_X\right)$$

with probability equal to 1. Thus if $\rho_{XY} = \pm 1$, Y is equal to a nonconstant linear function of X. On the other hand, if $Y = aX + b$, $a \neq 0$, then

$$\rho_{XY} = \rho_{X,aX+b} = \frac{\sigma_{X,aX+b}}{\sigma_X\sigma_{aX+B}} = \frac{a\sigma_X^2}{|a|\sigma_X^2} = \frac{a}{|a|},$$

so that $\rho_{XY} = 1$ if $a > 0$, and $\rho_{XY} = -1$ if $a < 0$. We conclude that $\rho_{XY} = \pm 1$ *if and only if* Y *is, with probability equal to* 1, *a nonconstant linear function of* X.

Of course, as we have already remarked in Section 2, two jointly continuous random variables X and Y have zero probability of exactly satisfying *any* linear function. For such variables, and in general for any pair (X, Y) of jointly distributed random variables, the foregoing interpretation of the extreme values of ρ_{XY} serves only to help interpret the intermediate values of ρ_{XY}. That is, this analysis of the meaning of the extreme values ± 1 of ρ_{XY} indicates that *the closer the probability mass or density concentrates around a line* $L = \{(x, y) : y = ax + b\}$ *with* $a > 0$, *the closer the value of* ρ_{XY} *is to* $+1$, *whereas the closer the probability mass or density concentrates around a line* L *with* $a < 0$, *the closer the value of* ρ_{XY} *is to* -1.

The contingency tables shown in Exhibit 3.4 give some possible bivariate probability distributions for the ranks X and Y on a three-point scale given by two raters. The marginal probability mass functions $p_X(x)$ and $p_Y(y)$ are kept constant from table to table; the differences between the contingency tables (and the associated values of the correlation coefficient ρ_{XY}) thus reflect differences in the interrelationships between the variables X and Y. Comparing these tables, we see that ρ_{XY} becomes more positive as there is more probability mass close to the rising line $L = \{(x, y) : y = x\}$, and ρ_{XY} becomes more negative as more probability mass is assigned to points (x, y) near the falling line $L = \{(x, y) : y = 4 - x\}$. Similar analyses of the correspondence between the values of ρ_{XY} and the strength of the probabilistic

EXHIBIT 3.4 *Joint Probability Model for the Ratings of Two People*

(a) $\rho_{XY} = 0$

x	y: 1	2	3	Total
3	1/9	1/9	1/9	1/3
2	1/9	1/9	1/9	1/3
1	1/9	1/9	1/9	1/3
Total	1/3	1/3	1/3	1

(b) $\rho_{XY} = \frac{1}{2}$

x	y: 1	2	3	Total
3	1/18	1/18	4/18	1/3
2	1/18	4/18	1/18	1/3
1	4/18	1/18	1/18	1/3
Total	1/3	1/3	1/3	1

(c) $\rho_{XY} = -\frac{1}{2}$

x	y: 1	2	3	Total
3	4/18	1/18	1/18	1/3
2	1/18	4/18	1/18	1/3
1	1/18	1/18	4/18	1/3
Total	1/3	1/3	1/3	1

(d) $\rho_{XY} = \frac{4}{9}$

x	y: 1	2	3	Total
3	1/27	2/27	6/27	1/3
2	2/27	5/27	2/27	1/3
1	6/27	2/27	1/27	1/3
Total	1/3	1/3	1/3	1

(e) $\rho_{XY} = -\frac{5}{9}$

x	y: 1	2	3	Total
3	6/27	2/27	1/27	1/3
2	2/27	5/27	2/27	1/3
1	1/27	2/27	6/27	1/3
Total	1/3	1/3	1/3	1

(f) $\rho_{XY} = \frac{2}{3}$

x	y: 1	2	3	Total
3	1/36	2/36	9/36	1/3
2	2/36	8/36	2/36	1/3
1	9/36	2/36	1/36	1/3
Total	1/3	1/3	1/3	1

(g) $\rho_{XY} = -\frac{1}{3}$

x	y: 1	2	3	Total
3	9/36	2/36	1/36	1/3
2	2/36	8/18	2/18	1/3
1	1/36	2/36	9/36	1/3
Total	1/3	1/3	1/3	1

relationship between X and Y are given in Chapter 12 for the bivariate normal distribution.

It is important to note that the correlation coefficient measures primarily the sign and strength of *linear* relationships between jointly distributed random variables X, Y. Consequently, *a correlation of* 0 *between X and Y need not imply that these variables are probabilistically unrelated* (*independent*). For example, consider the case of jointly distributed discrete variables X, Y whose probability mass function is

$$p(x, y) = \begin{cases} 1/3, & \text{if } (x, y) = (-1, 1) \text{ or } (1, 1), \\ 1/6, & \text{if } (x, y) = (-2, 4) \text{ or } (2, 4), \\ 0, & \text{otherwise.} \end{cases}$$

Note that for this probability distribution,

$$\begin{aligned} \mu_X &= E[X] = (-2)(\tfrac{1}{6}) + (-1)(\tfrac{1}{3}) + (1)(\tfrac{1}{3}) + 2(\tfrac{1}{6}) = 0, \\ \sigma_X^2 &= E[X^2] - \mu_X^2 = (-2)^2(\tfrac{1}{6}) + (-1)^2(\tfrac{1}{3}) + (1)^2(\tfrac{1}{3}) + (2)^2(\tfrac{1}{6}) - (0)^2 = 2, \end{aligned}$$

and similarly,

$$\mu_Y = E(Y) = \tfrac{4}{3}, \quad \sigma_Y^2 = 6 - (\tfrac{4}{3})^2 = \tfrac{38}{9}.$$

Also,

$$\begin{aligned} E[XY] &= \sum_x \sum_y xy\, p(x, y) \\ &= (-2)(4)(\tfrac{1}{6}) + (-1)(1)(\tfrac{1}{3}) + (1)(1)(\tfrac{1}{3}) + (2)(4)(\tfrac{1}{6}) = 0, \end{aligned}$$

so that

$$\sigma_{XY} = E[XY] - \mu_X \mu_Y = 0 - (0)(\tfrac{4}{3}) = 0.$$

Consequently,

$$\rho_{XY} = \frac{\sigma_{XY}}{\sigma_X \sigma_Y} = \frac{0}{\sqrt{2}\sqrt{38/9}} = 0.$$

On the other hand, there is a strong nonlinear relationship between X and Y because $y = x^2$ for every value of (x, y) for which $p(x, y)$ is not zero. That is, $P\{Y = X^2\} = 1$. Consequently, although $\rho_{XY} = 0$ indicates a lack of a linear relationship (probabilistically) between X and Y, it does not rule out the possibility that these variables are related nonlinearly.

4 Joint Moment Generating Function

Just as the moment generating function of a single random variable can be used to determine or characterize the distribution of that variable, the **joint moment generating function** of the pair of variables X, Y characterizes their joint distribution. The joint moment generating function of X, Y is defined by

$$M(s, t) = E[e^{sX+tY}] \tag{4.1}$$

provided that the expected value of the function e^{sx+ty} is finite. To characterize the joint distribution of X, Y, the moment generating function must be defined at least for all values (s, t) in a small enough circle containing the origin $(s, t) = (0, 0)$.

Note that

$$M(0, t) = E[e^{0X+tY}] = E[e^{tY}] = M_Y(t),$$

and similarly $M(s, 0) = M_X(s)$. Consequently, from the joint moment generating function, $M(s, t)$, of X, Y, we can easily obtain the (marginal) moment generating functions $M_X(s)$, $M_Y(t)$, of X, Y, respectively:

$$M_X(s) = M(s, 0), \qquad M_Y(t) = M(0, t). \tag{4.2}$$

Hence the moments of X and the moments of Y can be obtained by

$$E[X^k] = \left[\left(\frac{\partial}{\partial s}\right)^k M(s, t)\right]_{s=t=0}, \qquad E[Y^k] = \left[\left(\frac{\partial}{\partial t}\right)^k M(s, t)\right]_{s=t=0}, \tag{4.3}$$

for $k = 1, 2, \ldots$. It is also possible to obtain the expected values $E[X^k Y^l]$, $k, l = 0, 1, 2, \ldots$, from $M(s, t)$:

$$E[X^k Y^l] = \left[\frac{\partial^{k+l}}{(\partial s)^k(\partial t)^l} M(s, t)\right]_{s=t=0}. \tag{4.4}$$

To motivate this result, note that

$$\begin{aligned} \frac{\partial^{k+l}}{(\partial s)^k(\partial t)^l} M(s, t) &= \frac{\partial^{k+l}}{(\partial s)^k(\partial t)^l} E[e^{sX+tY}] \\ &= E\left[\frac{\partial^{k+l}}{(\partial s)^k(\partial t)^l} e^{sX+tY}\right] \\ &= E[X^k Y^l e^{sX+tY}] \end{aligned} \tag{4.5}$$

provided that the operations of partial differentiation and summation (or integration) can be interchanged. Setting $s = t = 0$ in (4.5) yields (4.4).

From the joint moment generating function, $M(s, t)$, of X, Y, it is possible to obtain the moment generating function, $M_{X+Y}(t)$, of the sum of X and Y. Indeed,

$$M_{X+Y}(t) = E[e^{t(X+Y)}] = E[e^{tX+tY}] = M(t, t). \tag{4.6}$$

If we can recognize $M(t, t)$ as being the moment generating function of a certain distribution, it follows that $X + Y$ has this distribution. Consequently, (4.6) provides a way of determining the distribution of the random variable $X + Y$.

EXAMPLE 4.1

Consider the Bernoulli random variables X, Y defined by

$$X = \begin{cases} 1, & \text{if event A occurs,} \\ 0, & \text{if event A does not occur;} \end{cases}$$

$$Y = \begin{cases} 1, & \text{if event B occurs,} \\ 0, & \text{if event B does not occur.} \end{cases}$$

Suppose that the probability of event A is 0.5, the probability of event B is 0.4, and the probability of $A \cap B$ (both events occur) is 0.1. Thus $P\{X = 1\} = 0.5$, $P\{Y = 1\} = 0.4$, and $P\{(X, Y) = (1, 1)\} = 0.1$. Further, $P\{X = 0\} = 1 - 0.5 = 0.5$ and $P\{Y = 0\} = 1 - 0.4 = 0.6$. The marginal totals in a contingency table for X, Y are thus determined, and also the entry in the cell corresponding to $(x, y) = (1, 1)$. The remaining entries in this table can now be filled in to satisfy the given marginal totals:

	y		
x	0	1	
0	0.2	0.3	0.5
1	0.4	0.1	0.5
	0.6	0.4	

The joint moment generating function of X, Y is

$$M(s, t) = \sum_{x=0}^{1} \sum_{y=0}^{1} e^{sx+ty} p(x, y) = 0.2 + (0.3)e^{t} + (0.4)e^{s} + (0.1)e^{s+t}.$$

Consequently, the moment generating function of X is

$$M_X(s) = M(s, 0) = 0.2 + 0.3e^0 + (0.4)e^s + (0.1)e^{s+0}$$
$$= 0.5 + (0.5)e^s.$$

This agrees with the moment generating function obtained from the marginal probability mass function for X.

To find the covariance σ_{XY} between X and Y, we compute

$$\mu_X = E[X] = \left[\frac{\partial}{\partial s} M(s, t)\right]_{s=t=0}$$
$$= [0 + 0 + (0.4)e^s + (0.1)e^{s+t}]_{s=t=0} = 0.4 + 0.1 = 0.5,$$
$$\mu_Y = E[Y] = \left[\frac{\partial}{\partial t} M(s, t)\right]_{s=t=0}$$
$$= [0 + (0.3)e^t + 0 + (0.1)e^{s+t}]_{s=t=0} = 0.3 + 0.1 = 0.4,$$

and

$$E[XY] = \left[\frac{\partial^2}{\partial s\, \partial t} M(s, t)\right]_{s=t=0} = [0 + 0 + 0 + (0.1)e^{s+t}]_{s=t=0} = 0.1.$$

Thus

$$\sigma_{XY} = E[XY] - \mu_X\mu_Y = 0.1 - (0.5)(0.4) = -0.1.$$

The moment generating function of $V = X + Y$ is

$$M_{X+Y} = M(t, t) = (0.2) + (0.3)e^t + (0.4)e^t + (0.1)e^{2t}$$
$$= (0.2) + (0.7)e^t + (0.1)e^{2t},$$

which is the moment generating function of the probability mass function

$$p(v) = \begin{cases} 0.2, & \text{if } v = 0, \\ 0.7, & \text{if } v = 1, \\ 0.1, & \text{if } v = 2, \\ 0 & \text{otherwise.} \end{cases}$$

Hence $V = X + Y$ has the probability mass function $p(v)$. ♦

EXAMPLE 4.2

Suppose that X, Y have the joint probability density function

$$f(x, y) = \begin{cases} |x - y|\, e^{-(x+y)}, & \text{if } x, y > 0, \\ 0 & \text{otherwise.} \end{cases}$$

The joint moment generating function of X, Y is

$$\begin{aligned}
M(s, t) &= \int_{-\infty}^{\infty} \int_{-\infty}^{\infty} e^{sx+ty} f(x, y)\, dx\, dy \\
&= \int_0^{\infty} \int_y^{\infty} (x - y) e^{-(1-s)x-(1-t)y}\, dx\, dy \\
&\quad + \int_0^{\infty} \int_y^{\infty} (y - x) e^{-(1-s)x-(1-t)y}\, dx\, dy \\
&= \left(\int_0^{\infty} v e^{-(1-s)v}\, dv\right)\left(\int_0^{\infty} e^{-(2-s-t)y}\, dy\right) \\
&\quad + \left(\int_0^{\infty} u e^{-(1-t)u}\, du\right)\left(\int_0^{\infty} e^{-(2-s-t)x}\, dx\right),
\end{aligned}$$

where the change of variables $v = x - y$, $u = y - x$ is made to simplify the integrals. It was shown in Chapter 10, Section 2, that for $c > 0$,

$$\int_0^{\infty} z^k e^{-cz}\, dz = \left(\frac{1}{c}\right)^k \Gamma(k + 1), \qquad k = 0, 1, 2, \ldots .$$

Thus we have

$$\begin{aligned}
M(s, t) &= \frac{\Gamma(2)}{(1 - s)^2} \frac{\Gamma(1)}{2 - s - t} + \frac{\Gamma(2)}{(1 - t)^2} \frac{\Gamma(1)}{2 - s - t} \\
&= \frac{1}{(1 - s)^2(2 - s - t)} + \frac{1}{(1 - t)^2(2 - s - t)} \qquad \text{if} \quad s, t < 1.
\end{aligned}$$

The marginal moment generating functions of X, Y are

$$\begin{aligned}
M_X(s) &= M(s, 0) = \frac{1}{(1 - s)^2(2 - s)} + \frac{1}{2 - s}, \qquad s < 1, \\
M_Y(t) &= M(0, t) = \frac{1}{2 - t} + \frac{1}{(1 - t)^2(2 - t)}, \qquad t < 1.
\end{aligned}$$

Although we do not recognize these moment generating functions as corresponding to known density functions, we can see that $M_X(u) = M_Y(u)$, all $u < 1$, so that X and Y have identical marginal distributions. Thus

$$\begin{aligned}
\mu_X = \mu_Y &= \left[\frac{\partial}{\partial t} M(s, t)\right]_{s=t=0} \\
&= \left[\frac{1}{(1 - s)^2(2 - s - t)^2} + \frac{2(2 - s - t) + (1 - t)}{(1 - t)^4(2 - s - t)^2}\right]_{s=t=0} \\
&= \frac{1}{4} + \frac{5}{4} = \frac{3}{2}.
\end{aligned}$$

Also,

$$E[XY] = \left[\frac{\partial}{\partial s}\frac{\partial}{\partial t} M(s, t)\right]_{s=t=0}$$
$$= \left[\frac{2(1-s)(2-s-t) + 2(1-s)^2}{(1-s)^4(2-s-t)^3} + \frac{2(2-s-t) + 2(1-t)}{(1-t)^4(2-s-t)^3}\right]_{s=t=0}$$
$$= \frac{6}{8} + \frac{6}{8} = \frac{3}{2},$$

so that

$$\sigma_{XY} = E[XY] - \mu_X\mu_Y = 3/2 - (3/2)^2 = -3/4.$$

Finally, the moment generating function of $V = X + Y$ is

$$M_{X+Y}(t) = M(t, t) = \frac{1}{(1-t)^3}, \qquad t < 1,$$

which is the moment generating function of the gamma distribution with parameters $r = 3$, $\theta = 1$. We conclude that $V = X + Y$ has a gamma distribution with parameters $r = 3$, $\theta = 1$. ◆

The joint moment generating function has other useful properties besides those stated above. In particular, it can be used to determine whether two jointly distributed random variables are independent (Exercise T.6), and it has a preservation of limits property similar to that for moment generating functions of a single random variable. Other properties of the joint moment generating function that are sometimes useful are given in Exercises T.7 and T.8.

5 Conditional Distributions

If two variables X, Y, are measured in a random experiment, we may learn the value of one random variable (say, X) before we obtain the value of the other (Y). Although the marginal probability distribution of Y gives probabilities for Y that are appropriate when only Y is measured, the information about X, which may be related to Y, may cause us to change these probabilities. That is, we should be interested in conditional probabilities for the values of Y given the observed value x of X.

If X and Y are both discrete random variables with joint probability mass function $p(x, y)$, a natural way to define the **conditional probability mass**

```
WtPct = (350*EPct+75*LPct+75*QPct)/500        Tot = E1A + E2A + .75*(
```

5.5 max order stat

5.1 $P(Y > X)$ joint pdf

5.4 Transform joint pdf

function of Y given that $X = x$ is

$$(5.1a) \qquad p_{Y|X}(y \mid x) = \frac{P\{X = x \text{ and } Y = y\}}{P\{X = x\}} = \frac{p(x, y)}{p_X(x)},$$

assuming that $p_X(x) > 0$. The conditional probability mass function is defined only for values x for which $p_X(x) > 0$ because only such values of X are observable. Correspondingly, the conditional probability mass function of X given $Y = y$ is defined by

$$(5.1b) \qquad p_{X|Y}(x \mid y) = \frac{p(x, y)}{p_Y(y)}, \qquad \text{if} \quad p_Y(y) > 0.$$

We assert that $p_{Y|X}(y \mid x)$ and $p_{X|Y}(x \mid y)$ are legitimate probability mass functions. For example, $p_{Y|X}(y \mid x)$ is clearly nonnegative for all values y and

$$\sum_y p_{Y|X}(y \mid x) = \sum_y \frac{p(x, y)}{p_X(x)} = \frac{1}{p_X(x)} \sum_y p(x, y) = \frac{p_X(x)}{p_X(x)} = 1.$$

Note that the value of x in $p_{Y|X}(y \mid x)$ is fixed in these calculations; it serves only to distinguish which conditional probability mass function for Y is under discussion. Thus there is a possibly different conditional probability mass function $p_{Y|X}(y \mid x)$ for Y for each value of x for which $p_X(x) > 0$.

EXAMPLE 5.1

In a study of voting habits, let $X = 1$ if a selected person voted in a presidential election, and $X = 0$ if the person did not vote in that election. Similarly, let $Y = 1$ if the person voted in the following congressional election, and $Y = 0$ if the person did not vote in that election. The following is a hypothesized joint probability mass function for X and Y:

	y		
x	0	1	$p_X(x)$
0	0.225	0.125	0.350
1	0.300	0.350	0.650
$p_Y(y)$	0.525	0.475	1.000

The conditional probability mass functions $p_{Y|X}(y \mid x)$ obtained from this joint distribution are displayed below in tabular form:

y	0	1
$p_{Y\mid X}(y \mid 0)$	0.643	0.357

y	0	1
$p_{Y\mid X}(y \mid 1)$	0.462	0.538

As an example of the calculations used to obtain these mass functions,

$$p_{Y|X}(y \mid 0) = \begin{cases} \dfrac{p(0, 0)}{p_X(0)} = \dfrac{0.225}{0.350} = 0.643, & y = 0, \\ \dfrac{p(1, 0)}{p_X(0)} = \dfrac{0.125}{0.350} = 0.357, & y = 1. \end{cases}$$

A comparison of the marginal and conditional probability distributions for Y shows that knowledge of whether or not a person voted in a presidential election ($X = 1$ or 0) changes our probabilities for Y. ◆

EXAMPLE 5.2

Example 2.1 presented a joint probability mass function for the ratings given to randomly selected wines by two wine tasters, I and II. Because the two wine tasters seemed to use similar standards for rating wines, we should be able to predict wine taster II's rating Y from knowledge of the rating $X = x$ given to that wine by wine taster I. From Exhibit 2.3, which gives the joint probability mass function $p(x, y)$ of X and Y, the conditional probability mass functions $p_{Y|X}(y \mid x)$ for Y given $X = 1, 2, 3, 4$, and 5 can be obtained. These are shown in Exhibit 5.1.

EXHIBIT 5.1 *Conditional Probability Mass Functions of Y Given X = x for the Wine-Tasting Experiment*

	y				
	1	2	3	4	5
$p_{Y\mid X}(y \mid 1)$	0.500	0.333	0.167	0.000	0.000
$p_{Y\mid X}(y \mid 2)$	0.111	0.444	0.278	0.111	0.056
$p_{Y\mid X}(y \mid 3)$	0.056	0.167	0.611	0.139	0.028
$p_{Y\mid X}(y \mid 4)$	0.034	0.069	0.207	0.621	0.069
$p_{Y\mid X}(y \mid 5)$	0.000	0.091	0.091	0.273	0.545

Before we knew wine taster I's rating, our best guess of wine taster II's rating Y would be the mode $y = 3$ of the marginal distribution of

Y (Exhibit 2.9) because this yields us the highest probability, $P\{Y = 3\} = 0.35$, of being correct. On the other hand, if we know taster I's rating x and predict that taster II's rating is also x, then from Exhibit 5.1 the conditional probability $p_{Y|X}(x \mid x)$ of a correct prediction is never less than 0.444 and can be as high as 0.621. ◆

For continuous random variables X, Y having a joint probability density function $f(x, y)$, the conditional probability mass function $f_{Y|X}(y \mid x)$ of Y given that $X = x$ and the conditional probability density function $f_{X|Y}(x \mid y)$ of X given that $Y = y$ are defined, analogous to the discrete bivariate case, as follows:

$$f_{Y|X}(y \mid x) = \frac{f(x, y)}{f_X(x)}, \quad \text{when } f_X(x) > 0, \tag{5.2a}$$

$$f_{X|Y}(x \mid y) = \frac{f(x, y)}{f_Y(y)}, \quad \text{when } f_Y(y) > 0. \tag{5.2b}$$

The proof that the functions in (5.2a) and (5.2b) are, in fact, probability density functions follows the lines of our early demonstration that (5.1a) is a probability mass function, replacing sums by integrals.

An idea of the shape of the graph of $f_{Y|X}(y \mid x^*)$ can be obtained by drawing a plane perpendicular to the (x, y) plane along the line $x = x^*$ and observing the curve cut by the plane on the surface of the graph of $f(x, y)$. Similarly, the shape of the graph of $f_{Y|X}(x \mid y^*)$ can be found by intersecting the plane perpendicular to the (x, y) plane along the line $y = y^*$ with the surface formed by the graph of $f(x, y)$ (see Exhibit 5.2).

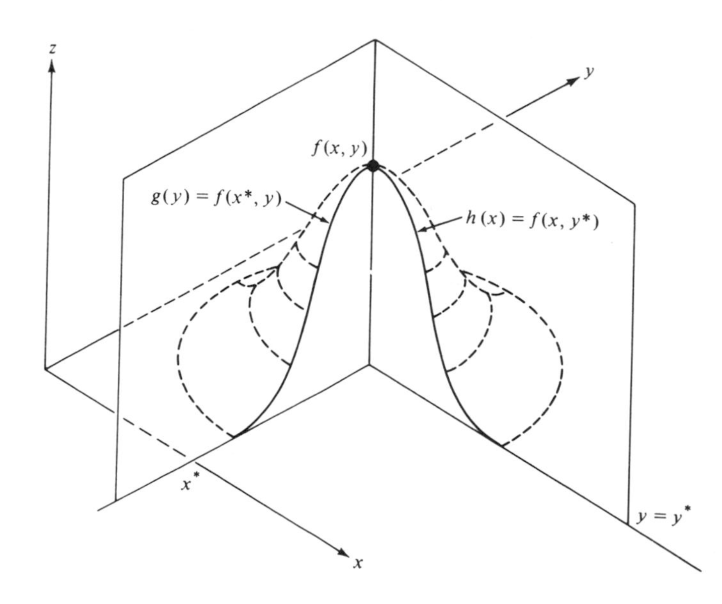

EXHIBIT 5.2 *Shapes of the graphs $g(y) = f(x^*, y)$ and $h(x) = f(x, y^*)$.*

EXAMPLE 5.3

Consider two continuous random variables with the joint probability density function

$$f(x, y) = \begin{cases} x + y, & 0 \leq x, y \leq 1, \\ 0, & \text{otherwise.} \end{cases}$$

The marginal density function for X is

$$f_X(x) = \int_0^1 (x + y)\, dy = x\left(y \Big|_0^1\right) + \frac{y^2}{2}\Big|_0^1 = x + \frac{1}{2}$$

for $0 \leq x \leq 1$, and $f_X(x) = 0$ otherwise. Note that a plane along $x = x^*$ perpendicular to the (x, y)-plane will intersect with the surface formed by $f(x, y)$, to form a straight line $ay + b$. Indeed, for $0 \leq x \leq 1$,

$$f_{Y|X}(y \mid x) = \frac{f(x, y)}{f_X(x)} = \begin{cases} \dfrac{x + y}{x + \frac{1}{2}}, & 0 \leq y \leq 1, \\ 0, & \text{otherwise.} \end{cases}$$

Recall that x is treated as a constant $f_{Y|X}(y \mid x)$, whereas y is the argument of the function; the graph of $f_{Y|X}(y \mid x)$ is a straight line with slope $a = 1/(x + \frac{1}{2})$ and intercept $b = x/(x + \frac{1}{2})$. ◆

Conditional Expected Values

To eliminate needless repetition, we continue to consider only the case of conditional distributions of Y given $X = x$. Results for the conditional distribution of X given $Y = y$ are analogous, and can be easily obtained by interchanging the roles of X and Y in the relevant formulas.

Let $h(Y)$ be any function of Y. Because $p_{Y|X}(y \mid x)$ and $f_{Y|X}(y \mid x)$ are legitimate probability mass functions and probability density functions, respectively, we can define the **conditional expected value of $h(Y)$ given $X = x$**, denoted $E[h(Y) \mid X = x]$, to be the expected value of $h(Y)$ with respect to these conditional probability distributions. That is,

$$(5.3) \qquad E[h(Y) \mid X = x] = \begin{cases} \sum_y h(y)\, p_{Y|X}(y \mid x), & \text{discrete case,} \\ \int_{-\infty}^{\infty} h(y) f_{Y|X}(y \mid x)\, dy, & \text{continuous case.} \end{cases}$$

for all x for which $p_X(x) > 0$ or $f_X(x) > 0$, respectively, and for which the sum or integral in (5.3) is well defined. Because $p_{Y|X}(y \mid x)$ and $f_{Y|X}(y \mid x)$ define legitimate univariate distributions, the conditional expected values $E[h(Y) \mid X = x]$ satisfy the preservation of inequality and linearity properties (Chapter 6, Section 4) of expected values. That is, for any functions

$h(y)$, $h_1(y)$, and $h_2(y)$ for which (5.3) is well defined and for any constants a and b,

(5.4) $\quad h(y) \geq 0 \quad$ all y implies that $\quad E[h(Y) \mid X = x] \geq 0,$

(5.5)
$$E[ah_1(Y) + bh_2(Y) \mid X = x] = aE[h_1(Y) \mid X = x] + bE[h_2(Y) \mid X = x].$$

Important special cases of the expected value computation are

$$\mu_{Y|X=x} = E[Y \mid X = x],$$

which is the **conditional mean of Y given that $X = x$,** and

$$\sigma^2_{Y|X=x} = E[(Y - \mu_{Y|X=x})^2 | X = x],$$

which is the **conditional variance of Y given that $X = x$.** The two indices $\mu_{Y|X=x}$ and $\sigma^2_{Y|X=x}$ play the same roles in describing properties of the conditional distribution of Y given $X = x$ that μ_Y and σ^2_Y play in describing properties of the unconditional (or marginal) distribution of Y. It follows from (5.3) that

$$(5.6) \qquad \mu_{Y|X=x} = \begin{cases} \sum_y y p_{Y|X}(y \mid x), & \text{discrete case,} \\ \int_{-\infty}^{\infty} y f_{Y|X}(y \mid x)\, dy, & \text{continuous case.} \end{cases}$$

and

(5.7)
$$\sigma^2_{Y|X=x} = \begin{cases} \sum_y (y - \mu_{Y|X=x})^2 p_{Y|X}(y \mid x), & \text{discrete case,} \\ \int_{-\infty}^{\infty} (y - \mu_{Y|X=x})^2 f_{Y|X}(y \mid x)\, dy, & \text{continuous case.} \end{cases}$$

It is often helpful to use the fact that

$$(5.8) \qquad \sigma^2_{Y|X=x} = E[Y^2 \mid X = x] - (\mu_{Y|X=x})^2,$$

analogous to the result proved for the unconditional variance σ^2_Y of Y in Chapter 6.

EXAMPLE 5.2 (continued)

In our example of two wine tasters, suppose wine taster I gives rating $x = 2$ to a wine. Using Exhibit 5.1, the conditional mean $\mu_{Y|X=2}$ of wine taster II's rating of this wine is

$$\begin{aligned}\mu_{Y|X=2} &= (1)(0.111) + (2)(0.444) + (3)(0.278) + (4)(0.111) + (5)(0.056)\\ &= 2.557,\end{aligned}$$

which differs from this wine taster's unconditional (average) rating

$$\mu_Y = (1)(0.08) + (2)(0.19) + (3)(0.35) + (4)(0.28) + (5)(0.10) = 3.13$$

calculated from the marginal probability mass function $p_Y(y)$ of Y. In a similar manner, the conditional variance

$$\begin{aligned}\sigma_{Y|X=2} &= E[Y^2 \mid X = 2] - \mu^2_{Y|X=2}\\ &= (1)^2(0.111) + (2)^2(0.444) + \cdots + (5)^2(0.056) - (2.557)^2\\ &= 7.565 - (2.557)^2 = 1.027\end{aligned}$$

differs from the marginal (unconditional) variance

$$\sigma^2_Y = E[Y^2] - \mu^2_Y = 10.97 - (3.13)^2 = 1.173. \quad ◆$$

EXAMPLE 5.3 (continued)

For the joint probability density function $f(x, y) = x + y$ for $0 \le x$, $y \le 1$ and $f(x, y) = 0$ otherwise, we found that the conditional density function of Y given $X = x$, $0 \le x \le 1$, was

$$f_{Y|X}(y \mid x) = \begin{cases} \dfrac{x + y}{x + \frac{1}{2}}, & \text{if } 0 \le y \le 1,\\ 0 & \text{otherwise.}\end{cases} \tag{5.9}$$

Given that $X = 0.25$, then

$$f_{Y|X}(y \mid 0.25) = \begin{cases} \frac{4}{3}y + \frac{1}{3}, & \text{if } 0 \le y \le 1,\\ 0, & \text{otherwise,}\end{cases}$$

and

$$\mu_{Y|X=0.25} = \int_0^1 y\left(\frac{4}{3}y + \frac{1}{3}\right) dy = \left[\frac{4y^3}{9} + \frac{1}{3}\frac{y^2}{2}\right]_0^1 = \frac{11}{18},$$

whereas

$$\mu_Y = \int_0^1 y\left(y + \frac{1}{2}\right) dy = \left[\frac{y^3}{3} + \frac{1}{2}\frac{y^2}{2}\right]_0^1 = \frac{7}{12}.$$

In similar fashion,

$$\sigma^2_{Y|X=0.25} = \int_0^1 y^2\left(\frac{4}{3}y + \frac{1}{3}\right)dy - \left(\frac{11}{18}\right)^2$$

$$= \left[\frac{4}{3}\frac{y^4}{4} + \frac{1}{3}\frac{y^3}{3}\right]_0^1 - \left(\frac{11}{18}\right)^2 = \frac{23}{324},$$

whereas

$$\sigma^2_Y = \int_0^1 y^2\left(y + \frac{1}{2}\right)dy - \left(\frac{7}{12}\right)^2 = \frac{11}{144}.$$

◆

Prediction and the Regression Function

When predicting a future value of a discrete random variable Y, we can use the mode of the distribution of Y if we want the highest probability of being *exactly* right. Alternatively, we can choose the predictor c of Y to minimize $E[(Y - c)^2]$, the expected squared error. In Chapter 6 we showed that under this last criterion, the mean μ_Y of Y is the best predictor of Y, in that $\sigma^2_Y = E[(Y - \mu_Y)^2]$ is the smallest expected squared error of prediction.

When we know that the random variables X and Y are jointly distributed and we have observed $X = x$, the probability model on which we base our predictions has changed, and $c = \mu_{Y|X=x}$ now provides the best predictor of Y in the sense that

$$\sigma^2_{Y|X=x} = E[(Y - \mu_{Y|X=x})^2 \mid X = x] \leq E[(Y - c)^2 \mid X = x],$$

for all values of c.

In general, the values of the conditional mean $\mu_{Y|X=x}$ of Y given $X = x$ vary over different values of x. The function that assigns to every possible value of x of X the corresponding value of $\mu_{Y|X=x}$ is called the **regression function** of Y on X. In practical applications, the regression function gives a rule for predicting Y when we know the value x of X. Further, the regression function expresses, at least "on the average," the mathematical relationship existing between the **dependent variable** Y and the **predictor variable** X.

EXAMPLE 5.2 *(continued)*

For our wine-tasting example, the regression function $\mu_{Y|X=x}$ is given by the expression

$$\mu_{Y|X=x} = \begin{cases} 1.667, & \text{if } x = 1, \\ 2.557, & \text{if } x = 2, \\ 2.919, & \text{if } x = 3, \\ 3.622, & \text{if } x = 4, \\ 4.272, & \text{if } x = 5, \end{cases} \tag{5.10}$$

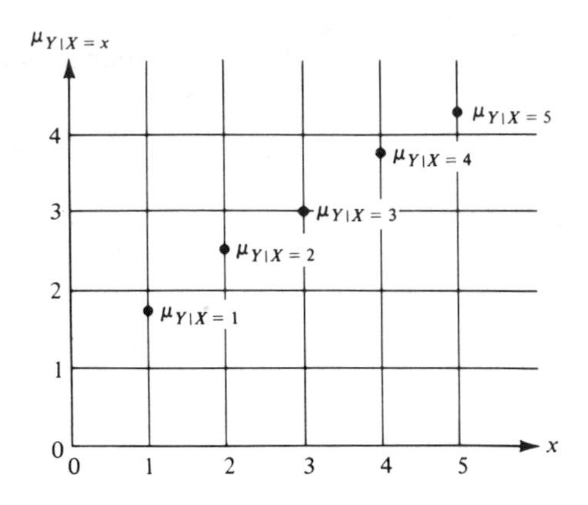

EXHIBIT 5.3 *The regression function $\mu_{Y|X=x}$ for the joint probability distribution of the ratings X, Y given to wines by two wine tasters.*

calculated in the manner previously shown for $\mu_{Y|X=2}$. The regression function (5.10) is graphed in Exhibit 5.3. ◆

EXAMPLE 5.3 *(continued)*

From the conditional density function (5.9) for Y given $X = x$, the regression function of Y on X is

$$\mu_{Y|X=x} = \int_0^1 (y)\left(\frac{x+y}{x+\frac{1}{2}}\right) dy = \frac{1}{x+\frac{1}{2}}\left[x\left(\frac{y^2}{2}\right) + \frac{y^3}{3}\right]_0^1$$
$$= \frac{\frac{1}{2}x + \frac{1}{3}}{x+\frac{1}{2}} = \frac{1}{2} + \frac{1}{12x+6}.$$

This is a decreasing nonlinear function of x. ◆

EXAMPLE 5.4

In Examples 2.3 and 3.1, we considered a joint probability density function $f(x, y) = 360\, xy^2(1 - x - y)$ for $0 \le x, y, x + y \le 1$ and $f(x, y) = 0$ otherwise, for the proportions X, Y of two metals in ore samples from a mine. In Section 2, we stated that the marginal density function of X is $f_X(x) = 30x(1 - x)^4$ for $0 \le x \le 1$, and 0 otherwise. Consequently, the conditional density function of Y given $X = x$ is

$$f_{Y|X}(y \mid x) = \frac{f(x, y)}{f_X(x)} = \begin{cases} \dfrac{12y^2(1 - x - y)}{(1 - x)^4}, & 0 \le y \le 1 - x, \\ 0, & \text{otherwise,} \end{cases}$$

for $0 \le x \le 1$. The regression function of Y on X is

$$\mu_{Y|X=x} = \int_{-\infty}^{\infty} y\, f_{Y|X}(y \mid x)\, dy = \int_0^{1-x} y\left[\frac{12y^2(1 - x - y)}{(1 - x)^4}\right] dy$$
$$= \frac{12}{(1 - x)^4}\left[(1 - x)\frac{y^4}{4} - \frac{y^5}{5}\right]_0^{1-x}$$
$$= (0.6)(1 - x), \qquad 0 \le x \le 1,$$

the graph of which is a decreasing straight line. The fact that the conditional mean of Y given $X = x$ decreases as x increases agrees with the result obtained in Section 3 that $\sigma_{XY} = -0.015$ (i.e., that there is a negative association between X and Y). If the proportion of the first metal in an ore sample is observed to be $X = 0.6$, we would expect the proportion of the second metal in that sample to be approximately $\mu_{Y|X=0.6} = 0.6(1 - 0.6) = 0.12$. In contrast, the expected value of Y in absence of knowledge about X is $\mu_Y = 3/7 = 0.43$.

If we predict Y to be equal to $\mu_{Y|X=x}$ when we observe $X = x$, the expected squared error of our prediction is

$$\begin{aligned}\sigma^2_{Y|X=x} &= E[Y^2 \mid X = x] - \mu^2_{Y|X=x} \\ &= \int_0^{1-x} y^2\left[\frac{12y^2(1 - x - y)}{(1 - x)^4}\right] dy - [0.6(1 - x)]^2 \\ &= (0.04)\,(1 - x)^2.\end{aligned}$$

Thus, when $X = 0.6$, we predict Y to be 0.12 and our expected squared error is $(0.04)(1 - 0.6)^2 = 0.0064$. Note that as x increases from $x = 0$ to $x = 1$, the expected squared error of prediction decreases monotonically to 0. ◆

Formulas Using Conditional Expectations

Both the regression function $\mu_{Y|X=x}$ and the conditional variance $\sigma^2_{Y|X=x}$ are functions of x. To emphasize this point, we temporarily write

$$M(x) = \mu_{Y|X=x}, \qquad S^2(x) = \sigma^2_{Y|X=x}.$$

These functions relate an observed value x of X to indices of the conditional distribution of Y given $X = x$. More generally, for any function $h(y)$ of Y for which $E[h(Y)]$ is well defined, let

$$H(x) = E[h(Y) \mid X = x].$$

Because $M(x)$, $S^2(x)$, and $H(x)$ are functions of x, we can consider the random variables $M(X)$, $S^2(X)$, and $H(X)$, the distributions of which are determined by the marginal distribution of X. In particular, we might want to determine the means (expected values) of these random variables.

Rather than add an extra notation, it is customary to use the notations $\mu_{Y|X} = E[Y \mid X]$, $\sigma^2_{Y|X}$, and $E[h(Y) \mid X]$ in place of $M(X)$, $S^2(X)$, and $H(X)$, respectively. That is, for example, if we write $E[h(Y) \mid X]$ we mean the random variable obtained by substituting X for x in the function $H(x) = E[h(Y) \mid X = x]$.

What is the expected value of $E[h(Y) \mid X]$? The answer may, or may not, surprise you. It is

$$E[E[h(Y)|X]] = E[h(Y)], \tag{5.11}$$

that is, the expected value (over X) of the *conditional* expected value of $h(Y)$ given X is equal to the *unconditional* expected value of $h(Y)$.

A special case of (5.11) is

$$E\{E[Y|X]\} = E[Y] = \mu_Y. \tag{5.12}$$

In other words, the "average" value of the regression function of Y on X is the unconditional mean of Y.

Equation (5.11) is easily verified. For example, when X and Y are both discrete variables, it follows from (5.1a) that

$$p(x, y) = p_X(x)p_{Y|X}(y|x).$$

Consequently,

$$\begin{aligned} E\{E[h(Y)|X]\} &= \sum_x E[h(Y)|X = x]p_X(x) \\ &= \sum_x \left[\sum_y h(y)p_{Y|X}(y|x)\right] p_X(x) = \sum_x \sum_y h(y)p_{Y|X}(y|x)p_X(x) \\ &= \sum_x \sum_y h(y)p(x, y) = E(h(Y)). \end{aligned}$$

The verification for the case when X, Y are both continuous is similar.

EXAMPLE 5.5 If the length X of the first jump of a broad jumper is exponentially distributed with mean μ and if, given $X = x$, the length Y of the second jump has conditionally an exponential distribution with mean $\mu_{Y|X=x} = E[Y|X = x] = x$, the unconditional mean of Y is

$$\mu_Y = E[Y] = E[E[Y|X]] = E[X] = \mu.$$ ◆

Having shown that the mean of Y is the expected value of the conditional mean of Y, we now can show that the same assertion is *not* true of the variance of Y. That is,

$$\sigma_Y^2 \neq E[\sigma_{Y|X}^2].$$

What is true is that *the variance of Y equals the expected value of the conditional variance plus the variance of the conditional expected value.* This is the formula

$$\sigma_Y^2 = E[\sigma_{Y|X}^2] + \text{variance of } E[Y|X]. \tag{5.13}$$

EXAMPLE 5.5 *(continued)*

The variance of an exponential distribution is the square of the mean. Thus because $E[Y|X = x] = x$, and Y is exponentially distributed given $X = x$, we have

$$\sigma_{Y|X=x}^2 = x^2.$$

Applying (5.13) yields

$$\begin{aligned}\sigma_Y^2 &= E[\sigma_{Y|X}^2] + \text{variance of } E[Y|X] \\ &= E(X^2) + \sigma_X^2.\end{aligned}$$

However, X has an exponential distribution with mean μ. Hence $\sigma_X^2 = \mu^2$. Because $\sigma_X^2 = E(X^2) - (\mu_X)^2$, it follows that

$$E(X^2) = \sigma_X^2 + (\mu_X)^2 = \mu^2 + \mu^2 = 2\mu^2.$$

We conclude that

$$\sigma_Y^2 = E(X^2) + \sigma_X^2 = 2\mu^2 + \mu^2 = 3\mu^2. \quad \blacklozenge$$

We now verify (5.13). Note that by (5.11) applied to $h(Y) = Y^2$, by (5.8), and by the linearity property of expected values,

$$\begin{aligned}\sigma_Y^2 &= E[Y^2] - \mu_Y^2 = E[E\{Y^2|X\}] - \mu_Y^2 \\ &= E\{E[Y^2|X] - (E[Y|X])^2 + (E[Y|X])^2\} - \mu_Y^2 \\ &= E\{E[Y^2|X] - (E[Y|X])^2\} + E(E[Y|X])^2 - \mu_Y^2 \\ &= E[\sigma_{Y|X}^2] + E(E[Y|X])^2 - \mu_Y^2.\end{aligned} \tag{5.14}$$

However, by (5.12),

$$E(E[Y|X]) = E[Y] = \mu_Y,$$

and thus

$$E(E[Y|X])^2 - \mu_Y^2 = E(E[Y|X])^2 - \{E(E[Y|X])\}^2$$
$$= \text{variance of } E[Y|X].$$

Substituting this last equation back into (5.14) yields (5.13).

Model Building Using Conditional Distributions

In many applications, variables X and Y are observed one at a time, with (say) X being the first observed. In such cases, an intuitively appealing way to construct a joint probability model for X and Y is to first determine a distribution for X, and then combine this with a distribution for Y conditional on the value x of X that has been observed.

In Example 5.5, a model of this type was constructed for the lengths X, Y of two consecutive jumps of a broad jumper. That is, the marginal distribution of the length X of the first jump was assumed to be exponentially distributed with mean μ, and given the length x of the first jump, the length Y of the second jump was taken to be exponentially distributed with mean x. A second example of this kind of stepwise modeling is the following.

EXAMPLE 5.6

In building a model for the sex distribution of children in a family, we have to account for the fact that families have different numbers, X, of children. Thus we first construct a probability mass function $p_X(x)$ for X. Given $X = x$, the number Y of male children might then be modeled as having a (conditional) binomial distribution (Chapter 8, Section 1) with parameters $n = x$, $p = 0.5$. Thus

$$p_{Y|X}(y|x) = \binom{x}{y}(0.5)^x, \qquad y = 0, 1, 2, \ldots, x$$

and otherwise

$$p_{Y|X}(y|x) = 0.$$ ◆

From the marginal distribution of X and the conditional distribution of Y given that $X = x$, the joint distribution of X, Y is easily determined as a direct consequence of (5.1a) and (5.2a):

(5.15)
$$p(x, y) = p_X(x)p_{Y|X}(y|x), \qquad \text{discrete case,}$$
$$f(x, y) = f_X(x)f_{Y|X}(y|x), \qquad \text{continuous case.}$$

EXAMPLE 5.6
(continued)

Suppose that the total number, X, of children in a randomly selected family has the binomial distribution with parameters $n = 10$ and $p = 0.2$. We assumed that the number Y of male children has, given $X = x$, a binomial distribution with parameters $n = x$ and $p = 0.5$. Thus the joint probability mass function of X, Y is

$$p(x, y) = p_X(x)p_{Y|X}(y|x) = \binom{10}{x}(0.2)^x(0.8)^{10-x}\binom{x}{y}(0.5)^x$$
$$= \frac{10!}{y!(x-y)!\,(10-x)!}(0.1)^x(0.8)^{10-x}$$

for $x, y = 0, 1, 2, \ldots, 10$, $y \leq x$. Otherwise, $p(x, y) = 0$. ◆

From (5.15) and (2.10), the marginal probability mass function of Y can be written

$$p_Y(y) = \sum_x p_{Y|X}(y|x)p_X(x). \tag{5.16}$$

Similarly, from (5.15) and (2.11),

$$f_Y(y) = \int_{-\infty}^{\infty} f(x, y)\, dx = \int_{-\infty}^{\infty} f_{Y|X}(y|x)f_X(x)\, dx. \tag{5.17}$$

Note. Because the events $\{X = x\}$ partition the sample space of (x, y) values, the result (5.16) can also be obtained from the law of partitions (Chapter 4). The result (5.17) is a natural extension of the law of partitions to density functions.

If we fix the value y in $p_{Y|X}(y|x)$ and $f_{Y|X}(y|x)$ and regard these as functions

$$g(x) = p_{Y|X}(y|x) \qquad \text{or} \qquad f_{Y|X}(y|x)$$

of x, we recognize (5.16) and (5.17) as being the expected value $E[g(X)]$ of $g(X)$. That is, for each y, we can regard the marginal mass function or density function of Y evaluated at y as being an expected value (average) of a certain function of X.

This same idea allows us to obtain the mass or density function for Y, even when X, Y are neither both discrete nor both continuous.

EXAMPLE 5.7

Suppose that X is a continuous random variable having an exponential distribution with parameter $\theta = 2$. Suppose that Y is a discrete random variable which, given $X = x$, has a conditional Poisson distribution with parameter $\lambda = x$. Thus for any fixed nonnegative integer y,

$$g(x) = P\{Y = y|X = x\} = \frac{x^y}{y!} e^{-x}.$$

The marginal probability mass function $p_Y(y)$ is then

$$\begin{aligned} p_Y(y) = E[g(X)] &= \int_{-\infty}^{\infty} \left[\frac{x^y e^{-x}}{y!}\right] 2e^{-2x}\, dx \\ &= \frac{2}{y!}\int_0^{\infty} x^{(y+1)-1} e^{-3x}\, dx = \frac{2}{y!}\frac{\Gamma(y+1)}{3^{y+1}} \\ &= (\tfrac{2}{3})(\tfrac{1}{3})^y, \qquad y = 0, 1, 2, \ldots, \end{aligned}$$

as follows from the definition and properties of the gamma function (Appendix A). We recognize $p_Y(y)$ as being the probability mass function of a negative binomial distribution with parameters $r = 1$, $p = \frac{2}{3}$. (Alternatively, $Y + 1$ has a geometric distribution with parameter $p = \frac{2}{3}$.)

◆

Conditional Distributions and Probabilistic Independence

The notion of probabilistic independence as applied to events, and to components of experiments, is discussed in Chapter 4. In Chapter 7 we discussed the independence of random variables but did not give a formal definition of this concept in terms of conditional distributions. Because probabilistic independence is inherently a conditional probability concept (Chapter 4), we need to remedy this omission.

Two jointly distributed random variables X and Y are said to be **probabilistically independent** if the conditional distribution of Y given $X = x$ is the same as the marginal (unconditional) distribution of Y for all possible values of x. That is, X and Y are probabilistically independent if

$$(5.18) \quad \begin{array}{lll} p_{Y|X}(y|x) = p_Y(y), & \text{all } y, \text{ all } x, & \text{discrete case,} \\ f_{Y|X}(y|x) = f_Y(y), & \text{all } y, \text{ all } x, & \text{continuous case.} \end{array}$$

Switching the roles of X and Y, we can also say that X and Y are independent if

$$(5.19) \quad \begin{array}{lll} p_{X|Y}(x|y) = p_X(x), & \text{all } x, \text{ all } y, & \text{discrete case,} \\ f_{X|Y}(x|y) = f_X(x), & \text{all } x, \text{ all } y, & \text{continuous case.} \end{array}$$

Indeed, (5.18) and (5.19) are equivalent definitions of independence. The essence of the concept of independence is in the fact that X and Y are independent if and only if knowledge of the values of one of these two random variables provides no new probabilistic information about the values of the other variable.

It follows directly from (5.18) and (5.19) that if $h(y)$ is any function of y, and X and Y are independent, then

$$E[h(Y)|X = x] = E[h(Y)], \qquad \text{all } x. \tag{5.20}$$

In particular,

$$\mu_{Y|X=x} = \mu_Y, \qquad \sigma^2_{Y|X=x} = \sigma^2_Y, \qquad \text{all } x, \tag{5.21}$$

so that the regression function is *constant* at the unconditional mean μ_Y of Y. Similar results hold for $E[h(X)|Y = y]$, $\mu_{X|Y=y}$, and $\sigma^2_{X|Y=y}$ when X and Y are independent.

In Chapter 7 we gave a computationally more convenient way of verifying whether two random variables X and Y are independent. If X and Y have joint probability mass function $p(x, y)$ or joint probability density function $f(x, y)$, then

X and Y are probabilistically independent if and only if

$$\begin{aligned} p(x, y) &= p_X(x)p_Y(y), \qquad \text{all } x, \text{ all } y, \qquad \text{discrete case,} \\ f(x, y) &= f_X(x)f_Y(y), \qquad \text{all } x, \text{ all } y, \qquad \text{continuous case.} \end{aligned} \tag{5.22}$$

To see that definitions (5.18) and (5.22) are equivalent in the discrete case, note first that if (5.18) holds, then

$$p_Y(y) = p_{Y|X}(y|x) = \frac{p(x, y)}{p_X(x)} \qquad \text{all } x, \text{ all } y,$$

from which, after multiplying all sides of this equality by $p_X(x)$, we obtain (5.22). On the other hand, if (5.22) holds, then

$$p_{Y|X}(y|x) = \frac{p(x, y)}{p_X(x)} = \frac{p_X(x)p_Y(y)}{p_X(x)} = p_Y(y), \qquad \text{all } x, \text{ all } y,$$

establishing (5.18). Interchanging the roles of X and Y, we can use the same steps to show that (5.19) and (5.22) are equivalent in the discrete case. Similar arguments establish the equivalence of (5.18), (5.19), and (5.22) in the continuous case.

EXAMPLE 5.8

Suppose that X and Y have the joint probability density function

$$f(x, y) = \begin{cases} e^{-(x+y)} & \text{if } x > 0 \text{ and } y > 0, \\ 0, & \text{otherwise.} \end{cases}$$

Then for $x > 0$,

$$f_X(x) = \int_{-\infty}^{\infty} f(x, y)\, dy = \int_0^{\infty} e^{-(x+y)}\, dy = e^{-x} \int_0^{\infty} e^{-y}\, dy = e^{-x},$$

whereas for $x < 0$, $f_X(x) = 0$. Thus

$$f_X(x) = \begin{cases} e^{-x}, & \text{if } x > 0, \\ 0, & \text{otherwise,} \end{cases}$$

and in similar fashion

$$f_Y(y) = \begin{cases} e^{-y}, & \text{if } y > 0, \\ 0, & \text{otherwise,} \end{cases}$$

so that the marginal distributions of X and Y are exponential distributions with parameter $\theta = 1$. Now when $x < 0$ or $y < 0$, $f_X(x) = f_Y(y) = 0 = f(x, y)$. If $x > 0$, $y > 0$, then

$$f_X(x) f_Y(y) = e^{-x} e^{-y} = e^{-(x+y)} = f(x, y).$$

Because $f(x, y) = f_X(x) f_Y(y)$ for all x, y, it follows that X and Y are independent. ◆

Two random variables that are not probabilistically independent are said to be **dependent.** Although the dependence of two variables can be shown by showing that one of the equalities given in (5.18), (5.19), or (5.22) does not hold, there are other, often more convenient ways to show that X and Y are dependent. For example, we can show that one of the equalities in (5.20) or (5.21) fails to hold for some value of x. As an example, the two broad jump lengths X, Y, in Example 5.5 are dependent because $\mu_{Y|X=x} = x$ differs from $\mu_Y = \mu$ whenever $x \neq \mu$; that is; the regression function is not constant. In Example 5.4 we could argue that the two proportions X, Y are dependent either because $\mu_{Y|X=x} = (0.6)(1 - x)$ is not a constant function of x or because $\sigma_{Y|X=x} = (0.04)(1 - x)^2$ is not constant in x. (Only one of these facts is needed to demonstrate dependence.) One useful index of dependence, the correlation coefficient ρ_{XY}, was introduced in Section 3.

Independence and Correlation

The usefulness of the correlation coefficient (and also the covariance σ_{XY}) as indices of probabilistic dependence between two jointly distributed random variables X and Y follows as a result of the following fact about independent random variables.

If X and Y are independent random variables, $g_1(x)$ is any function of x for which $E[g_1(x)]$ is well defined and $g_2(y)$ is any function of y for which $E[g_2(Y)]$ is well defined, then

$$E[g_1(X)g_2(Y)] = E[g_1(X)]E[g_2(Y)]. \tag{5.23}$$

To see this in the case when X and Y are continuous variables with joint probability density function $f(x, y)$, note that it follows from the independence of X and Y and equation (5.22) that

$$\begin{aligned} E[g_1(X)g_2(Y)] &= \int_{-\infty}^{\infty}\int_{-\infty}^{\infty} g_1(x)g_2(y)f(x, y)\,dx\,dy \\ &= \int_{-\infty}^{\infty}\int_{-\infty}^{\infty} g_1(x)g_2(y)f_X(x)f_Y(y)\,dx\,dy \\ &= \left[\int_{-\infty}^{\infty} g_1(x)f_X(x)\,dx\right]\left[\int_{-\infty}^{\infty} g_2(y)f_Y(y)\,dy\right] \\ &= E[g_1(X)]E[g_2(Y)]. \end{aligned}$$

An immediate consequence of this fact, with $g_1(x) = x$, $g_2(y) = y$, is that when X and Y are independent,

$$E[XY] = E[X]E[Y]$$

and consequently,

$$\sigma_{XY} = E[XY] - E[X]E[Y] = 0.$$

If X and Y are probabilistically independent, then

$$\sigma_{XY} = 0 \quad \text{and} \quad \rho_{XY} = 0.$$

Therefore, if we find that either $\sigma_{XY} \neq 0$ or $\rho_{XY} \neq 0$, it follows that X and Y are dependent random variables. Thus the number of visits X and number of sales Y in Example 3.2 are dependent random variables because we found that $\sigma_{XY} = 0.2327 \neq 0$. Similarly, the proportions X, Y of two metals in ore samples discussed in Example 3.1 are dependent because $\sigma_{XY} = -0.015 \neq 0$.

At this point, it is worth recalling that a covariance or correlation of 0 does not imply that X and Y are independent. This was illustrated at the end of Section 3 using the example of two discrete random variables X, Y having the joint probability mass function

$$p(x, y) = \begin{cases} 1/3, & \text{if } (x, y) = (-1, 1), (1, 1), \\ 1/6, & \text{if } (x, y) = (-2, 4), (2, 4), \\ 0, & \text{otherwise.} \end{cases}$$

There we showed that $\sigma_{XY} = 0$ and $\rho_{XY} = 0$, but that $Y = X^2$ with probability 1. Also note that

$$p_X(x) = \begin{cases} 1/3, & \text{if } x = -1, 1, \\ 1/6, & \text{if } x = -2, 2, \\ 0, & \text{otherwise,} \end{cases}$$

$$p_Y(y) = \begin{cases} 2/3, & y = 1, \\ 1/3, & y = 4, \\ 0, & \text{otherwise,} \end{cases}$$

but, for example,

$$p(1, 1) = 1/3 \neq (1/3)(2/3) = p_X(1)p_Y(1),$$

so that X and Y are probabilistically dependent random variables.

Notes and References

Multivariate Distributions

In this chapter we introduced the notion of a bivariate distribution to describe the joint variation of two variables. Extension of this concept to multivariate distributions that model the joint variation of several random variables, $X_1, \ldots, X_k$, is straightforward. If the variables are all discrete, the **multivariate probability mass function**

$$p(x_1, \ldots, x_k) = P\{X_1 = x_1 \text{ and } X_2 = x_2 \text{ and } \cdots \text{ and } X_k = x_k\}$$

can provide the needed information, whereas when all variables are continuous, the **joint probability density function** $f(x_1, \ldots, x_k)$ is used to model joint variation. Information about the joint variation of a subcollection of these variables can be obtained from the **marginal joint distributions** of just those variables of interest; these are obtained by summing $p(x_1, \ldots, x_k)$ or integrating $f(x_1, \ldots, x_k)$ over those variables not in the subcollection of interest. For example, if X, Y, Z are discrete random variables with trivariate probability mass function $p(x, y, z)$, the marginal bivariate probability mass function $p_{XY}(x, y)$ of X and Y is found by summing $p(x, y, z)$ over the values of z:

$$p_{XY}(x, y) = \sum_z p(x, y, z),$$

The marginal probability mass function $p_X(x)$ of X then can be found either by summing $p_{XY}(x, y)$ over y or as $p_X(x) = \sum_y \sum_z p(x, y, z)$.

The **expected value** $E[h(X_1, \ldots, X_k)]$ of any function $h(x_1, \ldots, x_k)$ of $x_1, \ldots, x_k$ is defined analogously to the expected value of a function of two random variables:

$$E[h(X_1, \ldots, X_k)] = \begin{cases} \sum_{x_1} \sum_{x_2} \cdots \sum_{x_k} h(x_1, \ldots, x_k) p(x_1, \ldots, x_k), \\ \int_{-\infty}^{\infty} \cdots \int_{-\infty}^{\infty} h(x_1, \ldots, x_k) f(x_1, \ldots, x_k) dx_1 \cdots dx_k, \end{cases}$$

and obeys the important linearity and preservation of inequalities properties:

$$E[a_1 h_1(X_1, \ldots, X_k) + a_2 h_2(X_1, \ldots, X_k)] = a_1 E[h_1(X_1, \ldots, X_k)] + a_2 E[h_2(X_1, \ldots, X_k)],$$

and

$$h(x_1, \ldots, x_k) \geq 0 \quad \text{all } x_1, \ldots, x_k \text{ implies that } E[h(X_1, \ldots, X_k)] \geq 0,$$

respectively. Important special cases of the expected value computation are the means $\mu_i = \mu_{X_i} = E[X_i]$, and variances $\sigma_i^2 = \sigma_{X_i}^2 = E[(X_i - \mu_i)^2]$, $i = 1, \ldots, k$, and covariances $\sigma_{X_iX_j} = E[(X_i - \mu_i)(X_j - \mu_j)]$, $i \neq j$. It follows directly from the definition of $\mu_1, \ldots, \mu_k$, the linearity property of the expected value computation and induction on k that the mean of $Y = \sum_{i=1}^k a_i X_i$ can be found from the means μ_i of the X_i, $i = 1, \ldots, k$; that is,

$$\mu_Y = E[Y] = E\left[\sum_{i=1}^k a_i X_i\right] = \sum_{i=1}^k a_i E[X_i] = \sum_{i=1}^k a_i \mu_i.$$

Similarly, induction and the linearity property can be used to show that

$$\sigma_Y^2 = E[(Y - \mu_Y)^2] = E\left[\left(\sum_{i=1}^{k} a_k X_i - \sum_{i=1}^{k} a_i \mu_i\right)^2\right]$$
$$= E\left[\sum_{i=1}^{k}\sum_{j=1}^{k} a_i a_j (X_i - \mu_i)(X_j - \mu_j)\right]$$
$$= \sum_{i=1}^{k}\sum_{j=1}^{k} a_i a_j E[(X_i - \mu_i)(X_j - \mu_j)]$$
$$= \sum_{i=1}^{k}\sum_{j=1}^{k} a_i a_j \sigma_{X_iX_j} = \sum_{i=1}^{k} a_i^2 \sigma_i^2 + 2\sum_{i<j} a_i a_j \sigma_{X_iX_j}.$$

The **mean vector** $(\mu_1, \mu_2, \ldots, \mu_k)$ serves to locate the joint probability distribution of $X_1, \ldots, X_k$ in k-dimensional space. The individual variances σ_i^2 serve as measures of dispersion of probability in the X_i-direction in k-dimensional space, $i = 1, 2, \ldots, k$, whereas the covariances and variances serve to identify dispersion in the $\sum_{i=1}^{k} a_i x_i$-direction through the formula for the variance of $Y = \sum_{i=1}^{k} a_i X_i$. Note that although derivation of the formulas above for μ_Y and σ_Y^2 required use of the expected value with respect to the entire joint distribution of $X_1, \ldots, X_k$, the means μ_i, variances σ_i^2, and covariances $\sigma_{X_iX_j}$ needed to compute μ_Y and σ_Y^2 can more easily be obtained from the individual marginal distributions of $X_1, \ldots, X_k$ or the marginal joint distribution of X_i and X_j, $i \neq j$. For these reasons, many important questions concerning the joint variation of $X_1, \ldots, X_k$ can be answered from their bivariate marginal distributions.

Conceptually, probabilities of any event concerning $X_1, X_2, \ldots, X_k$ of the form $\{(X_1, X_2, \ldots, X_k) \text{ is in } R\}$, where R is a region in k-dimensional space, can be computed by an appropriate summation or integration of probabilities over that region. However, such probabilities are generally quite difficult to compute. When such probabilities are needed, they are frequently worked out using methods of numerical approximation. One special case that is sometimes tabled is the **multivariate cumulative distribution function:**

$$F(x_1, x_2, \ldots, x_k) = \begin{cases} \displaystyle\sum_{t_1 \le x_1}\sum_{t_2 \le x_2} \cdots \sum_{t_k \le x_k} p(t_1, t_2, \ldots, t_k), & \text{discrete case,} \\ \displaystyle\int_{-\infty}^{x_1}\int_{-\infty}^{x_2} \cdots \int_{-\infty}^{x_k} f(t_1, t_2, \cdots, t_k)\, dt_1\, dt_2 \cdots dt_k, & \text{continuous case.} \end{cases}$$

Using the multivariate cumulative distribution function, probabilities of k-dimensional rectangles:

$$R = \{(x_1, x_2, \ldots, x_k): a_i \le x_i \le b_i, \quad i = 1, 2, \ldots, k\}$$

can be computed.

Conditional Distributions

Any random variable X_i or subcollection of random variables (e.g., X_i and X_j) can be predicted from knowledge of the remaining variables through the appropriate conditional probability distribution and associated regression function. For example, if X, Y, Z are continuous random variables with trivariate probability density function $f(x, y, z)$, and we want to predict Y from X and Z, we first form the conditional density function

$$f_{Y|X,Z}(y|x, z) = \frac{f(x, y, z)}{f_{XZ}(x, z)}$$

for Y given $X = x$ and $Z = z$. The conditional mean

$$\begin{aligned}\mu_{Y|X=x,Z=z} &= E[Y|X = x, Z = z] \\ &= \int_{-\infty}^{\infty} y f_{Y|X,Z}(y|x, z)\, dy\end{aligned}$$

viewed as a function of x and z is the **regression function** for Y given $X = x$ and $Z = z$. This function permits us to predict Y given observed values for X and Z, and also gives a formula that "on the average" relates the values of Y to those of X and Z. We can also predict Y and Z from X using the **bivariate conditional density function:**

$$f_{YZ|X}(y, z|x) = \frac{f(x, y, z)}{f_X(x)}.$$

The conditional mean vector $(\mu_{Y|X=x}, \mu_{Z|X=x})$ serves as a measure of location for this bivariate distribution, where

$$\mu_{Y|X=x} = \int_{-\infty}^{\infty}\int_{-\infty}^{\infty} y f_{YZ|X}(y, z|x)\, dy\, dz$$

and $\mu_{Z|X=x}$ is defined analogously. We can also find the variances (conditional variances) $\sigma^2_{Y|X=x}$, $\sigma^2_{Z|X=x}$ and covariance (conditional covariance) $\sigma_{YZ|X=x}$ to measure dispersion in any direction on the (y, z) plane given that $X = x$. From these, the **conditional (partial) correlation**

$$\rho_{YZ|X=x} = \frac{\sigma_{YZ|X=x}}{(\sigma^2_{Y|X=x}\sigma^2_{Z|X=x})^{1/2}}$$

can be computed to serve as a scale-free measure of the strength of the linear dependence between Y and Z after accounting for the relationship of these variables to X. The values of the partial correlation $\rho_{YZ|X=x}$ are interpreted precisely as we would interpret the values of the correlation coefficient ρ_{YZ}. If both Y and Z are linearly related to X, however, it is possible for ρ_{YZ} to be near 1 or -1 while $\rho_{YZ|X=x} = 0$, suggesting that the apparent linear relationship of Y and Z (shown by the value of ρ_{YZ} near ± 1) results from the relationship of these variables to X. It is also possible for ρ_{YZ} to be close to 0, while $\rho_{YZ|X=x}$ is near 1 or -1, suggesting that the relationships of Y and Z with X tend to "mask" the relationship between Y and Z.

Mutual (Probabilistic) Independence

One powerful structural assumption that is often made in constructing multivariate probability models is that of mutual independence among the random variables. If $X_1, X_2, \ldots, X_k$ have a multivariate probability mass function $p(x_1, x_2, \ldots, x_k)$, then we say that $X_1, X_2, \ldots, X_k$ are **mutually independent** if

$$p(x_1, x_2, \ldots, x_k) = p_{X_1}(x_1)p_{X_2}(x_2) \cdots p_{X_k}(x_k)$$

for all possible values $(x_1, x_2, \ldots, x_k)$ of $(X_1, X_2, \ldots, X_k)$. Similarly, if $X_1, X_2, \ldots, X_k$ have a multivariate probability density function $f(x_1, x_2, \ldots, x_k)$, then $X_1, X_2, \ldots, X_k$ are mutually probabilistically independent if

$$f(x_1, x_2, \ldots, x_k) = f_{X_1}(x_1)f_{X_2}(x_2) \cdots f_{X_k}(x_k).$$

The great advantage of the assumption of mutual independence is that we can determine the multivariate distribution from knowledge only of the marginal distributions; this provides a great simplification (see also Chapter 7).

In analogy to the case of independence of two random variables, mutual independence reflects the notion that knowledge of the values of any variable or combination of variables cannot give additional probabilistic information for predicting the values of any other variable, or other group of variables. For example, if we are interested in the three random variables X, Y, Z, it can be shown that X, Y, Z are mutually independent if and only if the conditional distribution of Y given $X = x$ and $Z = z$ is the same as the marginal distribution of Y *and also* the conditional distribution of Z given $X = x$ is the same as the marginal distribution of Z. Some consequences of the assumption of mutual independence are the following:

1. For any event of the form $\{X_1 \text{ in } A_1, X_2 \text{ in } A_2, \ldots, X_k \text{ in } A_k\}$, we have

$$\begin{aligned} &P\{X_1 \text{ in } A_1, X_2 \text{ in } A_2, \ldots, X_k \text{ in } A_k\} \\ &\quad = P\{X_1 \text{ in } A_1\}P\{X_2 \text{ in } A_2\} \cdots P\{X_k \text{ in } A_k\} \end{aligned}$$

and in particular

$$F(x_1, x_2, \ldots, x_k) = F_{X_1}(x_1)F_{X_2}(x_2) \cdots F_{X_k}(x_k).$$

2. $\sigma_{X_iX_j} = 0$ and $\rho_{X_iX_j} = 0$, for all $i \neq j$.
3. Every conditional distribution is identical to the corresponding unconditional distribution.
4. All regression functions are constant in value.

Violation of any one of these consequences of mutual independence can be used to show that $X_1, X_2, \ldots, X_k$ are not mutually independent, in which case we say that these variables are **mutually dependent.** We remark that the concepts of independence for two random variables (called **pairwise independence**) and of mutual independence are not equivalent. Although it is straightforward to show that *if* $X_1, X_2, \ldots, X_k$ *are mutually independent, then* X_i *and* X_j *are pairwise independent for all* $i \neq j$ (Exercise T.14), the converse of this assertion is false in general. That is, even if every pair X_i and X_j, $i \neq j$, of random variables is pairwise independent, it need not be the case that $X_1, X_2, \ldots, X_k$ are mutually independent [see Exercise T.15(b)].

From the brief outline given above, it should be apparent that multivariate distributions can be quite complex, and offer many avenues for analysis and interpretation. In dealing with multivariate distributions, a clear idea of the structure of the model is needed—both in terms of a specification of the functional form of the multivariate probability mass function or multivariate probability density function that determines the distribution, and also in terms of the types of relationships and interrelationships that will be looked for among the variables.

Glossary and Summary

Joint (multivariate) probability distribution: A probability model describing the joint variation of two or more random variables.

Bivariate distribution: A probability model describing the joint variation of two random variables, say X and Y.

Joint cumulative distribution function: The function $F(x, y)$, which assigns to each pair of numbers (x, y) the probability $P\{X \leq x \text{ and } Y \leq y\}$ of the region $\{(s, t): -\infty < s \leq x, -\infty < t \leq y\}$.

Joint probability mass function: For discrete random variables X and Y, the function $p(x, y)$ which for each pair of numbers (x, y) gives the probability of the event $\{X = x \text{ and } Y = y\}$.

Joint probability density function: For continuous random variables X and Y, the function

$$f(x, y) = \frac{\partial^2}{\partial x \, \partial y} F(x, y) = \frac{\partial^2}{\partial y \, \partial x} F(x, y),$$

which gives the simultaneous rate of accumulation of probability in x and y.

Contingency table: An alternative way of presenting the joint probability mass function for discrete random variables X and Y, each having a finite number of outcomes. The entries in the table are $p(x, y) = P\{X = x$ and $Y = y\}$.

Marginal probability mass functions: For discrete random variables X and Y, the univariate probability mass functions $p_X(x) = P\{X = x\}$ and $p_Y(y) = P\{Y = y\}$. If the joint probability mass function $p(x, y)$ of (X, Y) is given in a contingency table, these marginal probabilities are the marginal totals $p_X(x) = \sum_y p(x, y)$, $p_Y(y) = \sum_x p(x, y)$ of the entries in the table.

Marginal probability density functions: For continuous random variables X and Y, the univariate probability density functions

$$f_X(x) = \int_{-\infty}^{\infty} f(x, y) \, dy, \qquad f_Y(y) = \int_{-\infty}^{\infty} f(x, y) \, dx$$

obtained from the joint density function $f(x, y)$ in a fashion analogous to the way marginal probability mass functions are obtained from joint probability mass functions.

Joint mean or mean vector: The pair (μ_X, μ_Y), where μ_X is the mean of X and μ_Y is the mean of Y. This serves as an index of location (center of gravity) for the joint distribution of X and Y.

Expected value of a function $h(X, Y)$: The computation defined by

$$E[h(X, Y)] = \begin{cases} \sum_x \sum_y h(x, y)p(x, y) \\ \quad = \sum_y \sum_x h(x, y)p(x, y), & \text{discrete case,} \\ \int_{-\infty}^{\infty} \int_{-\infty}^{\infty} h(x, y)f(x, y) \, dx \, dy \\ \quad = \int_{-\infty}^{\infty} \int_{-\infty}^{\infty} h(x, y)f(x, y) \, dy \, dx, & \text{continuous case.} \end{cases}$$

Covariance: An index of the joint distribution of X, Y that describes the tendency of X and Y to either be **positively related** (large values of X tend to occur with large values of Y, small values of X tend to occur with small values of Y), **negatively** or **inversely related** (large values of Y tend to occur with small values of X and small values of Y tend to occur with large values of X), or unrelated. The covariance σ_{XY} is defined by

$$\sigma_{XY} = E[(X - \mu_X)(Y - \mu_Y)].$$

Positive values of σ_{XY} indicate that X and Y are positively related whereas negative values of σ_{XY} indicate that X and Y are negatively related.

Correlation: An index of the joint distribution of X and Y that conveys the same qualitative information about the positive or negative relationship of X and Y as the covariance, but whose values are not affected by scale changes in X and Y.

Joint moment generating function: A function

$$M(s, t) = E[e^{sX+tY}]$$

that plays a role for the joint distribution of X and Y analogous to the role played by the moment generating function of a single random variable. In particular, if $M(s, t)$ is well defined for all (s, t) within an open circle about $(0, 0)$, then $M(s, t)$ determines the joint distribution of X and Y. Further, using the formula

$$E(X^j Y^k) = \left[\frac{\partial^{j+k}}{\partial^j s \partial^k t} M(s, t)\right]_{s=t=0},$$

we can calculate all individual moments $E[X^r]$, $E[Y^r]$ and **joint moments** $E[X^j Y^k]$ of X and Y.

Conditional distribution: A probability model for the variation of one of two jointly distributed random variables when the value of the other random variable is known.

Conditional probability mass function: For two discrete random variables X and Y with probability mass function $p(x, y)$, the function

$$p_{Y|X}(y|x) = P\{Y = y|X = x\} = \frac{p(x, y)}{p_X(x)}$$

that assigns to each value y of Y its conditional probability of occurrence given that X is observed to be equal to x. This function is only defined when $p_X(x) > 0$. [The conditional probability mass function $p_{X|Y}(x|y)$ of X given $Y = y$ is defined analogously.]

Conditional probability density function: For two continuous random variables X and Y the function

$$f_{Y|X}(y|x) = \frac{f(x, y)}{f_X(x)}$$

that gives the conditional density of Y when $X = x$. This function is defined only for values x for which $f_X(x) > 0$.

Conditional expected value: For a function $h(y)$, the expected value

$$E[h(Y)|X = x] = \begin{cases} \sum_y h(y)p_{Y|X}(y|x), & \text{discrete case,} \\ \int_{-\infty}^{\infty} h(y)f_{Y|X}(y|x)\,dx, & \text{continuous case,} \end{cases}$$

of $h(Y)$ calculated with respect to the conditional distribution of Y given that $X = x$.

Conditional mean: The mean of the conditional distribution of Y given $X = x$; that is,

$$\mu_{Y|X=x} = E[Y|X = x].$$

Regression function: The function $M(x) = \mu_{Y|X=x}$ that assigns to every observable value x of X the mean of the conditional distribution of Y given that $X = x$. For each value x, this function provides the best predictor of Y, under squared-error loss, when it is known that X equals x.

Conditional variance: The variance

$$\sigma^2_{Y|X=x} = E[(Y - \mu_{Y|X=x})^2|X = x]$$

of the conditional distribution of Y given $X = x$. This function, for each value x, gives the expected squared-error loss for predicting Y by $\mu_{Y|X=x}$ when it is known that $X = x$.

Useful Facts

1. A function $p(x, y)$ is a joint probability mass function for some pair (X, Y) of discrete random variables if and only if (1) $p(x, y) \geq 0$ for all (x, y) pairs and is positive for only a countable number of such pairs, and (2) $\sum_x \sum_y p(x, y) = 1$.
2. Similarly, a function $f(x, y)$ is a joint probability density function for some pair (X, Y) of continuous random variables if and only if (1) $f(x, y) \geq 0$ for all (x, y) pairs, and (2) $\int_{-\infty}^{\infty} \int_{-\infty}^{\infty} f(x, y)\,dx\,dy = 1$.
3. For discrete random variables (X, Y) and any region R in the plane

$$P(R) \equiv P\{(X, Y) \text{ in } R\} = \sum_{(x,y) \text{ in } R} p(x, y).$$

4. For continuous random variables,

$$\begin{aligned} P(R) &= \iint_{\{(x,y) \text{ in } R} f(x, y)\,dx\,dy \\ &= \text{volume under the graph of } f(x, y) \text{ over the region } R. \end{aligned}$$

Consequently, for any region R having area equal to zero, $P(R) = 0$. For example, $P(X = Y) = 0$ because the region $R = \{(x, y): x = y\}$, which is a line in the plane, has zero area.

5. The same marginal probability mass functions $p_X(x), p_Y(y)$, for X and Y can be obtained from more than one joint probability mass function $p(x, y)$ for the pair (X, Y). Thus the marginal probability mass functions do not determine the joint probability mass function. Similarly, the joint probability density functions $f_X(x), f_Y(y)$ of (X, Y) do not determine the joint probability density function $f(x, y)$ of X and Y.
6. The expected value of a function of a pair (X, Y) of jointly distributed random variables satisfies the linearity property

$$E[ah_1(X, Y) + bh_2(X, Y)] = aE[h_1(X, Y)] + bE[h_2(X, Y)]$$

for any two functions $h_1(X, Y), h_2(X, Y)$ and any two numbers a and b. Also, if $h(x, y) \geq 0$ for all (x, y), then $E[h(X, Y)] \geq 0$ (preservation of inequalities property).

7. For all real numbers a, b, c, d,

$$\begin{aligned}
\mu_{aX+bY+c} &= a\mu_X + b\mu_Y + c, \\
\sigma_{aX+b,cY+d} &= ac\sigma_{XY}, \\
\rho_{aX+b,cY+d} &= \begin{cases} \rho_{XY}, & \text{if } ac > 0, \\ -\rho_{XY}, & \text{if } ac < 0; \end{cases} \\
\sigma^2_{aX+bY+c} &= a^2\sigma^2_X + b^2\sigma^2_Y + 2ab\sigma_{XY} \\
&= a^2\sigma^2_X + b^2\sigma^2_Y + 2ab\sigma_X\sigma_Y\rho_{XY}.
\end{aligned}$$

8. The correlation coefficient ρ_{XY} has the following properties:
   (a) $\rho_{XY} = 1$ if $P\{Y = aX + b\} = 1$, some $a > 0, b$.
   (b) $\rho_{XY} = -1$ if $P\{Y = aX + b\} = 1$, some $a < 0, b$.
   (c) $-1 \leq \rho_{XY} \leq 1$.
   (d) X and Y independent implies that $\rho_{XY} = 0$. However, if $\rho_{XY} = 0$, it is not necessarily the case that X and Y are independent random variables.
9. If (X, Y) have joint moment generating function $M(s, t)$, then

$$\begin{aligned}
M(s, 0) &= M_X(s) = \text{moment generating function of } X, \\
M(0, t) &= M_Y(t) = \text{moment generating function of } Y.
\end{aligned}$$

10. If $E[h(Y)]$ exists (i.e., $E|h(Y)| < \infty$), then

$$E[h(Y)] = E[E[h(Y)|X]].$$

Thus

$$E[Y] = E[E[Y|X]].$$

Here $E[h(Y)|X]$ is the random variable obtained by substituting X for x in the formula for the conditional expected value $E[h(Y)|X = x]$ of $h(Y)$ given that $X = x$.

11. $\sigma_Y^2 = E[\sigma_{Y|X}^2] + \text{variance}\,(E[Y|X])$.
12. If X and Y are independent, then
    (a) $p(x, y) = p_X(x)p_Y(y)$ all (x, y) (discrete case),
    $f(x, y) = f_X(x)f_Y(y)$ all (x, y) (continuous case).
    (b) $F(x, y) = F_X(x)F_y(y)$ all (x, y) where $F(x, y)$ is the joint cumulative distribution function of (X, Y) and $F_X(x)$, $F_Y(y)$ are the marginal cumulative distribution functions of X, Y respectively.
    (c) $p_{Y|X}(y|x) = p_Y(y)$, all (x, y) such that $p_X(x) > 0$,
    $p_{X|Y}(x|y) = p_X(x)$, all (x, y) such that $p_Y(y) > 0$,
    in the discrete case, and
    $f_{Y|X}(y|x) = f_Y(y)$, all (x, y) such that $f_X(x) > 0$,
    $f_{X|Y}(x|y) = f_X(x)$, all (x, y) such that $f_Y(y) > 0$,
    in the continuous case.
    (d) $E[h(Y)|X = x] = E[h(Y)]$, all observable x,
    $E[g(X)|Y = y] = E[g(X)]$, all observable y.
    (e) $E[g(X)h(Y)] = E[g(X)]E[h(Y)]$.

    Conversely, if any one of the results (a), (b), or (c) is true, then X and Y are independent.

Exercises

1. A sample of 455 married couples were asked to (i) rate the happiness of their marriage, and (ii) indicate the extent to which they agree on ways of dealing with their in-laws [Burgess and Cottrell (1955)]. Let Y be the rating of the happiness of the marriage of a given couple. Let X be a rating of the extent of agreement between the marriage partners on ways of dealing with in-laws. The joint distribution of Y and X is summarized in Exhibit E.1.
    (a) Find the marginal distributions of X and Y.
    (b) Find $P\{X + Y = 3\}$.
    (c) The lowest possible level of agreement on ways of dealing with in-laws is $X = 0$ and the highest possible level of agreement is $X = 5$. Find the conditional distributions of the happiness Y of a couple given that $X = 0$ and given that $X = 5$. Find $E[Y \mid X = 0]$ and $E[Y \mid X = 5]$.
    (d) Are X and Y independent? Support your assertion.
2. In Exercise 1, find the covariance σ_{XY} and correlation ρ_{XY} between X and Y. Is there a positive relationship between X and Y?

EXHIBIT E.1 *Comparison of Ratings of Happiness in Marriage with Extent of Agreement on Ways of Dealing with In-laws (Observed Relative Frequencies for 455 Couples)*

Extent of Agreement, x	*Rating of Happiness, y*				
	−2	−1	0	1	2
5	0.015	0.011	0.046	0.077	0.251
4	0.011	0.015	0.042	0.073	0.103
3	0.011	0.028	0.033	0.029	0.042
2	0.018	0.024	0.022	0.015	0.015
1	0.009	0.022	0.009	0.002	0.002
0	0.028	0.029	0.009	0.007	0.002

3. In an experiment to study the relationship between spatial perception and the ability to use a graphical method to solve algebra problems, subjects were asked to solve 3 puzzles and 3 algebra problems. Solutions of the puzzles involved identifying similar geometric objects when these were rotated in space relative to one another; solutions of the algebra problems involved determining where the graphs of linear and quadratic polynomials crossed the horizontal or vertical axes. Let X be the number of puzzles solved by a subject, and let Y be the number of algebra problems solved. Suppose that X and Y have the joint distribution displayed in Exhibit E.2.

EXHIBIT E.2 *Joint Distribution of the Number X of Spatial Puzzles Solved and the Number Y of Algebra Problems Correctly Answered*

x: Spatial Puzzles Solved	*y: Algebra Problems Correctly Answered*			
	0	1	2	3
3	0.00	0.01	0.03	0.06
2	0.01	0.09	0.20	0.10
1	0.04	0.16	0.10	0.10
0	0.05	0.04	0.01	0.00

(a) Find $p_X(x)$ and $p_Y(y)$.
(b) Find $P\{X = Y\}$ and $P\{Y > X\}$.

(c) Find the regression function $E[Y \mid X = x]$ for $x = 0, 1, 2, 3$. If a subject solves $X = 2$ spatial puzzles, how many algebra problems Y would you predict that the subject will answer correctly?

(d) Find ρ_{XY}. Is there a positive relationship between X and Y?

4. Suppose that X and Y are independent discrete random variables with the same marginal probability mass functions $p_X(x)$ and $p_Y(y)$ that were determined in part (a) of Exercise 3. Find the joint probability mass function $p^*(x, y)$ of X and Y and exhibit this in a contingency table similar to Exhibit E.2.

5. Let X be the number of successes in 4 independent Bernoulli trials with probability of success $p = 0.6$. Let Y be the length of the longest "run" in the 4 trials. A "run" is a consecutive sequence of 1 or more identical outcomes. For example, the outcome FSSF has 3 runs, and the longest run has length 2.

(a) Find the joint probability mass function $p(x, y)$ of X and Y.

(b) Find the marginal probability mass function $p_Y(y)$ of Y. Also find μ_Y and σ_Y^2.

(c) Find the conditional probability mass function $p_{X|Y}(x \mid 2)$ of X given that $Y = 2$.

(d) Find the conditional probability mass function $p_{Y|X}(y \mid 2)$ of Y given that $X = 2$.

(e) Find ρ_{XY}.

6. In families where both the husband and the wife work, and their combined income X (in thousands of dollars) exceeds 20, let Y be the proportion of this income earned by the husband. Although incomes are actually discrete variables, economists customarily treat income as a continuous random variable (because of the very large number of possible values). Thus regard X and Y as being continuous random variables and suppose that their joint probability density function is

$$f(x, y) = \begin{cases} \dfrac{(24{,}000)(x + 20)y^{20/x}}{x^5}, & \text{if } x \geq 20, \quad 0 \leq y \leq 1, \\ 0, & \text{otherwise.} \end{cases}$$

(a) Show that the marginal probability density function of X is

$$f_X(x) = \begin{cases} (24{,}000)x^{-4}, & \text{if } x \geq 20, \\ 0, & \text{otherwise.} \end{cases}$$

Find the mean μ_X of X.

(b) Find $\mu_Y = E[Y]$ directly from the joint density $f(x, y)$. (*Hint:* Integrate first with respect to y and then with respect to x.)

(c) Find the conditional probability density function $f_{Y|X}(y \mid x)$ of Y given $X = x$. For what values of x is this density defined?

(d) Find the regression function $E[Y \mid X = x]$ and show that it is a decreasing function of x. What does this result reveal about the relationship between the combined family income X and the proportion Y of that income earned by the husband?

7. In Exercise 6, based on the result in part (d), would you expect the covariance σ_{XY} between X and Y to be positive, negative, or zero? Verify your assertion by calculating the value of σ_{XY}.

8. Suppose that the time needed to complete a phone call has an exponential distribution with mean 1 minute. If two independent phone calls are made, let X be the time taken for the shorter phone call and Y be the time taken for the longer phone call. The joint probability density function of X and Y is

$$f(x, y) = \begin{cases} 2e^{-x-y}, & \text{if } 0 \le x \le y < \infty, \\ 0, & \text{otherwise.} \end{cases}$$

(a) Show that $P\{X = Y\} = 0$.
(b) Find $P\{Y > 2X\} = \int_0^\infty \int_0^{y/2} f(x, y)\, dx\, dy$.
(c) Show that the marginal distribution of X is the exponential distribution with mean $\frac{1}{2}$.
(d) Given that the shorter phone call takes $X = 1$ minute to complete, find the conditional density function $f_{Y|X}(y \mid 1)$ of the time taken for the longer phone call. Find the mean length $E[Y \mid X = 1]$ of that call.

9. In Exercise 8, are X and Y independent? Support your assertion.

10. Let X and Y have joint probability density function

$$f(x, y) = \begin{cases} ye^{-xy-y}, & \text{if } x > 0 \text{ and } y > 0. \\ 0, & \text{otherwise.} \end{cases}$$

(a) Find $f_X(x)$, $f_Y(y)$, μ_X, μ_Y, σ_X^2, and σ_Y^2.
(b) Find formulas for $f_{Y|X}(y \mid x)$, $\mu_{Y|X=x}$, and $\sigma_{Y|X=x}^2$.
(c) Find formulas for $f_{X|Y}(x \mid y)$, $\mu_{X|Y=y}$, and $\sigma_{X|Y=y}^2$.
(d) Find ρ_{XY}. Are X and Y independent?
(e) If you know that $X = 3$, what value would you predict for Y?

11. Let X and Y have joint probability density function

$$f(x, y) = \begin{cases} c, & \text{if } x^2 + y^2 \le 1, \\ 0, & \text{otherwise.} \end{cases}$$

(a) Find the value of c.
(b) Show that $\mu_X = \mu_Y = \rho_{XY} = 0$.
(c) Show that X and Y are dependent.

12. Two random variables X and Y have a joint probability distribution with $\mu_X = 0.5$, $\mu_Y = 2.0$, $\sigma_X^2 = 4.0$, $\sigma_Y^2 = 16.0$, $\sigma_{XY} = 4.0$. Let $Z = (0.5)X - 0.25$ and $W = (0.25)Y - 0.5$.

(a) Find μ_Z, μ_W, σ_Z^2, σ_W^2.

(b) Find σ_{ZY}, σ_{XW}, σ_{ZW}.

(c) Find ρ_{XY}, ρ_{ZY}, ρ_{ZW}.

(d) Find μ_{Z+W}, σ_{Z+W}^2.

13. Let X and Y have joint probability mass function:

$$p(x, y) = \begin{cases} p^2(1-p)^y, & \text{if } x, y = 0, 1, 2, \ldots, \text{ and } x \le y, \\ 0, & \text{otherwise,} \end{cases}$$

where $0 < p < 1$.

(a) Show that the marginal distribution of X is a negative binomial distribution with parameters $r = 1$ and p, while the marginal distribution of Y is a negative binomial distribution with parameters $r = 2$ and p.

(b) Find σ_{XY} and ρ_{XY}. Are X and Y probabilistically independent?

(c) The joint distribution of X and Y given above can arise when we observe independent Bernoulli trials with probability p of success, X is the total number of failures before the first success is observed, and Y is the total number of failures before two successes are observed. If we observe $X = 5$, what prediction should we make for the total number of failures observed before the second success is observed?

14. Find the joint moment generating function, $M(s,t)$, of:

(a) The discrete random variables X and Y having joint probability mass function

$$p(x, y) = \frac{4}{9}\left(\frac{1}{2}\right)^{x+y}, \qquad x, y = 0, 1.$$

(b) The discrete random variables X and Y having joint probability mass function

$$p(x, y) = \frac{4}{7}\left(\frac{1}{2}\right)^{x+y}, \qquad x, y = 0, 1, \quad x \le y.$$

(c) The continuous random variables X and Y having joint probability density function

$$f(x, y) = \begin{cases} 1, & 0 \le x, \quad y \le 1, \\ 0, & \text{otherwise.} \end{cases}$$

(d) The continuous random variables X and Y having the joint probability density function $f(x, y)$ given in Exercise 8.

Use $M(s, t)$ to find μ_X, μ_Y, σ_X^2, σ_Y^2, σ_{XY}, and ρ_{XY} in each of the cases above.

15. Use the moment generating functions $M(s, t)$ found in parts (a) and (b) of Exercise 14 to find the moment generating functions of $X + Y$ for these two joint distributions. Are the two moment generating functions for $X + Y$ in these situations the same?

Theoretical Exercises

T.1. (a) Let X, Y be jointly distributed random variables with joint cumulative distribution $F(x, y)$. Show that

$$P\{a < X \leq b, c < Y \leq d\}$$
$$= F(b, d) - F(b, c) - F(a, d) + F(a, c).$$

(*Hint:* Draw a picture.)

(b) If X and Y are discrete random variables with possible values $x_1 < x_2 < \cdots < x_r$ and $y_1 < y_2 < \cdots < y_s$, respectively, and joint cumulative distribution function $F(x, y)$ show that

$$p(x_i, y_j) = F(x_i, y_j) - F(x_{i-1}, y_j) - F(x_i, y_{j-1}) + F(x_{i-1}, y_{j-1}),$$
$$i = 1, \ldots, r, \quad j = 1, \ldots, s,$$

where $p(x, y)$ is the joint probability mass function of X and Y and $x_0 = y_0 = -\infty$.

T.2. Let X, Y be jointly distributed random variables, and let $W = \max(X, Y)$, $V = \min(X, Y)$. Suppose that (X, Y) has joint cumulative distribution function $F(x, y)$.

(a) Show that $F_W(w) = F(w, w)$.

(b) Show that $F_V(v) = F_X(v) + F_Y(v) - F(v, v)$.

(c) Show that

$$F_{V,W}(v, w) = \begin{cases} 0, & \text{if } v > w, \\ F(v, w) + F(w, v) - F(v, v), & \text{if } v \leq w. \end{cases}$$

(*Hint*: $W \leq w$ if and only if $X \leq w$ and $Y \leq w$, while $V > v$ if and only if $X > v$ and $Y > v$.)

T.3. Suppose that X and Y are independent random variables, and that both X and Y have a gamma distribution with parameters r and θ. Let $V = \min(X, Y)$, $W = \max(X, Y)$.

(a) Find the joint density $f(v, w)$ of V and W.

(b) Find $f_V(v)$.

(c) Find $f_W(w)$.

(d) Find $f_{V|W}(v \mid w)$.
(e) Find ρ_{VW}. Are V and W independent?

T.4. Let X and Y have joint cumulative distribution function $F(x, y)$ satisfying

$$F(x, y) = F_X(x)F_Y(y), \qquad \text{all } x, \text{ all } y.$$

Show separately (a) in the discrete case and (b) in the continuous case that X and Y are independent.

T.5. (a) Show that $E[XY] = E[XE[Y \mid X]]$. Hence show that

$$\sigma_{XY} = \sigma_{XE[Y|X]},$$

so that the covariance between X and Y is equal to the covariance between X and the expected value of Y given X.

(b) Suppose that the conditional distribution given that $X = x$ of the continuous variable Y is the normal distribution with mean x and variance 1. Find the correlation ρ_{XY} between X and Y when $P\{X = 1\} = \frac{1}{2} = P\{X = 0\}$.

T.6. If X and Y are independent random variables, show that

$$M(s, t) = M(s, 0)M(0, t)$$

for all values (s, t) for which the joint moment generating function $M(s, t)$ is well defined. (*Hint*: $M(s, t) = E[e^{sX+tY}] = E[e^{sX}\, e^{tY}]$.) Conversely, if $M(s, t) = M(s, 0)\, M(0, t)$ for all (s, t) in an open circle about (0, 0) show that X and Y must be independent by the following steps.

(a) Show that if X^*, Y^* are independent and have the same marginal distributions as X, Y, respectively, then the joint moment generating function of (X^*, Y^*) is equal to $M(s, t)$.

(b) Recall that if a joint moment generating function $M(s, t)$ of a pair of random variables exists for all (s, t) within an open circle about (0, 0), it determines the joint distribution. Thus argue that (X, Y) and (X^*, Y^*) have the same joint distribution, and thus X and Y must be independent.

T.7. Show that if $M(s, t)$ is the joint moment generating function of (X, Y), then $M_{aX+b,cY+d}(s, t) = e^{sb+td}M(as, ct)$.

T.8. Show that the joint moment generating function $M(s, t)$ of X and Y determines the moment generating function $M_{aX+bY}(u)$ of any linear combination $aX + bY$ of X and Y. On the other hand, show that if we know that $M(s, t)$ exists for all (s, t) within an open circle about (0, 0) and if we know the moment generating functions of all linear combinations $aX + bY$ of X and Y, we can then find $M(s, t)$. Consequently, in this case the joint distribution of X and Y both determines, and is determined by, the distributions of all linear combinations of X and Y.

T.9. Suppose that the conditional distribution of Y given $X = x$ is binomial with sample size $n = 10$ and probability of success x. The marginal distribution of X is the beta distribution with parameters r and s. Note that Y is a discrete variable and that X is a continuous variable.

(a) Using the method shown in Example 5.7, find the marginal probability mass function of Y.

(b) Find the conditional density function $f_{X|Y}(x \mid y)$ of X given that $Y = y$.

T.10. (Trinomial Distribution) As a generalization of the notion of a Bernoulli trial, consider a random experiment where there are three possible outcomes, A, B, and C, and where $P(A \text{ is observed}) = p_A$, $P\{B \text{ is observed}\} = p_B$, and consequently,

$$P(C \text{ observed}) = 1 - p_A - p_B.$$

Let there be $n = 4$ independent trials of this experiment and let $X =$ number of times outcome A is observed, $Y =$ number of times outcome B is observed.

(a) In these 4 trials, we might observe the outcome $ABAC$—that is, A is observed on the first and third trials, B on the second trial, and C on the fourth trial. What is the probability of $ABAC$?

(b) Note that for the outcome $ABAC$, $X = 2$ and $Y = 1$. What other outcomes yield $X = 2$ and $Y = 1$? Show that the total number of such outcomes is equal to $[4!/(2!)(1!)(1!)] = 12$ and that each such outcome has the probability found in part (a). Consequently, find $P\{X = 2 \text{ and } Y = 1\}$. (*Hint*: See Rule 3.4 of Chapter 3.)

(c) Extend your results in part (b) to find the joint probability mass function

$$p(x, y) = P\{X = x \text{ and } Y = y\}$$

for X and Y.

(d) In general, if n independent trials are made of the random experiment described above, argue that the joint probability mass function of $X =$ number of times outcome A is observed and $Y =$ number of times B is observed is the **trinomial distribution:**

$$p(x, y) = \binom{n}{x,\ y,\ n - x - y} p_A^x p_B^y (1 - p_A - p_B)^{n-x-y}$$

for $x, y = 0, 1, 2, \cdots, n$, $x + y \le n$, and $p(x, y) = 0$ otherwise.

(e) Argue that the marginal distribution of X is the binomial distribution with sample size n and probability of success p_A. What is the marginal distribution of Y?

T.11. For three numbers x, y, z, it is a fact that $x \geq y$ and $y \geq z$ implies that $x \geq z$. Suppose that the discrete random variables X, Y, Z have the following joint probability mass function:

(x, y, z)	$(1, 1, 1)$	$(1, 1, 0)$	$(0, 0, 1)$	$(0, 1, 1)$
$p(x, y, z)$	0.20	0.25	0.25	0.30

Find $P\{X \geq Y\}$, $P\{Y \geq Z\}$, $P\{X \geq Z\}$.

T.12.★ A "random walk" is the name given to a certain sequence of movements (steps) of a particle, where at each step the particle either moves 1 unit forward with probability p, or 1 unit backward with probability $1 - p$. The results of the various steps are mutually probabilistically independent. Let X_k represent the result of the kth step of the particle: $X_k = 1$ if the particle moves one step forward, and $X_k = -1$ if the article moves one step backward.

(a) Find the joint probability mass function of X_1, X_2, and X_3.

(b) Let U_k be the position relative to the starting point of the particle after k steps, so that $U_k = \Sigma_{i=1}^{k} X_i$. Find the joint probability mass function of U_1, U_2, and U_3.

(c) Find the marginal joint probability mass function of U_2 and U_3.

(d) Find the marginal probability mass function of U_3.

T.13.★ Let X, Y, and Z have joint probability density function

$$f(x, y, z) = \begin{cases} z^2 e^{-z(1+x+y)}, & \text{if } x, y, z \geq 0, \\ 0, & \text{otherwise.} \end{cases}$$

(a) Show that $\int_{-\infty}^{\infty} \int_{-\infty}^{\infty} \int_{-\infty}^{\infty} f(x, y, z)\, dx\, dy\, dz = 1$.

(b) Find ρ_{XY}. Are X and Y probabilistically independent?

(c) Find $f_Z(z)$.

(d) Find $f_{XZ}(x, z)$, $f_{X|Z}(x \mid z)$, $f_{YZ}(y, z)$, and $f_{Y|Z}(y \mid z)$,

(e) Find $f_{XY|Z}(x, y \mid z)$ and show that $f_{XY|Z}(x, y \mid z) = f_{X|Z}(x \mid z) f_{Y|Z}(y \mid z)$. What does this result imply about the relationship between X and Y? What is the value of $\rho_{XY|Z=z}$?

T.14. If X, Y, and Z are mutually independent random variables, show in (a) the discrete case and (b) the continuous case that X and Y, X and Z, and Y and Z are pairwise independent.

T.15. Let X, Y, and Z have the joint probability mass function $p(x, y, z)$ defined by

(x, y, z)	(1, 1, 0)	(0, 1, 0)	(0, 0, 1)	(1, 1, 1)
$p(x, y, z)$	0.25	0.25	0.25	0.25

(a) Find $p_{XY}(x, y)$, $p_{XZ}(x, z)$, and $p_{YZ}(y, z)$ and show that X and Y, X and Z, and Y and Z are pairwise independent.

(b) However, show that X, Y, and Z are *not* mutually independent.

(c) Find $p_{X|Y}(x \mid 1)$.

(d) Find $p_{XY|Z}(x, y \mid 1)$.

12

The Bivariate Normal Distribution

1 The Bivariate Normal Density

Joint normal distributions of two or more variables play such a central role in probability and statistics that it is appropriate to devote a chapter to discuss these distributions in some detail. Because joint normal distributions of two variables exhibit most of the characteristic properties of joint normal distributions for more than two variables, for conceptual simplicity we confine our present discussion entirely to the bivariate normal case.

Two continuous random variables X and Y have a bivariate normal distribution if the graph of their joint probability density function $f(x, y)$ has a shape similar to that shown in Exhibit 1.1. To be precise, the joint probability density function $f(x, y)$ of a bivariate normal distribution has the form

(1.1)

$$f(x, y) = \frac{\exp\left\{-\frac{1}{2}\left(\frac{1}{1-\rho^2}\right)\left[\frac{(x-\mu_1)^2}{\sigma_1^2} - 2\rho\frac{(x-\mu_1)(y-\mu_2)}{\sigma_1\sigma_2} + \frac{(y-\mu_2)^2}{\sigma_2^2}\right]\right\}}{2\pi\sigma_1\sigma_2\sqrt{1-\rho^2}}$$

for $-\infty < x, y < \infty$. The quantities $\mu_1, \mu_2, \sigma_1, \sigma_2, \rho$ are parameters of the bivariate normal distribution. The pair (μ_1, μ_2) tells us where the center of the "mountain" shown in Exhibit 1.1 is located in the (x, y)-plane, whereas σ_1^2 and σ_2^2 measure the spread of this "mountain" in the x and y directions, respectively. Finally, the parameter ρ, $-1 < \rho < 1$, determines the shape and orientation of the "mountain" shown in Figure 1.1. As may be inferred from the notation, μ_1 and σ_1^2 are the mean and variance of X, μ_2 and σ_2^2 are the mean and variance of Y, and ρ is the correlation coefficient between X and Y. If X and Y have the joint bivariate probability density function (1.1),

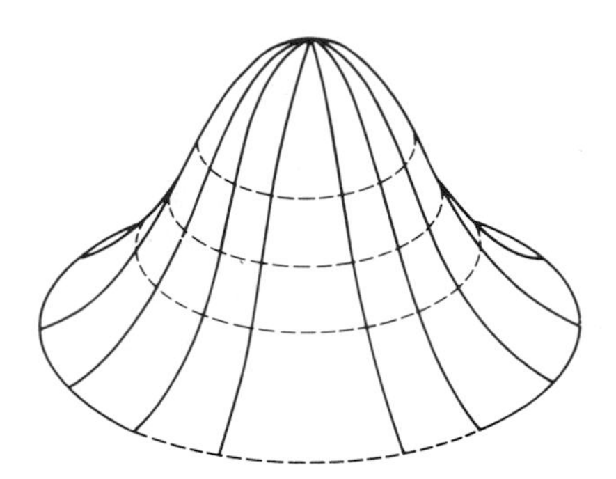

EXHIBIT 1.1 *The bivariate normal probability density function.*

we write $(X, Y) \sim \text{BVN}((\mu_1, \mu_2), (\sigma_1^2, \sigma_2^2; \rho))$, in a notation analogous to that used for the univariate normal distribution in Chapter 9.

Marginal Distributions

Consider the quadratic function

$$(1.2)\quad Q(x, y) = \frac{1}{1-\rho^2}\left[\frac{(x-\mu_1)^2}{\sigma_1^2} - \frac{2\rho(x-\mu_1)(y-\mu_2)}{\sigma_1\sigma_2} + \frac{(y-\mu_2)^2}{\sigma_2^2}\right],$$

which appears in the exponent in (1.1). Note that

(1.3)

$$\begin{aligned} Q(x, y) &- \frac{(x-\mu_1)^2}{\sigma_1^2} \\ &= \frac{1}{\sigma_2^2(1-\rho^2)}\left[(y-\mu_2)^2 - 2(y-\mu_2)\frac{\rho\sigma_2}{\sigma_1}(x-\mu_1) + \frac{\rho^2\sigma_2^2}{\sigma_1^2}(x-\mu_1)^2\right] \\ &= \frac{[y-\mu_2-(\rho\sigma_2/\sigma_1)(x-\mu_1)]^2}{\sigma_2^2(1-\rho^2)}. \end{aligned}$$

From (1.1) to (1.3), it follows that

$$f(x, y) = \frac{\exp[-\frac{1}{2}(x-\mu_1)^2/\sigma_1^2]}{\sqrt{2\pi\sigma_1^2}}\,\frac{\exp[-\frac{1}{2}(y-\theta)^2/\tau^2]}{\sqrt{2\pi\tau^2}},$$

where $\theta = \mu_2 + \rho\sigma_1^{-1}\sigma_2(x-\mu_1)$, $\tau^2 = \sigma_2^2(1-\rho^2)$. Consequently,

(1.4)

$$\begin{aligned} f_X(x) &= \int_{-\infty}^{\infty} f(x, y)\,dy = \frac{\exp[-\frac{1}{2}(x-\mu_1)^2/\sigma_1^2]}{\sqrt{(2\pi)\sigma_1^2}}\int_{-\infty}^{\infty}\frac{\exp[-\frac{1}{2}(y-\theta)^2/\tau^2]\,dy}{\sqrt{(2\pi)\tau^2}} \\ &= \frac{\exp[-\frac{1}{2}(x-\mu_1)^2/\sigma_1^2]}{\sqrt{(2\pi)\sigma_1^2}}, \end{aligned}$$

because we recognize the integrand of the integral in (1.4) as being the density function of a normal distribution. Thus the marginal probability density function $f_X(x)$ is a $\mathcal{N}(\mu_1, \sigma_1^2)$ probability density function, and hence X has mean μ_1 and variance σ_1^2, as asserted previously. In similar fashion, we can show that the margin distribution of Y is $\mathcal{N}(\mu_2, \sigma_2^2)$, so that $\mu_Y = \mu_2$, $\sigma_Y^2 = \sigma_2^2$.

We have shown that if $(X, Y) \sim \text{BVN}((\mu_1, \mu_2), (\sigma_1^2, \sigma_2^2; \rho))$, then X and Y each have marginal normal distributions. However, not only are X and Y each normally distributed, but also *any linear combination*, $aX + bY + c$, *of* X *and* Y *is normally distributed:*

$$aX + bY + c \sim \mathcal{N}(a\mu_1 + b\mu_2 + c, a^2\sigma_1^2 + 2ab\rho\sigma_1\sigma_2 + b^2\sigma_2^2). \tag{1.5}$$

The bivariate normal distribution is the *only* joint distribution of two random variables, X and Y, that has the property (1.5) for all constants a, b, c. This property indeed *characterizes* the bivariate normal distribution. On the other hand, many bivariate nonnormal distributions possess normal marginal distributions (see Exercises T.1 and T.2). Thus *the property of having normal marginal distributions does not characterize the bivariate normal distribution.*

The Correlation Coefficient

We have shown that the parameters $\mu_1, \mu_2, \sigma_1^2, \sigma_2^2$ of a bivariate normal distribution are the means of X and Y and the variances of X and Y, respectively. We now show that the covariance, σ_{XY}, between X and Y equals $\sigma_1\sigma_2\rho$. From the definition the covariance,

$$\begin{aligned}\sigma_{XY} &= \int_{-\infty}^{\infty}\int_{-\infty}^{\infty} (x - \mu_1)(y - \mu_2)f(x, y)\, dx\, dy \\ &= \int_{-\infty}^{\infty}\int_{-\infty}^{\infty} \frac{(x - \mu_1)(y - \mu_2)\exp[-\frac{1}{2}Q(x, y)]}{2\pi\sigma_1\sigma_2\sqrt{1 - \rho^2}}\, dx\, dy,\end{aligned} \tag{1.6}$$

where $Q(x, y)$ is defined by (1.2). Make the change of variables from x to $u = (x - \mu_1)/\sigma_1$, and then from y to $v = (y - \mu_2)/\sigma_2$ in (1.6) to obtain

(1.7)

$$\begin{aligned}\sigma_{XY} &= \sigma_1\sigma_2 \int_{-\infty}^{\infty}\int_{-\infty}^{\infty} \frac{uv\exp[-\frac{1}{2}(u^2 - 2\rho uv + v^2)/(1 - \rho^2)]}{2\pi\sqrt{1 - \rho^2}}\, du\, dv \\ &= \sigma_1\sigma_2 \int_{-\infty}^{\infty} \frac{v\exp(-\frac{1}{2}v^2)}{\sqrt{2\pi}} \left[\int_{-\infty}^{\infty} \frac{u\exp[-\frac{1}{2}(u - \rho v)^2/(1 - \rho^2)]}{\sqrt{2\pi(1 - \rho^2)}}\, du\right] dv.\end{aligned}$$

We recognize the inner integral (over u) as being that defining the mean of a $\mathcal{N}(\rho v, 1 - \rho^2)$ distribution, so that (1.7) becomes

$$\sigma_{XY} = \sigma_1\sigma_2 \int_{-\infty}^{\infty} \frac{v \exp(-\frac{1}{2}v^2)}{\sqrt{2\pi}} (\rho v)\, dv$$
$$= \sigma_1\sigma_2\rho \int_{-\infty}^{\infty} \frac{v^2 \exp(-\frac{1}{2}v^2)}{\sqrt{2\pi}}\, dv = \sigma_1\sigma_2\rho E[V^2],$$

where $V \sim \mathcal{N}(0, 1)$. Because $E[V^2] = \sigma_V^2 + \mu_V^2 = 1$, it follows that

$$\sigma_{XY} = \sigma_1\sigma_2\rho$$

and hence the correlation coefficient between X and Y is

$$\rho_{XY} = \frac{\sigma_{XY}}{\sigma_X\sigma_Y} = \frac{\sigma_1\sigma_2\rho}{\sigma_1\sigma_2} = \rho. \tag{1.8}$$

Probability Contours of a Bivariate Normal Distribution

If we slice the surface formed by the graph of the bivariate normal probability density function (see Exhibit 1.2) by a plane parallel to the (x, y)-plane, the cross section (**probability contour**) created by the intersection of the plane and the surface is ellipsoidal in shape, no matter what horizontal plane we use. Knowledge of this ellipsoidal shape is equivalent to knowledge

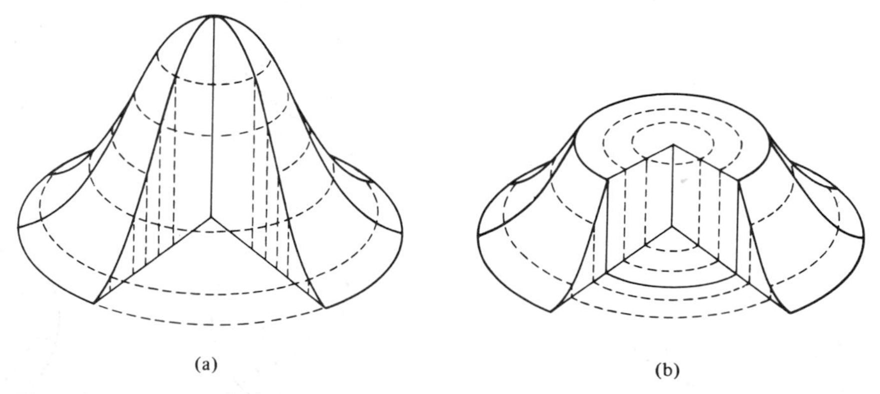

EXHIBIT 1.2 *Probability contours of the bivariate normal density function given in full view and in cross section.*

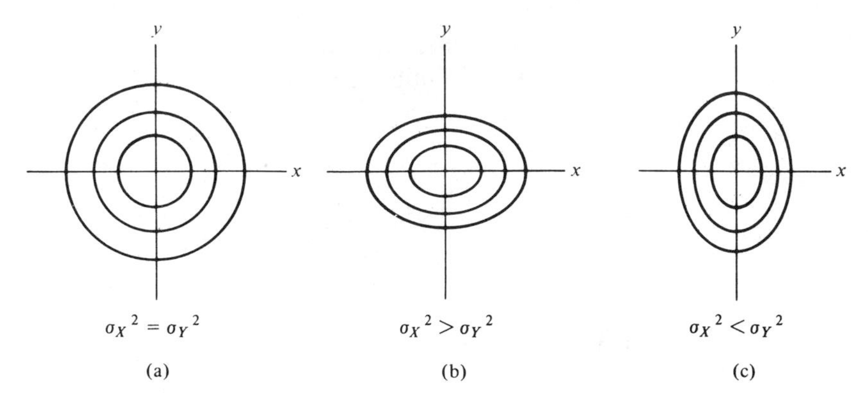

EXHIBIT 1.3 *Contours of the bivariate normal density function when X and Y are uncorrelated.*

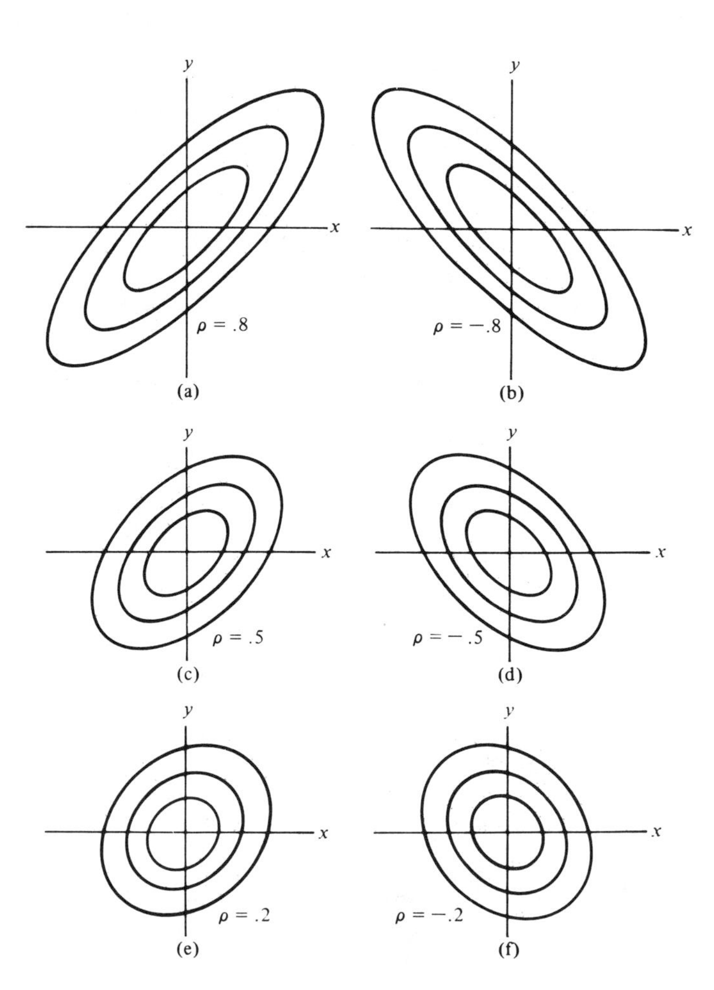

EXHIBIT 1.4 *Contours of the bivariate normal density function for different values of the correlation coefficient ρ when $\sigma_X^2 = \sigma_Y^2 = 1$.*

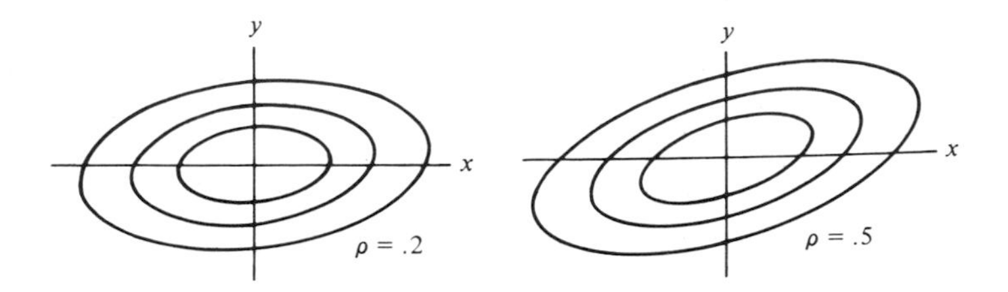

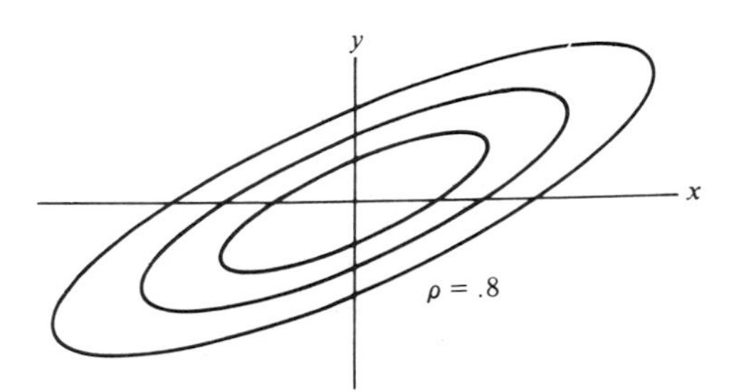

EXHIBIT 1.5 *Contours of the bivariate normal density function for different values of ρ when* $\sigma_X^2 = 4$ *and* $\sigma_Y^2 = 1$.

of how probability is distributed in the plane by the bivariate normal distribution.

Exhibits 1.3 to 1.5 represent different elliptical probability contours that define bivariate normal distributions. They illustrate how the correlation coefficient ρ determines the orientation and shape of the surface formed by a graph of the bivariate normal probability density function.

2 Conditional Distributions

When $(X, Y) \sim \text{BVN}((\mu_1, \mu_2), (\sigma_1^2, \sigma_2^2; \rho))$, it follows from (1.3) and (1.4) that the conditional probability density function of Y given $X = x$ is

$$f_{Y|X}(y|x) = \frac{f(x, y)}{f_X(x)} = \frac{\exp\{-\frac{1}{2}[y - \mu_2 - (\rho\sigma_2/\sigma_1)(x - \mu_1)]^2/\sigma_2^2(1 - \rho^2)\}}{\sqrt{(2\pi)\sigma_2^2(1 - \rho^2)}}, \tag{2.1}$$

which is the probability density function of a $\mathcal{N}(\mu_2 + (\rho\sigma_2/\sigma_1)(x - \mu_1), \sigma_2^2(1 - \rho^2))$ distribution. Hence the regression function of Y on X is

$$\mu_{Y|X=x} = \mu_2 + \frac{\rho\sigma_2}{\sigma_1}(x - \mu_1) = \mu_2 - \frac{\rho\sigma_2}{\sigma_1}\mu_1 + \frac{\rho\sigma_2}{\sigma_1}x, \tag{2.2}$$

which is a linear function, $ax + b$, of x with slope $a = \rho\sigma_2/\sigma_1$ and intercept $b = \mu_2 - (\rho\sigma_2/\sigma_1)\mu_1$. It also follows from (2.1) that the conditional variance of Y given $X = x$ is

$$\sigma^2_{Y|X=x} = \sigma^2_2(1 - \rho^2) = \sigma^2_Y(1 - \rho^2_{XY}), \tag{2.3}$$

which is the same quantity regardless of the value of x. Similarly, interchanging the roles of X and Y, the conditional distribution of X given $Y = y$ is normal with conditional mean

$$\mu_{X|Y=y} = \left(\mu_1 - \frac{\rho\sigma_1}{\sigma_2}\mu_2\right) + \frac{\rho\sigma_1}{\sigma_2}y \tag{2.4}$$

and conditional variance

$$\sigma^2_{X|Y=y} = \sigma^2_1(1 - \rho^2) = \sigma^2_X(1 - \rho^2_{XY}). \tag{2.5}$$

When $\rho_{XY} = \rho = 0$, the regression lines (2.2) and (2.4) have zero slopes and thus are constant.

Indeed, if (X, Y) has a bivariate normal distribution with parameters μ_1, μ_2, σ^2_1, σ^2_2 and ρ, and if $\rho_{XY} = \rho = 0$, the conditional distribution of Y given $X = x$ is a $\mathcal{N}(\mu_2, \sigma^2_2)$ distribution for all x. Because the marginal distribution of Y is also a $\mathcal{N}(\mu_2, \sigma^2_2)$ distribution, it follows that $f_{Y|X}(y \mid x) = f_Y(y)$ for all values of x and y, and hence X and Y are independent. On the other hand, we have seen that independent random variables X and Y have a zero correlation coefficient ($\rho_{XY} = 0$).

> When random variables X and Y have a bivariate normal distribution, the properties of being independent and uncorrelated are equivalent. If $(X, Y) \sim \text{BVN}((\mu_1, \mu_2), (\sigma^2_1, \sigma^2_2; \rho))$, then X and Y are independent if and only if $\rho = 0$.

Note that when X and Y have a $\text{BVN}((\mu_1, \mu_2), (\sigma^2_1, \sigma^2_2; \rho))$ distribution with $0 < |\rho| < 1$, then

$$\sigma^2_{Y|X=x} = \sigma^2_2(1 - \rho^2) < \sigma^2_2 = \sigma^2_Y.$$

That is, the expected squared error of prediction, $\sigma^2_{Y|X=x}$, for Y given that $X = x$ is always strictly less than the expected squared error of prediction σ^2_Y assuming ignorance of the value of X. Hence our predictions of Y are always more accurate when we know the value x of X than they are when we do not know x. A similar assertion is generally false for non-normal distributions (see Exercise T.3).

3 Joint Moment Generating Function

The joint moment generating function of a bivariate normal distribution is useful for deriving distributional results. The formula for this function is the following:

> If $(X, Y) \sim \text{BVN}((\mu_1, \mu_2), (\sigma_1^2, \sigma_2^2; \rho))$, then
>
> $$\begin{aligned} M(s, t) &= E[e^{sX+tY}] \\ &= \exp[s\mu_1 + t\mu_2 + \tfrac{1}{2}(s^2\sigma_1^2 + t^2\sigma_2^2 + 2st\rho\sigma_1\sigma_2)] \end{aligned} \tag{3.1}$$
>
> for $-\infty < s, t < \infty$.

Before verifying this result, we indicate some useful applications. Consider any linear combination $U = aX + bY + c$ of X and Y. The moment generating function of U is

$$\begin{aligned} M_U(t) &= E[e^{tU}] = E[e^{t(aX+bY+c)}] \\ &= e^{tc}E[e^{(ta)X+(tb)Y}] = e^{tc}M(ta, tb) \\ &= \exp\{tc + ta\mu_1 + tb\mu_2 + \tfrac{1}{2}[(ta)^2\sigma_1^2 + (tb)^2\sigma_2^2 + 2(ta)(tb)\sigma_1\sigma_2\rho]\} \\ &= \exp[t(a\mu_1 + b\mu_2 + c) + \tfrac{1}{2}t^2(a^2\sigma_1^2 + b^2\sigma_2^2 + 2ab\rho\sigma_1\sigma_2)], \end{aligned}$$

which is the moment generating function of a normal distribution with mean $a\mu_1 + b\mu_2 + c$ and variance $a^2\sigma_1^2 + b^2\sigma_2^2 + 2ab\rho\sigma_1\sigma_2$. This verifies the assertion (1.5) made in Section 1 that arbitrary linear combinations of bivariate normally distributed random variables X and Y are normally distributed.

Next, consider two linear combinations

$$U_1 = a_1X + b_1Y + c_1, \qquad U_2 = a_2X + b_2Y + c_2$$

of X and Y. The joint moment generating function of U_1 and U_2, after some algebraic simplification, is equal to

$$\begin{aligned} M_{U_1,U_2}(t_1, t_2) &= E[e^{t_1U_1+t_2U_2}] \\ &= E[e^{t_1(a_1X+b_1Y+c_1)+t_2(a_2X+b_2Y+c_2)}] \\ &= e^{t_1c_1+t_2c_2}M(t_1a_1 + t_2a_2, t_1b_1 + t_2b_2) \\ &= \exp[t_1\eta_1 + t_2\eta_2 + \tfrac{1}{2}(t_1^2\tau_1^2 + t_2^2\tau_2^2 + 2t_1t_2\gamma\tau_1\tau_2)], \end{aligned} \tag{3.2}$$

where

(3.3)

$$\eta_1 = a_1\mu_1 + b_1\mu_2 + c_1, \qquad \eta_2 = a_2\mu_1 + b_2\mu_2 + c_2,$$
$$\tau_1^2 = a_1^2\sigma_1^2 + b_1^2\sigma_2^2 + 2a_1b_1\rho\sigma_1\sigma_2, \qquad \tau_2^2 = a_2^2\sigma_1^2 + b_2^2\sigma_2^2 + 2a_2b_2\rho\sigma_1\sigma_2,$$
$$\gamma\tau_1\tau_2 = a_1a_2\sigma_1^2 + b_1b_2\sigma_2^2 + (a_1b_2 + a_2b_1)\rho\sigma_1\sigma_2.$$

This moment generating function has the form of (3.1), so we conclude that:

If $(X, Y) \sim \text{BVN}((\mu_1, \mu_2), (\sigma_1^2, \sigma_2^2; \rho))$, then for any linear combinations $U_1 = a_1X + b_1Y + c_1$, $U_2 = a_2X + b_2Y + c_2$,

$$(U_1, U_2) \sim \text{BVN}((\eta_1, \eta_2), (\tau_1^2, \tau_2^2; \gamma)),$$

where η_1, η_2, τ_1^2, τ_2^2, and γ are given by (3.3).

That is, pairs of linear combinations of bivariate normal random variables X, Y have a bivariate normal distribution.

It is often of interest to determine whether two linear combinations of X, Y are probabilistically independent. Because such linear combinations have a bivariate normal distribution, and because we have shown that bivariate normal random variables are independent if and only if they have zero correlation, it follows that:

If $(X, Y) \sim \text{BVN}((\mu_1, \mu_2), (\sigma_1^2, \sigma_2^2; \rho))$, then $U_1 = a_1X + b_1Y + c_1$ and $U_2 = a_2X + b_2Y + c_2$ are probabilistically independent if and only if

$$a_1a_2\sigma_1^2 + b_1b_2\sigma_2^2 + (a_1b_2 + a_2b_1)\sigma_1\sigma_2\rho = 0. \tag{3.4}$$

EXAMPLE 3.1

Suppose that X and Y are independent normally distributed random variables with $X \sim \mathcal{N}(\mu_1, \sigma_1^2)$, $Y \sim \mathcal{N}(\mu_2, \sigma_2^2)$. When is it the case that $X + Y$ and $X - Y$ are probabilistically independent?

To answer this question, we note first that (X, Y) has a bivariate normal distribution. This is shown by noting that by the independence of X, Y and the fact that X, Y are each normally distributed, the joint moment generating function of X, Y is

$$\begin{aligned} M_{X,Y}(s, t) &= M_X(s)M_Y(t) \\ &= \exp\{s\mu_1 + \tfrac{1}{2}s^2\sigma_1^2\} \exp\{t\mu_2 + \tfrac{1}{2}t^2\sigma_2^2\} \\ &= \exp\{s\mu_1 + t\mu_2 + \tfrac{1}{2}(s^2\sigma_1^2 + t^2\sigma_2^2)\} \end{aligned}$$

This is of the form (3.1) with $\rho = 0$. Thus independent normally distributed random variables always have a bivariate normal distribution with $\rho = 0$. Now, $X + Y = (1)X + (1)Y$ and $X - Y = (1)X + (-1)Y$, so that $a_1 = 1$, $b_1 = 1$, $a_2 = 1$, $b_2 = -1$ in (3.4) and

$$\begin{aligned} a_1a_2\sigma_1^2 + b_1b_2\sigma_2^2 + (a_1b_2 + b_1a_2)\sigma_1\sigma_2\rho &= (1)(1)\sigma_1^2 + (1)(-1)\sigma_2^2 + 0 \\ &= \sigma_1^2 - \sigma_2^2. \end{aligned}$$

Consequently, $X + Y$ and $X - Y$ are independent if and only if $\sigma_1^2 = \sigma_2^2$ (i.e., the variances of X and Y are equal). ◆

How can we simulate pairs (X, Y) of random variables having a bivariate normal distribution? From Section 4 of Chapter 10, we know how to simulate Z having a standard normal distribution. That is, we simulate a variable U having a uniform distribution on the interval [0, 1] and solve $F_Z(z) = U$ for z, where $F_Z(z)$ is the cumulative distribution function of the standard normal distribution. If we independently repeat this operation twice, we obtain two independent standard normal random variables Z_1, Z_2, which we know have a bivariate normal distribution with means 0 and 0, variances 1 and 1, and correlation equal to 0.[1]

Now form the linear combinations

$$V_1 = \sqrt{(1 - \rho^2)}\, Z_1 + \rho Z_2, \qquad V_2 = Z_2.$$

These random variables have a bivariate normal distribution with means

$$\mu_{V_1} = (\sqrt{1 - \rho^2})(0) + (\rho)(0) = 0, \qquad \mu_{V_2} = \mu_{Z_2} = 0,$$

variances

$$\begin{aligned} \sigma_{V_1}^2 &= (\sqrt{1 - \rho^2})^2(1) + (\rho^2)(1) + 2(\sqrt{1 - \rho^2})(\rho)(0) = 1, \\ \sigma_{V_2}^2 &= \sigma_{Z_1}^2 = 1, \end{aligned}$$

and covariance [see equation (3.4)]

$$\sigma_{V_1V_2} = (\sqrt{1 - \rho^2})(0)(1) + (\rho)(1)(1) + [(\sqrt{1 - \rho^2})(1) + (\rho)(0)](0) = \rho,$$

so that $\rho_{V_1V_2} = \rho$. That is,

$$(V_1, V_2) \sim \text{BVN}((0, 0), (1, 1; \rho)).$$

[1] Alternatively, many computer software packages have routines that directly simulate independent standard normal random variables Z_1, Z_2,

Finally, a further linear transformation

$$X^* = \sigma_1 V_1 + \mu_1, \qquad Y^* = \sigma_2 V_2 + \mu_2$$

can be made to give the simulated variables the correct means and variances. Recall from Chapter 11 that X^* has mean μ_1 and variance σ_1^2, Y^* has mean μ_2 and variance σ_2^2, and $\rho_{X^*Y^*} = \rho_{V_1V_2} = \rho$. Also, because X^* and Y^* are linear combinations of V_1 and V_2, (X^*, Y^*) has the desired bivariate normal distribution. Indeed, we have shown that:

> If $(V_1, V_2) \sim \text{BVN}((0, 0), (1, 1; \rho))$, and if $X = \sigma_1 V_1 + \mu_1$, $Y = \sigma_2 V_2 + \mu_2$, then $(X, Y) \sim \text{BVN}((\mu_1, \mu_2), (\sigma_1^2, \sigma_2^2, \rho))$.

Note. Of course, there is no need for two separate linear transformations. After obtaining the independent standard normal variables Z_1, Z_2, we can combine these transformations and simply calculate

$$X^* = \sigma_1 \sqrt{1 - \rho^2}\, Z_1 + \rho\sigma_1 Z_2 + \mu_1, \qquad Y^* = \sigma_2 Z_2 + \mu_2, \tag{3.5}$$

to obtain the simulated pair of variables (X^*, Y^*).

EXAMPLE 3.2

Suppose that we want to simulate

$$(X, Y) \sim \text{BVN}((1, 5), (4, 16; 0.6)).$$

First, we can independently simulate U_1, U_2 from the uniform distribution on [0, 1]. Suppose that we obtain $U_1 = 0.5438$ and $U_2 = 0.2337$. Using Exhibit B.4 of Appendix B to solve for the corresponding standard normal values, we obtain $Z_1 = 0.11$ and $Z_2 = -0.73$. Now

$$\begin{aligned} X^* &= \sqrt{(4)(1 - (0.6)^2)}\, Z_1 + (0.6)(\sqrt{4})\, Z_2 + 1 \\ &= (1.6)(0.11) + (1.2)(-0.73) + 1 = 0.124, \\ Y^* &= \sqrt{16} Z_2 + 5 = (4)(-0.73) + 5 = 2.080, \end{aligned}$$

so that the simulated observation of (X, Y) is (0.124, 2.080). ◆

As a last application of the moment generating function, we note that if $(X, Y) \sim \text{BVN}((\mu_1, \mu_2), (\sigma_1^2, \sigma_2^2, \rho))$, and we let

$$Z_1 = \frac{X - \mu_1}{\sigma_1}, \qquad Z_2 = \frac{Y - \mu_2}{\sigma_2}$$

be the standardizations of X and Y, then

$$M_{Z_1,Z_2}(t_1, t_2) = \exp\left[\frac{t_1(X-\mu_1)}{\sigma_1} + \frac{t_2(Y-\mu_2)}{\sigma_2}\right]$$
$$= \left\{\exp - \left[\frac{t_1\mu_1}{\sigma_1} + \frac{t_2\mu_2}{\sigma_2}\right]\right\} M\left(\frac{t_1}{\sigma_1}, \frac{t_2}{\sigma_2}\right)$$
$$= \exp[\tfrac{1}{2}(t_1^2 + t_2^2 + 2t_1t_2\rho)]$$

after some algebraic simplification. We recognize $M_{Z_1,Z_2}(t_1, t_2)$ as being the moment generating function (3.1) of a bivariate normal distribution with means 0 and 0, variances 1 and 1, and correlation equal to ρ. Thus

If $(X, Y) \sim \text{BVN}((\mu_1, \mu_2), (\sigma_1^2, \sigma_2^2; \rho))$, then if $Z_1 = (X - \mu_1)/\sigma_1$, $Z_2 = (Y - \mu_2)/\sigma_2$, we have

(3.6) $$(Z_1, Z_2) \sim \text{BVN}((0, 0), (1, 1; \rho)).$$

This result is useful for computing probabilities for bivariate normal distributions, as illustrated in Section 4.

We now return to verify the result (3.1). By definition

$$M(s, t) = E(e^{sX+tY}) = \int_{-\infty}^{\infty}\int_{-\infty}^{\infty} e^{sx+ty} f(x, y)\, dx\, dy,$$

where $f(x, y)$ is given by (1.1). Taking a hint from our last result, transform variables of integration one at a time from x, y to

$$z_1 = \frac{x - \mu_1}{\sigma_1}, \qquad z_2 = \frac{y - \mu_2}{\sigma_2}.$$

This yields

$$M(s, t)$$
$$= \int_{-\infty}^{\infty}\int_{-\infty}^{\infty} \frac{\exp\left[s(\sigma_1 z_1 + \mu_1) + t(\sigma_2 z_2 + \mu_2) - \dfrac{1}{2}\dfrac{z_1^2 + z_2^2 - 2\rho z_1 z_2}{1 - \rho^2}\right]}{2\pi\sqrt{1-\rho^2}}\, dz_1\, dz_2$$
$$= e^{s\mu_1 + t\mu_2}\int_{-\infty}^{\infty}\int_{-\infty}^{\infty} \frac{\exp[-\frac{1}{2}Q(z_1, z_2)]}{2\pi\sqrt{1-\rho^2}\, dz_1}\, dz_1\, dz_2,$$

where

$$Q(z_1, z_2) = \frac{z_1^2 + z_2^2 - 2\rho z_1 z_2 - 2(s^* z_1 + t^* z_2)}{1 - \rho^2},$$

and where $s^* = \sigma_1(1 - \rho^2)s$, $t^* = \sigma_2(1 - \rho^2)t$. If we let

$$a = \frac{\rho t^* + s^*}{1 - \rho^2} = \rho\sigma_2 t + \sigma_1 s, \qquad b = \frac{\rho s^* + t^*}{1 - \rho^2} = \rho\sigma_1 s + \sigma_2 t,$$

then some algebra shows that

$$Q(z_1, z_2) = \frac{(z_1 - a)^2 + (z_2 - b)^2 - 2\rho(z_1 - a)(z_2 - b)}{1 - \rho^2} - \frac{a^2 + b^2 - 2\rho ab}{1 - \rho^2}.$$

Consequently,

$$M(s, t) = \exp\left(s\mu_1 + t\mu_2 + \frac{1}{2}\frac{a^2 + b^2 - 2\rho ab}{1 - \rho^2}\right)$$

$$\times \int_{-\infty}^{\infty}\int_{\infty}^{\infty} \frac{\exp\left\{-\frac{1}{2}\left[\frac{(z_1 - a)^2}{1 - \rho^2} + \frac{(z_2 - b)^2}{1 - \rho^2} - \frac{2\rho(z_1 - a)(z_2 - b)}{1 - \rho^2}\right]\right\}}{2\pi\sqrt{1 - \rho^2}}\, dz_1\, dz_2. \tag{3.7}$$

We recognize the integral in (3.7) as the total probability for a bivariate normal distribution with means a and b, variances 1 and 1, and correlation equal to ρ. Hence this integral equals 1. Further algebraic simplification shows that

$$\frac{a^2 + b^2 - 2\rho ab}{1 - \rho^2} = \sigma_1^2 s^2 + \sigma_2^2 t^2 + 2\rho\sigma_1\sigma_2 ts,$$

so that the asserted result (3.1) is obtained.

4 Probability Calculations

For the univariate normal distribution, if $X \sim \mathcal{N}(\mu, \sigma^2)$, then $F_X(a) = F_Z((a - \mu)/\sigma)$, where $Z \sim \mathcal{N}(0, 1)$ has the standard normal distribution. It was this result that permitted us to compute the probabilities for any normal distribution from only one table, that of the cumulative distribution function of the standard normal distribution.

In the bivariate case, we have a similar key result: namely, if $(X, Y) \sim \text{BVN}((\mu_1, \mu_2), (\sigma_1^2, \sigma_2^2; \rho))$, then

$$F_{X,Y}(a, b) = F_{Z_1,Z_2}\left(\frac{a - \mu_1}{\sigma_1}, \frac{b - \mu_2}{\sigma_2}\right), \tag{4.1}$$

where $(Z_1, Z_2) \sim \text{BVN}((0, 0), (1, 1; \rho))$. Put another way, every pair of bivariate normal random variables X and Y can be transformed to standardized bivariate normal random variables

$$Z_1 = \frac{X - \mu_1}{\sigma_1}, \qquad Z_2 = \frac{Y - \mu_2}{\sigma_2},$$

so that $\mu_{Z_1} = \mu_{Z_2} = 0$, $\sigma_{Z_1} = \sigma_{Z_2} = 1$, and $\rho_{Z_1Z_2} = \rho_{XY} = \rho$. Although there is only one standard univariate normal distribution, there are many standard bivariate normal distributions, one for each value of the correlation coefficient ρ. Because tables of $F_{Z_1,Z_2}(h, k)$ have three variable arguments, h, k, and ρ, they require considerable space.

Let $A = \{Z_1 \le h\}$, $B = \{Z_2 \le k\}$, and note that

$$\begin{aligned} F_{Z_1,Z_2}(h, k) &= P(A \cap B) = 1 - P((A \cap B)^c) = 1 - P(A^c \cup B^c) \\ &= 1 - P(A^c) - P(B^c) + P(A^c \cap B^c) \\ &= 1 - (1 - F_{Z_1}(h)) - (1 - F_{Z_2}(k)) + P\{Z_1 > h \text{ and } Z_2 > k\} \\ &= F_{Z_1}(h) + F_{Z_2}(k) + P\{Z_1 > h \text{ and } Z_2 > k\} - 1. \end{aligned}$$

In summary,

$$(4.2) \quad \begin{aligned} &P\{Z_1 \le h \text{ and } Z_2 \le k\} \\ &\quad = P\{Z_1 \le h\} + P\{Z_2 \le k\} + P\{Z_1 > h \text{ and } Z_2 > k\} - 1. \end{aligned}$$

Thus if we have tables of $P\{Z_1 > h \text{ and } Z_2 > k\}$ for the standard bivariate normal distribution, these tables together with Exhibit B.4 of Appendix B allow us to calculate values of $F_{Z_1,Z_2}(h, k)$. Tables of probabilities of events of the form $P\{Z_1 > h \text{ and } Z_2 > k\}$ for $(Z_1, Z_2) \sim \text{BVN}((0, 0), (1, 1; \rho))$, $h, k = 0.0(0.1)4.0$, $\pm\rho = 0.00(0.05)0.95(0.01)1.00$ are provided in *Tables of the Bivariate Normal Distribution Function and Related Functions* [National Bureau of Standards (1959)]. (*Note:* In these tables, ρ is denoted by r.) Only nonnegative values of h and k appear in these tables, whereas we may need to calculate $F_{Z_1,Z_2}(h, k)$ when h or k are negative. For such calculations, it is helpful to note that: If $(Z_1, Z_2) \sim \text{BVN}((0, 0), (1, 1; \rho))$, then

$$(4.3) \quad \begin{aligned} &\text{(i) } (-Z_1, Z_2) \sim \text{BVN}((0, 0), (1, 1; -\rho)). \\ &\text{(ii) } (Z_1, -Z_2) \sim \text{BVN}((0, 0), (1, 1; -\rho)). \\ &\text{(iii) } (-Z_1, -Z_2) \sim \text{BVN}((0, 0), (1, 1; -\rho)). \end{aligned}$$

Let us adopt the notation

$$(4.4) \quad F(h, k; \rho) = F_{Z_1, Z_2}(h, k), \qquad \bar{F}(h, k; \rho) = P\{Z_1 > h \text{ and } Z_2 > k\},$$

EXHIBIT 4.1 *Formulas for Obtaining $F(h, k; \rho)$ and $\bar{F}(h, k; \rho)$ from Values of $\bar{F}(h, k; \rho)$ for $h, k \geq 0$*[a]

Case	$F(h, k; \rho)$	$\bar{F}(h, k; \rho)$
$h \geq 0, k \geq 0$	$F_Z(h) + F_Z(k) + \bar{F}(h, k; \rho) - 1$	$\bar{F}(h, k; \rho)$
$h \geq 0, k < 0$	$1 - F_Z(-k) - \bar{F}(h, -k; -\rho)$	$1 - \bar{F}(h, -k; -\rho) - F_Z(h)$
$h < 0, k \geq 0$	$1 - F_Z(-h) - \bar{F}(-h, k; -\rho)$	$1 - \bar{F}(-h, k; -\rho) - F_Z(k)$
$h < 0, k < 0$	$\bar{F}(-h, -k; -\rho)$	$F_Z(-h) + F_Z(-k) + \bar{F}(-h, -k; -\rho) - 1$

[a] $F_Z(a)$ is the cumulative distribution function of the $\mathcal{N}(0, 1)$ distribution.

where $(Z_1, Z_2) \sim \text{BVN}((0, 0), (1, 1; \rho))$. The function $\bar{F}(h, k; \rho)$ is the function tabled by the National Bureau of Standards for $h, k \geq 0$. Exhibit B.6 of Appendix B provides selected tables (for $\rho = \pm 0.5, \pm 0.85$) to illustrate our computations, and Exhibit 4.1 indicates how values of $F(h, k; \rho)$ and $\bar{F}(h, k; \rho)$ can be obtained from our tables and the more extensive National Bureau of Standards tables.

EXAMPLE 4.1

There are two methods available to make spectrographic determinations of zinc concentrations. One method yields the measurement X and the other yields the measurement Y. If both methods are used on the same specimen, X and Y have a joint distribution, which we here assume is the bivariate normal distribution:

$$(X, Y) \sim \text{BVN}((3.5, 3.5), (0.04, 0.16; 0.85)).$$

Suppose that we wish to find the probability, $P\{X > 3.9 \text{ and } Y > 3.9\}$, that both determinations exceed a critical threshold value of 3.9.

The desired probability is equal to

$$P\left(\frac{X - 3.5}{0.2} > \frac{3.9 - 3.5}{0.2} \quad \text{and} \quad \frac{Y - 3.5}{0.4} > \frac{3.9 - 3.5}{0.4}\right)$$
$$= P\{Z_1 > 2 \text{ and } Z_2 > 1\},$$

which in our present notation is $\bar{F}(2, 1: 0.85)$. This probability can be obtained directly from Exhibit B.6, Appendix B. That is, $P\{X > 3.9$ and $Y > 3.9) = \bar{F}(2, 1; 0.85) = 0.021838$.

On the other hand, if we wish to find the probability that both determinations yield measurements less than the threshold value of 3.9, we require $F(2, 1; 0.85)$. From (4.2) and Exhibits B.4 and B.6 of Appendix B,

$$\begin{aligned} F(2, 1; 0.85) &= P\{Z_1 \leq 2\} + P\{Z_2 \leq 1\} + \bar{F}(2, 1; 0.85) - 1 \\ &= 0.9772 + 0.8413 + 0.021838 - 1 \\ &= 0.8403. \end{aligned}$$

◆

To show how the entries in Exhibit 4.1 are obtained from (4.2) and (4.3), note first that the entries for $h \geq 0$, $k \geq 0$ follow directly from (4.4) and (4.2). For $h \geq 0$, $k < 0$, it follows from equation [4.3(i)] that

$$\begin{aligned} F(h, k; \rho) &= P\{Z_1 \leq h \text{ and } Z_2 \leq k\} = P\{Z_1 \leq h \text{ and } -Z_2 \geq -k\} \\ &= P\{-Z_2 > -k\} - P\{Z_1 > h \text{ and } -Z_2 > -k\} \\ &= 1 - F_Z(-k) - \bar{F}(h, -k; -\rho). \end{aligned}$$

Similar arguments establish the formulas for the case $h < 0$, $k \geq 0$, whereas when $h < 0$, $k < 0$, equation [4.3(iii)] is used to show that

$$\begin{aligned} F(h, k; \rho) &= P\{Z_1 \leq h \text{ and } Z_2 \leq k\} = P\{-Z_1 \geq -h \text{ and } -Z_2 \geq -k\} \\ &= P\{-Z_1 > -h \text{ and } -Z_2 > -k\} = \bar{F}(-h, -k; -\rho) \end{aligned}$$

and then (4.2) is used to obtain the formula for $\bar{F}(h, k; \rho)$ given in Exhibit 4.1.

Let $(X, Y) \sim \text{BVN}((\mu_1, \mu_2), (\sigma_1^2, \sigma_2^2; \rho))$, and suppose that we want to calculate the probability, $P\{(X, Y) \text{ in } R\}$, of the rectangle

$$R = \{(X, Y)\colon a \leq x \leq b \text{ and } c \leq y \leq d\}.$$

Because X and Y have a joint density function, the probability of any of the line segments making up the boundary of R is zero, and the probability of R equals the probability of any rectangle R' formed from R by deleting one or more of its boundaries. In particular,

$$\begin{aligned} P\{(X, Y) \text{ in } R\} &= P\{a < X \leq b \text{ and } c < Y \leq d\} \\ &= F_{X,Y}(b, d) + F_{X,Y}(a, c) - F_{X,Y}(a, d) - F_{X,Y}(b, c) \\ &= F\left(\frac{b - \mu_1}{\sigma_1}, \frac{d - \mu_2}{\sigma_2}; \rho\right) + F\left(\frac{a - \mu_1}{\sigma_1}, \frac{c - \mu_2}{\sigma_2}; \rho\right) \\ &\quad - F\left(\frac{a - \mu_1}{\sigma_1}, \frac{d - \mu_2}{\sigma_2}; \rho\right) - F\left(\frac{b - \mu_1}{\sigma_1}, \frac{c - \mu_2}{\sigma_2}; \rho\right), \end{aligned} \tag{4.5}$$

where the second equality in (4.5) is the result of Exercise T.1(a) of Chapter 11, and the last equality follows from (4.1) and (4.4). Also, substituting (4.2) into (4.5) gives the alternative formula

$$\begin{aligned} P\{(X, Y) \text{ in } R\} &= \bar{F}\left(\frac{b - \mu_1}{\sigma_1}, \frac{d - \mu_2}{\sigma_2}; \rho\right) + \bar{F}\left(\frac{a - \mu_1}{\sigma_1}, \frac{c - \mu_2}{\sigma_2}; \rho\right) \\ &\quad - \bar{F}\left(\frac{a - \mu_1}{\sigma_1}, \frac{d - \mu_2}{\sigma_2}; \rho\right) \\ &\quad - \bar{F}\left(\frac{b - \mu_1}{\sigma_1}, \frac{c - \mu_2}{\sigma_2}; \rho\right), \end{aligned} \tag{4.6}$$

enabling us to directly use the National Bureau of Standards tables.

EXHIBIT 4.2 *Mother's Stature and Son's Stature*

Son's stature (in.)	Mother's Stature (in.)																			Total
	52–53	53–54	54–55	55–56	56–57	57–58	58–59	59–60	60–61	61–62	62–63	63–64	64–65	65–66	66–67	67–68	68–69	69–70	70–71	
59–60	—	—	—	—	—	—	—	0.25	0.25	0.25	0.25	—	—	—	—	—	—	—	—	1
60–61	—	—	—	—	—	—	—	0.75	0.25	0.25	0.25	—	—	—	—	—	—	—	—	1.5
61–62	—	—	—	—	—	—	1	0.5	—	—	—	—	—	—	—	—	—	—	—	1.5
62–63	—	—	—	—	—	—	—	3.5	1.5	2	0.5	—	—	0.5	—	—	—	—	—	8
63–64	0.5	0.5	—	—	0.25	1.25	2.5	6.75	9	5	3.75	—	—	0.5	—	—	—	—	—	30
64–65	—	—	0.25	0.25	1	3.25	5.75	13	7.75	7	6.75	1.25	1.75	—	1	—	—	—	—	49
65–66	—	—	0.5	1	3.75	5.5	4.5	9	14.25	18.25	7.25	6.75	1.75	—	1.25	0.25	—	—	—	74
66–67	—	—	0.25	0.75	1.75	2.75	8.5	15	19.5	20	22.25	10	7.75	2	2.75	0.75	0.5	—	—	114.5
67–68	—	—	—	—	0.25	4.25	5	11.25	27.25	31	36	27	12.75	4	2.25	1.5	0.5	—	—	163
68–69	—	—	—	—	—	—	6.75	12.25	27.5	27.25	31.5	26.75	23.75	12.75	5	1.75	0.75	—	—	175.5
69–70	—	—	—	—	—	2	—	6	15.25	17	23.25	20.5	20	13.25	4.5	1	0.75	0.5	—	124
70–71	—	—	—	—	—	1	0.5	2.75	9.5	22	21.5	23.5	15.75	14.25	7.5	2.75	0.5	0.5	—	122
71–72	—	—	—	—	—	—	0.5	1	2.75	10.25	13.5	16	13.25	12.75	6	1	1	—	—	78
72–73	—	—	—	—	—	—	0.5	1	5	4.25	6	11	7.5	4	4.75	2.5	1	—	—	47.5
73–74	—	—	—	—	—	—	—	—	1.25	4	6.75	6.25	3.75	7	1.5	1.75	1.25	1.5	1	36
74–75	—	—	—	—	—	—	—	—	—	2.5	2.5	2.5	1	3	2	0.75	1.25	1.5	—	17
75–76	—	—	—	—	—	—	—	—	—	0.5	—	0.5	—	1	1.5	1	1.5	—	0.5	6.5
76–77	—	—	—	—	—	—	—	—	—	—	—	1	—	1	0.75	0.25	—	—	0.5	3.5
77–78	—	—	—	—	—	—	—	—	—	0.5	—	—	—	—	0.25	0.75	—	—	—	1.5
78–79	—	—	—	—	—	—	—	—	—	0.5	—	—	—	0.5	0.5	0.5	—	—	—	2
79–80	—	—	—	—	—	—	—	—	—	—	—	—	—	—	—	1	—	—	—	1
Total	0.5	0.5	1	2	7	20	35.5	83	141	172.5	182	153	108.5	76.5	41.5	17.5	9	4	2	1057

[a] The entries in this table are not always integers because of the rule Pearson and Lee (1903) used to assign observations (x, y) to cross-classified groups. If, for example, $(x, y) = (61.3, 64)$, one could assign the count for this observation either to (row 61–62, column 63–64), or to (row 61–62, column 64–65). Instead, Pearson and Lee gave half a count to both choices. If $(x, y) = (62, 64)$, then $\frac{1}{4}$ of a count was assigned to each of the row–column combinations having this point on its boundary.

Source: K. Pearson and A. Lee, On the laws of inheritance in man. I. Inheritance of physical characters. *Biometrika,* vol. 2, 1903, pp. 357–462

EXAMPLE 4.2

Exhibit 4.2 provides grouped data on the heights (in inches) of mothers (Y) and of their sons (X). To fit a bivariate normal distribution to these data, we need to calculate theoretical probabilities and theoretical frequencies of rectangles such as $\{(x, y): 68 \le x \le 69 \text{ and } 62 \le y \le 63\}$. To illustrate the computations, assume that X, Y have a BVN((68.65, 62.48), (7.3441, 5.7121; 0.50)) distribution. That is,

$$\mu_X = 68.65, \qquad \sigma_X = 2.71,$$
$$\mu_Y = 62.48, \qquad \sigma_Y = 2.39,$$

and $\rho_{XY} = 0.50$.

Suppose that we want to calculate the theoretical probability and theoretical frequency of the rectangle

$$R = \{(x, y): 68 \le x \le 69 \text{ and } 62 \le y \le 63\}.$$

Then

$$\begin{aligned} P\{(X, Y) \text{ in } R\} &= \bar{F}\left(\frac{69 - 68.65}{2.71}, \frac{63 - 62.48}{2.39}; 0.5\right) \\ &\quad + \bar{F}\left(\frac{68 - 68.65}{2.71}, \frac{62 - 62.48}{2.39}; 0.5\right) \\ &\quad - \bar{F}\left(\frac{69 - 68.65}{2.71}, \frac{62 - 62.48}{2.39}; 0.5\right) \\ &\quad - \bar{F}\left(\frac{68 - 68.65}{2.71}, \frac{63 - 62.48}{2.39}; 0.5\right) \\ &= \bar{F}(0.13, 0.22; 0.5) + \bar{F}(-0.24, -0.20; 0.5) \\ &\quad - \bar{F}(0.13, -0.20; 0.5) - \bar{F}(-0.24, 0.22; 0.5). \end{aligned} \tag{4.7}$$

Exhibit 4.3 gives the values of $\bar{F}(h, k; 0.5)$ and $\bar{F}(h, k; -0.5)$ taken from Exhibit B.6 of Appendix B, needed to calculate the quantities in (4.7).

EXHIBIT 4.3 *Values of* $\bar{F}(h, k; 0.5)$ *and* $\bar{F}(h, k; -0.5)$ *for* $h = 0.1(0.1)0.3$, $k = 0.2, 0.3$

$\rho = 0.5$				$\rho = -0.5$			
	h				h		
k	0.1	0.2	0.3	k	0.1	0.2	0.3
0.2	0.275161	0.257709	0.239718	0.2	0.1130216	0.0982164	0.0845419
0.3	0.255392	0.239718	0.223488	0.3	0.0976550	0.0845419	0.0724876

We must interpolate in Exhibit 4.3 to obtain the quantities in (4.7). For example, to obtain $\bar{F}(0.13, 0.22; 0.5)$, we first linearly interpolate by rows and then by columns:

	h		
k	0.1	0.13	0.2
0.2	0.275161	—	0.257709
0.22	0.271207	0.266078	0.254111
0.3	0.255392	—	0.239718

We thus obtain $\bar{F}(0.13; 0.22; 0.5) = 0.266078$. Proceeding similarly, we obtain

$$\begin{aligned}\bar{F}(0.24, 0.20; 0.5) &= 0.2505126,\\ \bar{F}(0.13, 0.20; -0.5) &= 0.1085801,\\ \bar{F}(0.24, 0.22; -0.5) &= 0.0901414,\end{aligned}$$

and from Exhibit B.4 of Appendix B:

$$\begin{aligned}F_Z(0.13) &= 0.55172, \qquad F_Z(0.20) = 0.57926\\ F_Z(0.22) &= 0.58706, \qquad F_Z(0.24) = 0.59483.\end{aligned}$$

Thus, from Exhibit 4.1,

$$\begin{aligned}\bar{F}(-0.24, -0.20; 0.5) &= F_Z(0.24) + F_Z(0.20) + \bar{F}(0.24, 0.20; 0.50) - 1\\ &= 0.59483 + 0.57926 + 0.2505126 - 1\\ &= 0.4246026,\\ \bar{F}(0.13, -0.20; 0.5) &= 1 - \bar{F}(0.13, 0.20; -0.5) - F_Z(0.13)\\ &= 1 - 0.1085801 - 0.55172 = 0.3396999,\end{aligned}$$

and

$$\begin{aligned}\bar{F}(-0.24, 0.22; 0.5) &= 1 - \bar{F}(0.24, 0.22; -0.5) - F_Z(0.22)\\ &= 1 - 0.0901414 - 0.58706 = 0.3227986.\end{aligned}$$

Substituting these results into (4.7) yields

$$\begin{aligned}&P\{68 \le X \le 69 \text{ and } 62 \le Y \le 63\}\\ &\qquad = 0.266078 + 0.4246026 - 0.3396999 - 0.3227986\\ &\qquad = 0.0281821.\end{aligned}$$

The theoretical frequency of the event $\{68 \leq X \leq 69$ and $62 \leq 63\}$ is equal to $(1057)(0.0281821) = 29.79$. The observed frequency of this event is, from Exhibit 4.2, equal to 31.5. ◆

The procedure illustrated above is tedious if we wish to fit a complete distribution, but is not too troublesome if we need only a few cells. If a complete distribution is needed, we may have to resort to a computer program.

Glossary and Summary

Bivariate normal distribution: Joint distribution of two continuous random variables whose joint density function has the shape shown in Exhibit 1.1 and the functional form (1.1). If (X, Y) have a bivariate normal distribution with $\mu_X = \mu_1$, $\mu_Y = \mu_2$, $\sigma_X^2 = \sigma_1^2$, $\sigma_Y^2 = \sigma_2^2$, and $\rho_{XY} = \rho$, we denote this fact by

$$(X, Y) \sim \text{BVN}((\mu_1, \mu_2), (\sigma_1^2, \sigma_2^2; \rho)).$$

Standard bivariate normal distribution: Distributions $\text{BVN}((0, 0), (1, 1; \rho))$ for choices of ρ in the interval $(-1, 1)$. These result from standardizing bivariate normal random variables; that is, if

$$(X, Y) \sim \text{BVN}((\mu_1, \mu_2), (\sigma_1^2, \sigma_2^2; \rho)),$$

then

$$\left(\frac{X - \mu_1}{\sigma_1}, \frac{Y - \mu_2}{\sigma_2}\right) \sim \text{BVN}((0, 0), (1, 1; \rho)).$$

Tables of standard bivariate normal distribution probabilities can be used to calculate probabilities for any bivariate normal distribution.

Probability contour: Cross section created by the intersection of a bivariate density function and a plane parallel to the (x, y)-plane (a horizontal plane). For any point (x, y) on this contour, the joint density function $f(x, y)$ has constant value. For the bivariate normal distribution, the probability contours are ellipses centered at $(x, y) = (\mu_1, \mu_2)$. The shape of such ellipses is determined by the values of σ_1^2, σ_2^2, and ρ (see Exhibits 1.3 to 1.5).

Facts About the Bivariate Normal Distribution

If $(X, Y) \sim \text{BVN}((\mu_1, \mu_2), (\sigma_1^2, \sigma_2^2; \rho))$, then:

1. The mean of X is μ_1, the mean of Y is μ_2, the variance of X is σ_1^2 and the variance of Y is σ_2^2. Also, the covariance between X and Y is $\sigma_1\sigma_2\rho$ and the correlation between X and Y is ρ.
2. The marginal distributions of X and Y are normal:

$$X \sim \mathcal{N}(\mu_1, \sigma_1^2), \qquad Y \sim \mathcal{N}(\mu_2, \sigma_2^2).$$

(*Note:* There exist bivariate density functions whose marginals are normal distributions, but which are not bivariate normal density functions—see Exercises T.1 and T.2.)

3. All linear combinations of X, Y have normal distributions:

$$aX + bY + c \sim \mathcal{N}(a\mu_1 + b\mu_2 + c,\ a^2\sigma_1^2 + b^2\sigma_2^2 + 2ab\sigma_1\sigma_2\rho).$$

4. All pairs of linear combinations of X, Y have bivariate normal distributions.
5. The conditional distribution of Y given $X = x$ is a normal distribution with conditional mean

$$\mu_{Y|X=x} = \left(\mu_2 - \rho\frac{\sigma_2}{\sigma_1}\mu_1\right) + \rho\frac{\sigma_2}{\sigma_1}x$$

and conditional variance $\sigma_{Y|X=x}^2 = \sigma_2^2(1 - \rho^2)$. Similarly, the conditional distribution of X given $Y = y$ is normal with conditional mean $\mu_{X|Y=y} = [\mu_1 - \rho(\sigma_1/\sigma_2)\mu_2] + \rho(\sigma_1/\sigma_2)y$ and conditional variance $\sigma_{X|Y=y}^2 = \sigma_1^2(1 - \rho^2)$.

6. X and Y are probabilistically independent if and only if $\rho = 0$. Further, $a_1X_1 + b_1Y + c_1$ and $a_2X + b_2Y + c_2$ are probabilistically independent if and only if

$$a_1a_2\sigma_1^2 + b_1b_2\sigma_2^2 + (a_1b_2 + a_2b_1)\sigma_1\sigma_2\rho = 0.$$

7. The moment generating function of (X, Y) is

$$M(s, t) = \exp[s\mu_1 + t\mu_2 + \tfrac{1}{2}(s^2\sigma_1^2 + t^2\sigma_2^2 + 2st\sigma_1\sigma_2\rho)].$$

8. $$F(x, y) = P\{X \le x \text{ and } Y \le y\} = P\left(Z_1 \le \frac{x - \mu_1}{\sigma_1} \text{ and } Z_2 \le \frac{y - \mu_2}{\sigma_2}\right)$$
$$= F_{Z_1,Z_2}\left(\frac{x - \mu_1}{\sigma_1}, \frac{y - \mu_2}{\sigma_2}\right),$$

where $(Z_1, Z_2) \sim \text{BVN}((0, 0), (1, 1; \rho))$. Further, if $\bar{F}_{Z_1,Z_2}(h, k) = P\{Z_1 > h \text{ and } Z_2 > k\}$, then

$$F_{Z_1,Z_2}(h, k) = F_{Z_1}(h) + F_{Z_2}(h) + \bar{F}_{Z_1,Z_2}(h, k) - 1.$$

9. If $X \sim \mathcal{N}(\mu_1, \sigma_1^2)$ and $Y \sim \mathcal{N}(\mu_2, \sigma_2^2)$ are independent, then $(X, Y) \sim \text{BVN}((\mu_1, \mu_2), (\sigma_1^2, \sigma_2^2; 0))$.
10. If Z_1, Z_2 are independent standard normal random variables, and

$$X = \sigma_1\sqrt{1 - \rho^2}\, Z_1 + \rho\sigma_1 Z_2 + \mu_1,$$
$$Y = \sigma_2 Z_2 + \mu_2,$$

then $(X, Y) \sim \text{BVN}((\mu_1, \mu_2), (\sigma_1^2, \sigma_2^2; \rho))$. This provides a way to simulate observations from the bivariate normal distribution.

Exercises

1. In Example 4.1, find:
(a) $P\{X > 3.9 \text{ and } Y \le 3.9\}$.
(b) $P\{X \le 3.9 \text{ and } Y > 3.9\}$.
(c) $P\{3.5 \le X \le 3.9 \text{ and } 3.1 \le Y \le 3.5\}$.
(d) $P\{3.1 \le X \le 3.5 \text{ and } 3.5 \le Y \le 3.9\}$.

2. Suppose that $(X, Y) \sim \text{BVN}((5, 3), (49, 16; 0.5))$. Use Exhibit 4.3 to find:
(a) $P\{X > 6 \text{ and } Y > 4\}$.
(b) $P\{X \le 6 \text{ and } Y \le 4\}$.
(c) $P\{X > 4 \text{ and } Y > 2\}$.
(d) $P\{X \le 4 \text{ and } Y \le 2\}$.
(e) $P\{X \ge 4 \text{ and } Y \ge 4\}$.
(f) $P\{X \le 6 \text{ and } Y \le 2\}$.
(g) $P\{4 \le X \le 6 \text{ and } 2 \le Y \le 4\}$.

3. Suppose that the heights of mothers (Y) and sons (X) followed the distribution fitted in Example 4.2; that is,

$$(X, Y) \sim \text{BVN}((68.65, 62.48), (7.3441, 5.7121; 0.5)).$$

(a) Find $\mu_{X|Y=y}$ and $\sigma^2_{X|Y=y}$. If a mother's height is $Y = 65$ inches, what value would you predict for her son's height? If you were not told the mother's height, what height would you predict for her son? Compare $\sigma^2_{X|Y=65}$ with σ^2_X.
(b) Find the probability, $P\{Y > X\}$, that a mother will be taller than her son.

4. The relationship, in German words, between the number X of syllables and the number Y of phonemes (letters) was studied for 20,453 headwords in Viëtor-Meyer's *Deutches Aussprachewörterbuch* [see Herdan (1966, p. 301)]. If we regard X and Y as if they are continuous variables that have been rounded to the nearest integer, a bivariate normal distribution with parameters

$$\mu_X = 2.780, \quad \sigma_X^2 = 1.2122, \quad \sigma_{XY} = 2.4040,$$
$$\mu_Y = 7.292, \quad \sigma_Y^2 = 6.6793,$$

provides a reasonably good fit to the data. Hence assume that (X, Y) has a BVN((2.780, 7.292), (1.2122, 6.6793; ρ_{XY})) distribution.

(a) Find ρ_{XY}.

(b) Find $\mu_{Y|X=x}$ and $\sigma_{Y|X=x}^2$. If a German headword has $X = 5$ syllables, how many letters would you predict that it had? If you were not told the number of syllables that the headword has, how many letters Y would you predict the word has? Compare $\sigma_{Y|X=5}^2$ with σ_Y^2.

5. An industrial process simultaneously produces two interlocking steel pipes. The lengths X and Y of the pipes have a bivariate normal distribution with parameters $\mu_X = \mu_Y = 3.25$ inches, $\sigma_X = \sigma_Y = 0.05$ inches, and ρ. After the pipes are produced, they are fitted together. Assuming that the pipes are joined without their combined length being lessened, find the probability that the combined length of the two joined pipes is less than 6.60 inches when:

(a) $\rho = 0.0$. (b) $\rho = 0.5$. (c) $\rho = -0.5$.
(d) $\rho = 1.0$. (e) $\rho = -1.0$.

6. Suppose that X and Y represent SAT mathematics and verbal scores, respectively, of a random individual, and that (X, Y) has a BVN((476, 423), (100, 100; 0.7)) distribution over the population of individuals taking the SAT. Some highly selective universities require that their applicants achieve a total SAT score, $X + Y$, of at least 1100 to be considered for admission.

(a) Find the distribution of the total SAT score $X + Y$.

(b) What proportion of all who take the SAT achieve the criterion, $X + Y \geq 1100$, for admission?

(c) A student who has applied to a highly selective university receives a SAT mathematics score of $X = 600$. This student's SAT verbal score is missing on the copy of the report sent by the SAT. Given that $X = 600$, what is the conditional probability that the student's total SAT score $X + Y$ exceeds 1100?

7. If $(X, Y) \sim$ BVN((2, 5), (4, 4: 0.3)), find the joint distribution of:

(a) $1 - Y$ and X.

(b) $2X + Y$ and X.

(c) $Y - 0.3X$ and X.

Be sure to state the means, variances, and correlation for each joint distribution. Which of these pairs of variables are probabilistically independent?

8. Let X be the Dow Jones average of common stocks and Y be the prime lending rate on a randomly chosen day. During a certain period, X and Y have a bivariate normal distribution with $\mu_X = 3250$, $\mu_Y = 6.12$, $\sigma_X^2 = 2500$, $\sigma_Y^2 = 0.36$, and $\rho_{XY} = -0.5$. Find the conditional probability,

$$P\{X > 3300 \mid Y \le 6.00\},$$

that the Dow Jones average exceeds 3300 when the prime lending rate is no greater than 6.00%. (*Hint:* You will need to find the probabilities $P\{Y \le 6.00\}$ and $P\{X > 3300 \text{ and } Y \le 6.00\}$.)

9. You wish to simulate an observation from a BVN((10, 0), (9, 25; 0.8)) distribution. Suppose that two independent simulations from the uniform distribution on [0, 1] are $U_1 = 0.82$ and $U_2 = 0.35$. What is the resulting observation (X^*, Y^*) from the bivariate normal distribution?

10. (a) If Z_1, Z_2 are independent standard normal random variables, what is the joint distribution of $W_1 = (Z_1 + Z_2)/\sqrt{2}$ and $W_2 = (Z_1 - Z_2)/\sqrt{2}$? Are W_1 and W_2 probabilistically independent?

(b) If X_1, X_2 are independent $\mathcal{N}(\mu, \sigma^2)$ random variables, what is the joint distribution of $W_1 = (X_1 + X_2)/\sqrt{2}$ and $W_2 = (X_1 - X_2)/\sqrt{2}$? Are W_1 and W_2 independent?

Theoretical Exercises

T.1. Suppose that (X, Y) have the joint probability density function

$$f(x, y) = \tfrac{1}{2}g_1(x, y) + \tfrac{1}{2}g_2(x, y),$$

where

$$g_1(x, y) = \frac{1}{2\pi\sqrt{1 - \rho_1^2}} \exp\left[-\frac{1}{2}\left(\frac{x^2 + y^2 - 2\rho_1 xy}{1 - \rho_1^2}\right)\right], \quad -\infty < x, y < \infty,$$

$$g_2(x, y) = \frac{1}{2\pi\sqrt{1 - \rho_2^2}} \exp\left[-\frac{1}{2}\left(\frac{x^2 + y^2 - 2\rho_2 xy}{1 - \rho_2^2}\right)\right], \quad -\infty < x, y < \infty,$$

and $\rho_1 \neq \rho_2$. Observe that $f(x, y)$ cannot be put into the form (1.1) of a bivariate normal density function.

(a) Show that $\int_{-\infty}^{\infty}\int_{-\infty}^{\infty} f(x, y)\, dx\, dy = 1$.

(b) Show that X and Y each have marginal $\mathcal{N}(0, 1)$ density functions.

T.2. Suppose that X and Y have the joint probability density function

$$f(x, y) = \begin{cases} \dfrac{\exp\{-\frac{1}{2}[(x^2 - 2\rho xy + y^2)/(1 - \rho^2)]\}}{\pi\sqrt{1 - \rho^2}}, & \text{if } xy \geq 0, \\ 0, & \text{if } xy < 0. \end{cases}$$

(a) Show that X and Y each have marginal $\mathcal{N}(0, 1)$ distributions, but that $f(x, y)$ is not the joint probability density function of a bivariate normal distribution.

(b) Find ρ_{XY}.

(c) Find $f_{X|Y}(x \mid y)$ and $\mu_{X|Y=y}$.

(d) Find $f_{Y|X}(y \mid x)$ and $\mu_{Y|X=x}$.

T.3. Suppose that (X, Y) has the joint probability density function

$$f(x, y) = \begin{cases} 1/2, & 0 \leq x, y, x + y \leq 1, \\ 0, & \text{otherwise.} \end{cases}$$

(a) Show that for all x, $0 \leq x < 1$,

$$f_{Y|X}(y|x) = \begin{cases} \dfrac{1}{1 - x}, & 0 \leq y \leq 1 - x, \\ 0, & \text{otherwise.} \end{cases}$$

(b) Show that $\sigma^2_{Y|X=x} = \frac{1}{12}(1 - x)^2$, $0 \leq x < 1$, and that $\sigma^2_Y = 1/18$. Conclude that $\sigma^2_Y < \sigma^2_{Y|X=x}$ for x sufficiently close to 0.

T.4. (a) Let $(X, Y) \sim \text{BVN}((0, 0), (1, 1; \rho))$. Find the moment generating function of

$$V = \frac{X^2 + Y^2 - 2\rho XY}{1 - \rho^2},$$

and hence show that $V \sim \chi^2_2$ (see Chapter 10, Section 2).

(b) Alternatively, show that

$$V = \frac{(X - \rho Y)^2}{1 - \rho^2} + Y^2,$$

and that $Z_1 = (X - \rho Y)/\sqrt{1 - \rho^2}$ and $Z_2 = Y$ are independent standard normal random variables. Use this result to prove that $V \sim \chi^2_2$.

(c) If $(X, Y) \sim \text{BVN}((\mu_1, \mu_2), (\sigma_1^2, \sigma_2^2; \rho))$, find the distribution of

$$V = \frac{(X - \mu_1)^2}{\sigma_1^2(1 - \rho^2)} + \frac{(Y - \mu_2)^2}{\sigma_2^2(1 - \rho^2)} - 2\rho \frac{(X - \mu_1)(Y - \mu_2)}{\sigma_1\sigma_2(1 - \rho^2)}.$$

T.5. If $(X, Y) \sim \text{BVN}((0, 0), (1, 1; \rho))$, find the joint moment generating function of X^2, Y^2.

T.6. It has been shown that if (X, Y) has a bivariate normal distribution, every linear combination of X, Y has a normal distribution. Supposing that (X, Y) has a joint moment generating function $M(s, t)$, prove the converse: *If every linear combination of two jointly distributed random variables is normally distributed,* (X, Y) *has a bivariate normal distribution.* [*Hint*: $sX + tY$ must have a $\mathcal{N}(s\mu_1 + t\mu_2, s^2\sigma_1^2 + t^2\sigma_2^2 + 2st\sigma_1\sigma_2\rho)$ distribution. Show that $M(s, t)$ equals the moment generating function $M(u)$ of $sX + tY$ evaluated at $u = 1$.]

13★

Transformations of Two Random Variables

In Chapter 5 we discussed transformations of a single random variable. In experiments where pairs (X, Y) of random variables are observed, we may be interested in two functions

$$U = g_1(X, Y), \qquad V = g_2(X, Y)$$

of these variables, or we might be interested in only one function, $W = g(X, Y)$. Both situations generalize the notion of a transformation. For example, we may have observed the initial size X of an organism and also its size Y after a defined period of time. We might then be interested in the relationship of the growth $V = Y - X$ of the organism to the ratio $U = X/X_0$ of its initial size to a known standard X_0, or might simply want to consider the percentage growth $W = 100(Y - X)/X$ of the organism. The variables U, V, and W are all transformations of the original observations X, Y into new variables.

In the present chapter we study methods for obtaining the distributions of transformed variables from the joint distribution of the original variables X, Y. In Section 1 we consider **bivariate transformations** $U = g_1(X, Y)$, $V = g_2(X, Y)$ that are one-to-one. This leads directly (Section 2) to methods for finding the distribution of a single transformation $W = g(X, Y)$, sometimes called a **many-to-one transformation.** Important special cases of many-to-one transformations are weighted sums $W = aX + bY$. Section 3 deals with such transformations and introduces the operation of **convolution** on distributions. Finally, Section 4 lists some special cases where the distributions of weighted sums of two or more random variables are known. That section, and also Section 3, demonstrate how distributions of transformations $W =$

$g(X_1, X_2, \ldots, X_p)$ of more than two random variables can often be found *recursively*, using the methods described in this chapter.

1 One-to-One Bivariate Transformations

Suppose that the random variables X and Y have a bivariate distribution defined either by a joint probability mass function $p(x, y)$ or a joint probability density function $f(x, y)$ and we wish to find the joint distribution of the bivariate transformation

$$U = g_1(X, Y), \qquad V = g_2(X, Y), \tag{1.1}$$

where this transformation is one-to-one. By **one-to-one** we mean that it is possible to solve (1.1) uniquely for X, Y as functions $h_1(U, V)$, $h_2(U, V)$ of U, V. This process is called **inverting** the transformation (1.1), and $h_1(u, v)$, $h_2(u, v)$ are said to be the **inverse functions** to $g_1(x, y)$ and $g_2(x, y)$. Consequently,

$$u = g_1(x, y),\ v = g_2(x, y) \quad \text{if and only if} \quad x = h_1(u, v),\ y = h_2(u, v). \tag{1.2}$$

For example, the transformation

$$U = X + Y, \qquad V = Y - X$$

can be inverted (solved for X, Y), yielding the inverse functions

$$X = \tfrac{1}{2}(U - V) = h_1(U, V), \qquad Y = \tfrac{1}{2}(U + V) = h_2(U, V).$$

Similarly, if X and Y are positive random variables, the transformation

$$U = XY, \qquad V = \frac{X}{Y}$$

can be inverted to yield

$$X = (UV)^{1/2} = h_1(U, V), \qquad Y = \left(\frac{U}{V}\right)^{1/2} = h_2(U, V).$$

Consequently, both of these bivariate transformations are one-to-one.

If X and Y are discrete random variables with joint probability mass function $p(x, y)$, it follows from (1.2) that the joint probability mass function of U and V is given by

$$p_{U,V}(u, v) = p(h_1(u, v), h_2(u, v)). \tag{1.3}$$

That is,

$$\begin{aligned} p_{U,V}(u, v) &= P\{g_1(X, Y) = u, g_2(X, Y) = v\} \\ &= P\{X = h_1(u, v), Y = h_2(u, v)\} \\ &= p(h_1(u, v), h_2(u, v)). \end{aligned}$$

EXAMPLE 1.1

In a clinical study of a new drug for treating skin rashes, the number X of subjects who are completely cured and the number Y of subjects who obtain relief from symptoms (but who are not completely cured) are observed. Suppose that n subjects have been given the drug, and that subjects respond independently with probability p_1 of being completely cured and p_2 of obtaining relief, $0 \le p_1, p_2, 1 - p_1 - p_2 \le 1$. Then X and Y have the trinomial distribution (Chapter 11, Exercise T.10), with joint probability mass function

$$p(x, y) = \frac{n!}{x!\, y!(n - x - y)!} p_1^x p_2^y (1 - p_1 - p_2)^{n-x-y}$$

for x, y nonnegative integers, $0 \le x + y \le n$.

The investigator may be more interested in the variables

$U = X + Y =$ total number of subjects who are improved (obtain relief or are cured),

$V = \dfrac{X}{X + Y} =$ proportion of improved subjects who are completely cured.

Because

$$X = UV = h_1(U, V), \qquad Y = U(1 - V) = h_2(U, V),$$

the joint probability mass function of U, V is

$$\begin{aligned} p_{U,V}(u, v) &= p(uv, u(1 - v)) \\ &= \frac{n!}{(uv)!\,(u(1 - v))!\,(n - u)!} p_1^{uv} p_2^{u(1-v)} (1 - p_1 - p_2)^{n-u} \end{aligned}$$

for uv, and $u(1 - v)$ nonnegative integers, and $0 \le u \le n$. ◆

If X, Y have joint probability density function $f(x, y)$ and if the first partial derivatives of $h_1(u, v)$ and $h_2(u, v)$ with respect to u and v exist, the joint probability density function of U, V is

$$f_{U,V}(u, v) = |J(u, v)| f(h_1(u, v), h_2(u, v)) \tag{1.4}$$

where

$$J(u, v) = \left(\frac{\partial h_1(u, v)}{\partial u}\right)\left(\frac{\partial h_2(u, v)}{\partial v}\right) - \left(\frac{\partial h_1(u, v)}{\partial v}\right)\left(\frac{\partial h_2(u, v)}{\partial u}\right) \tag{1.5}$$

is the *Jacobian* of the transformation $x = h_1(u, v)$, $y = h_2(u, v)$.

To verify formula (1.4), we first find the joint cumulative distribution function $F_{U,V}(u, v)$ of U, V. Thus

$$\begin{aligned} F_{U,V}(u, v) &= P\{U \le u, V \le u\} \\ &= P\{g_1(X, Y) \le u, g_2(X, Y) \le v\} \\ &= \iint\limits_{\substack{g_1(x,y)\le u \\ g_2(x,y)\le v}} f(x, y)\, dx\, dy \\ &= \int_{-\infty}^{u}\int_{-\infty}^{v} |J(s, t)| f(h_1(s, t), h_2(s, t))\, ds\, dt, \end{aligned}$$

where the last equality in (1.6) arises from changing the variables of integration from (x, y) to $s = g_1(x, y)$, $t = g_2(x, y)$ and making use of the theory of change of variables in multiple integration. Using the fact that

$$f_{U,V}(u, v) = \frac{\partial^2}{\partial u\, \partial v} F_{U,V}(u, v),$$

and two applications of the fundamental theorem of calculus, yields the desired result (1.4).

EXAMPLE 1.2

In Chapter 10 we remarked that waiting times often have gamma distributions. Suppose that X and Y are independent measurements of the times taken to complete two similar tasks. Assuming that X and Y each have a gamma distribution with parameters θ and r, the joint density function of X and Y is

$$\begin{aligned} f(x, y) &= f_X(x) f_Y(y) \\ &= \begin{cases} \dfrac{\theta^{2r} x^{r-1} y^{r-1} e^{-\theta x} e^{-\theta y}}{[\Gamma(r)]^2}, & \text{if } 0 < x < \infty, \quad 0 < y < \infty, \\ 0, & \text{otherwise.} \end{cases} \end{aligned}$$

Consider new variables $U = X + Y$, $V = X/Y$, which give the total time required and the ratio of the times required to do the tasks, respectively. The transformation $u = x + y$, $v = x/y$ is one-to-one, with

$$x = h_1(u, v) = \frac{uv}{1 + v}, \qquad y = h_2(u, v) = \frac{u}{1 + v}.$$

Applying (1.5), the Jacobian of this transformation is

$$J(u, v) = \left(\frac{v}{1 + v}\right)\left(\frac{-u}{(1 + v)^2}\right) - \left(\frac{u}{(1 + v)^2}\right)\left(\frac{1}{1 + v}\right) = -\frac{u}{(1 + v)^2}.$$

Hence, from equation (1.4), the joint density function of (U, V) is

$$f_{U,V}(u, v) = |J(u, v)| f(h_1(u, v), h_2(u, v))$$

$$= \begin{cases} \dfrac{u}{(1 + v)^2} \dfrac{\theta^{2r} u^{2r-2} v^{r-1} e^{-\theta u}}{(1 + v)^{2r-2}[\Gamma(r)]^2}, & \text{if } 0 < \dfrac{uv}{1 + v} < \infty, \\ & \quad 0 < \dfrac{u}{1 + v} < \infty, \\ 0, & \text{otherwise,} \end{cases}$$

$$= \begin{cases} \dfrac{\theta^{2r} u^{2r-1} e^{-\theta u}}{\Gamma(2r)} \dfrac{v^{r-1}}{B(r, r)(1 + v)^{2r}}, & \text{if } 0 < u < \infty, \\ & \quad 0 < v < \infty, \\ 0, & \text{otherwise.} \end{cases}$$

It is interesting to note that the joint density $f_{U,V}(u, v)$ of U and V is the product of a function of u only and a function of v only. Consequently, U and V are probabilistically independent. Further, the marginal density of U is

$$f_U(u) = \begin{cases} \dfrac{\theta^{2r} u^{2r-1} e^{-\theta u}}{\Gamma(2r)}, & \text{if } 0 < u < \infty, \\ 0, & \text{otherwise.} \end{cases}$$

which is a gamma distribution with scale parameter θ and shape parameter $2r$ (Chapter 10, Section 2). ◆

Bivariate Linear Transformation

Important special cases of one-to-one bivariate transformations are **nonsingular bivariate linear transformations:**

$$U = a_{11}X + a_{12}Y, \qquad V = a_{21}X + a_{22}Y, \tag{1.7}$$

for which $A \equiv a_{11}a_{22} - a_{12}a_{21} \neq 0$. Solving (1.7) for X, Y yields the inverse functions (inverse transformation):

$$(1.8)\quad X = b_{11}U + b_{12}V = h_1(U, V), \qquad Y = b_{21}U + b_{22}V = h_2(U, V),$$

where

$$b_{11} = \frac{a_{22}}{A}, \qquad b_{12} = \frac{-a_{12}}{A}, \qquad b_{21} = \frac{-a_{21}}{A}, \qquad b_{22} = \frac{a_{11}}{A}.$$

Notice that $A \neq 0$ is both necessary and sufficient for (1.7) to be uniquely solved for X, Y in terms of U, V. A special case of (1.7), namely $U = X + Y$, $V = Y - X$, was discussed earlier, for which $a_{11} = a_{12} = a_{22} = -a_{21} = 1$, $A = 2$, and $b_{11} = b_{22} = -b_{12} = b_{21} = \frac{1}{2}$.

A straightforward substitution in (1.5) using (1.8) yields

$$J(u, v) = b_{11}b_{22} - b_{12}b_{21} = \frac{a_{22}a_{11} - (-a_{12})(-a_{21})}{A^2} = \frac{1}{A}.$$

Consequently, it follows from (1.4) that

$$\begin{aligned} f_{U,V}(u, v) &= \frac{1}{|A|} f(b_{11}u + b_{12}v, b_{21}u + b_{22}v) \\ &= \frac{1}{|A|} f\left(\frac{a_{22}u - a_{12}v}{A}, \frac{a_{11}v - a_{21}u}{A}\right). \end{aligned}$$

is the joint probability density function of U, V when X, Y have joint probability density $f(x, y)$.

EXAMPLE 1.3

A student independently takes the Scholastic Aptitude Test (SAT) on two occasions, and obtains standardized scores X, Y, that are known to have marginal standard normal [$\mathcal{N}(0, 1)$] distributions. Thus the joint probability density function of X and Y is

$$\begin{aligned} f(x, y) = f_X(x)f_Y(y) &= \left(\frac{1}{\sqrt{2\pi}} e^{-(1/2)x^2}\right)\left(\frac{1}{\sqrt{2\pi}} e^{-(1/2)y^2}\right) \\ &= \frac{e^{-(1/2)(x^2+y^2)}}{2\pi}. \end{aligned}$$

The student is interested in the change $V = Y - X$ in his or her standardized score between the first and second tests, and also in the average $U = \frac{1}{2}(X + Y)$ of the two scores. The transformation from X, Y to U, V is of the linear form (1.7) with $a_{11} = a_{12} = \frac{1}{2}$, $a_{21} = -1$, $a_{22} = 1$. Thus $A = 1$, $b_{11} = 1$, $b_{12} = -\frac{1}{2}$, $b_{21} = 1$, $b_{22} = \frac{1}{2}$, and the joint probability density function of U, V is

$$f_{U,V}(u, v) = \frac{1}{|1|} \frac{\exp\left\{-\frac{1}{2}\left[\left(\frac{u - \frac{1}{2}v}{1}\right)^2 + \left(\frac{u + \frac{1}{2}v}{1}\right)^2\right]\right\}}{2\pi}$$

$$= \frac{\exp\{-\frac{1}{2}[u^2/(1/2)]\}}{\sqrt{2\pi(\frac{1}{2})}} \left[\frac{\exp\{-\frac{1}{2}[v^2/2]\}}{\sqrt{2\pi(2)}}\right], \qquad -\infty < u, v < \infty.$$

This joint probability density function is a product of two normal densities, so that U and V are probabilistically independent, and $U \sim \mathcal{N}(0, \frac{1}{2})$, $V \sim \mathcal{N}(0, 2)$. ◆

2 Many-to-One Transformations

Suppose that X, Y have a joint distribution and we want to find the distribution of only one function

$$V = g(X, Y). \tag{2.1}$$

The theory of one-to-one transformations does not hold here, because many values of (x, y) may yield the same value of $v = g(x, y)$. One method of obtaining the distribution of V is to find a second transformation

$$u = g_1(x, y)$$

so that this, together with $v = g(x, y)$, yields a one-to-one transformation from (x, y) to (u, v). Using (1.3) or (1.4), depending on whether the random variables X, Y are discrete or continuous, we can find the joint distribution $p_{U,V}(u, v)$ or $f_{U,V}(u, v)$ of (U, V). From this joint distribution, we can then find the marginal distribution $p_V(v)$ or $f_V(v)$ of V.

EXAMPLE 2.1

In Example 1.2, X and Y are continuous random variables arising from independent measurements of the lengths of time needed to complete two similar tasks. To obtain the distribution of the ratio $V = X/Y$, recall that

$$U = X + Y, \qquad V = \frac{X}{Y}$$

is a one-to-one transformation, for which the joint density function is

$$f_{U,V}(u, v) = \begin{cases} \dfrac{\theta^{2r} u^{2r-1} e^{-\theta u}}{\Gamma(2r)} \dfrac{v^{r-1}}{B(r, r)(1 + v)^{2r}}, & \text{if } 0 < u < \infty, \quad 0 < v < \infty, \\ 0, & \text{otherwise.} \end{cases}$$

The marginal density function $f_V(v)$ of V can be obtained directly from this joint density. Thus if $0 < v < \infty$,

$$f_V(v) = \int_0^\infty f_{U,V}(u, v)\, du = \frac{v^{r-1}}{B(r, r)(1+v)^{2r}} \int_0^\infty \frac{\theta^{2r} u^{2r-1} e^{-\theta u}\, du}{\Gamma(2r)}$$
$$= \frac{v^{r-1}}{B(r, r)(1+v)^{2r}},$$

whereas $f_V(v) = \int_0^\infty 0\, du = 0$ for $v \leq 0$.

In this example, we might instead have been interested in the total time $U = X + Y$ needed to complete the two tasks. We can obtain the distribution of U by finding its marginal density function

$$f_U(u) = \int_0^\infty f_{U,V}(u, v)\, dv = \begin{cases} \dfrac{\theta^{2r} u^{2r-1} e^{-\theta u}}{\Gamma(2r)}, & 0 < u < \infty, \\ 0, & \text{otherwise.} \end{cases}$$

We have already noted that U has a gamma distribution with parameters θ and $2r$. ◆

It is not always easy to find the remaining function $U = g_1(X, Y)$ that, together with $V = g(X, Y)$, will make up a workable one-to-one transformation. If in Example 2.1 we had not already obtained the joint distribution of $U = X + Y$ and $V = X/Y$, we might have had some trouble finding a transformation $U = g_1(X, Y)$ to pair with $V = X/Y$ in order to obtain a one-to-one transformation suitable for obtaining the marginal distribution of V.

One way to bypass this indirect approach is through the cumulative distribution function. If $V = g(X, Y)$, then

$$F_V(v) = P\{V \leq v\} = P\{g(X, Y) \leq v\} = \int \cdots \int_{g(x,y) \leq v} f(x, y)\, dx\, dy. \tag{2.2}$$

Suppose that there exists a function $h(v; y)$ such that

$$g(x, y) \leq v \quad \text{if and only if} \quad x \leq h(v; y). \tag{2.3}$$

Then

$$\begin{aligned} F_V(v) = \int \cdots \int_{g(x,y) \leq v} f(x, y)\, dx\, dy &= \int_{-\infty}^{\infty} \left[\int_{-\infty}^{h(v;y)} f(x, y)\, dx \right] dy \\ &= \int_{-\infty}^{\infty} q(v; y)\, dy, \end{aligned} \tag{2.4}$$

say, where

$$q(v; y) = \int_{-\infty}^{h(v;y)} f(x, y)\, dx. \tag{2.5}$$

Note that by the chain rule and the fundamental theorem of calculus,

$$\frac{\partial q(v; y)}{\partial v} = \frac{\partial h(v; y)}{\partial v} [f(h(v; y), y)]. \tag{2.6}$$

Thus if we can show that

$$\frac{\partial}{\partial v} \int_{-\infty}^{\infty} q(v; y)\, dy = \int_{-\infty}^{\infty} \frac{\partial}{\partial v} q(v; y)\, dy, \tag{2.7}$$

it follows from (2.4) and (2.6) that

$$f_V(v) = \frac{\partial}{\partial v} F_V(v) = \int_{-\infty}^{\infty} \frac{\partial h(v; y)}{\partial v} f(h(v; y), y)\, dy. \tag{2.8}$$

Of course, equality (2.7) need not always hold. However, if $q(v; y)$ and $\partial q(v; y)/\partial v$ are well defined and jointly continuous functions of v, y for $-\infty < v, y < \infty$, then equality (2.7) holds, and thus (2.8) will be valid.

EXAMPLE 2.2

A distributional problem of great importance to the theory of statistical inference is the following. Suppose that the random variables X and Y are independent, with $X \sim \mathcal{N}(0, 1)$ and Y having a gamma distribution with parameters $\theta = \frac{1}{2}$ and $r = n/2$, where $n \geq 1$ is an integer. We wish to find the distribution of

$$V = \frac{\sqrt{n} X}{\sqrt{Y}}.$$

Note that $g(x, y) = \sqrt{n}\, x/\sqrt{y}$, and that

$$\frac{\sqrt{n}\, x}{\sqrt{y}} \leq v \quad \text{if and only if} \quad x \leq v\sqrt{y}/\sqrt{n},$$

so that $h(v; y) = v\sqrt{y}/\sqrt{n}$. From the distributional assumptions given above, we have

$$\begin{aligned} f(x, y) &= f_X(x) f_Y(y) \\ &= \left(\frac{1}{\sqrt{2\pi}} e^{-(1/2)x^2}\right) \frac{y^{n/2-1} e^{-(1/2)y}}{2^{n/2}\Gamma(n/2)}, \qquad -\infty < x < \infty,\ 0 < y < \infty, \end{aligned}$$

and $f(x, y) = 0$, otherwise. Thus, from (2.5),

$$q(v; y) = \int_{-\infty}^{v\sqrt{y}/\sqrt{n}} \frac{1}{\sqrt{2\pi}} e^{-x^2/2} \frac{y^{(n/2)-1}e^{-y/2}}{2^{n/2}\Gamma(n/2)}\, dx,$$

and from (2.6),

$$\frac{\partial}{\partial v} q(v; y) = \left(\frac{\sqrt{y}}{\sqrt{n}}\right) \frac{1}{\sqrt{2\pi}} e^{-(v\sqrt{y}/\sqrt{n})^2/2} \frac{y^{(n/2)-1}e^{-y/2}}{2^{n/2}\Gamma(n/2)}.$$

Both of these functions are well-defined and continuous functions of v and y for all $-\infty < v < \infty$, $0 \le y < \infty$. From (2.8) it follows that

$$\begin{aligned} f_V(v) &= \frac{1}{\sqrt{2\pi n}\,\Gamma(n/2)2^{n/2}} \int_0^\infty y^{(n-1)/2} \exp\left\{-\tfrac{1}{2}\, y\left(1 + \frac{v^2}{n}\right)\right\} dy \\ &= \frac{\Gamma((n+1)/2)}{\Gamma(n/2)\sqrt{\pi n}\,(1 + v^2/n)^{(n+1)/2}}, \end{aligned} \tag{2.9}$$

for $-\infty < v < \infty$. ◆

Remark. The range of y can be taken to be $0 \le y < \infty$ rather than $-\infty < y < \infty$ in the preceding computation because $f_{X,Y}(x, y) = 0$ for $y < 0$.

The distribution corresponding to the density function $f_V(v)$ in (2.9) is called **Student's *t*-distribution with *n* degrees of freedom.** This distribution was originally studied in connection with a method of statistical inference for the mean of a normal distribution (called the "*t*-test") proposed by the English statistican William Sealy Gosset (1876–1937) writing under the pseudonym "Student."

Recall (Chapter 10, Section 2) that the distribution of the random variable Y in this example (namely, the gamma distribution with parameter $\theta = \frac{1}{2}$ and $r = n/2$, where n is a positive integer) is also called a **chi-squared distribution with *n* degrees of freedom,** and that the notation $Y \sim \chi_n^2$ is used to indicate that the random variable Y has a chi-squared distribution with n degrees of freedom. The chi-squared distribution originally arose in connection with work by the English statistican Karl Pearson (1857–1936) on the statistical analysis of sample contingency tables, and also arises in connection with statistical inferences about the variance σ^2 of a normal distribution.

Another distribution that is of statistical importance is the ***F*-distribution with *n* and *m* degrees of freedom,** which is used in connection with the statistical methodology called the analysis of variance. Suppose that X and Y are independent random variables, and that $X \sim \chi_n^2$, $Y \sim \chi_m^2$. Let

$$V = \frac{X/n}{Y/m}.$$

Then V is said to have the F-distribution with n and m degrees of freedom. The distribution of V can be obtained indirectly, using the one-to-one transformation $V = (X/n)/(Y/m)$, $U = X + Y$, by finding the joint probability density function $f_{U,V}(u, v)$ of U and V, and then integrating this joint density to find the marginal probability density function of V. This is the approach used in Examples 1.2 and 2.1. Alternatively, we can use the direct method of Example 2.2, making use of Equation (2.8), as illustrated below.

Note that

$$g(x, y) = \frac{x/n}{y/m} \le v \quad \text{if and only if} \quad x \le v\left(\frac{n}{m}\right)y,$$

so that $h(v; y) = v(n/m)y$. From the facts that X and Y are independent and that $X \sim \chi_n^2$, $Y \sim \chi_m^2$, we have

$$f(x, y) = f_X(x)f_Y(y) = \frac{x^{(n/2)-1}e^{-x/2}}{2^{n/2}\Gamma(n/2)}\frac{y^{(m/2)-1}e^{-y/2}}{2^{m/2}\Gamma(m/2)},$$

if $0 \le x, y < \infty$, and $f_{X,Y}(x, y) = 0$ otherwise. Thus

$$q(v; y) = \int_0^{(n/m)yv} \frac{x^{(n/2)-1}y^{(m/2)-1}e^{-(x+y)/2}}{2^{(n+m)/2}\Gamma(n/2)\Gamma(m/2)}\,dx$$

$$\frac{\partial}{\partial v}q(v; y) = \frac{v^{(n/2)-1}(n/m)^{n/2}y^{[(n+m)/2]-1}e^{-(y/2)(1+nv/m)}}{2^{(n+m)/2}\Gamma(n/2)\Gamma(m/2)}$$

are both well-defined and continuous functions of v and y for all $0 \le v, y < \infty$.

From (2.8), it follows that for $v \ge 0$,

$$\begin{aligned} f_V(v) &= \frac{(n/m)^{n/2}v^{(n/2)-1}}{2^{(n+m)/2}\Gamma(n/2)\Gamma(m/2)}\int_0^\infty y^{[(n+m)/2]-1}e^{-(y/2)y(1+nv/m)}\,dy \\ &= \frac{(n/m)^{n/2}\Gamma((n+m)/2)v^{(n/2)-1}}{\Gamma(n/2)\Gamma(m/2)(1+nv/m)^{(n+m)/2}}, \end{aligned} \tag{2.10}$$

whereas clearly $f_V(v) = 0$ for $v < 0$. The probability density function (2.10) thus defines the F-distribution with n and m degrees of freedom.

3 Sums and Weighted Sums of Random Variables

Perhaps the most frequently used many-to-one transformation of two random variables X, Y is their weighted sum,

$$W = aX + bY.$$

EXAMPLE 3.1

Suppose that X, Y are two independent observations from a population. Then the sample average $W = \frac{1}{2}(X + Y)$ has been previously suggested as an estimate of the population mean. ◆

EXAMPLE 3.2

In a "before–after" type of statistical study, measurements X and Y are made before and after the application of a treatment. The difference $W = Y - X$ is used to measure the change brought about by the treatment. ◆

EXAMPLE 3.3

A final grade for a course in which there was a midterm exam X and a final examination Y, might be determined by the total score $W = X + Y$ or by a weighted total score, say $W = 2X + 3Y$. ◆

To find the distribution of a weighted sum $W = aX + bY$ of the random variables X, Y, we first identify a coefficient a or b not equal to zero. For the sake of notational convenience, suppose that $a \neq 0$. Next, we consider the one-to-one transformation

$$W = aX + bY, \qquad V = Y,$$

for which $y = h_2(w, v) = v$, and

$$x = h_1(w, v) = \frac{w - bv}{a}.$$

We can now find the joint distribution of W, V and from this joint distribution find the marginal distribution of W. Hence when X, Y are discrete random variables with joint probability mass function $p(x, y)$, the steps above yield

$$p_W(w) = \sum_v p\left(\frac{w - bv}{a}, v\right). \tag{3.1}$$

Similarly, when X, Y are continuous random variables with joint probability density function $f(x, y)$, then

$$f_W(w) = \int_{-\infty}^{\infty} \left|\frac{1}{a}\right| f\left(\frac{w - bv}{a}, v\right) dv. \tag{3.2}$$

EXAMPLE 3.4

Suppose that (X, Y) have the joint probability mass function

$$p(x, y) = \frac{n!}{x!\, y!\, (n - x - y)!} p_1^x p_2^y (1 - p_1 - p_2)^{n-x-y}$$

if $x, y = 0, 1, \ldots, n$, $0 \leq x + y \leq n$, and $p(x, y) = 0$ otherwise (see Example 1.1). Then Equation (3.1) tells us that the probability mass function of $W = X + Y$ is

$$p_W(w) = \sum_{v=0}^{n} p(w - v, v) = \sum_{v=0}^{w} p(w - v, v)$$

if $w = 0, 1, \ldots, n$, and $p_W(w) = 0$ otherwise, because $p(x, y) = 0$ for x and y not equal to $0, 1, 2, \ldots, n$. Now,

$$\begin{aligned}\sum_{v=0}^{w} p(w - v, v) &= \frac{n!\,(1 - p_1 - p_2)^{n-w}}{(n - w)!\,w!}\left[\sum_{v=0}^{w} \frac{w!}{v!\,(w - v)!} p_1^{w-v} p_2^{v}\right] \\ &= \binom{n}{w} (1 - p_1 - p_2)^{n-w}(p_1 + p_2)^{w}.\end{aligned}$$

Thus

$$p_W(w) = \begin{cases} \binom{n}{w} (p_1 + p_2)^{w}(1 - p_1 - p_2)^{n-w}, & \text{if } w = 0, 1, \ldots, n. \\ 0, & \text{otherwise.} \end{cases}$$

which is the probability mass function of a binomial distribution with parameters n and $p = p_1 + p_2$. ♦

EXAMPLE 3.5

Let X and Y be statistically independent random variables, where X and Y both have exponential distributions with parameter $\theta = 1$. Thus

$$f(x, y) = \begin{cases} e^{-(x+y)}, & \text{if } x > 0,\ y > 0, \\ 0, & \text{otherwise.} \end{cases}$$

Let us find the distribution of $W = 2X + 3Y$. From equation (3.2), the probability density function, $f_W(w)$, of W is

$$f_W(w) = \int_{-\infty}^{\infty} \left|\frac{1}{2}\right| f\left(\frac{w - 3v}{2}, v\right) dv.$$

Because $f(x, y) = 0$ if $x < 0$, it follows that $f_W(w) = 0$ if $w < 0$. If $w > 0$, then

$$\begin{aligned} f_W(w) &= \frac{1}{2} \int_{0}^{w/3} \exp\left(-\frac{w - 3v}{2} - v\right) dv + \frac{1}{2} \int_{w/3}^{\infty} 0\, dv \\ &= \frac{1}{2} \exp\left(-\frac{1}{2} w\right) \int_{0}^{w/3} \exp\left(\frac{1}{2} v\right) dv \end{aligned}$$

$$= \frac{1}{2}\exp\left(-\frac{w}{2}\right)\left[2\exp\left(\frac{w}{6}\right) - 2\right]$$
$$= \exp\left(-\frac{w}{3}\right) - \exp\left(-\frac{w}{2}\right). \qquad \blacklozenge$$

Although equation (3.1) is sometimes useful in finding the distribution of a weighted sum of discrete random variables, in many cases a direct approach is easier.

EXAMPLE 3.6 Assume that the joint distribution of the discrete random variables, X, Y is given by

(x, y)	$(0, 0)$	$(1, 0)$	$(0, 1)$	$(1, 1)$
$p(x, y)$	0.20	0.30	0.35	0.15

Let $W = X - Y$ and note that

$$W = \begin{cases} 0, & \text{when } (x, y) = (0, 0) \text{ or } (1, 1), \\ 1, & \text{when } (x, y) = (1, 0), \\ -1, & \text{when } (x, y) = (0, 1). \end{cases}$$

Consequently,

$$P\{W = 0\} = p(0, 0) + p(1, 1) = 0.35,$$
$$P\{W = 1\} = p(1, 0) = 0.30,$$
$$P\{W = -1\} = p(0, 1) = 0.35,$$

and thus

$$p_W(w) = \begin{cases} 0.35, & \text{if } w = -1 \text{ or } 0, \\ 0.30, & \text{if } w = 1, \\ 0.0, & \text{otherwise.} \end{cases}$$

That is, for each value w of W we simply added the probabilities of all (x, y) pairs yielding that value of w. ◆

For the case of simple sums $W = X + Y$, formulas (3.1) and (3.2) become

(3.3) $$p_W(w) = \sum_v p(w - v, v)$$

in the discrete case [where $p(x, y)$ is the joint mass function of X, Y], and

(3.4)
$$f_W(w) = \int_{-\infty}^{\infty} f(w - v, v)\, dv$$

in the continuous case [where $f(x, y)$ is the joint probability density function of X, Y]. By relabeling variables, it is also the case that

(3.5)
$$p_W(w) = \sum_u p(u, w - u)$$

and

(3.6)
$$f_W(w) = \int_{-\infty}^{\infty} f(u, w - u)\, du.$$

EXAMPLE 3.7

Let $(X, Y) \sim \text{BVN}((0, 0), (1, 1; \rho))$. Then for $-\infty < x, y < \infty$, the joint density function of (X, Y) is

$$f(x, y) = \frac{1}{2\pi\sqrt{1 - \rho^2}} \exp\left[-\frac{x^2 - 2\rho xy + y^2}{2(1 - \rho^2)}\right],$$

and thus, applying (3.6), the probability density function of $W = X + Y$ is

$$\begin{aligned} f_W(w) &= \frac{1}{2\pi\sqrt{1 - \rho^2}} \int_{-\infty}^{\infty} \exp\left[-\frac{u^2 - 2\rho u(w - u) + (w - u)^2}{2(1 - \rho^2)}\right] du \\ &= \frac{\exp[-w^2/4(1 + \rho)]}{\sqrt{2\pi(2)(1 + \rho)}} \int_{-\infty}^{\infty} \frac{\exp\left[-\frac{1}{2}\frac{(u - \frac{1}{2}w)^2}{(1 - \rho)/2}\right]}{\sqrt{2\pi(1 - \rho)/2}}\, du \\ &= \frac{\exp[-\frac{1}{2}[w^2/2(1 + \rho)]}{\sqrt{2\pi(2)(1 + \rho)}}, \qquad -\infty < w < \infty, \end{aligned}$$

which is the probability density function of a $\mathcal{N}(0, 2(1 + \rho))$ random variable. ◆

When X and Y are independent discrete random variables, then $p(x, y) = p_X(x)p_Y(y)$ and

(3.7)
$$p_W(w) = \sum_u p_X(u)p_Y(w - u) = \sum_v p_X(w - v)p_Y(v).$$

Similarly, when X and Y are continuous independent random variables, then

$$(3.8) \quad f_W(w) \int_{-\infty}^{\infty} f_X(u)f_Y(w - u)\, du = \int_{-\infty}^{\infty} f_X(w - v)f_Y(v)\, dv.$$

Formulas (3.7) and (3.8) are said to be **convolutions** of the probability mass functions $p_X(x)$, $p_Y(y)$ or probability density functions $f_X(x), f_Y(y)$, respectively.

EXAMPLE 3.8

Suppose that X and Y are independent discrete random variables, and that X, Y, each have the probability mass function of a geometric distribution (number of failures) with parameter p:

$$p(z) = \begin{cases} p(1 - p)^z, & \text{if } z = 0, 1, 2, \ldots, \\ 0, & \text{otherwise.} \end{cases}$$

To obtain the distribution of $W = X + Y$, we can use the convolution formula (3.7). Thus for $w = 0, 1, 2, \ldots,$

$$\begin{aligned} p_W(w) &= \sum_{u=0}^{\infty} p_X(u)p_Y(w - u) = \sum_{u=0}^{w} p_X(u)p_Y(w - u) \\ &= \sum_{u=0}^{w} p^2(1 - p)^w = (w + 1)p^2(1 - p)^w, \end{aligned}$$

and, otherwise, $p_W(w) = 0$. Because

$$w + 1 = \binom{w + 2 - 1}{2 - 1},$$

$p_W(w)$ is the probability mass function of the negative binomial distribution (number of failures) with parameters $r = 2$ and p. ◆

EXAMPLE 3.9

Suppose that X and Y are independent random variables and that $X \sim \chi_n^2$ and $Y \sim \chi_m^2$. That is,

$$f_X(x) = \begin{cases} \dfrac{x^{n/2}e^{-x/2}}{2^{n/2}\Gamma(n/2)}, & \text{if } 0 < x < \infty, \\ 0, & \text{otherwise,} \end{cases}$$

$$f_Y(y) = \begin{cases} \dfrac{y^{m/2}e^{-y/2}}{2^{m/2}\Gamma(m/2)}, & \text{if } 0 < y < \infty, \\ 0, & \text{otherwise.} \end{cases}$$

To obtain the distribution of $W = X + Y$, use the convolution formula (3.8). When $w \le 0$, $f_W(w) = 0$; when $w > 0$,

$$\begin{aligned} f_W(w) &= \frac{e^{-w/2}}{2^{(n+m)/2}\Gamma(n/2)\Gamma(m/2)} \int_0^w (w - v)^{(1/2)n-1} v^{(1/2)m-1}\, dv \\ &= \frac{w^{[(n+m)/2]-1} e^{-w/2}}{2^{(n+m)/2}\Gamma[(n+m)/2]} \int_0^1 \frac{z^{(m/2)-1}(1-z)^{(n/2)-1}}{B(n/2, m/2)} \\ &= \frac{w^{[(n+m)/2]-1} e^{-w/2}}{2^{(n+m)/2}\Gamma[(n+m)/2]}, \end{aligned}$$

where we have made the change of variable from v to $z = v/w$. The density function of W is that of a gamma distribution with parameters $\theta = \frac{1}{2}$, $r = (n + m)/2$, and thus $W \sim \chi^2_{n+m}$. We have shown that the convolution of two chi-squared distributions is a chi-squared distribution with degrees of freedom equal to the sum of the degrees of freedom of the original chi-squared distributions. ◆

A Recursive Procedure

When $X_1, X_2, \ldots, X_p$ are mutually independent random variables, it is often possible to obtain the distribution of $W = X_1 + X_2 + \cdots + X_p$ recursively. That is, first we use a convolution formula [(3.7) or (3.8)] to obtain the distribution of $X_1 + X_2$ and then use it again to obtain the distribution of $(X_1 + X_2) + X_3$. Continuing in this fashion, we use the convolution formulas to obtain the distributions of $(X_1 + X_2 + X_3) + X_4$, $(X_1 + X_2 + X_3 + X_4) + X_5$, and so on, until at last, one final convolution gives us the distribution of $(X_1 + X_2 + \cdots + X_{p-1}) + X_p = W$.

Remark. In applying the foregoing iterative method for finding the distribution of $Y = X_1 + X_2 + \cdots + X_p$, we implicitly make use of the fact that if the random variables $X_1, X_2, \ldots, X_p$ are mutually independent, then $\sum_{i=1}^{j} X_i$ and $X_{j+1}, \ldots, X_p$ are independent. An intuitive justification of this assertion comes from noting that the definition of mutual independence of the X_i's states that no information about any subcollection (say, $X_1, \ldots, X_j$) of these variables can be used to predict the remaining variables $X_{j+1}, \ldots, X_p$. Because the value of $\sum_{i=1}^{j} X_i$ gives information about $X_1, \ldots, X_j$, it cannot be used to predict (change the odds of) $X_{j+1}, \ldots, X_p$, and hence $\sum_{i=1}^{j} X_i$ is independent of $X_{j+1}, \ldots, X_p$.

EXAMPLE 3.10

Suppose that $X_1, X_2, \ldots, X_p$ are independent, and that $X_i \sim \chi^2_{n_i}$, $i = 1, 2, \ldots, p$. We wish to find the distribution of $W = \sum_{i=1}^{p} X_i$. From Example 3.9 we know that $X_1 + X_2 \sim \chi^2_{n_1+n_2}$. Next, $\sum_{i=1}^{3} X_i = (X_1 + X_2) + X_3 \sim \chi^2_m$, where $m = (n_1 + n_2) + n_3 = \sum_{i=1}^{3} n_i$, and again

$\sum_{i=1}^4 X_i = \sum_{i=1}^3 X_i + X_4 \sim \chi_m^2$, where $m = \sum_{i=1}^3 n_i + n_4 = \sum_{i=1}^4 n_i$, and so on. Finally, $W = \sum_{i=1}^p X_i \sim \chi_m^2$ with $m = \sum_{i=1}^p n_i$; that is, *the convolution of independent random variables, each of which has a chi-squared distribution, is a chi-squared distribution with degrees of freedom equal to the sum of the individual degrees of freedom.* ◆

Use of Moment Generating Function

In Chapter 7 the moment generating function

$$M_X(t) = E[e^{tX}]$$

of a random variable X was introduced. One of the important properties of a moment generating function discussed in that chapter is that this function can be used to determine the distribution of the sum (convolution) of independent random variables $X_1, \ldots, X_p$. To do so, we proceed as follows:

1. Determine the moment generating function $M_{X_i}(t)$ of each X_i, $i = 1, \ldots, p$.
2. Form the product

$$M(t) = M_{X_1}(t)M_{X_2}(t) \cdots M_{X_p}(t)$$

of these moment generating functions. This product is the moment generating function of $Y = X_1 + X_2 + \cdots + X_p$.
3. Find (or recognize) the distribution corresponding to the moment generating function $M(t)$.

The distribution found in step 3, by the uniqueness property of moment generating functions, is then the distribution of W.

EXAMPLE 3.11

An alternative derivation of the result obtained in Example 3.10 starts by noting that if $X \sim \chi_m^2$, the moment generating function $M(t)$ of X is (for $1 - 2t > 0$):

$$\begin{aligned} M_X(t) &= \int_0^\infty e^{tx}\left(\frac{x^{(n/2)-1}e^{-(1/2)x}}{\Gamma(n/2)2^{n/2}}\right) dx \\ &= \int_0^\infty \frac{x^{(n/2)-1}e^{-(1/2)(1-2t)x}}{\Gamma(n/2)2^{n/2}}\, dx \\ &= \frac{1}{(1-2t)^{n/2}}\int_0^\infty \frac{u^{(n/2)-1}e^{-(1/2)/u}}{\Gamma(n/2)2^{n/2}}\, du, \end{aligned}$$

where we have made the change of variable $u = (1 - 2t)x$. The integral over u is recognizable as the total probability of a χ_n^2 random variable, and thus is equal to 1. Consequently,

$$M_X(t) = \frac{1}{(1 - 2t)^{n/2}}, \qquad t < \frac{1}{2}.$$

Now if $X_1, X_2, \ldots, X_p$ are independent random variables with $X_i \sim \chi_{m_i}^2$, $i = 1, 2, \ldots, p$, then for $t < \frac{1}{2}$,

$$M_{X_i}(t) = \frac{1}{(1 - 2t)^{m_i/2}}, \qquad i = 1, 2, \ldots, p,$$

and

$$M(t) = M_{X_1}(t)M_{X_2}(t) \cdots M_{X_p}(t) = \frac{1}{(1 - 2t)^{m/2}},$$

where $m = \sum_{i=1}^p m_i$. Because $M(t)$ is the moment generating function of $W = X_1 + X_2 + \cdots + X_p$ and also has the form of the moment generating function of a χ_m^2 random variable, it follows that $W \sim \chi_m^2$. ◆

The moment generating function also can be used to obtain the distributions of weighted sums of independent random variables. If $X_1, X_2, \ldots, X_p$ are mutually independent random variables with moment generating functions $M_{X_1}(t), \ldots, M_{X_p}(t)$, respectively, then the moment generating functions of $a_1X_1, a_2X_2, \ldots, a_pX_p$ are $M_{X_1}(a_1t), \ldots, M_{X_p}(a_pt)$, respectively (property 3 of moment generating functions; see Chapter 7, Section 3). Thus

$$M(t) = M_{X_1}(a_1t)M_{X_2}(a_2t) \cdots M_{X_p}(a_pt),$$

is the moment generating function of $W = a_1X_1 + a_2X_2 + \cdots + a_pX_p$.

EXAMPLE 3.12

If X has a normal distribution with mean μ and variance σ^2, then the moment generating function of X is (Chapter 9)

$$M_X(t) = e^{tu+(1/2)\sigma^2t^2}.$$

Consequently, if $X_1, X_2, \ldots, X_p$ are independent normally distributed random variables, with $X_i \sim \mathcal{N}(\mu_i, \sigma_i^2)$, $i = 1, \ldots, p$, then the movement generating function of $W = a_1X_1 + a_2X_2 + \cdots + a_pX_p$ is

$$\begin{aligned} M_W(t) &= M_{X_1}(a_1t)M_{X_2}(a_2t) \cdots M_{X_p}(a_pt) \\ &= [e^{(a_1t)\mu_1+(1/2)\sigma_1^2(a_1t)^2}][e^{(a_2t)\mu_2+(1/2)\sigma_2^2(a_2t)^2}] \cdots [e^{(a_pt)\mu_p+(1/2)\sigma_p^2(a_pt)^2}] \\ &= e^{t\mu+(1/2)\sigma^2t^2}, \end{aligned}$$

where $\mu = \sum_1^p a_i\mu_i$, $\sigma^2 = \sum_1^p a_i^2\sigma_i^2$. Note that $M_W(t)$ is the moment generating function of a $\mathcal{N}(\mu, \sigma^2)$ distribution. Hence $W \sim \mathcal{N}(\mu, \sigma^2)$. ◆

Note. If two random variables X_1, X_2 are not probabilistically independent, their joint moment generating function (Chapter 11)

$$M(s, t) = E[e^{sX_1+tX_2}]$$

can be used to find the distribution of the weighted sum $W = a_1X_1 + a_2X_2$. Indeed, the moment generating function of W is

$$M_W(t) = M(a_1t, a_2t),$$

as demonstrated in Chapter 11. Note that if the random variables X_1, X_2 are probabilistically independent, their joint moment generating function $M(s, t)$ is the product $M_{X_1}(s)M_{X_2}(t)$ of their marginal moment generating functions (Chapter 11), so that $M_W(t) = M_{X_1}(a_1t)M_{X_2}(a_2t)$.

4 Distributions of Sums (and Weighted Sums) in Special Cases

The following are special cases where the distributions of certain weighted sums are known.

1. **Normal distribution.** If $X_1, X_2, \ldots, X_m$ are mutually independent, and $X_i \sim \mathcal{N}(\mu_i, \sigma_i^2)$, $i = 1, 2, \ldots, m$, then $Y = \sum_{i=1}^m a_iX_i \sim \mathcal{N}(\sum_{i=1}^m a_i\mu_i, \sum_{i=1}^m a_i^2\sigma_i^2)$. In particular, if $\mu_i = \mu$, $\sigma_i^2 = \sigma^2$, $i = 1, 2, \ldots, m$, then

$$\bar{X} = \sum_{i=1}^m \frac{X_i}{m} \sim \mathcal{N}\left(\mu, \frac{\sigma^2}{m}\right).$$

2. **Binomial distribution.** If $X_1, X_2, \ldots, X_m$ are mutually independent, and if each X_i has a binomial distribution with parameters n_i and p, where p is the probability of success, then $W = \sum_{i=1}^m X_i$ also has a binomial distribution with parameters $n = \sum_{i=1}^m n_i$ and p.
3. **Poisson distribution.** If $X_1, X_2, \ldots, X_m$ are mutually independent, and if each X_i has a Poisson distribution with parameter λ_i, then $W = \sum_{i=1}^m X_i$ also has a Poisson distribution with parameter $\lambda = \sum_{i=1}^m \lambda_i$.
4. **Exponential distribution.** If $X_1, X_2, \ldots, X_m$ are mutually independent, and if each X_i has an exponential distribution with parameter θ, then $W = \sum_{i=1}^m X_i$ has a gamma distribution with parameters θ and m.

5. **Gamma distribution.** If $X_1, X_2, \ldots, X_m$ are mutually independent, and if each X_i has a gamma distribution with parameters θ and d_i, then $W = \sum_{i=1}^m X_i$ has a gamma distribution with parameters θ and $d = \sum_{i=1}^m d_i$.
6. **Cauchy distribution.** If $X_1, X_2, \ldots, X_m$ are mutually independent, and if each X_i has a Cauchy distribution with parameter, θ, then $\bar{X} = \sum_{i=1}^m X_i/m$ has a Cauchy distribution with the same parameter θ.

Case 5 has been demonstrated for the special case of the χ_n^2 distribution (a gamma distribution with $\theta = \frac{1}{2}$ and $r = n/2$) in Examples 3.9 to 3.11. Two different methods were illustrated: (i) use of the convolution formula (3.7) or (3.8) to verify the result for $m = 2$, followed by recursive arguments (Example 3.10) to extend the result to a situation where $m > 2$, and (ii) the use of moment generating functions. Because the exponential distribution is a special case of the gamma distribution (with parameters θ and $r = 1$), case 4 is a direct application of case 5. Case 1 was verified by moment generating function methods in Example 3.12; this case could also be established by recursive use of the convolution formula (3.8). Similarly, cases 2 and 3 follow either by recursive use of the convolution formula (3.7), or by use of moment generating functions.

Case 6 is more difficult. To verify case 6, we first establish a more general result for weighted sums of two independent and identically distributed Cauchy random variables: If X_1 and X_2 are independent random variables, each having a Cauchy distribution with parameter θ, then for any positive constants a_1 and a_2 with $a_1 + a_2 = 1$, $W = a_1X_1 + a_2X_2$ has a Cauchy distribution with parameter θ. It follows from this result that $\frac{1}{2}(X_1 + X_2)$ has a Cauchy distribution with parameter θ; that $\frac{1}{3}(X_1 + X_2 + X_3) = \frac{2}{3}[\frac{1}{2}(X_1 + X_2)] + \frac{1}{3}X_3$ has a Cauchy distribution with parameter θ; and in general that

$$\bar{X} = \frac{1}{m}\sum_{i=1}^{m} X_i = \frac{m-1}{m}\left(\frac{1}{m-1}\sum_{i=1}^{m-1} X_i\right) + \frac{1}{m}X_m$$

has a Cauchy distribution with parameter θ. The proof of the assertion that $a_1X_1 + a_2X_2$, where $a_1 > 0$, $a_2 > 0$, $a_1 + a_2 = 1$, has a Cauchy distribution with parameter θ involves straightforward, but complicated, algebra. The basic steps of the computation, however, can be illustrated in the special case where $\theta = 0$ and $a_1 = a_2 = \frac{1}{2}$.

EXAMPLE 4.1

Suppose that X and Y are independent random variables, where X and Y each have a Cauchy distribution with parameter $\theta = 0$. Let us find the distribution of $W = \frac{1}{2}X + \frac{1}{2}Y$. The joint probability density function of (X, Y) is

$$f(x, y) = f_X(x)f_Y(y) = \frac{1}{\pi^2(1 + x^2)(1 + y^2)}$$

for $-\infty < x, y < \infty$. Using equation (3.2), the probability density function $f_W(w)$ of W is

$$f_W(w) = \int_{-\infty}^{\infty} \frac{2\,dv}{\pi^2[1 + (2w - v)^2](1 + v^2)}.$$

Now

$$\frac{1}{[1 + (2w - v)^2](1 + v^2)} = \frac{Av + B}{(1 + v)^2} + \frac{D - Av}{1 + (2w - v)^2}.$$

where

$$A = \frac{1}{4w(w^2 + 1)}, \qquad B = Aw, \qquad D = 3Aw.$$

Thus

$$\begin{aligned} f_W(w) &= \frac{2}{\pi^2}\int_{-\infty}^{\infty} \frac{Av + B}{1 + v^2}\,dv + \int_{-\infty}^{\infty} \frac{-Av + D}{1 + (2w - v)^2}\,dv \\ &= \frac{2}{\pi^2}\int_{-\infty}^{\infty} \frac{Av + B}{1 + v^2}\,dv + \int_{-\infty}^{\infty} \frac{-Az - 2Aw + D}{1 + z^2}\,dz, \end{aligned}$$

where in the second integral we have made a change of variable from v to $z = v - 2w$. Combining integrals, we find that

$$\begin{aligned} f_W(w) &= \frac{2}{\pi}\int_{-\infty}^{\infty} \frac{B - 2Aw + D}{\pi(1 + z^2)}\,dz \\ &= \frac{4Aw}{\pi}\int_{-\infty}^{\infty} \frac{1}{\pi(1 + z^2)}\,dz = \frac{1}{\pi(1 + w^2)}, \end{aligned}$$

which is the probability density function of a Cauchy distribution with parameter $\theta = 0$. ◆

All of the special cases 1 to 6 listed above deal with weighted sums of independent random variables. In Chapter 12 we gave one example where the weighted sum of dependent random variables is known.

7. **Bivariate normal distribution.** If X and Y have a bivariate distribution with parameters μ_X, μ_Y, σ_X^2, σ_Y^2, and ρ, the weighted sum $aX + bY$ has a

normal distribution with mean

$$\mu_{aX+bY} = a\mu_X + b\mu_Y$$

and variance

$$\sigma^2_{aX+bY} = a^2\sigma^2_X + b^2\sigma^2_Y + 2ab\sigma_X\sigma_Y\rho.$$

A particular example of case 7 was verified in Example 3.7; there we showed that if $(X, Y) \sim \text{BVN}((0, 0), (1, 1; \rho))$, then $X + Y \sim \mathcal{N}(0, 2(1 + \rho))$. Similar arguments can be used to establish the more general result stated in case 7. However, it is considerably easier to use moment generating methods. If

$$(X, Y) \sim \text{BVN}((\mu_X, \mu_Y), (\sigma^2_X, \sigma^2_Y; \rho)),$$

it is shown in Chapter 12 that the joint moment generating function of (X, Y) is

$$M(t_1, t_2) = \exp[t_1\mu_X + t_2\mu_Y + \tfrac{1}{2}(\sigma^2_X t_1^2 + 2\sigma_X\sigma_Y\rho t_1 t_2 + \sigma^2_Y t_2^2)].$$

Thus the moment generating function of $Y = aX + bY$ is

$$\begin{aligned} M_Y(t) &= M(at, bt) \\ &= \exp[t(a\mu_X + b\mu_Y) + \tfrac{1}{2}t^2(a^2\sigma^2_X + 2ab\sigma_X\sigma_Y\rho + b^2\sigma^2_Y)], \end{aligned}$$

and we recognize this as being the moment generating function of the $\mathcal{N}(a\mu_X + b\mu_Y, a^2\sigma^2_X + 2ab\sigma_X\sigma_Y\rho + b^2\sigma^2_Y)$ distribution.

Glossary and Summary

Bivariate transformation: Functions $U = g_1(X, Y)$, $V = g_2(X, Y)$ of two jointly distributed random variables X and Y. A bivariate transformation is **one-to-one** if the equations $u = g_1(x, y)$, $v = g_2(x, y)$ have unique solutions $x = h_1(u, v)$, $y = h_2(u, v)$ for all possible values of u and v. The functions $h_1(u, v)$, $h_2(u, v)$ are said to be the **inverse functions** of $g_1(x, y)$, $g_2(x, y)$, and define the **inverse transformation** from U, V back to X, Y.

Many-to-one transformation: A function $W = g(X, Y)$ of two jointly distributed random variables X and Y. Of necessity, many (x, y)-pairs can yield the same value of $w = g(x, y)$.

Nonsingular bivariate linear transformation: A bivariate one-to-one transformation defined by linear functions $U = g_1(X, Y) = a_{11}X + a_{12}Y$,

$V = g_2(X, Y) = a_{21}X + a_{22}Y$ of X, Y for which $A = a_{11}a_{22} - a_{12}a_{21} \neq 0$. In this case, the inverse function of the transformation is

$$X = h_1(U, V) = \frac{a_{22}}{A}U - \frac{a_{12}}{A}V,$$

$$Y = h_2(U, V) = -\frac{a_{21}}{A}U + \frac{a_{11}}{A}V.$$

Jacobian of a inverse transformation: For $x = h_1(u, v)$, $y = h_2(u, v)$,

$$J(u, v) = \left(\frac{\partial h_1(u, v)}{\partial u}\right)\left(\frac{\partial h_2(u, v)}{\partial v}\right) - \left(\frac{\partial h_1(u, v)}{\partial v}\right)\left(\frac{\partial h_2(u, v)}{\partial u}\right).$$

The Jacobian of the inverse transformation of a nonsingular linear transformation $U = a_{11}X + a_{12}Y$, $V = a_{21}X + a_{22}Y$ is $(a_{11}a_{22} - a_{12}a_{21})^{-1} = A^{-1}$.

Convolution: Given probability mass functions $p_X(x)$, $p_Y(y)$, of independent discrete random variables X, Y, the function

$$p(u) = \sum_x p_X(x)p_Y(u - x) = \sum_y p_X(u - y)p_Y(y)$$

gives the probability mass function $p(u)$ of the sum $U = X + Y$. Similarly, given probability density functions $f_X(x)$, $f_Y(y)$ of two independent continuous random variables X and Y, the function

$$f(u) = \int_{-\infty}^{\infty} f_X(x)f_Y(u - x)\,dx = \int_{-\infty}^{\infty} f_X(u - y)f_Y(y)\,dy$$

gives the probability density function of the sum $U = X + Y$.

Student's t-distribution: The distribution of $T = \sqrt{n}\,X/\sqrt{Y}$, where X has a standard normal distribution, $Y \sim \chi^2_n$, and X and Y are independent. The positive integer n is the **degrees of freedom** of the t- distribution. T is a continuous random variable with probability density function

$$f(t) = \frac{\Gamma((n + 1)/2)}{\Gamma(n/2)\sqrt{\pi n}(1 + t^2/n)^{(1/2)(n+1)}}$$

***F*-distribution:** The distribution of $V = (m/n)(X/Y)$, where $X \sim \chi^2_n$, $Y \sim \chi^2_m$, and X and Y are independent. The positive constants n and m are called the numerator and denominator degrees of freedom, respectively, of the F-distribution. The F- distribution is a continuous distribution with probability density function given by equation (2.10).

Useful Facts

1. If $U = g_1(X, Y)$, $V = g_2(X, Y)$ is a one-to-one bivariate transformation of discrete random variables X, Y having joint probability mass function $p(x, y)$, the joint probability mass function of U, V is
$$p(u, v) = p(h_1(u, v), h_2(u, v)),$$
where $X = h_1(U, V)$, $Y = h_2(U, V)$ is the inverse transformation from (U, V) back to (X, Y).
2. If $U = g_1(X, Y)$, $V = g_2(X, Y)$ is a one-to-one bivariate transformation of continuous random variables X and Y having joint probability density function $f(x, y)$, the joint probability density function of U, V is
$$f(u, v) = |J(u, v)| f(h_1(u, v), h_2(u, v)),$$
where $J(u, v)$ is the Jacobian of the inverse transformation $X = h_1(U, V)$, $Y = h_2(U, V)$ from (U, V) back to (X, Y).
3. The probability mass function $p_U(u)$ of a many-to-one transformation $U = g(X, Y)$ of two discrete random variables X, Y with joint probability mass function $p(x, y)$ can be found by one of the following methods:
   (a) Finding a second function $V = g_2(X, Y)$ of (X, Y) such that the transformation from (X, Y) to (U, V) is one-to-one, finding the joint probability mass function $p(u, v)$ of U and V and then determining the marginal probability mass function $p_U(u)$ of U.
   (b) For each possible value u of U, summing the probabilities of all (x, y)-pairs that have the property that $g(x, y) = u$.
   (c) Finding the moment generating function $M_U(t)$ of U and recognizing this function as corresponding to a certain probability mass function [which then, by the uniqueness property of moment generating functions, must be equal to $p_U(u)$].
4. The probability density function $f_U(u)$ of a many-to-one transformation $U = g(X, Y)$ of two continuous variables X, Y with joint probability density function $f(x, y)$ can be found by one of the following methods:
   (a) Finding a second function $V = g_2(X, Y)$ of (X, Y) such that the transformation from (X, Y) to (U, V) is one-to-one, finding the joint probability density function $f(u, v)$ of U and V and then determining the marginal probability mass function $f_U(u)$ of U.
   (b) Finding a function $h(u; y)$ such that $g(x, y) \leq u$ if and only if $x \leq h(u; y)$ and such that $q(u; y) = \int_{-\infty}^{h(u; y)} f(x, y)\, dx$ and $\partial q(u; y)/\partial u$ are well defined and jointly continuous functions of u, y for $-\infty \leq u$, $y \leq \infty$. In this case
$$f_U(u) = \int_{-\infty}^{\infty} \frac{\partial h(u; y)}{\partial u} f(h(u; y), y)\, dy.$$

(c) Finding the moment generating function $M_U(t)$ of U and recognizing this as corresponding to a certain density function [which must then be equal to $f_U(u)$].

5. The distributions of sums and weighted sums of random variables having certain specified distributions (e.g., binomial, bivariate normal) are summarized in Section 4.

Exercises

1. Let X and Y be independent $\mathcal{N}(0, 1)$ random variables.
   (a) Find the joint probability density function of $V = X/Y$ and Y.
   (b) Find the (marginal) probability density function of V.
2. Let $X \sim \mathcal{N}(0, 1)$ and $Y \sim \chi_r^2$ be independent.
   (a) Find the joint probability density function of

$$U = \frac{X}{\sqrt{Y + X^2}}, \qquad V = Y + X^2.$$

   (b) Find the marginal probability density functions of U and V. Show that U and V are independent.
   (c) Show that U^2 has a beta distribution.
3. Let X and Y be independent random variables, where X has a chi-squared distribution with n degrees of freedom and Y has a chi-squared distribution with m degrees of freedom. Find the distribution of

$$W = \frac{X/n}{Y/m}$$

   as follows.
   (a) First, find the joint probability density function of $U = X_1/X_2$ and $V = X_1 + X_2$.
   (b) Then, find the marginal probability density function of U.
   (c) Finally, let $W = (m/n)U$ and find the probability density function of W from that of U.
   (d) What is the marginal distribution of V?
4. Let X have a negative binomial distribution (number of trials) with parameters r and p, where r is an integer. Let Y have a geometric distribution (number of trials) with parameter p. Let X and Y be independent. Find the joint probability mass function $p_{U,V}(u, v)$ of $U = X + Y$ and $V = Y$. Then find the marginal probability mass function of U.

5. Let $(X, Y) \sim \mathcal{N}((\mu_1, \mu_2), (\sigma_1^2, \sigma_2^2; \rho))$. Find the joint probability density function of

$$U = \frac{1}{\sqrt{2}}\left(\frac{X}{\sigma_1} + \frac{Y}{\sigma_2}\right), \qquad V = \frac{1}{\sqrt{2}}\left(\frac{X}{\sigma_1} - \frac{Y}{\sigma_2}\right).$$

Are U and V independent? Find the marginal distributions of U and V.

6. Let X and Y be independent random variables with X having a beta distribution with parameters r and s and Y having a gamma distribution with parameters $r + s$ and $\theta = 1$.
   (a) Find the joint probability density function of $U = XY$ and $V = (1 - X)Y$.
   (b) Find the marginal distributions of U and V, and show that U and V are independent.

7. Let X, Y be continuous random variables with joint probability density function

$$f(x, y) = \frac{\Gamma(a + b + c)}{\Gamma(a)\Gamma(b)\Gamma(c)} x^{a-1} y^{b-1} (1 - x - y)^{c-1}$$

for $0 \leq x, y, x + y \leq 1$; otherwise, $f(x, y) = 0$. Here the constants a, b, c are positive numbers. This joint distribution is known as the **bivariate Dirichlet distribution.**
   (a) Find the joint probability density function of $U = X + Y$ and $V = X/Y$.
   (b) Find the marginal distributions of U and $W = (b/a)V$. Are U and W independent?

8. Let X and Y be independent $U[0, 1]$ random variables. Let $U = \max(X, Y)$ and $V = \min(X, Y)$,
   (a) Show that

$$F_{V,U}(v, u) = \begin{cases} 1, & \text{if } 1 \leq v \leq u < \infty, \\ 2v - v^2, & \text{if } 0 \leq v \leq 1 < u < \infty, \\ 2vu - v^2, & \text{if } 0 \leq v \leq u \leq 1, \\ u^2, & \text{if } 0 \leq u \leq \min(v, 1) \\ 0, & \text{otherwise.} \end{cases}$$

   (b) Find the joint probability density function of V and $R = U - V$. Verify that V and R have a bivariate Dirichlet distribution (see Exercise 7).
   (c) Find the marginal probability density functions of V and R. Are V and R independent?

9. Let X_1, X_2 be independent random variables, each with an exponential

distribution with parameter $\theta = 1$. Let $U = \max(X_1, X_2)$, $V = \min(X_1, X_2)$.

(*a*) *Let* $Y_1 = 1 - \exp(-X_1)$, $Y_2 = 1 - \exp(-X_2)$. Show that Y_1 and Y_2 are independent $U[0, 1]$ random variables.

(b) Note that $\max(Y_1, Y_2) = 1 - \exp(-U)$ and $\min(Y_1, Y_2) = 1 - \exp(-V)$. Find the joint probability density function of U and V, and the marginal probability density functions of U and V.

10. Let X, Y be independent $U[0, 1]$ random variables. Find the probability density function of $W = XY$ in each of the following ways.

(a) By finding the joint probability density function of $W = XY$ and Y.

(b) By use of (2.8), verifying needed conditions.

(c) By showing that $U = -\log X$, $V = -\log Y$ are independent exponentially distributed random variables with parameter $\theta = 1$, verifying that $U + V$ has a gamma distribution with parameters $r = 2$ and $\theta = 1$, and then finding the probability density function of $W = \exp[-(U + V)] = XY$.

11. Let X and Y be independent random variables, each with a Cauchy distribution with parameter θ. Show that $W = aX + bY$ for general $a, b > 0$, $a + b = 1$, has a Cauchy distribution with parameter θ.

12. Show that when deriving the probability density function of the sum $W = X + Y$ of two jointly distributed random variables X, Y with joint probability density function $f(x, y)$, use of equation (2.8) yields the second integral in equation (3.8).

13. Let X and Y be independent random variables, each having a Weibull distribution with parameters $\alpha = 1$, $\beta = 2$, $\upsilon = 0$. Let $U = X^2 + Y^2$, $V = X + Y$.

(a) Find the probability density function of U, using (2.8).

(b) Find the probability distribution of U by first finding the joint probability density function of $S = X^2$ and $T = Y^2$, and then finding the probability density function of $U = S + T$.

(c) Find the probability density function of V.

14. Show that if $X_1, X_2, \ldots, X_r$ are mutually independent, and if X_i has a binomial distribution with parameters n_i and p, $i = 1, 2, \ldots, r$, then $\sum_{i=1}^{r} X_i$ has a binomial distribution with parameters $n = \sum_{i=1}^{r} n_i$ and p.

15. Show that if $X_1, X_2, \ldots, X_r$ are mutually independent, and if X_i has a Poisson distribution with parameters λ_i, $i = 1, 2, \ldots, r$, then $Y = \sum_{i=1}^{r} X_i$ has a Poisson distribution with parameter $\lambda = \sum_{i=1}^{r} \lambda_i$.

16. Show that if $X_1, X_2, \ldots, X_k$ are mutually independent random variables, and if X_i has a negative binomial distribution (number of trials) with parameters r_i and p, where r_i is an integer, $i = 1, 2, \ldots, k$, then $Y = \sum_{i=1}^{k} X_i$ has a negative binomial distribution (number of trials) with parameters $r = \sum_{i=1}^{k} r_i$ and p.

17. Let $X_1, X_2, \ldots, X_s$ be mutually probabilistically independent random variables and let X_i have a gamma distribution with parameters r_i and θ, $i = 1, 2, \ldots, s$.

(a) Show that $W = \sum_{i=1}^{s} X_i$ has a gamma distribution with parameters $r = \sum_{i=1}^{s} r_i$ and θ.

(b) Hence show that if $X_1, X_2, \cdots, X_s$ are mutually independent random variables, where X_i has an exponential distribution with parameter θ, then $W = \sum_{i=1}^{s} X_i$ has a gamma distribution with parameters s and θ.

18. Let $X_1, X_2, \ldots, X_r$ be mutually independent random variables, and let $X_i \sim U[0, 1]$, $i = 1, 2, \ldots, r$.

(a) For $r = 2$, find the probability density function of $W = X_1 + X_2$.

(b) For $r = 3$, find the probability density function of $W = X_1 + X_2 + X_3$.

19. The number of automobile accidents occurring on any given nonholiday weekend at a particular intersection has a Poisson distribution with parameter $\lambda = 2$. The number of accidents occurring on any one weekend is independent of the number of accidents occurring on all other weekends.

(a) What is the distribution of the total number of accidents that occur on weekends in a month (4 weekends) that has no holiday weekends?

(b) In a year there are 52 weekends, of which 6 are holiday weekends and 46 are nonholiday weekends. On any holiday weekend, the number of accidents that occur has a Poisson distribution with parameter $\lambda = 3.5$. What is the distribution of the total number of weekend accidents Y that occur in a year?

(c) Suppose that $X_1, X_2, \ldots, X_{52}$ are mutually independent random variables each of which has a Poisson distribution with parameter $\lambda = 113/52$. Show that $Z = X_1 + X_2 + \cdots + X_{52}$ has the same distribution as the random variable Y described in part (b). Thus show that $P\{Y \geq 120\}$ and $P\{Z \geq 120\}$ are equal.

(d) Use the result in part (c) and the Central Limit Theorem to find the probability that at least 120 automobile accidents occur on weekends at the given intersection in a year.

20. Let X and Y be independent random variables, with X having an exponential distribution with parameter θ_1 and Y having an exponential distribution with parameter θ_2. Let $W = X + Y$.

(a) Show, by finding the probability density function of W, that W does not have a gamma distribution unless $\theta_1 = \theta_2$.

(b) Find μ_W and σ_W^2.

(c) A common method for finding a distribution with known or easily calculable probabilities to approximate a given distribution, such as that of W in this problem, is to match moments. In the present case, we might look for a gamma distribution with parameters r and θ so that this distribution has the same mean and variance as W. Show that such a distribution has parameters

$$r = \frac{(\theta_1 + \theta_2)^2}{\theta_1^2 + \theta_2^2}, \qquad \theta = \frac{(\theta_1 + \theta_2)\theta_1 \theta_2}{\theta_1^2 + \theta_2^2}.$$

Show that if $\theta_1 = \theta_2$, then $r = 2$ and $\theta = \theta_1 = \theta_2$.

14

Fitting and Testing Goodness of Fit of Probability Models

A probability model for a random phenomenon may arise from a theoretical construct. The validity of this model (and thus of the theory) can then be tested by comparing probabilities based on the model with observed relative frequencies (Chapter 6). In comparing observed relative frequencies to theoretical probabilities, one does not expect perfect agreement to occur because the number of observations will usually not be large enough to force relative frequencies to stabilize. Discrepancies may exist by chance between the theoretical probabilities specified by the model and the observed relative frequencies, even if the model has been chosen correctly. Consequently, procedures are needed that can distinguish between discrepancies expected to occur by chance when the model holds, and discrepancies that indicate that the model is insufficiently accurate. Such procedures are called **goodness-of-fit tests** in the literature of statistics. One of the most frequently used goodness-of-fit tests, **Pearson's chi-squared goodness-of-fit test,** is presented in Section 1.

Often theoretical considerations lead not to a single probability model, but rather to a collection of similar probability models sharing a common functional form (shape), but differing in certain indices of location, dispersion, and so on. An example of such a collection of probability models is the collection (or *family*) of normal distributions (Chapter 9). All normal distributions have probability density functions of the same bell-like shape, but differ in their means μ and variances σ^2. The mean μ and variance σ^2 of a normal distribution are said to be the **parameters** of that distribution; specifying the values of μ and σ^2 tells us the particular normal distribution being considered. Without knowing the values of μ and σ^2, we cannot compute probabilities for a normal distribution.

Other examples of families (collections) of probability models are given in Chapters 8, 10, 11, and 12—for example, the family of binomial distributions, the family of exponential distributions, the family of bivariate normal distributions, and so on. A particular member of a family of distributions is identified by specifying values of certain parameters. These parameters are, unlike the case of the normal distribution, not always the mean or variance of the distribution. For example, the parameter θ of the exponential distribution is not the mean μ, but rather is $1/\mu$. In general, parameters are constants which, when given specified values, identify the particular distribution, and permit probabilities to be calculated. A collection of distributions of similar type or form for which each particular member of the collection is specified or identified by a particular value of a parameter (or parameters) is called a **parametric family.** Establishing that the probability distribution that describes a random variable X is one of a parametric family of distributions is a major step in modeling the random phenomenon described by X.

Once a family of distributions has been chosen, it still remains to identify that particular distribution within the family that yields probabilities for X. This can be done by estimating from data the corresponding parameter (or parameters) that specify or identify this distribution. Some commonly used methods for estimating parameters from data are briefly considered in Section 4. When using Pearson's chi-squared test to determine whether any member of a parametric family of distributions fits the observed relative frequencies, however, a special method of estimation, **minimum chi-squared estimation,** is used. This method, and the generalization of the chi-squared test of goodness of fit of a model needed when parameters must be estimated, is discussed in Section 2.

In some circumstances, investigators may agree on the parametric family of distributions that describes the probability model of a random variable X, but may disagree as to which member of that family is the correct one. Thus if there is a single parameter θ that identifies members of the family, some investigators may (on the basis of one theory) believe that $\theta = 5$ and other investigators (on the basis of a competing theory) believe that $\theta = 10$. To adjudicate these conflicting claims, data on X can be obtained. One can then use these data to test whether $\theta = 5$ or $\theta = 10$. Procedures for doing this are called **tests of hypotheses.** Such procedures are described in Section 3.

Before beginning our discussion, it should be noted that in this chapter the definition of a random sample is somewhat specialized (as compared to the general definition given in Chapter 3). In accordance with common statistical terminology, the term **random sample** will refer to mutually independent observations $X_1, X_2, \ldots, X_n$ of a random variable X.

1 The Chi-Squared Goodness-of-Fit Test

To verify that a particular probability model describes the variability of the outcomes of a random experiment, it is necessary to observe M repeated independent trials of the experiment and compare the observed relative frequencies r.f.(E_i), of various events E_i to the probabilities $P(E_i)$ given by the model to these events. If the number M of trials is large enough for the relative frequencies to stabilize (Chapter 1), and if the hypothesized probability model is correct, one expects r.f.(E_i) to be close to $P(E_i)$. When the number of trials is of moderate size, some discrepancies can be expected between the relative frequencies and corresponding probabilites obtained from the model.

For finite probability models, the events E_i used to make the comparison are the simple events, each of which contains exactly one outcome. Thus, in particular, for a random experiment yielding a discrete random variable X with possible values $1, 2, \ldots, k$ the events used are $E_i = \{X = i\}$, $i = 1, 2, \ldots, k$. For probability models having a very large, or infinite, number of possible values, such a choice is impractical. Because the number of outcomes might exceed the number M of trials, some simple events E_i could not be observed even if they had positive probability. In such cases a convenient partition $E_1, E_2, \ldots, E_k$ of the sample space (see Chapter 4) is used. For a continuous variable X, the events E_i are defined by the intervals of values used to construct the sample density function (histogram) of X.

Once a partition $E_1, E_2, \ldots, E_k$ has been chosen and data from repeated trials obtained, the observed relative frequencies, r.f.(E_i), of the events E_i can be compared to the probabilities $P(E_i)$ specified by the hypothesized model. One convenient summary of such comparisons is the **chi-squared statistic**

$$Q = M \sum_{i=1}^{k} \frac{[\text{r.f.}(E_i) - P(E_i)]^2}{P(E_i)} \tag{1.1}$$

proposed by Karl Pearson in 1900. Note that if there is perfect agreement between the observed relative frequencies and the hypothesized probabilities [that is, if r.f.$(E_i) = P(E_i)$, all i], then $Q = 0$. The larger the discrepancy between the observed relative frequencies and the hypothesized probabilities, the larger will be the value of Q. Thus large values of Q are an indication that the hypothesized probability model is not correct.

It must be remembered, however, that the observed relative frequencies are obtained from a random sample (the repeated trials), and thus are themselves random quantities. Hence even if the hypothesized probability model is true, some differences will occur by chance between the observed relative frequencies and the theoretical probabilities. A large value of Q could be due to sampling variation (chance) even when the hypothesized model is correct. If a large value of Q is observed, one needs to know how likely it is to obtain such a large value by chance under the hypothesized model before one can assert that such a value is evidence against this model.

Because Q is a function of the observed relative frequencies, it is a random variable. If M is large enough and the proposed model is correct, then Q has approximately a chi-squared distribution with $k - 1$ degrees of freedom. Experience shows that this chi-squared approximation is reasonably accurate for M as small as 50, particularly if the theoretical frequencies $T_i = M[P(E_i)]$ satisfy

$$T_i \geq 5 \text{ for all } i, \qquad i = 1, \ldots, k. \tag{1.2}$$

For this reason, investigators will often combine (form the union of) events E_i for which $T_i < 5$ in order to obtain a closer distributional approximation.

Using Pearson's approximation, one can find the probability

$$p = P\{\chi^2_{k-1} \geq Q_{\text{obs}}\} \tag{1.3}$$

that a chi-squared variable χ^2_{k-1} with $k - 1$ degrees of freedom is as large, or larger, than the observed value Q_{obs}. This probability p is called the *p-value* associated with the observed value Q_{obs}. If p is small, it is unlikely that such a large discrepancy from the model, as measured by Q_{obs}, could have occurred by chance, assuming that the hypothesized model is true. One thus has reason to doubt the model, because unlikely (low probability) events do not, by definition, usually happen. If, on the other hand, p is not small, the value Q_{obs} is not unlikely and one has no strong reason to disbelieve the model.

How does one know when a value of p is small? An analogy can be made with a fallible alarm system. Alarms sometimes are sounded when there is no emergency. If this can happen, one wants to be sure that false alarms do not happen very often. One might then specify what risk (probability) one can tolerate for false alarms. In scientific research, scientists have a low tolerance for false alarms—that is, for data that seem to conflict with an accepted and correct theory.

Thus the probability of false alarm is required to be very low in scientific research—say, 0.05, 0.01, or even 0.001. In medicine and in legal contexts, the value 0.05 has become standard. In other contexts, there is greater tolerance for false alarms, particularly if the theory or model challenged by the data is not an established one. Here one might allow probabilities of false alarm to be as high as 0.25.

Let the largest acceptable probability of a false alarm be denoted by α. In this case a formal rule for using the p-value to decide on the correctness of the hypothesized model is:

(1.4) assert that the model is incorrect if $p \leq \alpha$.

Equivalently, one can avoid calculating the p-value by using tables of the chi-squared distribution to determine percentiles of the chi-squared distribution with $k - 1$ degrees of freedom. Exhibit 1.1 displays the graph of the density of the chi-squared distribution with $k - 1$ degrees of freedom. The shaded area under this graph over the interval $[Q_{obs}, \infty)$ is the p-value corresponding to Q_{obs}. The $100(1 - \alpha)$ percentile $\chi^2_{k-1}(1 - \alpha)$ of the chi-squared distribution with $k - 1$ degrees of freedom is also shown on the horizontal axis. Note that the area under the graph over the interval $(-\infty, \chi^2_{k-1}(1 - \alpha))$ is by definition $1 - \alpha$, so that the area under the graph over the complimentary interval $[\chi^2_{k-1}(1 - \alpha), \infty)$ is α. It is easily seen from Exhibit 1.1 that the p-value (shaded area) for Q_{obs} is less than α if and only if

$$Q_{obs} \geq \chi^2_{k-1}(1 - \alpha). \tag{1.5}$$

Thus values of Q_{obs} obeying the inequality (1.5) lead to rejection of the hypothesized model.

To find $\chi^2_{k-1}(1 - \alpha)$, recall that $(\frac{1}{2})\chi^2_d$ has a gamma distribution with parameters $r = d/2$ and $\theta = 1$ (Chapter 10). Thus interpolation in Exhibit B.5 of Appendix B can be used to find the $100(1 - \alpha)$ percentile of a gamma distribution with parameters $r = (k - 1)/2$ and $\theta = 1$. Doubling this number yields $\chi^2_{k-1}(1 - \alpha)$.

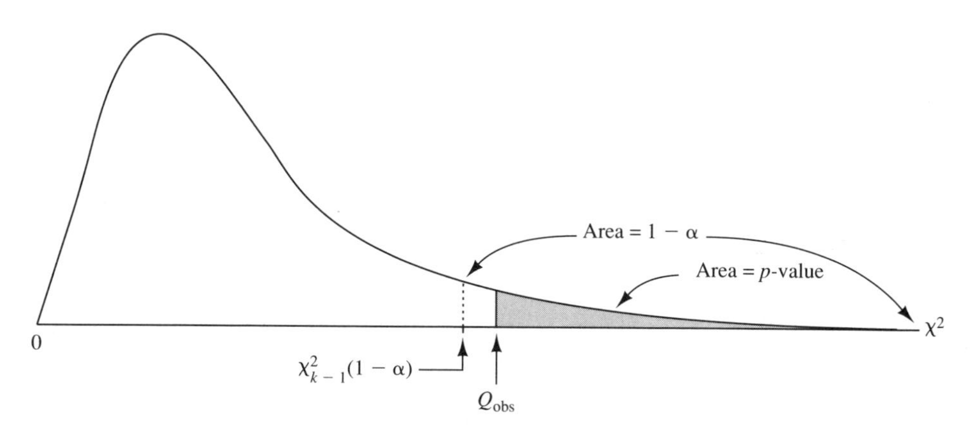

EXHIBIT 1.1 *Graph of the density function of χ^2_{k-1} showing the p-value corresponding to Q_{obs} and $\chi^2_{k-1}(1 - \alpha)$.*

EXAMPLE 1.1

Exhibit 4.9 of Chapter 5 compares relative frequencies and theoretical probabilities based on a hypothesized density function $f(x)$ for the lifetimes of $M = 417$ forty-watt incandescent lamps. This exhibit is reproduced below as Exhibit 1.2. In Exhibit 1.2, the first two intervals shown in Exhibit 4.9 of Chapter 5 have been combined into a single interval, 200–400, and the last three intervals have been similarly combined, so as to satisfy the requirement (1.2) for the accuracy of the chi-squared distributional approximation.

EXHIBIT 1.2 *Comparison of Observed Relative Frequencies and Theoretical Probabilities for the Lifetimes of Forty-Watt Incandescent Lamps*

Interval	*Relative Frequency*	*Theoretical Probability*	*Interval*	*Relative Frequency*	*Theoretical Probability*
200–400	0.0024	0.0064	900–1000	0.2182	0.1352
400–500	0.0000	0.0152	1000–1100	0.2038	0.2184
500–600	0.0072	0.0296	1100–1200	0.1918	0.1565
600–700	0.0240	0.0488	1200–1300	0.1055	0.1049
700–800	0.0504	0.0728	1300–1400	0.0552	0.0636
800–900	0.1079	0.1016	1400–1700	0.0336	0.0464

The observed value of Q is

$$\begin{aligned} Q_{\text{obs}} &= 417\left[\frac{(0.0024-0.0064)^2}{0.0064} + \frac{(0.0000-0.0152)^2}{0.0152}\right. \\ &\qquad \left. + \cdots + \frac{(0.0552-0.0636)^2}{0.0636} + \frac{(0.0336-0.0464)^2}{0.0464}\right] \\ &= 49.6538. \end{aligned}$$

Suppose that one is only willing to have a risk $\alpha = 0.01$ of false alarm. If the theoretical model is correct, Q has a $\chi^2_{12-1} = \chi^2_{11}$ distribution, because there are $k = 12$ intervals in Exhibit 1.2. The $100(1 - 0.01) = 99$th percentile of the χ^2_{12} distribution is found by first using Exhibit B.5 of Appendix B to find the 99th percentile of the gamma distribution with $r = 11/2$ and $\theta = 1$. The value of this percentile is 12.36, and doubling this value yields the value $2(12.36) = 24.72$ for $\chi^2_{11}(0.99)$. Because

$$Q_{\text{obs}} = 49.6538 \geq 24.72 = \chi^2_{11}(0.99),$$

one can assert that the hypothesized model is incorrect (insufficiently accurate) with a risk of false alarm equal to 0.01. (*Note:* The p-value corresponding to $Q_{\text{obs}} =$ is 0.000.) ◆

In the statistical literature, it is customary to compute Q by the formula

$$Q = \sum_{i=1}^{k} \frac{(O_i - T_i)^2}{T_i}, \tag{1.6}$$

where O_i is the observed frequency and $T_i = M[P(E_i)]$ is the theoretical frequency of event E_i, $i = 1, \ldots, k$.

In Exercise T.1, you are asked to show that formulas (1.1) and (1.6) for Q give the same result. Formula (1.6) is a little easier to use than formula (1.1) because the O_i's are integers. Yet another formula for Q is

$$Q = \sum_{i=1}^{k} \frac{O_i^2}{T_i} - M. \tag{1.7}$$

This formula is perhaps the most easy to use of all.

EXAMPLE 1.2

Exhibit 1.3 gives observed and theoretical frequencies for the number X of wrong connections in 267 trials (phone calls). The theoretical probability model asserts that X has a Poisson distribution with parameter $\lambda = 8.74$. As in Example, 1.1, we have combined some events to meet the requirement (1.2).

Using formula (1.6), we obtain

$$Q_{\text{obs}} = \frac{(6 - 6.68)^2}{6.68} + \frac{(11 - 10.41)^2}{10.41} + \cdots + \frac{(7 - 7.48)^2}{7.48} + \frac{(8 - 9.05)^2}{9.05}$$
$$= 7.698.$$

Because there are $k = 13$ intervals, the statistic Q has a chi-squared distribution with 12 degrees of freedom if the theoretical model is correct. The p-value of $Q_{\text{obs}} = 7.698$ is

$$p = P\{\chi_{12}^2 \geq 7.698\} = P\{Y \geq \tfrac{1}{2}(7.698)\}$$
$$= 1 - 0.1875 = 0.8125,$$

where Y has a gamma distribution with parameters $r = 6$, $\theta = 1$ (Exhibit B.5 of Appendix B). This p-value is large, so that there is no reason to doubt the proposed theoretical model. Alternatively, $\chi_{12}^2(0.50) = 11.32$ and $Q_{\text{obs}} = 7.698 < 11.32 = \chi_{12}^2(0.50)$. Thus even someone willing to tolerate a very large risk, $\alpha = 0.50$, of false alarm would find no reason to doubt the proposed model. ◆

EXHIBIT 1.3 *Observed Frequencies of x Wrong Connections in 267 Groups of Phone Calls over 5-Minute Intervals*

Number, x, of Wrong Connections	*Observed Frequency*	*Theoretical Frequency (Poisson Distribution,* $\lambda = 8.74$*)*
≤3	6	6.68
4	11	10.41
5	14	18.16
6	22	26.43
7	43	33.11
8	31	36.05
9	40	34.98
10	35	30.72
11	20	24.30
12	18	17.62
13	12	12.02
14	7	7.48
≥15	8	9.05

2 Testing Goodness of Fit of Parametric Models

How does one test whether one of a parametric family of probability models describes the variation of a random variable X? Suppose, for example, that a discrete random X is thought to have a probability model that is described by one of a family $p(x; \gamma)$ of probability mass functions parametrized by a parameter γ. If the value of γ is known, the model is specified completely, and one can use the chi-squared procedure to test whether this model fits the data. If γ is unknown, however, it must be estimated. Because the estimate $\hat{\gamma}$ used for γ is determined by the data, and is thus randomly chosen, the (large-sample) distribution of Q when the model is true may not be the chi-squared distribution described in Section 1. Consequently, naive use of this test with $\hat{\gamma}$ used in place of μ may be invalid.

Fortunately, there is a way to modify the Pearson chi-squared test to account for the fact γ must be estimated. This modification requires use of a particular method for estimating γ, called **minimum chi-squared estimation.** For every possible value of γ, the theoretical probabilities of the events E_i, $i = 1, 2, \ldots, k$, can be calculated. Thus let

$$P(E_i; \gamma), \qquad i = 1, 2, \ldots, k,$$

be these probabilities, and note that each such probability depends functionally on the value of γ. Consequently,

$$Q(\gamma) = M \sum_{i=1}^{k} \frac{[\text{r.f.}(E_i) - P(E_i; \gamma)]^2}{P(E_i; \gamma)} \tag{2.1}$$

also is a function of γ. Recall that the smaller Q is, the better the apparent fit to the data. This motivates estimating γ by the value $\hat{\gamma}$ that minimizes $Q(\gamma)$ over all possible values of γ. The rule that assigns to every data set the value $\hat{\gamma}$ of γ that minimizes $Q(\gamma)$ yields the **minimum chi-squared estimator of** γ and the minimum value

$$Q = \min_{\gamma} Q(\gamma) = Q(\hat{\gamma}) \tag{2.2}$$

is called the **minimum chi-squared test statistic.**

Although the description of the minimum chi-squared estimator $\hat{\gamma}$ and the minimum chi-squared test statistic Q in (2.2) has been given for the case where X is discrete and γ is a single parameter, an entirely similar approach yields minimum chi-squared estimators and test statistics when X is a continuous random variable or when more than one parameter $\gamma_1, \ldots, \gamma_r$ indexes the parametric family of distributions. The steps to be followed are:

1. Determine the probabilities of the partitioning events E_i as functions $P(E_i; \gamma_1, \ldots, \gamma_r)$ of the parameters $\gamma_1, \ldots, \gamma_r$, for $i = 1, 2, \ldots, k$.
2. Determine

$$Q(\gamma_1, \ldots, \gamma_r) = M \sum_{i=1}^{k} \frac{[\text{r.f.}(E_i) - P(E_i; \gamma_1, \ldots, \gamma_r)]^2}{P(E_i; \gamma_1, \ldots, \gamma_r)]}$$

as a function of $\gamma_1, \ldots, \gamma_r$ and minimize this function over all choices of $\gamma_1, \ldots, \gamma_r$. For this purpose, it is often easier to use the alternative formula [see equation (1.7)]

$$Q(\gamma_1, \ldots, \gamma_r) = \sum_{i=1}^{k} \frac{O_i^2}{MP(E_i; \gamma_1, \ldots, \gamma_r)]} - M. \tag{2.3}$$

The values $\hat{\gamma}_1, \ldots, \hat{\gamma}_r$ of $\gamma_1, \ldots, \gamma_r$ that minimize $Q(\gamma_1, \ldots, \gamma_r)$ are the minimum chi-squared estimators and $Q = Q(\hat{\gamma}_1, \ldots, \hat{\gamma}_r)$ is the minimum chi-squared test statistic.

If one model in the hypothesized family of parametric models indexed by $\gamma_1, \ldots, \gamma_r$ actually is the correct model for X, then R. A. Fisher showed that in large samples (i.e., when M is large) the statistic Q has approximately a chi-squared distribution with $k - r - 1$ degrees of freedom. Remember that k is the number of events E_i in the partition and that r is the number of parameters. Because the chi-squared distribution needs to have at least 1 degree of freedom, the number k of partitioning events used must be at least $r + 2$; we must have at least 2 more events in the partition than we have parameters in the model.

To test whether one member of the parametric family fits the data, the p-value of the observed statistic Q_{obs} can be calculated:

$$p = P\{\chi^2_{k-r-1} \geq Q_{\text{obs}}\}. \tag{2.4}$$

If p is small, there is reason to doubt that any member of the hypothesized parametric family of distributions provides an adequate model for X. Alternatively, if a prespecified probability α of false alarm is given, we can reject the hypothesized family of models either if $p \leq \alpha$, or equivalently if

$$Q \geq \chi^2_{k-r-1}(1 - \alpha). \tag{2.5}$$

Thus, apart from modifying the degrees of freedom from $k - 1$ to $k - r - 1$, the goodness-of-fit test is done in exactly the same way as the chi-squared test of a fully specified model given in Section 1. Note that if we do not reject the model, then $\hat{\gamma}_1, \hat{\gamma}_2, \ldots, \hat{\gamma}_r$ can serve as our estimates of the parameters.

A difficulty with the minimum chi-squared approach is that it rarely is possible to obtain explicit formulas for the estimates $\hat{\gamma}_1, \ldots, \hat{\gamma}_r$ or for the test statistic Q. Thus special computer software must be used. Actually, this is not a serious disadvantage given the number of excellent computer packages currently available to minimize functions, but it does prevent us from giving many practical examples of the calculation of χ^2. Instead, we must be content with illustrating how the test of fit is applied once the minimum chi-squared statistic has been calculated.

EXAMPLE 2.1

Exhibit 2.1 gives the results of a genetic experiment in which the number of "dominant" offspring in $M = 330$ litters of 5 mice were observed.

A parametric family of distributions that might fit these data is the family of binomial distributions with $n = 5$ and parameter θ, where θ is the probability of the dominant character for a single offspring. The $k = 6$ partitioning events $E_i = \{X = i - 1\}$ have probabilities

$$P(E_i;\theta) = \binom{5}{i-1}\theta^{i-1}(1-\theta)^{6-i}, \qquad i = 1, 2, 3, \ldots, 6,$$

and, using equation (2.3), we have

$$Q(\theta) = \sum_{i=1}^{6} \frac{O_i^2}{330\binom{5}{i-1}\theta^{i-1}(1-\theta)^{6-i}} - 330$$

$$= \frac{(9)^2}{330(1-\theta)^5} + \frac{(47)^2}{330(5)\theta(1-\theta)^4} + \cdots + \frac{(14)^2}{330\theta^5} - 330.$$

EXHIBIT 2.1 *Results of Observing the Number X of Mice with the Dominant Character in 330 Litters of 5 Mice*

Number X with Dominant Character	*Observed Frequency*
0	9
1	47
2	106
3	103
4	51
5	14
	330

A graph of $Q(\theta)$ is given in Exhibit 2.2. From a computer program to minimize $Q(\theta)$ over $0 < \theta < 1$ it is found that the minimum chi-squared estimate of θ is $\hat{\theta} = 0.511$ and that $Q_{obs} = Q(\hat{\theta}) = 1.187$. Because there are $k = 6$ partitioning events and $r = 1$ parameter, if a binomial distribution fits the data, Q has a chi-squared distribution with $k - r - 1 = 6 - 1 - 1 = 4$ degrees of freedom. The p-value of Q_{obs} is thus

$$\begin{aligned} p &= P\{\chi_4^2 \geq Q_{obs} = 1.187\} \\ &= P\{Y \geq \tfrac{1}{2}(1.187) = .59\} = 1 - 0.119 = 0.881, \end{aligned}$$

where Y has a standard gamma distribution with parameter $\frac{4}{2} = 2$. Because this p-value is quite large, there is no reason to doubt that a binomial distribution fits the data. ◆

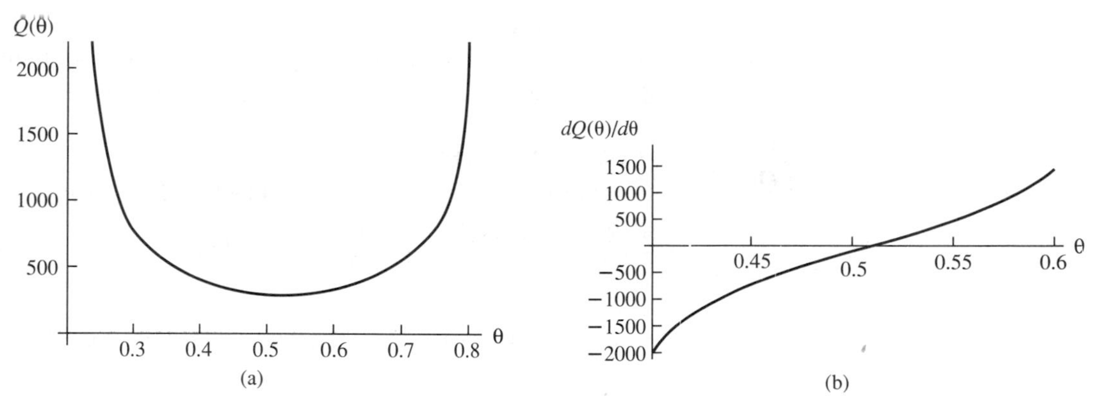

EXHIBIT 2.2 *Graphs of (a) $Q(\theta)$ and (b) $dQ(\theta)/d\theta$ as functions of θ.*

EXAMPLE 2.2

A study was made of the errors X made by medical technicians in counting cells on slides. Because the cells being counted are fairly large, however, technicians virtually never undercounted or overcounted by more than 2 cells. Thus the possible values of X were taken to be -2, -1, 0, 1, 2. (Here $X = -2$ means an undercount of 2 cells and $X = 2$ means an overcount of 2 cells; $X = 0$ means that the count was exactly correct.) Errors are commonly thought to have a symmetric distribution about 0; here this means that $P(X = -2) = P(X = 2) = \theta$, $P(X = -1) = P(X = 1) = \delta$. This is a parametric model with $r = 2$ parameters θ and δ. Note that $P\{X = 0\} = 1 - 2\theta - 2\delta$. A random sample of $M = 400$ counts by the technicians yielded the observed frequencies for the values of X given in Exhibit 2.3. Note that the observed frequencies do not have

EXHIBIT 2.3 *Frequency Distribution of Errors X Made in Counting Cells*

Error X	*Observed Frequency*	*Theoretical Probability*
−2	53	θ
−1	128	δ
0	90	$1 - 2\theta - 2\delta$
1	97	δ
2	32	θ
	400	

exactly the symmetry predicted by the parametric model, but it is possible that the deviations from symmetry may have occurred by chance.

For fixed values of θ and δ, using equation (2.3), we have

$$\begin{aligned}Q(\theta, \delta) &= \frac{(53)^2 + (32)^2}{400(\theta)} + \frac{(128)^2 + (97)^2}{400(\delta)} + \frac{(90)^2}{400(1 - 2\theta - 2\delta)} - 400 \\ &= \frac{9.5825}{\theta} + \frac{64.4825}{\delta} + \frac{20.25}{1 - 2\theta - 2\delta} - 400.\end{aligned}$$

The minimum of $Q(\theta, \delta)$ can be obtained by taking partial derivatives and solving

$$\frac{\partial}{\partial\theta} Q(\theta, \delta) = 0, \qquad \frac{\partial}{\partial\delta} Q(\theta, \delta) = 0.$$

for θ and δ. This yields the equations

$$\begin{aligned}&-\frac{9.5825}{\theta^2} + \frac{(20.25)(2)}{(1 - 2\theta - 2\delta)^2} = 0, \\ &-\frac{64.4825}{\delta^2} + \frac{(20.25)(2)}{(1 - 2\theta - 2\delta)^2} = 0.\end{aligned} \tag{2.6}$$

Substracting the second equation from the first, we find that

$$-\frac{9.5825}{\theta^2} + \frac{64.4825}{\delta^2} = 0$$

or

$$\frac{\delta}{\theta} = \left(\frac{64.4875}{9.5825}\right)^{1/2} = 2.594.$$

Substituting this result into the first equation in (2.6) yields

$$-\frac{9.5825}{\theta^2} + \frac{40.50}{(1 - 7.188\theta)^2} = 0$$

or

$$\frac{1 - 7.188\theta}{\theta} = \left(\frac{40.50}{9.5825}\right)^{1/2} = 2.0558,$$

from which

$$\hat{\theta} = \frac{1}{9.2438} = 0.108, \qquad \hat{\delta} = (2.594)(0.108) = 0.280.$$

are the minimum chi-squared estimators. Thus

$$Q_{\text{obs}} = Q(\hat{\theta}, \hat{\delta}) = \frac{9.5825}{0.108} + \frac{64.4825}{0.280} + \frac{20.25}{0.224} - 400$$
$$= 9.423.$$

If one of the hypothesized parametric family of symmetric distributions fits the data, then Q has a chi-squared distribution with $k - r - 1 = 5 - 2 - 1 = 2$ degrees of freedom. Because

$$Q_{\text{obs}} = 9.423 > \chi_2^2(0.99) = 9.20,$$

one can assert that the hypothesized family of symmetric error distributions is incorrect (fails to fit the data) with a risk of false alarm equal to 0.01. ◆

3 Statistical Tests of Hypotheses

The number X of bank failures in a year has been modeled by economists as having a Poisson distribution. The rate λ of such failures in a certain state has had the value $\lambda = 5$ for several years, but recent financial problems in the banking industry have led one economist to believe that this rate has increased to, say, $\lambda = 7$. The economist sets out to verify this fact by obtaining data. These data will be used to choose between two models for X. One model or hypothesis H_0, called the **null hypothesis,** reflects the status quo or past history. The second model H_a, called the **alternative hypothesis,** reflects the economist's belief (what the economist wants to prove). Thus the models compared are:

$$H_0 : X \text{ is Poisson distributed with parameter } \lambda = 5,$$
$$H_a : X \text{ is Poisson distributed with parameter } \lambda = 7.$$

At the end of the year, the economist will have a single observation X of the number of bank failures. The economist expects X to be close to λ; certainly, the mean of X equals λ. However, it is possible that the value of X will differ greatly from the true value of λ. Nevertheless, large values of X (say, values greater than or equal to 7, which is the mean of X under the economist's hypothesis H_a) intuitively support the truth of H_a, whereas small values of X (those less than 7) support the truth of H_0. The economist thus decides to say "H_a is true" if $X \geq 7$, and to say "H_0 is true" if $X < 7$. Because the economist does not know the truth, various scenarios are possible, as indicated in the following table:

	What Is Decided	
The Truth	*"H_0 is True"* ($X < 7$)	*"H_a is True"* ($X \geq 7$)
H_0 is true	Correct	Type I error
H_a is true	Type II error	Correct

A **type I error** is when one falsely contradicts the status quo (H_0), when this model is true. A **type II error** is when one fails to discredit the status quo when in fact H_a is true.

By setting up the rule

if $X \geq 7$, say "H_a is true,"
if $X < 7$, say "H_0 is true,"

the economist has defined a **test** of the hypotheses H_0 and H_a. This rule tells one what to do once one obtains data (here, the observation X), but a priori it is conceivable that the data may poorly reflect the true model. For example, if H_0 is true, it may still happen by chance that $X \geq 7$ (and thus one makes a type I error). On the other hand, even if H_a is true, by chance X can be less than 7, leading to a type II error. It is desirable to have a rule so that probabilities of both kinds of error,

$$\alpha = P\{\text{type I error}\} = P\{X \geq 7 \mid \lambda = 5.0\},$$
$$\beta = P\{\text{type II error}\} = P\{X < 7 \mid \lambda = 7.0\},$$

are small.[1] Using Table B.3 of Appendix B, the economist checks whether this is the case:

$$\begin{aligned}\alpha &= P\{X \geq 7 \mid \lambda = 5.0\}\\ &= p(7) + p(8) + \cdots\\ &= 1 - p(0) - p(1) - p(2) - \cdots - p(6)\\ &= 1 - 0.0067 - 0.0337 - \cdots - 0.1462\\ &= 1 - 0.7622 = 0.2378,\end{aligned}$$

[1] Here we abuse the conditional probability notation to indicate the model under which the probability is computed; thus $P\{X \geq 7 \mid \lambda = 5.0\}$ means that the probability of the count $\{X \geq 7\}$ is computed under the model where X is Poisson with parameter $\lambda = 5.00$.

$$\begin{aligned}\beta &= P\{X < 7 \mid \lambda = 7.0\} \\ &= p(0) + p(1) + \cdots + p(6) \\ &= 0.0009 + 0.0064 + \cdots + 0.1277 + 0.1490 \\ &= 0.4496.\end{aligned}$$

These probabilities are quite large, and the economist may want to do better. To do so, the economist can change the *critical region* $\{X \geq 7\}$ where "H_a is true" is decided. Making this region contain fewer values reduces α, but then β becomes larger.

Remark. Note that although α and β are probabilities of complementary events $\{X \geq 7\}$ and $\{X < 7\}$, it is *not true* that $\alpha = 1 - \beta$, because these probabilities are obtained from *different* models, so that the law of complements does not hold.

Usually, the critical region $\{X \geq c\}$ is chosen by making c large enough so that $\alpha = P\{X \geq c \mid H_0\}$ is a small number (e.g., $\alpha = 0.05$ or $\alpha = 0.10$). This protects against too frequently falsely throwing out the hypothesis (H_0) currently believed. Once c is chosen to make α small, the value of $\beta = P\{X < c \mid H_a\}$ is determined. If the value of β is too big, the data are unlikely to support the hypothesis H_a when this hypothesis is true. In this case one can try to find a better experiment (usually done by taking a larger number of observations). In the statistical literature, α is usually called the **level of significance** of the test. Researchers typically require that α be a small number (e.g., $\alpha = 0.05$).

In the example above, it was clear what information from the sample should be used to form the critical region (namely, the observed value of X). In general, it is not obvious what function

$$T(X_1, X_2, \ldots, X_n)$$

of a sample should serve to define the critical region for the test. That is, it is not clear what **test statistic** $T(X_1, X_2, \ldots, X_n)$ should define the critical region (the list of values that lead us to say "H_a is true"). If two possible test statistics T_1 and T_2 and corresponding critical regions, one based on T_1 and one on T_2, yield the same level of significance α, then T_1 is preferred to T_2 if the value of β for the test based on T_1 is less than the value of β for the test based on T_2. The quantity

$$\text{power} = 1 - \beta$$

for a test gives the probability tht the test correctly says that "H_a is true" when H_a is indeed true. Thus T_1 is preferred to T_2 if the power of T_1 is greater than the power of T_2, and we say that the test based on T_1 is **more powerful** than the test based on T_2.

EXAMPLE 3.1

A soda machine is supposed to dispense 8 ounces of soda per cup on the average. Let X be the amount of soda dispensed by the machine for a given cup. Then X is a (continuous) random variable. The company that owns the machine believes that X always has a normal distribution with variance $\sigma^2 = 0.01$ ounce^2. However, they are not certain whether or not the machine is filling the right amounts "on the average." If the machine is performing properly, the mean μ of X is 8.0, whereas if the machine is out of adjustment, $\mu \neq 8.0$. Thus two hypotheses are of interest:

$$H_0 : \mu = 8.0 \ (X \sim \mathcal{N}(8.0, 0.01)), \qquad \text{status quo,}$$
$$H_a : \mu \neq 8.0 \ (X \sim \mathcal{N}(\mu, 0.01)), \qquad \text{out of adjustment.}$$

The company plans to take a random sample $X_1, X_2, X_3, \ldots, X_9$ of the amounts of soda dispensed in $n = 9$ cups, and on the basis of this sample decide whether H_0 is true (in which case they leave the machine in operation) or H_a is true (in which case they repair the machine). If the machine is working properly (H_0 is true), the company does not want unnecessarily to send the machine for repair (have a "false alarm"). Thus they want to make the level of significance α of their test small. Indeed, they decide to let $\alpha = 0.05$, so that in only 5% of all samples taken from a properly working machine will they falsely send the machine for repair.

Intuitively, a reasonable test statistic for this problem is the sample mean

$$\hat{\mu} = \tfrac{1}{9}(X_1 + X_2 + \cdots + X_9).$$

If $\hat{\mu}$ is found to be too far away from $\mu = 8.0$, that is,

$$|\hat{\mu} - 8.0| > c, \qquad c \text{ is some constant to be chosen,} \tag{3.1}$$

there is reason to believe that H_a is true (the machine needs repair). The company thus decides on a critical region for their test of the form (3.1). If the null hypothesis H_0 is true, the sample average $\hat{\mu}$ has a normal distribution with mean 8.0 and variance (0.01)/9. Consequently,

$$\begin{aligned} \alpha &= P\{\text{type I error}\} = P\{|\hat{\mu} - 8.0| \geq c \mid H_0 \text{ true}\} \\ &= 1 - P\{-c \leq \hat{\mu} - 8.0 \leq c \mid H_0 \text{ true}\} \end{aligned}$$

$$= 1 - P\left\{-\frac{c}{\sqrt{0.01/9}} \le Z \le \frac{c}{\sqrt{0.01/9}}\right\}$$

$$= 1 - \left[2F_Z\left(\frac{c}{0.03\bar{3}}\right) - 1\right] = 2[1 - F_Z(30c)].$$

For example, if $c = 0.05$,

$$\alpha = 2[1 - F_Z(1.5)] = 2(1 - 0.93319) = 0.13362.$$

To make $\alpha = 0.05$, choose c so that $1 - F_Z(30c) = \alpha/2 = 0.025$. Equivalently, c must satisfy

$$F_Z(30c) = 1 - 0.025 = 0.975.$$

Note that $F_Z(1.96) = 0.975$. Thus

$$30c = 1.96.$$

Consequently, the company chooses

$$c = \frac{1.96}{30} = 0.0653$$

and adopts the critical region

$$\text{say ``}H_a \text{ is true'' if } |\hat{\mu} - 8.0| \ge 0.0653.$$

If the machine is working properly (H_0 is true), the probability that the company will get a sample that causes them falsely to say "H_a is true" is now predetermined to be $\alpha = 0.05$.

But what if the machine needs repair? Suppose that $\mu = 8.1$, so that the machine is overfilling the cups by 0.1 ounce on the average (a financial disaster for the company). What is the probability that the sample will correctly lead the company to send the machine for repair? That is, what is the *power*

$$\text{power} = 1 - \beta$$

of the company's test when $\mu = 8.1$? Because now $\mu = 8.1$ is assumed to be true,

$$\hat{\mu} \sim \mathcal{N}\left(8.1, \frac{0.01}{9}\right).$$

Thus

$$\begin{aligned}\text{power} = 1 - \beta &= P\{|\hat{\mu} - 8.0| \geq 0.0653 \mid \mu = 8.1\} \\ &= 1 - P\{8.0 - 0.0653 \leq \hat{\mu} \leq 8.0 + 0.0653 \mid \mu = 8.1\} \\ &= 1 - P\left\{\frac{7.9347 - 8.1}{\sqrt{(0.01/9)}} \leq Z \leq \frac{8.0653 - 8.1}{\sqrt{(0.01/9)}}\right\} \\ &= 1 - (F_Z(-1.04) - F_Z(-4.96)) \\ &= 1 - F_Z(-1.04) + 0 = F_Z(1.04) = 0.8508.\end{aligned}$$

Thus if the machine needs repair and $\mu = 8.1$, the correct decision to send the machine for repair (to say that "H_a is true") will be made 85% of the time.

The company is satisfied with these odds and proceeds with the experiment. They observe the amounts of soda $X_1, X_2, \ldots, X_9$ in 9 cups, and obtain the sample

$$8.2, 7.9, 8.0, 8.1, 8.2, 8.0, 7.9, 8.1, 8.2.$$

From this sample,

$$\hat{\mu} = \frac{8.2 + 7.9 + 8.0 + 8.1 + 8.2 + 8.0 + 7.9 + 8.1 + 8.2}{9} = 8.0667$$

and because

$$|\hat{\mu} - 8.0| = |8.0667 - 8.0| = 0.0667 > 0.0653,$$

the company must say "H_a is true" and send the machine for repair. ◆

EXAMPLE 3.2

It is commonly believed that in a human birth, the probability of a male birth is $p = 0.50$. Actually, there is evidence to suggest that $p > 0.50$ (indeed, that $p = 0.512$ for births to Caucasian parents). To prove that the probability of a male birth is larger than $\frac{1}{2}$, we randomly sample $n = 10$ births and count the number X of males born. Assuming independent births, X has a binomial distribution with parameters $n = 10$ and p, and it is the value of the parameter p that is in dispute. Because the current belief is that $p = 0.50$, the null hypothesis being tested is

$$H_0 : p = 0.50.$$

The alternative hypothesis is

$$H_a : p > 0.50.$$

A large number of male births will suggest that H_a is true, whereas a small number of male births suggests that H_0 is true. Thus we might use a critical region of the form

$$\text{say ``}H_a \text{ is true'' if } X \geq c,$$

where the constant c is chosen large enough so that the probability α of a type I error is small. Under H_0, the distribution of X is binomial with $n = 10$, $p = 0.50$; using Exhibit B.2 of Appendix B, we calculate the probability mass function:

Value, x	$p(x)$	*Value,* x	$p(x)$
0	0.00098	6	0.20503
1	0.00977	7	0.11719
2	0.04395	8	0.04395
3	0.11719	9	0.00977
4	0.20508	10	0.00098
5	0.24609		

from which we calculate

$$\alpha = P(X \geq c \mid H_0 \text{ true}) = p(c) + p(c + 1) + \cdots + p(10)$$

for various values of c:

c	$\alpha = P\{X \geq c \mid H_0\}$
5	0.62306
6	0.37697
7	0.17189
8	0.05470
9	0.01075
10	0.00098

From this table, $c = 8$ seems a good choice because the resulting probability of a type I error $\alpha = P\{X \geq 8 \mid H_0\} = 0.05470$ is close to $\alpha = 0.05$. Thus the decision rule is to

$$\text{say ``}H_a \text{ is true'' if } X \geq 8,$$
$$\text{say ``}H_0 \text{ is true'' if } X < 8.$$

This is a procedure that will satisfy other scientists that we are being "fair" to the currently held belief $H_0 : p = 0.50$, in the sense that if H_0 is true, there is a very low probability ($\alpha = 0.05470$) of observing a value X that will lead us to make the false assertion that H_a is true. Suppose, on the other hand, that H_a is true and $p = 0.512$. Then the probability of observing $X \geq 8$, and thus *correctly* asserting that H_a is true, is

$$\text{power} = 1 - \beta = P\{X \geq 8 \mid p = 0.512\},$$

where X has a binomial distribution with $n = 10$, $p = 0.512$. Hence

$$\text{power} = \binom{10}{8}(0.512)^8(0.488)^2 + \binom{10}{9}(0.512)^9(0.488) + \binom{10}{10}(0.512)^{10} = 0.0636.$$

This is a very low probability of being correct and suggests that we should not run the experiment as planned.

If more data are obtained (a larger sample size n), a critical region based on X can be constructed that has α near 0.05 and power $1 - \beta$ near (say) 0.90. This experiment would be worth running. The sample size needed to do this, however, is approximately 15,000. ◆

EXAMPLE 3.3

In cancer trials, it is sometimes reasonable to believe that the length of time X that a patient survives after being treated for cancer has an exponential distribution with parameter θ. Because

$$\mu_X = \frac{1}{\theta},$$

the lower is the value of θ, the longer is the expected lifetime for the patient (and thus the better is the treatment). The currently used treatment produces lifetimes with parameter $\theta = 0.02$. To verify that a new treatment improves survival, medical researchers give 15 patients the new treatment. On the basis of the observed lifetimes $X_1, X_2, \ldots, X_{15}$ of these patients, they have to show that the θ for the new treatment is less than 0.02 (i.e., is better). Thus, the null hypothesis H_0 is

$$H_0 : \theta = 0.02 \qquad \text{(new and current treatments do not differ)},$$

whereas the alternative hypothesis

$$H_a : \theta < 0.02 \qquad \text{(the new treatment is better)}$$

is the hypothesis that the researchers want to prove. The larger the average lifetime

$$\overline{X} = \tfrac{1}{15}(X_1 + X_2 + \cdots + X_{15})$$

of the patients treated by the new treatment, the better this treatment appears. Thus the researchers might adopt a critical region of the form

$$\text{say "}H_a \text{ is true" if } \overline{X} \geq c,$$

where c will be chosen so that

$$\alpha = P\{\text{type I error}\} = P\{\overline{X} \geq c \mid H_0 \text{ true}\}$$

is a small number (say, $\alpha = 0.05$). Note that

$$\overline{X} \geq c \quad \text{if and only if} \quad Y = 15\overline{X} \geq 15c,$$

and that $Y = X_1 + X_2 + \cdots + X_{15}$ has a gamma distribution with parameters $r = 15$ and θ. (Y is the sum of 15 independent exponential random variables with parameter θ.) When H_0 is true, $\theta = 0.02$ and

$$\alpha = P\{Y \geq 15c\} = 1 - P\{Y \leq 15c\} = 1 - I_{15}((0.02)(15)c).$$

From Exhibit B.5 of Appendix B, note that

$$I_{15}(21.5) = 0.94141,$$
$$I_{15}(22.0) = 0.95231.$$

By linear interpolation, the value of τ for which $I_{15}(\tau) = 0.95$ is

$$\tau = 21.5 + \left[\frac{0.95 - 0.94141}{0.95231 - 0.94141}\right](22.0 - 21.5) = 21.894.$$

Hence if

$$c = \frac{21.894}{(0.02)(15)} = 72.98,$$

then

$$\alpha = 1 - I_{15}((0.02)(15)(72.98)) = 1 - I_{15}(21.894) = 0.05.$$

The critical region is thus:

$$\text{say ``}H_a \text{ is true'' if } \overline{X} \geq 72.98.$$

If the researchers believe that the value of θ for their new treatment is $\theta = 0.01$, and if this is indeed the correct value of θ for the new treatment, then $Y = 15\overline{X}$ has a gamma distribution with parameters $r = 15$ and $\theta = 0.01$, and the power of the test is

$$\begin{aligned} \text{power} &= 1 - \beta = P\{\overline{X} \geq 72.98 \mid \theta = 0.01\} \\ &= P\{Y \geq (15)(72.98) \mid \theta = 0.01\} \\ &= 1 - I_{15}((0.01)(15)(72.98)) \\ &= 1 - I_{15}(10.947) \simeq 1 - I_{15}(11) = 0.85404. \end{aligned}$$

Thus if the true θ for the new treatment is $\theta = 0.01$, one will correctly say that "H_a is true" in 85.4% of all samples. This good risk supports running the experiment. ◆

4 *Estimation in Parametric Families*

As noted in the introduction to this chapter, it is often the case that theory and past experience lead to the assertion that the probability model describing a random experiment is a member of a parametric family of distributions. Individual members of such a family are distinguished by the value(s) of one or more parameters. For the sake of simplicity, discussion is confined here mostly to the case where a single parameter γ indexes family members.

The advantage of assuming that one of a parametric family of distributions (or probability models) describes the variation of outcomes of a random experiment is that attention can be concentrated on identifying the correct value of the parameter γ (a single number), rather than having to model probabilities for each of several events. This not only simplifies the modeling problem, it also means that smaller numbers of repeated trials are needed to identify or verify the correct model.

Typically, attention will be on a random variable X. If X is a discrete random variable, the probability mass function of X is denoted by $p(x; \gamma)$ to show the dependence of this mass function on the value of γ. Thus

$$p(x; \gamma) = P(X = x)$$

when γ is the value of the parameter that identifies the distribution of X.

EXAMPLE 4.1

If X has a geometric distribution with parameter θ, the mass function of X is

$$p(x; \theta) = \begin{cases} \theta(1-\theta)^{x-1}, & \text{if } x = 1, 2, \ldots, \\ 0, & \text{otherwise.} \end{cases}$$ ◆

EXAMPLE 4.2

If X has a Poisson distribution with parameter λ, the mass function of X is

$$p(x; \lambda) = \begin{cases} \dfrac{\lambda^x e^{-\lambda}}{x!}, & \text{if } x = 0, 1, 2, \ldots, \\ 0, & \text{otherwise.} \end{cases}$$ ◆

Similarly, if X has a continuous distribution that belongs to a family of continuous distributions parametrized by a parameter γ, the density function of X is denoted by $f(x; \gamma)$.

EXAMPLE 4.3

If X has an exponential distribution with parameter θ, the density function of X is

$$f(x; \theta) = \begin{cases} \theta e^{-\theta x}, & \text{if } x \geq 0, \\ 0, & \text{if } x < 0. \end{cases}$$ ◆

EXAMPLE 4.4

If X has a gamma distribution with parameters r and θ, the density function of X is

$$f(x; r, \theta) = \begin{cases} \dfrac{\theta^r x^{r-1} e^{-\theta x}}{\Gamma(r)}, & \text{if } x \geq 0, \\ 0, & \text{if } x < 0. \end{cases}$$

Note that here there are two parameters r and θ, but the notational principle is the same. ◆

To determine the value of the unspecified parameter γ of the distribution of X, a random sample $X_1, X_2, \ldots, X_n$ of observations on X is obtained. Recall that such a random sample consists of mutually independent random variables X_i, each having the same distribution as X. Thus if X has probability mass function $p(x; \gamma)$ or density function $f(x; \gamma)$, the joint distribution of the random sample $X_1, \ldots, X_n$ is determined by the joint probability mass function

$$p(x_1, x_2, \ldots, x_n; \gamma) = \prod_{i=1}^{n} p(x_i; \gamma), \tag{4.1}$$

or the joint density function

$$f(x_1, x_2, \ldots, x_n; \gamma) = \prod_{i=1}^{n} f(x_i; \gamma), \tag{4.2}$$

respectively, where $\prod_{i=1}^{n} a_i$ represents the product of the numbers $a_1, a_2, \ldots, a_n$. In the former case, $p(x_1, x_2, \ldots, x_n; \gamma)$ gives us the probability or **likelihood** of observing $X_1 = x_1, X_2 = x_2, \ldots, X_n = x_n$ when γ is the true value of the parameter. For this reason, (4.1) or (4.2) regarded as a function of the parameter γ and of the observed sample values is called the **likelihood function,** and is denoted by $L(\gamma; x_1, \ldots, x_n)$, or more simply by $L(\gamma; \mathbf{x})$. It is the likelihood function that relates the observed sample values $x_1, x_2, \ldots, x_n$ to the unknown parameter value γ.

EXAMPLE 4.1 (continued)

The likelihood function for a sample of n observations from the geometric distribution with parameter θ is

$$L(\theta; \mathbf{x}) = \prod_{i=1}^{n} p(x_i; \theta),$$
$$= \prod_{i=1}^{n} \theta(1-\theta)^{x_i - 1} = \left(\frac{\theta}{1-\theta}\right)^n (1-\theta)^{\sum_{i=1}^{n} x_i}.$$ ◆

EXAMPLE 4.4 (continued)

The likelihood function for a sample of n observations from a gamma distribution with parameters r and θ is

$$L(r, \theta; \mathbf{x}) = \prod_{i=1}^{n} f(x_i; r, \theta) = \prod_{i=1}^{n} \frac{\theta^r x_i^{r-1} e^{-\theta x_i}}{\Gamma(r)}$$
$$= \frac{\theta^{rn}(\prod_{i=1}^{n} x_i)^{r-1} \exp(-\theta \sum_{i=1}^{n} x_i)}{[\Gamma(r)]^n}.$$ ◆

A **point estimator** of a parameter γ based on a sample $X_1, X_2, \ldots, X_n$ is a rule that assigns a value $\hat{\gamma}$ (called **the estimate** of γ) to every sample $X_1, X_2, \ldots, X_n$.

EXAMPLE 4.2 (continued)

Because the parameter λ gives the mean μ_X of the Poisson distribution, this parameter might be estimated by the sample average

$$\hat{\lambda} = \hat{\lambda}(X_1, \ldots, X_n) = \frac{1}{n} \sum_{i=1}^{n} X_i.$$

For the sample 3, 1, 4, 7, 6 of $n = 5$ observations on X, the rule (estimator) given above yields

$$\hat{\lambda} = \frac{3 + 1 + 4 + 7 + 6}{5} = 4.2$$

as the estimate of λ. If, instead, the sample 4, 5, 9, 8, 3 were obtained, the estimate of λ yielded by the estimator $\hat{\lambda}$ would be $(4 + 5 + 9 + 8 + 3)/5 = 5.8$. ♦

Method of Moments

One can construct an estimator $\hat{\gamma}(X_1, \ldots, X_n)$ of a parameter γ in any way that seems reasonable for the problem at hand. There are, however, several general methods that have proven themselves as ways of constructing good point estimators. Perhaps the oldest of these methods is the **method of moments.** In this method, the moments $E[X]$, $E[X^2]$, $E[X^3]$, and so on, of X are written as functions of the unknown parameter(s) γ. Estimates of these theoretical moments are obtained by calculating corresponding sample moments

$$\hat{E}[X] = \frac{1}{n}\sum_{i=1}^{n} X_i, \qquad \hat{E}[X^2] = \frac{1}{n}\sum_{i=1}^{n} X_i^2, \quad \ldots .$$

Starting with the first moment, enough moments are chosen so that the set of equations obtained by setting the theoretical moments equal to their sample estimates yield a solution for the unknown parameter(s) γ. For a single parameter γ, it is usually sufficient to estimate γ using just the first sample moment $\hat{E}[X]$. The estimator $\hat{\lambda}$, already suggested for the parameter λ of the Poisson distribution illustrates this method. That is, because $E[X] = \lambda$ for the Poisson distribution, the method of moments would say to solve $\lambda = E[X] = \hat{E}[X]$ for λ, yielding

$$\hat{\lambda} = \frac{1}{n}\sum_{i=1}^{n} X_i$$

as the estimator. An estimator obtained by the method of moments is called a **method of moments estimator.**

For the exponential distribution (Example 4.3), the expression for the first moment $E[X]$ as a function of the parameter θ is

$$E[X] = \frac{1}{\theta}.$$

Thus the method of moments would solve

$$\frac{1}{\theta} = \hat{E}[X] = \frac{1}{n}\sum_{i=1}^{n} X_i$$

for θ, yielding the method of moments estimator

$$\hat{\theta} = \frac{1}{(1/n)\sum_{i=1}^{n} X_i}.$$

Note that because for the exponential distribution $E[X^2] = 2/\theta^2$, another possible method of moments estimator for θ is to solve the equation

$$\frac{2}{\theta^2} = \frac{1}{n}\sum_{i=1}^{n} X_i^2.$$

for θ. Generally, however, it is recommended to use the lowest possible moment to estimate a parameter by the method of moments.

EXAMPLE 4.4 *(continued)*

If X has a gamma distribution with parameters r and θ, one needs two equations (because there are two parameters) to estimate r and θ by the method of moments. It is shown in Chapter 10, Section 2, that $E[X] = r/\theta$ and $\sigma_X^2 = E[X^2] - (E[X])^2 = r/\theta^2$. Thus the method of moments requires solving

$$E[X] = \frac{r}{\theta} = \frac{1}{n}\sum_{i=1}^{n} X_i, \qquad E[X^2] = \frac{r}{\theta^2} + \left(\frac{r}{\theta}\right)^2 = \frac{1}{n}\sum_{i=1}^{n} X_i^2$$

for r and θ. This yields

$$\hat{r} = \frac{[(1/n)\sum_{i=1}^{n} X_i]^2}{(1/n)\sum_{i=1}^{n} X_i^2 - [(1/n)\sum_{i=1}^{n} X_i]^2},$$

$$\hat{\theta} = \frac{(1/n)\sum_{i=1}^{n} X_i}{(1/n)\sum_{i=1}^{n} X_i^2 - [(1/n)\sum_{i=1}^{n} X_i]^2}.$$

as method of moments estimators for r and θ. ◆

Maximum Likelihood Method

A second general method that which usually produces good estimators is the **maximum likelihood method.** In this method we regard the likelihood $L(\gamma; x_1, x_2, \ldots, x_n)$ as a function of γ, and use the value $\hat{\gamma}(x_1, x_2, \ldots, x_n)$

of γ that maximizes this function. Such an estimator is called a **maximum likelihood estimator.** In the discrete case, the maximum likelihood estimator of γ is that value of the parameter γ that would make the observed sample most likely.

EXAMPLE 4.4 *(continued)*

If X has a gamma distribution with parameters $r = 5$ and θ, we have seen that

$$L(\theta) = L(5, \theta; x_1, \ldots, x_n) = \frac{\theta^{5n}(\prod_{i=1}^n x_i)^4 \exp(-\theta \sum_{i=1}^n x_i)}{[\Gamma(5)]^n}.$$

To maximize this function over θ, we solve the equation that results from setting the derivative of $L(\theta)$ with respect to θ equal to 0:

$$0 = \frac{d}{d\theta} L(\theta) = \frac{(\prod_{i=1}^n x_i)^4}{[\Gamma(5)]^n} \left[5n\theta^{5n-1} e^{-\theta\sum_{i=1}^n x_i} - \left(\sum_{i=1}^n x_i\right)\theta^{5n} e^{-\theta\sum_{i=1}^n x_i} \right].$$

It is easily seen that the solution occurs for

$$\hat{\theta} = \frac{5n}{\sum_{i=1}^n x_i}.$$

This is the maximum likelihood estimator of θ. Note that in this case, the method of moments estimator of θ is obtained by solving

$$E[X] = \frac{5}{\theta} = \frac{1}{n}\sum_{i=1}^n x_i,$$

for θ, and the same estimator $\hat{\theta}$ is obtained. ◆

Comparing Estimators

Suppose that the maximum likelihood and method of moments had yielded different estimators in the previous example. How does one choose between them? Note that any estimator $\hat{\gamma}(X_1, X_2, \ldots, X_n)$ of a parameter γ varies as the sample $X_1, X_2, \ldots, X_n$ varies, and hence is a random variable. Thus it will rarely be the case that $\hat{\gamma} = \gamma$ exactly. Instead, it is reasonable to require that the probability

$$P\{-\varepsilon \leq \hat{\gamma} - \gamma \leq \varepsilon\}$$

that $\hat{\gamma}$ is within $\pm\varepsilon$ of γ (for some desired accuracy ε) will be as large as possible. If two estimators $\hat{\gamma}$ and $\tilde{\gamma}$ are available, then if

$$P\{-\varepsilon \le \hat{\gamma} - \gamma \le \varepsilon\} > P\{-\varepsilon \le \tilde{\gamma} - \gamma \le \varepsilon\} \tag{4.3}$$

for all values of γ, one would prefer $\hat{\gamma}$ to $\tilde{\gamma}$ because $\hat{\gamma}$ is most likely to yield estimates of γ that are sufficiently accurate. Unfortunately, the criterion (4.3) is difficult to use, because it requires one to find the distributions of $\hat{\gamma}$ and $\tilde{\gamma}$, and this can be a difficult task.

However, the mean and variance

$$\mu_{\hat{\gamma}} = E(\hat{\gamma}(x_1, x_2, \ldots, x_n)), \qquad \sigma^2_{\hat{\gamma}} = E(\hat{\gamma} - \mu_{\hat{\gamma}})^2$$

of an estimator $\hat{\gamma}$ are often more easily obtained. The Bienaymé–Chebychev inequality can then yield rough bonds to $P\{-\varepsilon \le \hat{\gamma} - \gamma \le \varepsilon\}$. In particular, if $\mu_{\hat{\gamma}} = \gamma$ (in which case we say that $\hat{\gamma}$ is **unbiased**),

$$P\{-\varepsilon \le \hat{\gamma} - \gamma \le \varepsilon\} \ge 1 - \frac{\sigma^2_{\hat{\gamma}}}{\varepsilon^2}.$$

Hence for an unbiased estimator $\hat{\gamma}$, the smaller the variance $\sigma^2_{\hat{\gamma}}$, the more likely that $\hat{\gamma}$ will be close to γ. This leads to the following criterion for comparing unbiased estimators:

Criterion. If $\hat{\gamma}$ and $\tilde{\gamma}$ are unbiased estimates of γ, that is,

$$\mu_{\hat{\gamma}} = E(\hat{\gamma}) = \gamma, \qquad \mu_{\tilde{\gamma}} = E(\tilde{\gamma}) = \gamma \qquad \text{for all } \gamma,$$

then $\hat{\gamma}$ **is preferred to** $\tilde{\gamma}$ if

$$\sigma^2_{\hat{\gamma}} < \sigma^2_{\tilde{\gamma}} \qquad \text{for all } \gamma. \tag{4.4}$$

EXAMPLE 4.2 (continued)

If X has a Poisson distribution with parameter λ, then

$$\hat{\lambda} = \frac{1}{n}\sum_{i=1}^{n} X_i$$

is both the maximum likelihood and method of moments estimator of λ. Because

$$E\left[\frac{1}{n}\sum_{i=1}^{n} X_i\right] = \mu_X = \lambda,$$

it follows that

$$\mu_{\hat{\lambda}} = E(\hat{\lambda}) = \lambda \qquad \text{for all } \lambda,$$

so that $\hat{\lambda}$ is an unbiased estimator of λ. Also, because $\sigma_X^2 = \lambda$ for the Poisson distribution,

$$\sigma_{\hat{\lambda}}^2 = \sigma_{(1/n)\Sigma X_i}^2 = \frac{\sigma_X^2}{n} = \frac{\lambda}{n}.$$

Now consider $\tilde{\lambda} = X_n$. Because X_n and X have the same distribution,

$$\mu_{\tilde{\lambda}} = \mu_{X_n} = \mu_X = \lambda, \qquad \sigma_{\tilde{\lambda}}^2 = \sigma_{X_n}^2 = \sigma_X^2 = \lambda.$$

Thus $\tilde{\lambda}$ is also an unbiased estimator of λ. When $n > 1$,

$$\sigma_{\hat{\lambda}}^2 = \frac{\lambda}{n} < \lambda = \sigma_{\tilde{\lambda}}^2, \qquad \text{all } \lambda > 0,$$

so that $\hat{\lambda}$ is a better estimator of λ than is $\tilde{\lambda}$. ◆

It may appear from the discussion above that the property of unbiasedness is an essential property for an estimator. Although earlier statisticians thought that this was the case, modern statistical research has shown that *biased* (i.e., not unbiased) estimators can sometimes be better than unbiased estimators in terms of either the criterion (4.3), or in terms of other reasonable criteria. Nevertheless, in general, users of statistics tend to feel more comfortable with, and thus prefer, unbiased estimators.

EXAMPLE 4.5

If X has a normal distribution with parameters μ and σ^2, both the method of moments and the method of maximum likelihood yield the estimators

$$\hat{\mu} = \frac{1}{n}\sum_{i=1}^{n} X_i = \overline{X}, \qquad \hat{\sigma}^2 = \frac{1}{n}\sum_{i=1}^{n} (X_1 - \hat{\mu})^2,$$

for μ and σ^2, respectively. In Chapter 7 it is shown that $\hat{\mu} = \overline{X}$ has mean μ and variance σ^2/n. Hence $\hat{\mu} = \overline{X}$ is an unbiased estimator of μ. On the other hand, $\hat{\sigma}^2$ is a biased estimator of σ^2. To see this, note that by the linearity property of expected values,

$$E(\hat{\sigma}^2) = E\left[\frac{1}{n}\sum_{i=1}^{n}(X_i - \hat{\mu})^2\right] = E\left[\frac{1}{n}\sum_{i=1}^{n}(X_i - \mu + \mu - \hat{\mu})^2\right]$$

$$= E\left[\frac{1}{n}\sum_{i=1}^{n}(X_i - \mu)^2 + 2(\mu - \hat{\mu})\frac{\sum_{i=1}^{n}(X_i - \mu)}{n} + (\hat{\mu} - \mu)^2\right]$$

$$= \frac{1}{n}\sum_{i=1}^{n}E(X_i - \mu)^2 - 2E(\hat{\mu} - \mu)^2 + E(\hat{\mu} - \mu)^2$$

$$= \frac{1}{n}\sum_{i=1}^{n}\sigma^2 - \frac{\sigma^2}{n} = \left(\frac{n-1}{n}\right)\sigma^2 \neq \sigma^2$$

because $(1/n)\sum_{i=1}^{n}(X_i - \mu) = \hat{\mu} - \mu$ and

$$E(\hat{\mu} - \mu)^2 = \sigma_{\hat{\mu}}^2 = \frac{\sigma^2}{n}.$$

Because statistical practitioners prefer to use unbiased estimators, they typically use

$$s^2 = \frac{n}{n-1}\hat{\sigma}^2 = \frac{1}{n-1}\sum_{i=1}^{n}(X_i - \hat{\mu})^2$$

in place of $\hat{\sigma}^2$ as an estimator of σ^2. That s^2 is an unbiased estimator of σ^2 follows from

$$E(s^2) = E\left(\frac{n}{n-1}\hat{\sigma}^2\right) = \frac{n}{n-1}E(\hat{\sigma}^2) = \frac{n}{n-1}\left(\frac{n-1}{n}\right)\sigma^2 = \sigma^2.$$

Note that the fact that $\hat{\mu} = \overline{X}$ and s^2 are unbiased estimators of $\mu_X = \mu$ and $\sigma_X^2 = \sigma^2$, respectively, is true whether or not X is normally distributed. ♦

Methods of Estimation and the Pearson Chi-Squared Test of Fit

Our discussion of methods of estimation has omitted the minimum chi-squared method (Section 2). The reason for this apparent oversight is that the minimum chi-squared estimators do not, in contrast to method of moments and maximum likelihood estimators, make use of the original data in ungrouped form (the raw data), but instead are based on relative frequencies of grouped data. Intuitively, the use of grouped data loses some information obtained in the raw data. Thus we would expect estimators obtained from grouped data not to have the accuracy of estimators based on the raw data. In general, this is the case. Consequently, when we are certain that a parametric family of distributions describes the data, we prefer to use estimators obtained from the raw data.

When we are uncertain about the fit of the parametric family, however, the minimum chi-squared estimators are useful for applying Pearson's chi-squared test of goodness of fit. Estimators based on the raw data use information not used by the test statistic Q, which is based only on grouped data. Consequently, using such estimators in place of the minimum chi-squared estimators to compute the chi-squared test statistic can change the distribution of this statistic, and thus invalidate p-values computed using the assumed χ^2_{k-r-1} distribution. For this reason *it is not advisable to use the method of moments or maximum likelihood estimators based on raw data to estimate the parameters when testing fit of a parametric family using the Pearson chi-squared test.*

Notes and References

The present chapter provides only a brief introduction to statistical methods for testing fit of probability models to data, and to concepts and methods used in estimating and testing hypotheses about the parameters of parametric families of distributions. We have concentrated here on only the most widely used test of fit, the Pearson chi-squared test. There are numerous other tests of fit (e.g., the Kolmogorov–Smirnov test based on the empirical cumulative distribution function). Each of these tests has advantages over the Pearson chi-squared test in certain contexts. The most important advantages of the Pearson chi-squared test are its adaptivity to a wide range of data types and models, and its simplicity of use. Readers interested in learning about other goodness-of-fit tests can consult one of the following references: Shapiro (1990), Rayner and Best (1989), and Conover (1980).

There are numerous textbooks that deal with concepts and methods for solving problems of statistical inference—far too many to list here. In this chapter we have confined our discussion to basic concepts and methods that are frequently used in statistical applications. Emphasis has been given to properties of methods of inference that can be expressed in terms of relative frequencies or averages over many repetitions of a type of problem. Thus, in hypothesis testing, probabilities of errors (type I or type II) or of correct inference (power) have been presented, whereas for estimation the focus has been on the means and variances of estimators. These **frequentist properties** reflect performance of a method over many uses rather than its success or failure in only one application.

In contrast, other approaches to statistical inference are concerned with the contribution of a particular data set to our belief about the truth of a certain hypothesis or the value of a parameter. In the **Bayesian** approach to statistical inference, one's strength of belief in various alternative hypotheses or parameter values is modeled by one's subjective or personal probabilities for such alternatives. As mentioned in Chapter 2, such subjective probabilities obey the axioms of probability theory, and thus mathematically can be treated on the same basis as probabilities that model the relative frequencies of the various possible outcomes for any data that may have been obtained. If one has a parametric family of probability distributions for the data, say $f(x_1, \ldots, x_n|\theta)$, and a subjective probability distribution $\pi(\theta)$ for the parameter θ of this parametric family, one can formally consider the joint probability distribution of $(X_1, \ldots, X_n)$ and θ and from this distribution obtain the conditional probability distribution of θ given the data $x_1, \ldots, x_n$:

$$\pi(\theta|x_1, \ldots, x_n) = \frac{f(x_1, \ldots, x_n|\theta)\pi(\theta)}{\int_{-\infty}^{\infty} \cdots \int_{-\infty}^{\infty} f(x_1, \ldots, x_n|\theta)\pi(\theta)\, dx_1 \cdots dx_n}.$$

This conditional (or **posterior**) distribution for θ now reflects what our new subjective belief (probability) should be about θ given the data $(x_1, \ldots, x_n)$, provided that our belief is consistent with the axioms of probability. For any hypothesis H that specifies that one of a collection Θ^* of values of θ is the correct one, our posterior subjective belief in this hypothesis H is now measured by

$$P(H|x_1, \ldots, x_n) = \int_{\Theta^*} \pi(\theta|x_1, \ldots, x_n)\, d\theta.$$

If we must choose between two such hypotheses H_0 and H_a, it is reasonable to choose the hypothesis for which our posterior subjective belief is the greatest. If an estimate of θ based on the data is desired, we can use any measure of location of the posterior distribution $\pi(\theta|x_1, \ldots, x_n)$ as our estimate; for example, the posterior mean

$$E[\theta|x_1, \ldots, x_n) = \int_{-\infty}^{\infty} \theta\pi(\theta|x_1, \ldots, x_n)\, d\theta.$$

Unlike the frequentist approach, however, we base our inference not on the long-run relative frequency properties of the method by which we arrived at an inference about θ, but rather on our subjective belief about θ based on the particular data we have gathered [as measured by the posterior probability distribution $\pi(\theta|x_1, \ldots, x_n)$]. From this brief description, it should be apparent that the frequentist and Bayesian approaches to inference are based on entirely different philosophies and goals. Both approaches are useful. However, it has been the frequentist approach that has dominated statistical applications over the last century. (On the other hand, the earliest attempts at statistical inference by Laplace and other scientists of the eighteenth and nineteenth centuries tended to be closer in spirit and philosophy to the Bayesian approach.) The Bayesian approach has recently begun to be used more widely, particularly in fields such as medicine and business. Consequently, it is now necessary to be familiar with both approaches to statistical inference: frequentist and Bayesian. Regardless of which of these approaches is adopted in a given situation, however, the probability models and methods for describing probability distributions that have been presented in the preceding chapters will be needed. Further, such models and descriptions are of importance and interest as scientific models of variability in their own right, apart from any potential role they may have as a background to statistical inference.

Glossary and Summary

Goodness-of-fit test: A statistical method for determining whether a hypothesized probability model, or family of probability models, adequately describes the observed relative frequencies of the outcomes (or events) of a random experiment.

Pearson chi-squared statistic: If a partition $E_1, E_2, \ldots, E_k$ of the sample space is chosen, if $p_1, p_2, \ldots, p_k$ are the probabilities of $E_1, \ldots, E_k$ specified by the probability model being tested, and $O_1, O_2, \ldots, O_k$ are the observed frequencies of $E_1, E_2, \ldots, E_k$ in M independent trials of the experiment, the Pearson chi-squared statistic Q is given by

$$Q = \sum_{i=1}^{k} \frac{(O_i - Mp_i)^2}{Mp_i} = M \sum_{i=1}^{k} \frac{[\text{r.f.}(E_i) - p_i]^2}{p_i}$$
$$= \sum_{i=1}^{k} \frac{O_i^2}{Mp_i} - M.$$

Here $\text{r.f.}(E_i) = O_i/M$. If the probability model being tested is correct and M is large (large enough so that $T_i = Mp_i \geq 5, i = 1, \ldots, k$), then Q has approximately a chi-squared distribution with $k - 1$ degrees of freedom. Large values of Q give evidence the the probability model being tested is not correct (does not adequately fit the observed relative frequencies).

***p*-Value:** The probability, assuming that the probability model being tested is correct, that a value of the statistic used to test that model is as extreme or more extreme than the observed value of the statistic. Small values of the p-value are evidence that the probability model being tested is incorrect, in that a small p-value shows that such an extreme value of the statistic is unlikely under the tested model and thus would be surprising if the model were correct. For the Pearson chi-square statistic Q, the corresponding p-value is (approximately)

$$p\text{-value} \simeq P\{\chi^2_{k-1} \geq Q_{\text{obs}}\},$$

where Q_{obs} is the observed value of Q.

Level of significance: The largest probability of "false alarm" allowed. (A "false alarm" is to declare a probability model to be incorrect when it is actually correct; *see* Type I error.) For tests of goodness of fit of a probability model, we can specify the level of significance (probability of "false alarm") to be a value α; a common choice is to let $\alpha = 0.05$. We can then declare the probability model being tested to be incorrect if the p-value corresponding to the test statistic used to test the model is less than or equal to α. Thus the probability model being tested is declared to be incorrect if

$$p\text{-level} \leq \alpha$$

with the probability of "false alarm" for this rule being equal to the specified value of α.

Critical value: For a specified level of significance α, the $100(1 - \alpha)$th percentile of the distribution of the test statistic used to test the probability model when the model is actually correct. For the Pearson chi-squared statistic, this critical value is $\chi^2_{k-1}(1 - \alpha)$. If the test statistic equals or exceeds this critical value, the p-value corresponding to the test statistic is less than or equal to α, and we would state that the probability model being tested is incorrect. Thus we say that the probability model being tested by the Pearson chi-squared test statistic is incorrect at level of significance $\alpha = 0.05$ if $Q \geq \chi^2_{k-1}(0.95)$.

Parametric family of distributions: A collection of probability distributions identified or distinguished from one another by the value(s) of one or more constants, called **parameters.** Examples are the family of Bernoulli distributions (parameter p = probability of success), the family of normal distributions (parameters μ = mean, σ^2 = variance), and the family of exponential distributions (parameter θ). To compute probabilities, the values of the parameters must be specified.

Minimum chi-squared test: To test whether one of a parametric family of distributions fits the data adequately, the chi-squared statistic Q is written as a function $Q(\theta_1, \theta_2, \ldots, \theta_r)$ of the values of the parameters $\theta_1, \theta_2, \ldots, \theta_r$ of the family. The minimum of this statistic over all possible values of the parameters is then found. (The values $\hat{\theta}_1, \ldots, \hat{\theta}_r$ of $\theta_1, \ldots, \theta_r$ that achieve this minimum Q are called **minimum chi-squared estimators** of $\theta_1, \ldots, \theta_r$.) The minimum value of Q is the observed test statistic Q_{obs}, and if some member of the parametric family of distributions is correct, has approximately a χ_{k-r-1} distribution. Thus the parametric family of distribution is said to be incorrect at level of significance α if

$$p\text{-level} \simeq P\{\chi^2_{k-r-1} > Q_{\text{obs}}\} \leq \alpha$$

or equivalently, if $Q_{\text{obs}} \geq \chi^2_{k-r-1}(1 - \alpha)$.

Test of hypothesis: A formal statistical procedure that uses the data to choose between one of two competing hypotheses concerning the unknown probability model that describes the data.

Null and alternative hypotheses H_0 and H_a: Two hypotheses that specify that one or more of a class of possible probability models fits the data. For a parametric family of models, such hypotheses specify one or more values of the parameter(s) of the family of models as being correct. The null hypothesis H_0 generally reflects the "status quo" (what is currently believed) and is the opposite of the alternative hypothesis H_a that the investigator believes is correct, and wishes to prove.

Test statistic: A function of the data that is used in testing H_0 versus H_a. A good test statistic will have a distribution under H_0 markedly different from its distribution under H_a.

Critical region: The collection of values of the test statistic that leads to the conclusion that the null hypothesis H_0 is false (and thus H_a is correct).

Type I error: Saying that the null hypothesis H_0 is false when this hypothesis is correct; thus a "false alarm" for the status quo.

Type II error: Saying that the null hypothesis H_0 is correct when this hypothesis is incorrect (and H_a is correct).

Probability of type I error: The probability that the observed value of the test statistic falls in the critical region (so that the null hypothesis H_0 is rejected) computed under a probability model specified by H_0.

Level of significance: If the null hypothesis H_0 specifies only one probability model, the level of significance α equals the probability of type I error. If H_0 specifies more than one model, the level of significance α is the largest probability of type I error over all models specified by H_0. The level of significance is usually required to be a small number (e.g., $\alpha = 0.05$).

Probability of type II error: The probability that the observed value of the test statistic does not fall in the critical region (so that H_0 is not rejected) computed under one of the probability models specified by the alternative hypothesis H_a. The probability β of type II error thus depends on which of these alternative models is used.

Power: The probability that the observed value of the test statistic falls in the critical region (so that H_0 is, correctly, rejected) computed under one of the probability models specified by the alternative hypothesis H_a.

Thus

$$\text{power} = 1 - \beta = 1 - \text{probability of type II error}.$$

Likelihood function: For a parametric family of distributions, the function that gives the probability or probability density of the observed data $X_1, \ldots, X_n$ as a function of the parameters $\theta_1, \ldots, \theta_r$ of this family.

Point estimator: A function of the data that provides an **estimate** of a parameter θ.

Method of moments estimator: A point estimator of a parameter (or parameters) obtained by setting sample moments $M_r = (1/n) \sum_{i=1}^{n} X_i^r$ equal to corresponding population moments $E_\theta(X^r)$, expressed as a function of the parameters θ, and solving the resulting equation(s) for θ.

Maximum likelihood estimator: The value of a parameter that maximizes the likelihood function. This value serves as a point estimator for the parameter.

Unbiased and biased estimators: A point estimator $\hat{\theta}$ of a parameter θ is said to be **unbiased** for θ if the population mean $\mu_{\hat{\theta}}$ of $\hat{\theta}$ equals θ for all possible values of θ. Otherwise, the estimator is said to be **biased.**

Frequentist approach to inference: Evaluates methods of statistical inference in terms of relative frequencies of good (or bad) performance over many repetitions of a type of inference problem.

Bayesian approach to inference: Models a person's strength of belief in various alternative models for the data in terms of subjective or personal probabilities. Because such subjective probabilities obey the same axioms as do frequentist probabilities for the data, they can be formally (mathematically) combined with these frequentist probabilities to yield joint probabilities for the models and the data.

Posterior probability distribution: The formal conditional probability distribution for models (or corresponding parameter values θ) given the observed values of the data. This conditional probability reflects the person's new strength of belief in the various models under consideration in light of the data that have been observed.

Exercises

1. Exhibit 3.4 of Chapter 5 compared observed relative frequencies $\hat{p}(x)$ for the values x of the number of letters in a random word from *Vanity Fair* to hypothesized theoretical probabilities

$$p_*(x) = (0.17307) \exp\left[-\frac{1}{2}\frac{(\log x - 1.224)^2}{0.588}\right]$$

for $x = 1, 2, \ldots$. This comparison is shown in Exhibit E.1.

EXHIBIT E.1 *Relative Frequencies of Number of Letters in a Random Word from Vanity Fair Compared to Hypothesized Theoretical Probabilities*

x	Relative Frequency $\hat{p}(x)$	Theoretical Probability $p_*(x)$	x	Relative Frequency $\hat{p}(x)$	Theoretical Probability $p_*(x)$
1	0.0290	0.0198	8	0.0500	0.0601
2	0.1575	0.1151	9	0.0315	0.0440
3	0.2400	0.1692	10	0.0215	0.0322
4	0.1755	0.1666	11	0.0080	0.0236
5	0.1220	0.1396	12	0.0075	0.0174
6	0.0770	0.1086	13	0.0020	0.0128
7	0.0760	0.0815	≥14	0.0025	0.0096

The relative frequencies are obtained from a sample of $M = 2000$ words.

(a) Compute the value of the chi-square statistic Q.

(b) Find the p-value associated with the observed value of Q.

(c) Which of the following is the best description of the p-value obtained in part (b)?

(i) The p-value is the probability that the theoretical model is correct.

(ii) The p-value is the probability that one would observe the value of Q obtained in part (a) when the theoretical model is correct.

(iii) The p-value is the probability that one would observe a value of Q as large, or larger, than the value found in part (a) when the theoretical model is correct.

(d) Suppose that the largest acceptable probability of false alarm is $\alpha = 0.05$. Is there sufficient evidence in the data to show that the theoretical probability model is incorrect? Show your reasoning.

2. Exercise 11 in Chapter 5 suggest that the probability mass function

$$p_{**}(x) = \frac{x^2 e^{-x}}{6}, \qquad x = 1, 2, 3, \ldots,$$

provides a better fit to the distribution of the number X of letters in a random word from *Vanity Fair* than does $p_*(x)$.

(a) Compute the theoretical probabilities for the values $x = 1, 2, \ldots,$ of X obtained from the model $p_{**}(x)$ and use these probabilities and the relative frequencies $\hat{p}(x)$ given in Exercise 1 to calculate the value of Q.

(b) Using a probability of false alarm of $\alpha = 0.05$, is there sufficient evidence in the data to indicate that the theoretical probability mass function $p_{**}(x)$ is incorrect?

3. A local radio station has a call-in game. Each caller selects one number from among the numbers 1, 2, 3, 4, and 5. The numbers correspond to the names of the station's advertisers. A wheel with these numbers on it is spun. When the wheel stops, a pointer will point at one of the numbers on the wheel. If this is the number the caller selected, he or she receives a prize from the advertiser corresponding to that number. In theory, each of the numbers 1 to 5 has an equal probability of occurring when the wheel is spun. A frequent listener has collected data on $M = 100$ spins of the wheel (Exhibit E.2) that leads her to doubt this theory.

EXHIBIT E.2 *Observed Frequencies in 100 Spins of a Wheel*

Number, x, Chosen	*Observed Frequency*
1	26
2	15
3	20
4	17
5	22
	$M = 100$

Use the chi-squared goodness-of-fit test to test the null hypothesis that the numbers 1 to 5 are equally likely to be chosen. Use $\alpha = 0.10$.

4. Exhibit E.3 gives observed frequencies for the values $x = 0, 1, 2, 3, 4$ of the number X of blue-eyed kittens in litters of four from pairs of non-blue-eyed Siamese cats known to each have one blue-eyed parent. If blue eyes is a recessive gene, genetic theory implies that the probability that a kitten will have blue eyes is $\frac{1}{4} = 0.25$ and, assuming independence of births, that X has a binomial distribution with parameters $n = 4$, $p = 0.25$.
   (a) Find the observed relative frequencies and theoretical probabilities $p(x)$ based on the binomial ($n = 4$, $p = 0.25$) model, for $x = 0, 1, 2, 3, 4$. Compute the value of Q.
   (b) Find the 0.95 quantile of the distribution of Q, assuming that the theoretical model is correct. What degrees of freedom does the chi-squared distribution have?
   (c) For a false-alarm probability of $\alpha = 0.05$, is there sufficient evidence in the data to show that blue-eyes is not a recessive gene (the genetic theory is incorrect)?

EXHIBIT E.3 *Observed Frequencies of the Number X of Blue-Eyed Kittens in M = 256 Litters of Four Siamese Kittens*

x	*Observed Frequency*
0	69
1	113
2	63
3	10
4	1
	256

5. In Exercise 25 of Chapter 9, a normal distribution with mean $\mu = 43$ and variance $\sigma^2 = 100$ was fit to data on $M = 370$ advanced placement composition scores X. After grouping intervals at the end to make theoretical frequencies acceptably large, the observed and theoretical frequencies in Exhibit E.4 were obtained.

EXHIBIT E.4 *Comparison of Observed and Theoretical Frequencies for 370 Advanced Placement Scores*

Scores x	*Interval*	*Observed Frequency*	*Theoretical Frequency*
≥68	[67.5, ∞)	2	2.64
64–67	[63.5, 67.5)	6	4.82
60–63	[59.5, 63.5)	13	10.84
56–59	[55.5, 59.5)	21	20.79
52–55	[51.5, 55.5)	35	34.04
48–51	[47.5, 51.5)	41	47.62
44–47	[43.5, 47.5)	58	56.87
40–43	[39.5, 43.5)	63	58.00
36–39	[35.5, 39.5)	46	50.52
32–35	[31.5, 35.5)	34	37.58
28–31	[27.5, 31.5)	28	23.87
24–27	[23.5, 27.5)	17	12.94
≤23	(−∞, 23.5)	6	9.47
		370	370.00

(a) Calculate Q for this table.
(b) Find the critical value $\chi_d^2\,(0.80)$ for a test of the theoretical model with false-alarm probability $\alpha = 0.20$. (How many degrees of free-

dom d does the test statistic distribution have when the theoretical model is correct?) Is there any statistical reason to doubt that the $\mathcal{N}(43,100)$ distribution provides a good fit to the data?

(c) The values of the mean μ and variance σ^2 of the theoretical normal distribution fit to the data were actually estimated from the data. If these estimates were minimum chi-squared estimators, find the correct degrees of freedom d and critical value $\chi_d^2(0.80)$ for the test of the model having false-alarm probability $\alpha = 0.20$. Is there any statistical reason to doubt that some normal distribution (note the change in hypothesis) provides a good fit to the data?

6. (a) How many degrees of freedom d does the large-sample distribution of Q have, assuming that some member of the given parametric family of models is correct, when:

(i) Relative frequencies for values $x = 0, 1, \ldots, 6$ are fit by a binomial distribution with $n = 6$ and unknown parameter p, $0 \leq p \leq 1$?

(ii) Relative frequencies for the 10 intervals [0.00–0.10), [0.10, 0.20), . . . , [0.90–1.00] are fit by a beta distribution with unknown parameters r and s ($r > 0$, $s > 0$)?

(iii) Relative frequencies for the intervals [0, 2.00), [2.00, 4.00), [4.00, 6.00), [6.00, 8.00), [8.00, 10.00), [10.00, 12.00), and [12.00, ∞) are fit by an exponential distribution with unknown parameter $\theta > 0$?

In each case, assume that the unknown parameters of the fitted model are estimated by minimizing Q (minimum chi-squared estimators).

(b) In each of cases (i) to (iii) in part (a), find the p-value corresponding to an observed minimum chi-squared statistic $Q_{obs} = 10.16$.

7. A manufacturer offers you a lot of $N = 100$ radios. The manufacturer claims that at most 4 of the radios in the lot are defective. You can tolerate 4 bad radios, because the manufacturer is giving you a very good deal, but if 10 or more radios in the lot are defective, customer returns will eat up your profits. The manufacturer agrees to let you inspect a random sample of $n = 8$ radios for defects. (Assume that you have a procedure which is certain to spot any defects in a radio if they exist.) Let X be the number of defective ratios found in your sample. Let p be the true proportion of defective radios in the lot.

(a) What is the name of the exact distribution of X?

(b) Let H_0 be the hypothesis that the manufacturer's claim is correct. State this hypothesis in terms of p, the proportion of radios in the lot that are defective.

(c) Suppose that you decide to reject H_0 (and thus not buy the lot) if you observe 2 or more defective radios in your sample of 8. Find the largest probability $P\{X \geq 2 | H_0 \text{ true}\}$ of type I error for your procedure. (*Note:* As p increases, the probability that $X \geq 2$ increases.)

(d) If 10 of the 100 radios are actually defective (so that you do not want to buy the lot), what is the probability of type II error for the procedure which rejects H_0 when $X \geq 2$? If you are being sold a bad lot, is this procedure likely to detect this fact?

8. A seismological experiment involves setting off a large explosion underground and measuring the times $X_1, X_2, \ldots, X_{10}$ between the first 10 shock waves. These times X_i are assumed to be independent exponential random variables with parameter θ. If the currently accepted theory of the structure of the earth's surface is true, $\theta = 100$. If the experimenter's new theory is true, then $\theta = 25$. Thus this is a hypothesis-testing problem with

$$H_0\colon \theta = 100, \qquad H_a\colon \theta = 25.$$

The experimenter decides to use a critical region (for rejecting H_0) of the form

$$Y = \sum_{i=1}^{10} X_i \geq 0.16.$$

(a) Find α, the probability of type I error, for this critical region. (*Hint:* What is the name of the exact distribution of Y when H_0 is true? See Chapter 10.)

(b) Find β, the probability of type II error, for the given critical region.

9. A cigarette manufacturer asserts that "on the average" its new cigarette "brand B" contains 1 milligram or less of tar per cigarette. The Office of the Surgeon General doubts that claim and decides to test it. They purchase 400 packs of cigarettes of brand B at random, and take one cigarette from each pack to test. Thus they have a sample $X_1, X_2, \ldots, X_{400}$ of independent observations on X = the amount of tar in one cigarette of brand B. Let μ be the population mean of X.

(a) In terms of μ, what is the null hypothesis H_0 being tested?

(b) The Surgeon General's office will calculate $\overline{X} = \frac{1}{400}\sum_{i=1}^{400} X_i$ and will reject H_0 when $\overline{X} > c$, where c is a constant chosen to make the probability α of type I error equal to 0.01 ($\alpha = 0.01$). The variance of the amount X of tar in brand B cigarettes is known to be $\sigma_X^2 = 0.36$. Use the Central Limit Theorem to find c.

(c) When the study was actually run, it was found that $\overline{X} = 1.13$ mg. Was this sufficient evidence to dispute the manufacturer's claim? Explain your answer.

10. It is claimed by a cereal manufacturer that their 16-ounce boxes of cereal have a mean dry weight of $\mu = 16.0$ ounces and a variance of $\sigma^2 = 4$. A consumers' organization believes that the true mean dry weight is less than 16.0 ounces (i.e., $\mu < 16.0$), and buys a random sample of $n = 100$ boxes of the manufacturer's cereal to test the claim.

(a) What is the null hypothesis being tested? What is the alternative hypothesis?

(b) If the sample average weight of the 100 boxes of cereal is $\bar{x} = 15.7$ ounces, is there sufficient evidence to reject the manufacturer's claim with a probability of type I error of $\alpha = 0.05$? (Use the Central Limit Theorem, and show explicitly what values of $\bar{x}$ would lead to rejection of the manufacturer's claim.)

11. The number X of new stars that are formed in a certain galaxy in a given year is a random variable that is assumed to have a Poisson distribution with parameter λ.

(a) If $\lambda = 2$, find the probability that 5 or more new stars will be formed in that galaxy this year.

(b) Actually, λ is unknown, as we wish to estimate the value of λ from a sample. We observe the galaxy for 8 years, and each year count the number X_i of new stars formed, $i = 1, 2, \ldots, 8$. We assume that $X_1, X_2, \ldots, X_8$ are independent, and that each X_i has a Poisson distribution with parameter λ. A statistician suggests that we use

$$\hat{\lambda} = \bar{X} = \tfrac{1}{8}(X_1 + X_2 + \cdots + X_8)$$

to estimate λ. Is $\hat{\lambda}$ an unbiased estimator of λ? Why, or why not?

(c) If $\lambda = 2$, what is the probability that $\hat{\lambda}$ gives exactly the right answer $\hat{\lambda} = \lambda$? (*Hint:* $P\{\hat{\lambda} = \lambda\} = P\{X_1 + X_2 + \cdots + X_8 = 8\lambda = 16\}$. Now, what is the distribution of $X_1 + X_2 + \cdots + X_8$?)

12. A newspaper at a state university wants to estimate the true proportion p of faculty who favor giving students a voting seat on the faculty senate. A random sample (without replacement) of $n = 50$ faculty members is selected from among the $N = 2110$ faculty at the university. Let X be the number of faculty in this sample who are in favor of a voting seat for students.

(a) Name the *exact* distribution of X and give formulas for the mean and variance of X in terms of the unknown value of p.

(b) The estimator of p to be used by the newspaper is $\hat{p} = X/81$. Show that $\hat{p}$ is an unbiased estimator of p and find the exact variance $\sigma^2_{\hat{p}}$ of $\hat{p}$ as a function of p.

(c) Show that $\sigma^2_{\hat{p}}$ is largest when $p = \frac{1}{2}$ and smallest when $p = 0$ or $p = 1$.

(d) Use the Bienaymé–Chebychev inequality to find a lower bound to the probability

$$P\{-0.05 \le \hat{p} - p \le 0.05\} = P\{|\hat{p} - p| \le 0.05\}$$

that $\hat{p}$ will be within ± 0.05 of the true value of p.

Theoretical Exercises

T.1. Show that equations (1.1) and (1.6) for Q give the same result and that $Q = \sum_{i=1}^{k} (O_i^2/T_i) - M$ as claimed in equation (1.7).

T.2. In each of the following cases, you are to find an estimator of the unknown parameter by the method of maximum likelihood. You are then to determine whether or not this estimator is unbiased.

(a) You have a random sample of 10 observations of X, where X has a Bernoulli distribution with parameter p (Chapter 6, Section 1).

(b) You have a random sample of 10 observations of X, where X has an exponential distribution with parameter θ.

(c) You have a random sample of 10 observations of X, where X has a normal distribution with unknown mean μ but known variance $\sigma^2 = 100$.

(d) You have a random sample of 10 observations of X, where X has a normal distribution with known mean $\mu = 0$, but unknown variance σ^2.

T.3. A random variable X has a probability mass function

$$p(x;\theta) = \begin{cases} \frac{1}{4}(1-\theta), & \text{if } x = 0, 1, 2, 3, \\ \theta, & \text{if } x = 4, \\ 0, & \text{otherwise,} \end{cases}$$

where θ, $0 \le \theta \le 1$, is an unknown parameter. We wish to estimate θ based on a sample of $n = 50$ independent observations $X_1, X_2, \ldots, X_{50}$ from the population of X.

(a) Let

$$Y = \text{number of } X_i\text{'s that are equal to 4.}$$

What is the name of the exact distribution of Y? Find the mean and variance of $\hat{\theta} = Y/50$ as a function of θ. Is $\hat{\theta}$ is an unbiased estimator of θ?

(b) Let $\overline{X} = (X_1 + X_2 + \cdots + X_{50})/50$. Show that $\overline{X}$ is a biased estimator of θ. Find numbers a and b such that $\tilde{\theta} = a\overline{X} + b$ is an unbiased estimator of θ.

(c) Find the variance of $\tilde{\theta}$ and compare this variance to the variance of $\hat{\theta}$. Is one of these estimators uniformly better than the other for all values of θ? If not, for what values of θ is each estimator best? (Here, by "best" we mean having the smallest variance.)

T.4. Look at Exercise 3 again and note that the test of goodness of fit used there made no use of the fact that callers-in to the radio program selected numbers. The same test could have been applied if the advertisers' names had been the outcomes. In fact, the chi-squared test can be used to test the fit of probability models for any kind of outcome of a random experiment. The steps for the use of the test are the same as those outlined in Section 1. If a parametric family of probability models is being tested for fit, the minimum chi-squared approach of Section 2 should be used. For example, consider multiple-choice questions that permit one of five possible responses: A, B, C, D, E. One probability model for how people answer such questions is that they guess; that is, select one response at random from among the five responses available. Call this model the *guessing model.*

(a) Use the chi-squared goodness-of-fit test to test the fit of the guessing model to the data shown in Exhibit E.5, obtained from the responses of $M = 300$ persons to a multiple-choice question. Use $\alpha = 0.05$.

EXHIBIT E.5 *Observed Frequencies of Responses to a Multiple-Choice Question*

Response	*Observed Frequency*
A	101
B	48
C	51
D	45
E	55
	300

(b) The question analyzed in part (a) concerned a date in history. The correct answer was response A. The data suggest that some of the people who answered the question knew the correct answer. Consequently, one might postulate a probability model in which a proportion θ of all people know the correct answer (and choose the correct response), whereas the remaining proportion, $1 - \theta$, of people guess. For this model, $P(A) = \theta + \frac{1}{5}(1 - \theta)$ and $P(B) = P(C) = P(D) = P(E) = \frac{1}{5}(1 - \theta)$. The parameter θ, $0 \leq \theta \leq 1$, is unknown and must be estimated. Use the minimum chi-squared test to test the fit of this model to the data in part (a). Again use $\alpha = 0.05$. What is the estimated value of θ?

T.5. The chi-squared test can also be used to test the hypothesis of independence between components of a composite experiment. For example, for each of $M = 500$ randomly selected insects, we might observe eye color (yellow, red, brown) and wing type (straight, curved, notched). Eye color and wing type are components of this composite experiment. Let p_1, p_2, p_3 be the marginal probabilities of yellow, red, and brown eye colors, respectively, and let q_1, q_2, q_3 be the marginal probabilities of straight, curved, and notched wing types. If eye color and wing type are independent, the probabilities of the outcomes of the composite experiment are products of these marginal probabilities. The theoretical probabilities of each of the $k = 9$ outcomes of the composite experiment and (in parentheses) also the observed relative frequencies are shown in Exhibit E.6. Note that $p_1 + p_2 + p_3 = 1$, $q_1 + q_2 + q_3 = 1$, so that only p_1, p_2, q_1, q_2 are free to vary (are parameters).

EXHIBIT E.6

	Wing Type		
Eye Color	Straight	Curved	Notched
Yellow	$p_1q_1(0.21)$	$p_1q_2(0.18)$	$p_1q_3(0.12)$
Red	$p_2q_1(0.11)$	$p_2q_2(0.15)$	$p_2q_3(0.09)$
Brown	$p_3q_1(0.04)$	$p_3q_2(0.07)$	$p_3q_3(0.03)$

(a) It can be shown that the minimum chi-squared estimates of the parameters p_1, p_2 are approximately the marginal relative frequencies $\hat{p}_1 = 0.51$, $\hat{p}_2 = 0.35$ of yellow and red, and that similarly $\hat{q}_1 = 0.36$, $\hat{q}_2 = 0.40$ are the marginal relative frequencies of straight and curved. Using this fact, compute the minimum chi-squared statistic Q for these data.

(b) Test the hypothesis of independence of eye color and wing type using $\alpha = 0.05$. (Note that the degrees of freedom of Q under this hypothesis is $9 - 4 - 1 = 4$.)

(c) If we test independence of two component characteristics having r and c possible outcomes, respectively, how many degrees of freedom will the minimum chi-squared statistic have?

(d) The assertion made in part (a) is generally correct for testing independence of two component characteristics; that is, the minimum chi-squared estimators of the marginal probabilities of the component outcomes are approximately the marginal relative frequencies of these outcomes. Show that

$$Q = \sum_{i=1}^{r} \sum_{j=1}^{c} \frac{(O_{ij} - M\hat{p}_i\hat{q}_j)^2}{M\hat{p}_i\hat{q}_j}$$

where O_{ij} is the observed frequency of the (i, j)th cell of the contingency table and $\hat{p}_i$, $\hat{q}_j$ are the marginal relative frequencies of the ith row and jth column, respectively, of this table.

A

The Gamma Function

The gamma function arises in many statistical contexts and serves as a useful tool for a variety of computations. It provides one way of extending the factorial $n!$ computation to all positive numbers z in a continuous fashion. Perhaps the most straightforward way to define the gamma function is through *Euler's integral* [named after the Swiss mathematician Leonhard Euler (1707–1783); actually, this integral is Euler's second integral]. Thus, define

$$\Gamma(z) = \int_0^\infty t^{z-1}e^{-t}\,dt, \tag{A.1}$$

where $z > 0$. To demonstrate the relationship of $\Gamma(z)$ to the factorial, integrate by parts in (A.1). Then, as long as $z \geq 1$,

$$\begin{aligned}\Gamma(z) &= \int_0^\infty t^{z-1}e^{-t}\,dt = -e^{-t}t^{z-1}\Big|_0^\infty + (z-1)\int_0^\infty t^{z-2}e^{-t}\,dt \\ &= (z-1)\Gamma(z-1),\end{aligned} \tag{A.2}$$

because when $z \geq 1$, $\lim_{t\to\infty} e^{-t}t^{z-1} = \lim_{t\to 0} e^{-t}t^{z-1} = 0$, as can be seen using L'Hospital's rule. Using (A.2) repeatedly, if $n \leq z < n + 1$ for some positive integer n, then

$$\begin{aligned}\Gamma(z) &= (z-1)\Gamma(z-1) \\ &= (z-1)(z-2)\Gamma(z-2) = \cdots \\ &= (z-1)(z-2)\cdots(z-n+1)\Gamma(z-n+1).\end{aligned} \tag{A.3}$$

Now, if $z = n$, then by (A.3),

$$\Gamma(n) = (n - 1)(n - 2) \cdot \cdot \cdot 2(1)\Gamma(1) = (n - 1)!\, \Gamma(1).$$

However,

$$\Gamma(1) = \int_0^\infty e^{-t}\, dt = -e^{-t} \Big|_0^\infty = 1.$$

Thus $\Gamma(n) = (n - 1)!$, or equivalently,

$$\Gamma(n + 1) = n! \tag{A.4}$$

In obtaining (A.2) and (A.3), we implicitly assumed that the definite integral in (A.1) is well defined and finite for positive values of z. This is, in fact, the case, and indeed it can be shown [see, e.g., Artin (1964)] that the integral in equation (A.1) is defined and finite for all real numbers z other than $0, -1, -2, -3, \ldots$; for $z = 0$ or negative integer values of z the integral in (A.1) is either $+\infty$ or $-\infty$. The graph of $\Gamma(z)$ for $-4 \le z \le 4$ is given in Exhibit A.1. (For $z > 4$, the graph continues to rise as shown. For $z < -4$, the graph continues to alternate between positive and negative curves of parabolic shape, where the minima of the positive curves and the maxima of the negative curves come closer and closer to 0 as $z \to -\infty$.) From Exhibit

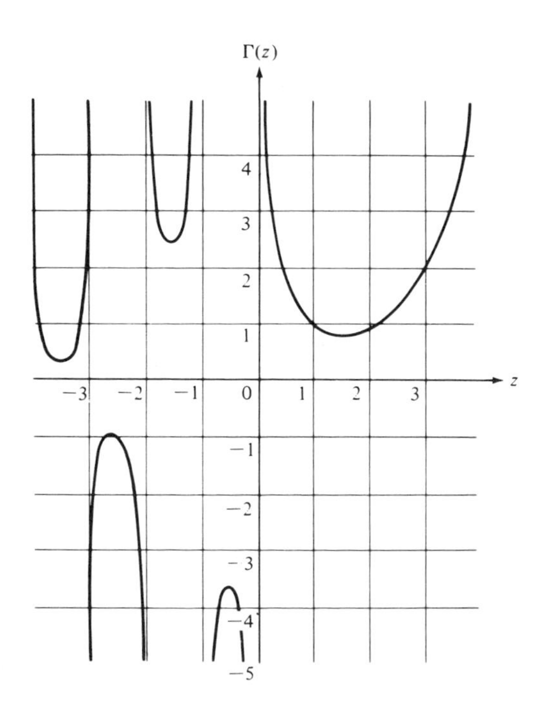

FIGURE A.1 *Graph of the gamma function $\Gamma(z)$ for $-4 \le z \le 4$. For $z > 4$, the graph continues to rise as shown. For $z < -4$, the graph continues to alternate between positive and negative curves of parabolic shape, where the minima of the positive curves and the maxima of the negative curves come closer and closer to 0 as $z \to -\infty$.*

A.1 it is apparent that $\Gamma(z)$ is continuous (and indeed differentiable) at all values z other than $z = 0, -1, -2, \ldots$.

Once we have values of $\Gamma(a)$ for $1 \le a < 2$, the value of $\Gamma(z)$ for any $z \ge 2$ can be obtained from (A.3). On the other hand, if $0 < z < 1$, then because

$$\Gamma(1 + z) = \int_0^\infty t^z e^{-t}\, dt = -e^{-t} t^z \Big|_0^\infty + z \int_0^\infty t^{z-1} e^{-t}\, dt,$$

and $\lim_{t \to \infty} e^{-t} t^z = \lim_{t \to 0} e^{-t} t^z = 0$, when $0 < z < 1$, it follows that

$$\Gamma(z) = \frac{\Gamma(1 + z)}{z}, \qquad \text{(A.5)}$$

so that again we can obtain $\Gamma(z)$ from a table of values for $\Gamma(a)$, $1 \le a < 2$. The relation (A.5) can also be shown to hold when z is negative, but not an integer. Thus, we conclude that the entire gamma function $\Gamma(z)$ can be computed [using (A.3) and (A.5)] once we know the values of $\Gamma(a)$ for $1 \le a < 2$. These values are provided in Exhibit A.2.

For example, to compute $\Gamma(3.45)$, we use (A.3) with $n = 3$ and Exhibit A.2 to obtain

$$\Gamma(3.45) = (2.45)(1.45)\Gamma(1.45) = (2.45)(1.45)(0.8857) = 3.1287.$$

Similarly, using (A.5) repeatedly,

$$\begin{aligned}\Gamma(-3.45) &= \frac{\Gamma(-2.45)}{-3.45} = \frac{\Gamma(-1.45)}{(-3.45)(-2.45)} = \frac{\Gamma(0.45)}{(-3.45)(-2.45)(-1.45)} \\ &= \frac{\Gamma(1.45)}{(-3.45)(-2.45)(-1.45)(0.45)} = -0.0016,\end{aligned}$$

$$\Gamma(0.67) = \frac{\Gamma(1.67)}{0.67} = \frac{0.9033}{0.67} = 1.3482.$$

EXHIBIT A.2 *Values of the Gamma Function $\Gamma(a)$, for $1 \le a < 2$*

a	*0.00*	*0.01*	*0.02*	*0.03*	*0.04*	*0.05*	*0.06*	*0.07*	*0.08*	*0.09*
1.00	1.0000	0.9943	0.9888	0.9835	0.9784	0.9735	0.9687	0.9642	0.9597	0.9555
1.10	0.9514	0.9474	0.9436	0.9399	0.9364	0.9330	0.9298	0.9267	0.9237	0.9209
1.20	0.9182	0.9156	0.9131	0.9108	0.9085	0.9064	0.9044	0.9025	0.9007	0.8990
1.30	0.8975	0.8960	0.8946	0.8934	0.8922	0.8912	0.8902	0.8893	0.8885	0.8879
1.40	0.8873	0.8868	0.8864	0.8860	0.8858	0.8857	0.8856	0.8856	0.8857	0.8859
1.50	0.8862	0.8866	0.8870	0.8876	0.8882	0.8889	0.8896	0.8905	0.8914	0.8924
1.60	0.8935	0.8947	0.8959	0.8972	0.8986	0.9001	0.9017	0.9033	0.9050	0.9068
1.70	0.9086	0.9106	0.9126	0.9147	0.9168	0.9191	0.9214	0.9238	0.9262	0.9288
1.80	0.9314	0.9341	0.9368	0.9397	0.9426	0.9456	0.9487	0.9518	0.9551	0.9384
1.90	0.9618	0.9652	0.9688	0.9724	0.9761	0.9799	0.9837	0.9877	0.9917	0.9948

By making various changes of variable in (A.1), we obtain other integral representations for $\Gamma(z)$ that frequently are useful. In (A.1) make the change of variable $u = \sqrt{t}$. then $dt = 2u\, du$, and

$$\Gamma(z) = 2\int_0^\infty u^{2z-1}e^{-u^2}\,du = \frac{1}{2^{z-1}}\int_0^\infty v^{2z-1}e^{-(1/2)v^2}\,dv, \tag{A.6}$$

where the last expression is obtained by a further change of variables to $v = (\sqrt{2})u$.

There is one value of z, namely $z = \frac{1}{2}$, which is a critical argument in the gamma function. We now show that

$$\Gamma(\tfrac{1}{2}) = \sqrt{\pi}. \tag{A.7}$$

There appears to be only one elementary proof of this fact. Using (A.6), we obtain

$$\Gamma\left(\frac{1}{2}\right) = 2\int_0^\infty e^{-u^2}\,du$$

so that

$$\begin{aligned}\left[\Gamma\left(\frac{1}{2}\right)\right]^2 &= \left(2\int_0^\infty e^{-u^2}\,du\right)^2 = 4\int_0^\infty e^{-u^2}\,du\int_0^\infty e^{-v^2}\,dv\\ &= 4\int_0^\infty\int_0^\infty e^{-(u^2+v^2)}\,du\,dv.\end{aligned} \tag{A.8}$$

Now let $u = r\sin\theta$, $v = r\cos\theta$ be a transformation to polar coordinates. Then $du\,dv = r\,dr\,d\theta$, and (A.8) becomes

$$\begin{aligned}\left[\Gamma\left(\frac{1}{2}\right)\right]^2 &= 4\int_0^{\pi/2}\int_0^\infty re^{-r^2}\,dr\,d\theta\\ &= 4\left(\int_0^{\pi/2}d\theta\right)\left(\int_0^\infty re^{-r^2}\,dr\right) = 4\left(\frac{\pi}{2}\right)\frac{\Gamma(1)}{2}\\ &= \pi,\end{aligned} \tag{A.9}$$

which completes the proof.

Stirling's Formula

The computation of $n!$ for large values of n [or of $\Gamma(z)$ for large values of z] is no easy task. We now try to find a more convenient mathematical expression to approximate $\Gamma(z)$.

For $x > 0$, a direct calculation shows that

$$\left(\frac{d}{dx}\right)^2\left[x \log\left(1 + \frac{1}{x}\right)\right] = \frac{d}{dx}\left[\log\left(1 + \frac{1}{x}\right) - \frac{1}{1 + x}\right]$$
$$= -\frac{1}{(x + 1)^2 x} < 0,$$
$$\left(\frac{d}{dx}\right)^2\left[(x + 1) \log\left(1 + \frac{1}{x}\right)\right] = \frac{d}{dx}\left[\log\left(1 + \frac{1}{x}\right) - \frac{1}{x}\right]$$
$$= \frac{1}{x^2(1 + x)} > 0,$$

so that

$$\frac{d}{dx}\left[x \log\left(1 + \frac{1}{x}\right)\right] = \log\left(1 + \frac{1}{x}\right) - \frac{1}{1 + x}$$

is decreasing in x, and

$$\frac{d}{dx}\left[(x + 1) \log\left(1 + \frac{1}{x}\right)\right] = \log\left(1 + \frac{1}{x}\right) - \frac{1}{x}$$

is increasing in x. It follows that

$$\frac{d}{dx}\left[x \log\left(1 + \frac{1}{x}\right)\right] \geq \lim_{x\to\infty}\left[\log\left(1 + \frac{1}{x}\right) - \frac{1}{1 + x}\right] = 0,$$
$$\frac{d}{dx}\left[(x + 1) \log\left(1 + \frac{1}{x}\right)\right] \leq \lim_{x\to\infty}\left[\log\left(1 + \frac{1}{x}\right) - \frac{1}{x}\right] = 0,$$

from which we conclude that the function $x \log (1 + 1/x)$ is increasing in x and $(x + 1) \log (1 + 1/x)$ is decreasing in x, for $x \geq 0$. Consequently,

$$\left(1 + \frac{1}{x}\right)^x = \exp\left[x \log\left(1 + \frac{1}{x}\right)\right] \leq \exp\left[\lim_{x\to\infty} x \log\left(1 + \frac{1}{x}\right)\right] = e,$$
$$\left(1 + \frac{1}{x}\right)^{x+1} = \exp\left[(x + 1) \log\left(1 + \frac{1}{x}\right)\right]$$
$$\geq \exp\left[\lim_{x\to\infty} (x + 1) \log\left(1 + \frac{1}{x}\right)\right] = e,$$

Putting these results together yields the well-known inequality

(A.10) $$\left(1 + \frac{1}{x}\right)^x \leq e \leq \left(1 + \frac{1}{x}\right)^{x+1}.$$

Substituting $x = 1, 2, 3, \ldots, n - 1$, in (A.11), we obtain

$$\left(1 + \frac{1}{j}\right)^j \leq e \leq \left(1 + \frac{1}{j}\right)^{j+1}, \qquad j = , \ldots, n - 1,$$

and multiplying these $(n - 1)$ inequalities together yields

$$\begin{aligned} &\left(1 + \frac{1}{1}\right)^1\left(1 + \frac{1}{2}\right)^2 \cdots \left(1 + \frac{1}{n-1}\right)^{n-1} \\ &\leq e^{n-1} \\ &\leq \left(1 + \frac{1}{1}\right)^2\left(1 + \frac{1}{2}\right)^3 \cdots \left(1 + \frac{1}{n-1}\right)^n, \end{aligned}$$

or

$$\text{(A.11)} \quad \left(\frac{2}{1}\right)^1\left(\frac{3}{2}\right)^2 \cdots \left(\frac{n}{n-1}\right)^{n-1} \leq e^{n-1} \leq \left(\frac{2}{1}\right)^2\left(\frac{3}{2}\right)^3 \cdots \left(\frac{n}{n-1}\right)^n.$$

Canceling terms in (A.11), we are able to simplify to

$$\text{(A.12)} \qquad \frac{n^{n-1}}{(n-1)!} \leq e^{n-1} \leq \frac{n^n}{(n-1)!}.$$

Finally, multiplying all sides of (A.13) by $n!\, e^{-(n-1)}$, we find that

$$\text{(A.13)} \qquad n^n e^{-(n-1)} \leq n! \leq n^{n+1} e^{-(n-1)}.$$

Thus we see that $n!$ is not free to increase with n at any rate, but must lie between $n^n e^{-(n-1)}$ and $n^{n+1} e^{-(n-1)}$.

Stirling's formula provides a compromise between these upper and lower bounds and approximates $n!$ by

$$\text{(A.14)} \qquad n! \cong \sqrt{2\pi}\, n^{n+1/2} e^{-n} \equiv S(n).$$

Exhibit A.3 shows the accuracy of the approximation $n! \simeq S(n)$ for $n = 1(1)15$.

Note from Exhibit A.3 that the difference $n! - S(n)$ is increasing as n increases. Indeed, $n! - S(n) \to \infty$ as $n \to \infty$. However, the *percent error* between $n!$ and $S(n)$ [i.e., $100(n! - S(n))/n!$] stays small, and in fact converges to 0 as $n \to \infty$. Thus, even for relatively small values of n, the Stirling approximation to $n!$ is proportionately quite accurate.

EXHIBIT A.3 Values of $n!$ and $S(n)$ for $n = 1(1)15$

n	$n!$	$S(n)$	Percent error $= 100\left[\frac{n! - S(n)}{n!}\right]$
1	1	0.9221370089	7.79
2	2	1.919004351	4.05
3	6	5.836209591	2.73
4	24	23.50617513	2.06
5	120	118.0191680	1.65
6	720	710.0781847	1.38
7	5,040	4,980.395833	1.18
8	40,320	39,902.39545	1.04
9	362,880	359,536.8729	0.92
10	3,628,800	3,598,695.619	0.83
11	39,916,800	39,615,625.05	0.75
12	479,001,600	475,687,486.5	0.69
13	6,227,020,800	6,187,239,475	0.64
14	87,178,291,200	86,661,001,78$\underline{0}$	0.59
15	1,307,674,368,000	1,300,430,722,$\underline{000}$	0.55

Remark. In Feller (1968) it is shown that $n!/S(n) = e^{r(n)/12n}$, where $(12n)/(12n + 1) < r(n) < 1$. Hence, as $n \to \infty$, $[n!/S(n)] \to 1$.

There is a similar Stirling approximation to $\Gamma(z + 1)$, namely, $S(z) = \sqrt{2\pi}\, z^{z+1/2}e^{-z}$. This approximation also has the property that the percentage error, $100(\Gamma(z + 1) - S(z))/\Gamma(z + 1)$, is small even for relatively small values of z, and converges to 0 as $z \to \infty$.

The Beta Function

Earlier, we mentioned that the integral (A.1) is known as Euler's second integral. Euler's first integral is

$$B(x, y) = \int_0^1 t^{x-1}(1 - t)^{y-1}\, dt, \tag{A.15}$$

and as a function of $x > 0$, $y > 0$, is known as the *beta function*.

The beta function has other representations. In (A.15) let $t = \sin^2\theta$, so that $dt = 2 \sin\theta \cos\theta\, d\theta$. Then

$$\mathrm{B}(x, y) = 2\int_0^{\pi/2} (\sin\theta)^{2x-1}(\cos\theta)^{2y-1}\, d\theta. \tag{A.16}$$

Equation (A.16) gives one alternative integral expression for $\mathrm{B}(x, y)$. Another expression can be obtained by making the change of variable from t to $s = t(1 - t)^{-1}$ in (A.15). Since $t = s/(1 + s)$, we have $dt = (1 + s)^{-2}\, ds$, and

(A.17)
$$B(x, y) = \int_0^\infty \frac{s^{x-1}}{(1+s)^{x+y}}\, ds.$$

Still a third expression is obtained by letting $u = 1 - t$ in (A.15). We obtain

(A.18)
$$B(x, y) = \int_0^1 t^{x-1}(1-t)^{y-1}\, dt = \int_0^1 (1-u)^{x-1}u^{y-1}\, du = B(y, x),$$

a relationship that could also have been inferred from the symmetry in the definition (A.15) of $B(x, y)$.

The beta function is related to the gamma function by the formula

(A.19)
$$B(x, y) = \frac{\Gamma(x)\Gamma(y)}{\Gamma(x+y)}.$$

To prove this relationship, we make use of (A.6):

$$\Gamma(x)\Gamma(y) = \left(2\int_0^\infty u^{2x-1}e^{-u^2}\, du\right)\left(2\int_0^\infty v^{2y-1}e^{-v^2}\, dv\right)$$
$$= 4\int_0^\infty \int_0^\infty u^{2x-1}v^{2y-1}e^{-(u^2+v^2)}\, du\, dv.$$

Let $u = r \sin\theta$, $v = r\cos\theta$. Then

(A.20)
$$\Gamma(x)\Gamma(y) = \left(2\int_0^\infty r^{2(x+y)-1}e^{-r^2}\, dr\right)\left(2\int_0^{\pi/2} (\sin\theta)^{2x-1}(\cos\theta)^{2y-1}\, d\theta\right)$$
$$= \Gamma(x+y)\left(2\int_0^{\pi/2} (\sin\theta)^{2x-1}(\cos\theta)^{2y-1}\, d\theta\right)$$
$$= \Gamma(x+y)B(x, y).$$

The beta function is closely related to the binomial coefficient $\binom{n}{k}$ through the expression

(A.21)
$$B(k+1, n-k+1) = \frac{1}{(n+1)\binom{n}{k}}.$$

Computation of $B(x, y)$ follows directly from (A.19) and computations with the gamma function. For example,

$$B(2.45, 2.67) = \frac{\Gamma(2.45)\Gamma(2.67)}{\Gamma(2.45 + 2.67)} = \frac{(1.45)(1.67)\Gamma(1.45)\Gamma(1.67)}{(4.12)(3.12)(2.12)\Gamma(1.12)}$$
$$= \frac{(1.45)(1.67)(0.8857)(0.9033)}{(4.12)(3.12)(2.12)(0.9436)} = 0.0753.$$

If one of the arguments is an integer n, then

$$\mathrm{B}(n, y) = \frac{\Gamma(n)\Gamma(y)}{\Gamma(n+y)} = \frac{(n-1)!\,\Gamma(y)}{(n-1+y)(n-2+y)\cdots(1+y)y\Gamma(y)}$$
$$= \frac{(n-1)!}{(n-1+y)(n-2+y)\cdots(1+y)(y)}.$$

If both of the arguments are integers ($x = n$, $y = m$), then

$$\mathrm{B}(n, m) = \frac{\Gamma(n)\Gamma(m)}{\Gamma(n+m)} = \frac{(n-1)!\,(m-1)!}{(n+m-1)!}.$$

Finally, the recursive formulas

$$\mathrm{B}(x, y) = \frac{y-1}{x}\mathrm{B}(x+1, y-1),$$
$$\mathrm{B}(x, y) = \frac{(x-1)(y-1)}{(x+y-1)(x+y-2)}\mathrm{B}(x-1, y-1),$$

often are useful.

B

Tables

Table Interpolation

Suppose that it is desired to evaluate a function $f(x)$ at a point x_0, and that a table of values of $f(x)$ is available for some, but not all, values of x. In particular, the table may not give the value $f(x_0)$ but may give values for $f(x_1)$ and $f(x_2)$ where $x_1 < x_0 < x_2$. Rather than evaluate $f(x_0)$ exactly by other more costly or difficult means, one can use the known values of $f(x)$ for $x = x_1, x_2$ to approximate the value of $f(x_0)$. This process is known as **interpolation.** Perhaps the most commonly used interpolation method is **linear interpolation.** If $f(x)$ is sufficiently smooth and not too curvilinear between $x = x_1$ and $x = x_2$, calculus tells us that $f(x)$ can be regarded as being nearly linear (see Exhibit B.0) over the interval $[x_1, x_2]$. That is,

$$f(x) \cong a + bx, \qquad x_1 \leq x \leq x_2.$$

Solving the equations

$$f(x_1) = a + bx_1, \qquad f(x_2) = a + bx_2$$

for a and b yields

$$b = \frac{f(x_2) - f(x_1)}{x_2 - x_1}, \qquad a = f(x_1) - \left[\frac{f(x_2) - f(x_1)}{x_2 - x_1}\right] x_1$$

so that

$$f(x_0) \cong a + bx_0 = f(x_1) + \left[\frac{f(x_2) - f(x_1)}{x_2 - x_1}\right] (x_0 - x_1).$$

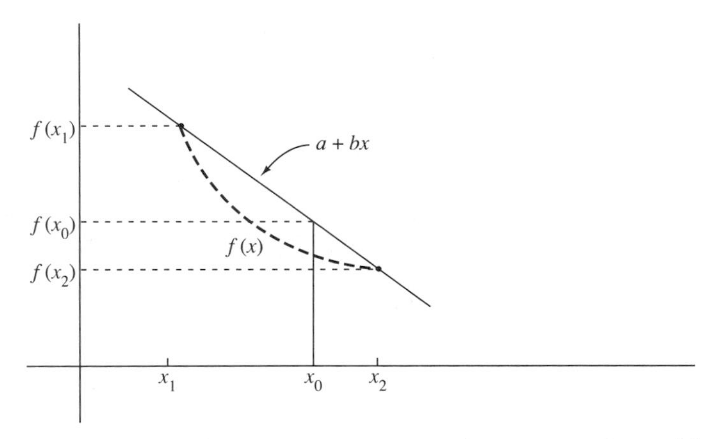

EXHIBIT B.0 *In linear interpolation, f(x) is approximated by a straight line; f(x₀) is the interpolated value.*

EXAMPLE B.1

Suppose that $f(x) = e^{-x}$ and that a table gives the values $f(3.0) = 0.04979$, $f(3.4) = 0.03337$. If the values of $f(3.2)$ and $f(3.3)$ are desired, linear interpolation yields

$$\begin{aligned} f(3.2) &\simeq f(3.0) + \left[\frac{f(3.4) - f(3.0)}{3.4 - 3.0}\right](3.2 - 3.0) \\ &= \tfrac{1}{2}[f(3.0) + f(3.4)] = \tfrac{1}{2}(0.04979 + 0.03337) \\ &= 0.04158, \\ f(3.3) &\simeq f(3.0) + \left[\frac{f(3.4) - f(3.0)}{(3.4 - 3.0)}\right](3.3 - 3.0) \\ &= \tfrac{1}{4}f(3.0) + \tfrac{3}{4}f(3.4) = \tfrac{1}{4}(0.04979) + \tfrac{3}{4}(0.03337) \\ &= 0.03748. \end{aligned}$$

These are fairly close to the exact values $f(3.2) = 0.04076$ and $f(3.3) = 0.03688$. ◆

If the curvilinearity of $f(x)$ appears too severe for linear interpolation, $f(x)$ can be approximated by polynomials of higher order, provided that more tabled values of $f(x)$ are available for x near x_0. For example, if values of $f(x_1), f(x_2), f(x_3)$ are available in the table, and all of $x_1 < x_2 < x_3$ are reasonably close to x_0, **quadratic interpolation** can be used. Here the approximation $f(x) \simeq a + bx + cx^2$ is utilized, with a, b, c found by solving the set of linear equations $f(x_i) = a + bx_i + cx_i^2$, $i = 1, 2, 3$. For most, if not all, uses of the tables in this book, however, linear interpolation will be sufficient.

Interpolation can also be used for functions $f(x, y)$ of two variables (two-way interpolation). In linear two-way interpolation, $f(x, y)$ is regarded as being approximately a bilinear function $a + bx + cy + dxy$ of x and y, $x_1 \le x \le x_2$, $y_1 \le y \le y_2$, where values of $f(x_i, y_j)$, $i, j = 1, 2$, are given in the table. The constants a, b, c, d can, as before, be obtained from the tabled values by solving a set of equations. It is easier, however, to first linearly interpolate on x and then linearly interpolate on y, or vice versa.

EXAMPLE B.2

In a table of the Poisson distribution, the values of $p(x; \lambda) = P\{X = x\}$ might be given for Poisson variables with $\lambda = 2.4$ and $\lambda = 2.6$, and for the values $x = 3$ and $x = 6$, but it may be desired to find $p(4; 2.5) = P\{X = 4\}$ for $\lambda = 2.5$. We can tabulate the known values as follows:

	λ	
x	2.4	2.6
3	0.2090	0.2176
6	0.0241	0.0319

Start by using linear interpolation columnwise to find $p(4; 2.4)$ and $p(4; 2.6)$:

$$\begin{aligned} p(4; 2.4) &\cong p(3; 2.4) + \left[\frac{p(6; 2.4) - p(3; 2.4)}{6 - 3}\right](4 - 3) \\ &= \tfrac{2}{3}\,p(3; 2.4) + \tfrac{1}{3}\,p(6; 2.4) \\ &= \tfrac{2}{3}(0.2090) + \tfrac{1}{3}(0.0241) = 0.1474, \\ p(4; 2.6) &\cong \tfrac{2}{3}\,p(3; 2.6) + \tfrac{1}{3}\,p(6; 2.6) = \tfrac{2}{3}(0.2176) + \tfrac{1}{3}(0.0319) \\ &= 0.1557. \end{aligned}$$

Then, linearly interpolate rowwise between $p(4; 2.4)$ and $p(4; 2.6)$ to find $p(4; 2.5)$:

$$\begin{aligned} p(4; 2.5) &\cong \tfrac{1}{2}[p(4; 2.4) + p(4; 2.6)] = \tfrac{1}{2}(0.1474 + 0.1557) \\ &= 0.1515. \end{aligned}$$

Alternatively, we could linearly interpolate rowwise,

$$\begin{aligned} p(3; 2.5) &\cong \tfrac{1}{2}[p(3; 2.4) + p(3; 2.6)] = \tfrac{1}{2}(0.2090 + 0.2176) = 0.2133, \\ p(6; 2.5) &\cong \tfrac{1}{2}[p(6; 2.4) + p(6; 2.6)] = \tfrac{1}{2}(0.0241 + 0.0319) = 0.0280, \end{aligned}$$

and then columnwise

$$p(4; 2.5) \cong \tfrac{2}{3}\,p(3; 2.5) + \tfrac{1}{3}\,p(6; 2.5) = \tfrac{2}{3}(0.2133) + \tfrac{1}{3}(0.0280) = 0.1515.$$

For the sake of comparison, the exact value of $p(4; 2.5)$ is 0.1336. The approximation would be improved if values of x closer to $x = 4$ had been used (e.g., $x = 3$, $x = 5$). ◆

Generalizations of this method of two-way interpolation can be used: for example, quadratic interpolation rowwise and then linear interpolation columnwise, or quadratic interpolation both rowwise and columnwise. Again, such more complicated interpolation methods will not be required.

EXHIBIT B.1 *2000 Random Numbers*

1	29280	39655	18902	92531	90374	07109	26627	59587	84340	98351
2	20123	82082	55477	22059	43168	12903	13436	25523	21090	73449
3	66405	35287	33248	67657	07702	01474	66068	01125	59258	30138
4	97299	83419	13069	17826	76984	48906	10567	17829	00723	46700
5	83923	92076	98880	33942	46841	58731	36513	16681	88722	61984
6	11258	92175	94894	97606	11134	51941	43733	00514	06694	27706
7	08522	48468	60789	47178	85587	78410	67050	41286	16545	22061
8	02114	89744	10115	39603	61089	79392	38945	77699	59054	07742
9	24580	05775	54677	04171	97815	35557	92626	29756	35289	97756
10	23937	25079	12306	23125	50842	51015	57436	71349	79397	06095
11	15529	38214	68209	82655	89015	73159	52233	38844	46373	69912
12	63113	89594	91488	84795	70688	85252	24903	18446	02349	69008
13	14510	74196	51065	89054	03642	05842	25909	85998	26482	01986
14	15632	45813	02378	87946	14803	55973	29915	55599	98152	14148
15	23522	94186	55528	02483	96313	87715	78336	18672	29260	53130
16	99919	54509	21242	84850	59905	28605	32816	04807	63554	54959
17	72159	69313	67810	64819	19314	70702	60778	48753	75775	75322
18	60172	73201	18902	71182	42158	01184	02708	21062	69869	87882
19	49787	07561	20928	86517	86489	58490	62550	38854	91070	28022
20	35948	92023	46949	38255	21897	99978	21935	36545	64959	74716
21	37693	73413	64495	82483	13357	87025	71542	93052	97979	47174
22	46037	36193	41518	01194	18296	35566	12194	21334	47829	52904
23	60812	80784	13284	11905	71124	62686	67589	73404	64680	03933
24	23331	07113	83029	84896	59786	27926	18360	05097	12401	09574
25	63632	20522	42599	94300	31586	95638	87504	39277	38156	01989
26	64002	03244	45331	22215	29861	61748	83109	55502	40330	24740
27	17068	03024	05730	27364	67220	55240	78238	63125	72740	50315
28	81672	87668	20124	88016	46143	27051	14797	48511	18819	14465
29	11939	77938	05824	55807	62458	67887	73792	16586	92715	46654
30	09132	50818	73756	68629	26017	39807	69961	64681	65147	57290
31	72504	07749	77006	48499	31761	25246	37820	88399	01792	75040
32	09987	95625	46237	37490	84790	58486	87957	21725	48499	72785
33	69552	66506	62946	07559	12931	32621	82019	62292	89331	43490
34	99314	14865	35477	28130	95554	19603	19912	97005	99278	00929
35	32335	45363	34713	94258	86756	42640	53419	56910	07923	72984
36	55081	08953	21330	11415	45284	97450	38929	12258	23491	48366
37	06433	68716	08164	19822	81748	85506	47619	80316	88505	66744
38	45349	00397	00007	59491	94383	39736	04154	23806	16753	10180
39	16739	22808	92714	50589	01966	86230	20874	86054	83081	74625
40	83498	41005	40147	55045	36609	10555	12859	08585	29829	55784

EXHIBIT B.2 *Individual Terms, $p_Z(k) = \binom{n}{k} p^k (1-p)^{n-k}$, of the Probability Mass Function of a Binomially Distributed Random Variable Z*

		p						
n	*k*	0.01	0.02	0.03	0.04	0.05	0.10	0.15
1	0	0.99000	0.98000	0.97000	0.96000	0.95000	0.90000	0.85000
	1	0.01000	0.02000	0.03000	0.04000	0.05000	0.10000	0.15000
2	0	0.98010	0.96040	0.94090	0.92160	0.90250	0.81000	0.72250
	1	0.01980	0.03920	0.05820	0.07680	0.09500	0.18000	0.25500
	2	0.00010	0.00040	0.00090	0.00160	0.00250	0.01000	0.02250
3	0	0.97030	0.94119	0.91267	0.88474	0.85738	0.72900	0.61413
	1	0.02940	0.05762	0.08468	0.11059	0.13538	0.24300	0.32513
	2	0.00030	0.00118	0.00262	0.00461	0.00712	0.02700	0.05737
	3	0.00000	0.00001	0.00003	0.00006	0.00013	0.00100	0.00337
4	0	0.96060	0.92237	0.88529	0.84935	0.81451	0.65610	0.52201
	1	0.03881	0.07530	0.10952	0.14156	0.17148	0.29160	0.36848
	2	0.00059	0.00230	0.00508	0.00885	0.01354	0.04860	0.09754
	3	0.00000	0.00003	0.00010	0.00025	0.00047	0.00360	0.01148
	4	0.00000	0.00000	0.00000	0.00000	0.00001	0.00010	0.00051
5	0	0.95099	0.90392	0.85873	0.81537	0.77378	0.59049	0.44371
	1	0.04803	0.09224	0.13279	0.16987	0.20363	0.32805	0.39150
	2	0.00097	0.00376	0.00821	0.01416	0.02143	0.07290	0.13818
	3	0.00001	0.00008	0.00025	0.00059	0.00113	0.00810	0.02438
	4	0.00000	0.00000	0.00000	0.00001	0.00003	0.00045	0.00215
	5	0.00000	0.00000	0.00000	0.00000	0.00000	0.00001	0.00008
6	0	0.94148	0.88584	0.83297	0.78276	0.73509	0.53144	0.37715
	1	0.05706	0.10847	0.15457	0.19569	0.23213	0.35429	0.39933
	2	0.00144	0.00553	0.01195	0.02038	0.03054	0.09842	0.17618
	3	0.00002	0.00015	0.00049	0.00113	0.00214	0.01458	0.04145
	4	0.00000	0.00000	0.00001	0.00004	0.00008	0.00121	0.00549
	5	0.00000	0.00000	0.00000	0.00000	0.00000	0.00005	0.00039
	6	0.00000	0.00000	0.00000	0.00000	0.00000	0.00000	0.00001
7	0	0.93207	0.86813	0.80798	0.75145	0.69834	0.47830	0.32058
	1	0.06590	0.12402	0.17492	0.21917	0.25728	0.37201	0.39601
	2	0.00200	0.00759	0.01623	0.02740	0.04062	0.12400	0.20965
	3	0.00003	0.00026	0.00084	0.00190	0.00356	0.02296	0.06166
	4	0.00000	0.00001	0.00003	0.00008	0.00019	0.00255	0.01088
	5	0.00000	0.00000	0.00000	0.00000	0.00001	0.00017	0.00115
	6	0.00000	0.00000	0.00000	0.00000	0.00000	0.00001	0.00007
	7	0.00000	0.00000	0.00000	0.00000	0.00000	0.00000	0.00000

			p					
0.20	0.25	0.30	0.35	0.40	0.45	0.50	n	k
0.80000	0.75000	0.70000	0.65000	0.60000	0.55000	0.50000	1	0
0.20000	0.25000	0.30000	0.35000	0.40000	0.45000	0.50000		1
0.64000	0.56250	0.49000	0.42250	0.36000	0.30250	0.25000	2	0
0.32000	0.37500	0.42000	0.45500	0.48000	0.49500	0.50000		1
0.04000	0.06250	0.09000	0.12250	0.16000	0.20250	0.25000		2
0.51200	0.42188	0.34300	0.27463	0.21600	0.16638	0.12500	3	0
0.38400	0.42188	0.44100	0.44362	0.43200	0.40838	0.37500		1
0.09600	0.14063	0.18900	0.23888	0.28800	0.33413	0.37500		2
0.00800	0.01563	0.02700	0.04287	0.06400	0.09113	0.12500		3
0.40960	0.31641	0.24010	0.17851	0.12960	0.09151	0.06250	4	0
0.40960	0.42188	0.41160	0.38447	0.34560	0.29948	0.25000		1
0.15360	0.21094	0.26460	0.31054	0.34560	0.36754	0.37500		2
0.02560	0.04688	0.07560	0.11148	0.15360	0.20048	0.25000		3
0.00160	0.00391	0.00810	0.01501	0.02560	0.04101	0.06250		4
0.32768	0.23730	0.16807	0.11603	0.07776	0.05033	0.03125	5	0
0.40960	0.39551	0.36015	0.31239	0.25920	0.20589	0.15625		1
0.20480	0.26367	0.30870	0.33642	0.34560	0.33691	0.31250		2
0.05120	0.08789	0.13230	0.18115	0.23040	0.27565	0.31250		3
0.00640	0.01465	0.02835	0.04877	0.07680	0.11277	0.15625		4
0.00032	0.00098	0.00243	0.00525	0.01024	0.01845	0.03125		5
0.26214	0.17798	0.11765	0.07542	0.04666	0.02768	0.01563	6	0
0.39322	0.35596	0.30253	0.24366	0.18662	0.13589	0.09375		1
0.24576	0.29663	0.32413	0.32801	0.31104	0.27795	0.23438		2
0.08192	0.13184	0.18522	0.23549	0.27648	0.30322	0.31250		3
0.01536	0.03296	0.05953	0.09510	0.13824	0.18607	0.23438		4
0.00154	0.00439	0.01021	0.02048	0.03686	0.06089	0.09375		5
0.00006	0.00024	0.00073	0.00184	0.00410	0.00830	0.01563		6
0.20972	0.13348	0.08235	0.04902	0.02799	0.01522	0.00781	7	0
0.36700	0.31146	0.24706	0.18478	0.13064	0.08719	0.05469		1
0.27525	0.31146	0.31765	0.29848	0.26127	0.21402	0.16406		2
0.11469	0.17303	0.22689	0.26787	0.29030	0.29185	0.27344		3
0.02867	0.05768	0.09724	0.14424	0.19354	0.23878	0.27344		4
0.00430	0.01154	0.02500	0.04660	0.07741	0.11722	0.16406		5
0.00036	0.00128	0.00357	0.00836	0.01720	0.03197	0.05469		6
0.00001	0.00006	0.00022	0.00064	0.00164	0.00374	0.00781		7

EXHIBIT B.2 *(continued)*

		p						
n	*k*	0.01	0.02	0.03	0.04	0.05	0.10	0.15
8	0	0.92274	0.85076	0.78374	0.72139	0.66342	0.43047	0.27249
	1	0.07457	0.13890	0.19392	0.24046	0.27933	0.38264	0.38469
	2	0.00264	0.00992	0.02099	0.03507	0.05146	0.14880	0.23760
	3	0.00005	0.00040	0.00130	0.00292	0.00542	0.03307	0.08386
	4	0.00000	0.00001	0.00005	0.00015	0.00036	0.00459	0.01850
	5	0.00000	0.00000	0.00000	0.00001	0.00002	0.00041	0.00261
	6	0.00000	0.00000	0.00000	0.00000	0.00000	0.00002	0.00023
	7	0.00000	0.00000	0.00000	0.00000	0.00000	0.00000	0.00001
	8	0.00000	0.00000	0.00000	0.00000	0.00000	0.00000	0.00000
9	0	0.91352	0.83375	0.76023	0.69253	0.63025	0.38742	0.23162
	1	0.08305	0.15314	0.21161	0.25970	0.29854	0.38742	0.36786
	2	0.00336	0.01250	0.02618	0.04328	0.06285	0.17219	0.25967
	3	0.00008	0.00060	0.00189	0.00421	0.00772	0.04464	0.10692
	4	0.00000	0.00002	0.00009	0.00026	0.00061	0.00744	0.02830
	5	0.00000	0.00000	0.00000	0.00001	0.00003	0.00083	0.00499
	6	0.00000	0.00000	0.00000	0.00000	0.00000	0.00006	0.00059
	7	0.00000	0.00000	0.00000	0.00000	0.00000	0.00000	0.00004
	8	0.00000	0.00000	0.00000	0.00000	0.00000	0.00000	0.00000
	9	0.00000	0.00000	0.00000	0.00000	0.00000	0.00000	0.00000
10	0	0.90438	0.81707	0.73742	0.66483	0.59874	0.34868	0.19687
	1	0.09135	0.16675	0.22807	0.27701	0.31512	0.38742	0.34743
	2	0.00415	0.01531	0.03174	0.05194	0.07463	0.19371	0.27590
	3	0.00011	0.00083	0.00262	0.00577	0.01048	0.05740	0.12983
	4	0.00000	0.00003	0.00014	0.00042	0.00096	0.01116	0.04010
	5	0.00000	0.00000	0.00001	0.00002	0.00006	0.00149	0.00849
	6	0.00000	0.00000	0.00000	0.00000	0.00000	0.00014	0.00125
	7	0.00000	0.00000	0.00000	0.00000	0.00000	0.00001	0.00013
	8	0.00000	0.00000	0.00000	0.00000	0.00000	0.00000	0.00001
	9	0.00000	0.00000	0.00000	0.00000	0.00000	0.00000	0.00000
	10	0.00000	0.00000	0.00000	0.00000	0.00000	0.00000	0.00000
12	0	0.88639	0.78472	0.69384	0.61271	0.54036	0.28243	0.14224
	1	0.10744	0.19218	0.25751	0.30636	0.34128	0.37657	0.30122
	2	0.00597	0.02157	0.04380	0.07021	0.09879	0.23013	0.29236
	3	0.00020	0.00147	0.00452	0.00975	0.01733	0.08523	0.17198
	4	0.00001	0.00007	0.00031	0.00091	0.00205	0.02131	0.06828
	5	0.00000	0.00000	0.00002	0.00006	0.00017	0.00379	0.01928
	6	0.00000	0.00000	0.00000	0.00000	0.00001	0.00049	0.00397
	7	0.00000	0.00000	0.00000	0.00000	0.00000	0.00005	0.00060
	8	0.00000	0.00000	0.00000	0.00000	0.00000	0.00000	0.00007
	9	0.00000	0.00000	0.00000	0.00000	0.00000	0.00000	0.00001
	10	0.00000	0.00000	0.00000	0.00000	0.00000	0.00000	0.00000
	11	0.00000	0.00000	0.00000	0.00000	0.00000	0.00000	0.00000
	12	0.00000	0.00000	0.00000	0.00000	0.00000	0.00000	0.00000

p								
0.20	0.25	0.30	0.35	0.40	0.45	0.50	n	k
0.16777	0.10011	0.05765	0.03186	0.01680	0.00837	0.00391	8	0
0.33554	0.26697	0.19765	0.13726	0.08958	0.05481	0.03125		1
0.29360	0.31146	0.29648	0.25869	0.20902	0.15695	0.10938		2
0.14680	0.20764	0.25412	0.27859	0.27869	0.25683	0.21875		3
0.04588	0.08652	0.13614	0.18751	0.23224	0.26266	0.27344		4
0.00918	0.02307	0.04668	0.08077	0.12386	0.17192	0.21875		5
0.00115	0.00385	0.01000	0.02175	0.04129	0.07033	0.10938		6
0.00008	0.00037	0.00122	0.00335	0.00786	0.01644	0.03125		7
0.00000	0.00002	0.00007	0.00023	0.00066	0.00168	0.00391		8
0.13422	0.07508	0.04035	0.02071	0.01008	0.00461	0.00195	9	0
0.30199	0.22525	0.15565	0.10037	0.06047	0.03391	0.01758		1
0.30199	0.30034	0.26683	0.21619	0.16124	0.11099	0.07031		2
0.17616	0.23360	0.26683	0.27162	0.25082	0.21188	0.16406		3
0.06606	0.11680	0.17153	0.21939	0.25082	0.26004	0.24609		4
0.01652	0.03893	0.07351	0.11813	0.16722	0.21276	0.24609		5
0.00275	0.00865	0.02100	0.04241	0.07432	0.11605	0.16406		6
0.00029	0.00124	0.00386	0.00979	0.02123	0.04069	0.07031		7
0.00002	0.00010	0.00041	0.00132	0.00354	0.00832	0.01758		8
0.00000	0.00000	0.00002	0.00008	0.00026	0.00076	0.00195		9
0.10737	0.05631	0.02825	0.01346	0.00605	0.00253	0.00098	10	0
0.26844	0.18771	0.12106	0.07249	0.04031	0.02072	0.00977		1
0.30199	0.28157	0.23347	0.17565	0.12093	0.07630	0.04395		2
0.20133	0.25028	0.26683	0.25222	0.21499	0.16648	0.11719		3
0.08808	0.14600	0.20012	0.23767	0.25082	0.23837	0.20508		4
0.02642	0.05840	0.10292	0.15357	0.20066	0.23403	0.24609		5
0.00551	0.01622	0.03676	0.06891	0.11148	0.15957	0.20508		6
0.00079	0.00309	0.00900	0.02120	0.04247	0.07460	0.11719		7
0.00007	0.00039	0.00145	0.00428	0.01062	0.02289	0.04395		8
0.00000	0.00003	0.00014	0.00051	0.00157	0.00416	0.00977		9
0.00000	0.00000	0.00001	0.00003	0.00010	0.00034	0.00098		10
0.06872	0.03168	0.01384	0.00569	0.00218	0.00077	0.00024	12	0
0.20616	0.12671	0.07118	0.03675	0.01741	0.00752	0.00293		1
0.28347	0.23229	0.16779	0.10885	0.06385	0.03385	0.01611		2
0.23622	0.25810	0.23970	0.19537	0.14189	0.09233	0.05371		3
0.13288	0.19358	0.23114	0.23669	0.21284	0.16996	0.12085		4
0.05315	0.10324	0.15850	0.20392	0.22703	0.22250	0.19336		5
0.01550	0.04015	0.07925	0.12810	0.17658	0.21239	0.22559		6
0.00332	0.01147	0.02911	0.05913	0.10090	0.14895	0.19336		7
0.00052	0.00239	0.00780	0.01990	0.04204	0.07617	0.12085		8
0.00006	0.00035	0.00149	0.00476	0.01246	0.02770	0.05371		9
0.00000	0.00004	0.00019	0.00077	0.00249	0.00680	0.01611		10
0.00000	0.00000	0.00002	0.00008	0.00030	0.00101	0.00293		11
0.00000	0.00000	0.00000	0.00000	0.00002	0.00007	0.00024		12

EXHIBIT B.2 *(continued)*

n	k	p = 0.01	0.02	0.03	0.04	0.05	0.10	0.15
15	0	0.86006	0.73857	0.63325	0.54209	0.46329	0.20589	0.08735
	1	0.13031	0.22609	0.29378	0.33880	0.36576	0.34315	0.23123
	2	0.00921	0.03230	0.06360	0.09882	0.13475	0.26690	0.28564
	3	0.00040	0.00286	0.00852	0.01784	0.03073	0.12851	0.21843
	4	0.00001	0.00018	0.00079	0.00223	0.00485	0.04284	0.11564
	5	0.00000	0.00001	0.00005	0.00020	0.00056	0.01047	0.04490
	6	0.00000	0.00000	0.00000	0.00001	0.00005	0.00194	0.01321
	7	0.00000	0.00000	0.00000	0.00000	0.00000	0.00028	0.00300
	8	0.00000	0.00000	0.00000	0.00000	0.00000	0.00003	0.00053
	9	0.00000	0.00000	0.00000	0.00000	0.00000	0.00000	0.00007
	10	0.00000	0.00000	0.00000	0.00000	0.00000	0.00000	0.00001
	11	0.00000	0.00000	0.00000	0.00000	0.00000	0.00000	0.00000
	12	0.00000	0.00000	0.00000	0.00000	0.00000	0.00000	0.00000
	13	0.00000	0.00000	0.00000	0.00000	0.00000	0.00000	0.00000
	14	0.00000	0.00000	0.00000	0.00000	0.00000	0.00000	0.00000
	15	0.00000	0.00000	0.00000	0.00000	0.00000	0.00000	0.00000
20	0	0.81791	0.66761	0.54379	0.44200	0.35849	0.12158	0.03876
	1	0.16523	0.27249	0.33637	0.36834	0.37735	0.27017	0.13680
	2	0.01586	0.05283	0.09883	0.14580	0.18868	0.28518	0.22934
	3	0.00096	0.00647	0.01834	0.03645	0.05958	0.19012	0.24283
	4	0.00004	0.00056	0.00241	0.00646	0.01333	0.08978	0.18212
	5	0.00000	0.00004	0.00024	0.00086	0.00225	0.03192	0.10285
	6	0.00000	0.00000	0.00002	0.00009	0.00030	0.00887	0.04537
	7	0.00000	0.00000	0.00000	0.00001	0.00003	0.00197	0.01601
	8	0.00000	0.00000	0.00000	0.00000	0.00000	0.00036	0.00459
	9	0.00000	0.00000	0.00000	0.00000	0.00000	0.00005	0.00108
	10	0.00000	0.00000	0.00000	0.00000	0.00000	0.00001	0.00021
	11	0.00000	0.00000	0.00000	0.00000	0.00000	0.00000	0.00003
	12	0.00000	0.00000	0.00000	0.00000	0.00000	0.00000	0.00000
	13	0.00000	0.00000	0.00000	0.00000	0.00000	0.00000	0.00000
	14	0.00000	0.00000	0.00000	0.00000	0.00000	0.00000	0.00000
	15	0.00000	0.00000	0.00000	0.00000	0.00000	0.00000	0.00000
	16	0.00000	0.00000	0.00000	0.00000	0.00000	0.00000	0.00000
	17	0.00000	0.00000	0.00000	0.00000	0.00000	0.00000	0.00000
	18	0.00000	0.00000	0.00000	0.00000	0.00000	0.00000	0.00000
	19	0.00000	0.00000	0.00000	0.00000	0.00000	0.00000	0.00000
	20	0.00000	0.00000	0.00000	0.00000	0.00000	0.00000	0.00000

p								
0.20	0.25	0.30	0.35	0.40	0.45	0.50	n	k
0.03518	0.01336	0.00475	0.00156	0.00047	0.00013	0.00003	15	0
0.13194	0.06682	0.03052	0.01262	0.00470	0.00157	0.00046		1
0.23090	0.15591	0.09156	0.04756	0.02194	0.00896	0.00320		2
0.25014	0.22520	0.17004	0.11096	0.06339	0.03177	0.01389		3
0.18760	0.22520	0.21862	0.17925	0.12678	0.07798	0.04166		4
0.10318	0.16515	0.20613	0.21234	0.18594	0.14036	0.09164		5
0.04299	0.09175	0.14724	0.19056	0.20660	0.19140	0.15274		6
0.01382	0.03932	0.08113	0.13193	0.17708	0.20134	0.19638		7
0.00346	0.01311	0.03477	0.07104	0.11806	0.16474	0.19638		8
0.00067	0.00340	0.01159	0.02975	0.06121	0.10483	0.15274		9
0.00010	0.00068	0.00298	0.00961	0.02449	0.05146	0.09164		10
0.00001	0.00010	0.00058	0.00235	0.00742	0.01914	0.04166		11
0.00000	0.00001	0.00008	0.00042	0.00165	0.00522	0.01389		12
0.00000	0.00000	0.00001	0.00005	0.00025	0.00099	0.00320		13
0.00000	0.00000	0.00000	0.00000	0.00002	0.00012	0.00046		14
0.00000	0.00000	0.00000	0.00000	0.00000	0.00006	0.00003		15
0.01153	0.00317	0.00080	0.00018	0.00004	0.00001	0.00000	20	0
0.05765	0.02114	0.00684	0.00195	0.00049	0.00011	0.00002		1
0.13691	0.06695	0.02785	0.00999	0.00309	0.00082	0.00018		2
0.20536	0.13390	0.07160	0.03226	0.01235	0.00401	0.00109		3
0.21820	0.18969	0.13042	0.07382	0.03499	0.01393	0.00462		4
0.17456	0.20233	0.17886	0.12720	0.07465	0.03647	0.01479		5
0.10910	0.16861	0.19164	0.17123	0.12441	0.07460	0.03696		6
0.05455	0.11241	0.16426	0.18440	0.16588	0.12207	0.07393		7
0.02216	0.06089	0.11440	0.16135	0.17971	0.16230	0.12013		8
0.00739	0.02706	0.06537	0.11584	0.15974	0.17706	0.16018		9
0.00203	0.00992	0.03082	0.06861	0.11714	0.15935	0.17620		10
0.00046	0.00301	0.01201	0.03359	0.07100	0.11852	0.16018		11
0.00009	0.00075	0.00386	0.01356	0.03550	0.07273	0.12013		12
0.00001	0.00015	0.00102	0.00450	0.01456	0.03662	0.07393		13
0.00000	0.00003	0.00022	0.00121	0.00485	0.01498	0.03696		14
0.00000	0.00000	0.00004	0.00026	0.00129	0.00490	0.01479		15
0.00000	0.00000	0.00001	0.00004	0.00027	0.00125	0.00462		16
0.00000	0.00000	0.00000	0.00001	0.00004	0.00024	0.00109		17
0.00000	0.00000	0.00000	0.00000	0.00001	0.00003	0.00018		18
0.00000	0.00000	0.00000	0.00000	0.00000	0.00000	0.00002		19
0.00000	0.00000	0.00000	0.00000	0.00000	0.00000	0.00000		20

EXHIBIT B.3 *Individual Terms, $p_X(x)$, of the Probability Mass Function of the Poisson Distribution with Parameter λ*

	λ									
x	0.1	0.2	0.3	0.4	0.5	0.6	0.7	0.8	0.9	1.0
0	.9048	.8187	.7408	.6703	.6065	.5488	.4966	.4493	.4066	.3679
1	.0905	.1637	.2222	.2681	.3033	.3293	.3476	.3595	.3659	.3679
2	.0045	.0164	.0333	.0536	.0758	.0988	.1217	.1438	.1647	.1839
3	.0002	.0011	.0033	.0072	.0126	.0198	.0284	.0383	.0494	.0613
4	.0000	.0001	.0003	.0007	.0016	.0030	.0050	.0077	.0111	.0153
5	.0000	.0000	.0000	.0001	.0002	.0004	.0007	.0012	.0020	.0031
6	.0000	.0000	.0000	.0000	.0000	.0000	.0001	.0002	.0003	.0005
7	.0000	.0000	.0000	.0000	.0000	.0000	.0000	.0000	.0000	.0001

	λ									
x	1.1	1.2	1.3	1.4	1.5	1.6	1.7	1.8	1.9	2.0
0	.3329	.3012	.2725	.2466	.2231	.2019	.1827	.1653	.1496	.1353
1	.3662	.3614	.3543	.3452	.3347	.3230	.3106	.2975	.2842	.2707
2	.2014	.2169	.2303	.2417	.2510	.2584	.2640	.2678	.2700	.2707
3	.0738	.0867	.0998	.1128	.1255	.1378	.1496	.1607	.1710	.1804
4	.0203	.0260	.0324	.0395	.0471	.0551	.0636	.0723	.0812	.0902
5	.0045	.0062	.0084	.0111	.0141	.0176	.0216	.0260	.0309	.0361
6	.0008	.0012	.0018	.0026	.0035	.0047	.0061	.0078	.0098	.0120
7	.0001	.0002	.0003	.0005	.0008	.0011	.0015	.0020	.0027	.0034
8	.0000	.0000	.0001	.0001	.0001	.0002	.0003	.0005	.0006	.0009
9	.0000	.0000	.0000	.0000	.0000	.0000	.0001	.0001	.0001	.0002

	λ									
x	2.1	2.2	2.3	2.4	2.5	2.6	2.7	2.8	2.9	3.0
0	.1225	.1108	.1003	.0907	.0821	.0743	.0672	.0608	.0550	.0498
1	.2572	.2438	.2306	.2177	.2052	.1931	.1815	.1703	.1596	.1494
2	.2700	.2681	.2652	.2613	.2565	.2510	.2450	.2384	.2314	.2240
3	.1890	.1966	.2033	.2090	.2138	.2176	.2205	.2225	.2237	.2240
4	.0992	.1082	.1169	.1254	.1336	.1414	.1488	.1557	.1622	.1680
5	.0417	.0476	.0538	.0602	.0668	.0735	.0804	.0872	.0940	.1008
6	.0146	.0174	.0206	.0241	.0278	.0319	.0362	.0407	.0455	.0504
7	.0044	.0055	.0068	.0083	.0099	.0118	.0139	.0163	.0188	.0216
8	.0011	.0015	.0019	.0025	.0031	.0038	.0047	.0057	.0068	.0081
9	.0003	.0004	.0005	.0007	.0009	.0011	.0014	.0018	.0022	.0027
10	.0001	.0001	.0001	.0002	.0002	.0003	.0004	.0005	.0006	.0008
11	.0000	.0000	.0000	.0000	.0000	.0001	.0001	.0001	.0002	.0002
12	.0000	.0000	.0000	.0000	.0000	.0000	.0000	.0000	.0000	.0001

EXHIBIT B.3 *(continued)*

	λ									
x	3.1	3.2	3.3	3.4	3.5	3.6	3.7	3.8	3.9	4.0
0	.0450	.0408	.0369	.0334	.0302	.0273	.0247	.0224	.0202	.0183
1	.1397	.1304	.1217	.1135	.1057	.0984	.0915	.0850	.0789	.0733
2	.2165	.2087	.2008	.1929	.1850	.1771	.1692	.1615	.1539	.1465
3	.2237	.2226	.2209	.2186	.2158	.2125	.2087	.2046	.2001	.1954
4	.1734	.1781	.1823	.1858	.1888	.1912	.1931	.1944	.1951	.1954
5	.1075	.1140	.1203	.1264	.1322	.1377	.1429	.1477	.1522	.1563
6	.0555	.0608	.0662	.0716	.0771	.0826	.0881	.0936	.0989	.1042
7	.0246	.0278	.0312	.0348	.0385	.0425	.0466	.0508	.0551	.0595
8	.0095	.0111	.0129	.0148	.0169	.0191	.0215	.0241	.0269	.0298
9	.0033	.0040	.0047	.0056	.0066	.0076	.0089	.0102	.0116	.0132
10	.0010	.0013	.0016	.0019	.0023	.0028	.0033	.0039	.0045	.0053
11	.0003	.0004	.0005	.0006	.0007	.0009	.0011	.0013	.0016	.0019
12	.0001	.0001	.0001	.0002	.0002	.0003	.0003	.0004	.0005	.0006
13	.0000	.0000	.0000	.0000	.0001	.0001	.0001	.0001	.0002	.0002
14	.0000	.0000	.0000	.0000	.0000	.0000	.0000	.0000	.0000	.0001

	λ									
x	4.1	4.2	4.3	4.4	4.5	4.6	4.7	4.8	4.9	5.0
0	.0166	.0150	.0136	.0123	.0111	.0101	.0091	.0082	.0074	.0067
1	.0679	.0630	.0583	.0540	.0500	.0462	.0427	.0395	.0365	.0337
2	.1393	.1323	.1254	.1188	.1125	.1063	.1005	.0948	.0894	.0842
3	.1904	.1852	.1798	.1743	.1687	.1631	.1574	.1517	.1460	.1404
4	.1951	.1944	.1933	.1917	.1898	.1875	.1849	.1820	.1789	.1755
5	.1600	.1633	.1662	.1687	.1708	.1725	.1738	.1747	.1753	.1755
6	.1093	.1143	.1191	.1237	.1281	.1323	.1362	.1398	.1432	.1462
7	.0640	.0686	.0732	.0778	.0824	.0869	.0914	.0959	.1002	.1044
8	.0328	.0360	.0393	.0428	.0463	.0500	.0537	.0575	.0614	.0653
9	.0150	.0168	.0188	.0209	.0232	.0255	.0280	.0307	.0334	.0363
10	.0061	.0071	.0081	.0092	.0104	.0118	.0132	.0147	.0164	.0181
11	.0023	.0027	.0032	.0037	.0043	.0049	.0056	.0064	.0073	.0082
12	.0008	.0009	.0011	.0014	.0016	.0019	.0022	.0026	.0030	.0034
13	.0002	.0003	.0004	.0005	.0006	.0007	.0008	.0009	.0011	.0013
14	.0001	.0001	.0001	.0001	.0002	.0002	.0003	.0003	.0004	.0005
15	.0000	.0000	.0000	.0000	.0001	.0001	.0001	.0001	.0001	.0002

	λ									
x	5.1	5.2	5.3	5.4	5.5	5.6	5.7	5.8	5.9	6.0
0	.0061	.0055	.0050	.0045	.0041	.0037	.0033	.0030	.0027	.0025
1	.0311	.0287	.0265	.0244	.0225	.0207	.0191	.0176	.0162	.0149
2	.0793	.0746	.0701	.0659	.0618	.0580	.0544	.0509	.0477	.0446
3	.1348	.1293	.1239	.1185	.1133	.1082	.1033	.0985	.0938	.0892
4	.1719	.1681	.1641	.1600	.1558	.1515	.1472	.1428	.1383	.1339
5	.1753	.1748	.1740	.1728	.1714	.1697	.1678	.1656	.1632	.1606
6	.1490	.1515	.1537	.1555	.1571	.1584	.1594	.1601	.1605	.1606
7	.1086	.1125	.1163	.1200	.1234	.1267	.1298	.1326	.1353	.1377
8	.0692	.0731	.0771	.0810	.0849	.0887	.0925	.0962	.0998	.1033
9	.0392	.0423	.0454	.0486	.0519	.0552	.0586	.0620	.0654	.0688
10	.0200	.0220	.0241	.0262	.0285	.0309	.0334	.0359	.0386	.0413
11	.0093	.0104	.0116	.0129	.0143	.0157	.0173	.0190	.0207	.0225
12	.0039	.0045	.0051	.0058	.0065	.0073	.0082	.0092	.0102	.0113
13	.0015	.0018	.0021	.0024	.0028	.0032	.0036	.0041	.0046	.0052
14	.0006	.0007	.0008	.0009	.0011	.0013	.0015	.0017	.0019	.0022
15	.0002	.0002	.0003	.0003	.0004	.0005	.0006	.0007	.0008	.0009
16	.0001	.0001	.0001	.0001	.0001	.0002	.0002	.0002	.0003	.0003
17	.0000	.0000	.0000	.0000	.0000	.0000	.0001	.0001	.0001	.0001

	λ									
x	6.1	6.2	6.3	6.4	6.5	6.6	6.7	6.8	6.9	7.0
0	.0022	.0020	.0018	.0017	.0015	.0014	.0012	.0011	.0010	.0009
1	.0137	.0126	.0116	.0106	.0098	.0090	.0082	.0076	.0070	.0064
2	.0417	.0390	.0364	.0340	.0318	.0296	.0276	.0258	.0240	.0223
3	.0848	.0806	.0765	.0726	.0688	.0652	.0617	.0584	.0552	.0521
4	.1294	.1249	.1205	.1162	.1118	.1076	.1034	.0992	.0952	.0912
5	.1579	.1549	.1519	.1487	.1454	.1420	.1385	.1349	.1314	.1277
6	.1605	.1601	.1595	.1586	.1575	.1562	.1546	.1529	.1511	.1490
7	.1399	.1418	.1435	.1450	.1462	.1472	.1480	.1486	.1489	.1490
8	.1066	.1099	.1130	.1160	.1188	.1215	.1240	.1263	.1284	.1304
9	.0723	.0757	.0791	.0825	.0858	.0891	.0923	.0954	.0985	.1014
10	.0441	.0469	.0498	.0528	.0558	.0588	.0618	.0649	.0679	.0710
11	.0245	.0265	.0285	.0307	.0330	.0353	.0377	.0401	.0426	.0452
12	.0124	.0137	.0150	.0164	.0179	.0194	.0210	.0227	.0245	.0264
13	.0058	.0065	.0073	.0081	.0089	.0098	.0108	.0119	.0130	.0142
14	.0025	.0029	.0033	.0037	.0041	.0046	.0052	.0058	.0064	.0071
15	.0010	.0012	.0014	.0016	.0018	.0020	.0023	.0026	.0029	.0033
16	.0004	.0005	.0005	.0006	.0007	.0008	.0010	.0011	.0013	.0014
17	.0001	.0002	.0002	.0002	.0003	.0003	.0004	.0004	.0005	.0006
18	.0000	.0001	.0001	.0001	.0001	.0001	.0001	.0002	.0002	.0002
19	.0000	.0000	.0000	.0000	.0000	.0000	.0000	.0001	.0001	.0001

EXHIBIT B.3 *(continued)*

	λ									
x	7.1	7.2	7.3	7.4	7.5	7.6	7.7	7.8	7.9	8.0
0	.0008	.0007	.0007	.0006	.0006	.0005	.0005	.0004	.0004	.0003
1	.0059	.0054	.0049	.0045	.0041	.0038	.0035	.0032	.0029	.0027
2	.0208	.0194	.0180	.0167	.0156	.0145	.0134	.0125	.0116	.0107
3	.0492	.0464	.0438	.0413	.0389	.0366	.0345	.0324	.0305	.0286
4	.0874	.0836	.0799	.0764	.0729	.0696	.0663	.0632	.0602	.0573
5	.1241	.1204	.1167	.1130	.1094	.1057	.1021	.0986	.0951	.0916
6	.1468	.1445	.1420	.1394	.1367	.1339	.1311	.1282	.1252	.1221
7	.1489	.1486	.1481	.1474	.1465	.1454	.1442	.1428	.1413	.1396
8	.1321	.1337	.1351	.1363	.1373	.1382	.1388	.1392	.1395	.1396
9	.1042	.1070	.1096	.1121	.1144	.1167	.1187	.1207	.1224	.1241
10	.0740	.0770	.0800	.0829	.0858	.0887	.0914	.0941	.0967	.0993
11	.0478	.0504	.0531	.0558	.0585	.0613	.0640	.0667	.0695	.0722
12	.0283	.0303	.0323	.0344	.0366	.0388	.0411	.0434	.0457	.0481
13	.0154	.0168	.0181	.0196	.0211	.0227	.0243	.0260	.0278	.0296
14	.0078	.0086	.0095	.0104	.0113	.0123	.0134	.0145	.0157	.0169
15	.0037	.0041	.0046	.0051	.0057	.0062	.0069	.0075	.0083	.0090
16	.0016	.0019	.0021	.0024	.0026	.0030	.0033	.0037	.0041	.0045
17	.0007	.0008	.0009	.0010	.0012	.0013	.0015	.0017	.0019	.0021
18	.0003	.0003	.0004	.0004	.0005	.0006	.0006	.0007	.0008	.0009
19	.0001	.0001	.0001	.0002	.0002	.0002	.0003	.0003	.0003	.0004
20	.0000	.0000	.0001	.0001	.0001	.0001	.0001	.0001	.0001	.0002
21	.0000	.0000	.0000	.0000	.0000	.0000	.0000	.0000	.0001	.0001

	λ									
x	8.1	8.2	8.3	8.4	8.5	8.6	8.7	8.8	8.9	9.0
0	.0003	.0003	.0002	.0002	.0002	.0002	.0002	.0002	.0001	.0001
1	.0025	.0023	.0021	.0019	.0017	.0016	.0014	.0013	.0012	.0011
2	.0100	.0092	.0086	.0079	.0074	.0068	.0063	.0058	.0054	.0050
3	.0269	.0252	.0237	.0222	.0208	.0195	.0183	.0171	.0160	.0150
4	.0544	.0517	.0491	.0466	.0443	.0420	.0398	.0377	.0357	.0337
5	.0882	.0849	.0816	.0784	.0752	.0722	.0692	.0663	.0635	.0607
6	.1191	.1160	.1128	.1097	.1066	.1034	.1003	.0972	.0941	.0911
7	.1378	.1358	.1338	.1317	.1294	.1271	.1247	.1222	.1197	.1171
8	.1395	.1392	.1388	.1382	.1375	.1366	.1356	.1344	.1332	.1318
9	.1256	.1269	.1280	.1290	.1299	.1306	.1311	.1315	.1317	.1318
10	.1017	.1040	.1063	.1084	.1104	.1123	.1140	.1157	.1172	.1186
11	.0749	.0776	.0802	.0828	.0853	.0878	.0902	.0925	.0948	.0970
12	.0505	.0530	.0555	.0579	.0604	.0629	.0654	.0679	.0703	.0728
13	.0315	.0334	.0354	.0374	.0395	.0416	.0438	.0459	.0481	.0504
14	.0182	.0196	.0210	.0225	.0240	.0256	.0272	.0289	.0306	.0324

EXHIBIT B.3 *(continued)*

	λ									
x	8.1	8.2	8.3	8.4	8.5	8.6	8.7	8.8	8.9	9.0
15	.0098	.0107	.0116	.0126	.0136	.0147	.0158	.0169	.0182	.0194
16	.0050	.0055	.0060	.0066	.0072	.0079	.0086	.0093	.0101	.0109
17	.0024	.0026	.0029	.0033	.0036	.0040	.0044	.0048	.0053	.0058
18	.0011	.0012	.0014	.0015	.0017	.0019	.0021	.0024	.0026	.0029
19	.0005	.0005	.0006	.0007	.0008	.0009	.0010	.0011	.0012	.0014
20	.0002	.0002	.0002	.0003	.0003	.0004	.0004	.0005	.0005	.0006
21	.0001	.0001	.0001	.0001	.0001	.0002	.0002	.0002	.0002	.0003
22	.0000	.0000	.0000	.0000	.0001	.0001	.0001	.0001	.0001	.0001

	λ									
x	9.1	9.2	9.3	9.4	9.5	9.6	9.7	9.8	9.9	10
0	.0001	.0001	.0001	.0001	.0001	.0001	.0001	.0001	.0001	.0000
1	.0010	.0009	.0009	.0008	.0007	.0007	.0006	.0005	.0005	.0005
2	.0046	.0043	.0040	.0037	.0034	.0031	.0029	.0027	.0025	.0023
3	.0140	.0131	.0123	.0115	.0107	.0100	.0093	.0087	.0081	.0076
4	.0319	.0302	.0285	.0269	.0254	.0240	.0226	.0213	.0201	.0189
5	.0581	.0555	.0530	.0506	.0483	.0460	.0439	.0418	.0398	.0378
6	.0881	.0851	.0822	.0793	.0764	.0736	.0709	.0682	.0656	.0631
7	.1145	.1118	.1091	.1064	.1037	.1010	.0982	.0955	.0928	.0901
8	.1302	.1286	.1269	.1251	.1232	.1212	.1191	.1170	.1148	.1126
9	.1317	.1315	.1311	.1306	.1300	.1293	.1284	.1274	.1263	.1251
10	.1198	.1210	.1219	.1228	.1235	.1241	.1245	.1249	.1250	.1251
11	.0991	.1012	.1031	.1049	.1067	.1083	.1098	.1112	.1125	.1137
12	.0752	.0776	.0799	.0822	.0844	.0866	.0888	.0908	.0928	.0948
13	.0526	.0549	.0572	.0594	.0617	.0640	.0662	.0685	.0707	.0729
14	.0342	.0361	.0380	.0399	.0419	.0439	.0459	.0479	.0500	.0521
15	.0208	.0221	.0235	.0250	.0265	.0281	.0297	.0313	.0330	.0347
16	.0118	.0127	.0137	.0147	.0157	.0168	.0180	.0192	.0204	.0217
17	.0063	.0069	.0075	.0081	.0088	.0095	.0103	.0111	.0119	.0128
18	.0032	.0035	.0039	.0042	.0046	.0051	.0055	.0060	.0065	.0071
19	.0015	.0017	.0019	.0021	.0023	.0026	.0028	.0031	.0034	.0037
20	.0007	.0008	.0009	.0010	.0011	.0012	.0014	.0015	.0017	.0019
21	.0003	.0003	.0004	.0004	.0005	.0006	.0006	.0007	.0008	.0009
22	.0001	.0001	.0002	.0002	.0002	.0002	.0003	.0003	.0004	.0004
23	.0000	.0001	.0001	.0001	.0001	.0001	.0001	.0001	.0002	.0002
24	.0000	.0000	.0000	.0000	.0000	.0000	.0000	.0001	.0001	.0001

EXHIBIT B.3 (continued)

	λ									
x	11	12	13	14	15	16	17	18	19	20
0	.0000	.0000	.0000	.0000	.0000	.0000	.0000	.0000	.0000	.0000
1	.0002	.0001	.0000	.0000	.0000	.0000	.0000	.0000	.0000	.0000
2	.0010	.0004	.0002	.0001	.0000	.0000	.0000	.0000	.0000	.0000
3	.0037	.0018	.0008	.0004	.0002	.0001	.0000	.0000	.0000	.0000
4	.0102	.0053	.0027	.0013	.0006	.0003	.0001	.0001	.0000	.0000
5	.0224	.0127	.0070	.0037	.0019	.0010	.0005	.0002	.0001	.0001
6	.0411	.0255	.0152	.0087	.0048	.0026	.0014	.0007	.0004	.0002
7	.0646	.0437	.0281	.0174	.0104	.0060	.0034	.0018	.0010	.0005
8	.0888	.0655	.0457	.0304	.0194	.0120	.0072	.0042	.0024	.0013
9	.1085	.0874	.0661	.0473	.0324	.0213	.0135	.0083	.0050	.0029
10	.1194	.1048	.0859	.0663	.0486	.0341	.0230	.0150	.0095	.0058
11	.1194	.1144	.1015	.0844	.0663	.0496	.0355	.0245	.0164	.0106
12	.1094	.1144	.1099	.0984	.0829	.0661	.0504	.0368	.0259	.0176
13	.0926	.1056	.1099	.1060	.0956	.0814	.0658	.0509	.0378	.0271
14	.0728	.0905	.1021	.1060	.1024	.0930	.0800	.0655	.0514	.0387
15	.0534	.0724	.0885	.0989	.1024	.0992	.0906	.0786	.0650	.0516
16	.0367	.0543	.0719	.0866	.0960	.0992	.0963	.0884	.0772	.0646
17	.0237	.0383	.0550	.0713	.0847	.0934	.0963	.0936	.0863	.0760
18	.0145	.0256	.0397	.0554	.0706	.0830	.0909	.0936	.0911	.0844
19	.0084	.0161	.0272	.0409	.0557	.0699	.0814	.0887	.0911	.0888
20	.0046	.0097	.0177	.0286	.0418	.0559	.0692	.0798	.0866	.0888
21	.0024	.0055	.0109	.0191	.0299	.0426	.0560	.0684	.0783	.0846
22	.0012	.0030	.0065	.0121	.0204	.0310	.0433	.0560	.0676	.0769
23	.0006	.0016	.0037	.0074	.0133	.0216	.0320	.0438	.0559	.0669
24	.0003	.0008	.0020	.0043	.0083	.0144	.0226	.0328	.0442	.0557
25	.0001	.0004	.0010	.0024	.0050	.0092	.0154	.0237	.0336	.0446
26	.0000	.0002	.0005	.0013	.0029	.0057	.0101	.0164	.0246	.0343
27	.0000	.0001	.0002	.0007	.0016	.0034	.0063	.0109	.0173	.0254
28	.0000	.0000	.0001	.0003	.0009	.0019	.0038	.0070	.0117	.0181
29	.0000	.0000	.0001	.0002	.0004	.0011	.0023	.0044	.0077	.0125
30	.0000	.0000	.0000	.0001	.0002	.0006	.0013	.0026	.0049	.0083
31	.0000	.0000	.0000	.0000	.0001	.0003	.0007	.0015	.0030	.0054
32	.0000	.0000	.0000	.0000	.0001	.0001	.0004	.0009	.0018	.0034
33	.0000	.0000	.0000	.0000	.0000	.0001	.0002	.0005	.0010	.0020
34	.0000	.0000	.0000	.0000	.0000	.0000	.0001	.0002	.0006	.0012
35	.0000	.0000	.0000	.0000	.0000	.0000	.0000	.0001	.0003	.0007
36	.0000	.0000	.0000	.0000	.0000	.0000	.0000	.0001	.0002	.0004
37	.0000	.0000	.0000	.0000	.0000	.0000	.0000	.0000	.0001	.0002
38	.0000	.0000	.0000	.0000	.0000	.0000	.0000	.0000	.0000	.0001
39	.0000	.0000	.0000	.0000	.0000	.0000	.0000	.0000	.0000	.0001

Source Reprinted with permission from William H. Beyer (Ed.), *Handbook of Tables for Probability and Statistics*, 2nd ed., pp. 207–11. Copyright 1964 The Chemical Rubber Co., CRC Press, Inc.

EXHIBIT B.4 *Cumulative Distribution Function of a Standard Normal Random Variable*

k	0.00	0.01	0.02	0.03	0.04	0.05	0.06	0.07	0.08	0.09
0.0	0.50000	0.50399	0.50798	0.51197	0.51595	0.51994	0.52392	0.52790	0.53188	0.53586
0.1	0.53983	0.54380	0.54776	0.55172	0.55567	0.55962	0.56356	0.56749	0.57142	0.57535
0.2	0.57926	0.58317	0.58706	0.59095	0.59483	0.59871	0.60257	0.60642	0.61026	0.61409
0.3	0.61791	0.62172	0.62552	0.62930	0.63307	0.63683	0.64058	0.64431	0.64803	0.65173
0.4	0.65542	0.65910	0.66276	0.66640	0.67003	0.67364	0.67724	0.68082	0.68439	0.68793
0.5	0.69146	0.69497	0.69847	0.70194	0.70540	0.70884	0.71226	0.71566	0.71904	0.72240
0.6	0.72575	0.72907	0.73237	0.73565	0.73891	0.74215	0.74537	0.74857	0.75175	0.75490
0.7	0.75804	0.76115	0.76424	0.76730	0.77035	0.77337	0.77637	0.77935	0.78230	0.78524
0.8	0.78814	0.79103	0.79389	0.79673	0.79955	0.80234	0.80511	0.80785	0.81057	0.81327
0.9	0.81594	0.81859	0.82121	0.82381	0.82639	0.82894	0.83147	0.83398	0.83646	0.83891
1.0	0.84134	0.84375	0.84614	0.84849	0.85083	0.85314	0.85543	0.85769	0.85993	0.86214
1.1	0.86433	0.86650	0.86864	0.87076	0.87286	0.87493	0.87698	0.87900	0.88100	0.88298
1.2	0.88493	0.88686	0.88877	0.89065	0.89251	0.89435	0.89617	0.89796	0.89973	0.90147
1.3	0.90320	0.90490	0.90658	0.90824	0.90988	0.91149	0.91309	0.91466	0.91621	0.91774
1.4	0.91924	0.92073	0.92220	0.92364	0.92507	0.92647	0.92785	0.92922	0.93056	0.93189
1.5	0.93319	0.93448	0.93574	0.93699	0.93822	0.93943	0.94062	0.94179	0.94295	0.94408
1.6	0.94520	0.94630	0.94738	0.94845	0.94950	0.95053	0.95154	0.95254	0.95352	0.95449
1.7	0.95543	0.95637	0.95728	0.95818	0.95907	0.95994	0.96080	0.96164	0.96246	0.96327
1.8	0.96407	0.96485	0.96562	0.96638	0.96712	0.96784	0.96856	0.96926	0.96995	0.97062
1.9	0.97128	0.97193	0.97257	0.97320	0.97381	0.97441	0.97500	0.97558	0.97615	0.97670

2.0	0.97725	0.97778	0.97831	0.97882	0.97932	0.97982	0.98030	0.98077	0.98124	0.98169
2.1	0.98214	0.98257	0.98300	0.98341	0.98382	0.98422	0.98461	0.98500	0.98537	0.98574
2.2	0.98610	0.98645	0.98679	0.98713	0.98745	0.98778	0.98809	0.98840	0.98870	0.98899
2.3	0.98928	0.98956	0.98983	0.99010	0.99036	0.99061	0.99086	0.99111	0.99134	0.99158
2.4	0.99180	0.99202	0.99224	0.99245	0.99266	0.99286	0.99305	0.99324	0.99343	0.99361
2.5	0.99379	0.99396	0.99413	0.99430	0.99446	0.99461	0.99477	0.99492	0.99506	0.99520
2.6	0.99534	0.99547	0.99560	0.99573	0.99585	0.99598	0.99609	0.99621	0.99632	0.99643
2.7	0.99653	0.99664	0.99674	0.99683	0.99693	0.99702	0.99711	0.99720	0.99728	0.99736
2.8	0.99744	0.99752	0.99760	0.99767	0.99774	0.99781	0.99788	0.99795	0.99801	0.99807
2.9	0.99813	0.99819	0.99825	0.99831	0.99836	0.99841	0.99846	0.99851	0.99856	0.99861
3.0	0.99865	0.99869	0.99874	0.99878	0.99882	0.99886	0.99889	0.99893	0.99896	0.99900
3.1	0.99903	0.99906	0.99910	0.99913	0.99916	0.99918	0.99921	0.99924	0.99926	0.99929
3.2	0.99931	0.99934	0.99936	0.99938	0.99940	0.99942	0.99944	0.99946	0.99948	0.99950
3.3	0.99952	0.99953	0.99955	0.99957	0.99958	0.99960	0.99961	0.99962	0.99964	0.99965
3.4	0.99966	0.99968	0.99969	0.99970	0.99971	0.99972	0.99973	0.99974	0.99975	0.99976
3.5	0.99977	0.99978	0.99978	0.99979	0.99980	0.99981	0.99981	0.99982	0.99983	0.99983
3.6	0.99984	0.99985	0.99985	0.99986	0.99986	0.99987	0.99987	0.99988	0.99988	0.99989
3.7	0.99989	0.99990	0.99990	0.99990	0.99991	0.99991	0.99992	0.99992	0.99992	0.99992
3.8	0.99993	0.99993	0.99993	0.99994	0.99994	0.99994	0.99994	0.99995	0.99995	0.99995
3.9	0.99995	0.99995	0.99996	0.99996	0.99996	0.99996	0.99996	0.99996	0.99997	0.99997
4.0	0.99997	0.99997	0.99997	0.99997	0.99997	0.99997	0.99998	0.99998	0.99998	0.99998

EXHIBIT B.5 *Values of The Incomplete Gamma Function $I_r(t)$*

	r									
t	0.5	1	1.5	2	2.5	3	3.5	4	4.5	5
0.2	0.47291	0.18127	0.05976	0.01752	0.00467	0.00115	0.00026	0.00006	0.00001	0.00000
0.4	0.62891	0.32968	0.15053	0.06155	0.02297	0.00793	0.00256	0.00078	0.00022	0.00006
0.6	0.72668	0.45119	0.24700	0.12190	0.05512	0.02312	0.00907	0.00336	0.00118	0.00039
0.8	0.79410	0.55067	0.34061	0.19121	0.09875	0.04742	0.02136	0.00908	0.00367	0.00141
1.0	0.84270	0.63212	0.42759	0.26424	0.15085	0.08030	0.04016	0.01899	0.00853	0.00366
1.2	0.87866	0.69881	0.50637	0.33737	0.20853	0.12051	0.06556	0.03377	0.01655	0.00775
1.4	0.90574	0.75340	0.57650	0.40817	0.26921	0.16650	0.09713	0.05372	0.02830	0.01425
1.6	0.92636	0.79810	0.63819	0.47507	0.33082	0.21664	0.13410	0.07881	0.04417	0.02368
1.8	0.94222	0.83470	0.69198	0.53716	0.39169	0.26938	0.17548	0.10871	0.06428	0.03641
2.0	0.95450	0.86466	0.73854	0.59399	0.45058	0.32332	0.22022	0.14288	0.08859	0.05265
2.2	0.96406	0.88920	0.77861	0.64543	0.50663	0.37729	0.26728	0.18065	0.11683	0.07250
2.4	0.97154	0.90928	0.81296	0.69156	0.55923	0.43029	0.31565	0.22128	0.14862	0.09587
2.6	0.97741	0.92573	0.84228	0.73262	0.60804	0.48157	0.36443	0.26400	0.18346	0.12258
2.8	0.98204	0.93919	0.86722	0.76892	0.65289	0.53055	0.41285	0.30806	0.22081	0.15232
3.0	0.98569	0.95021	0.88839	0.80085	0.69378	0.57681	0.46025	0.35277	0.26008	0.18474
3.2	0.98859	0.95924	0.90631	0.82880	0.73078	0.62010	0.50611	0.39748	0.30069	0.21939
3.4	0.99088	0.96663	0.92145	0.85316	0.76406	0.66026	0.55000	0.44164	0.34207	0.25582
3.6	0.99271	0.97268	0.93421	0.87431	0.79381	0.69725	0.59164	0.48478	0.38369	0.29356
3.8	0.99416	0.97763	0.94496	0.89262	0.82030	0.73110	0.63082	0.52652	0.42510	0.33216
4.0	0.99532	0.98168	0.95399	0.90842	0.84376	0.76190	0.66741	0.56653	0.46585	0.37116
4.2	0.99625	0.98500	0.96157	0.92202	0.86447	0.78976	0.70135	0.60460	0.50561	0.41017
4.4	0.99699	0.98772	0.96793	0.93370	0.88269	0.81486	0.73266	0.64055	0.54406	0.44882
4.6	0.99758	0.98995	0.97325	0.94371	0.89865	0.83736	0.76139	0.67429	0.58098	0.48677
4.8	0.99805	0.99177	0.97771	0.95227	0.91260	0.85746	0.78760	0.70577	0.61617	0.52374
5.0	0.99843	0.99326	0.98143	0.95957	0.92476	0.87535	0.81143	0.73497	0.64951	0.55951
5.2	0.99874	0.99448	0.98455	0.96580	0.93534	0.89121	0.83298	0.76193	0.68092	0.59387
5.4	0.99898	0.99548	0.98714	0.97109	0.94451	0.90524	0.85242	0.78671	0.71033	0.62669
5.6	0.99918	0.99630	0.98931	0.97559	0.95244	0.91761	0.86987	0.80938	0.73775	0.65785
5.8	0.99934	0.99697	0.99111	0.97941	0.95930	0.92849	0.88550	0.83004	0.76319	0.68728
6.0	0.99947	0.99752	0.99262	0.98265	0.96521	0.93803	0.89944	0.84880	0.78669	0.71494
6.2	0.99957	0.99797	0.99387	0.98539	0.97030	0.94638	0.91185	0.86577	0.80831	0.74082
6.4	0.99965	0.99834	0.99491	0.98770	0.97467	0.95368	0.92287	0.88108	0.82813	0.76493
6.6	0.99972	0.99864	0.99578	0.98966	0.97843	0.96003	0.93262	0.89485	0.84624	0.78730
6.8	0.99977	0.99889	0.99650	0.99131	0.98164	0.96556	0.94123	0.90719	0.86272	0.80797
7.0	0.99982	0.99909	0.99709	0.99270	0.98439	0.97036	0.94882	0.91823	0.87767	0.82701
7.2	0.99985	0.99925	0.99759	0.99388	0.98674	0.97453	0.95549	0.92808	0.89121	0.84448
7.4	0.99988	0.99939	0.99800	0.99487	0.98875	0.97813	0.96135	0.93685	0.90342	0.86047
7.6	0.99990	0.99950	0.99835	0.99570	0.99046	0.98124	0.96648	0.94463	0.91441	0.87506
7.8	0.99992	0.99959	0.99863	0.99639	0.99192	0.98393	0.97097	0.95152	0.92428	0.88833
8.0	0.99994	0.99966	0.99887	0.99698	0.99316	0.98625	0.97488	0.95762	0.93312	0.90037
9.0	0.99998	0.99988	0.99956	0.99877	0.99705	0.99377	0.98803	0.97877	.96483	0.94504
10.0	0.99999	0.99995	0.99983	0.99950	0.99875	0.99723	0.99443	0.98966	0.98209	0.97075
11.0	1.00000	0.99998	0.99993	0.99980	0.99948	0.99879	0.99746	0.99508	0.99112	0.98490
12.0	1.00000	0.99999	0.99998	0.99992	0.99978	0.99948	0.99886	0.99771	0.99570	0.99240
13.0	1.00000	1.00000	0.99999	0.99997	0.99991	0.99978	0.99950	0.99895	0.99796	0.99626
14.0	1.00000	1.00000	1.00000	0.99999	0.99996	0.99991	0.99978	0.99953	0.99905	0.99819
15.0	1.00000	1.00000	1.00000	0.10000	0.99999	0.99996	0.99991	0.99979	0.99956	0.99914

5.5	6	6.5	7	7.5	8	8.5	9	9.5	t
0.00000	0.00000	0.00000	0.00000	0.00000	0.00000	0.00000	0.00000	0.00000	0.2
0.00002	0.00000	0.00000	0.00000	0.00000	0.00000	0.00000	0.00000	0.00000	0.4
0.00013	0.00003	0.00001	0.00000	0.00000	0.00000	0.00000	0.00000	0.00000	0.6
0.00052	0.00018	0.00006	0.00002	0.00001	0.00000	0.00000	0.00000	0.00000	0.8
0.00150	0.00059	0.00023	0.00008	0.00003	0.00001	0.00000	0.00000	0.00000	1.0
0.00348	0.00150	0.00062	0.00025	0.00010	0.00004	0.00001	0.00000	0.00000	1.2
0.00689	0.00320	0.00144	0.00062	0.00026	0.00011	0.00004	0.00002	0.00001	1.4
0.01219	0.00604	0.00289	0.00134	0.00060	0.00026	0.00011	0.00005	0.00002	1.6
0.01981	0.01038	0.00525	0.00257	0.00122	0.00056	0.00025	0.00011	0.00005	1.8
0.03008	0.01656	0.00881	0.00453	0.00226	0.00110	0.00052	0.00024	0.00011	2.0
0.04328	0.02491	0.01386	0.00746	0.00390	0.00198	0.00098	0.00047	0.00022	2.2
0.05954	0.03567	0.02066	0.01159	0.00631	0.00334	0.00172	0.00086	0.00042	2.4
0.07891	0.04904	0.02948	0.01717	0.00971	0.00533	0.00285	0.00149	0.00076	2.6
0.10132	0.06511	0.04049	0.02441	0.01429	0.00813	0.00450	0.00243	0.00128	2.8
0.12664	0.08392	0.05385	0.03351	0.02025	0.01190	0.00681	0.00380	0.00207	3.0
0.15461	0.10541	0.06962	0.04462	0.02778	0.01683	0.00993	0.00571	0.00321	3.2
0.18496	0.12946	0.08784	0.05785	0.03704	0.02307	0.01401	0.00829	0.00479	3.4
0.21734	0.15588	0.10845	0.07327	0.04814	0.03079	0.01919	0.01167	0.00693	3.6
0.25138	0.18444	0.13135	0.09089	0.06118	0.04011	0.02563	0.01598	0.00974	3.8
0.28670	0.21487	0.15640	0.11067	0.07622	0.05113	0.03345	0.02136	0.01333	4.0
0.32291	0.24686	0.18340	0.13254	0.09325	0.06394	0.04277	0.02793	0.01783	4.2
0.35965	0.28009	0.21212	0.15635	0.11226	0.07858	0.05367	0.03580	0.02334	4.4
0.39656	0.31424	0.24232	0.18197	0.13317	0.09505	0.06622	0.04507	0.02999	4.6
0.43331	0.34899	0.27373	0.20920	0.15588	0.11333	0.08046	0.05582	0.03787	4.8
0.46961	0.38404	0.30607	0.23782	0.18026	0.13337	0.09639	0.06809	0.04705	5.0
0.53983	0.41909	0.33906	0.26761	0.20615	0.15508	0.11400	0.08193	0.05762	5.2
0.50519	0.45387	0.37243	0.29833	0.23336	0.17834	0.13323	0.09735	0.06962	5.4
0.57334	0.48814	0.40593	0.32974	0.26171	0.20302	0.15402	0.11432	0.08307	5.6
0.60555	0.52169	0.43932	0.36161	0.29098	0.22897	0.17627	0.13281	0.09800	5.8
0.63636	0.55432	0.47236	0.39370	0.32097	0.25602	0.19986	0.15276	0.11437	6.0
0.66566	0.58589	0.50485	0.42579	0.35147	0.28398	0.22467	0.17409	0.13218	6.2
0.69340	0.61626	0.53662	0.45767	0.38226	0.31268	0.25053	0.19669	0.15135	6.4
0.71955	0.64533	0.56752	0.48916	0.41315	0.34192	0.27730	0.22044	0.17182	6.6
0.74408	0.67302	0.59740	0.52008	0.44394	0.37151	0.30481	0.24523	0.19351	6.8
0.76701	0.69929	0.62616	0.55029	0.47447	0.40129	0.33290	0.27091	0.21631	7.0
0.78836	0.72410	0.65371	0.57964	0.50457	0.43106	0.36139	0.29733	0.24011	7.2
0.80816	0.74744	0.68000	0.60804	0.53408	0.46067	0.39012	0.32435	0.26478	7.4
0.82648	0.76932	0.70497	0.63538	0.56289	0.48996	0.41892	0.35181	0.29020	7.6
0.84336	0.78975	0.72859	0.66159	0.59088	0.51879	0.44765	0.37956	0.31622	7.8
0.85887	0.80876	0.75087	0.68663	0.61795	0.54704	0.47617	0.40745	0.34272	8.0
0.95466	0.88431	0.84248	0.79322	0.73733	0.67610	0.61116	0.54435	0.47756	9.0
0.91842	0.93291	0.90479	0.86986	0.82807	0.77978	0.72577	0.66718	0.60542	10.0
0.97563	0.96248	0.94464	0.92139	0.89220	0.85681	0.81528	0.76801	0.71574	11.0
0.98727	0.97966	0.96887	0.95418	0.93491	0.91050	0.88057	0.84497	0.80385	12.0
0.99351	0.98927	0.98300	0.97411	0.96198	0.94597	0.92554	0.90024	0.86981	13.0
0.99676	0.99447	0.99095	0.98577	0.97843	0.96838	0.95506	0.93794	0.91657	14.0
0.99842	0.99721	0.99529	0.99237	0.98808	0.98200	0.97365	0.96255	0.94820	15.0

EXHIBIT B.6 *Values of the Probability Content of an Upper Quadrant, P{X > h, Y > k}, for the Standard Bivariate Normal Distribution with Correlation ρ (decimal points omitted)*

$\rho = 0.500$

k \ h	0.0	0.1	0.2	0.3	0.4	0.5	0.6	0.7	0.8
0.0	333333	312961	291886	270344	248589	226878	205468	184605	164512
0.1	312961	294422	275161	255392	235345	215260	195377	175927	157126
0.2	291886	275161	257709	239718	221397	202965	184644	166650	149190
0.3	270344	255392	239718	223488	206888	190114	173370	156858	140769
0.4	248589	235345	221397	206888	191979	176847	161676	146649	131946
0.5	226878	215260	202965	190114	176847	163320	149694	136139	122816
0.6	205468	195377	184644	173370	161676	149694	137570	125451	113486
0.7	184605	175927	166650	156858	146649	136139	125451	114718	104071
0.8	164512	157126	149190	140769	131946	122816	113486	104071	094686
0.9	145388	139168	132448	125281	117733	109882	101819	093640	085448
1.0	127398	122215	116586	110550	104159	097477	090578	083546	076465
1.1	110671	106398	101733	096704	091350	085722	079882	073896	067839
1.2	095297	091814	087990	083844	079407	074718	069825	064784	059656
1.3	081329	078522	075421	072042	068404	064539	060485	056284	051988
1.4	068785	066546	064061	061337	058388	055237	051913	048451	044891
1.5	057646	055882	053912	051740	049377	046836	044142	041320	038401
1.6	047867	046493	044950	043238	041365	039340	037180	034905	032539
1.7	039379	038321	037126	035793	034325	032729	031018	029204	027307
1.8	032095	031290	030375	029348	028211	026968	025627	024198	022695
1.9	025912	025307	024615	023834	022963	022006	020968	019854	018677
2.0	020724	020274	019756	019169	018510	017782	016987	016130	015218
2.1	016417	016087	015704	015268	014776	014228	013627	012974	012276
2.2	012882	012642	012363	012043	011679	011272	010823	010332	009804
2.3	010011	009840	009638	009406	009141	008842	008510	008145	007750
2.4	007706	007585	007441	007275	007083	006867	006624	006357	006065
2.5	005875	005790	005689	005571	005435	005280	005105	004911	004698
2.6	004436	004377	004307	004225	004129	004019	003894	003755	003602
2.7	003317	003277	003229	003172	003105	003029	002941	002842	002733
2.8	002457	002430	002397	002358	002312	002259	002198	002130	002053
2.9	001802	001784	001762	001736	001705	001669	001627	001579	001526
3.0	001309	001297	001283	001265	001244	001220	001192	001159	001123
3.1	000942	000934	000925	000913	000899	000883	000864	000842	000817
3.2	000671	000666	000660	000652	000644	000633	000620	000606	000589
3.3	000473	000470	000466	000462	000456	000449	000441	000431	000420
3.4	000331	000329	000326	000323	000320	000315	000310	000304	000297
3.5	000229	000228	000226	000224	000222	000219	000216	000212	000207
3.6	000157	000156	000155	000154	000153	000151	000149	000146	000143
3.7	000107	000106	000106	000105	000104	000103	000102	000100	000098
3.8	000072	000071	000071	000071	000070	000069	000069	000068	000066
3.9	000048	000048	000047	000047	000047	000046	000046	000045	000045
4.0	000031	000031	000031	000031	000031	000031	000030	000030	000030

ρ = 0.500 (*continued*)

k \ h	0.9	1.0	1.1	1.2	1.3	1.4	1.5	1.6	1.7
0.0	145388	127398	110671	095297	081329	068785	057646	047867	039379
0.1	139168	122215	106398	091814	078522	066546	055882	046493	038321
0.2	132448	116586	101733	087990	075421	064061	053912	044950	037126
0.3	125281	110550	096704	083844	072042	061337	051740	043238	035793
0.4	117733	104159	091350	079407	068404	058388	049377	041365	034325
0.5	109882	097477	085722	074718	064539	055237	046836	039340	032729
0.6	101819	090578	079882	069825	060485	051913	044142	037180	031018
0.7	093640	083546	073896	064784	056284	048451	041320	034905	029204
0.8	085448	076465	067839	059656	051988	044891	038401	032539	027307
0.9	077344	069426	061786	054504	047649	041275	035421	030110	025349
1.0	069426	062514	055812	049394	043323	037651	032418	027648	023353
1.1	061786	055812	049991	044388	039062	034063	029428	025184	021345
1.2	054504	049394	044388	039545	034920	030556	026490	022749	019349
1.3	047649	043323	039062	034920	030942	027170	023639	020373	017391
1.4	041275	037651	034063	030556	027170	023944	020907	018085	015494
1.5	035421	032418	029428	026490	023639	020907	018323	015909	013681
1.6	030110	027648	025184	022749	020373	018085	015909	013864	011969
1.7	025349	023353	021345	019349	017391	015494	013681	011969	010372
1.8	021134	019534	017915	016297	014701	013146	011652	010233	008902
1.9	017447	016178	014888	013591	012304	011044	009826	008663	007566
2.0	014260	013266	012249	011221	010196	009186	008204	007261	006367
2.1	011539	010769	009977	009171	008363	007563	006780	006024	005304
2.2	009242	008654	008043	007420	006790	006163	005546	004947	004373
2.3	007328	006883	006419	005941	005457	004971	004490	004021	003568
2.4	005751	005418	005069	004708	004339	003968	003598	003234	002882
2.5	004468	004222	003962	003692	003415	003134	002852	002574	002303
2.6	003435	003255	003065	002865	002659	002449	002237	002027	001820
2.7	002613	002484	002346	002200	002049	001894	001736	001579	001424
2.8	001968	001876	001777	001672	001562	001449	001333	001217	001102
2.9	001467	001402	001332	001257	001178	001097	001013	000928	000843
3.0	001082	001037	000988	000935	000879	000821	000761	000700	000639
3.1	000789	000759	000725	000688	000649	000608	000566	000522	000478
3.2	000570	000549	000526	000501	000474	000446	000416	000385	000354
3.3	000408	000394	000378	000361	000343	000323	000303	000281	000260
3.4	000288	000279	000269	000257	000245	000232	000218	000203	000188
3.5	000202	000196	000189	000181	000173	000164	000155	000145	000135
3.6	000140	000136	000132	000127	000121	000115	000109	000102	000096
3.7	000096	000093	000091	000087	000084	000080	000076	000072	000067
3.8	000065	000064	000062	000060	000057	000055	000052	000049	000046
3.9	000044	000043	000042	000040	000039	000037	000036	000034	000032
4.0	000029	000028	000028	000027	000026	000025	000024	000023	000022

ρ = 0.500 (*continued*)

k \ h	1.8	1.9	2.0	2.1	2.2	2.3	2.4	2.5	2.6
0.0	032095	025912	020724	016417	012882	010011	007706	005875	004436
0.1	031290	025307	020274	016087	012642	009840	007585	005790	004377
0.2	030375	024615	019756	015704	012363	009638	007441	005689	004307
0.3	029348	023834	019169	015268	012043	009406	007275	005571	004225
0.4	028211	022963	018510	014776	011679	009141	007083	005435	004129
0.5	026968	022006	017782	014228	011272	008842	006867	005280	004019
0.6	025627	020968	016987	013627	010823	008510	006624	005105	003894
0.7	024198	019854	016130	012974	010332	008145	006357	004911	003755
0.8	022695	018677	015218	012276	009804	007750	006065	004698	003602
0.9	021134	017447	014260	011539	009242	007328	005751	004468	003435
1.0	019534	016178	013266	010769	008654	006883	005418	004222	003255
1.1	017915	014888	012249	009977	008043	006419	005069	003962	003065
1.2	016297	013591	011221	009171	007420	005941	004708	003692	002865
1.3	014701	012304	010196	008363	006790	005457	004339	003415	002659
1.4	013146	011044	009186	007563	006163	004971	003968	003134	002449
1.5	011652	009826	008204	006780	005546	004490	003598	002852	002237
1.6	010233	008663	007261	006024	004947	004021	003234	002574	002027
1.7	008902	007566	006367	005304	004373	003568	002882	002303	001820
1.8	007671	006546	005531	004626	003830	003138	002544	002041	001620
1.9	006546	005608	004758	003996	003322	002733	002225	001793	001429
2.0	005531	004758	004053	003418	002853	002358	001928	001560	001249
2.1	004626	003996	003418	002895	002427	002014	001654	001344	001081
2.2	003830	003322	002853	002427	002043	001703	001405	001146	000926
2.3	003138	002733	002358	002014	001703	001426	001181	000968	000785
2.4	002544	002225	001928	001654	001405	001181	000983	000809	000659
2.5	002041	001793	001560	001344	001146	000968	000809	000669	000548
2.6	001620	001429	001249	001081	000926	000785	000659	000548	000451
2.7	001273	001127	000989	000859	000740	000630	000532	000444	000367
2.8	000989	000879	000775	000676	000584	000500	000424	000355	000295
2.9	000760	000678	000600	000526	000457	000393	000334	000282	000235
3.0	000578	000518	000460	000405	000353	000305	000261	000221	000185
3.1	000434	000391	000349	000308	000270	000234	000201	000171	000144
3.2	000323	000292	000262	000232	000204	000178	000154	000131	000111
3.3	000238	000216	000194	000173	000153	000134	000116	000099	000084
3.4	000173	000157	000142	000127	000113	000099	000086	000075	000064
3.5	000124	000114	000103	000093	000083	000073	000064	000055	000047
3.6	000088	000081	000074	000067	000060	000053	000047	000040	000035
3.7	000062	000057	000052	000048	000043	000038	000034	000029	000025
3.8	000043	000040	000037	000033	000030	000027	000024	000021	000018
3.9	000030	000028	000025	000023	000021	000019	000017	000015	000013
4.0	000020	000019	000017	000016	000015	000013	000012	000010	000009

ρ = 0.500 (*continued*)

k \ h	2.7	2.8	2.9	3.0	3.1	3.2	3.3	3.4	3.5
0.0	003317	002457	001802	001309	000942	000671	000473	000331	000229
0.1	003277	002430	001784	001297	000934	000666	000470	000329	000228
0.2	003229	002397	001762	001283	000925	000660	000466	000326	000226
0.3	003172	002358	001736	001265	000913	000652	000462	000323	000224
0.4	003105	002312	001705	001244	000899	000644	000456	000320	000222
0.5	003029	002259	001669	001220	000883	000633	000449	000315	000219
0.6	002941	002198	001627	001192	000864	000620	000441	000310	000216
0.7	002842	002130	001579	001159	000842	000606	000431	000304	000212
0.8	002733	002053	001526	001123	000817	000589	000420	000297	000207
0.9	002613	001968	001467	001082	000789	000570	000408	000288	000202
1.0	002484	001876	001402	001037	000759	000549	000394	000279	000196
1.1	002346	001777	001332	000988	000725	000526	000378	000269	000189
1.2	002200	001672	001257	000935	000688	000501	000361	000257	000181
1.3	002049	001562	001178	000879	000649	000474	000343	000245	000173
1.4	001894	001449	001097	000821	000608	000446	000323	000232	000164
1.5	001736	001333	001013	000761	000566	000416	000303	000218	000155
1.6	001579	001217	000928	000700	000522	000385	000281	000203	000145
1.7	001424	001102	000843	000639	000478	000354	000260	000188	000135
1.8	001273	000989	000760	000578	000434	000323	000238	000173	000124
1.9	001127	000879	000678	000518	000391	000292	000216	000157	000114
2.0	000989	000775	000600	000460	000349	000262	000194	000142	000103
2.1	000859	000676	000526	000405	000308	000232	000173	000127	000093
2.2	000740	000584	000457	000353	000270	000204	000153	000113	000083
2.3	000630	000500	000393	000305	000234	000178	000134	000099	000073
2.4	000532	000424	000334	000261	000201	000154	000116	000086	000064
2.5	000444	000355	000282	000221	000171	000131	000099	000075	000055
2.6	000367	000295	000235	000185	000144	000111	000084	000064	000047
2.7	000300	000242	000194	000153	000120	000093	000071	000054	000040
2.8	000242	000197	000158	000126	000099	000077	000059	000045	000034
2.9	000194	000158	000128	000102	000081	000063	000049	000037	000028
3.0	000153	000126	000102	000082	000065	000051	000040	000030	000023
3.1	000120	000099	000081	000065	000052	000041	000032	000025	000019
3.2	000093	000077	000063	000051	000041	000032	000025	000020	000015
3.3	000071	000059	000049	000040	000032	000025	000020	000016	000012
3.4	000054	000045	000037	000030	000025	000020	000016	000012	000009
3.5	000040	000034	000028	000023	000019	000015	000012	000009	000007
3.6	000030	000025	000021	000017	000014	000011	000009	000007	000006
3.7	000022	000018	000015	000013	000011	000009	000007	000005	000004
3.8	000016	000013	000011	000009	000008	000006	000005	000004	000003
3.9	000011	000010	000008	000007	000006	000005	000004	000003	000002
4.0	000008	000007	000006	000005	000004	000003	000003	000002	000002

EXHIBIT B.6 *(continued)*

ρ = 0.500 (*continued*)

k \ h	3.6	3.7	3.8	3.9	4.0
0.0	000157	000107	000072	000048	000031
0.1	000156	000106	000071	000048	000031
0.2	000155	000106	000071	000047	000031
0.3	000154	000105	000071	000047	000031
0.4	000153	000104	000070	000047	000031
0.5	000151	000103	000069	000046	000031
0.6	000149	000102	000069	000046	000030
0.7	000146	000100	000068	000045	000030
0.8	000143	000098	000066	000045	000030
0.9	000140	000096	000065	000044	000029
1.0	000136	000093	000064	000043	000028
1.1	000132	000091	000062	000042	000028
1.2	000127	000087	000060	000040	000027
1.3	000121	000084	000057	000039	000026
1.4	000115	000080	000055	000037	000025
1.5	000109	000076	000052	000036	000024
1.6	000102	000072	000049	000034	000023
1.7	000096	000067	000046	000032	000022
1.8	000088	000062	000043	000030	000020
1.9	000081	000057	000040	000028	000019
2.0	000074	000052	000037	000025	000017
2.1	000067	000048	000033	000023	000016
2.2	000060	000043	000030	000021	000015
2.3	000053	000038	000027	000019	000013
2.4	000047	000034	000024	000017	000012
2.5	000040	000029	000021	000015	000010
2.6	000035	000025	000018	000013	000009
2.7	000030	000022	000016	000011	000008
2.8	000025	000018	000013	000010	000007
2.9	000021	000015	000011	000008	000006
3.0	000017	000013	000009	000007	000005
3.1	000014	000011	000008	000006	000004
3.2	000011	000009	000006	000005	000003
3.3	000009	000007	000005	000004	000003
3.4	000007	000005	000004	000003	000002
3.5	000006	000004	000003	000002	000002
3.6	000004	000003	000003	000002	000001
3.7	000003	000003	000002	000001	000001
3.8	000003	000002	000002	000001	000001
3.9	000002	000001	000001	000001	000001
4.0	000001	000001	000001	000001	000000

$\rho = -0.500$

k \ h	0.0	0.1	0.2	0.3	0.4	0.5	0.6	0.7	0.8
0.0	1666667	1472109	1288543	1117443	0959897	0816598	0687848	0573588	0473431
0.1	1472109	1295818	1130216	0976550	0835700	0708178	0594141	0493418	0405553
0.2	1288543	1130216	0982164	0845419	0720665	0608253	0508211	0420282	0343956
0.3	1117443	0976550	0845419	0724876	0615434	0517301	0430398	0354399	0288762
0.4	0959897	0835700	0720665	0615434	0520367	0435548	0360817	0295794	0239929
0.5	0816598	0708178	0608253	0517301	0435548	0362982	0299375	0244321	0197268
0.6	0687848	0594141	0508211	0430398	0360817	0299375	0245803	0199680	0160471
0.7	0573588	0493418	0420282	0354399	0295794	0244321	0199680	0161454	0129134
0.8	0473431	0405553	0343956	0288762	0239929	0197268	0160471	0129134	0102785
0.9	0386718	0329853	0278526	0232782	0192530	0157559	0127561	0102154	0080912
1.0	0312570	0265442	0223135	0185637	0152821	0124469	0100285	0079918	0062984
1.1	0249952	0211319	0176829	0146428	0119973	0097244	0077966	0061823	0048478
1.2	0197727	0166407	0138601	0114231	0093142	0075127	0059935	0047286	0036890
1.3	0154710	0129603	0107439	0088123	0071504	0057388	0045553	0035756	0027751
1.4	0119721	0099821	0082354	0067219	0054273	0043340	0034227	0026728	0020636
1.5	0091615	0076023	0062416	0050694	0040726	0032357	0025422	0019748	0015167
1.6	0069322	0057246	0046769	0037796	0030210	0023879	0018663	0014421	0011017
1.7	0051861	0042617	0034644	0027856	0022150	0017417	0013541	0010408	0007909
1.8	0038356	0031363	0025367	0020292	0016052	0012556	0009710	0007424	0005610
1.9	0028042	0022814	0018359	0014610	0011497	0008945	0006881	0005232	0003932
2.0	0020265	0016403	0013132	0010396	0008138	0006298	0004818	0003644	0002723
2.1	0014474	0011656	0009283	0007310	0005692	0004381	0003334	0002507	0001864
2.2	0010217	0008185	0006484	0005079	0003934	0003011	0002279	0001704	0001260
2.3	0007128	0005680	0004476	0003487	0002686	0002045	0001539	0001144	0000841
2.4	0004913	0003895	0003053	0002365	0001812	0001372	0001027	0000759	0000555
2.5	0003347	0002639	0002058	0001586	0001208	0000909	0000677	0000498	0000362
2.6	0002253	0001767	0001370	0001050	0000795	0000596	0000441	0000323	0000233
2.7	0001498	0001168	0000901	0000687	0000517	0000385	0000283	0000206	0000148
2.8	0000984	0000763	0000585	0000444	0000332	0000246	0000180	0000130	0000093
2.9	0000639	0000493	0000376	0000283	0000211	0000155	0000113	0000081	0000058
3.0	0000409	0000314	0000238	0000179	0000132	0000097	0000070	0000050	0000035
3.1	0000259	0000198	0000149	0000111	0000082	0000060	0000043	0000030	0000021
3.2	0000162	0000123	0000092	0000068	0000050	0000036	0000026	0000018	0000013
3.3	0000100	0000076	0000057	0000041	0000030	0000022	0000015	0000011	0000008
3.4	0000061	0000046	0000034	0000025	0000018	0000013	0000009	0000006	0000004
3.5	0000037	0000028	0000020	0000015	0000011	0000008	0000005	0000004	0000003
3.6	0000022	0000016	0000012	0000009	0000006	0000004	0000003	0000002	0000001
3.7	0000013	0000010	0000007	0000005	0000004	0000002	0000002	0000001	0000001
3.8	0000007	0000005	0000004	0000003	0000002	0000001	0000001	0000001	0000000
3.9	0000004	0000003	0000002	0000002	0000001	0000001	0000001	0000001	0000000
4.0	0000002	0000002	0000001	0000001	0000001	0000000	0000000	0000000	0000000

EXHIBIT B.6 *(continued)*

$\rho = -0.500$ (*continued*)

$k \backslash h$	0.9	1.0	1.1	1.2	1.3	1.4	1.5	1.6	1.7
0.0	0386718	0312570	0249952	0197727	0154710	0119721	0091615	0069322	0051861
0.1	0329853	0265442	0211319	0166407	0129603	0099821	0076023	0057246	0042617
0.2	0278526	0223135	0176829	0138601	0107439	0082354	0062416	0046769	0034644
0.3	0232782	0185637	0146428	0114231	0088123	0067219	0050694	0037796	0027856
0.4	0192530	0152821	0119973	0093142	0071504	0054273	0040726	0030210	0022150
0.5	0157559	0124469	0097244	0075127	0057388	0043340	0032357	0023879	0017417
0.6	0127561	0100285	0077966	0059935	0045553	0034227	0025422	0018663	0013541
0.7	0102154	0079918	0061823	0047286	0035756	0026728	0019748	0014421	0010408
0.8	0080912	0062984	0048478	0036890	0027751	0020636	0015167	0011017	0007909
0.9	0063376	0049085	0037587	0028455	0021294	0015751	0011515	0008319	0005940
1.0	0049085	0037823	0028814	0021699	0016152	0011884	0008641	0006209	0004409
1.1	0037587	0028814	0021836	0016357	0012111	0008863	0006409	0004580	0003234
1.2	0028455	0021699	0016357	0012188	0008975	0006532	0004698	0003339	0002345
1.3	0021294	0016152	0012111	0008975	0006574	0004758	0003403	0002405	0001680
1.4	0015751	0011884	0008863	0006532	0004758	0003425	0002436	0001712	0001189
1.5	0011515	0008641	0006409	0004698	0003403	0002436	0001723	0001204	0000831
1.6	0008319	0006209	0004580	0003339	0002405	0001712	0001204	0000837	0000574
1.7	0005940	0004409	0003234	0002345	0001680	0001189	0000831	0000574	0000392
1.8	0004190	0003093	0002257	0001627	0001159	0000815	0000567	0000389	0000264
1.9	0002921	0002144	0001556	0001115	0000790	0000553	0000383	0000261	0000176
2.0	0002012	0001469	0001059	0000755	0000532	0000370	0000254	0000173	0000116
2.1	0001369	0000994	0000713	0000505	0000354	0000245	0000167	0000113	0000075
2.2	0000920	0000664	0000474	0000334	0000232	0000160	0000108	0000073	0000048
2.3	0000611	0000439	0000311	0000218	0000151	0000103	0000070	0000046	0000031
2.4	0000401	0000286	0000201	0000140	0000097	0000066	0000044	0000029	0000019
2.5	0000260	0000184	0000129	0000089	0000061	0000042	0000028	0000018	0000012
2.6	0000166	0000117	0000082	0000056	0000039	0000026	0000017	0000011	0000007
2.7	0000105	0000074	0000051	0000035	0000024	0000016	0000010	0000007	0000004
2.8	0000066	0000046	0000031	0000021	0000014	0000010	0000006	0000004	0000003
2.9	0000040	0000028	0000019	0000013	0000009	0000006	0000004	0000002	0000002
3.0	0000025	0000017	0000012	0000008	0000005	0000003	0000002	0000001	0000001
3.1	0000015	0000010	0000007	0000005	0000003	0000002	0000001	0000001	0000001
3.2	0000009	0000006	0000004	0000003	0000002	0000001	0000001	0000000	0000000
3.3	0000005	0000003	0000002	0000002	0000001	0000001	0000000	0000000	0000000
3.4	0000003	0000002	0000001	0000001	0000001	0000000	0000000	0000000	0000000
3.5	0000002	0000001	0000001	0000000	0000000	0000000	0000000	0000000	0000000
3.6	0000001	0000001	0000000	0000000	0000000	0000000	0000000	0000000	0000000
3.7	0000001	0000000	0000000	0000000	0000000	0000000	0000000	0000000	0000000
3.8	0000000	0000000	0000000	0000000	0000000	0000000	0000000	0000000	0000000
3.9	0000000	0000000	0000000	0000000	0000000	0000000	0000000	0000000	0000000
4.0	0000000	0000000	0000000	0000000	0000000	0000000	0000000	0000000	0000000

EXHIBIT B.6 *(continued)*

$\rho =$ 0.500 (continued)

k \ h	1.8	1.9	2.0	2.1	2.2	2.3	2.4	2.5	2.6
0.0	0038356	0028042	0020265	0014474	0010217	0007128	0004913	0003347	0002253
0.1	0031363	0022814	0016403	0011656	0008185	0005680	0003895	0002639	0001767
0.2	0025367	0018359	0013132	0009283	0006484	0004476	0003053	0002058	0001370
0.3	0020292	0014610	0010396	0007310	0005079	0003487	0002365	0001586	0001050
0.4	0016052	0011497	0008138	0005692	0003934	0002686	0001812	0001208	0000795
0.5	0012556	0008945	0006298	0004381	0003011	0002045	0001372	0000909	0000596
0.6	0009710	0006881	0004818	0003334	0002279	0001539	0001027	0000677	0000441
0.7	0007424	0005232	0003644	0002507	0001704	0001144	0000759	0000498	0000323
0.8	0005610	0003932	0002723	0001864	0001260	0000841	0000555	0000362	0000233
0.9	0004190	0002921	0002012	0001369	0000920	0000611	0000401	0000260	0000166
1.0	0003093	0002144	0001469	0000994	0000664	0000439	0000286	0000184	0000117
1.1	0002257	0001556	0001059	0000713	0000474	0000311	0000201	0000129	0000082
1.2	0001627	0001115	0000755	0000505	0000334	0000218	0000140	0000089	0000056
1.3	0001159	0000790	0000532	0000354	0000232	0000151	0000097	0000061	0000039
1.4	0000815	0000553	0000370	0000245	0000160	0000103	0000066	0000042	0000026
1.5	0000567	0000383	0000254	0000167	0000108	0000070	0000044	0000028	0000017
1.6	0000389	0000261	0000173	0000113	0000073	0000046	0000029	0000018	0000011
1.7	0000264	0000176	0000116	0000075	0000048	0000031	0000019	0000012	0000007
1.8	0000177	0000117	0000077	0000050	0000032	0000020	0000012	0000008	0000005
1.9	0000117	0000077	0000050	0000032	0000020	0000013	0000008	0000005	0000003
2.0	0000077	0000050	0000032	0000021	0000013	0000008	0000005	0000003	0000002
2.1	0000050	0000032	0000021	0000013	0000008	0000005	0000003	0000002	0000001
2.2	0000032	0000020	0000013	0000008	0000005	0000003	0000002	0000001	0000001
2.3	0000020	0000013	0000008	0000005	0000003	0000002	0000001	0000001	0000000
2.4	0000012	0000008	0000005	0000003	0000002	0000001	0000001	0000000	0000000
2.5	0000008	0000005	0000003	0000002	0000001	0000001	0000000	0000000	0000000
2.6	0000005	0000003	0000002	0000001	0000001	0000000	0000000	0000000	0000000
2.7	0000003	0000002	0000001	0000001	0000000	0000000	0000000	0000000	0000000
2.8	0000002	0000001	0000001	0000000	0000000	0000000	0000000	0000000	0000000
2.9	0000001	0000001	0000000	0000000	0000000	0000000	0000000	0000000	0000000
3.0	0000001	0000000	0000000	0000000	0000000	0000000	0000000	0000000	0000000
3.1	0000000	0000000	0000000	0000000	0000000	0000000	0000000	0000000	0000000
3.2	0000000	0000000	0000000	0000000	0000000	0000000	0000000	0000000	0000000
3.3	0000000	0000000	0000000	0000000	0000000	0000000	0000000	0000000	0000000
3.4	0000000	0000000	0000000	0000000	0000000	0000000	0000000	0000000	0000000
3.5	0000000	0000000	0000000	0000000	0000000	0000000	0000000	0000000	0000000
3.6	0000000	0000000	0000000	0000000	0000000	0000000	0000000	0000000	0000000
3.7	0000000	0000000	0000000	0000000	0000000	0000000	0000000	0000000	0000000
3.8	0000000	0000000	0000000	0000000	0000000	0000000	0000000	0000000	0000000
3.9	0000000	0000000	0000000	0000000	0000000	0000000	0000000	0000000	0000000
4.0	0000000	0000000	0000000	0000000	0000000	0000000	0000000	0000000	0000000

EXHIBIT B.6 *(continued)*

$\rho = -0.500$ (*continued*)

$k \backslash h$	2.7	2.8	2.9	3.0	3.1	3.2	3.3	3.4	3.5
0.0	0001498	0000984	0000639	0000409	0000259	0000162	0000100	0000061	0000037
0.1	0001168	0000763	0000493	0000314	0000198	0000123	0000076	0000046	0000028
0.2	0000901	0000585	0000376	0000238	0000149	0000092	0000057	0000034	0000020
0.3	0000687	0000444	0000283	0000179	0000111	0000068	0000041	0000025	0000015
0.4	0000517	0000332	0000211	0000132	0000082	0000050	0000030	0000018	0000011
0.5	0000385	0000246	0000155	0000097	0000060	0000036	0000022	0000013	0000008
0.6	0000283	0000180	0000113	0000070	0000043	0000026	0000015	0000009	0000005
0.7	0000206	0000130	0000081	0000050	0000030	0000018	0000011	0000006	0000004
0.8	0000148	0000093	0000058	0000035	0000021	0000013	0000008	0000004	0000003
0.9	0000105	0000066	0000040	0000025	0000015	0000009	0000005	0000003	0000002
1.0	0000074	0000046	0000028	0000017	0000010	0000006	0000003	0000002	0000001
1.1	0000051	0000031	0000019	0000012	0000007	0000004	0000002	0000001	0000001
1.2	0000035	0000021	0000013	0000008	0000005	0000003	0000002	0000001	0000000
1.3	0000024	0000014	0000009	0000005	0000003	0000002	0000001	0000001	0000000
1.4	0000016	0000010	0000006	0000003	0000002	0000001	0000001	0000000	0000000
1.5	0000010	0000006	0000004	0000002	0000001	0000001	0000000	0000000	0000000
1.6	0000007	0000004	0000002	0000001	0000001	0000000	0000000	0000000	0000000
1.7	0000004	0000003	0000002	0000001	0000001	0000000	0000000	0000000	0000000
1.8	0000003	0000002	0000001	0000001	0000000	0000000	0000000	0000000	0000000
1.9	0000002	0000001	0000001	0000000	0000000	0000000	0000000	0000000	0000000
2.0	0000001	0000001	0000000	0000000	0000000	0000000	0000000	0000000	0000000
2.1	0000001	0000000	0000000	0000000	0000000	0000000	0000000	0000000	0000000
2.2	0000000	0000000	0000000	0000000	0000000	0000000	0000000	0000000	0000000

$\rho = -0.500$ (*continued*)

k \ h	3.6	3.7	3.8	3.9	4.0
0.0	0000022	0000013	0000007	0000004	0000002
0.1	0000016	0000010	0000005	0000003	0000002
0.2	0000012	0000007	0000004	0000002	0000001
0.3	0000009	0000005	0000003	0000002	0000001
0.4	0000006	0000004	0000002	0000001	0000001
0.5	0000004	0000002	0000001	0000001	0000000
0.6	0000003	0000002	0000001	0000001	0000000
0.7	0000002	0000001	0000001	0000000	0000000
0.8	0000001	0000001	0000000	0000000	0000000
0.9	0000001	0000001	0000000	0000000	0000000
1.0	0000001	0000000	0000000	0000000	0000000
1.1	0000000	0000000	0000000	0000000	0000000
1.2	0000000	0000000	0000000	0000000	0000000
1.3	0000000	0000000	0000000	0000000	0000000
1.4	0000000	0000000	0000000	0000000	0000000
1.5	0000000	0000000	0000000	0000000	0000000
1.6	0000000	0000000	0000000	0000000	0000000
1.7	0000000	0000000	0000000	0000000	0000000
1.8	0000000	0000000	0000000	0000000	0000000
1.9	0000000	0000000	0000000	0000000	0000000
2.0	0000000	0000000	0000000	0000000	0000000
2.1	0000000	0000000	0000000	0000000	0000000
2.2	0000000	0000000	0000000	0000000	0000000

	$\rho = 0.85$			$\rho = -0.85$		
k \ h	0.0	1.0	2.0	0.0	1.0	2.0
0.0	41170	15602	02275	08830	00264	00000
1.0	15602	10580	02184	00264	00001	00000
2.0	02275	02184	01140	00000	00000	00000

Bibliography

Abramowitz, M., and I. A. Stegun (eds.) (1965). *Handbook of Mathematical Functions*. New York: Dover Publications, Inc.

Aczél, J. (1966). *Lectures on Functional Equations and Their Applications*. New York: Academic Press, Inc.

Adams, J. D. (1962). Failure time distribution estimation. *Semiconductor Reliability,* vol. 2, pp. 41–52.

Ahrens, L. H. (1954a,b, 1957). The lognormal distribution of the elements. *Geochimica et Cosmochimica Acta,* vol. 5, pp. 49–73; vol. 6, pp. 121–31; vol. 11, pp. 205–12.

Arnold, B. C. (1983). *Pareto Distributions*. Fairland, MD: International Co-operative Publishing House.

Artin, E. (1964). *The Gamma Function*. New York: Holt, Rinehart, and Winston.

Ayer, A. J.(1965). Chance. *Scientific American,* vol. 213, October, p. 44.

Barlow, R. E., and F. Proschan (1981). *Statistical Theory of Reliability and Life Testing: Probability Models*. Silver Spring, MD: To Begin With.

Bessel, F. W. (1818). *Fundamenta Astronomiae*. Regiomonti: Frid. Nicolvium.

Birnbaum, Z. W., and S. C. Saunders (1958). A statistical model for the life-length of materials. *Journal of the American Statistical Association,* vol. 53, pp. 151–60.

Bliss, C. I. (1934). The method of probits. *Science,* vol. 79, pp. 38–39, 409–10.

Boswell, M. T., J. K. Ord, and P. P. Patil (1979). Chance mechanisms underlying univariate distributions. *Statistical Ecology, 4: Statistical Distributions in Ecological Work* (eds. J. K. Ord, G. P. Gatil, and C.

Taillie). Fairland, MD: International Co-operative Publishing House, pp. 1–156.

Burgess, E. W., and L. S. Cottrell (1955). The prediction of adjustment in marriage, in *The Language of Social Research* (ed. P. F. Lazarsfeld and M. Rosenberg). Glencoe, IL: The Free Press, pp. 267–76.

Chatfield, C. (1970). Distributions in market research, in *Random Counts in Physical Science, Geo Science, and Business* (ed. G. P. Patil). University Park, PA: Pennsylvania State University Press.

Chayes, F. (1954). The lognormal distribution of the elements: a discussion. *Geochimica et Cosmochimica Acta,* vol. 6, pp. 119–20.

Conover, W. J. (1980). *Practical Nonparametric Statistics* (2nd ed.). New York: John Wiley & Sons, Inc.

David, F. N. (1962). *Games, Gods, and Gambling*. New York: Hafner Publishing Co., Inc.

Davis, D. J. (1952). An analysis of some failure data. *Journal of the American Statistical Association,* vol. 47, pp. 113–50.

Davis, H. T. (1935). *Tables of Higher Mathematical Functions,* 2 vols. Bloomington, IN: Principia Press.

de Finetti, B. (1974, 1975). *Theory of Probability,* Vols. 1, 2. New York: John Wiley & Sons, Inc.

DeGroot, M. H. (1970). *Optimal Statistical Decisions*. New York: McGraw-Hill Book Company.

Dhrymes, P. J. (1962). On devising unbiased estimators for the parameters of the Cobb–Douglas production function. *Econometrica,* vol. 30, pp. 297–304.

Diaconis, P. (1988). *Group Representations in Probability and Statistics*. Lecture Notes–Monograph Series, Vol. 11. Hayward, CA: Institute of Mathematical Statistics.

Ebbinghaus, H. (1985). *Memory: A Contribution to Experimental Psychology* (translation). New York: Dover Press.

Efron, B., and R. Thisted (1976). Estimating the number of unseen species: how many words did Shakespeare know? *Biometrika,* vol. 63, pp. 435–47.

Efron, B., and R. Thisted (1987). Did Shakespeare write a newly-discovered poem? *Biometrika,* vol. 74, pp. 445–55.

Epstein, B. (1947). The mathematical description of certain breakage mechanisms leading to the logarithmico-normal distribution. *Journal of the Franklin Institute,* vol. 244, pp. 471–77.

Epstein, B. (1948). Statistical aspects of fracture problems. *Journal of Applied Physics,* vol. 19, pp. 140–49.

Fechner, G. T. (1897). *Kollektivmasslehre*. Leipzig: W. Engelmann.

Feinlieb, M. (1960). A method of analyzing log-normally distributed survival data with incomplete follow-up. *Journal of the American Statistical Association,* vol. 55, pp. 534–45.

Feller, W. (1968). *An Introduction to Probability Theory and Its Applications,* Vol. 1 (3rd ed.). New York: John Wiley & Sons, Inc.

Fieller, E. C. (1932). The distribution of the index in a normal bivariate population. *Biometrika,* vol. 24, pp. 428–40.

Fienberg, S. E. (1971). Randomization and social affairs: the 1970 draft lottery. *Science,* vol. 171, pp. 255–61.

Fisher, R. A. (1950). *Statistical Methods for Research Workers* (11th ed.). Edinburgh: Oliver & Boyd Ltd.

Fisher, R. A., and K. Mather (1936). A linkage test with mice. *Annals of Eugenics,* vol. 7, pp. 265–80.

Froggatt, P. (1970). Application of discrete distribution theory to the study of noncommunicable events in medical epidemiology, in *Random Counts in Biomedical and Social Sciences,* Vol. 2 (ed. G. P. Patil). University Park, PA: Pennsylvania State University Press.

Gaddum, J. H. (1945). Lognormal distributions. *Nature,* vol. 156, pp. 463–66.

Galton, F. (1879). The geometric mean in vital and social statistics. *Proceedings of the Royal Society, London,* vol. 29, pp. 365–67.

Galton, F. (1892). *Finger Prints*. London: Macmillan and Co. Ltd.

Gardner, M. J. (1989). Review of reported increases of childhood cancer rates in the vicinity of nuclear installations in the U.K. *Journal of the Royal Statistical Society,* Series A, vol. 152, pp. 307–25.

Gauss, C. F. (1809). *Theory of Motion of the Heavenly Bodies Moving About the Sun in Conic Sections* (translation). New York: Dover.

Glass, D. V., and J. R. Hall (1954). A study of intergeneration changes in status, in *Social Mobility in Britain* (ed. D. V. Glass). Glencoe, IL: The Free Press, pp. 177–241.

Goldthwaite, L. R. (1961). Failure rate study for the lognormal lifetime model. *Proceedings of the 7th National Symposium on Reliability and Quality Control in Electronics,* pp. 208–13.

Good, I. J. (1983). *Good Thinking: The Foundations of Probability and Its Applications*. Minneapolis, MN: University of Minnesota Press.

Greenwood, M., Jr. (1904). A first study of the weight, variability, and correlation of the human viscera, with special reference to the healthy and diseased heart. *Biometrika,* vol. 3, pp. 63–83.

Greenwood, M., and G. U. Yule (1920). An inquiry into the nature of frequency distributions representative of multiple happenings with particular reference to the occurrence of multiple attacks of disease or of repeated accidents. *Journal of the Royal Statistical Society,* Series A, vol. 83, pp. 259–79.

Gregory, S. (1963). *Statistical Methods and the Geographer*. London: Longmans, Green & Company Ltd.

Griffths, J. C. (1960). Frequency distributions in accessory mineral analysis. *Journal of Geology,* vol. 68, pp. 353–65.

Grundy, P. M. (1951). The expected frequencies in a sample of an animal population in which the abundances of species are log-normally distributed. I. *Biometrika,* vol. 38, pp. 427–34.

Gutenberg, B., and C. F. Richter (1944). Frequency of earthquakes in California. *Bulletin of the Seismological Society of America,* vol. 34, pp. 185–88.

Hacking, I. (1975). *The Emergence of Probability*. Cambridge: Cambridge University Press.

Hagstroem, K.-G. (1960). Remarks on Pareto distributions. *Skandinavisk Aktuarietidskrift,* vol. 43, pp. 59–71.

Hasofer, A. M. (1970). Random mechanisms in Talmudic literature, in *Studies in the History of Statistics and Probability* (ed. E. S. Pearson and M. G. Kendall). New York: Hafner Publishing Company, Inc. pp. 39–43.

Hatch, T., and S. P. Choute (1929). Statistical description of the size properties of non-uniform particules. *Journal of the Franklin Institute,* vol. 207, pp. 369–80.

Herdan, G. (1958). The relation between the dictionary distribution and the occurrence distribution of word length and its importance for the study of quantitative linguistics. *Biometrika,* vol. 45, pp. 222–28.

Herdan, G. (1960). *Small Particle Statistics* (2nd ed.). London: Butterworth & Company Ltd.

Herdan, G. (1964). *Quantitative Linguistics*. London: Butterworth & Company Ltd.

Herdan, G. (1966). *The Advanced Theory of Language as Choice and Chance*. New York: Springer-Verlag, New York, Inc.

Indow, T. (1971). Models for responses of customers with varying rate. *Journal of Marketing Research,* vol. 8, pp. 78–84.

Jaffee, J., and S. Feldstein (1970). *Rhythms of Dialogue*. New York: Academic Press, Inc.

James, J. (1953). The distribution of free-forming small group size. *American Sociological Review,* vol. 18, p. 569.

Jaynes, E. T. (1983). *Papers on Probability, Statistics and Statistical Physics* (ed. R. D. Rosenkrantz). Dordrecht, The Netherlands: D. Reidel Publishing Co.

Jeffreys, H. (1961). *Theory of Probability* (3rd ed.). London: Oxford University Press.

Johnson, N. L., and S. Kotz (1970). *Distributions in Statistics: Continuous Univariate Distributions,* Vols. 1, 2. Boston: Houghton Mifflin Company.

Johnson, N. L., S. Kotz, and A. W. Kemp (1992). *Univariate Discrete Distributions* (2nd ed.). New York: John Wiley & Sons, Inc.

Kac, M. (1964). Probability. *Scientific American,* vol. 211, September, p. 92.

Kendall, M. G. (1963). Isaac Todhunter's history of the mathematical theory of probability. *Biometrika,* vol. 50, pp. 204–5.

Kendall, M. G., and R. L. Plackett (1977). *Studies in the History of Statistics and Probability,* Vol. 2. London: Charles Griffin & Company Ltd.

Kitagawa, T. (1952). *Tables of the Poisson Distribution.* Tokyo: Baifukan.

Kline, M. (1962). *Mathematics: A Cultural Approach.* Reading, MA: Addison-Wesley Publishing Co., Inc.

Koch, G. S., Jr., and B. F. Link (1971). *Statistical Analysis of Geological Data,* Vol. 2. New York: John Wiley & Sons, Inc.

Kolmogorov, A. N. (1933/1950/1956/1965) *Foundations of the Theory of Probability* (translation of the original). New York: Chelsea Publishing Company, Inc.

Krumbein, W. C. (1936). Application of logarithmic moments to size frequency distributions of sediments. *Journal of Sedimentary Petrology,* vol. 6, pp. 35–47.

Laplace, P. S. de (1951). *A Philosophical Essay on Probabilities* (translation with an introduction by E. T. Bell). New York: Dover Publications, Inc.

Latter, O. H. (1902). The egg of *Cuculus canorus. Biometrika,* vol. 1, pp. 164–76.

Lazarsfeld, P. F., and W. Thielens, Jr. (1958). *The Academic Mind.* Glencoe, IL: The Free Press.

Lieberman, G. J., and D. B. Owen (1961). *Tables of the Hypergeometric Distribution.* Stanford, CA: Stanford University Press.

Ludwig, J. A., and J. F. Reynolds (1988). *Statistical Ecology: A Primer on Methods and Computing.* John Wiley & Sons, Inc.

Macdonnell, W. R. (1902). On criminal anthropometry and the identification of criminals. *Biometrika,* vol. 1, pp. 177–227.

Masuyama, M., and Y. Kuroiwa (1952). Table for the likelihood solutions of gamma distribution and its medical applications. *Reports of Statistical Application Research* (*JUSE*), vol. 1, pp. 18–23.

McAlister, D. (1879). The law of the geometric mean. *Proceedings of the Royal Society, London,* vol. 29, pp. 369–75.

McGill, W. J. (1963). Stochastic latency mechanisms, in *Handbook of Mathematical Psychology* (ed. R. D. Luce, R. R. Bush, and E. Galanter). New York: John Wiley & Sons, Inc.

Medley, G. F., R. M. Anderson, D. R. Cox, and L. Billard (1987). Incubation period of AIDS in patients infected via blood transfusion. *Nature,* vol. 6132, pp. 719–21.

Miller, R. L., and J. S. Kahn (1962). *Statistical Analysis in the Geological Sciences.* New York: John Wiley & Sons, Inc.

Molina, E. C. (1942) *Poisson's Exponential Binomial Limit.* New York: D. Van Nostrand Co., Inc.

Morgan, J. P., N. R. Chaganty, R. C. Dahiya, and M. J. Doviak (1991). Let's make a deal: the player's dilemma. *American Statistician,* vol. 45(4), pp. 284–89.

Mosteller, F. (1965). *Fifty Challenging Problems in Probability with Solutions*. New York: Dover.

Mosteller, F., and D. L. Wallace (1984). *Applied Bayesian and Classical Inference: The Case of "The Federalist Papers"* (2nd ed.). New York: Springer-Verlag.

Mueller, C. G. (1950). Theoretical relationships among some measures of conditioning. *Proceedings of the National Academy of Sciences,* vol. 56, pp. 123–34.

National Bureau of Standards (1950). *Tables of the Binomial Probability Distribution*. Applied Mathematics Series 6. Washington, D.C.: U.S. Government Printing Office.

National Bureau of Standards (1951). *Tables of* $n!$ *and* $\Gamma(n + \frac{1}{2})$ *for the First Thousand Values of* n. Applied Mathematics Series 16. Washington, DC: U.S. Government Printing Office.

National Bureau of Standards (1959). *Tables of the Bivariate Normal Distribution Function and Related Functions*. Applied Mathematics Series 50. Washington, D.C.: U.S. Government Printing Office.

Newman, J. R. (1956). *The World of Mathematics*. New York: Simon & Schuster, Inc.

Ore, O. (1953). *Cardano, The Gambling Scholar*. Princeton, NJ: Princeton University Press.

Ore, O. (1960). Pascal and the invention of probability theory. *American Mathematical Monthly,* vol. 67, pp. 409–19.

Pearl, R. (1905). Biometrical studies on man. I. Variation and correlation in brainweight. *Biometrika,* vol. 4, pp. 13–104.

Pearson, E. S., and M. G. Kendall (1970). *Studies in the History of Statistics and Probability*. New York: Hafner Publishing Company, Inc.

Pearson, K. (1924). On a certain double hypergeometrical series and its representation by continuous frequency surfaces. *Biometrika,* vol. 16, pp. 172–88.

Pearson, K. (1934). *Tables of the Incomplete Beta Function*. London: The "Biometrika" Office, University College.

Pearson, K., and A. Lee (1903). On the laws of inheritance in man. I. Inheritance of physical characters. *Biometrika,* vol. 2, pp. 357–462.

Pollard, R. (1973). Collegiate football scores and the negative binomial distribution. *Journal of the American Statistical Association,* vol. 68, pp. 351–52.

Proschan, F. (1963). Theoretical explanation of observed decreasing failure rate. *Technometrics,* vol. 5, pp. 375–84.

Protter, M. H., and C. B. Morrey, Jr. (1985). *Intermediate Calculus* (2nd ed.). New York: Springer-Verlag New York, Inc.

Rabinovitch, N. L. (1973). *Probability and Statistical Inference in Ancient and Medieval Jewish Literature*. Toronto: University of Toronto Press.

Rasch, G. (1960). *Probabilistic Models for Some Intelligence and Attainment Tests. Studies in Mathematical Psychology I*. Copenhagen: Nielson and Lydiche.

Rayner, J. W. C., and D. J. Best (1989). *Smooth Tests of Goodness of Fit*. Cary, NC: Oxford Press.

Richardson, L. F. (1944). Distribution of wars in time. *Journal of the Royal Statistical Society,* vol. 107, pp. 242–50.

Roberts, J. (1975). *Blood Pressure of Persons 18–74 Years, United States, 1971–72*. U.S. National Center for Health Statistics, Vital and Health Statistics Series 11, No. 150. Washington, DC: U.S. Department of Health, Education and Welfare.

Rogers, F. E. (1977). Fire losses and the effect of sprinkler protection of buildings in a variety of industries and trades. *Current Paper CP 9/77*. Garston, U.K.: Building Research Establishment.

Romig, H. G. (1953). *50–100 Binomial Tables*. New York: John Wiley & Sons, Inc.

Rutherford, E., and H. Geiger (1910). The probability variations in the distribution of particles. *Philosophical Magazine,* vol. 20, pp. 698–707.

Rutherford, E., J. Chadwick, and C. D. Ellis (1930). *Radiations from Radioactive Substances*. New York: Macmillan Publishing Co., Inc.

Sartwell, P. E. (1950). The distribution of incubation periods of infectious diseases. *American Journal of Hygiene,* vol. 51, pp. 310–18.

Sclove, S. L. (1981). Modeling the distribution of fingerprint characteristics, in *Statistical Distributions in Scientific Work*. Vol. 6: *Applications in Physical, Social and Life Sciences* (ed. C. Taillie, G. P. Patil, and B. A. Baldessari). Boston, MA: D. Reidel Publishing Company, pp. 111–130.

Selvin, S. (1975). A problem in probability. *American Statistician,* vol. 29(1), p. 67.

Shapiro, S. S. (1990). *How to Test Normality and Other Distributional Assumptions*. American Society for Quality Control. Milwaukee, WI: Quality Press.

Simon, H. A., and C. P. Bonini (1958). The size distribution of business firms. Reprint 20. Pittsburgh, PA: Graduate School of Business Administration, Carnegie Institute of Technology.

Simpson, G. G., A. Roe, and R. C. Lewontin (1960). *Quantitative Zoology*. New York: Harcourt, Brace & World, Inc.

Slack, H. A., and W. C. Krumbein (1955). Measurement and statistical evaluation of low-level radioactivity in rocks. *Transactions, American Geophysical Union,* vol. 36, pp. 460–64.

Spencer, H. (1877). *The Principles of Sociology*. New York: Appleton and Co.

Stigler, S. M. (1977). Do robust estimators work with *real* data? (with discussion). *Annals of Statistics,* vol. 5, pp. 1055–98.

Stigler, S. M. (1986). *The History of Statistics: The Measurement of Uncertainty Before 1900.* Cambridge, M.A.: Harvard University Press.

Svedberg, T. (1912). *Existenz der Moleküle.* Leipzig: Akademische Verlagsgesellschaft mbH.

Taylor, A. E. (1955). *Advanced Calculus.* Boston: Ginn & Company.

Terman, L. M. (1919). *The Intelligence of School Children.* Boston: Houghton Mifflin Company.

Todhunter, I. (1865/1949). *A History of the Mathematical Theory of Probability from the Time of Pascal to That of Laplace* (reprint). New York: Chelsea Publishing Company.

Trumpler, R. J., and H. F. Weaver (1953). *Statistical Astronomy.* Berkeley, CA: University of California Press.

Tversky, A., and D. Kahneman (1982). The psychology of preferences. *Scientific American,* vol. 246(1), January, pp. 160–173.

Urban, F. M. (1909). Die psychophysischen Massmethoden als Grundlage empirischer Messungen. *Archivfuer die Gesante Psychologie,* vol. 15, pp. 261–415.

Weaver, W. (1950). Probability. *Scientific American,* vol. 183, October, p. 44.

Weaver, W. (1952). Statistics. *Scientific American,* vol. 186, January, p. 60.

Weibull, W. (1951). A statistical distribution function of wide applicability. *Journal of Applied Mechanics,* vol. 18, pp. 293–97.

Williams, C. B. (1940). A note on the statistical analysis of sentence length as a criterion of literary style. *Biometrika,* vol. 31, pp. 356–61.

Williams, C. B. (1956). Studies in the history of probability and statistics. IV. A note on an early statistical study of literary style. *Biometrika,* vol. 43, pp. 248–56.

Williamson, E., and M. K. Bretherton (1963). *Tables of the Negative Binomial Probability Distribution.* New York: John Wiley & Sons, Inc.

Winkler, R. L. (1972). *Introduction to Bayesian Inference and Decision.* New York: Holt, Rinehart and Winston, Inc.

Yule, G. U. (1939). On sentence length as a statistical characteristic of style in prose. *Biometrika,* vol. 30, pp. 363–384.

Zipf, G. K. (1949). *Human Behavior and the Principle of Least Effort.* Reading, MA: Addison-Wesley Publishing Co., Inc.

Name Index

Subject Index

A

B

E

F

G

H

I

J

K

L

M

Q

R

S

T

U

V

W

Z